Lecture Notes in Physics

Monographs

Springer-Verlag Berlin Heidelberg GmbH

The Editorial Policy for Monographs

The series Lecture Notes in Physics reports new developments in physical research and teaching - quickly, informally, and at a high level. The type of material considered for publication in the monograph Series includes monographs presenting original research or new angles in a classical field. The timeliness of a manuscript is more important than its form, which may be preliminary or tentative. Manuscripts should be reasonably self-contained. They will often present not only results of the author(s) but also related work by other people and will provide sufficient motivation, examples, and applications.
The manuscripts or a detailed description thereof should be submitted either to one of the series editors or to the managing editor. The proposal is then carefully refereed. A final decision concerning publication can often only be made on the basis of the complete manuscript, but otherwise the editors will try to make a preliminary decision as definite as they can on the basis of the available information.
Manuscripts should be no less than 100 and preferably no more than 400 pages in length. Final manuscripts should be in English. They should include a table of contents and an informative introduction accessible also to readers not particularly familiar with the topic treated. Authors are free to use the material in other publications. However, if extensive use is made elsewhere, the publisher should be informed. Authors receive jointly 30 complimentary copies of their book. They are entitled to purchase further copies of their book at a reduced rate. No reprints of individual contributions can be supplied. No royalty is paid on Lecture Notes in Physics volumes. Commitment to publish is made by letter of interest rather than by signing a formal contract. Springer-Verlag secures the copyright for each volume.

The Production Process

The books are hardbound, and quality paper appropriate to the needs of the author(s) is used. Publication time is about ten weeks. More than twenty years of experience guarantee authors the best possible service. To reach the goal of rapid publication at a low price the technique of photographic reproduction from a camera-ready manuscript was chosen. This process shifts the main responsibility for the technical quality considerably from the publisher to the author. We therefore urge all authors to observe very carefully our guidelines for the preparation of camera-ready manuscripts, which we will supply on request. This applies especially to the quality of figures and halftones submitted for publication. Figures should be submitted as originals or glossy prints, as very often Xerox copies are not suitable for reproduction. For the same reason, any writing within figures should not be smaller than 2.5 mm. It might be useful to look at some of the volumes already published or, especially if some atypical text is planned, to write to the Physics Editorial Department of Springer-Verlag direct. This avoids mistakes and time-consuming correspondence during the production period.
As a special service, we offer free of charge LaTeX and TeX macro packages to format the text according to Springer-Verlag's quality requirements. We strongly recommend authors to make use of this offer, as the result will be a book of considerably improved technical quality.
For further information please contact Springer-Verlag, Physics Editorial Department II, Tiergartenstrasse 17, D-69121 Heidelberg, Germany.

Series homepage – http://www.springer.de/phys/books/lnpm

Francesco Calogero

Classical Many-Body Problems Amenable to Exact Treatments

(Solvable and/or Integrable and/or Linearizable...) in One-, Two- and Three-Dimensional Space

Springer

Author

Francesco Calogero
Department of Physics
University of Rome "La Sapienza"
p. Aldo Moro
00185 Roma, Italy
e-mail: francesco.calogero@roma1.infn.it
francesco.calogero@uniroma1.it

Library of Congress Cataloging-in-Publication Data.
Die Deutsche Bibliothek - CIP-Einheitsaufnahme

Calogero, Francesco:
Classical many body problems amenable to exact treatments : (solvable and, or integrable and, or linearizable ...) in one-, two- and three-dimensional space / Francesco Calogero.

(Lecture notes in physics : N.s. M, Monographs ; 66)
(Physics and astronomy online library)

ISSN 0940-7677 (Lecture Notes in Physics. Monographs)
ISBN 978-3-662-14344-5 ISBN 978-3-540-44730-6 (eBook)
DOI 10.1007/978-3-540-44730-6

http://www.springer.de

Originally published by Springer-Verlag Berlin Heidelberg New York in 2001

Typesetting: Camera-ready by perform, Heidelberg
Cover design: *design & production*, Heidelberg

Printed on acid-free paper
SPIN: 10644563 55/3141/du - 5 4 3 2 1 0

Foreword

This book focuses on *exactly treatable* many-body problems. This class does not include most physical problems. We are therefore reminded "of the story of the man who, returning home late at night after an alcoholic evening, was scanning the ground for his key under a lamppost; he knew, to be sure, that he had dropped it somewhere else, but only under the lamppost was there enough light to conduct a proper search" <C71>. Yet we feel the interest for such models is nowadays sufficiently widespread – because of their beauty, their mathematical relevance and their multifarious applicative potential – that no apologies need be made for our choice. In any case, whoever undertakes to read this book will know from its title what she is in for!

Yet this title may require some explanations: a gloss of it (including its extended version, see inside front cover) follows.

By "Classical" we mean *nonquantal* and *nonrelativistic* (although some consider the Ruijsenaars-Schneider models, which are indeed treated in this book, as relativistic versions of, previously known, nonrelativistic models; see below): our presentation is mainly focussed on many-body systems of point particles whose time evolution is determined by Newtonian equations of motion (*acceleration proportional to force*). The fact that we treat problems not only in one, but also in two, and even in three (and occasionally in an arbitrary number of), dimensions, is of course somewhat of a novelty: indeed the treatment of two-dimensional, and especially three-dimensional, (rotation-invariant!) models, is based on recent (sometimes very recent) findings. By "amenable to exact treatments" we mean that, to investigate the behavior of the many-body models identified and studied in this book, significant progress can be made by "exact" (i. e., not approximate) techniques. The extent to which one can thereby master the detailed behavior of these many-body systems varies from case to case: this is emphasized by the parenthetical part of our title, which perhaps requires some additional elaboration, to explain what we mean by our distinction – which is, to be sure, a *heuristic* one: quite useful, but not quite precise – among *solvable, integrable* and *linearizable* models.

Solvable models are characterized by the availability of a technique of solution which requires purely *algebraic operations* (such as inverting or diagonalizing finite matrices, or finding the zeros of known polynomials), and/or possibly solving *known* (generally linear, possibly nonautonomous) ODEs in terms of *known* special functions (say, of hypergeometric type), and/or perhaps the *inversion* of known functions (as in the standard *solution by quadratures*).

Integrable models are those for which some approach (for instance, a "Lax-pair", see below) is available, which yields an adequate supply of *constants of motion.* As a rule these models are also solvable, but generally this requires more labor. In the Hamiltonian cases, these models are generally *Liouville integrable.*

Thirdly, we refer to *linearizable* problems: their treatment generally requires, in addition to the operations mentioned above in the context of *solvable* models, the solution of *linear, generally nonautonomous, ODEs,* which, in spite of their being generally rather simple, might indeed give rise to quite complicated (chaotic?) motions. In the Hamiltonian cases these many-body models need not be integrable in the Liouville sense, although the linearity of the equations to be finally solved entails the possibility to introduce *constants of the motion* via the *superposition principle,* which guarantees that the general solution of a linear ODE can be represented as a linear combination with *constant* coefficients of an appropriate set of specific solutions. In any case a linearizable many-body problem is certainly much easier to treat than the generic (nonlinear!) many-body problem, inasmuch as its solution can be reduced to solving a *linear* first-order matrix ODE (indeed, in most cases, a *single linear second-order scalar* ODE – albeit a *nonautonomous* one – see below).

Clearly these three categories of problems – *solvable, integrable, linearizable* – are ordered in terms of increasing difficulty, so that (as indeed the title of this book indicates with its *and/or*'s), problems belonging to a lower category generally also belong to the following one(s).

But let us reemphasize that the distinction among *solvable, integrable* and *linearizable* models is imprecise: the boundaries among these categories are somewhat blurred, moreover we have been vague about what "solving" a problem really means: Finding the general solution? Solving the initial-value problem? For which class of initial data? And what about boundary conditions (which in some cases are essential to define the problem)? The final dots in the title underline the heuristic, and incomplete, character of this distinction among *solvable, integrable* and *linearizable* models (for instance, we shall also introduce below the notion of *partially solvable* models, whose initial-value problem can be solved only for a restricted subclass of initial data). Yet this distinction is convenient to convey synthetically the status of the various many-body problems treated in this book.

Two additional remarks.

(i). The genesis of exactly treatable models comes seldom from the discovery of a technique to solve a given problem; generally the actual development is the other way round, a suitable technique is exploited to discover all the models which can be treated (possibly solved) by it. Some disapprove of such an approach to research, in which, rather than trying to find the solution of a problem, one tries to find problems that fit a known (technique of) solution. Some, indeed, go as far as decrying "basic research," presumably because, in contrast to applied research, it does not solve specific problems: "Basic research is like shooting an arrow into the air and, where it lands, painting a target" (attributed to Homer Adkins (1984) <APS99>). This author, on the contrary, does not see anything wrong with this approach; it seems to me it is a normal way of making progress in science. For instance: occasionally an experimental device (say, a particle accelerator) is built for the specific purpose to discover something (say, a new elementary particle); but more often an experimental device is available (say, a particle accelerator), and the experimental activity is concentrated on whatever that particular device allows experimenters to do. And nobody sees anything wrong in this. Indeed there is a quotation from Carl Jacobi (which I am lifting from a classical treatise by Vladimir Arnold <A74>), that expresses this point of view in a context quite close to that of this book (although it refers specifically to an approach -- separation of variables -- we do not explicitly treat): "The main difficulty to integrate these differential equations is to find the appropriate change of variables. There is no rule to discover it. Hence we need to follow the inverse path, namely to introduce some convenient change of variables and investigate to which problems it can be successfully applied." And another quotation which expresses a point of view I sympathize with comes from Vladimir E. Zakharov: "A mathematician, using the dressing method to find a new integrable system, could be compared with a fisherman, plunging his net into the sea. He does not know what a fish he will pull out. He hopes to catch a goldfish, of course. But too often his catch is something that could not be used for any known to him purpose. He invents more and more sophisticated nets and equipments and plunges all that deeper and deeper. As a result he pulls on the shore after a hard work more and more strange creatures. He should not despair, nevertheless. The strange creatures may be interesting enough if you are not too pragmatic. And who knows how deep in the sea do goldfishes live? ". <Z90>

(ii). Models amenable to exact treatments are, of course, special. Why focus on them, rather than look at general cases, which capture many more problems, including the more "physical" ones? But again, this is to a large extent the essence of normal science. Pythagora' s theorem does not hold for all triangles, but only for rectangular ones. Should this be considered a shortcoming of this mathematical result, or instead its very essence? The answer is plain.

Finally, a few remarks on the presentation and the selection of the material.

The presentation is meant to facilitate the self-education of a reader who wishes to enter this research area. For instance, special cases are often presented in place or in advance of more general treatments, in order to introduce ideas and techniques in a simpler context. The division into a main text and a secondary part, separated by horizontal lines and distinguished by a slight difference in the size of the fonts, should also be helpful: in the secondary part we generally segregate remarks and arguments (often including proofs) which deviate from the main flow of the presentation (but the reader is well advised to read sequentially through these parts as well, which often contain material that is essential -- or at least helpful -- for the understanding of what follows; and this advise also applies to all *exercises*, which should all be read, even when there is no intention/possibility to invest immediately time in their solution). Almost all mentions of related references, historical remarks, due credits, etc., are also relegated elsewhere, to special sections ("Notes") located at the end of the chapters and of some appendices. Of course this book might also be used as background material for teaching a course (it actually emerged from such a context – indeed, it profited from such a test) .

The selection of the material presented in this book is unashamedly skewed towards research topics to which the author has personally contributed, or which he finds particularly congenial (such as the Ruijsenaars-Schneider model). The enormous amount of research on the topics treated in this book and/or on closely related areas that emerged in the last quarter century would have anyway doomed to failure any effort at providing a "complete" coverage; likewise any attempt to present a "complete" bibliographic record of the contributions on the topics treated would have been impossible, indeed perhaps futile given the great ease nowadays to retrieve relevant references via computer-assisted searches. These are admittedly lame excuses for the shortcomings of this book, whose worth (be it somewhat positive or largely negative) will in any case be best assessed by those who will use it as a (personal or didactic) teaching tool; but I like to express here my apologies to all those colleagues who contributed importantly to the development of this area of research and who will not find in this book any reference to their contribution.

The organization of the book into a rather detailed net of telescoped sections is meant to help the reader, both the first time he navigates through the book as well as when she might wish to retrieve some notion. Moreover, the table of contents provides a synthetic overview of the material covered in this book which might help the perplexed browser in deciding whether he wishes to become an engaged, or even a diligent, reader. Equations are numbered progressively within each section and appendix: equation (16) of Section 2.1.1 is referred to as (16) within that section, as (2.1.1.-16) elsewhere; and likewise (C.-10b) is equation (10b) of Appendix C (but within Appendix C it is referred to simply as (10b)).

Let me end this Foreword on a personal note. My father, Guido Calogero, was a philosopher who wrote many books (without formulas!), and he also had a great interest for, and much scholarship in, philology and archaeology (especially texts from ancient Greece). Hence, he always paid a keen attention to the appearance of any text; and he much disliked misprints. I inherited this attitude, but not his keen eye to weed out imperfections. Hence I must apologize for the many misprints and other defects this book certainly contains, and beg the reader to take the same benevolent attitude displayed by Hermann Weyl in his 1938 review <W38> of the second volume of the classic mathematical physics treatise by Richard Courant and David Hilbert <CH37>, when he wrote: "The author apologizes that lack of time prevented him from fitting out this book with a full sized index of literature and such paraphernalia. The same reason may be responsible for quite a few misprints on which the reader will occasionally stumble. But perhaps even these minor faults deserve praise rather than blame. Although I know that a craftsman's pride should be in having his work as perfect and shipshape as possible, even in the most minute and inessential details, I sometimes wonder whether we do not lavish on the dressing-up of a book too much time that would better go into more important things."

Yet I will be most grateful to whoever will take the trouble to bring to my attention shortcomings of this book (including misprints!), via an e-mail message sent to (both) these addresses: francesco.calogero@uniroma1.it, francesco.calogero@roma1.infn.it.

Preface

This book, as well as its title, are long, perhaps too long; and it took quite a long time to complete this project, well over three years of intense hard work. Throughout this period I sought and got advise from several colleagues and friends, and also from students to whom preliminary drafts were distributed and who helped me by spotting misprints and mistakes (letting them search for these turned indeed out to be a very efficient teaching technique!). For a special word of thanks I like to mention Mario Bruschi, Jean-Pierre Françoise, David Gomez Ullarte, Misha Olshanetsky, Orlando Ragnisco, Simon Ruijsenaars. But it is of course understood that I am solely responsible for all shortcomings of this book.

I also wish to thank: Alessandra Grussu and Matteo Sommacal for transforming my scribbled first draft into WORD files for me to work on; my Physics Department at the University of Rome I "La Sapienza" for supporting financially this typing job, and in particular the Administrator of my Department, Maria Vittoria Marchet, for organizing this arrangement, and the Director of my Department, Francesco Guerra, for encouraging me to undertake this project; and the staff at Springer, in particular Mrs. Brigitte Reichel-Mayer respectively Prof. Wolf Beiglboeck, for their cooperative attitude on the technical respectively substantive aspects of the production of this book.

This book is dedicated to the memory of Juergen Moser, whose seminal work was instrumental in opening up this field of research. Most regrettably, I never managed to meet him: I only spoke by telephone with him one time, more than twenty years ago, from JFK airport in New York, while he was in his office at the Courant Institute; then, through the years, various last minutes glitches postponed more than once our getting together. Alas, now it is too late to remedy this mistake.

September 2000 *Francesco Calogero*

Contents

1. Classical (nonquantal, nonrelativistic) many-body problems 1

1.1 Newton's equation in one, two and three dimensions 1

1.2 Hamiltonian systems - Integrable systems 6

1.N Notes to Chapter 1 16

2. One-dimensional systems. Motions on the line and on the circle 17

2.1 The Lax pair technique 17

2.1.1 A convenient representation. The functional equation (*) 23

2.1.2 A simple solution of the functional equation (*) 26

2.1.3 *N* particles on the line, interacting pairwise via repulsive forces inversely proportional to the cube of their mutual distance 27

2.1.3.1 Qualitative behavior 27

2.1.3.2 The technique of solution of Olshanetsky and Perelomov (OP) 30

2.1.3.3 Motion in the presence of an additional harmonic interaction. Extension of the OP technique of solution 37

2.1.4 General solution of the functional equation (*). Integrable many-body model with elliptic interactions 47

2.1.5 *N* particles on the line interacting pairwise via a repulsive hyperbolic force. Technique of solution OP 53

2.1.6 *N* particles on the circle interacting pairwise via a trigonometric force 61

2.1.7 Various tricks: changes of variables, particles of different types, duplications, infinite duplications (from rational to hyperbolic, trigonometric, elliptic forces), reductions (model with forces only among "nearest neighbors") 63

2.1.8 Another convenient representation for the Lax pair. The functional equation (**) 80

2.1.9 A simple solution of the functional equation (**) 84

2.1.9.1 Fake Lax pairs 86

2.1.10 N particles on the line, interacting pairwise via forces equal to twice the product of their velocities divided by their mutual distance 90
2.1.10.1 Technique of solution OP 92
2.1.10.2 Behavior of the solutions: mention of future developments 94
2.1.10.3 Can a fake Lax pair be used to solve a nontrivial many-body problem? 97
2.1.11 General solution of the functional equation (**) 99
2.1.12 The many-body problem of Ruijsenaars and Schneider (RS) 110
2.1.12.1 Hamiltonian and Newtonian equations for the RS model 113
2.1.12.2 Relativistic character of the RS model 115
2.1.12.3 Newtonian case. Complex extension presumably characterized by completely periodic motions 119
2.1.12.4 Solution via the OP technique in the rational, hyperbolic and trigonometric cases. Completely periodic character of the motion 122
2.1.13 Various tricks: changes of variables, duplications, infinite duplications, reductions to "nearest-neighbor" forces, elimination of velocity-dependent forces 126
2.1.14 Another Lax pair corresponding to a Hamiltonian many-body problem on the line. The functional equation (***) 135
2.1.15 A simple solution of the functional equation (***), and the corresponding Hamiltonian many-body problem on the line 138
2.1.15.1 Explicit solution 141
2.1.15.2 Reformulation via canonical transformations 147
2.1.16 A nonanalytic solution of the functional equation (***), and the corresponding Hamiltonian many-body problem 151
2.1.16.1 Proof of integrability. A new functional equation 154
2.2 Another exactly solvable Hamiltonian problem 159
2.3 Many-body problems on the line related to the motion of the zeros of solutions of linear partial differential equations in 1+1 variables (space + time) 163

2.3.1 A nonlinear transformation: relationships between the coefficients and the zeros of a polynomial 164
2.3.2 Some formulas for a polynomial and its derivatives, in terms of its coefficients and its zeros 165
2.3.3 Many-body problems on the line solvable via the identification of their motions with those of the zeros of a polynomial that evolves in time according to a linear PDE in 2 variables (space and time)............ 167
2.3.4 Examples 174
2.3.4.1 First-order systems.................................... 175
2.3.4.2 Second-order systems (Newtonian equations of motion)............... 188
2.3.5 Trigonometric extension... 197
2.3.6 Further extension 203
2.3.6.1 New solvable many-body problems via a new functional equation 207
2.3.6.2 General solution of the new functional equation.................. 213
2.3.6.3 A new solvable many-body problem with elliptic-type velocity-dependent forces 219
2.4 Finite-dimensional representations of differential operators, Lagrangian interpolation, and all that................................ 228
2.4.1 Finite-dimensional matrix representations of differential operators 229
2.4.2 Connection with Lagrangian interpolation................ 235
2.4.3 Algebraic approach... 240
2.4.4 The finite-dimensional (matrix) algebra of raising and lowering operators, and its realizations 253
2.4.5 Remarkable matrices and identities.......................... 264
2.4.5.1 Matrices with known spectrum.................... 265
2.4.5.2 Matrices with known inverse....................... 268
2.4.5.3 A remarkable matrix, and some related trigonometric identities 269
2.4.5.4 Matrices satisfying “fake” Lax equations..... 273
2.4.5.5 Determinantal representations of polynomials defined by ODEs or by recurrence relations 274
2.5 Many-body problems on the line solvable via techniques of exact Lagrangian interpolation.. 279
2.N Notes to Chapter 2 304

3. N-body problems treatable via techniques of exact Lagrangian interpolation in spaces of one or more dimensions 311
3.1 Generalized formulation of Lagrangian interpolation, in spaces of arbitrary dimensions.. 311
3.1.1 Finite-dimensional representation of the operator of differentiation.. 316
3.1.2 Examples 330
3.1.2.1 One-dimensional space ($S=1$) 330
3.1.2.2 Two-dimensional space ($S=2$).................... 341
3.1.2.3 Three-dimensional space ($S=3$) 352
3.2 N-body problems in spaces of one or more dimensions 355
3.2.1 One-dimensional examples.. 368
3.2.2 Two-dimensional examples (in the plane)................. 389
3.2.3 Few-body problems in ordinary (3-dimensional) space............................. 400
3.2.4 N-body problems in M-dimensional space, or M^2-body problems in one-dimensional space 404
3.3 First-order evolution equations and partially solvable N-body problems with velocity-independent forces............. 408
3.N Notes to Chapter 3 415

4. Solvable and/or integrable many-body problems in the plane, obtained by complexification 417
4.1 How to obtain by complexification rotation-invariant many-body models in the plane from certain many-body problems on the line... 418
4.2 Example: a family of solvable many-body problems in the plane.. 428
4.2.1 Origin of the model and technique of solution........... 429
4.2.2 The generic model; behavior in the remote past and future.. 432
4.2.3 Some special cases: models with a limit cycle, models with confined and periodic motions, Hamiltonian models, translation-invariant models, models featuring equilibrium and spiraling configurations, models featuring only completely periodic motions.............. 435
4.2.4 The simplest model: explicit solution (the game of musical chairs), Hamiltonian structure.. 448
4.2.5 The simplest model featuring only completely periodic motions 453
4.2.6 First-order evolution equations, and a partially solvable many-body problem with velocity-independent forces, in the plane 456

4.3 Examples: other families of solvable many-body problems in the plane ... 459
4.3.1 A rescaling-invariant solvable one-dimensional many-body problem ... 462
4.3.2 A rescaling- and translation-invariant solvable one-dimensional many-body problem ... 467
4.3.3 Another rescaling-invariant solvable one-dimensional many-body problem ... 469
4.4 Survey of solvable and/or integrable many-body problems in the plane obtained by complexification ... 471
4.4.1 Example one ... 472
4.4.2 Example two ... 474
4.4.3 Example three ... 476
4.4.4 Example four ... 479
4.4.5 Example five ... 481
4.4.6 Example six ... 482
4.4.7 Example seven ... 485
4.4.8 Example eight ... 489
4.4.9 Example nine ... 491
4.4.10 A Hamiltonian example ... 493
4.5 A many-rotator, possibly nonintegrable, problem in the plane, and its periodic motions ... 494
4.6 Outlook ... 509
4.N Notes to Chapter 4 ... 509

5. Many-body systems in ordinary (three-dimensional) space: solvable, integrable, linearizable problems ... 511
5.1 A simple example: a solvable matrix problem, and the corresponding one-body problem in three-dimensional space ... 512
5.2 Another simple example: a linearizable matrix problem, and the corresponding one-body problem in three-dimensional space ... 520
5.2.1 Motion of a magnetic monopole in a central electric field ... 535
5.2.2 Motion of a magnetic monopole in a central Coulomb field ... 543
5.2.3 Solvable cases of the (2×2)-matrix evolution equation $\ddot{\underline{U}} = 2a\dot{\underline{U}} + b\underline{U} + c[\dot{\underline{U}}, \underline{U}]$ ... 550
5.3 Association, complexification, multiplication: solvable few and many-body problems obtained from the previous ones . 553
5.4 A survey of matrix evolution equations amenable to exact treatments ... 569

5.4.1 A class of linearizable matrix evolution equations 570
5.4.2 Some integrable matrix evolution equations related to the non Abelian Toda lattice 585
5.4.3 Some other matrix evolution equations amenable to exact treatments 590
5.4.4 On the integrability of the matrix evolution equation $\ddot{\underline{U}} = f(\underline{U})$ 598
5.5 Parametrization of matrices via three-vectors 605
5.6 A survey of N-body systems in three-dimensional space amenable to exact treatments 613
5.6.1 Few-body problems of Newtonian type 614
5.6.2 Few-body problems of Hamiltonian type 626
5.6.3 Many-body problems of Newtonian type 628
5.6.4 Many-body problems of Hamiltonian type 641
5.6.5 Many-body problems in multidimensional space with velocity-independent forces: integrable unharmonic ("quartic") oscillators, and nonintegrable oscillators with lots of completely periodic motions 645
5.7 Outlook 661
5.N Notes to Chapter 5 662

Appendix A: Elliptic functions 663
A.N Notes to Appendix A 673

Appendix B: Functional equations 675
B.N Notes to Appendix B 684

Appendix C: Hermite polynomials: zeros, determinantal representations 685
C.N Notes to Appendix C 688

Appendix D: Remarkable matrices and related identities 689
D.N Notes to Appendix D 702

Appendix E: Lagrangian approximation for eigenvalue problems in one and more dimensions 703
E.N Notes to Appendix E 707

Appendix F: Some theorems of elementary geometry in multidimensions 708

Appendix G: Asymptotic behavior of the zeros of a polynomial whose coefficients diverge exponentially 723

Appendix H: Some formulas for Pauli matrices and three-vectors 732

References 735

1 CLASSICAL (NONQUANTAL, NONRELATIVISTIC) MANY-BODY PROBLEMS

In this introductory Chap. 1 we tersely review the basic notion of, and notation for, classical many-body problems in one-, two- and three-dimensional space, mainly by exhibiting the corresponding Newtonian equations of motion. We also tersely review the Hamiltonian formulation of such problems and we outline the notion of integrability associated with such Hamiltonian systems.

1.1 Newton's equation in one, two and three dimensions

The fundamental ("Newton's") equations characterizing a classical (i.e., nonquantal and nonrelativistic) N-body problem state that the *acceleration* of the n-th particle equals a *force* acting on it which depends, in an assumedly known manner, on the positions and velocities of all N particles:

$$\ddot{\vec{r}}_n(t) = \vec{f}_n[\vec{r}_m(t), \dot{\vec{r}}_m(t); t]. \tag{1a}$$

Here and generally below indices such as n, m label the different particles (they take the values $1,\ldots,N$, unless otherwise specified), t is the *time* and superimposed dots denote differentiation with respect to this ("independent") variable. $\vec{r}_n(t)$ identifies the position of the n-th particle at time t; in the following we often omit to indicate explicitly the time dependence. Likewise, $\vec{f}_n(\vec{r}_m, \dot{\vec{r}}_m; t)$ is the force acting on the n-th particle; as a rule it depends on the positions and velocities of the N particles. The force depends on the time via the positions and velocities; as we have indicated, it may also depend explicitly on the time variable, although hereafter we will almost exclusively focus on the *autonomous* case, characterized by the *absence* of such an *explicit* time-dependence. (Note that here we include the mass in the definition of the force: strictly speaking $\vec{f}_n$ is the force acting on the n-th particle divided by the mass of the n-th particle).

F. Calogero: LNPm 66, pp. 1 - 16, 2001

It is often notationally convenient to introduce an N-vector whose N components are labeled by an index, say n or m, that takes the N values $1,...,N$; hereafter such N-vectors are denoted by underlined lower-case symbols (upper-case underlined symbols are reserved for matrices, see below). Hence equivalent versions of (1a) read as follows:

$$\ddot{\vec{r}}_n(t) = \vec{f}_n[\underline{\vec{r}}(t), \underline{\dot{\vec{r}}}(t); t], \tag{1b}$$

$$\underline{\ddot{\vec{r}}}(t) = \underline{\vec{f}}[\underline{\vec{r}}(t), \underline{\dot{\vec{r}}}(t); t]. \tag{1c}$$

When the Newtonian equations (1) are *invariant under translations*, namely under the transformations

$$\vec{r}_n \rightarrow \tilde{\vec{r}}_n = \vec{r}_n + \vec{r}_0, \tag{2}$$

where $\vec{r}_0$ is arbitrary (but constant, $\dot{\vec{r}}_0 = 0$), they can conveniently be reformulated as follows:

$$\ddot{\vec{r}}_n = \vec{f}_n(\vec{r}_n - \vec{r}_m, \dot{\vec{r}}_m; t). \tag{3}$$

So far we have not specified the dimensionality d of space. Let us now do so by considering separately the 3 cases (S=1,2,3) on which we will hereafter focus, to introduce an appropriate notation as well as the important notion of rotation invariance (when applicable, namely for S=2,3).

In the *one-dimensional* case (S=1) we generally write, say, x_n in place of $\vec{r}_n$, so that Newton's equations read, say,

$$\ddot{x}_n = f_n(x_m, \dot{x}_m; t). \tag{4}$$

(In the Hamiltonian cases, on which we will largely focus, we often use the "canonical coordinates" $q_n(t)$ in place of $x_n(t)$). In this one-dimensional case we generally consider motions taking place on an (infinite, straight) line; although the case of motions on a circle will also be occasionally considered, see below.

In the *two-dimensional* case (S=2) the motion takes place *in a plane,* which we, for notational convenience, envisage to be embedded in ordinary (three-dimensional) space. Hence we introduce Cartesian and polar coordinates by setting

$$\vec{r}_n \equiv (x_n, y_n, 0) \equiv (\rho_n \cos\theta_n, \rho_n \sin\theta_n, 0). \tag{5}$$

We moreover introduce the unit vector $\hat{k}$ orthogonal to the plane,

$$\hat{k} \equiv (0,0,1), \tag{6}$$

and use the notation

$$\hat{k} \wedge \vec{r}_n \equiv (-y_n, x_n, 0) \equiv \left(\rho_n \cos(\theta_n + \frac{\pi}{2}), \rho_n \sin(\theta_n + \frac{\pi}{2}), 0\right) = (-\rho_n \sin\theta_n, \rho_n \cos\theta_n, 0). \tag{7}$$

The (symmetrical) *scalar product* is then defined in the standard manner:

$$\vec{r}_n \cdot \vec{r}_m \equiv x_n x_m + y_n y_m \equiv \rho_n \rho_m \cos(\theta_n - \theta_m) = \vec{r}_m \cdot \vec{r}_n, \tag{8}$$

while the (antisymmetrical) *pseudoscalar product* is defined as follows:

$$\hat{k} \cdot \vec{r}_n \wedge \vec{r}_m \equiv x_n y_m - x_m y_n \equiv \rho_n \rho_m \sin(\theta_m - \theta_n) = -\hat{k} \cdot \vec{r}_m \wedge \vec{r}_n \; . \tag{9}$$

Note that this latter quantity coincides, up to its sign, with twice the area of the plane triangle having as its 3 vertices the origin and the 2 vectors $\vec{r}_n$ and $\vec{r}_m$.

A *rotation in the plane* is defined as follows:

$$\vec{r}_n \to \tilde{\vec{r}}_n \equiv (\rho_n \cos(\theta_n + \theta_0), \rho_n \sin(\theta_n + \theta_0), 0), \tag{10}$$

where θ_0 is the *angle of rotation*. Clearly both the scalar and the pseudoscalar products are invariant under rotations, see (8) and (9); indeed, they are the only quantities having this property. Moreover, the scalar product (8) is invariant under the *inversion* transformations

$$\vec{r}_n \to \tilde{\vec{r}}_n \equiv (-x_n, y_n), \tag{11a}$$

$$\vec{r}_n \to \tilde{\vec{r}}_n \equiv (x_n, -y_n), \tag{11b}$$

while the pseudoscalar product (9) changes sign under such transformations. Since generally, in the following, we shall not consider the behavior under inversions, we will for simplicity often denote as *scalars* all quantities that are invariant under rotations, independently of their behavior under inversions (namely, independently of whether they are scalars, pseudoscalars or a combination of scalars and pseudoscalars). The requirement that the Newtonian equations of motion (1) be, in the *two-dimensional* case, *invariant* under (plane) rotations (by an arbitrary, but time-independent, angle θ_0) entails that they have the "covariant" form

$$\ddot{\vec{r}}_n = \sum_{m=1}^{N} \left\{ \vec{r}_m \varphi_{nm}^{(1)} + \dot{\vec{r}}_m \varphi_{nm}^{(2)} + \hat{k} \wedge \vec{r}_m \varphi_{nm}^{(3)} + \hat{k} \wedge \dot{\vec{r}}_m \varphi_{nm}^{(4)} \right\}, \tag{12}$$

where the $4N^2$ rotation-invariant (scalar) quantities $\varphi_{nm}^{(s)}$ are functions of the $\frac{1}{2}N(N+1)$ scalar products $\vec{r}_j \cdot \vec{r}_k$, of the N^2 scalar product $\vec{r}_j \cdot \dot{\vec{r}}_k$, of the $\frac{1}{2}N(N+1)$ scalar products $\dot{\vec{r}}_j \cdot \dot{\vec{r}}_k$, of the $\frac{1}{2}N(N-1)$ pseudoscalar products $\hat{k} \cdot \vec{r}_j \wedge \vec{r}_k$, of the N^2 pseudoscalar products $\hat{k} \cdot \vec{r}_j \wedge \dot{\vec{r}}_k$, and of the $\frac{1}{2}N(N-1)$ pseudoscalar products $\hat{k} \cdot \dot{\vec{r}}_j \wedge \dot{\vec{r}}_k$. The quantities $\varphi_{nm}^{(s)}$ may depend moreover on the time t (in the "nonautonomous" case), and on a number of scalar ("coupling") constants.

In the *three-dimensional* case (S=3) Cartesian and spherical coordinates are introduced in the standard manner:

$$\vec{r}_n \equiv (x_n, y_n, z_n) = (\rho_n \cos\varphi_n \cos\theta_n, \rho_n \cos\varphi_n \sin\theta_n, \rho_n \sin\varphi_n). \tag{13}$$

The (symmetrical) *scalar product* reads

$$\vec{r}_n \cdot \vec{r}_m \equiv x_n x_m + y_n y_m + z_n z_m = \vec{r}_m \cdot \vec{r}_n, \tag{14}$$

the (antisymmetrical) *vector product* reads

$$\vec{r}_n \wedge \vec{r}_m \equiv (y_n z_m - y_m z_n, z_n x_m - z_m x_n, x_n y_m - x_m y_n) = -\vec{r}_m \wedge \vec{r}_n, \tag{15}$$

and the (completely antisymmetrical) triple *pseudoscalar product* reads

$$\vec{r}_1 \cdot \vec{r}_2 \wedge \vec{r}_3 \equiv \begin{vmatrix} x_1 & y_1 & z_1 \\ x_2 & y_2 & z_2 \\ x_3 & y_3 & z_3 \end{vmatrix} = \vec{r}_2 \cdot \vec{r}_3 \wedge \vec{r}_1 = \vec{r}_3 \cdot \vec{r}_1 \wedge \vec{r}_2$$

$$= -\vec{r}_1 \cdot \vec{r}_3 \wedge \vec{r}_2 = -\vec{r}_2 \cdot \vec{r}_1 \wedge \vec{r}_3 = -\vec{r}_3 \cdot \vec{r}_2 \wedge \vec{r}_1. \tag{16}$$

This quantity, up to a sign, is 6 times the volume of the tetrahedron characterized by the 4 vertices $(\vec{0}, \vec{r}_1, \vec{r}_2, \vec{r}_3)$.

The scalar product (14) is invariant under the *space inversion* transformation which changes the sign of all vectors,

$$\vec{r}_n \rightarrow \tilde{\vec{r}}_n = -\vec{r}_n , \tag{17}$$

while the pseudoscalar product (16) changes sign under the space inversion (17). (Note that in the 3-dimensional case, in contrast to the 2-dimensional case, the transformation (17) -- which is of course supposed to hold for *all* vectors $\vec{r}_n$ -- is *not* a rotation). In the following, for simplicity of language, we occasionally neglect the difference between scalars and pseudoscalars, namely we term *scalar* any quantity that is *invariant under rotations*, irrespective of its behavior under space inversion. Note that the only scalar quantities that can be manufactured using 3-vectors are the (quadratic) scalar product (14) and the (cubic) pseudoscalar triple product (16).

The requirement that the Newtonian equations of motion (1) be, in the *three-dimensional* case, *invariant under rotations* (by an arbitrary, but time-independent, angle around an arbitrary, but time-independent, direction) entails that they have the "covariant" form

$$\ddot{\vec{r}}_n = \sum_{m=1}^{N} \left\{ \vec{r}_m \, \varphi_{nm}^{(1)} + \dot{\vec{r}}_m \, \varphi_{nm}^{(2)} \right\}$$

$$+ \sum_{m_1,m_2=1}^{N} \left\{ \vec{r}_{m_1} \wedge \vec{r}_{m_2} \varphi_{nm_1m_2}^{(3)} + \vec{r}_{m_1} \wedge \dot{\vec{r}}_{m_2} \varphi_{nm_1m_2}^{(4)} + \dot{\vec{r}}_{m_1} \wedge \dot{\vec{r}}_{m_2} \varphi_{nm_1m_2}^{(5)} \right\} , \tag{18}$$

where the $2N^2$ scalar (i.e., rotation invariant) quantities $\varphi_{nm}^{(s)}, s = 1,2$, as well as the $N^2(2N-1)$ quantities $\varphi_{nm_1m_2}^{(s)}, s = 3,4,5$, are functions of the $\frac{1}{2}N(N+1)$ scalars $\vec{r}_j \cdot \vec{r}_k$, of the N^2 scalars $\vec{r}_j \cdot \dot{\vec{r}}_k$, of the $\frac{1}{2}N(N+1)$ scalars $\dot{\vec{r}}_j \cdot \dot{\vec{r}}_k$, of the $\frac{1}{6}N(N-1)(N-2)$ pseudoscalars $\vec{r}_j \cdot \vec{r}_k \wedge \vec{r}_\ell$, of the $\frac{1}{2}N^2(N-1)$ pseudoscalars $\vec{r}_j \cdot \vec{r}_k \wedge \dot{\vec{r}}_\ell$, of the $\frac{1}{2}N^2(N-1)$ pseudoscalars $\vec{r}_j \cdot \dot{\vec{r}}_k \wedge \dot{\vec{r}}_\ell$, and of the $\frac{1}{6}N(N-1)(N-2)$ pseudoscalars $\dot{\vec{r}}_j \cdot \dot{\vec{r}}_k \wedge \dot{\vec{r}}_\ell$; the quantities φ may moreover depend on the time t (in the "nonautonomous" case), and on a number of scalar ("coupling") constants.

For instance, the classical problem of N gravitating (point, or spherical) bodies moving in 3-dimensional space corresponds to (18) with

$$\varphi_{nm}^{(2)} = \varphi_{nm_1m_2}^{(3)} = \varphi_{nm_1m_2}^{(4)} = \varphi_{nm_1m_2}^{(5)} = 0 \tag{19a}$$

and

$$\varphi_{nm}^{(1)} = -\delta_{nm} G \sum_{\ell=1,\ell\neq n}^{N} M_l r_{nl}^{-3} + (1-\delta_{nm}) G M_m r_{nm}^{-3} , \tag{19b}$$

where M_n is the mass of the n-th particle and G is the gravitational constant. Here, and often as well below, we use the short-hand notation

$$\vec{r}_{nm} \equiv \vec{r}_n - \vec{r}_m , \tag{20a}$$

entailing of course

$$r_{nm}^2 \equiv \vec{r}_{nm} \cdot \vec{r}_{nm} = r_n^2 + r_m^2 - 2\vec{r}_n \cdot \vec{r}_m . \tag{20b}$$

1.2 Hamiltonian systems - Integrable systems

In Sect. 1.2 we tersely review the basic notions of Hamiltonian dynamics. We restrict attention to the one-dimensional case, leaving as an elementary *exercise* for the diligent reader the reformulation of the following results in covariant form, in the two- and three-dimensional cases.

A Hamiltonian system is characterized by a Hamiltonian function $H(\underline{q},\underline{p})$, whose dependence on the N "canonical" coordinates q_n and momenta p_n determines the time-evolution of these quantities according to the Hamiltonian equations of motion

$$\dot{q}_n = \partial H(\underline{q},\underline{p}) / \partial p_n , \tag{1a}$$

$$\dot{p}_n = -\partial H(\underline{q},\underline{p}) / \partial q_n . \tag{1b}$$

Let us repeat that, here and throughout, an underlined lower-case letter denotes an N-vector: thus $\underline{q} = \underline{q}(t)$ is the N-vector of components $q_n = q_n(t)$, and so on. Note that, for simplicity, we assume that the Hamiltonian $H(\underline{q},\underline{p})$ does not depend explicitly on the time t. We also restrict attention only to this *standard* Hamiltonian formulation.

The corresponding Lagrangian reads

$$L(\underline{q},\underline{\dot{q}}) = -H(\underline{q},\underline{p}) + \sum_{m=1}^{N} \dot{q}_m p_m , \tag{2}$$

where the quantities p_m in the right hand side must be expressed in terms of $\underline{q}$ and $\underline{\dot{q}}$ by solving (for p_m) the (nondifferential) equations (1a) (assuming this can be done, namely that the Jacobian $\left|\partial^2 H / \partial p_m \partial p_n\right|$ does not vanish). It is then easily seen that the Hamiltonian equations (1) entail the Lagrangian evolution equation

$$d\left[\partial L(\underline{q},\underline{\dot{q}})/\partial \dot{q}_n\right]/dt = \partial L(\underline{q},\underline{\dot{q}})/\partial q_n \,. \tag{3}$$

The antisymmetrical *Poisson bracket* of 2 quantities $F(\underline{q},\underline{p})$ and $G(\underline{q},\underline{p})$ that depend on the canonical coordinates and momenta is defined as follows:

$$[F,G] \equiv \sum_{m=1}^{N} \left\{ \frac{\partial F}{\partial q_m}\frac{\partial G}{\partial p_m} - \frac{\partial F}{\partial p_m}\frac{\partial G}{\partial q_m} \right\} = -[G,F]. \tag{4}$$

This definition, together with the Hamiltonian equations of motion (1), entail that the time evolution of any quantity $F(\underline{q},\underline{p})$, that depends on the time t only via the canonical coordinates and momenta, evolves according to the equation

$$\dot{F} = [F,H]. \tag{5}$$

Proof:

$$\dot{F} = \sum_{m=1}^{N} \left\{ \frac{\partial F}{\partial q_m}\dot{q}_m + \frac{\partial F}{\partial p_m}\dot{p}_m \right\} = \sum_{m=1}^{N} \left\{ \frac{\partial F}{\partial q_m}\frac{\partial H}{\partial p_m} - \frac{\partial F}{\partial p_m}\frac{\partial H}{\partial q_m} \right\} = [F,H]. \tag{6}$$

Hence, any quantity $C(\underline{q},\underline{p})$ that "Poisson-commutes" with the Hamiltonian,

$$[C,H] = 0, \tag{7}$$

is a *constant of the motion,*

$$\dot{C} = 0 \tag{8}$$

(and, of course, viceversa: (7) implies (8), and (8) implies (7), see (5)). In particular, the Hamiltonian is itself a constant of the motion ("conservation of energy"):

$$\dot{H} = 0 . \qquad (9)$$

A *transformation* (namely, a change of variables), from the "old" canonical variables q_m, p_m to "new" canonical variables $\tilde{q}_n(\underline{q},\underline{p})$, $\tilde{p}_n(\underline{q},\underline{p})$, is *canonical* if the Poisson brackets of the new variables with respect to the old ones satisfy the rules

$$[\tilde{q}_n, \tilde{p}_m] = \delta_{nm}, \; [\tilde{q}_n, \tilde{q}_m] = [\tilde{p}_n, \tilde{p}_m] = 0 . \qquad (10)$$

Remark 1.2-1. The *identical* transformation, $\tilde{q}_n = q_n$, $\tilde{p}_n = p_n$, is *canonical.* It is a special case of the *canonical ("point") transformations*

$$\tilde{q}_n = \tilde{q}_n(q_n), \quad \tilde{p}_n = p_n / [\partial \tilde{q}_n(q_n) / \partial q_n] , \qquad (11)$$

or

$$\tilde{q}_n = q_n / [\partial \tilde{p}_n(p_n) / \partial p_n], \quad \tilde{p}_n = \tilde{p}_n(p_n) . \qquad (12)$$

Exercise 1.2-2. Verify that the transformations (11) and (12) are canonical.

The main property of canonical transformations is to leave invariant Hamilton's equations. Namely if a new Hamiltonian $\tilde{H}(\underline{\tilde{q}},\underline{\tilde{p}})$ is defined by setting

$$H(\underline{q},\underline{p}) = \tilde{H}(\underline{\tilde{q}},\underline{\tilde{p}}), \qquad (13)$$

and the transformation from the old variables $\underline{q},\underline{p}$ to the new variables $\underline{\tilde{q}},\underline{\tilde{p}}$ is canonical, then to the *Hamiltonian* equations (1) satisfied by $q_n(t)$, $p_n(t)$ there correspond the following standard *Hamiltonian* equations satisfied by the new variables $\tilde{q}_n(t)$, $\tilde{p}_n(t)$:

$$\dot{\tilde{q}}_n = \partial \tilde{H}(\underline{\tilde{q}},\underline{\tilde{p}}) / \partial \tilde{p}_n , \qquad (14a)$$

$$\dot{\tilde{p}}_n = -\partial \tilde{H}(\underline{\tilde{q}},\underline{\tilde{p}}) / \partial \tilde{q}_n . \qquad (14b)$$

Proof of (14a): from (5)

$$\dot{\tilde{q}}_n = [\tilde{q}_n, H] = \sum_{m=1}^{N} \left\{ \frac{\partial \tilde{q}_n}{\partial q_m} \frac{\partial H}{\partial p_m} - \frac{\partial \tilde{q}_n}{\partial p_m} \frac{\partial H}{\partial q_m} \right\}, \tag{15}$$

$$\frac{\partial H}{\partial p_m} = \sum_{\ell=1}^{N} \left\{ \frac{\partial \tilde{H}}{\partial \tilde{q}_\ell} \frac{\partial \tilde{q}_\ell}{\partial p_m} + \frac{\partial \tilde{H}}{\partial \tilde{p}_\ell} \frac{\partial \tilde{p}_\ell}{\partial p_m} \right\}, \tag{16a}$$

$$\frac{\partial H}{\partial q_m} = \sum_{\ell=1}^{N} \left\{ \frac{\partial \tilde{H}}{\partial \tilde{q}_\ell} \frac{\partial \tilde{q}_\ell}{\partial q_m} + \frac{\partial \tilde{H}}{\partial \tilde{p}_\ell} \frac{\partial \tilde{p}_\ell}{\partial q_m} \right\}, \tag{16b}$$

$$\dot{\tilde{q}}_n = \sum_{\ell=1}^{N} \left\{ \frac{\partial \tilde{H}}{\partial \tilde{q}_\ell} [\tilde{q}_n, \tilde{q}_\ell] + \frac{\partial \tilde{H}}{\partial \tilde{p}_\ell} [\tilde{q}_n, \tilde{p}_\ell] \right\} = \frac{\partial \tilde{H}}{\partial \tilde{p}_n}. \tag{17}$$

The proof of (14b) is analogous.

Action-angle variables. If the Hamiltonian $\tilde{H}$ is independent of one of the canonical coordinate, say $\tilde{q}_n$,

$$\partial \tilde{H} / \partial \tilde{q}_n = 0, \tag{18}$$

then the corresponding canonical momentum $\tilde{p}_n$ is a constant of motion (see (14b)):

$$\dot{\tilde{p}}_n = 0. \tag{19}$$

If the Hamiltonian $\tilde{H} = \tilde{H}(\underline{\tilde{p}})$ is independent of *all* canonical coordinates, then all canonical momenta are constants of the motion hence the quantities

$$v_n = \partial \tilde{H}(\underline{\tilde{p}}) / \partial \tilde{p}_n \tag{20}$$

are obviously constant as well, hence the canonical coordinate $\tilde{q}_n(t)$ evolve linearly (see (14a)):

$$\tilde{q}_n(t) = \tilde{q}_n(0) + v_n t. \tag{21}$$

Such variables $\tilde{p}_n$, $\tilde{q}_n$ are called "action-angle variables" (indeed, if the motion is confined, the "angle variables" $\tilde{q}_n(t)$ only vary on a finite range, which by appropriate rescaling can be reduced to the interval, say, $(0,2\pi)$, thereby fully justifying their characterization as "angles"). Clearly in terms of these variables the time evolution is trivially simple. Hence the identification of a canonical transformation from the original "physical" variables $\underline{q}(t)$, $\underline{p}(t)$ to *action-angle* variables $\underline{\tilde{q}}(t)$, $\underline{\tilde{p}}(t)$ provides a route to solve the Hamiltonian equations of motion (1). Whenever such a route to evince the time evolution of the Hamiltonian system is available, *with the corresponding canonical transformation being global and univalent*, the time evolution of the system cannot be too complicated ("chaotic"). Such Hamiltonian systems, whose time evolution does not exhibit chaotic features, are called *integrable.* They are *exceptional* (namely the chaotic behavior is in some sense *generic* for Hamiltonian systems with confined motions), yet they are of great importance from a mathematical/ theoretical, and also from an applicative, point of view.

Integrable systems. Assume that a Hamiltonian system possesses N constants of the motion, $C_n(\underline{q},\underline{p})$, globally defined by N univalent independent functions:

$$[C_n,H]=0, \qquad \dot{C}_n=0. \tag{22}$$

Assume moreover that these constants are "in involution", namely that they Poisson-commute,

$$[C_n,C_m]=0. \tag{23}$$

Then the Hamiltonian system is called "Liouville integrable".

Indeed, under such conditions, it is generally possible to identify the N quantities $C_n(\underline{q},\underline{p})$ as new canonical momenta $\tilde{p}_n$ (note that the validity of (23) is essential for this to be possible, see (10)), and then to identify corresponding canonical coordinates $\tilde{q}_n(\underline{q},\underline{p})$. Since the new canonical momenta are constants of the motion, one has thereby succeeded to reformulate the Hamiltonian problem in terms of action-angle variables.

Remark 1.2-3. It is of course essential that the constants of motion in involution $C_n(\underline{q},\underline{p})$ depend nontrivially on the canonical variables q_m, p_m (for instance, they cannot be numerical constants independent of these variables!) and moreover that different C_n's be functionally independent of each other. The fact that their number N coincide with the number of degrees of freedom of the Hamiltonian system is also essential.

While it can be proven that, if these conditions prevail, together with (22) and (23), then the assignment position $\tilde{p}_n=C_n(\underline{q},\underline{p})$ to identify new (constant) canonical momenta can indeed be supplemented by the introduction of appropriate canonical coordinates $\tilde{q}_n(\underline{q},\underline{p})$, the *explicit* implementation of this program cannot be generally carried out (the proof that quantities $\tilde{q}_n(\underline{q},\underline{p})$ do exist is not constructive).

Hence, after the integrability of a Hamiltonian system has been demonstrated by exhibiting N constants of the motion $C_n(\underline{q},\underline{p})$ with all the required properties, the job to obtain an explicit expression of the action-angle variables - or, equivalently, to actually get the solution of the Hamiltonian equations of motion - remains as a nontrivial task. We generally call *solvable* the systems for which this additional step can be performed in explicit form, or at least reduced to purely algebraic operations (see the Foreword).

A Hamiltonian system describing a (one-dimensional) many-body problem is called *normal* if the Hamiltonian function $H(\underline{q},\underline{p})$ is separated into kinetic and potential energy parts as follows:

$$H(\underline{q},\underline{p})=T(\underline{p})+V(\underline{q}), \tag{24a}$$

$$T(p)=\frac{1}{2}\sum_{m=1}^{N} p_m^2/\mu_m . \tag{24b}$$

Note the special form of the kinetic energy, $T(\underline{p})$, as well as the independence of the potential energy $V(\underline{q})$ from the canonical momenta.

The corresponding Hamiltonian equations read

$$\dot{q}_n = p_n/\mu_n , \tag{25a}$$

$$\dot{p}_n = -\partial V(q_m)/\partial q_n , \tag{25b}$$

entailing Newton's equations of motion

$$\mu_n \ddot{q}_n = f_n(q_m), \tag{26a}$$

$$f_n(q_m) = -\partial V(\underline{q})/\partial q_n . \tag{26b}$$

These Newtonian equations of motion ("mass times acceleration equal force") can also be obtained directly, see (3), from the Lagrangian (see (2), (24) and (25a))

$$L(\underline{q},\dot{\underline{q}})=\frac{1}{2}\sum_{m=1}^{N}\mu_m\dot{q}_m^2-V(\underline{q}). \tag{27}$$

Of special interest is the case with only one- and two-body interactions:

$$V(\underline{q})=\sum_{m=1}^{N} V_m^{(1)}(q_m)+\frac{1}{2}\sum_{n,m=1;m\neq n}^{N} V_{nm}^{(2)}(q_n,q_m) \tag{28}$$

entailing

$$\mu_n\ddot{q}_n=-\frac{\partial}{\partial q_n}\left[V_n^{(1)}(q_n)+\frac{1}{2}\sum_{m=1,m\neq n}^{N}\left\{V_{nm}^{(2)}(q_n,q_m)+V_{mn}^{(2)}(q_m,q_n)\right\}\right]. \tag{29}$$

The requirement that the Newtonian equations of motion be invariant under translations $(q_n(t)\to\tilde{q}_n(t)=q_n(t)+q_0,\dot{q}_0=0)$ implies the following restrictions on the one-body and two-body potentials:

$$V_n^{(1)}(q_n)=-a_nq_n+b_n, \tag{30a}$$

$$V_{nm}^{(2)}(q_n,q_m)=V_{nm}^{(2)}(q_n-q_m), \tag{30b}$$

entailing

$$\mu_n\ddot{q}=a_n-\frac{1}{2}\sum_{m=1}^{N}\left\{V_{nm}^{(2)\prime}(q_n-q_m)-V_{mn}^{(2)\prime}(q_m-q_n)\right\}. \tag{31}$$

Here and below of course $V_{nm}^{(2)\prime}(q)\equiv\partial V_{nm}^{(2)}(q)/\partial q$.

Remark 1.2-4. The requirement that not only the Newtonian equations of motion (31), but the Hamiltonian itself, see (24) with (28), be invariant under translations entails the additional condition $a_n=0$, see (30a).

Remark 1.2-5. The equations of motion (31) are as well invariant under the *Galileian transformation*

$$q_n(t)\to\tilde{q}_n(t)=q_n(t)+v_0\,t\ ,\quad \dot{v}_0=0. \tag{32}$$

The two-body potential is generally *even*, in the following sense:

$$V_{nm}^{(2)}(q)=V_{mn}^{(2)}(-q)\ , \tag{33a}$$

entailing

$$\partial V_{nm}^{(2)}(q)/\partial q=-\partial V_{mn}^{(2)}(x)/\partial x\,\big|_{x=-q}\ . \tag{33b}$$

This corresponds to the *action-reaction* principle: "the force exerted by the n-th particle on the m-th particle is equal (in modulus) and opposite (in direction) to that exerted by the m-th particle on the n-th particle" (see (26) and (33b)). In this case the Newtonian equations of motion (31) become

$$\mu_n \ddot{q}_n = a_n - \sum_{m=1, m\neq n}^{N} V_{nm}^{(2)\prime}(q_n - q_m), \tag{34}$$

and the center-of-mass coordinate $\bar{q}(t)$,

$$\bar{q}(t) = M^{-1} \sum_{n=1}^{N} \mu_n q_n(t), \tag{35a}$$

$$M = \sum_{n=1}^{N} \mu_n, \tag{35b}$$

satisfies the simple evolution equation (see (34) and(33b))

$$M\ddot{\bar{q}} = A, \qquad A = \sum_{n=1}^{N} a_n, \tag{36}$$

entailing

$$\bar{q}(t) = \bar{q}(0) + \dot{\bar{q}}(0)\, t + \frac{1}{2}(A/M)\, t^2. \tag{37}$$

When the Hamiltonian is itself invariant under translations (namely, when $a_n = 0$; see (24) and (30)) the center-of-mass moves freely, $\ddot{\bar{q}} = 0$ (see (36)), and the total momentum P,

$$P = \sum_{n=1}^{N} p_n, \tag{38}$$

is a constant of motion,

$$\dot{P} = 0. \tag{39}$$

Proof. Use (25a), $p_n = \mu_n \dot{q}_n$, and sum (34) over n from 1 to N: the right hand side then vanishes due to the antisymmetry of the summand, see (33b).

Exercise 1.2-6. Verify that the (*not* normal!) Hamiltonian

$$H(\underline{q},\underline{p})=\sum_{n=1}^{N}\varphi_n(p_n)\exp\left[-\frac{1}{2}\sum_{m=1,m\neq n}^{N}W_{nm}(q_n-q_m)\right] \tag{40a}$$

with

$$W_{nm}(-q)=W_{mn}(q)\ , \tag{40b}$$

yields the Newtonian equations of motion

$$\ddot{q}_n=\frac{1}{2}\dot{q}_n\sum_{m=1\ m\neq n}^{N}\{\ w_{nm}(q_n-q_m)[\ \dot{q}_m\{1+\varphi_n''(p_n)\varphi_m(p_m)/[\varphi_n'(p_n)\varphi_m'(p_m)]\}$$

$$-\dot{q}_n\{1-\varphi_n''(p_n)\varphi_n(p_n)/[\varphi_n'(p_n)]^2\}\ ]\ \} \tag{41a}$$

with

$$w_{nm}(q)=\partial W_{nm}(q)/\partial q\ , \tag{41b}$$

entailing of course (see (40b))

$$w_{nm}(-q)=-w_{mn}(q)\ . \tag{41c}$$

Exercise 1.2-7. Verify that the Hamiltonian

$$H(\underline{q},\underline{p})=\sum_{n=1}^{N}\exp(s_n\,p_n)\exp\left[-\frac{1}{2}\sum_{m=1,m\neq n}^{N}W_{nm}(q_n-q_m)\right]\ , \tag{42}$$

with (40b) and where the N constants s_n are *arbitrary*, yields the Newtonian equations of motion

$$\ddot{q}_n=\frac{1}{2}\sum_{m=1,m\neq n}^{N}(1+s_n/s_m)\dot{q}_n\,\dot{q}_m\,w_{nm}(q_n-q_m) \tag{43}$$

with (41b). *Hint*: use (41a).

Exercise 1.2-8. Verify that the Hamiltonian

$$H(\underline{q},\underline{p})=\sum_{n=1}^{N}\cosh(s_n\,p_n)\exp\left[-\frac{1}{2}\sum_{m=1,m\neq n}^{N}W_{nm}(q_n-q_m)\right]\ , \tag{44}$$

with (40b) and where the N constants s_n are *arbitrary*, yields the Newtonian equations of motion

$$\ddot{q}_n = \frac{1}{2} \sum_{m=1\, m\neq n}^{N} \{ w_{nm}(q_n - q_m) \{ \dot{q}_n \dot{q}_m + s_n^2 u_n^2$$

$$+ s_n^2 \left[u_n^2 + (\dot{q}_n / s_n)^2\right]^{1/2} \left[u_m^2 + (\dot{q}_m / s_m)^2\right]^{1/2} \} \} \tag{45a}$$

with (41b) and

$$u_n \equiv u_n(\underline{q}) \equiv \exp\left[-\frac{1}{2} \sum_{m=1, m\neq n}^{N} W_{nm}(q_n - q_m) \right] . \tag{45b}$$

Hint: use (41a).

Exercise 1.2-9. Verify that the (one-body) Hamiltonian

$$\tilde{H}(q,p) = 2^{1/2} (i/s) \sinh(sp) [V(q)]^{1/2} \tag{46}$$

with s an arbitrary constant, yields the *same* Newtonian equation of motion,

$$\ddot{q} = -V'(q) , \tag{47}$$

as the (normal) Hamiltonian

$$H(q,p) = p^2/2 + V(q) . \tag{48}$$

In conclusion, we have seen that the essential condition in order that a Hamiltonian system be *integrable* is that it possess N constants of motion. Hence all Hamiltonian systems (of the type considered herein) are integrable if $N = 1$ (because the Hamiltonian H is itself a constant of the motion), and those for which the total momentum P is a constant of motion, see (39), are also all integrable for $N = 2$.

Finally let us note that, in the special case of equal particles, all masses are equal, $\mu_n = \mu$ ($=1$ for notational simplicity), and the potentials do not depend on the particle indices, for instance in (28) $V_m^{(1)}(q_m) = V^{(1)}(q_m)$ and $V_{nm}^{(2)}(q_n, q_m) = V^{(2)}(q_n, q_m)$. The diligent reader will note the simplified form that the various equations written above take in this special case, on which our attention will be mainly focussed.

1.N. Notes to Chapter 1

The material surveyed in Chapter 1. is standard and can be retrieved from any textbook, see for instance <A74>, <AM78>, <G83>. The diligent reader will thereby realize that the presentation given in this Chapter 1 is sketchy and will be able to complement it with the many details, generalizations and proofs that have been omitted herein.

2 ONE-DIMENSIONAL SYSTEMS - MOTIONS ON THE LINE AND ON THE CIRCLE

In Chap. 2 we discuss integrable many-body problems in one-dimensional space (on a straight line or on a circle). We begin by introducing the idea of a Lax pair, and we show how it can be used to identify integrable Hamiltonian systems and in some cases to solve them. Several instances of such integrable many-body problems are studied; their treatment involves the discussion of various functional equations. We then introduce another technique to identify solvable many-body problems, we demonstrate its effectiveness by exhibiting several examples, and we outline the connections of this technique with the classical problem of Lagrangian interpolation for functions of one variable.

2.1 The Lax pair technique

Let $\underline{L}(\underline{q},\underline{p})$ and $\underline{M}(\underline{q},\underline{p})$ be two $(N\times N)$-matrices, which depend in some conveniently assigned manner (see below) on the N canonical coordinates q_m and on the N canonical momenta p_m.

Notation. We hereafter denote $(N\times N)$-matrices by upper case underlined Latin letters, N-vectors by lower case underlined Latin letters, and use the standard notation and rules for matrix-vector algebra. Hence, for instance,

$$(\underline{A}\underline{v})_j=\sum_{k=1}^{N}A_{jk}v_k\,, \tag{1a}$$

$$(\underline{u}\underline{A})_j=\sum_{k=1}^{N}u_kA_{kj}\,, \tag{1b}$$

$$(\underline{A}\underline{B})_{jk}=\sum_{\ell=1}^{N}A_{j\ell}B_{\ell k}\,, \tag{1c}$$

F. Calogero: LNPm 66, pp. 17 - 309, 2001

$$(\underline{u},\underline{v}) = \sum_{j=1}^{N} u_j v_j . \tag{1d}$$

Such a pair of matrices $\underline{L}(\underline{q},\underline{p})$, $\underline{M}(\underline{q},\underline{p})$ is called a "Lax pair" whenever it can be exploited according to the following ("Lax" [L68]) technique.

Let the two matrices $\underline{L}$ and $\underline{M}$ satisfy the (matrix) "Lax" evolution equation

$$\dot{\underline{L}} = [\underline{L},\underline{M}] \tag{2}$$

Notation. The *commutator* $[\underline{A}, \underline{B}]$ of the two square matrices $\underline{A}$, $\underline{B}$ is defined in the standard manner:

$$[\underline{A},\underline{B}] \equiv \underline{A}\,\underline{B} - \underline{B}\,\underline{A} = -[\underline{B},\underline{A}], \tag{3a}$$

entailing

$$[\underline{A},\underline{B}]_{jk} = \sum_{\ell=1}^{N} \{A_{j\ell}B_{\ell k} - B_{j\ell}A_{\ell k}\}. \tag{3b}$$

Beware: do not confuse the commutator of 2 (square) matrices with the Poisson bracket of 2 functions of the canonical variables (see (1.2-4)).

The time-dependence of the matrices $\underline{L}(\underline{q},\underline{p})$, $\underline{M}(\underline{q},\underline{p})$ (see for instance the left hand side of (2), where the superimposed dot denotes of course the time-derivative) obtains from their dependence on the canonical variables $q_m(t), p_m(t)$. Hence, for any specific *ansatz* for the matrices $\underline{L}(\underline{q},\underline{p})$, $\underline{M}(\underline{q},\underline{p})$, the Lax evolution equation (2) entails evolution equations for the canonical variables $q_n(t), p_n(t)$ -- provided of course such an *ansatz* is compatible with the time evolution. This last condition is far from trivial, since the matrix evolution equation (2) corresponds *a priori* to N^2 scalar evolution equations, for the $2N$ quantities q_m, p_m; hence for $N > 2$ there are more equations than variables ("overdetermined" problem). In the following subsections we show that there exist nevertheless certain *ansaetze* for the Lax pair $\underline{L}(\underline{q},\underline{p})$, $\underline{M}(\underline{q},\underline{p})$ which are compatible with the Lax time-evolution (2). But firstly, in the remaining part of Sect. 2.1, we derive an important consequence of the Lax evolution equation (2), namely the existence of N constants of the motion.

To this end let us assume the matrix $\underline{L}$ to be *diagonalizable*, and denote by $\lambda^{(m)}$ its N eigenvalues and by $\underline{v}^{(m)}$, $\underline{u}^{(m)}$ its right- and left-eigenvectors:

$$\left[\underline{L}-\lambda^{(n)}\right]\underline{v}^{(n)}=0, \tag{4a}$$

$$\underline{u}^{(m)}\left[\underline{L}-\lambda^{(m)}\right]=0. \tag{4b}$$

Let us also assume, for convenience (see below), that these eigenvectors are *orthonormal*,

$$(\underline{u}^{(m)},\underline{v}^{(n)})=\delta_{nm}. \tag{5a}$$

Here and throughout the symbol δ_{nm} denotes the Kronecker delta, $\delta_{nm}=1$ if $n=m$, $\delta_{nm}=0$ if $n\neq m$.

The fact that $(\underline{u}^{(m)},\underline{v}^{(n)})=0$ if $\lambda^{(m)}\neq\lambda^{(n)}$ is a consequence of (4). (*Proof*: multiply (4a) from the left by $\underline{u}^{(m)}$, (4b) from the right by $\underline{v}^{(n)}$, and subtract). Note that, in writing (5a), we are implicitly assuming that the eigenvalues $\lambda^{(n)}$ of $\underline{L}$ are all different, $\lambda^{(n)}\neq\lambda^{(m)}$ if $n\neq m$. This will generally be the case in the following. But of course one can define the eigenvectors $\underline{v}^{(n)}$ and $\underline{u}^{(m)}$ so that (5a) hold even if the matrix $\underline{L}$ is degenerate, namely if some of its eigenvalues coincide.

The fact that $(\underline{u}^{(n)},\underline{v}^{(n)})=1$ corresponds to the (standard, convenient) choice of normalization for the eigenvectors, which are *a priori* defined, by (4), up to an arbitrary (nonvanishing) multiplicative (scalar) constant.

The "orthonormality relation" (5a) entails the following "completeness relation":

$$\sum_{m=1}^{N}u_j^{(m)}v_k^{(m)}=\delta_{jk}. \tag{5b}$$

(*Proof*: call the left-land-side of (5b) γ_{jk}, multiply it by $v_j^{(n)}$, sum over j from 1 to N, and use (5a) to get $\sum_{j=1}^{N}v_j^{(n)}\gamma_{jk}=v_k^{(n)}$, which clearly entails $\gamma_{jk}=\delta_{jk}$).

Note that we are not assuming the matrix $\underline{L}$ to be *symmetrical* (in which case the eigenvectors $\underline{v}^{(n)}$ and $\underline{u}^{(n)}$ would coincide) nor *Hermitian* (in which case $\underline{v}^{(n)}$ and $\underline{u}^{(n)}$ would be complex-conjugate of each other), but merely that it be *diagonalizable.* However in most of the following applications all the eigenvalues $\lambda^{(m)}$ of $\underline{L}$ shall be real -- as it would automatically be the case if $\underline{L}$ were Hermitian.

It is easy to prove that, as a consequence of the Lax equation (2), the eigenvalues $\lambda^{(m)}$ are time-independent,

$$\dot{\lambda}^{(m)} = 0. \tag{6}$$

Proof. From (4) and (5) we infer

$$\lambda^{(n)} = \left(\underline{u}^{(n)}, \underline{L}\,\underline{v}^{(n)}\right). \tag{7}$$

Hence

$$\dot{\lambda}^{(n)} = \left(\underline{u}^{(n)}, \dot{\underline{L}}\,\underline{v}^{(n)}\right) + \left(\dot{\underline{u}}^{(n)}, \underline{L}\,\underline{v}^{(n)}\right) + \left(\underline{u}^{(n)} \underline{L}, \dot{\underline{v}}^{(n)}\right), \tag{8a}$$

$$\dot{\lambda}^{(n)} = \left(\underline{u}^{(n)}, [\underline{L}, \underline{M}]\, v^{(n)}\right) + \lambda^{(n)} \left[\left(\dot{\underline{u}}^{(n)}, \underline{v}^{(n)}\right) + \left(\underline{u}^{(n)}, \dot{\underline{v}}^{(n)}\right) \right], \tag{8b}$$

$$\dot{\lambda}^{(n)} = \left[\lambda^{(n)} - \lambda^{(n)}\right] \left(\underline{u}^{(n)}, \underline{M}\,\underline{v}^{(n)}\right) + \lambda^{(n)} (d/dt) \left(\underline{u}^{(n)}, \underline{v}^{(n)}\right), \tag{8c}$$

$$\dot{\lambda}^{(n)} = 0. \tag{8d}$$

(To go from (8a) to (8b) we used (2) and (4); to go from (8b) to (8c), we used (4); and to go from (8c) to (8d), we used (5a)).

We have therefore seen that, whenever the evolution of the canonical coordinates $q_m(t), p_m(t)$ is of Lax type, see (2), one gets as a bonus N constants of the motion, in the guise of the N eigenvalues $\lambda^{(n)}$ of the Lax matrix $\underline{L}(q_m, p_m)$, which are of course functions of the canonical variables q_m, p_m.

A way is thereby open to invent/discover *integrable* Hamiltonian systems. This can be done (see below) by choosing appropriate *ansaetze* for the Lax pair (as functions of the canonical variables), which are compatible with the Lax time evolution (2), and are moreover identifiable with a *Hamiltonian* evolution of the canonical variables $q_n(t), p_n(t)$, indeed one that is interpretable as describing an N-body problem.

This part of the program requires imagination and luck (the main components of original research). One must then check that the N constants of motion obtained in this manner are nontrivial (namely that the Lax pair is not a "fake" one, see Sect. 2.1.9.1), that they are functionally independent of each other, and that they Poisson-commute (see (1.2-23)).

It is sometimes convenient to focus attention, rather than on the N eigenvalues $\lambda^{(n)}$ of the Lax matrix $\underline{L}(\underline{q},\underline{p})$, on some other equivalent set of N constants of the motion. Two such sets which are often used are the traces T_n of the (first) N powers of the Lax matrix $\underline{L}(\underline{q},\underline{p})$,

$$T_n \equiv \text{trace}\left[\underline{L}^n\right] = \sum_{m=1}^{N} \left[\lambda^{(m)}\right]^n, \qquad n=1,\dots,N, \tag{9}$$

or the symmetric invariants J_n of the matrix $\underline{L}$,

$$\det\left[\lambda \underline{1} - \underline{L}\right] \equiv \lambda^N + \sum_{n=1}^{N} \lambda^{N-n} J_n. \tag{10}$$

The trace of a matrix is of course defined as the sum of its diagonal elements, and the expression (10) of course vanishes whenever λ coincides with one of the N zeros of the polynomial of degree N appearing in the right hand side of (10), namely with an eigenvalue $\lambda^{(m)}$ of $\underline{L}$.

Exercise 1.2-1: prove (9). *Hint:* use (4), (5b) and (5a).

It is indeed well known that the sets, $\{\lambda^{(n)}; n=1,\dots,N\}$, $\{T_n; n=1,\dots,N\}$ and $\{J_n; n=1,\dots,N\}$ are in bi-univocal correspondence (namely, each one of these sets determines uniquely the other two), and of course the properties to be constant (time-independent), to depend nontrivially on the canonical variables q_m, p_m, to be functionally independent among themselves, and to Poisson-commute, if valid for (all) the elements of any one of these three sets, are also automatically valid for the elements of the other two sets.

The transitivity of these properties is too obvious to require a proof, except perhaps for the last one. Hence let us prove that, if $F(\lambda^{(m)})$, $G(\lambda^{(m)})$ are functions of N quantities $\lambda^{(n)}(q_m, p_m)$ which Poisson-commute among themselves,

$$[\ \lambda^{(n)}, \lambda^{(m)}\ ] = 0, \tag{11}$$

then F and G also Poisson-commute among themselves,

$$[F, G] = 0. \tag{12}$$

Indeed (see (1.2-4))

$$[F,G]=\sum_{m=1}^{N}\left\{\frac{\partial F}{\partial q_m}\frac{\partial G}{\partial p_m}-\frac{\partial F}{\partial p_m}\frac{\partial G}{\partial q_m}\right\}$$

$$=\sum_{m,m_1,m_2=1}^{N}\frac{\partial F}{\partial \lambda^{(m_1)}}\frac{\partial G}{\partial \lambda^{(m_2)}}\left\{\frac{\partial \lambda^{(m_1)}}{\partial q_m}\frac{\partial \lambda^{(m_2)}}{\partial p_m}-\frac{\partial \lambda^{(m_1)}}{\partial p_m}\frac{\partial \lambda^{(m_2)}}{\partial q_m}\right\}$$

$$=\sum_{m_1,m_2=1}^{N}\frac{\partial F}{\partial \lambda^{(m_1)}}\frac{\partial G}{\partial \lambda^{(m_2)}}\left[\lambda^{(m_1)},\lambda^{(m_2)}\right]. \tag{13}$$

Hence (11) entails (12).

Exercise 2.1-2. Prove directly from the definition

$$T_n \equiv \mathrm{trace}[\ \underline{L}^n\ ] \tag{14}$$

(namely, without using the equality in (9)) that, if $\underline{L}$ satisfies the Lax equation (2),

$$\dot{T}_n = 0\ . \tag{15}$$

Hint: use the *identity*

$$\mathrm{trace}\left[\underline{A}\,\underline{B}\right]=\mathrm{trace}\left[\underline{B}\,\underline{A}\right]. \tag{16}$$

Let us end Sect. 2.1 by emphasizing that *different* Lax pairs may correspond to the *same* equations of motion: we will see instances of this phenomenon in the following. It is moreover clear that, if $\underline{L}$ and $\underline{M}$ constitute a Lax pair, namely they satisfy the Lax evolution equation (2), then the new pair $\underline{\tilde{L}},\underline{\tilde{M}}$,

$$\underline{\tilde{L}}=\underline{U}\,\underline{L}\,\underline{U}^{-1}+c\underline{1}\ , \tag{17}$$

$$\underline{\tilde{M}}=\underline{U}\left(\underline{M}+f(t)\underline{L}\right)\underline{U}^{-1}-\underline{\dot{U}}\,\underline{U}^{-1}+g(t)\underline{1}\ , \tag{18}$$

also satisfies the Lax evolution equation. Here $\underline{U}$ is an arbitrary (invertible) matrix, c is an arbitrary constant, and $f(t),g(t)$ are two arbitrary functions.

Exercise 2.1-3. Verify!

2.1.1 A convenient representation. The functional equation (*)

We now make a convenient *ansatz* for a Lax pair $\underline{L}(\underline{q},\underline{p})$, $\underline{M}(\underline{q},\underline{p})$, and we then ascertain under which conditions it is compatible with the Lax (matrix) evolution equation (2.1-2). The main one of these conditions is the functional equation (*) (see (16) below). When these conditions are satisfied the resulting dynamical system can be identified with a Hamiltonian system of normal type (see (1.2-25)), describing a one-dimensional many-body problem whose Newtonian equations of motion feature (velocity-independent) forces whose functional form is determined by the solution of the functional equation (*).

Our starting *ansatz* for the Lax pair:

$$L_{nm} = p_n \text{ if } m = n, \tag{1a}$$

$$L_{nm} = \alpha(q_n - q_m) \text{ if } m \neq n, \tag{1b}$$

$$M_{nm} = \sum_{\ell=1,\ell\neq n}^{N} \beta(q_n - q_\ell) \text{ if } m = n, \tag{2a}$$

$$M_{nm} = \gamma(q_n - q_m) \text{ if } m \neq n, \tag{2b}$$

where $\alpha(q), \beta(q)$ and $\gamma(q)$ are 3 functions to be determined, see below.

Let us now insert this *ansatz* in the Lax evolution equation (2.1-2), namely into

$$\dot{L}_{nm} = \sum_{\ell=1,\ell\neq n}^{N} \{L_{n\ell} M_{\ell n} - L_{\ell n} M_{n\ell}\} \text{ if } m = n, \tag{3a}$$

$$\dot{L}_{nm} = (L_{nn} - L_{mm}) M_{nm} + L_{nm}(M_{mm} - M_{nn}) + \sum_{\ell=1;\ell\neq m,n}^{N} \{L_{n\ell} M_{\ell m} - L_{\ell m} M_{n\ell}\} \text{ if } m \neq n. \tag{3b}$$

The diagonal terms $(m = n)$ then yield

$$\dot{p}_n = \sum_{l=1,l\neq n}^{N} \{\alpha(q_n - q_l)\,\gamma(q_l - q_n) - \alpha(q_l - q_n)\,\gamma(q_n - q_l)\}, \tag{4}$$

while the off-diagonal $(m \neq n)$ terms yield

$$\alpha'(q_n - q_m)\,[\dot{q}_n - \dot{q}_m] = (p_n - p_m)\,\gamma(q_n - q_m)$$

$$+\alpha(q_n - q_m)\left[\sum_{\ell=1,\ell\neq m}^{N}\beta(q_m - q_\ell) - \sum_{\ell=1,\ell\neq n}^{N}\beta(q_n - q_\ell)\right]$$

$$+\sum_{\ell=1;\ell\neq m,n}^{N}\{\alpha(q_n - q_\ell)\gamma(q_\ell - q_m) - \alpha(q_\ell - q_m)\gamma(q_n - q_\ell)\}. \tag{5}$$

This latter equation is clearly satisfied if the following equations hold:

$$p_n = \dot{q}_n, \tag{6}$$

$$\gamma(q) = \alpha'(q), \tag{7}$$

$$\beta(q) = \beta(-q), \tag{8}$$

$$\alpha(q_n - q_m)[\beta(q_m - q_\ell) - \beta(q_n - q_\ell)]$$

$$+\alpha(q_n - q_\ell)\gamma(q_\ell - q_m) - \alpha(q_\ell - q_m)\gamma(q_n - q_\ell) = 0\ . \tag{9}$$

The insertion of (7) into (4) yields

$$\dot{p}_n = -\sum_{m=1,m\neq n}^{N} v'(q_n - q_m) \tag{10}$$

with

$$v(q) = \alpha(q)\alpha(-q). \tag{11}$$

It is now clear that (6) and (10) are precisely the Hamiltonian equations of motion produced by a Hamiltonian of normal type (see (1.2-25)) with pair interactions,

$$H = \frac{1}{2}\sum_{n=1}^{N} p_n^2 + \sum_{n,m=1;m<n}^{N} V^{(2)}(q_n - q_m), \tag{12}$$

with

$$V^{(2)}(q) = v(q) \tag{13}$$

given by (11). Note that this definition, see (11), entails that $V^{(2)}(q)$ is even,

$$V^{(2)}(-q) = V^{(2)}(q), \tag{14}$$

a fact of which we already took advantage in writing the right hand side of (12). Let us also recall that this fact entails the validity of the *action-reaction principle* (see Sect. 1.2).

The corresponding equations of motion take of course the Newtonian form

$$\ddot{q}_n = -\sum_{m=1,m\neq n}^{N} v'(q_n - q_m), \tag{15}$$

characteristic of the one-dimensional N-body problem describing N identical particles (whose mass, without loss of generality, is set to unity) which interact pairwise via the (even) potential energy $v(q)$ entailing the interparticle force $f(q) = -v'(q)$, where q is of course the interparticle distance.

There remains to take account of (8) and (9). The main condition is encoded in the latter equation, which using (8) and via the positions $x = q_n - q_\ell$, $y = q_\ell - q_m$, can be reformulated as the following *functional equation (*)*:

$$(*)\quad [\,\alpha(x)\,\alpha'(y) - \alpha(y)\,\alpha'(x)\,] \,/\, \alpha(x+y) = \beta(x) - \beta(y), \tag{16}$$

(complemented, for our purposes, with the condition that $\beta(x)$ be even, $\beta(-x) = \beta(x)$, see (8)).

Solutions of this functional equation are discussed in the following sections. They correspond to the Hamiltonian many-body problem (12), via (13) with (11). Such a many-body problem is then generally *integrable*, since the N eigenvalues of the Lax matrix, see (1), provide N constants of motion, which generally turn out to be functionally independent and in involution.

Let us recall that equivalent sets of N constants of motion, which are sometimes more convenient to handle, are provided by the N traces T_n, see (2.1-9), or by the N symmetric invariants J_n, see (2.1-10). It is rather obvious that these sets generally satisfy the additional conditions (required to guarantee integrability) to depend non-trivially on the canonical variables q_m, p_m and to be functionally independent: for instance the trace T_n is clearly a polynomial of degree n in the canonical momenta p_m, see (1) and (2.1-9). It is less trivial to show that these constants of the motion Poisson-commute among themselves (they of course all Poisson-commute with the Hamiltonian (12), since they are constants of the motion; see (1.2-5,7,8)). In specific cases, see below, this can be easily proven. General proofs based on the functional equation (*), see (16), are also available. The diligent reader may try to construct such a proof, or look it up in the literature (see Sect. 2.N below for references).

Note that the *ansatz* (1) entails

$$T_1 = \sum_{n=1}^{N} p_n \equiv P , \tag{17}$$

$$T_2 = 2H , \tag{18}$$

where P is clearly the total momentum and H is of course the Hamiltonian (12). Hence the constancy in time of the first 2 traces, T_1 and T_2, corresponds to the conservation of the (total) momentum P and the (total) energy H.

The formula (17) follows immediately from (1) via (2.1-9), and (18) also obtains easily from (1) using (2.1-9), (12), (13) and (11).

It is also easily seen that the Poisson-commutativity of all the traces T_n, see (2.1-9), with the total momentum P, see (17), is an immediate consequence of the translation-invariant character of these quantities (indeed, the Lax matrix itself, see (1), is invariant under the translation $q_n \to q_n + q_0$, with q_0 an arbitrary constant).

2.1.2 A simple solution of the functional equation (*)

The functional equation (*) (see (2.1.1-16)) admits the solution

$$\alpha(x) = b/x , \tag{1}$$

$$\beta(x) = b/x^2 , \tag{2}$$

where b is an arbitrary constant.

Proof.

$$[\alpha(x)\alpha'(y) - \alpha(y)\alpha'(x)] / \alpha(x+y) = -b(xy)^{-2}(x-y)(x+y) =$$

$$= -b(xy)^{-2}(x^2 - y^2) = b(x^{-2} - y^{-2}) = \beta(x) - \beta(y). \tag{3}$$

It is convenient to set

$$b = ig , \tag{4}$$

so that the Hamiltonian (2.1.1-12) become

$$H = \frac{1}{2}\sum_{n=1}^{N} p_n^2 + g^2 \sum_{n,m=1;m<n}^{N} (q_n - q_m)^{-2}, \tag{5}$$

and the corresponding Newtonian equations of motion (2.1.1-15) read

$$\ddot{q}_n = 2g^2 \sum_{m=1,m\neq n}^{N} (q_n - q_m)^{-3}. \tag{6}$$

These equations describe the motion on the line of N equal particles of unit mass interacting pairwise via a two-body repulsive force inversely proportional to the cube of the interparticle distance.

We postpone to the following sections an investigation of this Hamiltonian system, that has played a seminal role in the study of integrable dynamical systems over the last quarter century. Here we report the expressions of the corresponding Lax pair:

$$L_{nm} = \delta_{nm} p_n + (1 - \delta_{nm})\, i g (q_n - q_m)^{-1}, \tag{7}$$

$$M_{nm} = \delta_{nm}\, i g \sum_{\ell=1,\ell\neq n}^{N} (q_n - q_\ell)^{-2} - (1 - \delta_{nm})\, i g (q_n - q_m)^{-2}. \tag{8}$$

2.1.3 N particles on the line, interacting pairwise via repulsive forces inversely proportional to the cube of their mutual distance

In this Section, which is conveniently subdivided into 3 subsections, we investigate the many-body problem described in the title, which, as demonstrated in the preceding Sect. 2.1.2, is an integrable Hamiltonian system (see (2.1.2-5)) whose time evolution (see (2.1.2-6)) coincides, via the *ansatz* (2.1.2-7), with the Lax (matrix) evolution equation (2.1-2).

2.1.3.1 Qualitative behavior

In Sect. 2.1.3.1 we discuss the qualitative behavior of the system on the line characterized by the Hamiltonian (2.1.2-5) and by the Newtonian equations of motion (2.1.2-6).

The force acting among every pair of particles is repulsive, singular at zero separation and vanishing as the separation diverges (see (2.1.2-6)).

Hence this N-body system cannot be bound nor contain any bound subsystem (due to the repulsive nature of the forces): asymptotically, i.e. in the remote past and future, the particles necessarily separate from each other and eventually move freely,

$$q_n(t) = p_n^{(\pm)} t + q_n^{(\pm)} + o(1) \text{ as } t \to \pm\infty . \tag{1}$$

Moreover, the ordering of the particles on the line cannot change throughout the motion, due to the singular (and repulsive) character of the forces at zero separation. It is therefore convenient to label the particles according to their ordering on the line, say from left to right:

$$q_n(t) < q_{n+1}(t), \qquad n = 1,2,\ldots,N-1 . \tag{2}$$

Note that, via (1), this entails

$$p_n^{(-)} > p_{n+1}^{(-)}, \qquad n = 1,2,\ldots,N-1 , \tag{3a}$$

$$p_n^{(+)} < p_{n+1}^{(+)}, \qquad n = 1,2,\ldots,N-1 , \tag{3b}$$

corresponding to the intuitive picture that sees in the remote past the particles coming in from far away and in the remote future the particles moving out far away from each other. Of course the center-of-mass,

$$\overline{q}(t) = N^{-1} \sum_{n=1}^{N} q_n(t), \tag{4a}$$

moves uniformly,

$$\overline{q}(t) = \overline{q}(0) + Pt , \tag{4b}$$

where P is the total momentum,

$$P = \sum_{n=1}^{N} p_n(t) = \sum_{n=1}^{N} p_n^{(\pm)}, \tag{4c}$$

which is of course a constant of the motion (see (2.1.1-17)).

Exercise 2.1.3.1-1. Verify!

The classical scattering problem has the following formulation: given the quantities $p_n^{(-)}, q_n^{(-)}$, which characterize the behavior of the system in the remote past, find the quantities $p_n^{(+)}, q_n^{(+)}$ which characterize its behavior in the remote future (see (1)). In this formulation the quantities $p_n^{(-)}, q_n^{(-)}$, can be arbitrarily assigned, except for the restriction (3a); they determine the subsequent evolution of the system for all time, and in particular the values of the quantities $p_n^{(+)}$ and $q_n^{(+)}$ that characterize the behavior of the system in the remote future, see (1) (of course the quantities $p_n^{(+)}$ shall automatically satisfy the inequalities (3b)).

Note that the "scattering problem", as formulated above, is different from the "initial-value problem", where the quantities $q_n(0)$ and $p_n(0)$ are assigned (arbitrarily, except for the restriction (2)) and the subsequent (or previous) evolution of the system is then computed, including the parameters $q_n^{(\pm)}$ and $p_n^{(\pm)}$ characterizing the behavior of the system in the remote future and past, see (1). Of course in the "initial-value" case, as a consequence of the initial data $q_n(0)$ satisfying the restriction (2), the asymptotic momenta $p_n^{(-)}$ and $p_n^{(+)}$ turn out to satisfy the inequalities (3a) and (3b).

For the particular system under consideration the outcome of the scattering process is exceedingly simple, being specified by the following simple rules:

$$p_n^{(+)} = p_{N+1-n}^{(-)}, \tag{5}$$

$$q_n^{(+)} = q_{N+1-n}^{(-)}. \tag{6}$$

The rule (5), which is clearly compatible with the simultaneous validity of (3a) and (3b), is a simple consequence of the Lax pair formulation.

Proof. In the remote past and future the Lax matrix (2.1.2-7) becomes diagonal, since the off-diagonal terms vanish asymptotically proportionally to $|t|^{-1}$, see (1):

$$L_{nm}(t) \xrightarrow[t\to\pm\infty]{} = L_{nm}(\pm\infty) = \delta_{nm} p_n^{(\pm)}. \tag{7}$$

Hence at $t = +\infty$ the set of the N eigenvalues $\lambda^{(n)}$ (which are time-independent!) coincides with the set $\{p_n^{(+)}; n = 1,..., N\}$ of eigenvalues of the (diagonal!) matrix $\underline{L}(+\infty)$, and likewise at $t = -\infty$ $\{\lambda^{(n)}; n = 1,..., N\}$ coincides with $\{p_n^{(-)}; n = 1,..., N\}$. This entails that the two sets $\{p_n^{(+)}; n = 1,..., N\}$ and $\{p_n^{(-)}; n = 1,..., N\}$ coincide; via the ordering rules (3a) and (3b) this immediately entails (5), namely $p_1^{(+)} = p_N^{(-)}$,

$p_2^{(+)} = p_{N-1}^{(-)}$ and so on (note the difference among the two cases with N odd and N even: only in the former case the central particle re-acquires in the remote future the *same* momentum it had in the remote past).

Let us also note that the argument given above entails the Poisson-commutativity of the N constants of the motion yielded by the Lax pair approach: since these constants of the motion are time-independent, they can be evaluated at the asymptotic times, when the Lax matrix becomes diagonal, see (7), hence its eigenvalues coincide with the canonical momenta, which of course Poisson–commute among themselves.

A proof of the rule (6) is given below (see Sect. 2.1.3.2).

2.1.3.2 The technique of solution of Olshanetsky and Perelomov (OP)

In Sect. 2.1.3.2 we show how the explicit solution of the initial-value problem for the one-dimensional many-body system characterized by the Newtonian equations of motion (2.1.2-6),

$$\ddot{q}_n = 2g^2 \sum_{m=1, m\neq n}^{N} (q_n - q_m)^{-3}, \tag{1}$$

can be reduced to the purely algebraic task of finding the eigenvalues of a (time-dependent) $(N \times N)$-matrix explicitly given in terms of the initial data $q_n(0)$ and $p_n(0) = \dot{q}_n(0)$.

Let us introduce the diagonal $(N \times N)$-matrix $\underline{Q}(t)$,

$$\underline{Q}(t) = \mathrm{diag}[q_n(t); n = 1, \dots, N], \quad Q_{nm}(t) = \delta_{nm}\, q_n(t), \tag{2}$$

as well as the matrix $\underline{\tilde{Q}}(t)$ obtained from $\underline{Q}(t)$ via a similarity transformation,

$$\underline{\tilde{Q}}(t) = \underline{U}(t)\, \underline{Q}(t)\, [\underline{U}(t)]^{-1}. \tag{3}$$

The properties required of the matrix $\underline{U}(t)$ will be specified below; although in the end we shall, remarkably, find out that this matrix plays no explicit role in determining the solution.

It is clear from the definitions (2) and (3) that the canonical coordinates $q_n(t)$ are the N eigenvalues of the matrix $\underline{\tilde{Q}}(t)$, which is, by construction, diagonalizable, see (3). The strategy of solution that we now pursue is to obtain an evolution equation for $\underline{\tilde{Q}}(t)$, to solve it (yes, it turns out that this equation will be explicitly solvable!) and to thereby obtain an explicit expression for $\underline{\tilde{Q}}(t)$: the computation of the canonical coordinates $q_n(t)$, namely of the solutions of the Newtonian equations of motion (1), is then reduced to evaluating the N eigenvalues of the known matrix $\underline{\tilde{Q}}(t)$.

Let us now introduce a matrix $\underline{M}(t)$ related to $\underline{U}(t)$, see (3), as follows:

$$\underline{M} = \underline{U}^{-1}\underline{\dot{U}}. \tag{4}$$

Then, from (3),

$$\dot{\underline{\tilde{Q}}} = \underline{U}\,\underline{L}\,\underline{U}^{-1} \tag{5}$$

with

$$\underline{L} = \underline{\dot{Q}} - [\,\underline{Q}, \underline{M}\,], \tag{6}$$

and from (5)

$$\ddot{\underline{\tilde{Q}}} = \underline{U}\left\{\underline{\dot{L}} - [\underline{L}, \underline{M}]\right\}\underline{U}^{-1}. \tag{7}$$

Proof of (5) with (6), and of (7). Indeed, quite generally, if

$$\underline{A} = \underline{U}\,\underline{B}\,\underline{U}^{-1}, \tag{8}$$

then

$$\underline{\dot{A}} = \underline{U}\left\{\underline{\dot{B}} - [\underline{B}, \underline{M}]\right\}\underline{U}^{-1} \tag{9}$$

with $\underline{M}$ related to $\underline{U}$ by (4). This can be verified as follows: from (8)

$$\underline{\dot{A}} = \underline{U}\,\underline{\dot{B}}\,\underline{U}^{-1} + \underline{\dot{U}}\,\underline{B}\,\underline{U}^{-1} - \underline{U}\,\underline{B}\,\underline{U}^{-1}\underline{\dot{U}}\,\underline{U}^{-1}, \tag{10a}$$

$$\underline{\dot{A}} = \underline{U}\left\{\underline{\dot{B}} + \underline{U}^{-1}\underline{\dot{U}}\,\underline{B} - \underline{B}\,\underline{U}^{-1}\underline{\dot{U}}\right\}\underline{U}^{-1}. \tag{10b}$$

The last equation, via (4), reproduces (9). And clearly the rule (9), applied to (3) respectively to (5), yields (5) with (6) respectively (7).

We now make, consistently with (6), for $\underline{L}$ and $\underline{M}$ the choices (2.1.2-7,8), namely

$$L_{nm} = \delta_{nm}\,\dot{q}_n + (1-\delta_{nm})\, i\, g\, (q_n - q_m)^{-1}, \tag{11}$$

$$M_{nm} = \delta_{nm} M_{nn} - (1-\delta_{nm})(q_n - q_m)^{-1} L_{nm}. \tag{12}$$

The diligent reader will check that (11) and (12) coincide with (2.1.2-7,8). Of course to get (11) from (2.1.2-7) we also used (2.1.1-6). As for (12), we wrote it in the most convenient manner to check that (11) and (12) are consistent with (6). To this end the explicit expression of the diagonal part of $\underline{M}$ is irrelevant: since $\underline{Q}$ is diagonal, see (2), the diagonal part of $\underline{M}$ commutes with $\underline{Q}$ hence does not contribute to the right hand side of (6). Moreover, since $\underline{Q}$ is diagonal, the commutator in the right hand side of (6) has no diagonal component. The consistency of the diagonal part of (6) with (2) and (11) is therefore clear. As for the off-diagonal part of (6), it reads (see (2))

$$L_{nm} = -(q_n - q_m) M_{nm}, \quad m \neq n, \tag{13}$$

so that its consistency with (12) is obvious.

We now use the fact that $\underline{L}$ and $\underline{M}$ satisfy the Lax evolution equation (2.1-2), to infer from (7) that $\underline{\tilde{Q}}(t)$ satisfies the (amazingly simple) evolution equation

$$\ddot{\underline{\tilde{Q}}} = 0 . \tag{14}$$

The general (matrix) solution of this equation reads

$$\underline{\tilde{Q}}(t) = \underline{\tilde{Q}}(0) + \dot{\underline{\tilde{Q}}}(0)\, t . \tag{15}$$

To get explicit expressions of $\underline{\tilde{Q}}(0)$ and $\underline{\dot{\tilde{Q}}(0)}$ we now make the convenient assumption

$$\underline{U}(0) = \underline{1}. \tag{16}$$

The matrix $\underline{U}(t)$, which characterizes the similarity transformation (3), is defined by (4), namely by the evolution equation

$$\dot{\underline{U}} = \underline{U}\,\underline{M}, \tag{17}$$

which can be supplemented by an arbitrary initial condition assigning $\underline{U}(0)$ (with the only restriction that this matrix be invertible, namely that $\det[\ \underline{U}(0)\ ]\neq 0$). Different choices for $\underline{U}(0)$ would yield different matrices $\underline{\tilde{Q}}(t)$, all of them however having the same eigenvalues, namely yielding the same canonical coordinates $q_n(t)$. The choice (16) is the most convenient one to get explicit results.

Let us re-emphasize that, to obtain our final result, namely the quantities $q_n(t)$, we need not evaluate the matrix $\underline{U}(t)$ by integrating (17), nor indeed do we need to evaluate the matrix $\underline{M}(t)$.

From (16) and (3) we get

$$\underline{\tilde{Q}}(0)=\underline{Q}(0)=\mathrm{diag}\left[q_n(0);\ n=1,\ldots,N\right], \tag{18}$$

and likewise from (16) and (5) we get

$$\dot{\underline{\tilde{Q}}}(0)=\underline{L}(0), \tag{19}$$

where (see (11))

$$L_{nm}(0)=\delta_{nm}\,\dot{q}_n(0)+\left(1-\delta_{nm}\right)i\,g\left[q_n(0)-q_m(0)\right]^{-1}. \tag{20}$$

Hence, from (15), we get

$$\underline{\tilde{Q}}(t)=\underline{Q}(0)+\underline{L}(0)\,t, \tag{21a}$$

namely (see (20))

$$\tilde{Q}_{nm}(t)=\delta_{nm}\left[q_n(0)+\dot{q}_n(0)\,t\right]+\left(1-\delta_{nm}\right)i\ g\ \left[q_n(0)-q_m(0)\right]^{-1}t, \tag{21b}$$

a completely explicit expression of the matrix $\underline{\tilde{Q}}(t)$ in terms of the initial data, $q_n(0)$ and $\dot{q}_n(0)$, of the many-body problem (1).

In conclusion, as promised, the solution of the initial-value problem for (1) has now being reduced to the purely algebraic task of finding the N eigenvalues $q_n(t)$ of the $(N\times N)$-matrix (21).

Let us now outline how to recover the results (2.1.3.1-5,6). It is clear from (21a) and (2.1.3.1-1) that, in the asymptotic $t\rightarrow\pm\infty$ limits, the N quantities $p_n^{(+)}$, as well as the N quantities $p_n^{(-)}$, are the N eigenvalues of the (same!) matrix $\underline{L}(0)$. Besides providing the connection among the asymptotic momenta $p_n^{(\pm)}$ and the data $q_n(0)$ and $p_n(0)$ of the initial-value problem (which define the matrix $\underline{L}(0)$, see (20)), this fact entails the coincidence of the two sets $\{p_n^{(+)}; n=1,\ldots,N\}$ and $\{p_n^{(-)}; n=1,\ldots,N\}$, and this together with (2.1.3.1-3a,b) entails (2.1.3.1-5). (This proof is of course closely analogous to that given in the preceding Sect. 2.1.3.1 (after (2.1.3.1-6)).

Let us then indicate with $\underline{v}^{(n)(\pm)}$ and $\underline{u}^{(n)(\pm)}$ the right- and left-eigenvectors of the matrix $\underline{L}(0)$ corresponding to the eigenvalues $p_n^{(\pm)}$,

$$\underline{L}(0)\underline{v}^{(n)(\pm)} = p_n^{(\pm)}\underline{v}^{(n)(\pm)}, \tag{22a}$$

$$\underline{u}^{(n)(\pm)}\underline{L}(0) = p_n^{(\pm)}\underline{u}^{(n)(\pm)}, \tag{22b}$$

orthonormalized so that

$$\left(\underline{u}^{(n)(\pm)}, \underline{v}^{(m)(\pm)}\right) = \delta_{nm}. \tag{23}$$

(*Beware*: we use here an abbreviated notation. Do not be misled to think that, say, $\underline{v}^{(n)(+)}$ is the eigenvector of $\underline{L}(t)$ as $t \to \infty$).

It is then clear from standard (first-order) perturbation theory (for the evaluation of the eigenvalues of a matrix) that, in the asymptotic limit $(t \to \pm\infty)$, (2.1.3.1-1) and (21a) entail

$$q_n^{(\pm)} = \left(\underline{u}^{(n)(\pm)}, \underline{Q}(0)\,\underline{v}^{(n)(\pm)}\right). \tag{24}$$

In all these equations, of course, whenever a double sign is featured one should systematically take either the upper or the lower choice. But from (2.1.3.1-5) (which was proven in Sect. 2.1.3.1 and has again been proven just above), and from (22), we infer

$$\underline{v}^{(n)(+)} = \underline{v}^{(N+1-n)(-)}, \tag{25a}$$

$$\underline{u}^{(n)(+)} = \underline{u}^{(N+1-n)(-)}, \tag{25b}$$

and, via (24), this entails (2.1.3.1-6).

Let us compute (from (21)) the solution in the $N=2$ case:

$$\begin{vmatrix} q_1(0) + \dot{q}_1(0)\,t - q_{1,2}(t) & ig\left[q_1(0) - q_2(0)\right]^{-1} t \\ -ig\left[q_1(0) - q_2(0)\right]^{-1} t & q_2(0) + \dot{q}_2(0)\,t - q_{1,2}(t) \end{vmatrix} = 0, \tag{26a}$$

$$[q_{1,2}(t)]^2 - q_{1,2}(t)\{q_1(0)+q_2(0)+[\dot{q}_1(0)+\dot{q}_2(0)]\,t\}$$

$$+[q_1(0)+\dot{q}_1(0)\,t]\,[q_2(0)+\dot{q}_2(0)\,t]-g^2[q_1(0)-q_2(0)]^{-2}t^2=0, \tag{26b}$$

$$\bar{q}(t)\equiv\frac{1}{2}[q_1(t)+q_2(t)],\quad q(t)\equiv q_2(t)-q_1(t), \tag{27a}$$

$$q_1(t)\equiv\bar{q}(t)-\frac{1}{2}q(t),\quad q_2(t)\equiv\bar{q}(t)+\frac{1}{2}q(t), \tag{27b}$$

$$\bar{q}(t)=\bar{q}(0)+\dot{\bar{q}}(0)\,t, \tag{28}$$

$$q(t)=\{q(0)+2\,t\;q(0)\;\dot{q}(0)+t^2([\dot{q}(0)]^2+4g^2[q(0)]^{-2})\}^{1/2}. \tag{29}$$

A comparison with (2.1.3.1-1) entails

$$p_1^{(\pm)}=p_2^{(\mp)}=\frac{1}{2}\left([\dot{q}_1(0)+\dot{q}_2(0)]\pm\{[\dot{q}_1(0)-\dot{q}_2(0)]^2+4g^2[q_1(0)-q_2(0)]^{-2}\}^{1/2}\right), \tag{30}$$

$$q_1^{(\pm)}=q_2^{(\mp)}=\frac{1}{2}(\;[q_1(0)+q_2(0)]$$

$$\mp[q_1(0)-q_2(0)][\dot{q}_1(0)-\dot{q}_2(0)]\{[\dot{q}_1(0)-\dot{q}_2(0)]^2+4g^2[q_1(0)-q_2(0)]^{-2}\}^{-1/2}\;). \tag{31}$$

Exercise 2.1.3.2-1. Verify!

Let us recall that, in the *two-body* case $(N=2)$, the fact that $p_1^{(+)}=p_2^{(-)}$, $p_2^{(+)}=p_1^{(-)}$ is a general result (valid for a large class of *two-body* problems), being a consequence of momentum and energy conservation,

$$p_1^{(+)}+p_2^{(+)}=p_1^{(-)}+p_2^{(-)}, \tag{32}$$

$$[p_1^{(+)}]^2+[p_2^{(+)}]^2=[p_1^{(-)}]^2+[p_2^{(-)}]^2. \tag{33}$$

Indeed the latter equation (energy conservation) holds in this form for *any* two-body problem (with velocity-independent forces) in which the particles separate asymptotically and the forces vanish at large separation, with the quantities $p_n^{(\pm)}$ defined by (2.1.3.1-1), or equivalently by the identification $p_n^{(\pm)}=p_n(\pm\infty)$.

This solution, see (27b) with (28) and (29), could have been easily obtained by solving directly the equations of motion (1) via the position (27), corresponding to the separation of the motions of the center-of-mass coordinate $\overline{q}(t)$ and of the relative coordinate $q(t)$. But already for $N=3$ a direct solution of (1) is difficult, and for $N>3$ only the technique described above solves the problem (in the sense of reducing it to a purely algebraic task).

The results we have just obtained for the $N=2$ case entail an important observation: the outcome of the scattering process in the general case (namely, for arbitrary N), as given by the simple rules (2.1.3.1-5,6), is the same that would be produced if the scattering were the results of a sequence of two-body encounters. This phenomenon is often referred to as the property of "factorization".

Exercise 2.1.3.2-2. Investigate the relevance of this remark by drawing schematically, for instance for $N=3$ and $N=4$, the trajectories of the particles as functions of time (say, in the (q,t) plane,), for a few cases characterized by the same asymptotic parameters but by different values of the coupling constant g^2, including the limiting case of almost vanishing g^2 (when the particles move freely except when they collide).

The property of factorization has a deeper significance and import than is for the moment apparent. Let us in any case emphasize that it only applies to the *integrable* model (1); indeed the remarkable properties (2.1.3.1-5,6) of the scattering process do not hold, for $N>2$, if the equations of motion (1) were generalized by allowing different coupling constants, so as to read

$$\ddot{q}_n = 2 \sum_{m=1, m\neq n}^{N} g_{nm}^2 (q_n - q_m)^{-3} , \qquad (34)$$

with generic values of the (positive) coupling constants g_{nm}^2.

Exercise 2.1.3.2-3. Prove this statement, for $N=3$. *Hint:* see <KL72>.

Note that, provided $g_{nm}^2 = g_{mn}^2$, the equations of motion (34) obtain from the Hamiltonian

$$H = \frac{1}{2}\sum_{n=1}^{N} p_n^2 + \sum_{m=1, m<n}^{N} g_{nm}^2 (q_n - q_m)^{-2} . \qquad (35)$$

2.1.3.3 Motion in the presence of an additional harmonic interaction. Extension of the OP technique of solution

In Sect. 2.1.3.3 we modify the model considered in the preceding Sect. 2.1.3.2, by adding to the Newtonian equations of motion (2.1.3.2-1) or (2.1.2-6) a harmonic interaction, so that they read

$$\ddot{q}_n = -\omega^2 q_n + 2g^2 \sum_{m=1, m\neq n}^{N} (q_n - q_m)^{-3}. \tag{1}$$

These equations of motion are of course obtained from the Hamiltonian

$$H = \frac{1}{2}\sum_{n=1}^{N} \left(p_n^2 + \omega^2 q_n^2\right) + g^2 \sum_{n,m=1;m<n}^{N} (q_n - q_m)^{-2}. \tag{2}$$

Of course both the equations of motion (1) and the Hamiltonian (2) reduce to those of Sect. 2.1.2 and 2.1.3.2 (see (2.1.2-6) or (2.1.3.2-1), and (2.1.2-5)) if the "circular frequency" ω vanishes.

The Hamiltonian (2) (in contrast to the Hamiltonian (2.1.2-5)), as well as the equations of motion (1) (in contrast to (2.1.3.2-1), (2.1.2-6)), are *not* invariant under translations $(q_n \to \tilde{q}_n = q_n + q_0, \dot{q}_0 = 0)$. It is however well-known that the *non-translation-invariant* model (1), (2) is closely related to the *translation-invariant* model characterized by a harmonic interaction that, rather than acting as an external potential (pulling every particle towards the origin, see (1)), acts between every pair of particles. Such a model is characterized by the equations of motion

$$\ddot{x}_n = -\Omega^2 \sum_{m=1}^{N} (x_n - x_m) + 2g^2 \sum_{m=1, m\neq n}^{N} (x_n - x_m)^{-3}, \tag{3}$$

which obtain from the Hamiltonian

$$H = \frac{1}{2}\sum_{n=1}^{N} p_n^2 + \sum_{n,m=1;m<n}^{N} \left\{\Omega^2 (x_n - x_m)^2 + g^2 (x_n - x_m)^{-2}\right\}, \tag{4}$$

where x_n are of course now the canonical coordinates and p_n the corresponding canonical momenta. As can be easily verified, the connection between these two models, (1), (2) respectively (3), (4), is given by the relations

$$\Omega^2 = \omega^2 / N, \tag{5}$$

$$q_n = x_n - \bar{x}, \tag{6}$$

$$\bar{x} = N^{-1} \sum_{n=1}^{N} x_n \,. \tag{7}$$

Note that the equations of motion (3) entail that the center-of-mass $\bar{x}(t)$ of the translation-invariant model (3, 4) moves freely,

$$\ddot{\bar{x}} = 0 \,, \tag{8}$$

while (6) and (7) entail that the center-of-mass $\bar{q}(t)$,

$$\bar{q}(t) \equiv N^{-1} \sum_{n=1}^{N} q_n(t) \tag{9}$$

of the non-translation-invariant model (1, 2) is fixed at the origin,

$$\bar{q}(t) = 0 \,. \tag{10}$$

This is of course compatible with the equations of motion (1), which clearly yield

$$\ddot{\bar{q}} + \omega^2 \bar{q} = 0 \,, \tag{11}$$

entailing

$$\bar{q}(t) = \bar{q}(0) \cos(\omega t) + \dot{q}(0)\, \omega^{-1} \sin(\omega t) \,. \tag{12}$$

We focus hereafter on the *non-translation-invariant* model characterized by the equations of motion (1) and the Hamiltonian (2); the above formulas indicate how to translate any result valid for this model, into a corresponding result for the *translation-invariant* model characterized by (3) and (4).

We now proceed exactly as in the preceding Sect. 2.1.3.2, see (2.1.3.2-2,3,4,5,6,7) as well as (2.1.3.2-11,12) or equivalently (2.1.2-7,8). The only novelty is that, while previously the *ansatz* (2.1.2-7,8) corresponded via the equations of motion (2.1.3.2-1) to the Lax evolution equation (2.1.-2), now the equations of motion (1) yield instead the modified Lax equation

$$\dot{\underline{L}} - [\underline{L}, \underline{M}\,] = -\omega^2 \underline{Q} \,, \tag{13}$$

of course with $\underline{Q}$ defined by (2.1.3.2-2).

Exercise 2.1.3.3-1. Verify!

Hence now (2.1.3.2-7) gives, instead of (2.1.3.2-14), the evolution equation

$$\ddot{\underline{\tilde{Q}}}+\omega^2\underline{\tilde{Q}}=0\,, \tag{14}$$

the *general* solution of which can again be exhibited explicitly:

$$\underline{\tilde{Q}}(t)=\underline{\tilde{Q}}(0)\cos(\omega t)+\dot{\underline{\tilde{Q}}}(0)\,\omega^{-1}\sin(\omega t)\,. \tag{15}$$

Of course (14), (15) reduce to (2.1.3-14,15) for $\omega=0$.

Now we can again proceed in close analogy to the treatment of the preceding Sect. 2.1.3.2, see (2.1.3.2-16,19,20), getting (in place of (2.1.3.2-21))

$$\underline{\tilde{Q}}(t)=\underline{Q}(0)\cos(\omega t)+\underline{L}(0)\,\omega^{-1}\sin(\omega t)\,, \tag{16a}$$

namely

$$\tilde{Q}_{nm}(t)=\delta_{nm}\left[q_n(0)\cos(\omega t)+\dot{q}_n(0)\,\omega^{-1}\sin(\omega t)\right]$$

$$+\left(1-\delta_{nm}\right)i\;g\,\left[q_n(0)-q_m(0)\right]^{-1}\omega^{-1}\sin(\omega t)\,, \tag{16b}$$

which is again a completely explicit expression of the matrix $\underline{\tilde{Q}}(t)$ in terms of the initial data, $q_n(0)$ and $\dot{q}_n(0)$, of the many-body problem (1). And of course, as implied by the above treatment, the solution $q_n(t)$ of the initial-value problem for the system (1) is now given by the N eigenvalues of this matrix $\underline{\tilde{Q}}(t)$.

For $\omega=0$ this matrix reduces to (2.1.3.2-21) and the results of the previous Section (2.1.3.2) are recovered. But there is a qualitative difference among the two cases. If $\omega=0$ the motion is *not* confined, and the phenomenological behavior of the system is characterized by the scattering process described in the preceding Sect. 2.1.3.2. In the case (with $\omega\neq 0$) considered in Sect. 2.1.3.3, the motion is instead confined to some neighborhood of the origin, due to the elastic force pulling back each particle towards the origin (see the first term in the right hand side of the Newtonian equations of motion (1)). In fact it turns out that, in this case with $\omega\neq 0$, the motion is *completely periodic*, with period

$$T=2\pi/\omega\,, \tag{17}$$

for any arbitrary set of initial data:

$$q_n(t+T)=q_n(t)\,. \tag{18}$$

This is implied by (16), which shows that the matrix $\underline{\tilde{Q}}(t)$ is periodic with period T, see (17): $\underline{\tilde{Q}}(t+T)=\underline{\tilde{Q}}(t)$. Hence the set $S(t)=\{q_n(t);1,...,N\}$ of the N eigenvalues of $\underline{\tilde{Q}}(t)$ is periodic, $S(t+T)=S(t)$. This by itself does not imply that each eigenvalue is periodic with period T: if the eigenvalues could be reshuffled through the motion, the periodicity of the (unordered) set $S(t)$ with period T would only imply that each eigenvalue is periodic with period (at most) $\tilde{T}=T\cdot N!$, since there are (at most) $N!$ way to reshuffle N objects. But in the case under consideration the singular character at zero separation of the repulsive pair force (see the sum in the right hand side of (1)) excludes any such reshuffling (see (2.1.3.1-2)); this entails the periodicity with period T of each canonical coordinate, see (18).

The completely periodic character, see (18), of all solutions of (1) is a characteristic property of the integrable model (1). It would not obtain if, for instance, (1) were replaced by the more general equations of motion

$$\ddot{q}_n=-\omega^2 q_n+2\sum_{m=1,m\neq n}^{N} g_{nm}^2 (q_n-q_m)^{-3}, \tag{19}$$

with a generic choice of the (*positive*: to avoid any singularity possibly caused by forces which are infinitely *attractive* at zero separation) coupling constants $g_{nm}^2=g_{mn}^2$; note that this symmetry condition is not essential for the validity of the previous statement, but it is implied by the requirement that (19) follow from the Hamiltonian

$$H=\frac{1}{2}\sum_{n=1}^{N}\left(p_n^2+\omega^2 q_n^2\right)+\sum_{n,m=1;m<n}^{N} g_{nm}^2 (q_n-q_m)^{-2}. \tag{20}$$

The property of periodicity (18) is therefore the counterpart, for the model treated in Sect. 2.1.3.3, of the factorization property (2.1.3.1-5) of the model treated in the preceding Sect. 2.1.3.2.

There is indeed a connection among the model treated in the preceding Sect. 2.1.3.2, see (2.1.3.2-1), and that treated in this Section, see (1); this also entails a connection among the factorization property (2.1.3.1-5) and the periodicity property (18).

To demonstrate these connections let us first of all note the following remarkable fact: if $q_n(t)$ satisfies (2.1.3.2-34),

$$\ddot{q}_n=2\sum_{m=1,m\neq n}^{N} g_{nm}^2 (q_n-q_m)^{-3}, \tag{21}$$

then $\tilde{q}_n(t)$,

$$\tilde{q}_n(t) = \cos(\omega t)\, q_n(\tau), \tag{22a}$$

$$\tau = \omega^{-1} \tan(\omega t), \tag{22b}$$

satisfies (19), namely

$$\ddot{\tilde{q}}_n = -\omega^2 \tilde{q}_n + 2 \sum_{m=1, m\neq n}^{N} g_{nm}^2 (\tilde{q}_n - \tilde{q}_m)^{-3}. \tag{23}$$

Exercise 2.1.3.3-2. Verify!

Remark 2.1.3.3-3. Since both models, see (21) and (23), are invariant under a shift of the time variable $(t \to \tilde{t} = t + t_o)$, it is clear that the above treatment would apply equally if (22a) and (22b) were replaced by the more general formulas

$$\tilde{q}_n(t) = \cos[\omega (t - t_0)]\, q_n(\tau), \tag{24a}$$

$$\tau = \tau_0 + \omega^{-1} \tan[\omega (t - t_0)], \tag{24b}$$

with τ_0 and t_0 arbitrary constants. This would merely replace with a more complicated formula the simple relation among the "initial" values of $q_n(t)$ and $\tilde{q}_n(t)$ implied by (22),

$$\tilde{q}_n(0) = q_n(0). \tag{25}$$

Note that we have established a connection between the (more general, nonintegrable) models with different coupling constants; it holds of course *a fortiori* for the *integrable* models characterized by the restriction $g_{nm}^2 = g^2$ (see (2.1.3.2-1) and (1)).

From (22) it might appear that $\tilde{q}_n(t)$ is generally periodic in t with period T, see (17); indeed the prefactor $\cos(\omega t)$ in the right hand side of (22a) certainly possesses this property, and the new variable τ is clearly, see (22b), itself a periodic function of t (in fact, with period $T/2$). This suggests that all solutions of (23) (or, equivalently, of (19)) are periodic in t with period T. But this conclusion is wrong: it only holds in the *integrable* case (1), as we now explain.

Exercise 2.1.3.3-4. Find the solution to this riddle on your own, *before* reading the explanation given below.

The point is, that the change of (dependent and independent) variables from $q_n(t)$ to $\tilde{q}_n(t)$, see (22), is *not* global. When the variable t in (22b) spans the interval, say, from $-T/4$ to $T/4$ (see (17); or, say, from $T/4$ to $3T/4$, etc.), the variable τ spans the entire interval from $-\infty$ to $+\infty$ (see (22b)). Accordingly, the quantity $q_n(\tau)$, see the right hand side of (22a), spans its entire trajectory, which is not confined, so that, as $\tau \to \pm\infty$, $q_n(\tau)$ diverges (to the right or to the left, as the case may be). But when $\tau \to \pm\infty$, namely when $t \to \pm T/4$ (mod $(T/2)$), the prefactor $\cos(\omega t)$, see (22a), vanishes; hence, in these limits, $\tilde{q}_n(t)$, see (22a), remains finite. Indeed it is easily seen, from (17), (22) and (2.1.3.1-1), that

$$\lim_{t \to \pm T/4 + \text{mod}(T)} [\tilde{q}_n(t)] = \pm \omega^{-1} p_n^{(\pm)}, \tag{26a}$$

$$\lim_{t \to \pm T/4 + T/2 + \text{mod}(T)} [\tilde{q}_n(t)] = \mp \omega^{-1} p_n^{(\pm)}. \tag{26b}$$

Proof:

$$\lim_{t \to \pm T/4 + sT/2 + \text{mod}(T)} [\tilde{q}_n(t)] = \omega^{-1} p_n^{(\pm)} \lim_{t \to \pm T/4 + sT/2 + \text{mod}(T)} \cos(\omega t) \tan(\omega t)$$

$$= \pm \sin\left(\frac{\pi}{2} + s\pi\right) \omega^{-1} p_n^{(\pm)}, s = 0,1. \tag{27}$$

Hence the behavior of the canonical coordinates $\tilde{q}_n(t)$ of the confined system (23) over the interval from $t = -T/4$ to $t = T/4$ (see (26a) and (17)) corresponds to the entire trajectory of the canonical coordinates $q_n(\tau)$ of the unconfined system (21) from $\tau = -\infty$ to $\tau = +\infty$, with the values of the canonical coordinates $\tilde{q}_n(t)$ at the beginning and at the end of the time interval from $-T/4$ to $T/4$ related to the asymptotic momenta of the trajectories of the unconfined model, see (1.1.3.1-1), by the rule

$$\tilde{q}_n(\pm T/4) = \pm \omega^{-1} p_n^{(\pm)}. \tag{28a}$$

In the spirit of evincing, via the transformation (22), the solutions $\tilde{q}_n(t)$ of the confined model from the solutions $q_n(\tau)$ of the unconfined model, one would assign the "initial" values $\tilde{q}_n(-T/4)$, obtain via (28a) the corresponding values $p_n^{(-)}$, let the unconfined model run its entire history, from $\tau = -\infty$ (with the asymptotic momenta $p_n^{(-)}$ in the remote past, see (2.1.3.1-1)) to $\tau=+\infty$, determine thereby the asymptotic momenta $p_n^{(+)}$ in the remote future (see (2.1.3.1-17)), and thus obtain, again via (28a), the values $\tilde{q}_n(T/4)$ at the end of the interval (as well, of course, as the values $\tilde{q}_n(t)$ for $-T/4\le t<T/4$ via (22)).

One would then repeat the process in the subsequent interval, from $T/4\le t\le 3T/4$, starting from the "initial" values $\tilde{q}_n(T/4)$ just obtained and determining the "final" values $\tilde{q}_n(3T/4)$ via a completely analogous procedure, except for the fact that, in this interval, (28a) should be replaced, see (26b), by the relation

$$\tilde{q}_n(\pm T/4+T/2)=\mp\omega^{-1}p_n^{(\pm)}. \tag{28b}$$

And so on.

In the nonintegrable (unconfined) case with different coupling constants g_{nm}^2, see (21), there is (for $N>2$) no general rule to connect the asymptotic momenta $p_n^{(+)}$ in the remote future to the asymptotic momenta $p_n^{(-)}$ in the remote past (see (3.1.3.1-1)). Hence there is no justification to expect, for the corresponding confined model, see (23), any general rule connecting the values of the canonical coordinates $\tilde{q}_n(t)$ at the end of one of the time intervals considered above to those at the beginning of that time interval. In particular, therefore, there is no justification to expect the motions of this system to be periodic, in spite of what might have been naively inferred from (22).

The situation is different in the integrable case ($g_{nm}^2=g^2$), see (2.1.3.1-1)), where we have the simple rule (2.1.3.2-5) connecting the asymptotic momenta in the remote past and future, see (2.1.3.1-1). Via (28a) this entails

$$\tilde{q}_n(T/4)=-\tilde{q}_{N+1-n}(-T/4), \tag{29a}$$

and likewise, via (28b),

$$\tilde{q}_n(3T/4)=-\tilde{q}_{N+1-n}(T/4). \tag{29b}$$

But these two formulas yield

$$\tilde{q}_n(-T/4+T)=\tilde{q}_n(-T/4), \tag{30}$$

which in fact corresponds to the property of periodicity (18).

This is a consequence of the above treatment and of the invariance of both models, (21) and (23), under shifts of the respective time variables (see the *Remark 2.1.3.3-3*).

However to complete the above analysis and thereby make more cogent the conclusion that (18) is implied, via the transformation (22), by (2.1.3.1-5), one should prove that, in addition to (30), there also holds the relation

$$\dot{\tilde{q}}_n(-T/4+T)=\dot{\tilde{q}}_n(-T/4)\,. \tag{31}$$

This is left as an *exercise* for the diligent reader.

We also leave, as another *exercise* for the diligent reader, to clarify the consistency of the above treatment with the validity of the inequalities (2.1.3.1-2,3), and in particular the role played, in this connection, by the (sign) difference among (28a) and (28b).

In contrast to the system without harmonic potential considered in the two preceding Sects. 2.1.3.1 and 2.1.3.2, which only features repulsive forces, the system with an additional harmonic potential considered here, see (1) and (2), possesses an *equilibrium configuration.* Indeed it is clear, see (1), that an initial configuration characterized by the initial data

$$q_n(0)=r_n\,,\quad \dot{q}_n(0)=0\,, \tag{32}$$

yields the static solution

$$q_n(t)=r_n\,, \tag{33}$$

if the N quantities r_n satisfy the N relations

$$\omega^2 r_n = 2g^2 \sum_{m=1;m\neq n}^{N} (r_n - r_m)^{-3}\,. \tag{34}$$

This set of N algebraic relations for the N quantities r_n defines the equilibrium configuration in which the (attractive and repulsive) forces acting on every particle are exactly balanced (see the right hand side of (1)). Physical intuition suggests that this set of algebraic equations admit one, and only one, *real* solution, up to the ambiguity entailed by the possibility to reshuffle the particles among themselves (there are of course $N!$ different permutations). This ambiguity can be lifted by sticking to some ordering convention, say (see (2.1.3.1-2))

$$r_n < r_{n+1}\,,\; n=1,\ldots,N-1\,. \tag{35}$$

The correctness of this physical intuition will be proven later: indeed we will see below (and report in Appendix C) that the N quantities z_n characterized by the N algebraic relations

$$z_n = 2 \sum_{m=1, m \neq n}^{N} (z_n - z_m)^{-3} \tag{36}$$

coincide with the N zeros of the Hermite polynomial of order N,

$$H_N(z_n) = 0 . \tag{37}$$

It is on the other hand obvious, see (33) and (35), that up to a rescaling the quantities r_n coincide with these quantities z_n, namely

$$r_n = (g/\omega)^{1/2} z_n . \tag{38}$$

The fact that the equilibrium configuration of the many-body problem considered herein, see (1) and (2), coincide essentially with the N zeros of the Hermite polynomial of order N is remarkable (see comments in Sect. 2.N). We now show that an additional remarkable property of the zeros of Hermite polynomials can be evinced from the study of our many-body problem.

Let us indeed look at the behavior of the system (1) in the neighborhood of the equilibrium configuration (33) with (34), by applying the standard theory of the *small oscillations* of a dynamical system in the neighborhood of a (stable) equilibrium configuration. Hence we set

$$q_n(t) = r_n + \varepsilon \, (g/\omega)^{1/2} \xi_n(t) , \tag{39}$$

where ε is a "small parameter". Insertion of this "small oscillations" *ansatz* in (1) yields, by expanding in ε and using the equilibrium condition (34), and the definition (38), the *linear* evolution equations

$$\ddot{\xi}_n + \omega^2 \sum_{m=1}^{n} A_{nm} \, \xi_m = 0 \tag{40}$$

with

$$A_{nm} = \delta_{nm} \left[1 + 6 \sum_{\ell=1, \ell \neq n}^{N} (z_n - z_\ell)^{-4} \right] - (1 - \delta_{nm}) \, 6 \, (z_n - z_m)^{-4} . \tag{41}$$

Proof. From (1) and (38)

$$\varepsilon(g/\omega)^{1/2}\ddot{\xi}_n = -\omega^2\left[r_n + \varepsilon(g/\omega)^{1/2}\xi_n\right] + 2g^2 \sum_{m=1,m\neq n}^{N} \left[r_n - r_m + \varepsilon\,(g/\omega)^{1/2}(\xi_n - \xi_m)\right]^{-3}, \tag{42a}$$

$$\varepsilon(g/\omega)^{1/2}\ddot{\xi}_n = -\omega^2[\, r_n + \varepsilon(g/\omega)^{1/2}\xi_n \,]$$

$$+2g^2 \sum_{m=1,m\neq n}^{N} (r_n - r_m)^{-3}\left[1 + \varepsilon(g/\omega)^{1/2}(\xi_n - \xi_m)/(r_n - r_m)\right]^{-3}. \tag{42b}$$

Expanding to first order in ε and using (34) we get

$$\ddot{\xi}_n + \omega^2\xi_n + 6g^2 \sum_{m=1,m\neq n}^{N} (r_n - r_m)^{-4}(\xi_n - \xi_m) = 0, \tag{43}$$

and, via (38), this yields (40) with (41).

The $(N\times N)$-matrix $\underline{A}$ is defined, see (41), in terms of the N zeros z_n of the Hermite polynomial $H_N(x)$, see (37). Let α_n indicate its N eigenvalues. It is then well known that (40) entails that, in the neighborhood of the equilibrium configuration, the system oscillates with the N circular frequencies

$$\tilde{\omega}_n = \omega\alpha_n^{1/2}, \tag{44}$$

namely

$$\xi_n(t) = \sum_{m=1}^{N} \left[a_{nm}\cos(\tilde{\omega}_m t) + b_{nm}\sin(\tilde{\omega}_m t)\right]. \tag{45}$$

But we known that the *general* solution of (1) is completely periodic with period $T = 2\pi/\omega$, see (17) and (18). This must be *a fortiori* true of the solution characterizing the behavior of the system in the neighborhood of its equilibrium configuration. But this is compatible with (44) iff all the eigenvalues α_n of the matrix $\underline{A}$, see (41) and (37), are the square of a (nonvanishing) integer. Indeed (see Appendix C)

$$\alpha_n = n^2,\ n = 1,2,\ldots,N. \tag{46}$$

Another remarkable property of the zeros of Hermite polynomials is evidenced by the following

Exercise 2.1.3.3-5. Prove that the ($N\times N$)-matrix

$$M_{nm}(\varphi)=\delta_{nm}\cos(\varphi)\,z_n+i(1-\delta_{nm})\sin(\varphi)(z_n-z_m)^{-1} \quad , \tag{47}$$

where the numbers z_n are the N zeros of the Hermite polynomial of order N, see (37), has, for *all* values of the "angle" φ, the (same) zeros z_n as its eigenvalues (the result is of course trivial if $\varphi=0$). *Hint*: insert the *equilibrium* configuration (32) in (16), recall the key properties of the eigenvalues of this matrix, (16), and use (33) with (38) (for another, "less physical" hence "more mathematical" proof, see Sect. 2.4.5.5).

2.1.4 General solution of the functional equation (*). Integrable many-body model with elliptic interactions

Let us return to the functional equation (*) (see (2.1.1-16)),

$$[\ \alpha(x)\,\alpha'(y)-\alpha(y)\,\alpha'(x)\ ]\,/\,\alpha(x+y)=\beta(x)-\beta(y)\,, \tag{1}$$

with the additional constraint (see (2.1.1-8))

$$\beta(-x)=\beta(x)\,. \tag{2}$$

Let us recall that from the solution $\alpha(x)$ of (1) with (2) one obtains the (even) potential (see (2.1.1-11))

$$v(x)=\alpha(x)\alpha(-x)\,, \tag{3}$$

and that the corresponding Hamiltonian many-body problem, see (2.1.1-12,13,15), is then generally integrable. In Sect. 2.1.4 we obtain the most general expression of $v(x)$ compatible with (1), (2) and (3).

Clearly the functional equation (1) is invariant under the transformations

$$\beta(x)\to\tilde{\beta}(x)=ab\,\beta(ax)+\beta_0 \tag{4}$$

$$\alpha(x)\to\tilde{\alpha}(x)=b\exp(cx)\,\alpha(ax)\,, \tag{5}$$

with a,b,c and β_0 arbitrary constants. The (only) effect of this transformation on the potential $v(x)$, see (3), is to multiply it by the constant b^2 and to rescale its argument,

$$v(x) \to \tilde{v}(x) = b^2\, v(ax)\,. \tag{6}$$

Exercise 2.1.4-1. Verify all these assertions.

The functional equation (1) admits the solutions

$$\alpha(x) = \mathrm{cn}\,(x,k)/\,\mathrm{sn}\,(x,k)\,, \tag{7a}$$

$$\alpha(x) = \mathrm{dn}\,(x,k)/\,\mathrm{sn}\,(x,k)\,, \tag{7b}$$

in both cases with

$$\beta(x) = -\alpha(x)\,\alpha(-x)\,. \tag{8}$$

In (7) k is an arbitrary constant, $0 \le k \le 1$. Note that here and throughout Sect. 2.1.4 we use the notation of Appendix A.

Proof. From (7a) and (A-6a),

$$\alpha'(x) = -d\,(x)/s^2\,(x)\,. \tag{9}$$

Here and below we use the short-hand notation (see (A-1))

$$c(x) \equiv \mathrm{cn}\,(x,k),\; s\,(x) \equiv \mathrm{sn}\,(x,k),\; d\,(x) \equiv \mathrm{dn}\,(x,k)\,. \tag{10}$$

Hence, from (7a) and (9)

$$\alpha(x)\alpha'(y) - \alpha(y)\alpha'(x) = -[s(x)\; s(y)]^{-2}\,[c(x)s(x)d(y) - c(y)s(y)d(x)]\,. \tag{11}$$

From (7a) and (A-10b,c),

$$\alpha(x+y) = D(x,y)/[s(x)c(y)d(y) + s(y)c(x)d(x)]\,, \tag{12}$$

$$D(x,y) = c(x)c(y) - s(x)d(x)s(y)d(y)\,. \tag{13}$$

Hence, from (11) and (12),

$$[\alpha(x)\,\alpha'(y) - \alpha(y)\alpha'(x)]/\,\alpha(x+y)$$

$$= -[s(x)s(y)]^{-2}\,[c(x)s(x)d(y) - c(y)s(y)d(x)]\cdot$$

$$[s(x)c(y)d(y) + s(y)c(x)d(x)]/\,D(x,y)\,. \tag{14}$$

But

$$[c(x)s(x)d(y) - c(y)s(y)d(x)][s(x)c(y)d(y) + s(y)c(x)d(x)] =$$

$$=c(x)c(y)[\ s^2(x)d^2(y)-s^2(y)d^2(x)\ ]+s(x)d(x)s(y)d(y)[\ c^2(x)-c^2(y)\ ]$$

$$=c(x)c(y)[\ s^2(x)-s^2(y)\ ]+s(x)d(x)s(y)d(y)[\ -s^2(x)+s^2(y)\ ]$$

$$=[\ s^2(x)-s^2(y)\ ]D(x,y). \tag{15}$$

In the second step we used (A-3).

Insertion of (15) into (14) yields

$$[\alpha(x)\,\alpha'(y)-\alpha(y)\,\alpha'(x)]/\alpha(x+y)=[s(x)]^{-2}-[s(y)]^{-2}. \tag{16}$$

The fact that the right hand side of this equation separates into the difference of a function of x minus the same function of y is the crucial property guaranteeing that $\alpha(x)$ satisfy the functional equation (1), of course now with

$$\beta(x)=[s(x)]^{-2}. \tag{17}$$

It is easy to verify, via (A-8) and (A-3), that this expression corresponds to (8) with (7a), up to an irrelevant additive constant.

The analogous proof that (7b) with (8) satisfy (1) is left as an *exercise* for the diligent reader.

Up to an irrelevant additive constant, (7a) and (7b) both yield, via (3) and (A-53) with (A-49), the same expression for the potential $v(x)$:

$$v(x)=A\wp(ax|\omega,\omega') \tag{18}$$

with

$$A=-a^2,\ a=(e_1-e_3)^{-1/2} \tag{19}$$

(for the definition of e_1 and e_3, see (A-19,21)).

Remark 2.1.4-2. The two semiperiods ω,ω' of the Weierstrass function $\wp(ax|\omega,\omega')$ in (18) can be chosen arbitrarily, but to get a real potential one should make the standard choice, $\omega=$ real and $\omega'=$imaginary. The other 2 constants, A and a, in (18) can eventually also be chosen arbitrarily, see (6). But the freedom to chose arbitrarily all 3 constants a,ω,ω' is apparent, since only the ratio of these constants plays a role (see the remark after (A-35)). Hence the freedom of choice in (18) is in fact restricted to the choice of only 2 constants, in addition to A, i.e. altogether 3 constants.

We now show that the solution we obtained for the potential $v(x)$, see (18), is the most general one that is compatible, via (3), with the functional equation (1) with (2). The method of proof also suggests the route that was originally used <C75> to discover the solutions (7), which were given above "out of the blue".

We set

$$y=-x+\varepsilon \tag{20}$$

in (1), and consider the limit of vanishing ε. Consistency requires that in this limit

$$\alpha(\varepsilon)=c_{-1}/\varepsilon+c_0+c_1\varepsilon+o(\varepsilon)\,; \tag{21a}$$

we moreover hereafter set

$$c_{-1}=1,\ \ c_0=0\,, \tag{21b}$$

since these two conditions can be enforced by taking advantage of (5).

Exercise 2.1.4-3. Verify, via (A-9), that the solutions (7) are consistent with (21).

We now insert (20) and (21) in (1) and, using (2), we equate the first 3 terms that obtain by expanding (1) (or rather, more conveniently, the equation that obtains multiplying (1) by $\alpha(x+y)$) around $\varepsilon=0$:

$$\beta'(x)=\alpha(x)\alpha'(-x)-\alpha(-x)\alpha'(x)\ \ , \tag{22a}$$

$$\beta''(x)=2[\alpha'(x)\alpha'(-x)-\alpha''(x)\alpha(-x)]\ \ , \tag{22b}$$

$$\beta'''(x)+6c_1\,\beta'(x)=3\left[\alpha''(x)\alpha'(-x)-\alpha'''(x)\alpha(-x)\right]. \tag{22c}$$

Exercise 2.1.4-4. Verify!

The first of these equations can be immediately integrated and, up to an irrelevant additive constant (see (4)), it yields (8).

Insertion of (8) into (22b) yields

$$\alpha''(x)\alpha(-x)=\alpha(x)\alpha''(-x)\ \ , \tag{23}$$

and from this equation and (21) one concludes that $\alpha(x)$ must be odd,

$$\alpha(-x)=-\alpha(x) \quad , \tag{24}$$

so that (23) becomes an identity.

Let $\alpha(x)=\alpha_e(x)+\alpha_o(x)$, with $\alpha_e(x)$ even and $\alpha_0(x)$ odd, $\alpha_e(-x)=\alpha_e(x)$, $\alpha_o(-x)=-\alpha_o(x)$. Then the odd part of (23) reads $\alpha_e(x)\alpha_o''(x)=\alpha_o(x)\alpha_e''(x)$, and this equation is consistent with (21) only if $\alpha_e(x)=0$. Note that the conclusion that $\alpha(x)$ is odd holds up to the transformation (5), which breaks this property, but has been now frozen by the requirement that the constant c_0 vanish, see (21).

Using (24) we see that (8) yields

$$\beta(x)=\alpha^2(x) \tag{25}$$

while (3) and (8) of course entail

$$v(x)=-\beta(x). \tag{26}$$

It is now easy to show that $v(x)$ must satisfy the second order nonlinear ODE

$$2v''(x)\,v(x)-3[v'(x)]^2-24c_1\,[v(x)]^2=0. \tag{27}$$

To get this ODE we note first of all that (22c) can be integrated once to yield

$$\beta''(x)+6c_1\,\beta(x)+3\alpha''(x)\;\alpha(-x)=0. \tag{28}$$

Here we set to zero the integration constant, since it amounts merely to the addition of an (irrelevant) arbitrary constant to $\beta(x)$ (see (4)).

We now note that, via (3) and (24) (or, equivalently, (25) and (26)),

$$[\alpha(x)]^2=-v(x) \quad , \tag{29a}$$

entailing

$$2\;\alpha'(x)\;\alpha(x)=-v'(x) \quad , \tag{29b}$$

$$2\;\alpha''(x)\alpha(x)+2[\alpha'(x)]^2=-v''(x) \quad , \tag{29c}$$

hence, via (29b),

$$\alpha''(x)\,\alpha(x) = -\{2v''(x) - [v'(x)]^2 / v(x)\}/4, \tag{29d}$$

namely, via (24),

$$\alpha''(x)\alpha(-x) = \{2v''(x) - [v'(x)]^2/v(x)\}/4. \tag{29e}$$

Inserting this expression in (28) and using (26) we get (27).

If $v(x)$ is a special solution of (27), then

$$\tilde{v}(x) = A\,v(x - x_0) \tag{30}$$

is clearly also a solution, and, since it depends on 2 arbitrary constants (A and x_0), it is the *general* solution of the second–order ODE (27). On the other hand it is easy to verify, using (A.–23a) and (A.–24), that

$$v(x) = \wp(ax; g_2, g_3) + \gamma, \tag{31a}$$

$$a^2 = 2c_1/\gamma,\ g_2 = 12\gamma^2,\ g_3 = 8\gamma^3, \tag{31b}$$

is such a solution.

The last 2 of the 3 equations (31b) entail a relation among g_2 and g_3. This has emerged from some of the special choices we have made, see (21b) and the remark after (28). But this fact does not contradict our conclusions, see below.

The ODE (27) is a consequence of (1), (2) and (3); hence any function $v(x)$ consistent with (1), (2) and (3) must satisfy (27) (although the converse statement need not, *a priori*, hold). Hence, from (31) and (30) we can conclude that, up to an (irrelevant) additive constant, the potential (18) (with A, a, ω and ω' arbitrary constants; see, however, *Remark 2.1.4-2*) is the most general one consistent with (1), (2) and (3).

Note that, to reach this conclusion, one must exclude that x_0, see (30), take any other value than $x_0 = 0$. But this condition is indeed implied, see (30), (31) and (A-13), by the requirement that $v(x)$ be even, see (3) (or, equivalently, see (2) and (26)). Note that to the uniqueness, as described above, of the potential $v(x)$ yielded by the functional equation (1) with (2) via (3), there does not correspond an analogous uniqueness of the function $\alpha(x)$ yielded by (1) with (2), as indeed demonstrated by the existence of the 2 different solutions (7a) and (7b), as well as by the invariance property (5).

In conclusion we see that the most general Hamiltonian many-body problem consistent with the *ansatz* (2.1.1-1,2) for the Lax pair reads as follows:

$$H = \frac{1}{2}\sum_{n=1}^{N} p_n^2 + g^2 a^2 \sum_{n,m=1;m<n}^{N} \wp[a(q_n - q_m)|\omega,\omega'] \ , \tag{32}$$

where $\wp(z|\omega,\omega')$ is the (doubly-periodic) Weierstrass elliptic function. This Hamiltonian depends on 4 arbitrary *real* constants, say $g^2 > 0$ and a, b, c, with $\omega = b$ and $\omega' = ic$ (however, of the 3 constants $a,b,c,$ only 2 really play a role, see (A-34a); hence the Hamiltonian (32) depends effectively only on 3 real constants). The corresponding Newtonian equations of motion read

$$\ddot{q}_n = -g^2 a^3 \sum_{m=1,m\neq n}^{N} \wp'[a(q_n - q_m)|\omega,\omega'] \ . \tag{33}$$

For special ("degenerate") values of the semiperiods ω,ω' one gets, from this N-body model with *elliptic* forces, the more special (but perhaps "physically" more interesting – to the extent any one-dimensional model can be "physical"!) models with inverse-cube forces treated in Sect. 2.1.3 (this corresponds to $\omega = \infty$, $\omega' = i\infty$, see (A-37)), as well as those with hyperbolic, or trigonometric, forces treated in the following two Sects. 2.1.5 and 2.1.6 (this corresponds to $\omega = \infty$, $\omega' = i\pi/(2a)$, or to $\omega = \pi/(2a)$, $\omega' = i\infty$, see (A-36b)).

2.1.5 N particles on the line interacting pairwise via a repulsive hyperbolic force. Technique of solution OP

Another simple solution of the functional equation (*), see (2.1.1-16), reads

$$\alpha(x) = i\,g\,a/\sinh(ax), \tag{1a}$$

$$\beta(x) = i\,g\,a^2/\sinh^2(ax), \tag{1b}$$

where a and g are 2 arbitrary (real) constants.

Proof:

$$\left[\alpha(x)\alpha'(y)-\alpha(y)\alpha'(x)\right]/\alpha(x+y)=$$

$$=-i\,g\,a^2\left[\sinh(ax)\sinh(ay)\right]^{-2}\cdot$$

$$\cdot\left[\sinh(ax)\cosh(ay)-\sinh(ay)\cosh(ax)\right]\left[\sinh(ax)\cosh(ay)+\sinh(ay)\cosh(ax)\right]$$

$$=-i\,g\,a^2\left[\sinh(ax)\sinh(ay)\right]^{-2}\left[\sinh^2(ax)\cosh^2(ay)-\sinh^2(ay)\cosh^2(ax)\right]$$

$$=-i\,g\,a^2\left[\sinh(ax)\sinh(ay)\right]^{-2}\left[\sinh^2(ax)-\sinh^2(ay)\right]$$

$$=-i\,g\,a^2\left[\sinh^2(ay)-\sinh^2(ax)\right]$$

$$=\beta(x)-\beta(y). \tag{1c}$$

The diligent reader will also check how this solution can be obtained from the general solution given in the preceding Sect. 2.1.4. *Hint:* use (2.1.4 –5)), (2.1.4 –7b)) and (A-11b) with $u=ax$ (as well as (2.1.4 – 4), (2.1.4 –17)), and again (A-11b) with $u=ax$.

The diligent reader will also verify that another simple solution of the functional equation (*), see (2.1.1–16), reads

$$\alpha(x)=i\,g\,a\,\text{cotanh}(ax), \tag{2}$$

with the same expression (1b) for $\beta(x)$ (recall that $\beta(x)$ is always defined up to an arbitrary additive constant). Hereafter we focus on the choice (1a), which has the advantage to yield a function $\alpha(x)$ that vanishes as $x\to\pm\infty$, a feature that simplifies some of the arguments made below. The Hamiltonian yielded by this choice, see below, is of course the same (up to an irrelevant additive constant) as that which obtains from the choice (2), as the diligent reader will easily verify.

The corresponding Hamiltonian, see (2.1.1.–12), reads

$$H=\frac{1}{2}\sum_{n=1}^{N}p_n^2+g^2a^2\sum_{n,m=1;m<n}^{N}\{\sinh[a(q_n-q_m)]\}^{-2}, \tag{3}$$

yielding the Hamiltonian equations of motion

$$\dot{q}_n=p_n, \tag{4a}$$

$$\dot{p}_n = -g^2 a^2 (\partial/\partial q_n) \sum_{j,k=1;k<j}^{N} \{\sinh[a(q_j - q_k)]\}^{-2}, \tag{4b}$$

and the Newtonian equations of motion

$$\ddot{q}_n = 2g^2 a^3 \sum_{m=1,m\neq n}^{N} \cosh[a(q_n - q_m)] \{\sinh[a(q_n - q_m)]\}^{-3}. \tag{5}$$

Hence this model describes the motion of N equal particles that interact pairwise via a *repulsive* force which is *singular at zero separation* (thereby preventing the particles from crossing over, so that their ordering on the line remains unchanged throughout the motion) and which *vanishes exponentially at large separation* ("short range force", such as, say, the strong interaction acting among nucleons).

Except for this last feature ("short range"), these characteristics of the interaction are analogous to those of the simple model discussed in Sect. 2.1.3.1, to which the present model indeed reduces for $a = 0$. Hence all the conclusions of Sect. 2.1.3.1, with the sole exception of the rule (2.1.3.1-6) (which is replaced by a different formula, see (44) below), remain valid.

The diligent reader will check this fact, using, if need be, the expression of the Lax matrix given immediately below.

The Lax pair for this model reads

$$L_{nm} = \delta_{nm} p_n + (1 - \delta_{nm})\, i g a / \sinh[a(q_n - q_m)] \ , \tag{6}$$

$$M_{nm} = \delta_{nm}\, i g a^2 \sum_{l=1,l\neq n} \{\sinh[a(q_n - q_\ell)]\}^{-2}$$

$$- (1 - \delta_{nm})\, i g a^2 \cosh[a(q_n - q_m)] \{\sinh[a(q_n - q_m)]\}^{-2} \ . \tag{7}$$

Note that the off-diagonal terms of these matrices, $\underline{M}$ and $\underline{L}$, are related as follows:

$$M_{nm} = -a \operatorname{cotanh}[a(q_n - q_m)] L_{nm}, \quad n \neq m \quad . \tag{8}$$

We now reduce the solution of the equations of motion for the N-body problem under consideration, see (3), (4) and (5), to a purely algebraic task, via the OP technique. Our starting point are the following two matrix evolution equations:

$$\dot{\underline{L}} = [\underline{L}, \underline{M}], \tag{9}$$

$$\dot{\underline{E}} = [\underline{E}, \underline{M}] + a\{\underline{E}, \underline{L}\}, \tag{10}$$

where $\underline{E}(t)$ is the diagonal matrix having $\exp[2aq_m(t)]$ as (diagonal) elements,

$$\underline{E}(t) = \mathrm{diag}\{\ \exp[2aq_m(t)]\ \}\ . \tag{11}$$

In (10) and always below the notation $\{\underline{A}, \underline{B}\}$ denotes the *anticommutator* of the two matrices $\underline{A}$ and $\underline{B}$,

$$\{\underline{A}, \underline{B}\} \equiv \underline{A}\,\underline{B} + \underline{B}\,\underline{A}. \tag{12}$$

The matrix equation (9) is of course the Lax evolution equation, which we know to be equivalent to the Hamiltonian evolution equations (4) entailed by (3), namely to the Newtonian equations of motion (5).

The diagonal part of the matrix evolution equation (10) is an immediate consequence of the definition (11), and of (4a) (note that the first term in the right hand side of (10) does not contribute to the diagonal part). As for the nondiagonal part of (10), it follows from (11) and (8), as the diligent reader will readily verify.

Let us now introduce the similarity transformation

$$\tilde{\underline{L}} = \underline{U}\,\underline{L}\,\underline{U}^{-1}, \tag{13a}$$

$$\tilde{\underline{M}} = \underline{U}\,\underline{M}\,\underline{U}^{-1}, \tag{13b}$$

$$\tilde{\underline{E}} = \underline{U}\,\underline{E}\,\underline{U}^{-1}, \tag{13c}$$

where the (invertible) matrix $\underline{U}(t)$ is characterized by the evolution equation

$$\dot{\underline{U}} = \underline{U}\,\underline{M}, \tag{14a}$$

entailing of course

$$\underline{U}^{-1}\dot{\underline{U}} = \underline{M}, \tag{14b}$$

$$\dot{\underline{U}}\,\underline{U}^{-1} = \tilde{\underline{M}}, \tag{14c}$$

with the convenient initial condition

$$\underline{U}(0) = \underline{1}. \tag{15}$$

The evolution equation (14a), together with the initial condition (15), define uniquely the matrix $\underline{U}(t)$, in terms of the matrix $\underline{M}(t)$. In the following, however, we shall not need to ascertain the explicit time evolution of neither $\underline{M}(t)$ nor $\underline{U}(t)$.

Note that the simple initial condition (15) entails, via (13a) and (13c), that, at the "initial" time $t=0$,

$$\tilde{\underline{L}}(0)=\underline{L}(0), \tag{16}$$

$$\tilde{\underline{E}}(0)=\underline{E}(0). \tag{17}$$

The definitions (13c) and (11) entail that *the quantities* $\exp[2aq_n(t)]$ *coincide with the N eigenvalues of the matrix* $\tilde{\underline{E}}(t)$. Hence our strategy will be to obtain an evolution equation for $\tilde{\underline{E}}(t)$, to solve this evolution equation, and to thereby obtain an explicit expression for this matrix. The computation of the canonical coordinates $q_n(t)$ is thereby reduced, up to taking logarithms (see (11)), to the purely algebraic task of evaluating the eigenvalues of the (explicitly known) matrix $\tilde{\underline{E}}(t)$.

Let us therefore time-differentiate (13a) and (13c). We get

$$\dot{\tilde{\underline{L}}}=0, \tag{18a}$$

$$\dot{\tilde{\underline{E}}}=a\{\tilde{\underline{E}},\tilde{\underline{L}}\}. \tag{19}$$

The first of these 2 equations, (18), follows immediately from (13a) via the Lax matrix equation (9) and (14):

$$\dot{\tilde{\underline{L}}}=\underline{U}\{\dot{\underline{L}}+\underline{U}^{-1}\dot{\underline{U}}\,\underline{L}-\underline{L}\,\underline{U}^{-1}\dot{\underline{U}}\}\underline{U}^{-1}=\underline{U}\{[\underline{L},\underline{M}]+[\underline{M},\underline{L}]\}\,\underline{U}^{-1}=0. \tag{18b}$$

The second equation, (19), is likewise entailed by (13c) via (10), (14) and (13a,c).

But (18a), together with (16), entails

$$\tilde{\underline{L}}(t)=\underline{L}(0), \tag{20}$$

and this equation allows to integrate (19), yielding, via (17), the following explicit expression for $\tilde{E}(t)$:

$$\tilde{E}(t)=\exp[a\underline{L}(0)\,t]\,\tilde{\underline{E}}(0)\,\exp[a\underline{L}(0)t]. \tag{21}$$

Of course the matrices $\underline{L}(0)$ and $\underline{E}(0)$ are given, in terms of the initial data $q_m(0)$, $p_m(0) = \dot{q}_m(0)$, by the explicit formulas (6) and (11).

We have thereby achieved our goal, to reduce the solution of the N-body problem, see (3), (4) and (5), to the purely algebraic task of computing the eigenvalues of this matrix, which then yield the coordinates $q_n(t)$ via (11).

The fact that there is a one-to-one correspondence between the eigenvalues of the matrix $\underline{\widetilde{E}}(t)$, see (21), and the canonical coordinates $q_n(t)$, is implied by the fact that the ordering of the quantities $q_n(t)$ on the line, corresponding to the rank ordering of the eigenvalues of $\underline{\widetilde{E}}(t)$ (N positive integers), does not change throughout the motion. Note moreover that, using the well-known identities

$$\det[\exp(\underline{A})] = \exp[\ \text{trace}(\underline{A})\ ], \tag{22}$$

$$\det[\underline{A}\ \underline{B}] = \det[\underline{A}]\det[\underline{B}], \tag{23}$$

valid for diagonalizable matrices, one obtains from (21), using (6), the relation

$$\overline{q}(t) = \overline{q}(0) + Pt, \tag{24}$$

where $\overline{q}(t)$ is the center of mass of the system,

$$\overline{q}(t) = N^{-1}\sum_{n=1}^{N} q_n(t), \tag{25}$$

and

$$P = P(0) = N^{-1}\sum_{n=1}^{N} p_n(0) \tag{26}$$

is the total momentum. Note that (24) is consistent with the fact that the center of mass moves freely,

$$\ddot{\overline{q}} = 0, \tag{27}$$

an equation of motion which is an obvious consequence, using (25), of the equations of motion (5).

For $N=2$, the solution can be written in explicit form (also by integrating the equations of motion (5), see below). It reads:

$$q_1(t)=\overline{q}(t)-q(t)/2\ ,\quad q_2(t)=\overline{q}(t)+q(t)/2\ ,\quad q_2(t)>q_1(t)\ , \tag{28a}$$

$$\overline{q}(t)=[q_1(t)+q_2(t)]/2,\quad q(t)=q_2(t)-q_1(t)>0\ , \tag{28b}$$

$$\overline{q}(t)=Pt+\overline{q}(0), \tag{29}$$

$$P=[p_1(t)+p_2(t)]/2=[\dot{q}_1(t)+\dot{q}_2(t)]/2\ , \tag{30}$$

$$q(t)=a^{-1}\log[\ b\cosh[2a(pt+c)]+\{b^2\cosh^2[2a(pt+c)]-1\}^{1/2}\ ], \tag{31}$$

$$b=[1+g^2a^2/p^2]^{1/2}, \tag{32}$$

with P, $\overline{q}(0)$, c and p arbitrary constants $(p>0)$.

Proof. Of course (29) with (30) is an obvious consequence of the free motion, $\ddot{\overline{q}}(t)=0$, of the center of mass coordinate $\overline{q}(t)$, while (30) corresponds to the conservation of momentum. On the other hand from (3) and (4a) one gets

$$H=\{[\dot{q}_1(t)]^2+[\dot{q}_2(t)]^2\}/2+g^2a^2/\sinh^2\{a[q_2(t)-q_1(t)]\}, \tag{33}$$

hence, via (28) and (30),

$$\dot{q}^2=p^2-g^2a^2/\sinh^2(2aq) \tag{34}$$

with

$$p^2=H-P^2. \tag{35}$$

The differential equation (35) is easily integrated (*hint:* introduce the new dependent variable $u=\cosh(2aq)$, then set $u=b\cosh(v)$ with b given by (32)), to yield

$$q(t)=a^{-1}\mathrm{arccosh}\{b\cosh[2a(pt+c)]\}\ , \tag{36}$$

which coincides with (31) since

$$\mathrm{arccosh}(u)=\log[\ u+(u^2-1)^{1/2}\ ]. \tag{37}$$

Exercise 2.1.5-1. Reobtain this result by computing the eigenvalues of (21).

Let us now look at the behavior as $t \to \pm\infty$, setting (see (2.1.3.1-1), as well as (2.1.3.1-2,3)):

$$q_{1,2}(t) = p_{1,2}^{(-)}\, t + q_{1,2}^{(-)} + o(1) \text{ as } t \to -\infty, \tag{38a}$$

$$q_{1,2}(t) = p_{1,2}^{(+)}\, t + q_{1,2}^{(+)} + o(1) \text{ as } t \to +\infty. \tag{38b}$$

It is then clear that these results, (28)-(32), yield

$$p_1^{(-)} = P + p,\ p_2^{(-)} = P - p, \tag{39a}$$

$$q_1^{(-)} = \overline{q}(0) + c - (2a)^{-1} \log b,\ q_2^{(-)} = \overline{q}(0) - c + (2a)^{-1} \log b, \tag{40a}$$

$$p_1^{(+)} = P - p,\ p_2^{(+)} = P + p, \tag{39b}$$

$$q_1^{(+)} = \overline{q}(0) - c - (2a)^{-1} \log b,\ q_2^{(+)} = \overline{q}(0) + c + (2a)^{-1} \log b, \tag{40b}$$

implying

$$p_1^{(+)} = p_2^{(-)},\ p_2^{(+)} = p_1^{(-)}, \tag{41}$$

$$q_1^{(+)} = q_2^{(-)} + \Delta\left(p_2^{(-)} - p_1^{(-)}; g, a\right),\ q_2^{(+)} = q_1^{(-)} + \Delta\left(p_1^{(-)} - p_2^{(-)}; g, a\right), \tag{42}$$

with the following definition of the "shift" $\Delta(p; g, a)$:

$$\Delta(p; g, a) = \operatorname{sign}(p)\,(2a)^{-1} \log\left(1 + g^2 a^2 / p^2\right). \tag{43}$$

Note that, as expected, (41) reproduces the rule (2.1.3.1-5), while (42) with (43), as indicated above, differs from (2.1.3.1-6) (but it reproduces it, as indeed it should, for $a = 0$, since clearly (43) entails $\Delta(p; g, 0) = 0$).

For $N > 2$, the rule (2.1.3.1-5) continues of course to hold, while (2.1.3.1-6) is replaced by the remarkably neat formula

$$q_n^{(+)} = q_{N+1-n}^{(-)} + \sum_{m=1, m \neq n}^{N} \Delta\left(p_m^{(-)} - p_n^{(-)}; g, a\right), \tag{44}$$

which reflects the factorized character of the scattering process: its outcome is the same as if it consisted of a sequence of isolated two-body encounters, each of which causes (due to energy and momentum conservation) an exchange of momenta, see (41), leading to (2.1.3.1-5), as well as a coordinate shift, see (42) with (43), which sum up to yield (44).

Exercise 2.1.5-2. Draw a qualitative picture of the motion (in the (q,t) plane) for $N=2,3,4$. *Hint:* see *Exercise 2.1.3.2-2,* but keep in mind the difference among (2.1.3.1-6) and (44).

Exercise 2.1.5-3. Prove (44). *Hint:* see Sect. 2.N.

2.1.6 N particles on the circle interacting pairwise via a trigonometric force

In the preceding Sect. 2.1.5 we discussed and essentially solved the problem characterized by the Hamiltonian (2.1.5-3) and by the equations of motion (2.1.5-4,5). In this section we consider the model that obtains from that one via the formal replacement of the (real) constant a with the imaginary constant ia. The Hamiltonian, and the corresponding equations of motion, for this model read

$$H=\frac{1}{2}\sum_{n=1}^{N} p_n^2+g^2 a^2 \sum_{n,m=1;m<n}^{N} \{\sin[a(q_j-q_k)]\}^{-2}, \tag{1}$$

$$\dot{q}_n=p_n, \tag{2a}$$

$$\dot{p}_n=-g^2 a^2(\partial/\partial q_n) \sum_{j,k=1;k<j}^{N} \{\sin[a(q_j-q_k)]\}^{-2} \tag{2b}$$

$$\ddot{q}_n=2g^2 a^3 \sum_{m=1,m\neq n}^{N} \cos[a(q_n-q_m)]\{\sin[a(q_n-q_m)]\}^{-3}. \tag{3}$$

Likewise, the solution of the initial-value problem is given by the following prescription: the eigenvalues $\eta_n(t)$ of the matrix $\underline{\tilde{E}}(t)$ defined in terms of the initial data $q_n(0)$, $\dot{q}(0)=p_n(0)$ by the formulas

$$\underline{\tilde{E}}(t)=\exp[ia\underline{L}(0)t]\,\underline{E}(0)\,\exp[ia\underline{L}(0)\,t], \tag{4}$$

$$L_{nm}(0)=\delta_{nm}\,\dot{q}_n(0)+(1+\delta_{nm})\,iga\,/\sin\{a[q_n(0)-q_m(0)]\}, \tag{5}$$

$$E_{nm}(0)=\delta_{nm}\,\exp[2ia\,q_n(0)]\,, \tag{6}$$

yield the canonical coordinates $q_n(t)$ via the relation

$$\eta_n(t)=\exp[2ia\,q_n(t)]. \tag{7}$$

These results are clearly implied by those of the preceding Sect. 2.1.5. The fact that the canonical coordinates $q_n(t)$ obtained in this manner are real and satisfy, throughout their motion, ordering constrains such as (2.1.3.1-2) is not immediately apparent from (7), but it is of course implied by (3).

Clearly in this model the particle coordinates $q_n(t)$ are defined $\mathrm{mod}(\pi/a)$, see (1), (2) and (3). This suggest introducing new rescaled variables

$$\theta_n(t) = 2a\, q_n(t)\,, \tag{8}$$

so that the equations of motion read

$$\ddot{\theta}_n = G^2 \sum_{m=1, m\neq n}^{N} \cos[(\theta_n - \theta_m)/2]\, \{\sin[(\theta_n - \theta_m)/2]\}^{-3} \tag{9}$$

with

$$G = 2ga^2\,. \tag{10}$$

The new variables $\theta_n(t)$ are then defined $\mathrm{mod}(2\pi)$ and therefore can be interpreted as "angles", detailing the angular positions of the particles, which are then interpretable as moving on a circle rather than on an straight line. This interpretation justifies the title of Sect. 2.1.6.

Exercise 2.1.6-1. For clear reasons of symmetry, the system (9) admits the equilibrium configuration

$$\theta_n = 2\pi n/N\ , \tag{11a}$$

as well as the (more general) rotating configuration

$$\theta_n = vt + 2\pi n/N \tag{11b}$$

with v an arbitrary constant. Verify the validity of the corresponding trigonometric identities (see the right hand side of (9)):

$$\sum_{m=1, m\neq n}^{N} \cos[\pi(n-m)/N]\, \{\sin[\pi(n-m)/N]\}^{-3} = 0,\ \ n = 1,2,\ldots,N;\ N = 2,3,\ldots\ . \tag{12}$$

2.1.7 Various tricks: changes of variables, particles of different types, duplications, infinite duplications (from rational to hyperbolic, trigonometric, elliptic forces), reductions (model with forces only among "nearest neighbors")

In Sect. 2.1.7 we illustrate various tricks, by displaying how they work in specific instances.

Changes of dependent and independent variables. Consider the equations of motion

$$q_n''(\tau) = f_n[\ \underline{q}(\tau)\ ]\,. \tag{1}$$

Here and below the primes denote differentiations with respect to the variable τ, while we continue to indicate with superimposed dots differentiations with respect to the time t (see below), and of course the notation $\underline{q}(\tau)$ denotes the N-vector of components $q_n(\tau)$.

Let us now introduce the following *change of dependent and independent variables:*

$$\underline{q}(\tau) = \varphi(t)\ \underline{x}(t),\ \ \tau = \tau(t)\,, \tag{2}$$

where we keep for the moment open the option to assign the two functions $\varphi(t)$ and $\tau(t)$.

The transformation (2) entails that (1) become the following equations for $x_n(t)$:

$$\ddot{x}_n + [2(\dot{\varphi}/\varphi) - (\ddot{\tau}/\dot{\tau})]\,\dot{x}_n + [(\ddot{\varphi}/\varphi) - (\dot{\varphi}/\varphi)(\ddot{\tau}/\dot{\tau})]\,x_n = (\dot{\tau}^2/\varphi)\ f_n[\varphi\underline{x}]\,. \tag{3}$$

Proof. Time-differentiations of (2) yields

$$q_n'\,\dot{\tau} = \dot{\varphi}\,x_n + \varphi\,\dot{x}_n\,, \tag{4}$$

$$q_n''\,\dot{\tau}^2 + q_n'\,\ddot{\tau} = \ddot{\varphi}\,x_n + 2\dot{\varphi}\,\dot{x}_n + \varphi\,\ddot{x}_n\,. \tag{5}$$

From (5), using (4), we get

$$\varphi\ddot{x}_n + [2\,\dot{\varphi} - \varphi(\ddot{\tau}/\dot{\tau})\ ]\,\dot{x}_n + [\ddot{\varphi} - \dot{\varphi}(\ddot{\tau}/\dot{\tau})]\,x_n = q_n''\,\dot{\tau}^2\,, \tag{6}$$

and, via (1) and (2), this yields (3).

Let us pursue the analysis by considering two options, corresponding respectively to the requirements that the second, or the third, term in the left hand side of (3) vanish.

Let us start from the first one of these two possibilities, by setting

$$\dot{\tau}(t) = [\varphi(t)]^2 \tag{7a}$$

entailing

$$\ddot{\tau}(t)/\dot{\tau}(t) = 2\dot{\varphi}(t)/\varphi(t), \tag{7b}$$

so that (3) become

$$\ddot{x}_n + [(\ddot{\varphi}/\varphi) - 2(\dot{\varphi}/\varphi)]\, x_n = [\varphi(t)]^3\, f_n[\varphi \underline{x}], \tag{8a}$$

or equivalently, via the convenient position

$$\varphi(t) = 1/\psi(t). \tag{9}$$

$$\ddot{x}_n - (\ddot{\psi}/\psi)\, x_n = [\psi(t)]^{-3}\, f_n[\underline{x}/\psi]. \tag{8b}$$

Proof: time-differentiation of the logarithm of (9) yields

$$\dot{\varphi}/\varphi = -\dot{\psi}/\psi, \tag{10a}$$

and a further differentiation yields

$$(\ddot{\varphi}/\varphi) - (\dot{\varphi}/\varphi)^2 = -(\ddot{\psi}/\psi) + (\dot{\psi}/\psi)^2 \tag{10b}$$

entailing, via (10a),

$$(\ddot{\varphi}/\varphi) - 2(\dot{\varphi}/\varphi)^2 = -(\ddot{\psi}/\psi). \tag{10c}$$

If moreover the forces $f_n(\underline{q})$, see (1), satisfy the scaling property

$$f_n(\lambda \underline{q}) = \lambda^{-3}\, f_n(\underline{q}), \tag{11}$$

then (8b) read

$$\ddot{x}_n - (\ddot{\psi}/\psi)\, x_n = f_n(\underline{x}). \tag{12}$$

Hence, if the equations of motion (1) are solvable with a force satisfying the rescaling property (11), the more general equations of motion (12), with a function $\psi(t)$ which we are still free to choose, is also solvable, via the transformation (2) with (7) and (9).

Note that the forces appearing in the right hand side of the equations of motion (2.1.3.3-21) do satisfy the scaling property (11). Hence the equations of motion

$$\ddot{x}_n(t) + \chi(t)x_n(t) = 2 \sum_{m=1,m\neq n}^{N} g_{nm}^2 \left[x_n(t) - x_m(t)\right]^{-3}, \tag{13}$$

with an essentially arbitrary choice of the function

$$\chi(t) = -\ddot{\psi}(t)/\psi(t), \tag{14a}$$

entailing of course

$$\ddot{\psi}(t) + \chi(t)\psi(t) = 0, \tag{14b}$$

are no more difficult to solve than (2.1.3.3-21) (except for the need to solve the *linear* ODE (14b) for $\psi(t)$); and in particular, if $g_{nm}^2 = g^2$, they are *solvable*, see Sect. 2.1.3.2.

The special choice

$$\psi(t) = \cos\left[\omega\left(t - t_0\right)\right] \tag{15}$$

yields (see (14a))

$$\chi(t) = \omega^2 \tag{16}$$

and, via (7a) and (9)

$$\tau(t) = \tau_0 + \omega^{-1}\tan\left[\omega\left(t - t_0\right)\right]. \tag{17}$$

The results of Sect. 2.1.3.3, see (2.1.3.3-21)-(2.1.3.3-24), are thereby reproduced, with $\tilde{q}_n(t) = x_n(t)$.

Another interesting choice is

$$\psi(t) = \left(1 + bt\right)/a, \ \dot{\psi}(t) = b/a, \ \ddot{\psi}(t) = 0, \tag{18}$$

entailing (see (9) and (7a))

$$\varphi(t)=a/(1+bt)^{-1},\ \tau(t)=\tau_0-a^2/\left[b(1+bt)\right]. \tag{19}$$

Note that (18), together with (13) and (14), implies that the equations of motion (2.1.3.3-21) are *invariant* under the transformation (2) with (19) (as the diligent reader will verify by direct computation).

More generally, (8b) together with (18) entail that the equations of motion

$$\ddot{x}_n(t)=a^3(1+bt)^{-3}\ f_n\left[a\underline{x}(t)/(1+bt)\right] \tag{20}$$

are transformable into (1) (via (2) with (19)). Hence the many body-problems characterized by the (nonautonomous) Newtonian equations of motion

$$\ddot{x}_n=2g^2a^3(1+bt)^{-3}\sum_{m=1,m\neq n}^{N}\cosh[a(x_n-x_m)/(1+bt)]\{\sinh[a(x_n-x_m)/(1+bt)]\}^{-3} \tag{21}$$

or, more generally,

$$\ddot{x}_n=g^2a^3(1+bt)^{-3}\sum_{m=1,m\neq n}^{N}\wp'[a(x_n-x_m)/(1+bt)] \tag{22}$$

are amenable to exact treatment ((21) is transformed into (2.1.5-5) via (2) with (9), (18) and (19); likewise (22) is transformed into (2.1.4-33)).

Let us now return to (3) to explore the second option, corresponding to the position

$$\tau(t)=c\varphi(t), \tag{23}$$

whose insertion in (3) clearly yields

$$\ddot{x}_n+\left[2(\dot{\varphi}/\varphi)-(\ddot{\varphi}/\dot{\varphi})\right]\dot{x}_n=c^2(\dot{\varphi}^2/\varphi)\ f_n[\varphi\underline{x}]. \tag{24}$$

If we moreover assume the forces f_n to satisfy the scaling law (11), then (24) read

$$\ddot{x}_n+\left[2(\dot{\varphi}/\varphi)-(\ddot{\varphi}/\dot{\varphi})\right]\dot{x}_n=c^2(\dot{\varphi}/\varphi^2)^2\ f_n[\underline{x}], \tag{25a}$$

or equivalently, via (9) (see (10a) and (10c))

$$\ddot{x}_n-(\ddot{\psi}/\dot{\psi})\dot{x}_n=(c\dot{\psi})^2\ f_n[\underline{x}]. \tag{25b}$$

We may therefore conclude that, for a largely arbitrary choice of the function $\psi(t)$, the equations of motion (see (2.1.2-5))

$$\ddot{x}_n-(\ddot{\psi}/\psi)\dot{x}_n=(c\,g\,\dot{\psi})^2\sum_{m=1,\,m\neq n}^{N}(x_n-x_m)^{-3} \tag{26}$$

are *solvable*. In particular the choice

$$\psi(t)=b\exp(-at) \tag{27}$$

with a,b arbitrary constants, yields

$$\ddot{x}_n=a^2\dot{x}_n+[G\exp(-at)]^2\sum_{m=1,m\neq n}^{N}(x_n-x_m)^{-3}\ ,\ G=abcg. \tag{28}$$

The diligent reader will enjoy exploring the behavior of the solutions of (28), using the results of Sect. 2.1.3; as well as other options resulting from the possibility to transform (1) into (3).

Particles of different types. To illustrate this trick, let us return to the N-body model of Sect. 2.1.5, characterized by the Hamiltonian (see (2.1.5-3))

$$H=\frac{1}{2}\sum_{n=1}^{N}p_n^2+\frac{1}{2}g^2a^2\sum_{n,m=1;\,m\neq n}^{N}\{\sinh[a(q_n-q_m)]\}^{-2}, \tag{29}$$

which, let us recall, describes N particles on the line interacting pairwise via a short-range repulsive force singular at zero separation.

We now set

$$N=N_1+N_2, \tag{30a}$$

$$q_n(t)=q_n^{(1)}(t),\ \ p_n(t)=p_n^{(1)}(t),\ n=1,\ldots,N_1, \tag{30b}$$

$$q_{n+N_1}(t)=q_n^{(2)}(t)+i\pi/(2a),\ \ p_{n+N_1}(t)=p_n^{(2)}(t),\ n=1,\ldots,N_2. \tag{30c}$$

Thereby (29) becomes

$$H=\frac{1}{2}\sum_{n=1}^{N_1}\left[p_n^{(1)}\right]^2+\frac{1}{2}\sum_{n=1}^{N_2}\left[p_n^{(2)}\right]^2+$$

$$\frac{1}{2}\sum_{n,m=1;m\neq n}^{N_1}V_e(q_n^{(1)}-q_m^{(1)})+\frac{1}{2}\sum_{n,m=1,m\neq n}^{N_2}V_e(q_n^{(2)}-q_m^{(2)})+\frac{1}{2}\sum_{n=1}^{N_1}\sum_{m=1}^{N_2}V_d(q_n^{(1)}-q_m^{(2)})\,, \quad (31a)$$

with

$$V_e(q)=g^2\,a^2/\sinh^2(aq)\,, \quad (31b)$$

$$V_d(q)=-g^2\,a^2/\cosh^2(aq)\,. \quad (31c)$$

Clearly this Hamiltonian describes $N=N_1+N_2$ particles on the line, N_1 of them of one type and N_2 of another, all having the same (unit) mass, and interacting pairwise via the *repulsive singular* potential $V_e(q)$ (see (31b)) acting among *equal* particles, and via the *attractive nonsingular* potential $V_d(q)$ (see (31c)) acting among *different* particles. This system is of course *integrable* indeed *solvable*: its Lax matrices, as well as the reduction of its solution to a purely algebraic task, can be immediately obtained, via the position (30), from the results of Sect. 2.1.5 for the Hamiltonian (29).

Exercise 2.1.7-1. (i) Draw a graph of the potentials $V_e(q)$ and $V_d(q)$, see (31b) and (31c); *(ii)* write the Hamiltonian and Newtonian equations of motion entailed by (31); *(iii)* verify that they (obviously!) coincide with the equations of motion that obtain from (2.1.5-4, 5) via (30); *(iv)* ponder on the extent to which the results of Sect. 2.1.5 continue to hold. *Hint*: the main phenomenological changes are that different particles can now go through each other, and that bound states can now exist (see below).

Two different particles may now form a bound state, with negative total energy (aside from the kinetic energy associated to the free motion of the center-of-mass). The phenomenology associated to the N-body model (31) includes therefore scattering processes involving bound "molecules" in addition to single particles. The diligent reader will, for instance, show that in a scattering process characterized (say, in the center of mass frame of reference) by particle 1, say, of the first kind coming, in the remote past, say, from the left, while a bound state composed of a couple of particles of different kinds (say, particle 2 of the first kind and particle 3 of the second kind) comes from the right, the phenomenon of partner swapping generally occurs: namely, in the remote future, particle 2 returns alone to the right, while a bound pair formed by particles 1 and 3 escapes to the left.

There is a special configuration of this many body problem with

$$N_1 = N_2 = \tilde{N} \tag{32a}$$

which is worth a special mention. It corresponds to the assignment

$$q_n^{(1)}(t) = q_n^{(2)}(t) = q_n(t), \quad p_n^{(1)}(t) = p_n^{(2)}(t) = p_n(t), \quad n = 1,...,\tilde{N}, \tag{32b}$$

which clearly describes $\tilde{N}$ "molecules", each composed of 2 particles of different types bound together maximally. Clearly such a configuration is compatible with the time evolution entailed by the Hamiltonian H; namely, if the conditions (32b) hold initially, say at $t = 0$, for *all* values of n, $n = 1,...,\tilde{N}$, they remain always valid. It is then sufficient to consider the time evolution of the $\tilde{N}$ quantities $q_n(t)$, $p_n(t)$.

Hence one has thereby manufactured a (new?) many-body problem, whose Hamiltonian is given by the following formula:

$$H = \frac{1}{2}\sum_{n=1}^{\tilde{N}} p_n^2 + \sum_{n,m=1;\, m<n}^{N} W(q_n - q_m), \tag{33a}$$

with

$$W(q) = V_e(q) + V_d(q). \tag{33b}$$

The skeptical reader will verify that the Hamiltonian (33) produces indeed the same equations of motion that (31a) yields for, say, $q_n^{(1)}(t)$ (or, equivalently, for $q_n^{(2)}(t)$) whenever the configuration (32) prevails.

One might believe to have thereby discovered a *new* integrable many-body problem. But it is easy to verify that (33b) with (31b,c) entail

$$W(q) = g^2 (2a)^2 / \sinh^2(2aq). \tag{33c}$$

Proof:

$$g^2 a^2 \left\{ [\sinh(aq)]^{-2} - [\cosh(aq)]^{-2} \right\}$$

$$= g^2 a^2 [\sinh(aq)\cosh(aq)]^{-2} [\cosh^2(aq) - \sinh^2(aq)]$$

$$= g^2 (2a)^2 [\sinh(2aq)]^{-2} . \tag{34}$$

Hence one has in this manner re-obtained the original system (29), except for the replacement of the constant a with $2a$.

There is, however, an amusing twist. Consider the $[(2N)\times(2N)]$-Lax matrices for the $(2N)$-body problem (29) with (30) and (32); and replace in it a with $a/2$. This yields a *new* Lax pair, composed of $(2N \times 2N)$-matrices, for the original system (29)! Moreover, this procedure can be repeated again and again, getting thereby, always for the system (29), new Lax pairs composed of square matrices of size $4N, 8N, 16N$ and so on. But of course these new Lax matrices will only yield N independent constants of the motion, as the diligent reader will verify.

Duplication. For a first illustration of this idea, let us consider the $\tilde{N}$-body problem characterized by the following Hamiltonian of normal type with one-and two-body interactions:

$$H = \frac{1}{2}\sum_{n=1}^{\tilde{N}} \tilde{p}_n^2 + \sum_{n=1}^{\tilde{N}} V^{(1)}(\tilde{q}_n) + \frac{1}{2}\sum_{n,m=1;m\neq n}^{\tilde{N}} V^{(2)}(\tilde{q}_n - \tilde{q}_m), \tag{35a}$$

$$V^{(1)}(-\tilde{q}) = V^{(1)}(\tilde{q}),\ V^{(2)}(-\tilde{q}) = V^{(2)}(\tilde{q}). \tag{35b}$$

The corresponding Newtonian equations of motion read

$$\ddot{\tilde{q}}_n = f^{(1)}(\tilde{q}_n) + \sum_{m=1,m\neq n}^{\tilde{N}} f^{(2)}(\tilde{q}_n - \tilde{q}_m)\ ,\ n = 1,\ldots,\tilde{N}, \tag{36a}$$

with

$$f^{(1)}(\tilde{q}) = -dV^{(1)}(\tilde{q})/d\tilde{q},\quad f^{(2)}(\tilde{q}) = -dV^{(2)}(\tilde{q})/d\tilde{q}\ , \tag{36b}$$

entailing, via (35b),

$$f^{(1)}(-\tilde{q}) = -f^{(1)}(\tilde{q}), \quad f^{(2)}(-\tilde{q}) = -f^{(2)}(\tilde{q}). \tag{36c}$$

Consider now the following special (symmetrical) configuration of this $\tilde{N}$-body system:

$$\tilde{N} = 2N + M, \tag{37a}$$

$$\tilde{q}_n(t) = q_n(t), \quad \tilde{p}_n(t) = p_n(t), \quad n = 1,\ldots,N, \tag{37b}$$

$$\tilde{q}_{n+N}(t) = -q_n(t), \quad \tilde{p}_{n+N}(t) = -p_n(t), \quad n = 1,\ldots,N, \tag{37c}$$

$$\tilde{q}_{n+2N}(t) = 0, \quad \tilde{p}_{n+2N}(t) = 0, \quad n = 1,\ldots,M. \tag{37d}$$

It is easily seen that this configuration is compatible with the equations of motion (36a) with (36c): if it holds at any one time, say at $t = t_0$, it holds for all time. It is also clear that, without loss of generality, at any one time $t = t_0$ we can supplement the *ansatz* (37) with the prescription, say,

$$0 < q_1(t_0) < q_2(t_0) < \ldots < q_N(t_0), \tag{38}$$

although the persistence of this rule during the time evolution depends then on the nature of the forces: a sufficient condition is that the two-body force be infinitely repulsive at zero separation, so as to prevent the particles from crossing over each other.

We can now fix our attention only on the behavior of the first N particles, namely on the canonical coordinates $q_n(t)$ and momenta $p_n(t)$, see (37b); the behavior of the other N particles, see (37c), duplicates faithfully that of the first N; and the last M particles simply sit at the origin.

Clearly the equations of motion for the coordinates $q_n(t)$ now read

$$\ddot{q}_n = f^{(1)}(q_n) + M f^{(2)}(q_n) + f^{(2)}(2q_n) + \sum_{n=1, m\neq n}^{N} \left[f^{(2)}(q_n - q_m) + f^{(2)}(q_n + q_m)\right], \tag{39}$$

and it is easily seen that they are obtainable from the Hamiltonian

$$H = \frac{1}{2}\sum_{n=1}^{N} p_n^2 + \sum_{n=1}^{N} \left[V^{(1)}(q_n) + M V^{(2)}(q_n) + \frac{1}{2} V^{(2)}(2q_n)\right]$$

$$+ \frac{1}{2} \sum_{n,m=1; m\neq n}^{N} \left[V^{(2)}(q_n - q_m) + V^{(2)}(q_n + q_m)\right]. \tag{40}$$

These equations of motion, (39), follow from (36a) (with $n=1,...,N$) via (37); their applicability is implied by the compatibility of the *ansatz* (37) with (36a,c), a fact which is rather obvious (a graphical representation may be helpful in this respect; the diligent reader may wish to go through a formal analytic proof).

The Hamiltonian (40) is obtained, again via (37), from (35a). It is wise, when making such derivations, to actually check the consistency of the equations of motion (in our case, (39)) with the Hamiltonian (in our case, (40)), to make sure that the two are indeed consistent (trouble might originate from "double-counting" when deriving the new Hamiltonian). Note that, in deriving the Hamiltonian, there may be a constant term, that can of course be neglected: for instance, in our case, the term $(M-1)\, V^{(2)}(0)$ arising from the part of the last sum in the right hand side of (35a) with the indices n and m in the range from $2N+1$ to $2N+M$). If the two-body potential $V^{(2)}(q)$ is singular at $q=0$, this term is actually infinite; but this fact can be safely ignored, as the diligent reader will verify by checking the consistency of (39) with (40).

It is also clear that, if the problem we started from, see (35) and (36), was *solvable* and/or *integrable*, these same properties hold for the problem (39) and (40). Hence from the *integrable* and *solvable* models treated in the preceding Sections one can obtain new models by using this kind of "duplication".

If $V^{(1)}=0$, the model (35, 36) is translation-invariant; but this invariance property is no more featured by the new model (39, 40).

Exercise 2.1.7-2. The diligent reader will write out explicitly the models that obtain via this trick from those discussed in preceding Sections, and will analyze qualitatively their behavior.

Other *duplications* are possible, which however entail some excursions into the complex plane. The basic idea is again to identify a special configuration that is preserved throughout the motion. Let us illustrate this kind of trick by exhibiting one specific example.

We take as starting point the simple model of Sect. 2.1.3.3, characterized by the Hamiltonian (see (2.1.3.3-2))

$$H=\frac{1}{2}\sum_{n=1}^{N}(p_n^2+\omega^2 q_n^2)+g^2\sum_{n,m=1,m<n}^{N}(q_n-q_m)^{-2}, \tag{41}$$

and correspondingly by the equations of motion (see (2.1.3.3-1))

$$\ddot{q}_n + \omega^2 q_n = 2g^2 \sum_{m=1, m \neq n}^{N} (q_n - q_m)^{-3}, \quad n = 1, \ldots, N. \tag{42}$$

Let us now set

$$N = N_1 + 2N_2 \ , \tag{43a}$$

$$q_n(t) = z_n(t) \ , \quad n = 1, \ldots, N_1 \ , \tag{43b}$$

$$q_{n+N_1}(t) = x_n(t) + i\, y_n(t) \ , \quad n = 1, \ldots, N_2 \ , \tag{43c}$$

$$q_{n+N_1+N_2}(t) = x_n(t) - i\, y_n(t) \ , \quad n = 1, \ldots, N_2 \ . \tag{43d}$$

Note that we are assuming here the quantities x_n, y_n, z_n to be *real;* it is easy to see that this is compatible with (42), provided g^2 and ω^2 are also *real* (indeed, we hereafter assume that they are nonnegative).

It is now a matter of trivial algebra to obtain from (42), via (43), the following equations:

$$\ddot{z}_n + \omega^2 z_n = 2g^2 \sum_{m=1, m \neq n}^{N_1} (z_n - z_m)^{-3}$$

$$+ 4g^2 \sum_{m=1}^{N_1} (z_n - x_m) \left[(z_n - x_m)^2 - 3y_m^2 \right] \left[(z_n - x_m)^2 + y_m^2 \right]^{-3} , \quad n = 1, \ldots, N_1 \ , \tag{44a}$$

$$\ddot{x}_n + \omega^2 x_n = 2g^2 \sum_{m=1}^{N_1} (x_n - z_m) \left[(x_n - z_m)^2 - 3y_n^2 \right] \left[(x_n - z_m)^2 + y_n^2 \right]^{-3}$$

$$+ 2g^2 \sum_{m=1, m \neq n}^{N_2} (x_n - x_m) \left\{ \left[(x_n - x_m)^2 - 3(y_n - y_m)^2 \right] \left[(x_n - x_m)^2 + (y_n - y_m)^2 \right]^{-3} \right.$$

$$\left. + \left[(x_n - x_m)^2 - 3(y_n + y_m)^2 \right] \left[(x_n - x_m)^2 + (y_n + y_m)^2 \right]^{-3} \right\} , \quad n = 1, \ldots, N_2 \ , \tag{44b}$$

$$\ddot{y}_n + \omega^2 y_n = \frac{1}{4} g^2 y_n^{-3} - 2g^2 \sum_{m=1}^{N_1} y_n \left[3(x_n - z_m)^2 - y_n^2 \right] \left[(x_n - z_m)^2 + y_n^2 \right]^{-3}$$

$$- 2g^2 \sum_{m=1, m \neq n}^{N_2} \left\{ (y_n - y_m) \left[(y_n - y_m)^2 - 3(x_n - x_m)^2 \right] \left[(x_n - x_m)^2 + (y_n - y_m)^2 \right]^{-3} \right.$$

$$\left. + (y_n + y_m) \left[(y_n + y_m)^2 - 3(x_n - x_m)^2 \right] \left[(x_n - x_m)^2 + (y_n + y_m)^2 \right]^{-3} \right\} , \quad n = 1, \ldots, N_2 . \tag{44c}$$

Clearly these equations can be interpreted as the Newtonian equations of motion of an N-body problem (see (43a)) composed of N_1 particles of one kind (coordinates $z_n(t)$, $n=1,...,N_1$), N_2 of a second kind (coordinates $x_n(t)$, $n=1,...,N_2$) and (again) N_2 of a third kind (coordinates $y_n(t), n=1,...,N_2$), all having the same (unit) mass and moving on the (straight) line. The forces acting among these particles can be read from (44). They include forces of one-body type (see the first term in the right hand side of (44c); in addition of course to the "elastic" force, see the second term in the left hand sides of (44a,b,c)), of two-body type (see the first sum in the right hand side of (44a)), of three-body type (see the second sum in the right hand side of (44a) and the first sum in the right hand sides of (44b) and (44c)), and of four-body type (see the second sum in the right hand side of (44b) and (44c)).

Note the lack of translation invariance (even in the $\omega=0$ case).

The solutions of these equations of motion, (44), can be obtained from the solutions of (42); hence the phenomenology discussed in Sects. 2.1.3.1, 2.1.3.2 and 2.1.3.3 is, to a large extent, also featured by this model, as well as the possibility to reduce the solution of the initial-value problem to the purely algebraic task of computing the eigenvalues of a matrix explicitly known in terms of the initial data and the time. In particular, for $\omega^2>0$ the model features a *completely periodic* behavior, and for $\omega^2=0$ the scattering process gives rise to *no new momenta* (the set of those characterizing the behavior in the remote future coincides with the set of those characterizing the behavior in the remote past).

There is, however, a nontrivial difference with respect to the models of Sects. 2.1.3.1, 2.1.3.2 and 2.1.3.3: while (as in the previous models) the N_1 particles of the first kind cannot overtake each other (see the first term in the right hand side of (44a)), now (in contrast to the previous models) no analogous restriction applies to the motion of the $2N_2$ particles of the second and third kind (except for the fact that the particles of the third kind cannot cross the origin, see the first term in the right hand side of (44c)). The diligent reader will investigate the qualitative modifications, relative to the treatments of Sects. 2.1.3.1, 2.1.3.2 and 2.1.3.3, caused by these differences.

Exercise 2.1.7-3. Understand the origin of these differences, in terms of motions in the complex plane.

As the diligent reader will readily verify, the Newtonian equations of motion (44) are yielded by the Hamiltonian

$$H=\frac{1}{2}\sum_{n=1}^{N_1}(\zeta_n^2+\omega^2 z^2)+\sum_{n=1}^{N_2}(\xi_n^2+\omega^2 x_n^2)-\sum_{n=1}^{N_2}(\eta_n^2+\omega^2 y_n^2+\frac{1}{4}g^2 y_n^{-2})$$

$$+\frac{1}{2}g^2\sum_{n,m=1;m\neq n}^{N_1}(z_n-z_m)^{-2}+\sum_{n=1}^{N_1}\sum_{m=1}^{N_2}\left\{\left[(z_n-x_m)^2-y_m^2\right]/\left[(z_n-x_m)^2+y_m^2\right]^2\right\}$$

$$+g^2\sum_{n,m=1;m\neq n}^{N_2}\left\{\left[(x_n-x_m)^2-(y_n-y_m)^2\right]/\left[(x_n-x_m)^2+(y_n-y_m)^2\right]^2\right.$$

$$\left.+\left[(x_n-x_m)^2-(y_n+y_m)^2\right]/\left[(x_n-x_m)^2+(y_n+y_m)^2\right]^2\right\}. \tag{45}$$

Here of course ζ_n is the canonical momentum conjugated to the canonical coordinate z_n, and likewise for ξ_n relative to x_n, and η_n relative to y_n.

This Hamiltonian, (45), is obtained from (41) via the assignments positions (43), supplemented by analogous assignments positions for the momenta:

$$N=N_1+2N_2\ , \tag{46a}$$

$$p_n(t)=\zeta_n(t)\ ,\ n=1,\dots,N_1\ , \tag{46b}$$

$$p_{n+N_1}(t)=\xi_n(t)-i\eta_n(t),\ n=1,\dots,N_2\ , \tag{46c}$$

$$p_{n+N_1+N_2}(t)=\xi_n(t)-i\eta_n(t),\ n=1,\dots,N_2\ . \tag{46d}$$

Note however the sign difference of the imaginary term in the right hand side, among (46c) and (41c), and likewise among (46d) and (41d) (although it should be noted that this choice is motivated by philosophical reasons, see Sect. 4.1, but it is really of no consequence here: clearly the opposite sign choice yields the same result!).

Also note the negative sign in front of the third sum in the right hand side of (45).

The diligent reader is advised to reflect on the significance, in terms of motions in the complex plane, of the *ansatz* (43). There clearly are various other *ansaetze* which are also compatible with the evolution (42), and the diligent reader is also advised to explore them, before looking at the literature where additional instances are given <CF92>.

Infinite duplications. To introduce this trick we start from the simple integrable system of Sect. 2.1.3, whose Hamiltonian function and equations of motion we write here as follows:

$$H=\frac{1}{2}\sum_J\tilde{p}_J^2+\frac{1}{2}g^2\sum_{J,K;J\neq K}(\tilde{q}_J-\tilde{q}_K)^{-2}\quad , \tag{47}$$

$$\ddot{\tilde{q}}_J = -g^2 (\partial / \partial \tilde{q}_J) \sum_{K \neq J} (\tilde{q}_J - \tilde{q}_K)^{-2} \quad , \tag{48a}$$

$$\ddot{\tilde{q}}_J = 2 g^2 \sum_{K \neq J} (\tilde{q}_J - \tilde{q}_K)^{-3} \quad . \tag{48b}$$

We now assume that there are in fact an *infinite* number of particles, arranged (in the *complex* plane) according to the following configuration: N particles are located on the real axis, at the N positions $q_n(t)$; to each of them there correspond in addition an infinity of other particles, located equispaced on a vertical straight line (in the complex plane), at the positions

$$\tilde{q}_J(t) = q_n(t) + i \pi s / a, \; s = \pm 1, \pm 2, \pm 3, \ldots , \tag{49a}$$

where a is a *real* constant (independent of the index n). It is clear that such a configuration is compatible with the equations of motion (48): indeed, the symmetry of this configuration, and the antisymmetrical character of the pair force (see the right hand side of (48b)) entail that the vertical component of the forces acting on each particle cancels out, while the horizontal components are exactly the same for all the particles located on each vertical line. This of course entails that, throughout the motion characterized by the configuration (49a), all the particles sitting equispaced on a vertical line move horizontally with the same velocity or, equivalently, have the same momentum: namely (49a) is complemented by

$$\tilde{p}_J(t) = \dot{\tilde{q}}_J(t) = \dot{q}_n(t), \tag{49b}$$

which is clearly compatible (indeed implied) by (49a).

The evolution of this configuration is completely characterized by the motion on the N particles on the real line, namely by the time-evolution of the N coordinates $q_n(t)$, see (49a). The corresponding equations of motion read

$$\ddot{q}_n = -g^2 a^2 (\partial / \partial q_n) \sum_{m=1, m \neq n}^{N} \{\sinh [a(q_n - q_m)]\}^{-2}, \tag{50a}$$

$$\ddot{q}_n = 2 g^2 a^3 \sum_{m=1, m \neq n}^{N} \cosh[a(q_n - q_m)] \{\sinh[a(q_n - q_m)]\}^{-3}, \tag{50b}$$

which are clearly obtainable from the N-body Hamiltonian

$$H = \frac{1}{2} \sum_{n=1}^{N} p_n^2 + g^2 \sum_{n,m=1; m<n}^{N} \{\sinh[a(q_n - q_m)]\}^{-2}. \tag{51}$$

Proof: The equations of motion (50) obtain from (48) (and likewise the Hamiltonian (51) from (47)) via the *identity*

$$\sum_{s=\infty}^{+\infty}(q+i\pi s/a)^{-2}=a^2/\sinh^2(aq). \tag{52}$$

(This well-known identity is implied by the following statements: both sides of (52) are meromorphic functions of q, both are periodic in q with period $i\pi/a$, both have, as their only singularities, double poles, with the same strength, at the same locations, $q=i\pi s/a$, $s=0,\pm1,\pm2,\ldots$, and both remain finite (or vanish) as $q\to\infty$ in any direction in the complex plane away from the poles). This identity holds of course for arbitrary (complex) values of both q and a.

We have thereby obtained the (integrable) many-body model of Sect. 2.1.5, featuring hyperbolic interactions, from the (integrable) model of Sect. 2.1.3, featuring rational interactions.

Exercise 2.1.7-4. (i) Reflect on the significance of these findings, in terms of the motion in the complex plane of the infinite number of particles of the model (48) corresponding to the configuration (49), a motion which is, as we just saw, equivalent to (50). *(ii)* Ditto, but replacing a with ia, so that (50) gets replaced by (2.1.6-3). *(iii)* Ditto, but for the model with particles of 2 different kinds, see (31). *(iv)* Ditto, but for the phenomenon discussed in connection with the assignment (32) (indeed, what is the interpretation of (34) with (32) and (31) in terms of symmetrical configurations of infinite points in the complex plane, and what is the relation of (34) to (52) ?). *(v)* Try and derive via (49) the Lax pair for the system with hyperbolic interactions, see (2.1.5-6,7), from the Lax pair for the system with rational interactions, see (2.1.2-6,7).

It is easily seen that, in an analogous manner, but adding to every particle on the real axis, characterized by the coordinate $q_n(t)$, a *double infinity* of particles, according to the rule

$$\tilde{q}_J(t)=q_n(t)+s\omega/(2a)+s'\omega'/(2a);\ s=\pm1,\pm2,\ldots;\ s'=\pm1,\pm2,\ldots\ , \tag{53}$$

it is possible to pass from the integrable Hamiltonian many-body problem characterized by the rational two-body potential g^2q^{-2}, see (2.1.2-4), to the integrable Hamiltonian many-body problem characterized by the "elliptic" potential $g^2\wp(aq|\ \omega,\omega')$, see (2.1.4-32). Here the key formula is (A-12). Note that ω and ω' are 2 *a priori* arbitrary constants, but a natural prescription is that ω be *real* and ω' be *imaginary*.

The diligent reader will go through the appropriate derivation, proceeding in close analogy to the previous treatment. In fact this can be done, as indicated above (see (53)), starting from the rational case and adding to each particle a *double* infinity of replicas; or one can start from the hyperbolic or trigonometric cases of Sect. 2.1.5 or 2.1.6 and add appropriately a *single* infinity of replicas to either of those cases, getting thereby again the "elliptic" model of Sect. 2.1.4

The diligent reader is also advised to repeat the series of *Exercise 2.1.7-4,* to the extent they are applicable, in the doubly-periodic elliptic context. It is also of interest to ask oneself why the process of infinite duplication cannot be pursued any further. (*Hint:* there are *simply periodic* – i.e., hyperbolic or trigonometric – and *doubly periodic* – i.e., elliptic – functions of a complex variable; there do not exist *triple periodic* functions).

Before closing the topic of infinite duplications we report one amusing observation, which we discuss for simplicity in the context of the first model discussed above, see after (47). Indeed let us return to the argument presented after (49a). Clearly it entails that the configuration (49a) would be compatible with the equations of motion (48) even if the quantity π/a, which characterizes the interparticle separation of the equispaced particles on each vertical line, varies *linearly in time* rather than being *constant* (both behaviors are indeed compatible with the lack, or rather the balancing off, of forces in the vertical direction). This suggests replacing a with $a/(1+bt)$ (where a and b are now 2 real constants). This entails replacing, say, (50b) with

$$\ddot{x}_n = 2g^2a^3(1+bt)^{-3}\sum_{m=1,m\neq n}^{N}\cosh[a(x_n-x_m)/(1+bt)]\{\sinh[a(x_n-x_m)/(1+bt)]\}^{-3}. \tag{54}$$

The argument we just made suggests that these equations of motion, in spite of their nonautonomous character, should be amenable to exact treatments. But this is a forgone finding: see (21)!

We end this tricky section by reporting a technique based on an appropriate limiting process, whereby from the model of Sect. 2.1.5 one obtains a new many-body problem featuring only "nearest neighbor" interactions.

Let us start from the equations of motion (2.1.5-5),

$$\ddot{\tilde{q}}_n = 2g^2a^3\sum_{m=1,m\neq n}^{N}\cosh[a(\tilde{q}_n-\tilde{q}_m)]\{\sinh[a(\tilde{q}_n-\tilde{q}_m)]\}^{-3}\ ; \tag{55}$$

for definiteness, we assume hereafter the constant a to be positive, $a>0$.

We now set in these equations

$$\tilde{q}_n(t)=q_n(t)+\lambda\, n,\ \ n=1,\dots,N\ , \tag{56a}$$

$$g^2=c(2a)^{-3}\exp(2a\lambda), \tag{56b}$$

and we let the parameter λ diverge to positive infinity, $\lambda\to\infty$. It is then clear that (55) yield

$$\ddot{q}_n=c\{\exp[2a(q_{n-1}-q_n)]-\exp[2a(q_{n+1}-q_n)]\},\ \ n=2,\dots,N-1, \tag{57a}$$

$$\ddot{q}_1=-c\exp[2a(q_2-q_1)], \tag{57b}$$

$$\ddot{q}_N=c\exp[2a(q_{N-1}-q_N)]. \tag{57c}$$

Proof. As $\lambda\to\infty$, for $n>m$

$$2g^2a^3\cosh[a(\tilde{q}_n-\tilde{q}_m)]/\sinh^3[a(\tilde{q}_n-\tilde{q}_m)]$$

$$=c\exp(2a\lambda)\{\exp[a\lambda(n-m)+a(q_n-q_m)]+O(\exp[-a\lambda(n-m)])\}$$

$$/\{\exp[a\lambda(n-m)+a(q_n-q_m)]+O(\exp[-a\lambda(n-m)])\}^3$$

$$=c\exp\{2a\lambda[1-(n-m)]\}\exp[-2a(q_n-q_m)]\{1+O(\exp[-2a\lambda(n-m)])\}. \tag{58}$$

Hence, for $m<n-1$, the limit vanishes, while for $m=n-1$, in the $\lambda\to\infty$ limit one gets

$$2g^2a^3\cosh[a(\tilde{q}_n-\tilde{q}_m)]/\sinh^3[a(\tilde{q}_n-q_m)]=c\exp[2a(q_{n-1}-q_n)]. \tag{59a}$$

Likewise, for $m>n+1$ one gets a vanishing limit, while for $m=n+1$ in the $\lambda\to\infty$ limit one gets

$$2g^2a^3\cosh[a(\tilde{q}_n-\tilde{q}_m)]/\sinh^3[a(\tilde{q}_n-\tilde{q}_m)]=-c\exp[2a(q_{n+1}-q_n)]. \tag{59b}$$

It is easily seen that these equations of motion obtain from the Hamiltonian

$$H=\frac{1}{2}\sum_{n=1}^{N}p_n^2+[c/(2a)]\sum_{n=2}^{N}\exp[2a(q_n-q_{n-1})]. \tag{60}$$

The N-body problem on the line defined by this Hamiltonian, hence by the Newtonian equations of motion (59), is characterized by an interaction acting only among "nearest neighbors"; namely, the n-th particle interacts only with the particles identified by the labels $n-1$ and $n+1$ (independently, of course, from where these particles actually happen to be through the time evolution, be it far away, or close to each other). Note moreover that the two-body potential, $V^{(2)}(q)=-(c^2/2a)\exp(2aq)$, see (60), is not even, $V^{(2)}(-q)\neq V^{(2)}(q)$, and that it vanishes at one end ($V^{(2)}(-\infty)=0$) but diverges at the other, namely if $q\to+\infty$.

The way this model has been obtained suggests that it should also be amenable to exact treatment: *integrable*, indeed *solvable*. This is indeed the case, but we do not elaborate on this aspect here, except as material for *exercises*, see below.

Exercise 2.1.7-5. (i) Derive from the model characterized be (57) and (60) an analogous model containing N different "coupling" constants. *Hint*: replace $q_n(t)$ with $q_n(t)+\log(c_n)$. *(ii)* Find a Lax pair for this model. *Hint:* apply (56) to (2.1.5-6,7). *(iii)* Perform a qualitative analysis of the "scattering" behavior of this model. *(iv)* Find a Lax pair for the model characterized by the validity of the equations of motion (57a) for $n=1,...,N$ (i.e., also for $n=1$ and $n=N$), with the "periodicity" prescription $q_0(t)\equiv q_N(t)$, $q_{N+1}(t)\equiv q_1(t)$. *(v)* Find techniques whereby the solution of these models is reduced to a purely algebraic task. *Solution*: see, for instance, <P90>.

2.1.8 Another convenient representation for the Lax pair. The functional equation (**)

In this Section we introduce another convenient representation for a Lax pair. It reads:

$$L_{nm}=\dot{q}_n \quad \text{if} \quad m=n\ , \tag{1a}$$

$$L_{nm}=(\dot{q}_n\,\dot{q}_m)^{1/2}\ \alpha(q_n-q_m) \quad \text{if} \quad m\neq n\ , \tag{1b}$$

$$M_{nm}=\sum_{l=1,l\neq n}^{N}\dot{q}_l\ \beta(q_n-q_l) \quad \text{if} \quad m=n\ , \tag{2a}$$

$$M_{nm}=(\dot{q}_n\,\dot{q}_m)^{1/2}\ \gamma(q_n-q_m) \quad \text{if} \quad m\neq n\ . \tag{2b}$$

Let us pause to analyze the similarities and differences of these expressions, (1) and (2), with respect to (2.1.1-1) and (2.1.1-2).

The two matrices $\underline{L}$ and $\underline{M}$ are written in terms of $q_n(t)$ and $\dot{q}_n(t)$, rather than $q_n(t)$ and $p_n(t)$ (but note that (1a) coincides with (2.1.1-1a) via (2.1.1-6)). Hence we shall require, see below, that the Lax equation, see (2.1-2), with (1) and (2), correspond to equations of motion of Newtonian, rather than Hamiltonian, type. (We will eventually also find a Hamiltonian formulation, see below).

Let us also note that, from (1) and (2), via the positions

$$q_n(t) = t + \varepsilon \tilde{q}_n(t), \tag{3a}$$

$$\alpha(q) = \varepsilon \tilde{\alpha}(\tilde{q}/\varepsilon),\ \beta(q) = \tilde{\beta}(\tilde{q}/\varepsilon),\ \gamma(q) = \tilde{\gamma}(\tilde{q}/\varepsilon), \tag{3b}$$

one obtains

$$\underline{L} = \underline{1} + \varepsilon \underline{\tilde{L}} \ , \tag{4a}$$

$$\underline{M} = \underline{\tilde{M}} \ , \tag{4b}$$

with

$$\tilde{L}_{nm} = \dot{\tilde{q}}_n \quad \text{if} \quad m = n \tag{5a}$$

$$\tilde{L}_{nm} = [\ (1+\varepsilon\dot{\tilde{q}}_n)(1+\varepsilon\dot{\tilde{q}}_m)\ ]^{1/2}\ \tilde{\alpha}(\tilde{q}_n - \tilde{q}_m) \quad \text{if} \quad m \neq n \ , \tag{5b}$$

$$\tilde{M}_{nm} = \sum_{l=1, l\neq n}^{N} (1+\varepsilon\, \dot{\tilde{q}}_l)\ \tilde{\beta}(\tilde{q}_n - \tilde{q}_l) \quad \text{if} \quad m = n \ , \tag{6a}$$

$$\tilde{M}_{nm} = [\ (1+\varepsilon\dot{\tilde{q}}_n)(1+\varepsilon\dot{\tilde{q}}_m)\ ]^{1/2}\ \tilde{\gamma}(\tilde{q}_n - \tilde{q}_m) \quad \text{if} \quad m \neq n \ . \tag{6b}$$

But (4) clearly entails that, if $\underline{L}$ and $\underline{M}$ satisfy the Lax equation $\dot{\underline{L}} = [\underline{L}, \underline{M}]$, $\underline{\tilde{L}}$ and $\underline{\tilde{M}}$ obey the same Lax equation, $\dot{\underline{\tilde{L}}} = [\underline{\tilde{L}}, \underline{\tilde{M}}]$. On the other hand it is also clear that, up to trivial notational changes, for $\varepsilon = 0$ (5) respectively (6) coincide, via (2.1.1-6), with (2.1.1-1) respectively (2.1.1-2). This indicates that the results of this section contain, as a special (limiting) case, the results of Sect. 2.1.1.

Exercise 2.1.8-1. Verify!

Let us now insert this *ansatz,* see (1) and (2), in the Lax equation, which is to this end conveniently written in the form (2.1.1-3).

From the diagonal terms (see (2.1.1-3a)) we now get the Newtonian equations of motion

$$\ddot{q}_n = \sum_{m=1, m\neq n}^{N} \dot{q}_n \dot{q}_m w(q_n - q_m) , \tag{7}$$

with

$$w(x) = \alpha(x)\gamma(-x) - \alpha(-x)\gamma(x) , \tag{8}$$

entailing that $w(q)$ is odd,

$$w(-x) = -w(x) . \tag{9}$$

Note that the "forces" appearing in the right hand side of (7) have two-body character and are translation-invariant; they do however depend (in contrast to the previous case) on the velocities (quadratically), hence (7) is *not* invariant under the Galilei transformation $q_n(t) \to \tilde{q}_n(t) = q_n(t) + v_0 t$.

Next, we look at the off-diagonal terms of the Lax equation, see (2.1.1-3b), again of course with the *ansatz* (1) and (2). We get:

$$\beta(x) = \beta(-x) , \tag{10}$$

$$\gamma(x) = \alpha'(x) - \alpha(x)\, \eta(x) , \tag{11}$$

$$\eta(x) = \beta(x) + \frac{1}{2} w(x) , \tag{12}$$

$$\alpha(x)\alpha'(y) - \alpha(y)\alpha'(x) = [\alpha(x+y) - \alpha(x)\alpha(y)][\eta(x) - \eta(y)] , \tag{13}$$

of course with $w(q)$, see (12), defined by (8).

Proof. From (2.1.1-3b)

$$\dot{L}_{nm} / L_{nm} = (L_{nn} - L_{mm}) M_{nm} / L_{nm} - M_{nn} + M_{mm} + \sum_{l=1; l\neq n,m}^{N} [L_{nl} M_{lm} - M_{nl} L_{lm}] / L_{nm} . \tag{14}$$

Here, of course, $m \neq n$. Now, using (1) and (2), we get

$$\frac{1}{2}[(\ddot{q}_n / \dot{q}_n) + (\ddot{q}_m / \dot{q}_m)] + [\alpha'(q_n - q_m) / \alpha(q_n - q_m)](\dot{q}_n - \dot{q}_m)$$

$$= (\dot{q}_n - \dot{q}_m)[\gamma(q_n - q_m) / \alpha(q_n - q_m)] - \sum_{l=1, l\neq n}^{N} \dot{q}_l \beta(q_n - q_l) + \sum_{l=1, l\neq m}^{N} \dot{q}_l \beta(q_m - q_l)$$

$$+\sum_{l=1;l\neq n,m}^{N} \dot{q}_l \left[\alpha(q_n-q_l)\gamma(q_l-q_m)-\gamma(q_n-q_l)\alpha(q_l-q_m)\right]/\alpha(q_n-q_m)\,. \qquad (15)$$

We now use (7) and re-write this equation as follows (by taking out of the sums all terms proportional to $\dot{q}_n$ or $\dot{q}_m$):

$$\dot{q}_n\left[\frac{1}{2}w(q_m-q_n)+\alpha'(q_n-q_m)/\alpha(q_n-q_m)-\gamma(q_n-q_m)/\alpha(q_n-q_m)-\beta(q_m-q_n)\right]$$

$$+\dot{q}_m\left[\frac{1}{2}w(q_n-q_m)-\alpha'(q_n-q_m)/\alpha(q_n-q_m)+\gamma(q_n-q_m)/\alpha(q_n-q_m)+\beta(q_n-q_m)\right]$$

$$+\sum_{l=1;l\neq n,m}^{N} \dot{q}_l\left\{\,[w(q_n-q_l)+w(q_m-q_l)]/2+\beta(q_n-q_l)-\beta(q_m-q_l)\right.$$

$$\left.-\left[\alpha(q_n-q_l)\gamma(q_l-q_m)-\gamma(q_n-q_l)\alpha(q_l-q_m)\right]/\alpha(q_n-q_m)\,\right\}=0\,. \qquad (16)$$

Equating to zero the factors that multiply $\dot{q}_n$ and $\dot{q}_m$ (and setting, for notational convenience, $q_n-q_m=x$), and using (9) and the definition (12), one gets

$$\alpha'(x)-\gamma(x)-\eta(x)\,\alpha(x)+[\beta(x)-\beta(-x)]\,\alpha(x)=0\ , \qquad (17a)$$

$$\alpha'(x)-\gamma(x)-\eta(x)\,\alpha(x)=0\ , \qquad (17b)$$

which clearly yield (10) and (11).

Finally, one equates to zero the factor multiplying $\dot{q}_l$ in the sum in the left hand side of (16). Setting, for notational convenience, $q_n-q_l=x$, $q_l-q_m=y$ (entailing $q_n-q_m=x+y$), and using (9), (10) and (12), one gets

$$\alpha(x)\,\gamma(y)-\gamma(x)\,\alpha(y)=[\eta(x)-\eta(y)]\alpha(x+y)\ , \qquad (18)$$

which, via (11), yields (13).

Clearly the main constraint is provided by (13), namely by the functional equation (**):

$$(**)\ \ [\alpha(x)\,\alpha'(y)-\alpha(y)\,\alpha'(x)]\,/\,[\alpha(x+y)-\alpha(x)\,\alpha(y)]=\eta(x)-\eta(y)\,. \qquad (19)$$

This is essentially a functional equation for the function $\alpha(x)$, corresponding to the requirement that the combination appearing in the left hand side of (19) separate additively into two terms, one depending only on x and the other only on y. The fact that these two terms are the difference of a function of x only, $\eta(x)$, minus the *same* function of y only,

$\eta(y)$ (see the right hand side of (19)) is then automatically implied by the antisymmetry of the left hand side of (19). Once a function $\alpha(x)$ having this property is found, it determines the function $\eta(x)$ up to an (irrelevant, see below) additive constant via (19). Then $\alpha(x)$ and $\eta(x)$ yield $\gamma(x)$ via (11), and $\eta(x)$ yields $\beta(x)$ and $w(x)$ via (12) with (9) and (10), which clearly entail

$$\beta(x) = [\eta(x) + \eta(-x)]/2 = \text{ even part of } \eta(x) \ , \tag{20}$$

$$w(x) = \eta(x) - \eta(-x) = \text{ odd part of } 2\eta(x) \ . \tag{21}$$

Note that, if the function $\alpha(x)$ has a finite value $\alpha(0)$ at $x=0$ (as we shall indeed find, see below), by setting $y=-x$ in (19) and using (21) we get

$$w(x) = v'(x)/[v(x) - \alpha(0)] = (d/dx)\log[v(x) - \alpha(0)], \tag{22}$$

with

$$v(x) = \alpha(x)\alpha(-x) = v(-x) . \tag{23}$$

Let us end this section by noting the explicit form of the first 2 traces, T_1 and T_2 (see (2.1-19)), of the Lax matrix (1), which are of course two constants of motion for the system whose time evolution is determined by the Newtonian equations (7):

$$T_1 = \sum_{n=1}^{N} \dot{q}_n(t) \ , \tag{24}$$

$$T_2 = \sum_{n=1}^{N} [\dot{q}_n(t)]^2 + \sum_{n,m=1;m\neq n}^{N} \dot{q}_n(t)\dot{q}_m(t) v[q_n(t) - q_m(t)] \ , \tag{25a}$$

$$T_2 = T_1^2 + \sum_{n,m=1;m\neq n}^{N} \dot{q}_n(t)\, \dot{q}_m(t) \ \{ \ v[q_n(t) - q_m(t)] - 1 \ \} \ . \tag{25b}$$

The fact that T_1 is a constant of motion is also an immediate consequence of (7) and of the odd character of w, see (9).

As for T_2, (see (25), which has been obtained from (1) via (23)), the fact that (7) entail its time-independence is less obvious (although it is of course guaranteed by our treatment).

2.1.9 A simple solution of the functional equation (**)

In Sect. 2.1.9 we introduce a simple solution of the functional equation (**), see (2.1.8-19):

$$[\alpha(x)\,\alpha'(y)-\alpha(y)\alpha'(x)]\,/\,[\alpha(x+y)-\alpha(x)\,\alpha(y)]=\eta(x)-\eta(y)\,. \tag{1}$$

We then introduce and discuss (but in the separate Sect. 2.1.9.1) the notion of "fake" Lax pairs.

A simple solution of the functional equation (**), see (1), reads

$$\alpha(x)=(1+a\,x)/(1+b\,x)\,, \tag{2}$$

with a and b two arbitrary constants. The corresponding expression for $\eta(x)$ reads

$$\eta(x)=1/[x\,(1+b\,x)\,]\,. \tag{3}$$

Proof.

$$\alpha'(x)=(a-b)/(1+bx)^2\;, \tag{4}$$

$$\alpha(x)\alpha'(y)-\alpha(y)\alpha'(x)=(a-b)[\;(1+bx)(1+by)]^{-2}\,[(1+ax)(1+bx)-(1+ay)(1+by)]$$

$$=(a-b)\big[(1+bx)(1+by)\big]^{-2}\,(x-y)[(a+b)+ab\,(x+y)\,]\,, \tag{5}$$

$$\alpha(x+y)-\alpha(x)\alpha(y)=[1+a(x+y)]\,/\,[1+b(x+y)]-(1+ax)(1+ay)/[(1+bx)(1+by)]$$

$$=\{[1+b(x+y)](1+bx)(1+by)\}^{-1}\cdot$$

$$\cdot\{(1+bx)\,(1+by)[1+a(x+y)]-(1+ax)(1+ay)[1+b(x+y)]\}$$

$$=\{[1+b(x+y)](1+bx)(1+by)\}^{-1}\,(b-a)\,xy\,[a+b+ab(x+y)]\,, \tag{6}$$

$$[\alpha(x)\alpha'(y)-\alpha(y)\alpha'(x)]/[\alpha(x+y)-\alpha(x)\alpha(y)]=$$

$$-[xy(1+bx)(1+by)]^{-1}\,\{(x-y)\,[1+b\,(x+y)\,]\}$$

$$=-[xy(1+bx)(1+by)]^{-1}\,\{x(1+bx)-y(1+by)\}$$

$$=[x(1+bx)]^{-1}-[y(1+by)]^{-1} \tag{7}$$

The corresponding expressions of $\beta(x)$, $\gamma(x)$ and $w(x)$ read:

$$\beta(x) = -b/(1-b^2 x^2), \tag{8}$$

$$\gamma(x) = -1/[x(1-bx)], \tag{9}$$

$$w(x) = 2/[x(1-b^2 x^2)]. \tag{10}$$

Exercise 2.1.9-1. Verify that these expressions follow, respectively, from (2.1.8-20) with (3), from (2.1.8-11) with (2), (3) and (4), and from (2.1.8-21) with (3).

Remark 2.1.9-2. These expressions, (8), (9) and (10), of $\beta(x)$, $\gamma(x)$ and $w(x)$, as well as the expression (3) of $\eta(x)$, are independent of the quantity *a*, which does instead appear in $\alpha(x)$, see (2). One may therefore set $a=0$, with the advantage of getting, at least for $b \neq 0$, a function $\alpha(x)$, see (2), that has the convenient feature to vanish as $x \to \infty$. On the other hand if both a and b vanish, $a=b=0$, the Lax matrix becomes exceedingly simple, since then

$$\alpha(x) = 1. \tag{11}$$

The peculiarities of this case are discussed in the following Sect. 2.1.9.1. Note that with this assignment, (11), of $\alpha(x)$, the relation (2.1.8-13) is identically satisfied, implying no condition on $\eta(x)$.

2.1.9.1 Fake Lax pairs

In this Section we focus on the very special case of the Lax pair of type (2.1.8-1,2) with $\alpha(x)=1$, which we write as follows:

$$L_{nm} = \delta_{nm}\,\dot{q}_n + (1-\delta_{nm})(\dot{q}_n\,\dot{q}_m)^{1/2} = (\dot{q}_n\,\dot{q}_m)^{1/2} \quad , \tag{1}$$

$$M_{nm} = -\frac{1}{2}(1-\delta_{nm})(\dot{q}_n\,\dot{q}_m)^{1/2}\,w(q_n-q_m) \quad , \tag{2}$$

where however we assume now that $w(q)$ is an *arbitrary* function, except for the restriction to be odd,

$$w(-x) = -w(x). \tag{3}$$

Indeed it is easy to verify that the corresponding Lax equation,

$$\dot{\underline{L}} = [\underline{L}, \underline{M}],\tag{4}$$

is satisfied if the quantities $q_n(t)$ evolve according to the Newtonian equations of motion (see (2.1.8-7))

$$\ddot{q}_n = \sum_{m=1, m\neq n}^{N} \dot{q}_n \dot{q}_m w(q_n - q_m).\tag{5}$$

The ansatz (1), (2) is obtained by setting

$$\alpha(x) = 1\tag{6}$$

in (2.1.8-1,2). Indeed, as noted at the end of the preceding Sect. 2.1.9, when (6) holds, (2.1.8-13) is identically satisfied, entailing no restriction on $\eta(x)$. The conditions that must still be implemented are (2.1.8-9,10,11,12), which yield (from (2.1.8-11) and (6))

$$\gamma(x) = -\eta(x) \ ,\tag{7}$$

as well as

$$\eta(x) = \beta(x) + \frac{1}{2} w(x)\tag{8}$$

or, more precisely, (2.1.8-20,21). To get (2) we have, for simplicity, made the additional choice

$$\beta(x) = 0 .\tag{9}$$

Let us reemphasize that we now have no restriction on the function $w(q)$, other than (3). Does this imply that (5) is integrable for any arbitrary (odd) choice of the function $w(q)$?

Not so. The Lax pair (1), (2) is a *fake* Lax pair. Indeed it is clear that (1) yields, for the traces T_n, see (2.1-9), of the Lax matrix (1), the simple identity

$$T_p = (T_1)^p \ , \ p = 1,2,3,... \ .\tag{10}$$

Hence the time-independence of the traces of the Lax matrix $\underline{L}$, see (1), yields in this case *only one* conserved quantity, the trace T_1, see (2.1.8-24), whose constancy in time is indeed an obvious consequence of (5) with (3).

Many other instances of fake Lax pairs can be manufactured. For instance, let $p_n(\underline{q})$ be N arbitrarily given functions of the N coordinates $q_m(t)$, and let these N coordinates evolve according to the N "equations of motion"

$$\dot{q}_n = p_n(\underline{q}) . \tag{11}$$

Introduce then the following Lax pair:

$$L_{nm} = \delta_{nm}\, q_n \sum_{l=1, l\neq n}^{N} (q_n - q_l)^{-1} + (1-\delta_{nm})\, q_n (q_n - q_m)^{-1} \;, \tag{12}$$

$$M_{nm} = -\delta_{nm} \sum_{l=1, l\neq n}^{N} p_l(\underline{q})\,(q_n - q_l)^{-1} - (1-\delta_{nm})\; p_n(\underline{q})\,(q_n - q_m)^{-1} . \tag{13}$$

Here $\underline{q}$ indicates of course the N-vector of components q_n.

Statement (i): the evolution (11) corresponds to the Lax equation (4). *Statement (ii)*: this Lax pair is a fake one. There indeed holds the following *Statement (iii)*: the matrix $\underline{L}$, see (12), has as eigenvalues the numbers *1, 2, ...,N* for any arbitrary choice of the N quantities q_n. Hence the constancy in time of the eigenvalues of $\underline{L}$ provides no information on the time evolution (11) of the coordinates $q_n(t)$, which is actually largely arbitrary, given the assumed arbitrariness of the functions $p_n(\underline{q})$.

The diligent reader will ponder, and perhaps try to prove, these statements, whose validity will be shown later (see Sect. 2.4.5.1).

Let us end Sect. 2.1.9.1 by reporting a result that generalizes that mentioned at its beginning. The Lax equation (4) with $\underline{L}$ given by (1) and $\underline{M}$ defined as follows,

$$M_{nm} = -\frac{1}{2}(1-\delta_{nm})(\dot{q}_n \dot{q}_m)^{1/2}\, w_{nm}(q_n - q_m) \;, \tag{14}$$

where $w_{nm}(q)$ is an *arbitrary* $(N \times N)$-matrix-valued function only subject to the "oddness" restriction

$$w_{nm}(-x) = -w_{mn}(x) \;, \tag{15}$$

corresponds to the Newtonian equations of motion

$$\ddot{q}_n = \sum_{m=1, m\neq n}^{N} \dot{q}_n \dot{q}_m\, w_{nm}(q_n - q_m) . \tag{16}$$

Moreover, these equations of motion obtain from the following Hamiltonian (not of normal type):

$$H(\underline{q},\underline{p})=\sum_{n=1}^{N} h_n(s\,p_n;\underline{q}) \ , \tag{17a}$$

where s is an arbitrary (nonvanishing) constant and

$$h_n(p;\underline{q})=\exp\left[\, p-\frac{1}{2}\sum_{l=1,l\neq n}^{N} W_{nl}(q_n-q_l)\right] , \tag{17b}$$

$$w_{nm}(x)=[\; W'_{nm}(x)-W'_{mn}(-x)\;]/2 \ . \tag{17c}$$

Note that this formula, (17c), implies (15).

Proof. It is easily seen that (16) corresponds, via (15), to the diagonal terms of the Lax equation (4). As for the off-diagonal terms of (4), they yield (see (2.1.8-15), (1) and (14))

$$(\ddot{q}_n/\dot{q}_n)+(\ddot{q}_m/\dot{q}_m)=-(\dot{q}_n-\dot{q}_m)\,w_{nm}(q_n-q_m)$$

$$+\sum_{l=1;l\neq n,m}^{N} \dot{q}_l\left[-w_{lm}(q_l-q_m)+w_{nl}(q_n-q_l)\right] , \tag{18}$$

and it is easily seen that (16) and (15) entail that this equation is indeed satisfied.

The Hamiltonian equations, see (1.2-1), corresponding to (17a) read

$$\dot{q}_n=\partial h_n(s\,p_n;\underline{q})/\partial p_n \ , \tag{19a}$$

$$\dot{p}_n=-\sum_{l=1}^{N}\partial h_l(s\,p_l;\underline{q})/\partial q_n \ . \tag{19b}$$

From (19a), using (17b), we get

$$\dot{q}_n=s\exp\left[\, s\,p_n-\frac{1}{2}\sum_{l=1,l\neq n}^{N} W_{nl}(q_n-q_l)\right] , \tag{20a}$$

$$\dot{q}_n=s\,h_n(s\,p_n;\underline{q}) \ . \tag{20b}$$

Logarithmic t-differentiation of (20a) yields

$$\ddot{q}_n/\dot{q}_n=s\,\dot{p}_n-\frac{1}{2}\sum_{l=1,l\neq n}^{N}(\dot{q}_n-\dot{q}_l)\,W'_{nl}(q_n-q_l) \ . \tag{21}$$

One the other hand logarithmic differentiation with respect to q_n of h_l, see (17b), yields

$$[\ \partial h_l(s\,p_l;\underline{q})/\partial q_n\ ]/h_l(s\,p_l;\underline{q})$$

$$=-\delta_{nl}\frac{1}{2}\sum_{l'=1,l'\neq n}^{N}W'_{nl'}(q_n-q_{l'})+(1-\delta_{nl})\frac{1}{2}W'_{ln}(q_l-q_n)\ , \tag{22}$$

hence, using (20b),

$$\partial h_n(sp_l;\underline{q})/\partial q_n$$

$$=(\dot{q}_\ell/s)\{-\delta_{n\ell}\frac{1}{2}\sum_{\ell'=1,\ell'\neq n}^{N}W'_{n\ell'}(q_n-q_\ell)+(1-\delta_{n\ell})\frac{1}{2}W'_{\ell n}(q_\ell-q_n)\}\ . \tag{23}$$

Insertion of this into (19b) yields

$$\dot{p}_n=s^{-1}\frac{1}{2}\sum_{\ell=1,\ell\neq n}^{N}\{\dot{q}_n W'_{n\ell}(q_n-q_\ell)-\dot{q}_\ell W'_{\ell n}(q_\ell-q_n)\}\ , \tag{24}$$

and insertion of this into (21) yields

$$\ddot{q}_n/\dot{q}_n=\frac{1}{2}\sum_{\ell=1,\ell\neq n}^{N}\dot{q}_\ell\left[W'_{n\ell}(q_n-q_\ell)-W'_{\ell n}(q_\ell-q_n)\right], \tag{25}$$

which coincides, via (17c), with (16).

2.1.10 N particles on the line, interacting pairwise via forces equal to twice the product of their velocities divided by their mutual distance

In Sect. 2.1.10, and in the following Sects. 2.1.10.1,2, we discuss the N-body problem characterized by the Newtonian equations of motion

$$\ddot{q}_n=2\sum_{m=1,m\neq n}^{N}\dot{q}_n\,\dot{q}_m/(q_n-q_m)\,. \tag{1}$$

These equations of motion contain no ("coupling") constant. They are invariant under ("space") translations $(q_n(t)\rightarrow\tilde{q}_n(t)=q_n(t)+q_0, \dot{q}_0=0)$; they are not invariant under Galilei transformations $(q_n(t)\rightarrow\tilde{q}_n(t)=q_n(t)+v_0\,t)$; they are autonomous, hence invariant under translation of the time variable $(q_n(t) \to \tilde{q}_n(t) = q_n(t+t_0))$; and they are moreover invariant under rescaling of both the independent ("time") variable and the dependent ("space") variables $(q_n(t) \to \tilde{q}_n(t) = a q_n(bt))$, with a and b two arbitrary constants). These latter two invariance properties underline the remarkable nature of these equations of motion.

This model, (1), is discussed in some detail in Sect. 2.1.10 and the following Sects. 2.1.10.1,2; but, for a much richer understanding of its dynamics one must pass from the *real* to the *complex*: see Chap. 4, and in particular Sect. 4.2.4.

The equations of motion (1) correspond to (2.1.8-7) with

$$w(x) = 2/x. \tag{2}$$

Hence the evolution (1) is of Lax type, $\dot{\underline{L}} = [\underline{L}, \underline{M}]$, the matrices $\underline{L}$ and $\underline{M}$ being given by the representation (2.1.8-1,2) with

$$\alpha(x) = 1 + ax \ , \tag{3}$$

$$\beta(x) = 0 \ , \tag{4}$$

$$\gamma(x) = -1/x \ . \tag{5}$$

Exercise 2.1.10-1. Verify that these expressions, (2), (3), (4) respectively (5), correspond to (2.1.9-10), (2.1.9-2), (2.1.9-8) respectively (2.1.9-9), with $b = 0$.

Hence the explicit expressions of the matrices $\underline{L}$ and $\underline{M}$ read (see (2.1.8-1,2))

$$L_{nm} = \delta_{nm}\dot{q}_n + (1-\delta_{nm})(\dot{q}_n \dot{q}_m)^{1/2}[1 + a(q_n - q_m)] = (\dot{q}_n \dot{q}_m)^{1/2}[1 + a(q_n - q_m)] \ , \tag{6}$$

$$M_{nm} = -(1-\delta_{nm})(\dot{q}_n \dot{q}_m)^{1/2}(q_n - q_m)^{-1} \ . \tag{7}$$

We have seen in the previous Sect. 2.1.9.1 that, for $a = 0$, this is a fake Lax pair: the time-independence of the traces of the powers of the Lax matrix (6) with $a = 0$ (which coincides with (2.1.9.1-1)) yields only one constant of motion, see (2.1.9.1-10). This is as well (or perhaps as badly ?) the case for the Lax matrix (6) with $a \neq 0$, except that in this case one gets 3 constants of the motion rather than only one, namely

$$c_p = \sum_{n=1}^{N} \dot{q}_n (q_n)^p, \quad p = 0,1,2 \ . \tag{8}$$

It is indeed clear from (6) that the traces T_n, see (2.1.9), can all be expressed in terms of c_0, c_1 and c_2, see (8).

Proof: from (2.1-9) and (6)

$$T_n = \sum_{m_1,m_2,\ldots,m_n=1}^{N} \dot{q}_{m_1}\dot{q}_{m_2}\ldots\dot{q}_{m_n}\left[1+a(q_{m_1}-q_{m_2})\right]\left[1+a(q_{m_2}-q_{m_3})\right]$$

$$\ldots\left[1+a(q_{m_{n-1}}-q_{m_n})\right]\left[1+a(q_{m_n}-q_{m_1})\right] \tag{9a}$$

hence, see (8),

$$T_n = F_n(c_0, c_1, c_2) , \tag{9b}$$

where F_n is a function of its 3 arguments whose computation is left as an (irrelevant) exercise for the very diligent reader.

One might therefore doubt that the system (1) is *integrable*. But in the following Sect. 2.1.10.1 we show that this system is actually *solvable*, via an (appropriate) application of the OP technique.

2.1.10.1 Technique of solution OP

To apply this technique we return, for simplicity, to the Lax pair (2.1.10-6) with $a=0$, which can be written as follows:

$$L_{nm} = (\dot{q}_n\,\dot{q}_m)^{1/2} , \tag{1}$$

$$M_{nm} = -(1-\delta_{nm})(q_n - q_m)^{-1} L_{nm} . \tag{2}$$

Note the close similarity of the last equation to (2.1.3.2-12), or, equivalently, to (2.1.5-8) with $a=0$. This immediately suggests that we introduce, as in Sect. 2.1.3.2, the matrix

$$\underline{Q}(t) = \mathrm{diag}\left[q_n(t)\right],\quad \underline{Q}_{nm}(t) = \delta_{nm}\, q_n(t) , \tag{3}$$

and we take as starting point for the application of the OP technique, in addition to the Lax evolution equation

$$\dot{\underline{L}} = \left[\underline{L}, \underline{M}\right], \tag{4}$$

the matrix evolution equation

$$\dot{\underline{Q}} = [\,\underline{Q}, \underline{M}\,] + \underline{L} \,, \tag{5}$$

which is clearly entailed by (2) and (1).

Proof. The diagonal terms of (5) yield the identities $\dot{q}_n = \dot{q}_n$, and the off-diagonal terms yield the equations

$$0 = (q_n - q_m) M_{nm} + L_{nm}, \quad m \neq n, \tag{5a}$$

which are obviously satisfied, see (2).

We now proceed in close analogy to the previous treatments, see Sect. 2.1.3.2 and 2.1.5, introducing the similarity transformation

$$\underline{\tilde{L}} = \underline{U}\,\underline{L}\,\underline{U}^{-1} \,, \tag{6a}$$

$$\underline{\tilde{M}} = \underline{U}\,\underline{M}\,\underline{U}^{-1} \,, \tag{6b}$$

$$\underline{\tilde{Q}} = \underline{U}\,\underline{Q}\,\underline{U}^{-1} \,, \tag{6c}$$

with the invertible matrix $\underline{U}(t)$ defined by the evolution equation

$$\dot{\underline{U}} = \underline{U}\,\underline{M} \,, \tag{7a}$$

which entails of course

$$\underline{U}^{-1}\dot{\underline{U}} = \underline{M} \,, \tag{7b}$$

$$\dot{\underline{U}}\,\underline{U}^{-1} = \underline{\tilde{M}} \,, \tag{7c}$$

and by the convenient initial condition

$$\underline{U}(0) = \underline{1} \,. \tag{8}$$

We thereby obtain

$$\dot{\underline{\tilde{L}}} = 0 \,, \tag{9}$$

$$\dot{\underline{\tilde{Q}}} = \underline{\tilde{L}} \,, \tag{10}$$

entailing (see (8) and (6))

$$\underset{\sim}{\tilde{L}}(t)=\underset{\sim}{\tilde{L}}(0)=\underset{\sim}{L}(0) \quad , \tag{11}$$

$$\underset{\sim}{\tilde{Q}}(t)=\underset{\sim}{\tilde{Q}}(0)+\underset{\sim}{\tilde{L}}(0)t=\underset{\sim}{Q}(0)+\underset{\sim}{L}(0)t \, . \tag{12}$$

Hence we conclude, see (6c), that the initial-value problem for (2.1.10-1) is solved by the following recipe: the coordinates $q_n(t)$ coincide with the eigenvalues of the matrix $\underset{\sim}{\tilde{Q}}(t)$, whose explicit expression reads

$$[\ \underset{\sim}{\tilde{Q}}(t)\ ]_{nm}=\delta_{nm}\left[q_n(0)+\dot{q}_n(0)t\right]+(1-\delta_{nm})\left[\dot{q}_n(0)\dot{q}_m(0)\right]^{1/2}t \quad , \tag{13a}$$

$$[\ \underset{\sim}{\tilde{Q}}(t)\ ]_{nm}=\delta_{nm}\,q_n(0)+\left[\dot{q}_n(0)\dot{q}_m(0)\right]^{1/2}t \, . \tag{13b}$$

2.1.10.2 Behavior of the solutions: mention of future developments

The eigenvalues of the matrix $\underset{\sim}{\tilde{Q}}(t)$ coincide with the N roots of the polynomial of degree N in q defined by the formula

$$\det\left[q\underset{\sim}{1}-\underset{\sim}{\tilde{Q}}(t)\right]=q^N+\sum_{m=1}^{N}c_m(t)q^{N-m} \, . \tag{1}$$

Here $\underset{\sim}{\tilde{Q}}(t)$ is defined by (2.1.10.1-13), and this formula, (1), defines the N coefficients $c_m(t)$.

Hence we can write

$$\prod_{n=1}^{N}\left[q-q_n(t)\right]=q^N+\sum_{m=1}^{N}c_m(t)q^{N-m} \quad . \tag{2}$$

As it is well known, the mapping between the set $\{q_n(t); n=1,...,N\}$ and the set $\{c_m(t); m=1,...,N\}$ is one-to-one (it is the mapping between the N zeros and the N coefficients of a monic polynomial): of course the set $\{q_n(t); n=1,...,N\}$ is not ordered (namely, the $N!$ sets obtained by permuting the zeros $q_n(t)$ among themselves must not be considered as different sets), while the set $\{c_m(t); m=1,...,N\}$ is of course ordered.

The fact that the matrix $\underset{\sim}{\tilde{Q}}(t)$ is linear in t, see (2.1.10.1-13), clearly entails that the left hand side of (1) is a polynomial in t, of degree at most N; hence the coefficients $c_m(t)$, see (1), are also polynomials in t, of degree at most N. But in fact the special nature of the matrix $\underset{\sim}{\tilde{Q}}(t)$, which is the sum of a time-independent diagonal matrix and of a dyadic matrix linear in t, see (2.1.10.1-13), entails that the left hand side of (1), hence all the quantities $c_m(t)$, are *linear* in *t*, so that

$$\ddot{c}_m(t)=0 \ . \tag{3}$$

Proof. Let us rewrite (2.1.10.1-13b) as follows:

$$\underline{\tilde{Q}}(t)=\mathrm{diag}[q_n(0)]+t\,V(0)\underline{v}^{(1)}\otimes\underline{v}^{(1)} \ , \tag{4a}$$

$$[\ \underline{\tilde{Q}}(t)\ ]_{nm}=\delta_{nm}\,q_n(0)+t\,V(0)v_n^{(1)}\,v_m^{(1)} \ , \tag{4b}$$

with

$$V(0)=\sum_{n=1}^{N}q_n(0) \ , \tag{5}$$

$$v_n^{(1)}=\{[\dot{q}_n(0)]/V(0)\}^{1/2} \ , \tag{6}$$

so that the vector $\underline{v}^{(1)}$ is normalized,

$$(\underline{v}^{(1)},\underline{v}^{(1)})=1 \ . \tag{7}$$

Let us now introduce $N-1$ other vectors, $\underline{v}^{(j)}$, $j=2,...,N$, which, together with $\underline{v}^{(1)}$ form an orthonormal set, so that

$$(\underline{v}^{(n)},\underline{v}^{(m)})\equiv\sum_{j=1}^{N}v_j^{(n)}\,v_j^{(m)}=\delta_{nm} \ , \tag{8a}$$

as well as

$$\sum_{n=1}^{N}v_j^{(n)}\,v_k^{(n)}=\delta_{jk} \ . \tag{8b}$$

Note that all these vectors are time-independent.

Now define the (time-independent) matrix $\underline{V}$ whose matrix elements are given by the following prescription:

$$(\underline{V})_{nm}=v_m^{(n)} . \tag{9a}$$

Then clearly (8) entails

$$(\underline{V}^{-1})_{nm}=v_n^{(m)} . \tag{9b}$$

Let now

$$\underline{\tilde{\tilde{Q}}}(t) = \underline{V}\,\underline{\tilde{Q}}(t)\underline{V}^{-1} . \tag{10}$$

Clearly

$$\det\left[\underline{\tilde{\tilde{Q}}}(t)\right] = \det\left[\underline{\tilde{Q}}(t)\right] . \tag{11}$$

But (4a), (8a) and (9b) imply that only the first diagonal element of the matrix $\underline{\tilde{\tilde{Q}}}(t)$ depends (linearly) on t, while all the other elements are t-independent:

$$\left[\underline{\tilde{\tilde{Q}}}(t)\right]_{nm} = \sum_{j=1}^{N} v_j^{(n)}\, q_j(0)\, v_j^{(m)} + \delta_{n1}\,\delta_{m1}\, V(0)\,\mathrm{t} . \tag{12}$$

Exercise 2.1.10.2-1. Verify!

Hence the determinant of $\underline{\tilde{\tilde{Q}}}(t)$, and as well the determinant of $\underline{\tilde{Q}}(t)$, see (11), is *linear* in t: the determinant of a matrix is a sum of terms, each of which is a product (of elements of the matrix) which contains only one element belonging to the first line (or to the first column).

This proof requires that the initial conditions entail $V(0) \neq 0$, see (5); if instead $V(0) = 0$, the above treatment has to be modified. Since in any case we will return to this topic below, see Sect. 2.3.4.2, we leave this adjustment as an *exercise* for the diligent reader.

The result (3) is remarkable, and it leads to the following neat solution of the initial-value problem for the equations of motion (2.1.10-1): the N roots of the following equation in q,

$$\sum_{n=1}^{N} \dot{q}_n(0)/\left[q - q_n(0)\right] = t^{-1} , \tag{13}$$

yield the N coordinates $q_n(t)$. Note the neat way in which the initial data, $q_n(0)$ and $\dot{q}_n(0)$, are encoded in this formula.

Exercise 2.1.10.2-2. Prove this statement, and analyze its implications, namely the motions entailed by (2.1.10-1). *Solution*: see Sect. 2.3.4.2.

2.1.10.3 Can a fake Lax pair be used to solve a nontrivial many-body problem?

The development reported in Sect. 2.1.10.1 demonstrates that the question posed by the title of the present Sect. 2.1.10.3 must be answered positively. It is therefore justified to wonder how much further such an approach can be pushed.

In particular we saw in Sect. 2.1.9.1 that the Newtonian equations of motion (2.1.9.1-16),

$$\ddot{q}_n = \sum_{m=1,m\neq n}^{N} \dot{q}_n \dot{q}_m \, w_{nm}(q_n - q_m), \tag{1}$$

correspond to the Lax evolution equation

$$\dot{\underline{L}} = [\underline{L}, \underline{M}] , \tag{2}$$

with $\underline{L}$ and $\underline{M}$ given by (2.1.9.1-1) and (2.1.9.1-14),

$$L_{nm} = (\dot{q}_n \dot{q}_m)^{1/2} , \tag{3}$$

$$M_{nm} = -\frac{1}{2}(1-\delta_{nm})(\dot{q}_n \dot{q}_m)^{1/2} \, w_{nm}(q_n - q_m), \tag{4}$$

under the sole “oddness” condition (2.1.9.1-15),

$$w_{nm}(x) = -w_{mn}(x) . \tag{5}$$

On the other hand we have learned, from repeated applications of the OP technique (see Sects. 2.1.3.2, 2.1.3.3, 2.1.5 and 2.1.10.1), that key to the applicability of this technique of solution is the availability of a second matrix evolution equation, in addition to the Lax equation (2). Hence we propose now to explore whether, by assuming the existence of such a second equation, we can discover some new *solvable* model. To this end we introduce the matrix

$$\underline{G} = \mathrm{diag}[g(q_n)] , \tag{6}$$

with $g(q)$ a function yet to be determined, and we require that it satisfy the evolution equation

$$\dot{\underline{G}} = [\underline{G}, \underline{M}] + a\{\underline{G}, \underline{L}\} + b[\underline{G}, \underline{L}] + c\,\underline{G} + d\,\underline{L} + h , \tag{7}$$

with a,b,c,d,h five arbitrary (scalar) constants. The justification for this *ansatz* for the right hand side of this equation is that it entails the possibility to perform all subsequent steps in the OP technique.

The compatibility of (7) with (6), (3), (4) and (5) yields the following constraints:

$$b=c=h=0 \ , \tag{8}$$

$$g(x)=C\exp(2ax)-d/(2a) \ , \tag{9}$$

$$w_{nm}(x)=2a\,\mathrm{cotanh}(ax) \ , \tag{10}$$

with C another arbitrary constant. Note however that neither C, nor d, enter in the expression (10) of $w_{nm}(x)$, which is moreover independent of the indices n and m.

Proof. Insertion of (3), (4) and (6) in the diagonal part of (7) yields

$$g'(q_n)\,\dot{q}_n=2a\,g(q_n)\,\dot{q}_n+c\,g(q_n)+d\,\dot{q}_n+h \tag{11}$$

implying

$$c=h=0 \tag{12}$$

and

$$g'(x)=2a\,g(x)+d \tag{13}$$

which yields (9).

Likewise, and using (12), the off-diagonal part of (7) yields

$$[g(q_n)-g(q_m)]\,w_{nm}(q_n-q_m)=2a[g(q_n)+g(q_m)]+2b[g(q_n)-g(q_m)]+2d \tag{14}$$

which, via (9), yields

$$w_{nm}(x)=2a\,\mathrm{cotanh}(ax)+2b\,. \tag{15}$$

Finally the requirement (5) entails

$$b=0\,. \tag{16}$$

This concludes the proof. The diligent reader will check that, if one had started from the more general *ansatz* for $\underline{G}$ that obtains by replacing in (6) $g(q_n)$ with $g_n(q_n)$, the more general result obtained would have merely corresponded to that entailed by the trivial freedom to shift every coordinate $q_n(t)$ by an arbitrary (different) constant,

$$q_n(t) \to \tilde{q}_n(t) = q_n(t) + c_n, \; \dot{c}_n = 0 .$$

We therefore conclude that the Newtonian equations of motion

$$\ddot{q}_n = 2a \sum_{m=1, m \neq n}^{N} \dot{q}_n \, \dot{q}_m \operatorname{cotanh}[a(q_n - q_m)] \tag{17}$$

are *solvable*. Note that, for $a = 0$, they reduce to (2.1.10-1). We will see in the following Sect. 2.1.11 that this model, (17), is also contained in the class of many-body problems whose integrability is entailed by a Lax pair of type (2.1.8-1,2), leading to the functional equation (**), see (2.1.8-19).

Exercise 2.1.10.3-1. Obtain, via the OP technique, the formulas that provide the solution of the initial-value problem for (17). *Solution*: see Sects. 2.1.12.4 and 2.3.4.2.

Exercise 2.1.10.3-2. Obtain for the *solvable* many-body problem (17) a result in some way analogous to that yielded, for (2.1.10-1), by (2.1.10.2-2,3). *Solution*: see Sect. 2.3.4.2.

2.1.11 General solution of the functional equation (**)

The functional equation (**), see (2.1.8-19), reads

$$[\alpha(x)\alpha'(y) - \alpha(y)\alpha'(x)]/[\alpha(x+y) - \alpha(x)\alpha(y)] = \eta(x) - \eta(y) . \tag{1}$$

In Sect. 2.1.11 we discuss its general solution.

Let us recall that the Newtonian equations of motion (2.1.8-7),

$$\ddot{q}_n = \sum_{m=1, m \neq n}^{N} \dot{q}_n \, \dot{q}_m \, w(q_n - q_m) , \tag{2}$$

are associated with the functional equation (1), via the relation

$$w(x) = v'(x)/[v(x) - 1] = (d/dx)\log[v(x) - 1] \tag{3}$$

where

$$v(x) = \alpha(x)\alpha(-x) . \tag{4}$$

This formula, (3) with (4), coincides with (2.1.8-22,23) via the relation

$$\alpha(0) = 1 \, , \tag{5}$$

which is easily seen to follow from (1) (see below). Note that (3) with (5) entail that $w(x)$ is singular at $x = 0$ (see below). We also recall that $w(x)$ coincides with the odd part of $2\eta(x)$, see (2.1.8-21),

$$w(x) = \eta(x) - \eta(-x) = -w(-x) \, . \tag{6}$$

Let us emphasize that the simultaneous validity of (3) with (4) and (5), and of (6), is nontrivial.

The functional equation (1), or rather the equation (2.1.8-13), admits the trivial solution

$$\alpha(x) = \exp(\rho x) \tag{7}$$

with arbitrary $\eta(x)$. The corresponding Lax matrix, up to the similarity transformation from $\underline{L}$ to $\underline{\tilde{L}}$ (see (2.1-17)),

$$\underline{\tilde{L}} = \underline{R} \, \underline{L} \, \underline{R}^{-1} \tag{8}$$

with

$$\underline{R} = \mathrm{diag}\left[\exp(\rho q_n)\right] , \tag{9}$$

coincides with the case considered in Sect. 2.1.9.1, see (2.1.9.1-1). In the following we ignore this anomalous (and trivial) case.

The functional equation (1) with (5) is clearly invariant under the transformation

$$\alpha(x) \to \tilde{\alpha}(x) = \alpha(ax) \exp(bx) \, , \tag{10a}$$

$$\eta(x) \to \tilde{\eta}(x) = a\eta(ax) + \eta_0 \, , \tag{10b}$$

with a, b and η_0 arbitrary constants $(a \neq 0)$, which entails

$$w(x) \to \tilde{w}(x) = a \, w(ax) \, . \tag{10c}$$

It is also invariant under the (somewhat less trivial) transformation

$$\alpha(x) \to \tilde{\alpha}(x) = 1/\alpha(x), \tag{11a}$$

$$\eta(x) \to \tilde{\eta}(x) = \eta(x) - \alpha'(x)/\alpha(x), \tag{11b}$$

which entails (via (3) and (4))

$$w(x) \to \tilde{w}(x) = w(x)/v(x) = w(x)/[\alpha(x)\alpha(-x)]. \tag{11c}$$

Of course these two transformations, (10) and (11), can also be combined, i.e. performed sequentially.

The next step is to derive from (1), or rather from (2.1.8-13), an expression of $\eta(x)$ in terms of $\alpha(x)$ and its derivatives, as well as a differential equation for $\alpha(x)$. To this end we firstly parametrize the behavior of $\alpha(\varepsilon)$ and $\eta(\varepsilon)$ as $\varepsilon \to 0$, by setting

$$\alpha(\varepsilon) = \alpha_0 + \alpha_1\,\varepsilon + \frac{1}{2}\alpha_2\,\varepsilon^2 + \frac{1}{6}\alpha_3\,\varepsilon^3 + O(\varepsilon^4)\ , \tag{12}$$

$$\eta(\varepsilon) = \eta_{-1}\varepsilon^{-1} + \eta_0 + \eta_1\,\varepsilon + O(\varepsilon^2)\ . \tag{13}$$

(The validity of this parameterization is implied by the consistency of the following developments). Then, by setting $y = \varepsilon$ in (1), or rather in (2.1.8-13), we get

$$\alpha_0 = 1, \tag{14}$$

$$\eta_{-1} = 1, \tag{15}$$

$$\eta(x) = \eta_0 + \frac{1}{2}[\alpha''(x) - 2\alpha_1\,\alpha'(x) + \alpha_2\,\alpha(x)]/[\alpha'(x) - \alpha_1\,\alpha(x)]\ , \tag{16}$$

and

$$2[\alpha'(x) - \alpha_1\,\alpha(x)]\,\alpha'''(x) - 3[\alpha''(x) - 2\alpha_1\,\alpha'(x)]\,\alpha''(x)$$

$$+ a[\alpha'(x)]^2 + b\,\alpha(x)\alpha'(x) + c[\alpha(x)]^2 = 0\ , \tag{17a}$$

with

$$a = 6(2\eta_1 - \alpha_2)\ , \tag{17b}$$

$$b = 4(\alpha_3 - 6\alpha_1\,\eta_1)\ , \tag{17c}$$

$$c = 3\alpha_2^2 - 4\alpha_1\alpha_3 + 12\alpha_1^2\eta_1 \ . \tag{17d}$$

Note that (14) corresponds to (5), that is thereby proven.

Proof. From (2.1.8-13) with $y = \varepsilon$ we get, in the $\varepsilon \to 0$ limit, via (12) and (13),

$$\alpha(x)\left[\alpha_1 + \alpha_2\varepsilon + \frac{1}{2}\alpha_3\,\varepsilon^2\right] - \left[\alpha_0 + \alpha_1\,\varepsilon + \frac{1}{2}\alpha_2\,\varepsilon^2\right]\alpha'(x)$$

$$= \left\{\left[\alpha(x) + \varepsilon\,\alpha'(x) + \frac{1}{2}\varepsilon^2\,\alpha''(x) + \frac{1}{6}\varepsilon^3\,\alpha'''(x)\right] - \alpha(x)\left[\alpha_0 + \alpha_1\,\varepsilon + \frac{1}{2}\alpha_2\,\varepsilon^3 + \frac{1}{6}\alpha_3\,\varepsilon^3\right]\right\}\cdot$$

$$\cdot\left[\eta(x) - \eta_{-1}\,\varepsilon^{-1} - \eta_0 - \eta_1\,\varepsilon\right] + O(\varepsilon^3) \ . \tag{18}$$

This yields, to order $\varepsilon^p, p = -1,0,1,2,$ the following relations:

$$(1-\alpha_0)\eta_{-1} = 0 \ , \tag{19a}$$

$$\left[\alpha_1\,\alpha(x) - \alpha'(x)\right](1-\eta_{-1}) = (1-\alpha_0)\,\alpha(x)\left[\eta(x) - \eta_0\right], \tag{19b}$$

$$\alpha_2\,\alpha(x) - \alpha_1\,\alpha'(x) = \frac{1}{2}\left[\alpha_2\,\alpha(x) - \alpha''(x)\right]\eta_{-1} - \left[\alpha_1\,\alpha(x) - \alpha'(x)\right]\left[\eta(x) - \eta_0\right], \tag{19c}$$

$$\frac{1}{2}\left[\alpha_3\,\alpha(x) - \alpha_2\,\alpha'(x)\right]$$

$$= \frac{1}{6}\left[\alpha_3\,\alpha(x) - \alpha'''(x)\right]\eta_{-1} - \frac{1}{2}\left[\alpha_2\,\alpha(x) - \alpha''(x)\right]\left[\eta(x) - \eta_0\right] + \left[\alpha_1\,\alpha(x) - \alpha'(x)\right]\eta_1, \tag{19d}$$

and using these equations one gets (16) from (19c). Finally, using (14), (15) and (19c), with a little (trivial) labor, one gets (17).

The 7 coefficients $\alpha_0, \alpha_1, \alpha_2, \alpha_3, \eta_{-1}, \eta_0, \eta_1$ in (12) and (13) were *a priori* arbitrary. Now 2 of these, α_0 and η_{-1}, have been fixed to unity, see (14) and (15), and another one, η_0, is clearly irrelevant (see (10b)). There remain 4, so far undetermined, coefficients, which correspond to the 4 coefficients, α_1, a, b, c, appearing in (17a).

The consistency of (17) with (12) and (14) is obviously implied by the way (17) has been derived, yet the diligent reader might wish to verify it directly.

The ODE (17a) is a consequence of the functional equation (1) with (5); hence any solution $\alpha(x)$ of (1) must satisfy (17a) (note that the converse need not be true, there is no *a priori* guarantee that a solution of (17a) also satisfy (1)). But the most general solution of the third order ODE (17a) can contain at most 6 arbitrary constants: indeed $6=4+3-1$, 4 being the number of *a priori* arbitrary constants that appear in (17a), 3 being the order of this ODE, and 1 being the number of constraints the solution $\alpha(x)$ is required to satisfy (see (5)); of course the 6 parameters characterizing a solution of (17a) and (5) determine the values of the 4 constants $\alpha_1,a,b,c,$ appearing in (17a), in addition to, say, the values of $\alpha'(0)$ and $\alpha''(0)$ $(\alpha(0)=1$, see (5)). Hence the most general solution of the functional equation (1) can contain at most 6 arbitrary parameters.

We now exhibit such a solution of (1), containing 6 arbitrary parameters. It reads

$$\alpha(x)=\exp(\rho x)\frac{\sigma(\mu|\omega,\omega')\sigma(\lambda x+\nu|\omega,\omega')}{\sigma(\nu|\omega,\omega')\sigma(\lambda x+\mu|\omega,\omega')}, \tag{20}$$

where σ is the "sigma" Weierstrass function, see Appendix A.

The 6 arbitrary parameters in this expression, (20), of $\alpha(x)$ are $\rho,\mu,\nu,\lambda,\omega,\omega'$. It is plain that (20) entails $\alpha(0)=1$, see (5).

The diligent reader, using the appropriate formulas from Appendix A, will readily verify that this expression of $\alpha(x)$ is invariant under the transformations (10a) and (11a), whose only effect is to cause a redefinition of (some of) the 6 parameters in (20).

The verification that $\alpha(x)$, see (20), satisfies the ODE (17a) is a cumbersome task, that can be left as an *exercise* for the diligent reader. This result is of course implied by the property of (20) to satisfy (1), which is proved below.

To prove that the expression (20) of $\alpha(x)$, with no (nontrivial) restriction on the 6 parameters $\rho,\mu,\nu,\lambda,\omega,\omega'$, satisfies (1) (yielding in the process an appropriate expression for $\eta(x)$), is a cumbersome task, that can however be eased thanks to the following two remarks.

In the first place, by taking advantage of the invariance property (10a) one can simplify (20) to read

$$\alpha(x)=\frac{\sigma(\mu)\sigma(x+\nu)}{\sigma(\nu)\sigma(x+\mu)}. \tag{21}$$

Here and below, as it is also often done in Appendix A, we omit to indicate explicitly the dependence on the 2 "semiperiods" ω and ω'.

The expression of $\eta(x)$ that corresponds to (21) is, up to an irrelevant additive constant,

$$\eta(x) = \zeta(x) - \zeta(x+\mu) \ , \tag{22}$$

where ζ is the “zeta” Weierstrass function, see Appendix A.

The validity of this expression of $\eta(x)$ is proven below; the diligent reader might try and derive it immediately from (21) via (16).

The second remark that simplifies our task to verify that (20) satisfies (1), but has in fact an interest of its own, is contained in the following

Proposition 2.1.11-1. If $\alpha(x)$ satisfies the functional equation

$$\alpha(x+y) = \alpha(x)\alpha(y) + \varphi(x)\varphi(y)\psi(x+y) \ , \tag{23}$$

with $\varphi(x)$ and $\psi(x)$ *a priori* arbitrary functions, then it also satisfies the functional equation (1).

Of course in (1) the function $\eta(x)$ is also, *a priori*, arbitrary. But it actually turns out that the two *a priori* arbitrary functions $\eta(x)$ and $\varphi(x)$ are related to each other as follows:

$$\eta(x) = \varphi'(x) / \varphi(x) . \tag{24}$$

Before proving this Proposition 2.1.11-1, as well as (24), let us interject 3 *remarks.*

Remark 2.1.11-2. The functional equation (23), in contrast to the functional equation (1), contains no differentiations.

Remark 2.1.11-3. By taking advantage of the *a priori* arbitrariness of the 3 functions $\alpha(x)$, $\varphi(x)$ and $\psi(x)$ appearing in (23), this functional equation can be presented in many different guises, for instance

$$\alpha(x)\alpha(y) / \alpha(x+y) = 1 + \varphi(x)\varphi(y) / \chi(x+y) \ , \tag{25}$$

which corresponds to (23) via the position

$$\chi(x) = -\alpha(x) / \psi(x) . \tag{26}$$

The diligent reader will obtain, by appropriate substitutions, several other avatars of the functional equation (23) (see Appendix B).

Remark 2.1.11-4. It is plain that if $\alpha(x)$, $\varphi(x)$ and $\psi(x)$ satisfy (23), so do

$$\tilde{\alpha}(x) = \alpha(ax)\exp(bx) \ , \tag{27a}$$

$$\tilde{\varphi}(x) = c\,\varphi(ax)\exp[(b+d)x] \ , \tag{27b}$$

$$\tilde{\psi}(x) = c^{-2}\psi(ax)\exp(-d\,x) \ , \tag{27c}$$

with a,b,c,d arbitrary constants, as well as

$$\tilde{\alpha}(x) = 1/\alpha(x) \ , \tag{28a}$$

$$\tilde{\varphi}(x) = \varphi(x)/\alpha(x) \ , \tag{28b}$$

$$\tilde{\psi}(x) = -\psi(x)/\alpha(x) \ . \tag{28c}$$

Clearly these invariance properties correspond to (10) respectively (11).

Let us then proceed to prove the *Proposition 2.1.11-1* stated above. To this end we define

$$F(x,y) = \log[1-\alpha(x)\alpha(y)/\alpha(x+y)] \ , \tag{29}$$

and we note that (1) implies

$$F_x(x,y) - F_y(x,y) = \eta(x) - \eta(y) \ , \tag{30}$$

where the subscripts denote partial differentiation. This first-order linear PDE has the general solution

$$F(x,y) = \mathrm{H}(x) + \mathrm{H}(y) + g(x+y) \ , \tag{31}$$

with $g(x)$ an *arbitrary* function and

$$\mathrm{H}'(x) = \eta(x) \ . \tag{32}$$

But (29) and (31) entail precisely (23) with (24), via the positions

$$\psi(x) = \alpha(x)\exp[g(x)] \ , \tag{33}$$

$$\varphi(x) = \exp[\mathrm{H}(x)] \ . \tag{34}$$

The last two equations, (33) and (34), correspond to the arbitrariness of $\psi(x)$ in (23) and, via (32), to the fact that $\varphi(x)$ is related to $\eta(x)$ by (24), which is thereby proven.

To prove that $\alpha(x)$, see (21), satisfies (23) we now conveniently set

$$N \equiv [\sigma(\nu)\sigma(x+\mu)\sigma(y+\mu)\sigma(x+y+\nu)] - [\mu \leftrightarrow \nu] , \tag{35}$$

$$D \equiv \sigma(\mu)\sigma(x+\nu)\sigma(y+\nu)\sigma(x+y+\mu) \quad , \tag{36}$$

so that (23) with (21) read

$$\alpha(x+y)/[\alpha(x)\alpha(y)] - 1 = N/D \ . \tag{37}$$

Of course the symbol $[\mu \leftrightarrow \nu]$ in the right hand side of (35) means: "the same expression within the square bracket preceding it, except for the interchange of μ and ν".

The main tool one must now use is (A-58a),

$$\sigma(z_1+z_2)\sigma(z_1-z_2) = \sigma^2(z_1)\sigma^2(z_2)[\wp(z_2) - \wp(z_1)] \ , \tag{38}$$

which entails the following relations:

$$\sigma(\nu)\sigma(x+y+\nu) = \sigma^2[\nu+(x+y)/2]\sigma^2[(x+y)/2]\{\wp[(x+y)/2] - \wp[\nu+(x+y)/2]\}, \tag{39a}$$

$$\sigma(x+\mu)\sigma(y+\mu) = \sigma^2[\mu+(x+y)/2]\sigma^2[(x-y)/2]\{\wp[(x-y)/2] - \wp[\mu+(x+y)/2]\}, \tag{39b}$$

and of course analogous ones with ν exchanged with μ.

Hence from (35), via (39), one gets

$$N = \tilde{N}\tilde{\tilde{N}} \tag{40}$$

$$\tilde{N} = \{\sigma[\mu+(x+y)/2]\,\sigma[\nu+(x+y)/2]\,\sigma[(x-y)/2]\,\sigma[(x+y)/2]\}^2 \quad , \tag{41}$$

$$\tilde{\tilde{N}} = \{\wp[(x+y)/2] - \wp[\nu+(x+y)/2]\}\,\{\wp[(x-y)/2] - \wp[\mu+(x+y)/2]\}$$
$$-\{\wp[(x+y)/2] - \wp[\mu+(x+y)/2]\}\,\{\wp[(x-y)/2] - \wp[\nu+(x+y)/2]\} , \tag{42a}$$

$$\tilde{\tilde{N}} = \{\wp[(x+y)/2] - \wp[(x-y)/2]\}\,\{\wp[\nu+(x+y)/2] - \wp[\mu+(x+y)/2]\} \ . \tag{42b}$$

Now one uses again (38), to get from (42b)

$$\tilde{\tilde{N}} = \sigma(x)\ \sigma(y)\ \sigma(\nu-\mu)\ \sigma(x+y+\mu+\nu)/\tilde{N} \quad . \tag{43}$$

Hence, from (40),

$$N = \sigma(x)\ \sigma(y)\ \sigma(\nu-\mu)\ \sigma(x+y+\mu+\nu)\ , \tag{44}$$

and, via (37) and (36), this yields precisely (23) with

$$\varphi(x) = \alpha(x)\ \sigma(x) / \sigma(x+\nu)\ , \tag{45}$$

$$\psi(x) = \sigma(\nu-\mu)\ \sigma(x+\mu+\nu) / \left[\sigma(\mu)\ \sigma(x+\mu)\right]\ . \tag{46}$$

From (45) and (21) one gets

$$\varphi(x) = \sigma(\mu)\ \sigma(x)\ /\ \left[\sigma(\nu)\ \sigma(x+\mu)\right]\ . \tag{47}$$

The last formula, (47), yields (22) via (24) and (A-39).

In conclusion let us report the most general expression (up to a trivial rescaling of the variable x) for the function $\alpha(x)$, as well as corresponding expressions for $\beta(x)$ and $\gamma(x)$, see (2.1.8-1,2), for $\eta(x)$, see (1), for $\varphi(x)$ and $\psi(x)$, see (23), for $\nu(x)$, see (4), and finally, and most importantly, for $w(x)$, see (2), (3) and (6):

$$\alpha(x) = \exp(\rho x)\ \sigma(\mu)\ \sigma(x+\nu) / \left[\sigma(\nu)\ \sigma(x+\mu)\right]\ , \tag{48a}$$

$$\beta(x) = c + \frac{1}{2}\wp'(\mu) / \left[\wp(x) - \wp(\mu)\right]\ , \tag{48b}$$

$$\gamma(x) = \alpha(x)\ \left[\zeta(x+\nu) - \zeta(x)\right]\ , \tag{48c}$$

$$\eta(x) = \eta_0 + \zeta(x) - \zeta(x+\mu)\ , \tag{48d}$$

$$\varphi(x) = \exp(\eta_0 x)\ \sigma(\mu)\ \sigma(x)\ / \left[\sigma(\nu)\ \sigma(x+\mu)\right]\ , \tag{48e}$$

$$\psi(x) = \exp\left[(\rho - \eta_0)x\right]\ \sigma(\nu-\mu)\ \sigma(x+\mu+\nu) / \left[\sigma(\mu)\ \sigma(x+\mu)\right]\ , \tag{48f}$$

$$\nu(x) = \left[\wp(x) - \wp(\nu)\right] / \left[\wp(x) - \wp(\mu)\right]\ , \tag{48g}$$

$$w(x) = \wp'(x) / \left[B - \wp(x)\right] \tag{48h}$$

with

$$B = \wp(\mu)\ . \tag{48i}$$

The expression (48g) of $v(x)$ follows from (48a) and (4) via (A-58); likewise, the expression (48h) with (48i) of $w(x)$ follows from (48d) and (6) via (A-59) and (A-41), or, more directly, from (48g) and (3).

The diligent reader will note again the obvious consistency of (48a) with (5), as well as the fact that the transformation (28) entails merely an exchange among the parameters μ and ν, as well as a change of sign of the (largely irrelevant) parameter ρ.

The diligent reader may also verify, using the appropriate formulas of Appendix A, that by taking advantage of the transformations (10) and (11) several alternative definitions of the function $\alpha(x)$ can be introduced, all of which however lead to the same expression (48h) for $w(x)$, possibly up to a rescaling of the independent variable x and a redefinition of the constant B. For instance the following 3 expressions of $\alpha(x)$, in terms of Jacobian elliptic functions, which also provide solutions of the functional equation (1),

$$\alpha(x) = \mathrm{sn}(\mu)/\mathrm{sn}(x+\mu) \ , \tag{49a}$$

$$\alpha(x) = \mathrm{sn}(\mu)\,\mathrm{cn}(x+\mu)/[\,\mathrm{cn}(\mu)\,\mathrm{sn}(x+\mu)] \ , \tag{49b}$$

$$\alpha(x) = \mathrm{sn}(\mu)\,\mathrm{dn}(x+\mu)/[\,\mathrm{dn}(\mu)\,\mathrm{sn}(x+\mu)] \ , \tag{49c}$$

all yield the same expression (48h) of $w(x)$, while the following 3 expressions of $\alpha(x)$,

$$\alpha(x) = \mathrm{cn}(x+\mu')/\mathrm{cn}(\mu'), \quad \mu' = \mu - \omega' \tag{50a}$$

$$\alpha(x) = \mathrm{dn}(x+\mu')/\mathrm{dn}(\mu'), \quad \mu' = \mu - \omega' \tag{50b}$$

$$\alpha(x) = \mathrm{dn}(\mu'')\,\mathrm{cn}(x+\mu'')/[\mathrm{cn}(\mu'')\,\mathrm{dn}(x+\mu'')], \quad \mu'' = \mu + \omega + \omega' \tag{50c}$$

which also provide solutions of the functional equation (1), all yield the same rescaled version of (48h), obtained via the replacement of x with ax where (see (A-19))

$$a = [\wp(\omega) - \wp(\omega')]^{-1/2} = (e_1 - e_3)^{-1/2} \ . \tag{51}$$

Let us emphasize that these developments entail that the most general function $w(x)$ for which (2) can be recast in Lax form via the *ansatz* (2.1.8-1,2) is given by the expression (48h), which features the 3 arbitrary constants B, ω and ω'. Note moreover that, quite generally, this function $w(x)$ has a simple pole at $x = 0$ with residue 2:

$$\lim_{x\to 0}[x\,w(x)] = 2 \ . \tag{52}$$

This follows from (A-22a).

The diligent reader, using (A-34,35), will check that the possibility to rescale the variables q_n in (2) ($q_n \to \tilde{q}_n = a q_n$) does not introduce an additional arbitrary constant, but corresponds merely to a rescaling of the 3 constants B, ω and ω'.

Likewise, no arbitrary constant is produced in (2) by a rescaling of the time variable $t \to \tilde{t} = bt$, since (2) (in contrast to, say, (2.1.1-15)) is clearly invariant under such a transformation.

Finally let us exhibit the special expressions that obtain, in place of (48), in the degenerate cases in which one, or both, of the semiperiods ω and ω' diverge.

For $\omega = \infty$, $\omega' = i\pi/2$ one obtains from (48) (setting moreover, for simplicity, $\rho = 0$, $\eta_0 = 0$, $c = -\operatorname{cotanh}(\mu)$), via (A-36) and (A-54),

$$\alpha(x) = \exp[(\mu - \nu)x/3] \sinh(\mu) \sinh(x+\nu)/[\sinh(\nu) \sinh(x+\mu)] \ , \tag{53a}$$

$$\beta(x) = \frac{1}{2} \sinh(2\mu)/[\sinh^2(x) - \sinh^2(\mu)] \ , \tag{53b}$$

$$\gamma(x) = -\alpha(x)\{\nu/3 + \sinh(\nu)/[\sinh(x) \sinh(x+\nu)]\} \ , \tag{53c}$$

$$\eta(x) = \sinh(\mu)/[\sinh(x) \sinh(x+\mu)] \ , \tag{53d}$$

$$\varphi(x) = \exp(\mu x/3) \sinh(\mu) \sinh(x)/[\sinh(\nu) \sinh(x+\mu)] \ , \tag{53e}$$

$$\psi(x) = \exp(-\nu x/3) \sinh(\nu - \mu) \sinh(x+\mu+\nu)/[\sinh(\mu) \sinh(x+\mu)] \ , \tag{53f}$$

$$v(x) = [\sinh(\mu)/\sinh(\nu)]^2 [\sinh^2(x) - \sinh^2(\nu)]/[\sinh^2(x) - \sinh^2(\mu)] \ , \tag{53g}$$

$$w(x) = 2\sinh^2(\mu) \operatorname{cotanh}(x)/[\sinh^2(\mu) - \sinh^2(x)] \ . \tag{53h}$$

Likewise, for $\omega = \infty$, $\omega' = i\infty$, setting for simplicity $\rho = 0$, $\eta_0 = 0$, $c = -b$, $\mu = b^{-1/2}$, $\nu = a^{-1/2}$, one gets via (A-37) and (A-55),

$$\alpha(x) = (1+ax)/(1+bx) \ , \tag{54a}$$

$$\beta(x) = -b/(1+b^2x^2) \ , \tag{54b}$$

$$\gamma(x) = -1/[x(1+bx)] \ , \tag{54c}$$

$$\eta(x) = 1/[x(1+bx)] \ , \tag{54d}$$

$$\varphi(x) = \mu x / [\nu(x+\mu)] \ , \tag{54e}$$

$$\psi(x) = (\nu - \mu)(x + \mu + \nu) / [\mu(x+\mu)] \ , \tag{54f}$$

$$v(x) = (1 - a^2 x^2)/(1 - b^2 x^2) \ , \tag{54g}$$

$$w(x) = 2/[x(1 - b^2 x^2)] \ . \tag{54h}$$

The diligent reader will note that (54a), (54b), (54c), (54d) respectively (54h) reproduce (2.1.9-2), (2.1.9-8), (2.1.9-9), (2.1.9-3) respectively (2.1.9-10).

2.1.12 The many-body problem of Ruijsenaars and Schneider (RS)

The results of the previous Sects. 2.1.8 and 2.1.11 entail that the many-body problems characterized by the equations of motion

$$\ddot{q}_n = \sum_{m=1, m \neq n}^{N} \dot{q}_n \dot{q}_m w(q_n - q_m) \ , \tag{1}$$

with $w(q)$ an appropriate, odd,

$$w(-x) = -w(x) \ , \tag{2}$$

function belonging to a certain class, see below, are reducible to the Lax form, see (2.1-2),

$$\dot{\underline{L}} = [\underline{L}, \underline{M}] \ , \tag{3}$$

via the *ansatz*, see (2.1.8-1,2),

$$L_{nm} = \delta_{nm} \dot{q}_n + (1 - \delta_{nm})(\dot{q}_n \dot{q}_m)^{1/2} \alpha(q_n - q_m), \tag{4}$$

$$M_{nm} = \delta_{nm} \sum_{l=1, l \neq n}^{N} \dot{q}_l \beta(q_n - q_l) + (1 - \delta_{nm})(\dot{q}_n \dot{q}_m)^{1/2} \gamma(q_n - q_m) . \tag{5}$$

The most general version of the function $w(x)$ reads

$$w(x) = \wp'(x)/[\wp(\mu) - \wp(x)] = -(d/dx)\log[\wp(\mu) - \wp(x)] \ , \tag{6a}$$

and corresponding expressions of the functions $\alpha(x)$, $\beta(x)$ and $\gamma(x)$ read

$$\alpha(x) = \sigma(\mu)\,\sigma(x+\nu)/[\sigma(\nu)\,\sigma(x+\mu)] \ , \tag{6b}$$

$$\beta(x) = \frac{1}{2}\wp'(x)/[\wp(x) - \wp(\mu)] \ , \tag{6c}$$

$$\gamma(x) = \alpha(x)\,[\zeta(x+\nu) - \zeta(x)] \ . \tag{6d}$$

Here we are of course using the notation of Appendix A.

The following special cases are particularly interesting:

Case (i):

$$w(x) = 2/x \ , \tag{7a}$$

$$\alpha(x) = 1 \ , \tag{7b}$$

$$\beta(x) = 0 \ , \tag{7c}$$

$$\gamma(x) = -1/x \ ; \tag{7d}$$

case (ii):

$$w(x) = 2/[x(1+x^2/g^2)] = 2g^2/[x(g^2+x^2)] \ , \tag{8a}$$

$$\alpha(x) = 1/(1+i\,x/g) \ , \tag{8b}$$

$$\beta(x) = [i\,g\,(1+x^2/g^2)]^{-1} \ , \tag{8c}$$

$$\gamma(x) = -[x(1+i\,x/g)]^{-1} = -x^{-1}\,\alpha(x) \ ; \tag{8d}$$

case (iii):

$$w(x) = 2\,a\,\mathrm{cotanh}(ax) \ , \tag{9a}$$

$$\alpha(x) = \cosh(ax) \ , \tag{9b}$$

$$\beta(x) = 0 \ , \tag{9c}$$

$$\gamma(x) = -a/\sinh(a\,x) = -a\,\mathrm{cotanh}(a\,x)\,\alpha(x) \ ; \tag{9d}$$

case (iv):

$$w(x) = 2a/\sinh(ax) \quad , \tag{10a}$$

$$\alpha(x) = 1/\cosh(ax/2) \quad , \tag{10b}$$

$$\beta(x) = 0 \quad , \tag{10c}$$

$$\gamma(x) = -(a/2)/\sinh(ax/2) \quad ; \tag{10d}$$

case (v):

$$w(x) = 2g^2 a \text{ cotanh}(ax)/[g^2 + \sinh^2(ax)] \quad , \tag{11a}$$

$$\alpha(x) = \sinh(a\mu) \; / \sinh[a(x+\mu)] \quad , \tag{11b}$$

$$\beta(x) = (a/2) \sinh(2a\mu)/[g^2 + \sinh^2(ax)] \quad , \tag{11c}$$

$$\gamma(x) = -a\sinh(a\mu) \text{ cotanh}(ax)/\sinh[a(x+\mu)] = -a \text{ cotanh}(ax)\, \alpha(x) \quad , \tag{11d}$$

$$g = i \sinh(a\mu) \quad . \tag{11e}$$

The neater notation has been chosen in each case; the diligent reader will have no difficulty to relate these formulas with those of Sect. 2.1.11, by identifying appropriately the constants appearing in those formulas, and if need be by rescaling the coordinates $q_n(t)$.

Let us also emphasize that these 5 cases are all special cases of (6); moreover, clearly *case (iv)* obtains from *case (v)* by setting $g^2 = 1$ and moreover by replacing a with $a/2$, *case (iii)* obtains from *case (v)* via the limit $g^2 \to \infty$, *case (ii)* obtains from *case (v)* by first replacing in it g with ga and then letting $a \to 0$; and finally *case (i)* obtains from *case (ii)* via $g^2 \to \infty$.

Case (i) and *case (ii)* are hereafter referred to as *rational* models; *case (iii), case (iv)* and *case (v),* as *hyperbolic* models. The replacement of a with ia in these last 3 cases turns the *hyperbolic* functions into *trigonometric* functions, without negating the property of $w(x)$ to be real (provided g^2 is also real, as we generally assume); the corresponding cases are then referred to as trigonometric models.

All the models of this class are referred to as Ruijsenaars-Schneider (for short, RS) models (see Sect. 2.N).

Let us finally note that, as shown below in Sect. 2.3.6.3, the more general model that obtains by adding the term λq, with λ an arbitrary constant, to $w(q)$, see (1), can be reduced to the case without this extra term via a change of the independent ("time") variable, if attention is restricted to the subclass of solutions satisfying the single condition

$$\sum_{n=1}^{N} \dot{q}_n(t) = 0 \ , \tag{12}$$

a constraint which is clearly compatible with (1) via (2).

2.1.12.1 Hamiltonian and Newtonian equations for the RS model

Let us recall that the Newtonian equations of motion (2.1.12-1) with (2.1.12-2) are Hamiltonian (see the last part of Sect. 2.1.9.1). The corresponding Hamiltonian function reads (see 2.1.9.1-17))

$$H(\underline{q},\underline{p};s) = \sum_{n=1}^{N} h_n(s\,p_n;\underline{q}) \ , \tag{1}$$

$$h_n(p;\underline{q}) = \exp\left[p - \frac{1}{2}\sum_{l=1,l\neq n}^{N} W(q_n - q_l)\right] \ , \tag{2}$$

with (see (2.1.9.1-17c))

$$W'(x) = w(x) \ , \tag{3}$$

entailing (see (2.1.12-2)) that $W(x)$ is even,

$$W(-x) = W(x) \ . \tag{4}$$

As emphasized by the notation employed in the left hand side of (1) the Hamiltonian H contains the arbitrary constant s, which however does not appear in the Newtonian equations of motion (2.1.12-1); it does instead appear in the Hamiltonian equations of motion, that read (see (2.1.9.1-20), (2.1.9.1-24), (3) and (2.1.12-2))

$$\dot{q}_n = s\,h_n(s\,p_n;\underline{q}) \quad , \tag{5a}$$

$$\dot{p}_n = \sum_{l=1,l\neq n}^{N} w(q_n - q_l)\left[h_n(s\,p_n;\underline{q}) - h_l(s\,p_l;\underline{q})\right]/2 \ . \tag{5b}$$

The formulas written above apply for any choice of the functions $W(x)$ and $w(x)$, provided they are related by (3) and are, respectively, even and odd, see (4) and (2.1.12-2).

Let us now specialize to the RS models, see Sect. 2.1.12; but before doing so, let us rewrite (2) in the form

$$h_n(p;\underline{q}) = \exp(p)\left[\prod_{l=1, l\neq n}^{N} Y(q_n - q_l) \right]^{1/2} , \tag{6}$$

where of course

$$Y(x) = \exp[-W(x)] . \tag{7}$$

Then from (2.1.12-6a), (3) and (7) we get

$$Y(x) = | \wp(x) - \wp(\mu) | , \tag{8}$$

and the special versions of this formula corresponding to the various degenerate cases considered in Sect. 2.1.12 read as follows:

case (i):

$$w(x) = 2/x, \quad Y(x) = x^{-2} ; \tag{9a}$$

case (ii):

$$w(x) = 2g^2/[x(g^2 + x^2)], \quad Y(x) = 1 + g^2/x^2 ; \tag{9b}$$

case (iii):

$$w(x) = 2a\,\mathrm{cotanh}(ax), \quad Y(x) = [a/\sinh(ax)]^2 ; \tag{9c}$$

case (iv):

$$w(x) = 2a/\sinh(ax), \quad Y(x) = [(a/2)/\tanh(ax/2)]^2 ; \tag{9d}$$

case (v):

$$w(x) = 2g^2 a\,\mathrm{cotanh}(ax)/[g^2 + a^{-2}\sinh^2(ax)], \; Y(x) = 1 + g^2 a^2/\sinh^2(ax) . \tag{9e}$$

2.1.12.2 Relativistic character of the RS model

Let us formally replace, in the basic *ansatz* (2.1.8-1,2) for the Lax pair (that underlines all subsequent developments on the RS model) $q_n(t)$ with $ct+q_n(t)$ (hence $\dot{q}_n(t)$ with $c+\dot{q}_n(t)$) and let us simultaneously rescale the 3 functions $\alpha(q)$, $\beta(q)$, $\gamma(q)$ by dividing them by the constant c. One obtains thereby the following version of (2.1.8-1,2):

$$L_{nm} = c + \dot{q}_n \qquad \text{if} \quad m = n \,, \tag{1a}$$

$$L_{nm} = \left[(1+c^{-1}\dot{q}_n)(1+c^{-1}\dot{q}_m)\right]^{1/2} \alpha(q_n - q_m) \quad \text{if } m \neq n \,, \tag{1b}$$

$$M_{nm} = \sum_{l=1,\, l\neq n}^{N} (1+c^{-1}\dot{q}_m)\, \beta(q_n - q_l) \qquad \text{if} \quad m = n \,, \tag{2a}$$

$$M_{nm} = \left[(1+c^{-1}\dot{q}_n)(1+c^{-1}\dot{q}_m)\right]^{1/2} \gamma(q_n - q_m) \quad \text{if } m \neq n \,. \tag{2b}$$

It is then clear, by comparing this *ansatz*, (1,2), with the *ansatz* (2.1.1-1,2), that the RS models (as treated in the Sections from 2.1.8. to 2.1.13) can be considered as (*integrable!*) deformations of the *integrable* models treated in Sect. from 2.1.1 to 2.1.7, with the constant c (or rather its inverse!) playing the role of deformation parameter.

It is indeed clear that, in the limit $c \to \infty$, (1) respectively (2) essentially coincide with (2.1.1-1) respectively (2.1.1-2).

Exercise 2.1.12.2-1. Verify! *Hint:* recall that, as implied by the very structure of the Lax equation (2.1-2), the Lax matrix $\underline{L}$ is always defined up to addition of a constant multiple of the unit matrix, say $a\underline{1}$, with a an *arbitrary* constant; and also take note of (2.1.1-6).

The parameter c, as introduced above, has clearly the dimensions of a *velocity*. It is therefore appealing to interpret the RS models as *relativistic* models, inasmuch as they yield, in the limit in which the parameter c (now interpreted as the *speed of light*) diverges, the many-body models whose time evolution is characterized by classical *nonrelativistic* Hamiltonian dynamics, see (2.1.2-4), (2.1.4-32), (2.1.5-3), (2.1.6-1). This very appealing interpretation is made more cogent by the following considerations.

The group-theoretical structure underlying nonrelativistic classical dynamics (in $1+1$ dimensions) is associated with the Galilei algebra,

$$[H,P]=0, \tag{3a}$$

$$[Q,H]=P, \tag{3b}$$

$$[Q,P]=N. \tag{3c}$$

In the algebraic context these square brackets must be read as commutators; while in the context of Hamiltonian dynamics, these same relations, (3), are read in terms of the Poisson brackets (1.2-4), with the following definitions (and interpretations) of the 4 quantities H, P, Q, N: H is the Hamiltonian (total energy – generator of time translations)

$$H=T+V, \tag{4a}$$

$$T=\frac{1}{2}\sum_{n=1}^{N} p_n^2, \tag{4b}$$

$$V=\sum_{n,m=1;\,n\neq m}^{N} V(q_n-q_m), \tag{4c}$$

(note that we set to unity the mass of the particles, see (4b)); P is the total momentum (generator of space translations),

$$P=\sum_{n=1}^{N} p_n; \tag{5}$$

$Q=N\overline{q}$ with $\overline{q}$ the center-of-mass coordinate,

$$Q=\sum_{n=1}^{N} q_n ; \tag{6}$$

and N is the number of particles (which is of course independent of $\underline{p}$ and $\underline{q}$, hence it Poisson-commutes with *all* physical quantities, and in particular with H, P, Q).

Exercise 2.1.12.2-2. Verify that (4), (5) and (6) satisfy (3). *Hint*: note that, via (1.2-4), the definitions (5) respectively (6) entail

$$[F(\underline{q}),P]=\sum_{m=1}^{N}\partial F(\underline{q})/\partial q_m, \tag{7a}$$

$$[F(\underline{p}),P]=0\ , \tag{7b}$$

respectively

$$[F(\underline{q}),Q]=0\ , \tag{8a}$$

$$[F(\underline{p}),Q]=-\sum_{m=1}^{N}\partial F(\underline{p})/\partial p_m, \tag{8b}$$

where F indicates a generic function of its arguments; and recall that $V(q)$ (see the right hand side of (4c)) is even, $V(-q)=V(q)$ (hence $V'(-q)=-V'(q)$).

In the relativistic ($(1+1)$-dimensional) context, the Galilei algebra (3) is replaced by the following Poincaré algebra:

$$[\tilde{H},\tilde{P}]=0, \tag{9a}$$

$$[Q,\tilde{H}]=\tilde{P}, \tag{9b}$$

$$[Q,\tilde{P}]=\tilde{H}/c^2, \tag{9c}$$

with $\tilde{H}$, $\tilde{P}$ respectively $B=-Q$ (see (6)) interpreted as the Hamiltonian (total energy), the total momentum (generator of space translations) respectively the "boost" generator. Here again, in the algebraic context, the square brackets in (9) are commutators. But the following remarkable fact was discovered by Ruijsenaars and Schneider <RS86>: the relations (9), with the square brackets interpreted as Poisson brackets, see (1.2-4), are satisfied by the following expressions of $\tilde{H}$, $\tilde{P}$ (and Q, see (6)):

$$\tilde{H}(\underline{q},\underline{p})=c^2\sum_{n=1}^{N}\cosh(p_n/c)\,u_n(\underline{q}), \tag{10a}$$

$$\tilde{P}(\underline{q},\underline{p})=c\sum_{n=1}^{N}\sinh(p_n/c)\,u_n(\underline{q}), \tag{10b}$$

with

$$u_n(\underline{q}) = \left\{ \prod_{m=1, m\neq n}^{N} \left[1 + 2(g\,a/c)^2 \wp[a(q_n - q_m)|\omega,\omega'] \right] \right\}^{1/2} . \tag{10c}$$

The verification that (10) and (6) entail (9b) and (9c) is easy: it follows immediately from the definition (1.2-4) of the Poisson bracket, quite independently of the specific expression (10c) of $u_n(q)$.

The verification that (10) entail (9a) is highly non trivial; in fact the discovery that the requirement (9a), dictated by relativistic invariance, could be satisfied by the *ansatz* (10) and that it would yield an *integrable* many-body problem (see below) is quite amazing.

Exercise 2.1.12.2-3. Verify that (10) satisfies (9a). *Solution*: see <RS86>.

The alert reader will have noticed that, up to a multiplicative constant, the Hamiltonian (10a) is just the sum of the two (Poisson commuting!) RS Hamiltonians (2.1.12.1-1) with $s = \pm 1/c$:

$$\tilde{H}(\underline{q},\underline{p}) = k\left[H(\underline{q},\underline{p};1/c) + H(\underline{q},-\underline{p};1/c)\right] , \tag{11a}$$

with $H(\underline{q},\underline{p};s)$ given by (2.1.12.1-1) and

$$k = c\,g\,a , \tag{11b}$$

$$\wp(\mu) = -(c\,g\,a)^{-2} \tag{11c}$$

(see (2.1.12-6,8)). And it is moreover clear that, in the *nonrelativistic limit* $c \to \infty$, $\tilde{H}(\underline{q},\underline{p})$, see (10a), yields precisely the nonrelativistic Hamiltonian (2.1.4-32) (up to the addition of the rest mass contribution):

$$\tilde{H}(\underline{q},\underline{p}) = N c^2 + H(\underline{q},\underline{p}) + O(c^{-2}) , \tag{12a}$$

with $H(\underline{q},\underline{p})$ given by (2.1.4-32), and likewise the total relativistic momentum $\tilde{P}(\underline{q},\underline{p})$, see (10b), yields the nonrelativistic total momentum,

$$\tilde{P}(\underline{q},\underline{p}) = P + O(c^{-2}) , \tag{12b}$$

with P defined by (5).

Exercise 2.1.12.2-4. Verify all these statements; and trace in detail how analogous results apply for the special cases of the Weierstrass functions (see (2.1.12.1-9), and identify the corresponding nonrelativistic Hamiltonians).

Exercise 2.1.12.2-5. Compare the Newtonian equations of motions yielded by the Hamiltonians $\tilde{H}(\underline{q},\underline{p})$, see (10a), and $H(\underline{q},\underline{p})$, see (2.1.12.1-1,2). *Hint:* see *Exercise 1.1-7* and *1.1-8.*

In our treatment above, and below as well, we focus on the subclass of Hamiltonian evolutions of RS type that are described in a more straightforward manner by *Newtonian* equations of motions: hence we generally restrict our attention to Hamiltonians of type (2.1.12.1-1,2) rather than (10a).

Let us end Sect. 2.1.12.2 by noting that the "relativistic" character of the RS models is further evidenced by the possibility to identify in some cases the motion of RS particles with that of the "solitons" of a *relativistically invariant* integrable PDE in $(1+1)$ dimensions, the so-called Sine-Gordon equation (this type of connections is one of many topics we have not treated in this book). But it is on other hand well known that no many-body dynamics of Newtonian type, as considered in this book, can be fully consistent with Einsteinian (special) relativity, inasmuch as it is fundamentally based on the notion of a single universal ("absolute") time.

2.1.12.3 Newtonian case. Complex extension presumably characterized by completely periodic motions

In Sect. 2.1.12.3 three *remarks* are offered, the third of which led to a *conjecture* which is also reported. A fourth *remark* that considerably strengthens the plausibility of that *conjecture* is then given. A proof of the *conjecture* is provided in the subsequent Sect. 2.1.12.4, for the *cases (i)-(v)* (see Sect. 2.1.12).

Remark 2.1.12.3-1. Addition of a term $\sum_{n=1}^{N} g_n(q_n)$ to the Hamiltonian (2.1.12.1-1), so that it read

$$H(\underline{q},\underline{p};s)=\sum_{n=1}^{N}\left[h_n(s\,p_n;\underline{q})+g_n(q_n)\right] , \tag{1}$$

with $h_n(p;\underline{q})$ defined by (2.1.12.1-2,3,4), leads to the Newtonian equations of motion

$$\ddot{q}_n + s\, g_n'(q_n)\dot{q}_n = \sum_{m=1,m\neq n}^{N} \dot{q}_n\, \dot{q}_m\, w(q_n - q_m) \ . \qquad (2)$$

Proof. It is closely analogous to that given above, see Sect. 2.1.9.1. Indeed the addition of a term $\sum_{n=1}^{N} g_n(q_n)$ in the right hand side of (2.1.9.1-17a) only entails the presence of an additional term $g_n'(q_n)$ in the left hand sides of (2.1.9.1-19b) hence of (2.1.9.1-24), leading, see (2.1.9.1-21), merely to the addition of a term $s\, g_n'(q_n)\dot{q}_n$ to the left hand side of (2.1.9.1-16).

Hereafter we focus on the special case

$$g_n(x) = -(\lambda / s)\, x \ , \qquad (3)$$

and correspondingly on the Newtonian equations

$$\ddot{q}_n - \lambda \dot{q}_n = \sum_{m=1,m\neq n}^{N} \dot{q}_n\, \dot{q}_m\, w(q_n - q_m) \ . \qquad (4)$$

Remark 2.1.12.3-2. The equations of motion (4) correspond to the (modified) Lax equation

$$\dot{\underline{L}} - \lambda \underline{L} = [\underline{L}, \underline{M}] \ . \qquad (5)$$

This statement applies of course only in the context of the models under consideration, see the formulas of Sect. 2.1.12. It is easily proven by retracing the treatment of Sect. 2.1.8, a task that is left as an easy *exercise* for the diligent reader.

Remark 2.1.12.3-3. The modified Lax equation (5) entails that the traces of the p-th powers of the Lax matrix,

$$T_p = \mathrm{trace}[\ \underline{L}^p\ ],\ p = 1,2,\dots\ , \qquad (6)$$

all evolve as follows

$$T_p(t) = T_p(0)\ \exp(p\, \lambda\ t) \ . \qquad (7)$$

In particular, if λ is imaginary,

$$\lambda = i\omega \ , \qquad (8)$$

with ω *real* and *nonvanishing,* the traces $T_p(t)$ are *all* periodic in t with period

$$T = 2\pi / \omega \ . \tag{9}$$

Proof. Time-differentiation of (6) yields (using (5) and (2.1-16))

$$\dot{T}_p - p\lambda\, T_p = 0 \ ,$$

which immediately entails (7).

Conjecture 2.1.12.3-4. All solutions of the many-body problem characterized by the Newtonian equation of motion

$$\ddot{q}_n - i\omega\dot{q}_n = \sum_{m=1, m\neq n}^{N} \dot{q}_n \dot{q}_m\, w(q_n - q_m) \ , \tag{10}$$

with ω an arbitrary (*nonvanishing, real*) constant and $w(x)$ as given in Sect. 2.1.12, are *completely periodic.*

Note that this *Conjecture 2.1.12.3-4* refers to a *complex* extension of the RS model, see (10), as well as (1) with (3) and (8).

Remark 2.1.12.3-5. If $q_n(t)$ is a solution of the equations of motion (2.1.12-1) (namely, of (10) with $\omega = 0$), then $\tilde{q}_n(t)$,

$$\tilde{q}_n(t) = q_n(\tau) \ , \tag{11a}$$

$$\tau(t) = [\exp(i\omega t) - 1]/(i\omega) \ , \tag{11b}$$

is a solution of the equations of motion (10).

Proof.

$$\dot{q}_n(t) = \tilde{q}_n'(\tau)\, \dot{\tau}(\tau) = \tilde{q}'(\tau) \exp(i\omega t) \ , \tag{12a}$$

$$\ddot{q}_n(t) - i\omega\dot{q}_n(t) = \tilde{q}_n''(\tau)\, [\dot{\tau}(\tau)]^2 = \tilde{q}''(\tau) \exp(2i\omega t) \ . \tag{12b}$$

Insertion of (12a) and (12b) in (10) yields, up to a trivial notational change, (2.1.12-1).
Note that $\tau(t)$, see (11b), is periodic in t with period T, see (9).

Obviously this *Remark 2.1.12.3-5* strengthens the plausibility of the *Conjecture 2.1.12.3-4*, although it does not quite prove it. Indeed the transformation (11), entailing a relation among the equations of motion (10) and (2.1.12-1), plays an important role in several subsequent developments related to models featuring *completely periodic* motions (see, for instance, Sect. 4.5).

2.1.12.4 Solution via the OP technique in the rational, hyperbolic and trigonometric cases. Completely periodic character of the motion

In Sect. 2.1.12.4 we provide the solution of the RS many-body problem of Sect. 2.1.12 in the *rational, hyperbolic* and *trigonometric* cases. The *Conjecture 2.1.12.3-4* is thereby proven, for these cases, via the *Remark 2.1.12.3-5.*

We focus on the more general *hyperbolic* case *(v)*, see (2.1.12-11); the treatment also covers the *trigonometric* case, since to obtain the solution no assumption needs to be made, see below, on the constant a (see (2.1.12-11), which may therefore be *imaginary* as well as *real* (indeed, arbitrarily complex). The other cases *(i)-(iv)* are then easily taken care of, see below, by appropriate specializations of the values of the two *a priori* arbitrary constants a and g^2 (see (2.1.12-11)), as discussed in Sect. 2.1.12 (after (2.1.12-11e)).

The procedure is actually quite close to that of Sect. 2.1.5. Indeed the starting point are the two matrix evolution equations

$$\dot{\underline{L}} = [\underline{L}, \underline{M}] \ , \tag{1}$$

$$\dot{\underline{E}} = [\underline{E}, \underline{M}] + a\, \{\underline{E}, \underline{L}\} \ , \tag{2}$$

with the diagonal matrix $\underline{E}(t)$ defined as follows:

$$\underline{E}(t) = \mathrm{diag}\{\exp[2 a q_n(t)]\} \ . \tag{3}$$

The first of these two matrix evolution equations, (1), is of course the Lax equation, see (2.1.12-1,4,11); the second, (2) with (3), can be shown to hold by exactly the same argument as given in Sect. 2.1.5 (after (2.1.5-11); the key point is that (2.1.5-8) holds as well in the present case, see (2.1.12-4,5,11d)).

Since (1), (2) and (3) coincide with (2.1.5-9,10,11), one can directly jump to the conclusion of the treatment of Sect. 2.1.5, namely to the following

Proposition 2.1.12.4-1. The quantities $\exp[2aq_n(t)]$ coincide with the N eigenvalues of the matrix $\tilde{\underline{E}}(t)$,

$$\tilde{\underline{E}}(t) = \exp[a\underline{L}(0)\,t]\,\underline{E}(0)\,\exp[a\underline{L}(0)\,t]\ , \tag{4}$$

where of course now

$$\underline{E}(0) = \text{diag}\{\exp[2aq_n(0)]\},\ E_{nm}(0) = \delta_{nm}\exp[2aq_n(0)]\ , \tag{5}$$

$$L_{nm}(0) = \delta_{nm}\,\dot{q}_n(0) + (1-\delta_{nm})[\dot{q}_n(0)\,\dot{q}_m(0)]^{1/2}\,\alpha[q_n(0) - q_m(0)]\ , \tag{6}$$

(see (2.1.12-4)), and $\alpha(x)$ is given by (2.1.12-11b),

$$\alpha(x) = \sinh(a\mu)/\sinh[a(x+\mu)] \tag{7a}$$

with

$$g = i\,\sinh(a\mu)\ . \tag{7b}$$

As mentioned above, the *trigonometric* cases obtains by the simple replacement $a \to ia$; and the (*hyperbolic* or *trigonometric*) cases *(iii)* respectively *(iv)*, by replacing the definition (7a) of $\alpha(x)$ with the appropriate definitions as given in Sect. 2.1.12, see (2.1.12-9b) respectively (2.1.12-10b). As for the *rational* cases *(i)* respectively *(ii)*, the formula (4) must be replaced by

$$\tilde{\underline{Q}}(t) = \underline{Q}(0) + \underline{L}(0)\,t\ , \tag{8}$$

where

$$\underline{Q}(0) = \text{diag}[q_n(0)],\ Q_{nm}(0) = \delta_{nm}\,q_n(0)\ , \tag{9}$$

while $\underline{L}(0)$ is still given by (6), but with $\alpha(x)$ given by (2.1.12-7b) respectively (2.1.12-8b); and in these 2 cases, the preceding *Proposition 2.1.12.4-1* is replaced by the following

Proposition 2.1.12.4-2. The coordinates $q_n(t)$ coincide directly with the eigenvalues of the matrix $\tilde{\underline{Q}}(t)$, see (8).

These results for the rational case obtain easily from those for the hyperbolic case by taking the limit $a \to 0$; otherwise, they could be obtained directly, indeed this is precisely what was done in Sect. 2.1.10.1 for case *(i)*.

From the explicit formulas (4) or (8) it is now easy to prove the *Conjecture 2.1.12.3-4*. Indeed the *Remark 2.1.12.3-5* entails that the solution of the equations of motion (see (2.1.12.3-4) with (2.1.12.3-8))

$$\ddot{q}_n - i\omega\dot{q}_n = \sum_{m=1, m\neq n}^{N} w(q_n - q_m) \quad , \tag{10}$$

which obtain from the Hamiltonian (see (2.1.12.3-1) with (2.1.12.3-3,8))

$$H = \sum_{n=1}^{N} [h_n(s\, p_n; \underline{q}) - i(\omega/s) q_n] \quad , \tag{11}$$

with h_n defined by (2.1.12.1-6), are given, in the cases *(iii)*, *(iv)* and *(v)*, by *Proposition 2.1.12.4-1*, and in cases *(i)* and *(ii)* by *Proposition 2.1.12.4-2*, both however modified via the replacement of the time t, in (4) or (8) (as the case may be), by (see (2.1.12.3-11b))

$$\tau = [\exp(i\omega t) - 1]/(i\omega) \ . \tag{12}$$

This entails that the matrices $\underline{\tilde{E}}$ respectively $\underline{\tilde{Q}}$, see (4) and (8), are now periodic in t with period T, see (2.1.12.3-9):

$$\underline{\tilde{E}}(t+T) = \underline{\tilde{E}}(t) \quad , \tag{13a}$$

$$\underline{\tilde{Q}}(t+T) = \underline{\tilde{Q}}(t) \ . \tag{13b}$$

Hence the (unordered) set of the (complex) eigenvalues of these matrices is also periodic with period T, and this entails that each eigenvalue, considered as a continuous function of t, is also periodic,

$$q_n(t+\tilde{T}) = q_n(t) \tag{14}$$

with period (at most)

$$\tilde{T} = T \cdot N! \ . \tag{15}$$

This argument guarantees that the coordinates $q_n(t)$ have periodicity $\widetilde{T}$ (rather than T), because of the possibility that they get reshuffled through the motion (which now takes place in the complex plane, see (10)). The mechanism whereby this may happen is analyzed in Sect. 4.5, in the context of a, possibly even *nonintegrable*, generalization of the simplest *rational* case *(i)*.

Note that the complete periodicity of the motion holds in spite of the (evident !) translation-invariant character of the equations of motion (10) (which, incidentally, are instead, in contrast to (2.1.12-1), *not* invariant under a rescaling $t \to \tilde{t} = bt$ of the time variable). Indeed it is clear (see (2.1.12-2)) that (10) entail, for the center of mass

$$\overline{q}(t) = N^{-1} \sum_{n=1}^{N} q_n(t) \quad , \tag{16}$$

the equation of motion

$$\ddot{\overline{q}} - i\omega \dot{\overline{q}} = 0 \quad , \tag{17}$$

whose solution $\overline{q}(t)$,

$$\overline{q}(t) = \overline{q}(0) + \dot{\overline{q}}(0) \left[\exp(i\omega t) - 1 \right] / (i\omega) \quad , \tag{18}$$

is indeed periodic with period T, see (9).

Exercise 2.1.12.4-3. We saw in Sect. 2.1.9.1 that the Lax pair (2.1.12-4,5) with (2.1.12-7) ("case *(i)*") was a fake Lax pair. Show that (2.1.12-4,5) with (2.1.12-9) ("case *(iii)*") is also a fake Lax pair (although a little less so: in this case *(iii)* the traces T_1, T_2 and T_3 are functionally independent, but T_p with $p > 3$ is a function of T_1, T_2 and T_3).

Exercise 2.1.12.4-4. The equations of motion (2.1.12-1) with (2.1.12-9) ("case *(iii)*") have been shown in Sect. 2.1.10.3 to be solvable (see (2.1.10.3-17)). Is the technique of solution identical to that of this Section? *Hint*: compare (2.1.9.1-1) with (2.1.12-4,9b)). Do the two techniques yield the same solution? *Reply:* of course !

Exercise 2.1.12.4-5. The separable character of the Lax matrix (2.1.12-4,7) ("case *(i)*") was taken advantage of in Sect. 2.1.12.2 to obtain the result (2.1.12.2-3), leading to (2.1.12.2-13). Try and obtain analogous results for case *(iii)*.

Exercise 2.1.12.4-6. Obtain (17) from (4) and (12).

2.1.13 Various tricks: changes of variables, duplications, infinite duplications, reductions to "nearest-neighbor" forces, elimination of velocity-dependent forces

In Sect. 2.1.7 various tricks were introduced and illustrated by showing how they work in a few exemplary cases. In Sect. 2.1.13 we revisit some of those tricks, and we introduce a few new ones. Our treatment is again based on the discussion of specific examples, this being the appropriate way to show how tricks work. And we naturally select now these examples from the material presented above, after Sect. 2.1.7.

Changes of variables. Let us review here an approach already utilized repeatedly above. Consider the class of Newtonian equations of motion

$$q_n''(\tau) = \sum_{m,\ell=1}^{N} q_m'(\tau) q_\ell'(\tau) f_{nm\ell}\left[\underline{q}(\tau)\right] , \tag{1}$$

where the primes indicate of course derivatives with respect to the independent variable τ. Here and below we use the notation $f_{nm\ell}(\underline{q})$ to indicate that the functions $f_{nm\ell}$ may depend on all the coordinates q_j, $j = 1,\ldots,N$. Note that these equations of motion, (1), are invariant under rescaling of the independent variable $(\tau \to c\tau)$. (Verify!).

Let us now introduce the following *change of dependent and independent variables* (see (2.1.7-2)):

$$q_n(\tau) = \varphi(t)\, x_n(t) , \qquad \tau = \tau(t) , \tag{2a}$$

where we keep for the moment open the option to assign the two functions $\varphi(t)$ and $\tau(t)$.

These transformations, (2), entail the following relations (see (2.1.7-4,5)):

$$q_n'(\tau) = \left[\, \varphi(t)\, \dot{x}_n(t) + \dot{\varphi}(t)\, x_n(t)\right] / \dot{\tau}(t) , \tag{2b}$$

$$q_n''(\tau) = \{\varphi(t)\, \ddot{x}_n(t) + \left[2\,\dot{\varphi}(t) - \varphi(t)\, \ddot{\tau}(t)/\dot{\tau}(t)\right] \dot{x}_n(t)$$

$$+\left[\, \ddot{\varphi}(t) - \dot{\varphi}(t)\, \ddot{\tau}(t)/\dot{\tau}(t)\right] x_n(t) \} / \left[\dot{\tau}(t)\right]^2 . \tag{2c}$$

The Newtonian equations (1) take then the following form:

$$\ddot{x}_n + \left[2(\dot{\varphi}/\varphi) - (\ddot{\tau}/\dot{\tau})\right] \dot{x}_n + \left[(\ddot{\varphi}/\varphi) - (\ddot{\tau}/\dot{\tau})\,(\dot{\varphi}/\varphi)\right] x_n$$

$$= \varphi \sum_{m,\ell=1}^{N} \left[\dot{x}_m + (\dot{\varphi}/\varphi)\, x_m\right] \left[\dot{x}_\ell + (\dot{\varphi}/\varphi) x_\ell\right] f_{nm\ell}\left[\varphi\, \underline{x}\right] . \tag{3}$$

Two cases deserve to be singled out. For

$$\varphi = 1 \tag{4a}$$

(and unrestricted $f_{nm\ell}(\underline{q})$), (3) take the neat form

$$\ddot{x}_n(t) - [\ddot{\tau}(t)/\dot{\tau}(t)]\,\dot{x}_n(t) = \sum_{m,\ell=1}^{N} \dot{x}_m(t)\,\dot{x}_\ell(t)\,f_{nm\ell}[\underline{x}(t)] \tag{4b}$$

The close similarity among (4b) and (1) should be noted. If we moreover set

$$\tau(t) = [\exp(at) - 1]/a \tag{5a}$$

(with a an arbitrary constant), which clearly entails $\tau(0) = 0,\ \dot{\tau}(0) = 1$, $\ddot{\tau}(t)/\dot{\tau}(t) = a$, then (4b) reads

$$\ddot{x}_n - a\dot{x}_n = \sum_{m,\ell=1}^{N} \dot{x}_m\,\dot{x}_\ell\,f_{nm\ell}[\underline{x}]\ . \tag{5b}$$

Note that these results imply the following

Proposition 2.1.13-1. To every solution $q_n(\tau)$ of (1) that is *analytic* in τ inside a disk centered at $\tau = i/\omega$ with radius $1/\Omega$, there corresponds (via (2a), (4a) and (5a)) a solution of (5b) with

$$a = i\omega\,, \tag{6a}$$

ω being a *positive* constant larger than Ω,

$$\omega > \Omega\,, \tag{6b}$$

which is periodic in t with period

$$T = 2\pi/\omega\ , \tag{6c}$$

see (5a) with (6a). And of course if *all* solutions of (1) possess such an analyticity property -- a fact which happens, presumably, only for special choices of the functions $f_{nm\ell}(\underline{q})$ (if at all) -- then all solutions of (5b) with (6a,b) are periodic with period T, see (6c).

The second case we like to single out is characterized by functions $f_{nm\ell}(\underline{q})$ which are homogeneous of degree -1 in their arguments,

$$f_{nm\ell}(\lambda \underline{q}) = \lambda^{-1} f_{nm\ell}(\underline{q}) \tag{7a}$$

(for an example, see (2.1.10-1); note that (7a) entails that (1) is invariant under rescaling of the coordinates $(q_n \to c\, q_n)$). Then of course (3) can be replaced by

$$\ddot{x}_n + [2(\dot{\varphi}/\varphi) - (\ddot{\tau}/\dot{\tau})]\, \dot{x}_n + [(\ddot{\varphi}/\varphi) - (\ddot{\tau}/\dot{\tau})\, (\dot{\varphi}/\varphi)]\, x_n$$

$$= \sum_{m,\ell=1}^{N} [\dot{x}_m + (\dot{\varphi}/\varphi) x_m]\, [\dot{x}_\ell + (\dot{\varphi}/\varphi) x_\ell]\, f_{nm\ell}[\underline{x}] \ . \tag{7b}$$

Let us also record the simple form taken by these equations of motion if the choice (5a) is made for $\tau(t)$, and moreover the following simple choice is made for $\varphi(t)$:

$$\varphi(t) = \exp(b\, t) \ , \tag{8a}$$

entailing of course $\varphi(0) = 1$. Then (7b) reads simply

$$\ddot{x}_n + (2b - a)\dot{x}_n + b(b-a) x_n = \sum_{m,\ell=1}^{N} (\dot{x}_m + b\, x_m)(\dot{x}_\ell + b\, x_\ell) f_{nm\ell}(\underline{x}) \ . \tag{8b}$$

Exercise 2.1.13-2. Show that the solution of the initial-value problem for the generalized version of the equations of motion (1) that obtains by replacing there $f_{nml}[\underline{q}(\tau)]$ with $f_{nml}[\underline{q}(\tau)] + \lambda\, \delta_{nm}\, [q_n(\tau) - q_l(\tau)]$, where λ is an *arbitrary* constant, can be reduced, via an appropriate change of the independent variable, to the solution of the initial-value problem for (1), provided attention is restricted to the class of solutions whose center-of-mass does not move, $\sum_{n=1}^{N} q_n'(\tau) = 0$ (and provided of course this constraint is compatible with the equations of motion (1)); and write the relevant change of independent variable in the context of the initial-value problem. *Hint*: see Sect. 2.3.6.3.

This ends our discussion here of the implications of the "trick" based on changes of (dependent and independent) variables.

We do not repeat here, but leave it as an instructive task for the diligent reader, a treatment of the results (for instance, in the context of the RS models treated above, see Sect. 2.1.12) entailed by the possibility to introduce *particles of different kinds* via shifts of the corresponding variables, see (2.1.7-30).

Likewise, we do not repeat here the analysis of the results implied by the consideration of *duplications* based on *symmetrical configurations*, see for instance (2.1.7-39). The diligent reader will have no difficulty to obtain, via such an approach, generalized versions of the models considered above, or below.

As an example of *finite duplications*, we start from the solvable system (see (2.1.10-1))

$$\ddot{q}_n = 2 \sum_{m=1, m\neq n}^{\tilde{N}} \dot{q}_n \dot{q}_m / (q_n - q_m) \quad , \tag{9}$$

where we assume of course the indices n and m to go from 1 to $\tilde{N}$. We then set

$$\tilde{N} = N + 2M \quad , \tag{10a}$$

$$q_n(t) = z_n(t), \qquad n = 1, \ldots, N \quad , \tag{10b}$$

$$q_{n+N}(t) = x_n(t) + i\, y_n(t), \qquad n = 1, \ldots, M \quad , \tag{10c}$$

$$q_{n+N+M}(t) = x_n(t) - i\, y_n(t), \qquad n = 1, \ldots, M \quad . \tag{10d}$$

It is easily seen that this *ansatz*, (10), is compatible with the equations of motion (9), and it yields the following (new, *real*) equations of motion:

$$\ddot{z}_n = 2 \sum_{m=1, m\neq n}^{N} \dot{z}_n \dot{z}_m / (z_n - z_m)$$

$$+ 4 \sum_{m=1}^{M} \dot{z}_m \left[\dot{x}_m (z_n - x_m) - \dot{y}_m y_m \right] / \left[(z_n - x_m)^2 + y_m^2 \right], \quad n = 1, \ldots, N \; , \tag{11a}$$

$$\ddot{x}_n = 4 \sum_{m=1, m\neq n}^{M} A_{nm} / D_{nm}$$

$$+ 4 \sum_{m=1}^{N} \dot{z}_n \left[\dot{x}_n (x_n - z_m) - \dot{y}_n y_n \right] / \left[(x_n - z_m)^2 + y_n^2 \right], \quad n = 1, \ldots, M \; , \tag{11b}$$

$$\ddot{y}_n = -(\dot{x}_n^2 + \dot{y}_n^2)/y_n + 4\sum_{m=1,m\neq n}^{M} B_{nm}/D_{nm}$$

$$+4\sum_{m=1}^{N} \dot{z}_m\left[\dot{y}_n(x_n - z_n) - \dot{x}_n y_n\right]/\left[(x_n - z_m)^2 + y_n^2\right], \quad n = 1,\ldots,M \ , \tag{11c}$$

$$A_{nm} \equiv \dot{x}_n\,\dot{x}_m\,(x_n - x_m)[\,(x_n - x_m)^2 + y_n^2 + y_m^2\,] - 2\,\dot{y}_n\,\dot{y}_m\,(x_n - x_m)\,y_n\,y_m$$

$$-\dot{x}_n\,\dot{y}_m\,y_m[\,(x_n - x_m)^2 - y_n^2 + y_m^2\,] + \dot{y}_n\,\dot{x}_m\,y_n[\,(x_n - x_m)^2 + y_n^2 - y_m^2\,]\ , \tag{11d}$$

$$B_{nm} \equiv -\dot{x}_n\,\dot{x}_m\,y_n[\,(x_n - x_m)^2 + y_n^2 - y_m^2\,] + \dot{y}_n\,\dot{y}_m\,y_m[\,(x_n - x_m)^2 - y_n^2 + y_m^2\,]$$
$$+2\dot{x}_n\,\dot{y}_m\,(x_n - x_m)\,y_n\,y_m + \dot{y}_n\,\dot{x}_m\,(x_n - x_m)\,[\,(x_n - x_m)^2 + y_n^2 + y_m^2\,]\ , \tag{11e}$$

$$D_{nm} \equiv [\,(x_n - x_m)^2 + (y_n - y_m)^2\,][\,(x_n - x_m)^2 + (y_n + y_m)^2\,]\ . \tag{11f}$$

These equations, (11), may be interpreted as the Newtonian equations of motion of a many-body problem on the line featuring $N+2M$ particles of three different kinds, N of one kind represented by the coordinates z_n, $n=1,\ldots,N$, and M each of two other, different, kinds, represented by the coordinates x_m and y_m, $m=1,\ldots,M$.

The original system, (9), features only two-body forces and is invariant under the translation $q_n \to q_n + c$; the new system is invariant under the (partial) translation $z_n \to z_n + c$, $x_m \to x_m + c$, $(n=1,\ldots,N;\ m=1,\ldots,M)$, but not under the (overall) translation $z_n \to z_n + c$, $x_m \to x_m + c$, $y_m \to y_m + c$, $(n=1,\ldots,N;\ m=1,\ldots,M)$, affecting the coordinates of *all* particles. It is however possible to recreate a fully translation-invariant system via the change of dependent variables

$$z_n(t) = \exp[a\,\zeta_n(t)]\ , \qquad n=1,\ldots,N\ , \tag{12a}$$

$$x_m(t) = \exp[a\,\xi_m(t)],\quad y_m(t) = \exp[a\,\eta_m(t)]\ , \qquad m=1,\ldots,M\ . \tag{12b}$$

Exercise 2.1.13-3. Write the equations of motion satisfied by the "new particle coordinates" ζ_n, ξ_m, η_m , and, for $N=M=1$, find the explicit solution of these equations of motion, or equivalently of the equations of motion (11). *Hint*: see (2.1.10.2-13).

Exercise 2.1.13-4. Modify the equations of motion (11), and the equations of motion obtained from (11) via (12), so that *all* their solutions are periodic. *Hint*: note that the system (11) belongs to the class (1), and see *Proposition 2.1.13-1*.

Infinite duplications. This trick has been described in Sect. 2.1.7, see the treatment beginning with (2.1.7-47). It will be instructive for the diligent reader to use this approach to obtain, in the context of the equations of motion (2.1.12-1), for instance, (2.1.12-9a) from (2.1.12-7a) and likewise (2.1.12-11a) from (2.1.12-8a).

Hints: see (2.1.7-49), and take advantage of the *identities*

$$\sum_{s=-\infty}^{+\infty} (x+i\, s\pi/a)^{-1} = a \,\text{cotanh}(a x), \tag{13a}$$

$$2g^2\left[x(g^2+x^2)\right]^{-1} = 2x^{-1} - (x+ig)^{-1} - (x-ig)^{-1}, \tag{14}$$

$$\text{cotanh}(x+i\,y) = \left[\text{cotanh}(x)\,\text{cotan}(y)+i\right] / \left[\text{cotan}(y)+i\,\text{cotanh}(x)\right] \tag{15}$$

(actually (13a) should, more rigorously, be written in the following form, see eq. 1.421.3 of <GRJ 94>:

$$x^{-1} + 2x\sum_{s=1}^{\infty} (x^2+s^2\pi^2/a^2)^{-1} = a \,\text{cotanh}(a\,x), \tag{13b}$$

to eliminate any doubt about the convergence of the infinite sum in the left-hand side of (13a)). Let us mention that the version of (2.1.12-11a) that will be obtained in this manner from (2.1.12-8a) will feature the quantity $\sin(g)$ in place of g, a modification that amounts merely to a notational change.

The diligent reader is also advised to revisit more generally the treatment of Sect. 2.1.7 (from (2.1.7-49) to (2.1.7-55)), in the context of the RS many-body problem, see Sect. 2.1.12.

Next, we discuss again the trick, see (2.1.7-56), whereby models involving only "nearest-neighbor" interactions are obtained from certain models with pair forces. To illustrate how this works in the context of RS models, see Sect. 2.1.12, we rewrite (2.1.12-2) as follows:

$$\ddot{\tilde{q}}_n = \sum_{m=1, m\neq n}^{N} \dot{\tilde{q}}_n \dot{\tilde{q}}_m \, w(\tilde{q}_n - \tilde{q}_m) \ , \tag{16}$$

and we focus on the assignment (2.1.12-11a),

$$w(x) = 2\,g^2\, a \,\text{cotanh}(ax) / \left[g^2 + \sinh^2(a\,x)\right] \ . \tag{17a}$$

Here the sign of the *real* constant a is irrelevant. Hereafter, for definiteness we assume that a is positive,

$$a > 0 \ . \tag{17b}$$

We then set

$$\tilde{q}_n(t) = q_n(t) + n\lambda \ , \tag{18a}$$

$$g^2 = (2/c)^2 \exp(2a\lambda) \ , \tag{18b}$$

and take the limit $\lambda \to \infty$, obtaining thereby

$$\ddot{q}_n = 2a\left[\dot{q}_n\,\dot{q}_{n-1} / \{1 + c^2 \exp[2a(q_n - q_{n-1})]\}\right.$$

$$\left. - \dot{q}_n\,\dot{q}_{n+1} / \{1 + c^2 \exp[-2a(q_n - q_{n+1})]\}\ \right], \qquad n = 2,\ldots,N-1 \ , \tag{19a}$$

$$\ddot{q}_1 = -2a\dot{q}_1\,\dot{q}_2 / \{1 + c^2 \exp[-2a(q_1 - q_2)]\} \ , \tag{19b}$$

$$\ddot{q}_N = 2a\dot{q}_N\,\dot{q}_{N-1} / \{1 + c^2 \exp[2a(q_N - q_{N-1})]\} \ . \tag{19c}$$

Proof. From (16), (17) and (18) we get

$$\ddot{q}_n = \sum_{m=1, m\neq n}^{N} \dot{q}_n\,\dot{q}_m\,\tilde{w}(q_n - q_m; \lambda) \ , \tag{20}$$

with

$$\tilde{w}_{nm}(x;\lambda) = 2a \ \text{cotan}\{a[x + \lambda(n-m)]\}\cdot$$

$$\cdot\{1 + c^2 [\exp\{a[x + \lambda(n-m-1)]\} - \exp\{-a[x + \lambda(n-m+1)]\}]^2\}^{-1} \ . \tag{21}$$

In the limit $\lambda \to \infty$ (with (17b)), this yields

$$\tilde{w}_{nm}(x;\infty) = 2a \ \text{sign}(n-m)\cdot$$

$$\cdot\{\delta_{m,n-1}[1 + c^2 \exp(2ax)]^{-1} + \delta_{m,n+1}[1 + c^2 \exp(-2ax)]\} \ , \tag{22}$$

and the insertion of this expression in (20) clearly yields (19).

The "Ruijsenaars-Toda" (RT) system (19) is of course integrable, indeed, as the system from which it has now been derived, its solution can be reduced to purely algebraic operations.

Let us recall that the solution of (2.1.12-1) with (2.1.12-11a) is detailed in Sect. 2.1.12.4. We leave it as an instructive (unnumbered!) *Exercise* for the diligent reader to provide an analogous treatment for the system (19). Such a procedure, as well as the Lax pair associated to (19), can of course be derived from the corresponding results for the system (2.1.12-1) with (2.1.12-11) (conveniently rewritten by replacing $q_n(t)$ by $\tilde{q}_n(t)$, which amounts merely to a notational change), via (18) followed by a careful scrutiny of the limit $\lambda \to \infty$.

There are two other interesting many-body systems with nearest-neighbor interactions that are easily obtained from the RT system (19). They read:

$$\ddot{u}_n = \dot{u}_n^2 / u_n - \dot{u}_n \dot{u}_{n-1} / (u_n - u_{n-1}) - \dot{u}_n \dot{u}_{n+1} u_n / [u_{n+1} (u_n - u_{n+1})] \quad, n = 2, \ldots, N-1, \tag{23a}$$

$$\ddot{u}_1 = \dot{u}_1^2 / u_1 - \dot{u}_1 \dot{u}_2 u_1 / [u_2 (u_1 - u_2)] \quad, \tag{23b}$$

$$\ddot{u}_N = \dot{u}_N^2 / u_N - \dot{u}_N \dot{u}_{N-1} / (u_N - u_{N-1}) \quad, \tag{23c}$$

and

$$\ddot{v}_n = -\dot{v}_n [\dot{v}_{n-1} / (v_n - v_{n-1}) + \dot{v}_{n+1} / (v_n - v_{n+1})] \quad, \quad n = 2, \ldots, N-1, \tag{24a}$$

$$\ddot{v}_1 = -\dot{v}_1 \dot{v}_2 / (v_1 - v_2) \quad, \tag{24b}$$

$$\ddot{v}_N = -\dot{v}_N \dot{v}_{N-1} / (v_N - v_{N-1}) \quad . \tag{24c}$$

To derive the system (23) from (19) we set

$$u_n(t) = \exp[2 a q_n(t)] \quad, \tag{25a}$$

entailing

$$\dot{u}_n / u_n = 2 a \dot{q}_n \quad, \tag{25b}$$

$$\ddot{u}_n / u_n = (\dot{u}_n / u_n)^2 + 2 a \ddot{q}_n \quad, \tag{25c}$$

hence, via (19a) and (25),

$$\ddot{u}_n = \dot{u}_n^2 / u_n + \dot{u}_n \dot{u}_{n-1} / (u_{n-1} + c^2 u_n) - \dot{u}_n \dot{u}_{n+1} u_n / [u_{n+1} (u_n + c^2 u_{n+1})] \quad, \tag{26}$$

which coincides with (23a) for

$$c^2 = -1 \ . \tag{27}$$

The derivation of (23b,c) from (19b,c) is completely analogous.

To obtain (24), one can set, in (23),

$$v_n = u_n + \lambda \ , \tag{28}$$

and then take the limit $\lambda \to \infty$. Alternatively, and equivalently, one can set (27) in (19) and then take the limit $a \to 0$. This yields again (24), except for the notational replacement of v_n by q_n.

Note that (19) and (24), but not (23), are translation-invariant; moreover, (23) and (24), but not (19), are invariant under rescaling of the dependent variables (q_n, u_n, or v_n, as the case may be); and all three systems, (19), (23) and (24), are invariant under rescaling of the independent ("time") variable. Also note that all these systems, (19), (23) and (24), belong to the class (1) (up to trivial notational changes); hence addition to these (integrable !) equations of motion, (19), (23) or (24), of a term $i\omega\,\dot{q}_n$ (or $i\omega\dot{u}_n$, or $i\omega\dot{v}_n$, as appropriate), with ω a nonvanishing real constant, yields new (complex) equations of motion featuring only completely periodic solutions (see the discussion above, after (5a)).

Let us finally introduce a trick whereby *certain equations of motion featuring velocity-dependent forces are reduced to equations without such forces*. As usual, rather than presenting a “general” approach, we illustrate the procedure by applying it to a specific example.

Let us take as starting point the equations of motion (2.1.12-1) with (2.1.12-8a), which we write here in the convenient form

$$\ddot{\tilde{q}}_n = 2 \sum_{m=1, m\neq n}^{N} \dot{\tilde{q}}_n \dot{\tilde{q}}_m / \left\{ (\tilde{q}_n - \tilde{q}_m) \left[1 + (\tilde{q}_n - \tilde{q}_m)^2 / \varepsilon^2 \right] \right\} \ , \tag{29}$$

(for notational convenience we replaced g with ε).

We now set

$$\tilde{q}_n(t) = q_n(t) + (g/\varepsilon)\, t \ , \tag{30a}$$

entailing

$$\dot{\tilde{q}}_n = g/\varepsilon + \dot{q}_n \ , \qquad \ddot{\tilde{q}}_n = \ddot{q}_n \ , \tag{30b}$$

so that we get from (29)

$$\ddot{q}_n = 2g^2 \sum_{m=1, m\neq n}^{N} \left[1 + (\varepsilon/g)\dot{q}_n \right] \left[1 + (\varepsilon/g)\dot{q}_m \right] / \left\{ (q_n - q_m) \left[(q_n - q_m)^2 + \varepsilon^2 \right] \right\}. \tag{31}$$

It is then clear that, in the limit $\varepsilon \to 0$, (31) become

$$\ddot{q}_n = 2g^2 \sum_{m=1, m\neq n}^{N} (q_n - q_m)^{-3} \quad , \tag{32}$$

which coincide with (2.1.2-5).

This result should not come as a surprise: see (2.1.8-3a) and the treatment following this equation.

Exercise 2.1.13-5. By an analogous treatment obtain (2.1.5-5) from (2.1.12-1) with (2.1.12-11a) and, more generally, (2.1.4-33) from (2.1.12-1) with (2.1.12-6a).

Exercise 2.1.13-6. Explore the connection entailed by this trick among the Lax pairs, as well as the techniques of explicit solution, of the RS models, see Sections 2.1.12 and 2.1.12.4, and those of the models featuring no velocity-dependent forces, see Sections 2.1.1, 2.1.3.2 and 2.1.5.

Exercise 2.1.13-7. To what extent can this connection (see *Exercise 2.1.13-6*) be interpreted as the nonrelativistic limit of a relativistic model ? *Hint:* see Sect. 2.1.12.2.

Exercise 2.1.13-8. Repeat the analysis of the three preceding *exercises* in the context of models with nearest-neighbor interactions only: in particular, obtain (2.1.7-57) from (19).

2.1.14 Another Lax pair corresponding to a Hamiltonian many-body problem on the line. The functional equation (***)

In Sect. 2.1.1 a convenient representation of a Lax pair was introduced, see (2.1.1-1,2), and in subsequent sections results that obtain from this *ansatz* were reported. Likewise, another (in fact more general) representation of a Lax pair was introduced in Sect. 2.1.8, see (2.1.8-1, 2), and its implications were treated in subsequent sections. We now introduce a third representation of a Lax pair and, in the following Sect. 2.1.15 and its subsections, we investigate results that flow from it.

The Lax pair we now introduce reads

$$L_{nm} = (p_n \, p_m)^{1/2} \, \alpha(q_n - q_m) \quad , \tag{1}$$

$$M_{nm} = (p_n \, p_m)^{1/2} \, \gamma(q_n - q_m) \quad , \tag{2}$$

where we are of course assuming that the two functions $\alpha(x)$ and $\gamma(x)$ are finite at $x = 0$.

Remark. These equations, (1) and (2), are rather similar to (2.1.8-1,2): indeed the resemblance among (1) and (2.1.8-1) is reinforced by (2.1.11-5), which entails that (1) obtains from (2.1.8-1) via the replacement of $\dot{q}_j$ with p_j. But this replacement corresponds in fact to an important change, since for the system we are now considering, as indeed for the system that obtained from the *ansatz* (2.1.8-1,2), there does *not* hold the relation (2.1.1-6) (nor (1.2-25a), with $\mu_n = 1$): see (5a) below, and (2.1.12.1-5a) with (2.1.12.1-2).

One might also consider a more general *ansatz* for $\underline{M}$, including an additional *diagonal* term of type

$$\delta_{nm} \sum_{\ell=1}^{N} p_\ell \, \beta(q_n - q_\ell) \quad , \tag{3}$$

in the right-hand side of (2). But this does not entail a significant generalization. We leave an exploration of its implications as an *exercise* for the diligent reader, who can usefully consult in this connection the relevant papers referred to in Sect. 2.N.

It is now easily seen that the Lax equation,

$$\underline{\dot{L}} = [\underline{L}, \underline{M}] \quad , \tag{4}$$

with the ansatz (1) and (2), corresponds to the Hamiltonian equations

$$\dot{q}_n = 2 \sum_{m=1}^{N} p_m \, f(q_n - q_m) \ , \tag{5a}$$

$$\dot{p}_n = -2 p_n \sum_{m=1}^{N} p_m \, f'(q_n - q_m) \ , \tag{5b}$$

entailed by the Hamiltonian

$$H = \sum_{n,m=1}^{N} p_n \, p_m \, f(q_n - q_m) \ , \tag{6}$$

provided the *even* (see (6)) function $f(x)$,

$$f(-x) = f(x) \quad , \tag{7a}$$

satisfies, together with the two functions $\alpha(x)$ and $\gamma(x)$, see (1) and (2), the following *functional equation (***)*:

(***) $2\alpha'(x+y)[f(x)-f(y)]-\alpha(x+y)[f'(x)-f'(y)]=\alpha(x)\gamma(y)-\alpha(y)\gamma(x)$. (8)

Proof. The diagonal term of the Lax matrix equation (4) with (1) and (2) yields

$$\dot{p}_n\alpha(0)=\sum_{m=1,m\neq n}^{N} p_n p_m[\alpha(q_n-q_m)\,\gamma(q_m-q_n)-\alpha(q_m-q_n)\,\gamma(q_n-q_m)]\,. \tag{9}$$

But the functional equation (8), for $y=-x$, yields, using (7a) (or rather the relation

$$f'(-x)=-f'(x) \tag{7b}$$

entailed by (7a)),

$$-2\alpha(0)\,f'(x)=\alpha(x)\,\gamma(-x)-\alpha(-x)\,\gamma(x)\,. \tag{10}$$

Insertion of this expression in the right-hand side of (9) demonstrates that this formula coincides with (5b).

Let us then look at the off-diagonal terms of the Lax equation (4) with (1) and (2). One gets

$$\frac{1}{2}[(\dot{p}_n/p_n)+(\dot{p}_m/p_m)]\,\alpha(q_n-q_m)+(\dot{q}_n-\dot{q}_m)\,\alpha'(q_n-q_m)=$$

$$=\sum_{\ell=1}^{N} p_\ell\,[\alpha(q_n-q_\ell)\,\gamma(q_\ell-q_m)-\alpha(q_\ell-q_m)\,\gamma(q_n-q_\ell)]\,. \tag{11}$$

We now use, in the right hand side of this equation, (11), the formula (see (8), with $x=q_n-q_\ell$, $y=q_\ell-q_m$)

$$\alpha(q_n-q_\ell)\,\gamma(q_\ell-q_m)-\alpha(q_\ell-q_m)\,\gamma(q_n-q_\ell)$$

$$=2\alpha'(q_n-q_m)\,[f(q_n-q_\ell)-f(q_\ell-q_m)]-\alpha(q_n-q_m)\,[f'(q_n-q_m)-f'(q_\ell-q_m)], \tag{12}$$

and then equate the terms which multiply $\alpha(q_n-q_m)$ respectively $\alpha'(q_n-q_m)$. We thus obtain (5b) respectively (5a): in each case twice, namely once for the index n and once for the index m.

The proof is thereby completed: the matrix Lax equation (4) with (1), (2), (8) and (7) is equivalent to the Hamiltonian equations of motion (5).

2.1.15 A simple solution of the functional equation (***), and the corresponding Hamiltonian many-body problem on the line

A simple solution of the functional equation (***), see (2.1.14-8), reads

$$\alpha(x) = c \sin(ax/2) , \tag{1}$$

$$\gamma(x) = 0 , \tag{2}$$

$$f(x) = \lambda + \mu \cos(ax) . \tag{3}$$

Proof.

$$2\alpha'(x+y)[f(x)-f(y)] - \alpha(x+y)[f'(x)-f'(y)]$$

$$= ac\mu\{\cos[a(x+y)/2][\cos(ax)-\cos(ay)] + \sin[a(x+y)/2][\sin(ax)-\sin(ay)]\}$$

$$= ac\mu\{\cos[a(y-x)/2] - \cos[a(x-y)/2]\} = 0 . \tag{4}$$

Here the first step has been made by using the expression (1) and (3) of $\alpha(x)$ and $f(x)$, and the second step by using (twice) the trigonometric identity $\cos(A)\cos(B)+\sin(A)\sin(B)=\cos(A-B)$. The final result demonstrates the validity of (2.1.14-8) with (1), (2) and (3).

We now note that, in this case, thanks to (2.1.14-2) and (2), the Lax equation (2.1.14-4) entails

$$\dot{\underline{L}} = 0 . \tag{5}$$

Hence in this case, not only the eigenvalues of the Lax matrix $\underline{L}$, but every matrix element of $\underline{L}$ is a constant of the motion. Let us focus, for notational convenience, on twice the squares of these matrix elements:

$$c_{nm} = 2 p_n p_m \sin^2[a(q_n - q_m)/2] = p_n\, p_m \{1 - \cos[a(q_n - q_m)]\} . \tag{6}$$

Note that there are $\frac{1}{2}N(N-1)$ constants c_{nm}, since clearly $c_{nm} = c_{mn}$ and $c_{nn} = 0$.

Not all these $\frac{1}{2}N(N-1)$ constants of the motion are in involution, but it is easily seen that

$$[c_{nm},c_{n'm'}]=0 \quad \text{if} \qquad n'\neq n, n'\neq m, m'\neq n, m'\neq m \ , \tag{7}$$

and moreover that, if one defines

$$h_M=\sum_{n,m=1}^{M} c_{nm} \ , \qquad\qquad M=2,3,\dots,N, \tag{8}$$

then

$$[h_M,h_{M'}]=0 \ . \tag{9}$$

Proofs. From (6)

$$\partial c_{nm}/\partial q_\ell=(\delta_{n\ell}-\delta_{m\ell})\,p_n\,p_m\sin[a(q_n-q_m)] \ , \tag{10}$$

$$\partial c_{nm}/\partial p_\ell=2(\delta_{n\ell}\,p_m+\delta_{m\ell}\,p_n)\,\sin^2[a(q_n-q_m)/2] \ . \tag{11}$$

Hence, see (1.2-4),

$$[c_{nm},c_{n'm'}]=2\sum_{\ell=1}^{N}\{(\delta_{n\ell}-\delta_{m\ell})(\delta_{n'\ell}\,p_{m'}+\delta_{m'\ell}\,p_{n'})\,p_n\,p_m\cdot$$

$$\cdot\sin[a(q_n-q_m)]\sin^2[a(q_{n'}-q_{m'})/2]\}-(n\leftrightarrow n',m\leftrightarrow m')$$

$$=2\,(\delta_{nn'}\,p_{m'}-\delta_{mn'}\,p_{m'}+\delta_{nm'}\,p_{n'}-\delta_{mm'}\,p_{n'})\,p_n\,p_m\cdot$$

$$\cdot\sin[a(q_n-q_m)]\sin^2[a(q_{n'}-q_{m'})/2]-(n\leftrightarrow n',m\leftrightarrow m'), \tag{12}$$

and this proves (7).

To prove (8), let us assume, without loss of generality, that $M'>M$, and let us note that the definition (8) entails

$$h_{M'}=h_M+\sum_{n=1}^{M}\ \sum_{m=M+1}^{M'}c_{nm}+\sum_{n=M+1}^{M'}\sum_{m=1}^{M}c_{nm}+\sum_{n=M+1}^{M'}\ \sum_{m=M+1}^{M'}c_{nm} \ . \tag{13}$$

Hence, see (8) and (7),

$$[h_{M'},h_M]=\sum_{n,m=1}^{M}\ \sum_{m'=M+1}^{M'}[c_{nm},c_{nm'}]+\sum_{n,m=1}^{M}\ \sum_{n'=M+1}^{M'}[c_{nm},c_{n'm}]$$

$$=2\sum_{n,m=1}^{M}\sum_{m'=M+1}^{M'}\left[c_{nm},c_{nm'}\right] . \quad (14)$$

To perform the last step we used the symmetry of c_{nm}, $c_{nm}=c_{mn}$, see (6).

We now use (12) (as well as $c_{mm}=0$, see (6)):

$$\left[h_{M'},h_{M}\right]=4\sum_{n,m=1}^{M}\sum_{m'=M+1}^{M'}p_n\,p_m\,p_{m'}\cdot$$

$$\cdot\left\{\sin\left[a(q_n-q_m)\right]\sin^2\left[a(q_n-q_{m'})/2\right]-\sin\left[a(q_n-q_{m'})\right]\sin^2\left[a(q_n-q_m)/2\right]\right\}$$

$$=8\sum_{n,m=1}^{M}\sum_{m'=M+1}^{M'}p_n\,p_m\,p_{m'}\sin\left[a\,(q_n-q_m)/2\right]\sin\left[a\,(q_n-q_{m'})/2\right]\cdot$$

$$\cdot\left\{\cos\left[a(q_n-q_m)/2\right]\sin\left[a(q_n-q_{m'})/2\right]-\cos\left[a(q_n-q_{m'})/2\right]\sin\left[a(q_n-q_m)/2\right]\right\}$$

$$=8\sum_{n,m=1}^{M}\sum_{m'=M+1}^{M'}p_n\,p_m\,p_{m'}\sin\left[a\,(q_n-q_m)/2\right]\sin\left[a\,(q_n-q_{m'})/2\right]\sin\left[a(q_m-q_{m'})/2\right]. \quad (15)$$

To make the second step we replaced $\sin\left[a(q_n-q_m)\right]$ with $2\sin\left[a(q_n-q_m)/2\right]\cos\left[a(q_n-q_m)/2\right]$, and likewise for $\sin\left[a(q_n-q_{m'})\right]$; while in the last step we used the identity $\cos(A)\sin(B)-\cos(B)\sin(A)=\sin(B-A)$. Now the summand in the right hand side of the last equation is antisymmetrical under the exchange of the two dummy indices n and m, which are summed over the same range. Hence the sum vanishes, and (9) is proven.

It is thus seen that the many-body system defined by the Hamiltonian (see (2.1.14-6) and (3))

$$H=\lambda P^2+\mu\sum_{n,m=1}^{N}p_n\,p_m\cos\left[a(q_n-q_m)\right] , \quad (16)$$

where P is the total momentum,

$$P=\sum_{n=1}^{N}p_n , \quad (17)$$

is integrable: it possesses the $N-1$ constants of motion in involution (8). An additional constant of motion is the total momentum P,

$$\dot{P}=0 , \quad (18)$$

or the Hamiltonian itself, see (16); note that these two constants are related to the constant of motion h_N, see (8), as follows:

$$H=(\lambda+\mu)P^2-\mu\, h_N \ . \tag{19}$$

This relation is an immediate consequence of (8), (6), (16) and (17).

Note that the translation-invariance of all the constants of motion h_M, see (8) and (6), and of the Hamiltonian H, see (16), implies that all these quantities Poisson-commute with P.

Let us also emphasize that the term $\lambda\, P^2$ in the right hand side of (16) could be replaced by an arbitrary function of P, say $g(P)$, without modifying, other than trivially, the evolution of the system. Indeed the additive presence of a term $g(P)$ in the Hamiltonian has the only effect to induce the same, additive, linear time-dependence $g'(P)\, t$ on all the canonical coordinates $q_n(t)$, and it has no effect on the canonical momenta $p_n(t)$. In particular, see below, the only effect of the term $\lambda\, P^2$ appearing in the right-hand side of (16) is to add the linear term $2\lambda\, Pt$ to all $q_n(t)$'s (hence one could, without any significant loss of generality, set $\lambda=0$, although we refrain from doing so).

The equations of motion of this many-body problem read (see (2.1.14.5) and (3))

$$\dot{q}_n=2\sum_{m=1}^{N} p_m\left\{\lambda+\mu\cos\left[a(q_n-q_m)\right]\right\} \ , \tag{20a}$$

$$\dot{p}_n=2a\mu\, p_n\sum_{m=1}^{N} p_m\sin\left[a(q_n-q_m)\right] \ . \tag{20b}$$

There is no straightforward way to write them in Newtonian form.

This system is not only *integrable*: it is *solvable*, indeed, as we show in the next Sect. 2.1.15.1, its general solution can be exhibited in *completely explicit* form.

2.1.15.1 Explicit solution

In Sect. 2.1.15.1 we report the general solution of the Hamiltonian equations of motion (2.1.15-20). It reads

$$q_n(t)=(\alpha/a)+2(\lambda+\mu)Pt$$

$$-(2/a)\arctan\{2a\mu[(P-A)/\omega]\,\tan[\omega(t-t_n)/2]\}\ , \tag{1a}$$

$$p_n(t)=p_n(0)\{P+A\cos[\omega(t-t_n)]\}\,/\,\{P+A\cos(\omega\,t_n)\}\ , \tag{1b}$$

$$\omega=2a\mu(P^2-A^2)^{1/2}\ . \tag{1c}$$

In these formulas a, λ and μ are the constants that appear in the Hamiltonian (2.1.15-16,17) and in the equations of motion (2.1.15-20), P is the total momentum, see (2.1.15-17), and the two constants α and A are defined by the expressions

$$\alpha=\arctan(S/C)-2a\mu Pt\ , \tag{2a}$$

$$A=(C^2+S^2)^{1/2}\ , \tag{2b}$$

where

$$C=\sum_{n=1}^{N} p_n\cos(aq_n)\ , \tag{3a}$$

$$S=\sum_{n=1}^{N} p_n\sin(aq_n)\ . \tag{3b}$$

These two quantities, C and S, evolve in time (unless $\lambda+\mu=0$, see (6) below), but the two quantities α and A, defined by (2) and (3), are time-independent (see (12) and (13) below); hence they, as well as the total momentum P, see (2.1.15-17), can be evaluated, via (2) and (3), at the initial time $t=0$, namely in terms of the initial data $q_n(0)$, $p_n(0)$. Finally, the quantities $p_n(0)$, appearing in the right-hand-side of (1b), are of course the initial values of the canonical momenta $p_n(t)$; while the N constants t_n are determined, in terms of the initial coordinates $q_n(0)$, by the requirement that (1a) hold at $t=0$.

Note that the time-evolution of the coordinates $q_n(t)$, except for the *linear* contribution displayed by the second term in the right-hand side of (1a), is given by a "universal" function of t (namely, the same function of the time t, independent of the index n, for all the q_n's), evaluated, for each q_n, at an appropriately shifted time $t-t_n$, see (1a). Also note that, as indeed entailed by the equations of motion (2.1.15-20), the coordinates $q_n(t)$ are defined $\mathrm{mod}(2\pi/a)$; as they evolve over time, this ambiguity is of course lifted, for each of them, by the requirement of continuity.

We have written the formulas (1) in the manner which is most appropriate to the case characterized by the inequality

$$P^2 - A^2 = h_N > 0 \ . \tag{4}$$

The validity of this inequality depends on the initial conditions; it is clearly satisfied if all the initial momenta, $p_n(0)$, have the same sign (see (2.1.15-8) and (2.1.15-6), as well as (2.1.15-17), (2b) and (3) which entail the equality in (4)). When the inequality (4) holds, clearly the canonical coordinates $q_n(t)$, except for the linear drift represented by the second term in the right-hand side of (1a) (absent if $\lambda + \mu = 0$), are all periodic with period

$$T = 2\pi / \omega \ , \tag{5}$$

see (1c); likewise, if the inequality (4) holds, the canonical momenta $p_n(t)$ are all periodic with the same period, and never vanish (so that their signs are determined by their initial values).

If instead the inequality in (4) does *not* hold, then the solution (1) is perhaps better written in the following (equivalent) form:

$$q_n(t) = (\alpha / a) + 2(\lambda + \mu) P t$$

$$-(2/a) \arctan\{2 a \mu [(P - a)/\nu] \tanh[\nu (t - t_n)/2]\} \ , \tag{6a}$$

$$p_n(t) = p_n(0) \{P + A \cosh[\nu (t - t_n)]\} / \{P + A \cosh(\nu t_n)\} \ , \tag{6b}$$

$$\nu = 2 a \mu (A^2 - P^2)^{1/2} \ . \tag{6c}$$

In this case neither the canonical coordinates $q_n(t)$, nor the canonical momenta $p_n(t)$, are periodic; note, however, that also in this case the canonical momenta never vanish (this general property is indeed already apparent from (2.1.14-5 b)).

Here and below we assume that the constant a is *real*, and of course that the canonical coordinates and momenta, $q_n(t)$ and $p_n(t)$, are *real* as well. The reality of the canonical coordinates and momenta, $q_n(t)$ and $p_n(t)$, is also compatible with a purely imaginary choice for the constant a, namely with the replacement of a with ia, which amounts essentially to a replacement everywhere (in the Hamiltonian, in the equations of motion, in the expression of the solution: see (2.1.15-16), (2.1.15-20) and (1)) of the trigonometric functions with their hyperbolic counterparts. We leave the exploration of the model obtained in this manner as an interesting *exercise* for the diligent reader.

Let us now prove that (1) provides a solution to the equations of motion (2.1.15-20). The following three *remarks* are instrumental to this end.

Remark 2.1.15.1-1. The Hamiltonian (2.1.15-16) can be rewritten, via (3) and (2b), in the simple form

$$H = \lambda P^2 + \mu (C^2 + S^2) = \lambda P^2 + \mu A^2 \ . \tag{7}$$

Remark 2.1.15.1-2. The evolution equations (2.1.15-20) can be recast in the following form:

$$\dot{q}_n = 2\lambda P + 2\mu [C \cos(a q_n) + S \sin(a q_n)] \ , \tag{8a}$$

$$\dot{p}_n = 2 a \mu p_n [-S \cos(a q_n) + C \sin(a q_n)] \ , \tag{8b}$$

with P, C and S defined by (2.1.15-17) and (3).

Remark 2.1.15.1-3. The quantities $C(t)$ and $S(t)$, see (3), evolve according to the following *linear* evolution equations:

$$\dot{C} = -\rho S \ , \tag{9a}$$

$$\dot{S} = \rho C \ , \tag{9b}$$

$$\rho = 2(\lambda + \mu) a P \ . \tag{9c}$$

Proofs. (7) follows from (2.1.15-16) and (3) (as well as (2b)), via the trigonometric identity

$$\cos[a(q_n - q_m)] = \cos(a q_n) \cos(a q_m) + \sin(a q_n) \sin(a q_m) \ . \tag{10a}$$

This identity, and the analogous one,

$$\sin[a(q_n - q_m)] = \sin(a q_n) \cos(a q_m) - \cos(a q_n) \sin(a q_m) \ , \tag{10b}$$

are also instrumental to show, by inserting the definitions (2.1.15-17) and (3) in the right hand side of (8), that these equations of motion coincide with (2.1.15-20).

Finally, to prove (9), let us time-differentiate the definitions (3):

$$\dot{C}=\sum_{n=1}^{N}\{\dot{p}_n\cos(aq_n)-a\,p_n\,\dot{q}_n\sin(aq_n)\}\ , \tag{11a}$$

$$\dot{S}=\sum_{n=1}^{N}\{\dot{p}_n\sin(aq_n)+a\,p_n\,\dot{q}_n\cos(aq_n)\} \tag{11b}$$

We then use (8) in the right-hand sides of (11a), getting thereby

$$\dot{C}=\sum_{n=1}^{N}p_n\{2a\mu[-S\cos^2(aq_n)+C\sin(aq_n)\cos(aq_n)]-2\lambda Pa\sin(aq_n)$$

$$-2a\mu\left[C\sin(aq_n)\cos(aq_n)+S\sin^2(aq_n)\right]\} \tag{12a}$$

$$=-2a\mu PS-2a\lambda PS\ . \tag{12b}$$

To get the last equation, (12b), we used the definitions (2.1.15-17) and (3b) of P and S. It is immediately seen, via (9c), that (12b) coincides with (9a).

The proof of (9b) is completely analogous.

The solution of (9a) and (9b) is easy:

$$C(t)=C(0)\cos(\rho t)-S(0)\sin\rho t)\ , \tag{13a}$$

$$S(t)=S(0)\cos(\rho t)+C(0)\sin(\rho t)\ , \tag{13b}$$

namely

$$C(t)=A\cos(\gamma t+\alpha)\ , \tag{14a}$$

$$S(t)=A\sin(\gamma t+\alpha)\ , \tag{14b}$$

with the 3 constants ρ, A and α defined by (9c), (2a) and (2b).

Proofs. The verification that (10b) satisfy (9) is too trivial to require any elaboration.

Summing the squares of (13a) and (13b) one gets

$$A^2=C^2(t)+S^2(t)=C^2(0)+S^2(0)\ , \tag{15}$$

which proves that A, see (2b), is indeed time-independent.

Finally the ratio of (14b) over (14a) yields precisely (2a), while comparison of (13a) with (14a), and of (13b) with (14b), entails

$$\alpha = \arctan[S(0)/C(0)] \ , \tag{16}$$

which demonstrates the time independence of α.

Insertion of (14) in (8) yields

$$\dot{q}_n = 2\lambda P + 2\mu A \cos(a q_n - \rho t - \alpha) \ , \tag{17a}$$

$$\dot{p}_n = 2 a \mu A \sin(a q_n - \rho t - \alpha) \ . \tag{17b}$$

It is now easily seen that these (decoupled !) ODEs can be integrated to yield (1).

Proofs. Let us set

$$q_n(t) = \tilde{q}_n(t) + (\rho t + \alpha)/a \tag{18}$$

so that (17a) become (using (9c))

$$\dot{\tilde{q}}_n = -2\mu[P - A\cos(a\tilde{q}_n)] \ , \tag{19}$$

entailing

$$\int_{\tilde{q}_n(0)}^{\tilde{q}_n(t)} dy/[P - A\cos(a y)] = -2\mu t \ . \tag{20}$$

The integral on the left hand side can be performed explicitly (see for instance eq. 2.553.3. of < GRJ 94>) and, via (18) and (1c), there obtains

$$\tilde{q}_n(t) = (2/a) \arctan\{2 a \mu[(P-A)/\omega] \tan[\omega(t_n - t)/2]\} \ , \tag{21}$$

which, via (18), yields (1a).

To obtain (1b) one notes that (17b), (18) and (19) entail

$$\dot{p}_n / p_n = -(d/dt) \log[P - A\cos(a\tilde{q}_n)] \tag{22}$$

hence

$$p_n(t) = p_n(0) \ \{P - A\cos[a\tilde{q}_n(0))]\}/\{P - A\cos[a\tilde{q}_n(t))]\} \ . \tag{23}$$

We then use (twice) the trigonometric *identity*

$$\cos(z) = [1 - \tan^2(z/2)]/[1 + \tan^2(z/2)] \tag{24}$$

to obtain, from (21),

$$P - A\cos(a\tilde{q}_n) = (P^2 - A^2)/\{P + A\cos[\omega(t - t_n)]\} . \tag{25}$$

The insertion of this expression in (23) yields (1b).

Let us end Sect. 2.1.15.1 with the following

Exercise 2.1.15.1-4. Investigate the many-body system defined by the Hamiltonian

$$H = \sum_{n,m=1}^{N} p_n\, p_m [\lambda + \mu\,(q_n - q_m)^2] . \tag{26}$$

Hint: Note that this Hamiltonian corresponds to (2.1.14-6) with

$$f(q) = \lambda + \mu\, q^2 . \tag{27}$$

Then in (2.1.15-1) replace c with $2\tilde{c}/a$, in (2.1.15-3) replace λ with $\lambda + 2\tilde{\mu}/a^2$, set $\mu = -2\tilde{\mu}/a^2$, take the limit $a \to 0$ and then eliminate all tildes. Or, more directly, perform this same transformation on the Hamiltonian (2.1.15-16).

2.1.15.2 Reformulation via canonical transformations

Another approach to the solution of the problem characterized by the Hamiltonian (2.1.15-16,17) is via appropriate canonical transformations. We indicate two of them in Sect. 2.1.15.2.

Firstly, let us set

$$\xi_n = \exp(i a q_n), \qquad \eta_n = -i(p_n / a)\exp(-i a q_n) , \tag{1a}$$

$$q_n = -i a^{-1} \log(\xi_n), \qquad p_n = i a \xi_n \eta_n . \tag{1b}$$

This transformation, from the ("old") canonical coordinates and momenta q_n, p_n, to the ("new") coordinates and momenta ξ_n, η_n, is *canonical.*

Proof.

$$\partial \xi_n / \partial q_\ell = \delta_{n\ell}\, i a \xi_n\,, \quad \partial \xi_n / \partial p_\ell = 0\ , \tag{2a}$$

$$\partial \eta_m / \partial q_\ell = -\delta_{m\ell}\, i a \eta_m\,, \quad \partial \eta_m / \partial p_\ell = \delta_{m\ell} (i a \xi_m)^{-1}\ . \tag{2b}$$

Hence (see (1.2-4))

$$[\xi_n, \eta_m] = \delta_{nm}\ , [\xi_n, \xi_m] = [\eta_n, \eta_m] = 0\ . \tag{3}$$

It is now convenient to introduce the quantities

$$E_s = \sum_{n=1}^{N} \eta_n (\xi_n)^s\,, \qquad s = 0,1,2\ , \tag{4}$$

and to note that there then hold the following equations:

$$P = i a E_1\ , \tag{5}$$

$$H = -a^2 (\lambda E_1^2 + \mu E_0 E_2)\ ; \tag{6}$$

$$\dot{\xi}_n = -a^2 (\mu E_2 + 2\lambda E_1 \xi_n + \mu E_0 \xi_n^2)\ , \tag{7a}$$

$$\dot{\eta}_n = 2 a^2 \eta_n (\lambda E_1 + \mu E_0 \xi_n)\ ; \tag{7b}$$

$$\dot{E}_s = (1-s) \beta E_s\ , \qquad s = 0,1,2\ , \tag{8a}$$

$$\beta = 2 a^2 (\lambda + \mu) E_1 = -i\gamma\ , \tag{8b}$$

$$E_s(t) = E_s(0) \exp[(1-s)\beta t]\ , \quad s = 0,1,2\ . \tag{8c}$$

The expressions (5) and (6) of the total momentum P, see (2.1.15-17), and of the Hamiltonian H, see (2.1.15-16), in terms of the quantities E_s, see (4), are immediate consequences of the second (1b), and of the formula

$$p_n p_m \cos[a(q_n - q_m)] = -(a^2/2) \eta_n \eta_m (\xi_n^2 + \xi_m^2)\ , \tag{9}$$

which is itself an immediate consequence of (1). Note that these expressions, (5) and (6), suggest that E_1 and $E_0\, E_2$ are time-independent, see indeed (8c).

The evolution equations (7) are the Hamiltonian equations, see (1.2-1), entailed by (6) and by the relations

$$\partial E_s / \partial \xi_n = s \eta_n \xi_n \,, \qquad \partial E_s / \partial \eta_n = \xi_n^s \tag{10}$$

entailed by (4).

The evolution equations (8a) with (8b) obtain by time-differentiation of the definitions (4) (separately, for $s=0$, $s=1$ and $s=2$), using (7); note that (8a) with $s=1$ entails that E_1, hence β, see (8b), are time-independent; and this entails (8c) (from (8a)). Finally, the consistency of (8b) with (2.1.15.1-9c) is guaranteed by (5).

The structures of (7a) and (8c) suggest the position

$$\xi_n(t) = \tilde{\xi}_n(t) \exp(-\beta t) \quad , \tag{11}$$

entailing

$$\dot{\tilde{\xi}}_n(t) = -\mu a^2 \{E_2(0) - 2E_1(0)\tilde{\xi}_n(t) + E_0(0)[\tilde{\xi}_n(t)]^2\} \,, \tag{12}$$

an ODE with *constant* coefficients that is immediately integrated (by quadratures), yielding

$$\tilde{\xi}_n(t) = \{E_1 + i[\omega/(2\mu a^2)] \operatorname{cotan}[\omega(t-\tau_n)/2)]\}/E_0(0) \,, \tag{13a}$$

with

$$\omega = 2\mu a [E_0(0)E_2(0) - E_1^2]^{1/2} \,, \tag{13b}$$

where the (integration) constants τ_n are determined by the initial conditions, see (11),

$$\tilde{\xi}_n(0) = \xi_n(0) = \{E_1 + [\omega/(2\mu a^2)] \operatorname{cotan}[-\omega\tau_n/2]\}/E_0(0) \ . \tag{13c}$$

(The diligent reader will check the consistency of the definitions (13b) and (2.1.15.1-1c)).

Via (11), (13c) yields

$$\xi_n(t) = \exp(-\beta t)\{E_1 + [\omega/(2\mu a^2)] \operatorname{cotan}[\omega(t-\tau_n)/2]\}/E_0(0) \ . \tag{14}$$

Hence, via (7b) and (8c),

$$\dot{\eta}_n = \eta_n\{2(\lambda+\mu)a^2 E_1 + \omega \operatorname{cotan}[\omega(t-\tau_n)/2]\} \,, \tag{15}$$

and this can be immediately integrated by quadratures, to yield

$$\eta_n(t)=\eta_n(0)\,\exp\left[2\,(\lambda+\mu)\,a^2\,E_1 t\right]\left\{\sin\left[\omega(\tau_n-t)/2\right]/\sin(\tau_n/2)\right\}^2 \,. \tag{16}$$

Exercise 2.1.15.2-2. Verify that, via (1), these expressions of $\xi_n(t)$ and $\eta_n(t)$ yield (2.1.15.1-1). *Hint*: use the following two *identities*:

$$2\,\log\left[i\,x+y\,\mathrm{cotan}(\theta)\right]=\log(y^2-x^2)+4\,i\,\arctan\left[z\tan(\theta-\theta_0)\right]\ , \tag{17a}$$

$$z=(x/y)\left\{\left[1-(y/x)^2\right]^{1/2}-1\right\}\ , \tag{17b}$$

$$\theta_0=-i\,(x/y)\left\{\left[1-(y/x)^2\right]+1\right\}\ ; \tag{17c}$$

$$2\left[x\sin(\theta)-y\cos(\theta)\right]\sin\theta=x+(x^2+y^2)^{1/2}\cos\left[2\,(\theta-\theta_0)\right]\ , \tag{18a}$$

$$\tan(2\,\theta_0)=y/x\ . \tag{18b}$$

As we just saw, the canonical transformation (1) provides an alternative, perhaps more elegant, route for the solution of the problem characterized by the Hamiltonian (2.1.15-16,17). But this approach appears in fact more suitable to treat the case in which the parameter a is purely *imaginary*, because only then the transform (1) is *real*, namely it maps *real* variables into *real* variables. We introduce now another canonical transformation, which is instead *real* when the parameter a is *real.* However, in this case, we only report some of the formulas relevant to this new approach, leaving their proofs, as well as the derivation (for the third time !) of the explicit solution of the problem, as an *exercise* for the diligent reader.

The canonical transformation among the "old" variables q_n, p_n and the "new" variables x_n, y_n reads

$$x_n=2\,(p_n/a)^{1/2}\sin(aq_n/2)\ ,\qquad y_n=2\,(p_n/a)^{1/2}\cos(aq_n/2)\ , \tag{19a}$$

$$p_n=(a/4)(x_n^2+y_n^2)\ ,\qquad q_n=(2/a)\arctan(x_n/y_n)\ . \tag{19b}$$

And there hold then the following formulas:

$$\left[x_n,y_m\right]=\delta_{nm}\,,\ \left[x_n,x_m\right]=\left[y_n,y_m\right]=0\ ; \tag{20}$$

$$H=\lambda P^2+\mu(C^2+S^2)\ , \tag{21}$$

with

$$P=(a/4)\sum_{n=1}^{N}(x_n^2+y_n^2) \ , \tag{22}$$

$$C=(a/4)\sum_{n=1}^{N}(-x_n^2+y_n^2) \ , \tag{23a}$$

$$S=(a/2)\sum_{n=1}^{N}x_n\,y_n \ ; \tag{23b}$$

$$\dot{x}_n=a\left[\mu S x_n+(\lambda P+\mu C)\,y_n\right] \ , \tag{25a}$$

$$\dot{y}_n=-a\left[\mu\,S\,y_n+(\lambda\,P-\mu\,C)\,x_n\right] \ , \tag{25b}$$

which are easily seen to entail (2.1.15-18) as well as (2.1.15.1-9), hence (2.1.15.1-13,14).

2.1.16 A nonanalytic solution of the functional equation (***), and the corresponding Hamiltonian many-body problem

The functional equation (***), see (2.1.14-8), admits a more general solution than that presented at the beginning of Sect. 2.1.15, see (2.1.15-1,2,3). It reads as follows:

$$\alpha(x)=\cos(a\,x/2)+M\,\mathrm{sign}(x)\,\sin(a\,x/2) \ , \tag{1}$$

$$\gamma(x)=a\,\mu'\left[\mathrm{sign}(x)\,\cos(ax/2)-M^{-1}\sin(a\,x/2)\right] \ , \tag{2}$$

$$f(x)=\lambda+\mu\cos(a\,x)+\mu'\,\mathrm{sign}(x)\,\sin(a\,x) \ , \tag{3}$$

$$M=\left\{\left[1+(\mu'/\mu)^2\right]^{1/2}-1\right\}^{1/2}/\left\{\left[1+(\mu'/\mu)^2\right]^{1/2}+1\right]\right\}^{1/2} \ . \tag{4}$$

Without loss of generality we assume throughout Sect. 2.1.16 that a is a *positive* constant, $a>0$.

A comparison of (3) with (2.1.15-3) clarifies in which sense this solution of the functional equation (***) is more general than that given at the beginning of Sect. 2.1.15. However, for $\mu'=0$, while (3) becomes (2.1.15-3), (1) and (2) yield

$$\alpha(x)=\cos(a\,x/2) \ , \tag{5}$$

$$\gamma(x)=-2\,\mu\,a\,\sin(a\,x/2) \ , \tag{6}$$

which differ from (2.1.15-1,2). The diligent reader may verify that these expressions, (5) and (6), together with (2.1.15-3), do satisfy the functional equation (***) (this is of course implied by the result proved below).

To verify that (1), (2), (3) and (4) satisfy the functional equation (***), see (2.1.14-8), we must show that the following expression vanishes:

$$a^{-1}\{2\alpha'(x+y)[f(x)-f(y)]-\alpha(x+y)[f'(x)-f'(y)]-[\alpha(x)\gamma(y)-\alpha(y)\gamma(x)]\}$$

$$=\{-\sin[a(x+y)/2]+M\,\mathrm{sign}(x+y)\,\cos[a(x+y)/2]\}\cdot$$
$$\cdot\{\mu[\cos(ax)-\cos(ay)]+\mu'[\mathrm{sign}(x)\,\sin(ax)-\mathrm{sign}(y)\,\sin(ay)]\}$$

$$-\{\cos[a(x+y)/2]+M\,\mathrm{sign}(x+y)\,\sin[a(x+y)/2]\}\cdot$$

$$\cdot\{-\mu[\sin(ax)-\sin(ay)]+\mu'\,[\mathrm{sign}(x)\,\cos(ax)-\mathrm{sign}(y)\,\cos(ay)]\}$$

$$-\mu'\{[\cos(ax/2)+M\,\mathrm{sign}(x)\,\sin(ax/2)][\mathrm{sign}(y)\,\cos(ay/2)-M^{-1}\sin(ay/2)]$$

$$-[\cos(ay/2)+M\,\mathrm{sign}(y)\,\sin(ay/2)][\mathrm{sign}(x)\,\cos(ax/2)-M^{-1}\sin(ax/2)]\}$$

$$=2\mu\,\sin[a(x+y)/2]-\mu'\,[\mathrm{sign}(x)-\mathrm{sign}(y)]\cos[a(x+y)/2]$$

$$+M\,\mu'\,\mathrm{sign}(x+y)\,[\mathrm{sign}(x)+\mathrm{sign}(y)]\,\sin[a(x-y)/2]$$

$$-\mu'\,[\,M^{-1}+M\,\mathrm{sign}(x)\,\mathrm{sign}(y)]\,\sin[a(x-y)/2]$$

$$+\mu'[\mathrm{sign}(x)-\mathrm{sign}(y)]\cos[a(x+y)/2]=\sin[a(x-y)/2]\,Z\ , \tag{7}$$

$$Z=2\mu-\mu'[M^{-1}+M\,\mathrm{sign}(x)\,\mathrm{sign}(y)]+\mu'M\,\mathrm{sign}(x+y)\,[\mathrm{sign}(x)+\mathrm{sign}(y)]\,. \tag{8}$$

To obtain this result we used several times the trigonometric identities $\sin(A)\sin(B)+\cos(A)\cos(B)=\cos(A-B)$, $\sin(A)\cos(B)-\cos(A)\sin(B)=\sin(A-B)$.

We now note that the definition (4) entails, as can be easily verified, the relation

$$M^{-1}=M+2\mu/\mu'\ , \tag{9}$$

hence (8) becomes

$$Z=\mu' M z\ , \tag{10}$$

$$z=-[1+\mathrm{sign}\,(x)\,\mathrm{sign}\,(y)]+\mathrm{sign}\,(x+y)[\mathrm{sign}\,(x)+\mathrm{sign}\,(y)]\ . \tag{11}$$

But z clearly vanishes: if x and y have opposite signs, obviously each square bracket in the right hand side of (11) vanishes, if x and y have the same sign, then $x+y$ also has the same sign, hence (11) yields $z=-2+2=0$.

The many-body problem corresponding to the Lax equation (2.1.14-4) with (2.1.14-1), (2.1.14-2) and (1), (2), (4) is therefore characterized, see (2.1.14-6) and (3), by the Hamiltonian

$$H=\lambda P^2+\sum_{n,m=1}^{N} p_n\, p_m\{\mu\cos[a(q_n-q_m)]+\mu'\sin|a(q_n-q_m)|\ \}\ , \tag{12}$$

where as usual P is the total momentum, see (2.1.15-17). The corresponding equations of motion read

$$\dot{q}_n=2\sum_{m=1}^{N} p_m\{\lambda+\mu\cos[a(q_n-q_m)]+\mu'\sin|a(q_n-q_m)|\ \}\ , \tag{13a}$$

$$\dot{p}_n=2a\,p_n\sum_{m=1}^{N} p_m\{\mu\sin[a(q_n-q_m)]-\mu'\,\mathrm{sign}(q_n-q_m)\cos[a(q_n-q_m)]\}\ . \tag{13b}$$

Of course for $\mu'=0$ these formulas, (12) respectively (13), reduce to (2.1.15-16) respectively (2.1.15-20).

The possibility to recast these equations of motion, (13), in Lax form, entails the existence of N integrals of motion; let us focus on the traces of the powers of the Lax matrix $\underline{L}$,

$$T_n=\mathrm{trace}[\ \underline{L}^n\ ]\ ,\qquad n=1,2,\ldots,N\ . \tag{14}$$

The fact that these quantities are constants of motion is implied by the general treatment of Sect. 2.1 (see for instance (2.1-9, 10)); to conclude that the corresponding Hamiltonian system, see (12) and (13), is *completely integrable* one must moreover show that these N quantities are *in involution*. We prove this in the following Sect. 2.1.16.1; here we note that (14), with (2.1.14-1) and (1), entail

$$T_1=P\ , \tag{15}$$

$$T_2=(M/\mu')[H+(\mu-\lambda+M\mu')P^2\ ]\ . \tag{16}$$

Proofs. (15) is an immediate consequence of (14), (2.1.14-1) and (1). To prove (16) we note that (14) and (2.1.14-1) yield

$$T_2=\sum_{n,m=1}^{N} L_{nm}\,L_{mn}=\sum_{n,m=1}^{N} p_n\, p_m[\alpha(q_n-q_m)]^2\ , \tag{17}$$

where we also used the fact that α is even, $\alpha(-q)=\alpha(q)$, see (1). We now use (1) to get

$$[\alpha(x)]^2 = \cos^2(ax/2) + M^2 \sin^2(ax/2) + 2M\cos(ax/2)\,\sin(a|x|/2) \;, \tag{18a}$$

$$[\alpha(x)]^2 = \frac{1}{2}(1+M^2) + \frac{1}{2}\,(1-M^2)\,\cos(ax) + M\sin(a|x|) \;, \tag{18b}$$

$$[\alpha(x)]^2 = (M/\mu')[\mu + M\mu' + \mu\,\cos(ax) + \mu'\sin(a|x|)] \;. \tag{18c}$$

To get (18b) from (18a) we used the trigonometric identities $\cos^2(A/2) = [\cos(A)+1]/2$, $\sin^2(A/2) = [-\cos(A)+1]/2$; to get (18c) from (18b), we used (9).

We now insert (18c) in (17), and use the definitions of H and P, see (12) and (2.1.15-17); this yields (16), which is thereby proven.

Of course the treatment given in Sect. 2.1.16, as well as in the following Sect. 2.1.16.1, holds equally if the constant a is replaced by ia, entailing generally the replacement of trigonometric with hyperbolic functions. A particularly interesting instance in this case obtains for the special choice $\mu' = i\mu$, since the Hamiltonian (12) takes then the neat form

$$H = \lambda P^2 + \mu \sum_{n,m=1}^{N} p_n\, p_m \exp[-a|q_n - q_m|] \;. \tag{19}$$

2.1.16.1 Proof of integrability. A new functional equation

In Sect. 2.1.16.1 we prove that the traces T_n defined by (2.1.16-14) with (2.1.14-1) and (2.1.16-1) are *in involution,*

$$[T_n, T_{n'}] = 0; \qquad n, n' = 1, \ldots, N \tag{1}$$

(see (1.2-4)).

Since T_1 coincides with the total momentum P, see (2.1.16-15) and (2.1.15-17), and all the traces T_n, as indeed the Lax matrix $\underline{L}$ itself, see (2.1.14-1), are translation-invariant, the validity of (1) for $n=1$, and for $n'=1$, is trivial. Hence in the following we assume that both n and n' are larger than unity, say $1 < n' < n$.

The result (1) is proven via the following two steps.

Proposition 2.1.16.1-1. The quantities T_n defined by (2.1.16-14) with (2.1.14-1) are *in involution,* see (1), if the even function $\alpha(x)$,

$$\alpha(-x) = \alpha(x) \;, \tag{2}$$

satisfies the following functional equation:

$$\alpha'(x)\alpha(x+y+z)-\alpha'(x+y+z)\alpha(x)+\alpha'(y)\alpha(z)+\alpha'(z)\alpha(y)=$$

$$=\alpha(y+z)\left[\beta_1(x)+\beta_2(y)+\beta_3(z)-\beta_4(x+y+z)\right] \, , \tag{3}$$

where the 4 functions $\beta_s(x)$ are all *odd*,

$$\beta_s(-x)=-\beta_s(x), \qquad s=1,2,3,4 \, , \tag{4}$$

but are otherwise arbitrary.

Proposition 2.1.16.1-2. The (clearly even) function $\alpha(x)$ defined by (2.1.16-1) satisfies the functional equation (3) with

$$\beta_s(x)=(a/2)\, M \,\text{sign}(x), \qquad s=1,2,3,4 \, , \tag{5}$$

which clearly satisfy (4).

Proofs. The proof of *Proposition 2.1.16.1-1* is somewhat cumbersome. We note first of all that the definition (2.1.16-14) of the traces T_n entails

$$\partial T_n/\partial q_j = n \sum_{j_1,\dots,j_n=1}^{N} (\partial L_{j_1 j_2}/\partial q_j) L_{j_2 j_3} \dots L_{j_n j_1} \, , \tag{6a}$$

$$\partial T_n/\partial p_j = n \sum_{j_1,\dots,j_n=1}^{N} (\partial L_{j_1 j_2}/\partial p_j) L_{j_2 j_3} \dots L_{j_n j_1} \, , \tag{6b}$$

and that the definition (2.1.14-1) of the Lax matrix $\underline{L}$ entails

$$\partial L_{jk}/\partial q_n = (p_j\, p_k)^{1/2} (\delta_{jn}-\delta_{kn})\, \alpha'(q_j-q_k) \, , \tag{7a}$$

$$\partial L_{jk}/\partial p_n = (p_j\, p_k)^{1/2} (\delta_{jn}+\delta_{kn})\, \alpha(q_j-q_k)/(2\, p_n) \tag{7b}$$

Hence, via (1.2-4) and after a little elementary algebra,

$$\left[T_n, T_{n'}\right] = (n n'/2) \sum_{j_1,\dots,j_n,j'_1,\dots,j'_{n'}=1}^{N} p(2,3)(3,4)\dots(n,1)(2',3')(3',4')\dots(n',1') \cdot$$

$$\{\delta_{j_1 j'_1}\left[(1,2)'(1',2')-(1',2')'(1,2)\right]/p_{j_1} + (1 \leftrightarrow 2) + (1' \leftrightarrow 2') + (1 \leftrightarrow 2, 1' \leftrightarrow 2')\} \, . \tag{8}$$

Here we have introduced the following convenient notation:

$$p \equiv p_{j_1} \cdots p_{j_n}\, p_{j'_1} \cdots p_{j'_n} \, , \tag{9}$$

$$(1,2) \equiv \alpha(q_{j_1} - q_{j_2}),\ (1',2') \equiv \alpha(q_{j'_1} - q_{j'_2}) \quad \text{and so on} \ , \tag{10a}$$

$$(1,2)' \equiv \alpha'(q_{j_1} - q_{j_2}),\ (1',2')' \equiv \alpha'(q_{j'_1} - q_{j'_2}) \quad \text{and so on} \ . \tag{10b}$$

The notation $(1 \leftrightarrow 2)$ indicates a repetition of the same expression, with the indices j_1 and j_2 exchanged; likewise $(1' \leftrightarrow 2')$, for the indices j'_1 and j'_2, and so on.

The product p, see (9), is clearly invariant under any exchange of summation indices. The product $(2,3)(3,4)\cdots(n,1)$ is invariant under the transformation $(1 \leftrightarrow 2, n \leftrightarrow 3, n-1 \leftrightarrow 4, \ldots)$, thanks to the even character of $\alpha(x)$, see (2.1.16-1), which entails, see (9), $(1,2) = (2,1)$ and so on. Likewise the product $(2',3')(3',4')\cdots(n',1')$ is invariant under the transformation $(1' \leftrightarrow 2', n' \leftrightarrow 3', n'-1 \leftrightarrow 4', \ldots)$. Hence

$$[T_n, T_{n'}] = 2nn' \sum_{j_1,\ldots,j_n,j'_2,\ldots,j'_{n'}}^{N} (p/p_{j'_1})(2,3)(3,4)\cdots(n,1)(1,n')(n',n'-1)\cdots(3'2')\cdot$$

$$\cdot[(1,2)'(1,2') - (1,2')'(12)] \ . \tag{11}$$

Note that the transition from (8) to (11) has involved *two* steps: the elimination of the 3 additional terms inside the curly bracket (compensated by the replacement of the factor (1/2) with 2 in front of the sum), and then the sum over the index j'_1 (using the Kronecker $\delta_{j_1 j'_1}$). Of course the ratio $(p/p_{j'_1})$ is independent of $p_{j'_1}$, see (9). Note that, as indicated, now (and below) the sum operates on all the indices j_ℓ and j'_ℓ except j'_1.

Let us now consider the following transformation, under which the product $(2,3)(3,4)\cdots(m,1)(1,m')(m',m'-1)\cdots(3',2')$ is clearly invariant: $(2' \leftrightarrow 2, 3' \leftrightarrow 3, \ldots, 1 \leftrightarrow h, \ldots)$. It is obvious how this transformation is identified, in terms of the above mentioned product it leaves invariant; it is also obvious that this transformation does not affect the ratio $p/p_{j'_1}$, since it does not involve the index j'_1. It would be easy to identify the integer h, namely the corresponding summation index j_h; but this identification is not needed in the following.

Using this transformation on the summation indices we get from (11)

$$[T_n, T_{n'}] = nn' \sum_{j_1,\ldots,j_n,j'_2,\ldots,j'_{n'}=1}^{N} (p/p_{j'_1})(2,3)(3,4)\cdots(n,1)(1,n')(n',n'-1)\cdots(3'2')\cdot$$

$$\cdot[(1,2)'(1,2') - (1,2')'(1,2) + (h2')'(h2) - (h2)'(h2')] \ . \tag{12}$$

We now use the formula

$$[(1,2)'(1,2') - (1,2')'(1,2) + (h2')'(h2) - (h2)'(h2')] = (2,2')[\beta_{12} + \beta_{2h} + \beta_{h2'} + \beta_{2'1}], \tag{13}$$

where

$$\beta_{12} \equiv \beta_1 (q_{j_1} - q_{j_2}), \beta_{2h} \equiv \beta_2 (q_{j_2} - q_{j_h}), \beta_{h2'} \equiv \beta_3 (q_{j_h} - q_{j_2'}), \beta_{2'1} \equiv \beta_4 (q_{j_2'} - q_{j_1}) . \tag{14}$$

This formula, via (10), coincides with (3) with $x = q_{j_1} - q_{j_2}, y = q_{j_h} - q_{j_2'}, z = -q_{j_h} + q_{j_2}$ hence $x + y + z = q_{j_1} - q_{j_2'}$ (recall that $\alpha(z)$ is even, see (2), hence $\alpha'(z)$ is odd, $\alpha'(-z) = -\alpha'(z)$).

The proof is now completed, since the sum in the right hand side of (12) with (13) can be written as follows:

$$[T_n, T_{n'}] = n n' \sum_{j_1,\ldots,j_n,j_2',\ldots,j_{n'}'}^{N} S_{j_1,j_2,\ldots,j_n,j_2',\ldots,j_{n'}'} [\beta_{12} + \beta_{2h} + \beta_{h2'} + \beta_{2'1}] , \tag{15}$$

with

$$S_{j_1,\ldots,j_n,j_2',\ldots,j_{n'}'} = (p / p_{j_1'}) (2,3)(3,4) \cdots (n,1)(1,n')(n',n'-1) \cdots (3'2')(2',2) , \tag{16}$$

where of course p is defined by (9). Hence $S_{j_1,\ldots,j_n,j_2',\ldots,j_{n'}'}$ is symmetrical under the exchange of any two of the summation indices, while each of the 4 terms in the square bracket in the right hand side of (15), $\beta_{12}, \beta_{2h}, \beta_{h2'}, \beta_{2'1}$, is antisymmetrical under the exchange of its two summation indices, see (14) and (4): and this of course entails that the sum vanishes.

The proof of *Proposition 2.1.16.1-2* is easier. The left hand side of (3) with (2.1.16-1) reads

$$\begin{aligned}
&(a/2)(\{-\sin(ax/2) + M\,\mathrm{sign}(x)\cos(ax/2)\}\{\cos[a(x+y+z)/2] \\
&+ M\,\mathrm{sign}(x+y+z)\,\sin[a(x+y+z)/2]\} - \{-\sin[a(x+y+z)/2] \\
&+ M\,\mathrm{sign}(x+y+z)\,\cos[a(x+y+z)/2]\}\,\{\cos(ax/2) + M\,\mathrm{sign}(x)\,\sin(ax/2)\} \\
&+ \{-\sin(ay/2) + M\,\mathrm{sign}(y)\,\cos(ay/2)\}\,\{\cos(az/2) + M\,\mathrm{sign}(z)\sin(az/2)\} \\
&+ \{-\sin(az/2) + M\,\mathrm{sign}(z)\,\cos(az/2)\}\,\{\cos(ay/2) + M\,\mathrm{sign}(y)\,\sin(ay/2)\}) \\
&= (a/2)\, M\, \{M \sin[a(y+z)/2]\,[\mathrm{sign}(x)\,\mathrm{sign}(x+y+z) + \mathrm{sign}(y)\,\mathrm{sign}(z)] \\
&+ \cos[a(y+z)/2]\,[\mathrm{sign}(x) + \mathrm{sign}(y) + \mathrm{sign}(z) - \mathrm{sign}(x+y+z)]\} .
\end{aligned} \tag{17}$$

To obtain this equality we used repeatedly the trigonometric *identities*
$\sin(A)\,\cos(B) - \cos(A)\,\sin(B) = \sin(A-B)$,
$\cos(A)\,\cos(B) + \sin(A)\,\sin(B) = \cos(A-B)$.

We now use the *identity* (proven below)

$$\text{sign}(x)\ \text{sign}(x+y+z)+\text{sign}(y)\ \text{sign}(z)$$

$$=\text{sign}(y+z)\ [\text{sign}(x)+\text{sign}(y)+\text{sign}(z)-\text{sign}(x+y+z)]\ , \qquad (18)$$

and the definition (2.1.16-1) of $\alpha(x)$, to rewrite the right hand side of (17) as follows:

$$(a/2)\,M\,\alpha(y+z)\ [\text{sign}(x)+\text{sign}(y)+\text{sign}(z)-\text{sign}(x+y+z)]\ , \qquad (19)$$

which coincides, via (5), with the right-hand side of (3).

This completes the proof of *Proposition 2.1.16.1-2*, except that there remains to ascertain the validity of the identity (18). This is a trivial task that can be left to the reader, who needs to verify that (17) holds for any *compatible* choice of the signs of $x, y, z, y+z$ and $x+y+z$. There are $2^5=32$ such choices, of which however only 20 are compatible: for instance if y and z are positive, $y+z$ cannot be negative. Moreover, there are two symmetries of (17) which lessen the burden of checking: this identity is obviously invariant under the simultaneous change of sign of x, y and z, as well as under the exchange of y and z. Hence one can restrict attention only to cases with, say, $x>0$, and among those, one need consider, of the cases with y and z having different signs, only those in which, say, $y>0, z<0$: altogether 12 cases, of which 8 with $x>0, y<0$ and 4 with $x>0, y>0, z<0$. Of these 12 cases, 4 are incompatible, 5 yield the equality $0=0$, 2 the equality $-2=-2$, and 1 the equality $2=2$.

Let us end Sect. 2.1.16.1 by emphasizing that *Proposition 2.1.16.1-1* entails that any one of the traces T_n defined by (2.1.16-14) with (2.1.14-1), or, more generally, any nontrivial function of them, can be chosen as the Hamiltonian of a system, which turns then out to be *completely integrable* provided the *even* function $\alpha(x)$, see (2), satisfies the functional equation (3) with *odd* $\beta_s(x)$, see (4). *Proposition 2.1.16.1-2* guarantees that such a function is provided by (2.1.16-1,4) with μ and μ' arbitrary constants, hence that the system characterized by the Hamiltonian (2.1.16-12) is *completely integrable*: indeed this Hamiltonian is a linear combination of T_1 and T_1^2, see (2.1.16-16) and (2.1.16-15). The question of the existence of other even functions $\alpha(x)$ which satisfy the functional equation (3) with (4) is open.

Finally note that the Hamiltonian model treated in Sect. 2.1.16 and 2.1.16.1, see (2.1.16-12), has been shown to be *completely integrable*, but has *not* been explicitly solved, in contrast to the model (to which it reduces in the special case $\mu'=0$) treated in Sections 2.1.15, 2.1.15.1 and 2.1.15.2, see (2.1.15-16), which has instead been explicitly solved, see (2.1.15.1-1).

2.2 Another exactly solvable Hamiltonian problem

In Sect. 2.1.14, and in the sections following it, many-body models are discussed which are characterized by Hamiltonians of type (2.1.14-6). In particular, the solution is reported, see (2.1.15.1-1), of the model characterized by the Hamiltonian (2.1.15-16). In Sect. 2.2 we report the explicit solution of a somewhat analogous model, characterized by the Hamiltonian

$$H=\sum_{n,m=1}^{N}(p_n\,p_m)^{1/2}\cos(q_n-q_m)\ . \tag{1}$$

Since this Hamiltonian is translation-invariant (i.e., invariant under $q_n \to q_n+q_0$), it commutes with the total momentum $P=\sum_{n=1}^{N} p_n$. Therefore this quantity is a constant of the motion. Addition of a function $g(P)$ to the Hamiltonian (1) would only, trivially, add the term $g'(P)\,t$ to each of the canonical coordinates $q_n(t)$, with no effect on the canonical momenta $p_n(t)$. We therefore forsake here any such addition. Note moreover that we have introduced no constants in (1); of course an arbitrary constant can be inserted by the trivial rescaling transformation $q_n(t)\to a\,q_n(t)$, while multiplication of the Hamiltonian by μ amounts merely to an analogous rescaling, $t\to\mu t$, of time.

Hereafter we assume the momenta p_n to be all positive,

$$p_n>0\ , \tag{2}$$

and we take the positive determination of the square roots, see (1). It is easily seen, for instance directly from the equations of motion (3b), see below, that validity of this condition, (2), at $t=0$, guarantees its validity for all time $t>0$.

The equations of motion entailed by (1) read

$$\dot q_n=p_n^{-1/2}\sum_{m=1}^{N} p_m^{1/2}\cos[a\,(q_n-q_m)], \tag{3a}$$

$$\dot p_n=2\,a\,p_n^{1/2}\sum_{m=1}^{N} p_m^{1/2}\sin[a\,(q_n-q_m)]\ . \tag{3b}$$

The general solution of these equations of motion reads

$$q_n(t) = q_n(0) + \arctan\{\sin(Nt/2)\cos(q_n(0) - \alpha - Nt/2)$$

$$/[\sin(Nt/2)\sin(q_n(0) - \alpha - Nt/2) + [p_n(0)]^{1/2}/b]\} \ , \tag{4a}$$

$$p_n(t) = p_n(0) + 2b\,[p_n(0)]^{1/2}\sin(Nt/2)\sin[q_n(0) - \alpha - Nt/2] + b^2\sin^2(Nt/2), \tag{4b}$$

with the two constants α and b given, in terms of the initial data, as follows:

$$b = (2/N)\{\sum_{n,m=1}^{N}[p_n(0)\,p_m(0)]^{1/2}\cos[q_n(0) - q_m(0)]\}^{1/2} = 2H^{1/2}/N \ , \tag{4c}$$

$$\alpha = \arctan[\{\sum_{n=1}^{N}[p_n(0)]^{1/2}\sin[q_n(0)]\}/\{\sum_{n=1}^{N}[p_n(0)]^{1/2}\cos[q_n(0)]\}] \ . \tag{4d}$$

Note that the coordinates and momenta, $q_n(t)$ and $p_n(t)$, of the n-th particle depend on the initial data of the other particles (namely, on the $q_m(0)$'s and $p_m(0)$'s with $m \neq n$) only via these two quantities, b and α. Let us also emphasize that these formulas entail that the solutions, see (4), are completely periodic with period $T = 4\pi/N$ (of course the coordinates $q_n(t)$ are defined $\mathrm{mod}(2\pi)$, see (1)).

It is trivially easy to verify that (4) hold at $t = 0$. The diligent reader will verify that (4) satisfy (3). We prefer to indicate below a route whereby the solution (4) is *obtained* from (1).

The stating point of our treatment is the following *canonical transformation:*

$$u_n = (2p_n)^{1/2}\sin(q_n), \qquad v_n = (2p_n)^{1/2}\cos(q_n) \ , \tag{5a}$$

$$p_n = (u_n^2 + v_n^2)/2, \qquad q_n = \arctan(u_n/v_n) \ . \tag{5b}$$

In terms of the new variables u_n, v_n the Hamiltonian (1) reads

$$H = (U^2 + V^2)/2 \tag{6}$$

with

$$U = \sum_{n=1}^{N} u_n, \quad V = \sum_{n=1}^{N} v_n \ , \tag{7}$$

and the corresponding equations of motion read

$$\dot{u}_n = V, \quad \dot{v}_n = -U \ , \tag{8}$$

entailing

$$\dot{U} = NV, \quad \dot{V} = -NU \ , \tag{9}$$

$$U(t) = B \sin(Nt + \alpha), \quad V(t) = B \cos(Nt + \alpha) \ , \tag{10}$$

$$u_n(t) = u_n(0) + (2B/N) \sin(Nt/2) \cos(\alpha + Nt/2) \ , \tag{11a}$$

$$v_n(t) = v_n(0) - (2B/N) \sin(Nt/2) \sin(\alpha + Nt/2) \ . \tag{11b}$$

Proof. To verify that the transformation (5) is canonical one must check (see (1.2-10)) the properties

$$[u_n, v_n] = \delta_{nm} \ , [u_n, u_m] = [v_n, v_m] = 0 \tag{12}$$

(see (1.2-4)), and this is a trivial task. It is likewise trivial, from the definitions (7) and (5a), using the trigonometric identity $\cos(q_n)\cos(q_m) + \sin(q_n)\sin(q_m) = \cos(q_n - q_m)$, to show the coincidence of the definitions (6) and (1) of the Hamiltonian H. The equations of motion (8) are then, again trivially, the Hamiltonian equations, see (1.2-1), entailed by (6) and (7), and by summing them over n from 1 to N one immediately gets (9) via (7). The verification that (10) satisfy (9) is trivial; moreover (10) imply

$$U(0) = B \sin(\alpha), \quad V(0) = B \cos(\alpha) \ , \tag{13a}$$

hence

$$B = \{[U(0)]^2 + [V(0)]^2\}^{1/2}, \quad \alpha = \arctan[U(0)/V(0)] \ . \tag{13b}$$

Via (7) and (5a) it is immediate to see that this definition of the constant α coincides with (4d), and that

$$B = (2H)^{1/2} = 2^{-1/2} N b \tag{13c}$$

with b defined by (4c).

Finally, clearly (11) hold at $t=0$, and to verify that they satisfy (8) with (10) one may conveniently use, in the right-hand sides of (11), the trigonometric *identities* $\sin(A)\cos(B)=[\sin(A+B)+\sin(A-B)]/2$,
$\sin(A)\sin(B)=[-\cos(A+B)+\cos(A-B)]/2$.

The last task is to show that (5) entail (4) via (11). Indeed, from the second of the (5b) and (11) one gets, using (13c),

$$q_n(t)=\arctan\{[u_n(0)+2^{1/2}b\,\sin(Nt/2)\,\cos(\alpha+Nt/2)]$$

$$/[v_n(0)-2^{-1/2}b\,\sin(Nt/2)\,\sin(\alpha+Nt/2)]\}\ , \tag{14a}$$

hence, via (5a),

$$q_n(t)=\arctan\{[\sin[q_n(0)]+b[p_n(0)]^{-1/2}\sin(Nt/2)\,\cos(\alpha+Nt/2)]$$

$$/[\cos[q_n(0)]-b[p_n(0)]^{-1/2}\sin(Nt/2)\,\sin(\alpha+Nt/2)]\}\ . \tag{14b}$$

We now use the trigonometric *identity*

$$\arctan\{[\sin(A)+C]/[\cos(A)+D]\}=A+\arctan(E)\ , \tag{15a}$$

$$E=[C\cos(A)-D\sin(A)]/[1+C\sin(A)+D\cos(A)]\ , \tag{15b}$$

to get from (14b)

$$q_n(t)=q_n(0)\arctan\{[b\,[p_n(0)]^{-1/2}\sin(Nt/2)\,\cos[q_n(0)-\alpha-Nt/2]]$$

$$/\ [1+b\,[p_n(0)]^{-1/2}\sin(Nt/2)\,\sin[q_n(0)-\alpha-Nt/2)]]\}\ , \tag{14c}$$

which clearly coincides with (4a).

To get (4b) one starts from the first of the (5b) and gets, via (11),

$$p_n(t)=p_n(0)+2(B/N)^2\sin^2(Nt/2)+(2B/N)[2\,p_n(0)]^{1/2}\sin(Nt/2)\cdot$$

$$\cdot\{\sin[q_n(0)]\cos(\alpha+Nt/2)-\cos[q_n(0)]\sin(\alpha+Nt/2)\} \tag{15a}$$

which, via (13c), clearly yields (4b).

Note that (4b) can be rewritten in the form

$$p_n(t)=\{[p_n(0)]^{1/2}+b\,\sin(Nt/2)\,\sin[q_n(0)-\alpha-Nt/2]\}^2$$

$$+b^2\sin^2(Nt/2)\cos^2[q_n(0)-\alpha-Nt/2]\ , \tag{15b}$$

which displays the fact that $p_n(t)$ is always positive.

2.3 Many-body problems on the line related to the motion of the zeros of solutions of linear partial differential equations in 1+1 variables (space + time)

In Sect. 2.3, which contains several subsections, see below, the following idea is exploited.

Let $\psi(x,t)$ be a function that satisfies a linear Partial Differential Equation (PDE) in x and t; in particular, let us assume that $\psi(x,t) \equiv p_N(x,t)$ is a (monic) polynomial of degree N in x,

$$\psi(x,t) \equiv p_N(x,t) = x^N + \sum_{m=1}^{N} c_m(t)\, x^{N-m} = \prod_{n=1}^{N} [x - x_n(t)] \,, \tag{1}$$

that evolves in time according to a linear PDE, say

$$[\ A_0 + A_1 x + A_2 x^2 + A_3 x^3\ ]\psi_{xx} + [\ B_0 + B_1 x - 2(N-1) A_3 x^2\ ]\psi_x$$

$$+ C\psi_{tt} + [\ E - (N-1) D_2\, x\ ]\psi_t + [\ D_0 + D_1 x + D_2 x^2\ ]\psi_{xt}$$

$$-[\ N(N-1)(A_2 - A_3 x) + N B_1\ ]\psi = 0 \ . \tag{2}$$

A polynomial is called "monic" if the coefficient of its term of highest degree is unity.

Exercise 2.3-1. Verify that the evolution PDE (2) is satisfied by functions $\psi(x,t)$ that are (for all time) a monic polynomial of degree N in x. *Solution:* see Sect. 2.3.3.

The formula (1) represents a monic polynomial of degree N in two different ways: via its N coefficients c_m, and via its N zeros x_n. This entails a bi-univocal (nonlinear) mapping among the two sets $C = \{c_m, m = 1,...,N\}$ and $X = \{x_n, n = 1,...,N\}$; note however that the elements c_m of the set C are uniquely defined, while the elements x_n of the set X are defined up to permutations. If the zeros x_n are all real, this ambiguity might be lifted by an ordering convention, say

$$x_1 \le x_2 \le ... \le x_N \ . \tag{3}$$

The *nonlinear mapping* among the two sets C and X is the main tool for the developments discussed in Sect. 2.3 (including its subsections, see below). We will see that certain simple, *linear* time-evolutions of the coefficients $c_m(t)$ correspond to more complicated *nonlinear* time-evolutions of the zeros $x_n(t)$. We will moreover see that these latter time-evolutions of the N zeros $x_n(t)$ are rather naturally interpretable as many-body problems on the line (and, in some cases, via complexification, also as many-body problems in the plane: see Chap. 4). This will open the way to solving such many-body problems; hence this approach provides a technique to identify *solvable* many-body problems, and then to study their behavior.

2.3.1 A nonlinear transformation: relationships between the coefficients and the zeros of a polynomial

Let $p_N(x)$ be a monic polynomial of degree N in x,

$$p_N(x) = x^N + \sum_{m=1}^{N} c_m \, x^{N-m} = \prod_{n=1}^{N} (x - x_n) \ . \tag{1}$$

In Sect. 2.3.1 formulas are reported which exhibit some (well-known!) expressions of the coefficients c_m in terms of the zeros x_n.

$$c_1 = -\sum_{n=1}^{N} x_n = -s_1 \ , \tag{2a}$$

$$c_2 = \frac{1}{2} \sum_{n_1, n_2 = 1, n_1 \neq n_2}^{N} x_{n_1} \, x_{n_2} = \frac{1}{2}(s_1^2 - s_2) \ , \tag{2b}$$

$$c_3 = -\frac{1}{6} \sum_{n_1, n_2, n_3 = 1; n_1 \neq n_2, n_2 \neq n_3, n_3 \neq n_1}^{N} x_{n_1} \, x_{n_2} \, x_{n_3} = -\frac{1}{6}(s_1^3 - 3 s_1 \, s_2 + 2 s_3) \ , \tag{2c}$$

and so on, where

$$s_p = \sum_{n=1}^{N} (x_n)^p \ , \quad p = 1,2,\dots \ ; \tag{3}$$

as well as

$$c_N = (-)^N \prod_{n=1}^{N} x_n \ . \tag{4}$$

These expressions follow in an obvious way from (1). The inverse problem, to express the zeros x_n of the polynomial $p_N(x)$ in terms of its coefficients c_m, can of course be solved in explicit form only for $N=1,2,3$ and 4.

2.3.2 Some formulas for a polynomial and its derivatives, in terms of its coefficients and its zeros

Let $p_N(x,t)$ be a monic polynomial of degree N in x, whose N coefficients $c_m(t)$ are time-dependent, as well as its N zeros $x_n(t)$:

$$\psi(x,t) \equiv p_N(x,t) = x^N + \sum_{m=1}^{N} c_m(t)\, x^{N-m} \quad , \tag{1}$$

$$\psi(x,t) \equiv p_N(x,t) = \prod_{n=1}^{N} \left[x - x_n(t)\right] . \tag{2}$$

In Sect. 2.3.2 we report several formulas that express the (partial) derivatives of $\psi(x,t) \equiv p_N(x,t)$, first in terms of its coefficients $c_m(t)$, see (1), then in terms of its zeros $x_n(t)$, see (2).

$$x\psi_x(x,t) = N x^N + \sum_{m=1}^{N-1} (N-m)\, c_m(t)\, x^{N-m} \quad , \tag{3}$$

$$x^2\psi_{xx}(x,t) = N(N-1)x^N + \sum_{m=1}^{N-2} (N-m)\,(N-1-m)\, c_m(t)\, x^{N-m} \quad , \tag{4}$$

and so on;

$$\psi_t(x,t) = \sum_{m=1}^{N} \dot{c}_m(t)\, x^{N-m} \quad , \tag{5}$$

$$\psi_{tt}(x,t) = \sum_{m=1}^{N} \ddot{c}_m(t)\, x^{N-m} \quad , \tag{6}$$

and so on.

$$\psi_x(x,t) = \psi(x,t) \sum_{n=1}^{N} \left[x - x_n(t)\right]^{-1} \quad , \tag{7}$$

$$\psi_t(x,t)=\psi(x,t)\sum_{n=1}^{N}\left[x-x_n(t)\right]^{-1}\left\{-\dot{x}_n(t)\right\}\ , \tag{8}$$

$$\psi_{xx}(x,t)=\psi(x,t)\sum_{n=1}^{N}\left[x-x_n(t)\right]^{-1}\left\{2\sum_{m=1,m\neq n}^{N}\left[x_n(t)-x_m(t)\right]^{-1}\right\}\ , \tag{9}$$

$$\psi_{xt}(x,t)=\psi(x,t)\sum_{n=1}^{N}\left[x-x_n(t)\right]^{-1}\left\{-\sum_{m=1,m\neq n}^{N}\left[\dot{x}_n(t)+\dot{x}_m(t)\right]\left[x_n(t)-x_m(t)\right]^{-1}\right\}\ , \tag{10}$$

$$\psi_{tt}(x,t)=\psi(x,t)\sum_{n=1}^{N}\left[x-x_n(t)\right]^{-1}\left\{-\ddot{x}_n(t)+2\sum_{m=1,m\neq n}^{N}\dot{x}_n(t)\dot{x}_m(t)\left[x_n(t)-x_m(t)\right]^{-1}\right\}, \tag{11}$$

$$x\,\psi_x(x,t)-N\,\psi(x,t)=\psi(x,t)\sum_{n=1}^{N}\left[x-x_n(t)\right]^{-1}\left\{x_n(t)\right\}\ , \tag{12}$$

$$x\psi_{xx}(x,t)=\psi(x,t)\sum_{n=1}^{N}\left[x-x_n(t)\right]^{-1}\left\{2\sum_{m=1,m\neq n}^{N}x_n(t)\left[x_n(t)-x_m(t)\right]^{-1}\right\}\ , \tag{13}$$

$$x\psi_{xt}(x,t)=\psi(x,t)\sum_{n=1}^{N}\left[x-x_n(t)\right]^{-1}\left\{-\sum_{m=1,m\neq n}^{N}x_n(t)\left[\dot{x}_n(t)+\dot{x}_m(t)\right]\left[x_n(t)-x_m(t)\right]^{-1}\right\}, \tag{14}$$

$$x^2\,\psi_{xx}(x,t)-N(N-1)\psi(x,t)=\psi(x,t)\sum_{n=1}^{N}\left[x-x_n(t)\right]^{-1}\left\{2\sum_{m=1,m\neq n}^{N}\left[x_n(t)\right]^2\left[x_n(t)-x_m(t)\right]^{-1}\right\}, \tag{15}$$

$$x\left[\ x^2\,\psi_{xx}(x,t)-2(N-1)\,x\psi_x(x,t)+N(N-1)\psi(x,t)\ \right]$$

$$=\psi(x,t)\sum_{n=1}^{N}\left[x-x_n(t)\right]^{-1}\left\{2\sum_{m=1,m\neq n}^{N}\left[x_n(t)\right]^2x_m(t)\left[x_n(t)-x_m(t)\right]^{-1}\right\}\ , \tag{16}$$

$$x\left[x\psi_{xt}(x,t)-(N-1)\,\psi_t(x,t)\right]$$

$$=\psi(x,t)\sum_{n=1}^{N}\left[x-x_n(t)\right]^{-1}\left\{-\sum_{m=1,m\neq n}^{N}x_n(t)\left[x_n(t)\dot{x}_m(t)+x_m(t)\dot{x}_n(t)\right]\left[x_n(t)-x_m(t)\right]^{-1}\right\}. \tag{17}$$

Proof. The formulas (3, 4, 5, 6) follow immediately from (1). Likewise (7) and (8) obtain immediately by differentiation of the logarithm of (2).

To obtain (9) one differentiates (7) with respect to x, and uses again (7). This yields

$$\psi_{xx} = \psi \left\{ \left[\sum_{n=1}^{N} (x - x_n)^{-1} \right]^2 - \sum_{n=1}^{N} (x - x_n)^{-2} \right\} , \tag{18a}$$

$$\psi_{xx} = \psi \sum_{n,m=1;m\neq n}^{N} (x - x_n)^{-1} (x - x_m)^{-1} , \tag{18b}$$

$$\psi_{xx} = \psi \sum_{n,m=1;m\neq n}^{N} (x_n - x_m)^{-1} \left[(x - x_n)^{-1} - (x - x_m)^{-1} \right] , \tag{18c}$$

$$\psi_{xx} = 2\psi \sum_{n=1}^{N} (x - x_n)^{-1} \sum_{m=1,m\neq n}^{N} (x_n - x_m)^{-1} . \tag{18d}$$

Exercise 2.3.1-1. Prove (10) and (11). *Hint*: proceed as above.

To prove (12), one starts from (7) multiplied by x:

$$x\psi_x = \psi \sum_{n=1}^{N} x(x - x_n)^{-1} = \psi \left[N + \sum_{n=1}^{N} x_n (x - x_n)^{-1} \right] . \tag{19}$$

Exercise 2.3.2-2. Prove (13)-(17). *Hint*: to prove (13), multiply (9) by x and, in the right hand side, replaces x with $x - x_n + x_n$; and so on for the other equations.

2.3.3 Many-body problems on the line solvable via the identification of their motions with those of the zeros of a polynomial that evolves in time according to a linear PDE in 2 variables (space and time)

It is now clear, by taking a linear combinations of the formulas (2.3.2-7 $\div$ 17), that to the *linear* PDE (see (2.3-2))

$$[\ A_0 + A_1 x + A_2 x^2 + A_3 x^3\ ]\psi_{xx} + [\ B_0 + B_1 x - 2(N-1) A_3 x^2\ ]\psi_x$$

$$+ C\psi_{tt} + [\ E - (N-1) D_2 x\ ]\psi_t + [\ D_0 + D_1 x + D_2 x^2\ ]\psi_{xt}$$

$$-[\ N(N-1)(A_2 - A_3 x) + N B_1\ ]\psi = 0 , \tag{1}$$

there corresponds the *nonlinear* "equations of motion"

$$C\ddot{x}_n + E\dot{x}_n = B_0 + B_1 x_n - 2(N-1)A_3 x_n^2 + \sum_{m=1,m\neq n}^{N} (x_n - x_m)^{-1} \cdot$$

$$\cdot \left[2C\dot{x}_n\dot{x}_m - (\dot{x}_n + \dot{x}_m)(D_0 + D_1 x_n) - D_2 x_n (\dot{x}_n x_m + \dot{x}_m x_n) \right.$$

$$\left. + 2(A_0 + A_1 x_n + A_2 x_n^2 + A_3 x_n^3) \right] . \qquad (2)$$

The correspondence works obviously both ways: if the N "particle coordinates" $x_n(t)$ evolve according to the equations of motion (2), the (monic) polynomial $\psi(x,t)$ identified by the N zeros $x_n(t)$, see (2.3.2-2), satisfies the linear PDE (1); if a (monic) polynomial satisfies the PDE (1) (which because of the way it has been obtained, is guaranteed to admit such a polynomial solution), then its zeros $x_n(t)$ evolve according to the equations of motion (2).

The equations of motion (2) are naturally interpretable as the Newtonian equations of motion of a N-body problem on the line with one- and two-body forces. In some cases, see below, these equations of motion are Hamiltonian.

Before discussing the solvability of the equations of motion (2) let us interject several *remarks* and an *exercise*.

Remark 2.3.3-1. The PDE (1), as well as the ODEs (2), feature linearly the 11 quantities $C, E, B_0, B_1, D_0, D_1, D_2, A_0, A_1, A_2, A_3$; only 10 of these, however, play a role, since any one (but only one!) of these 11 quantities could be replaced by unity, by dividing (1) and/or (2) by it.

Remark 2.3.3-2. As implied by the way it has been established, the correspondence among (1) and (2) does not require the 11 quantities $C, E, B_0, B_1, D_0, D_1, D_2, A_0, A_1, A_2, A_3$ to be constant (time-independent); however, for simplicity's sake, we hereafter assume these quantities to be indeed *all* constant.

Exercise 2.3.3-3. Verify that the simple change of variables

$$x_n(t) = a\tilde{x}_n(\tau) + c, \qquad \tau = bt, \qquad (3)$$

with a, b, c arbitrary constants, when inserted in (2), yields a new system which has the same form as (2) except for a redefinition of the constants, and evaluate the new constants in terms of the old ones and of a, b, c. Of course by taking advantage of such transformations one can in some cases reduce a many-body problem of type (2) to a somewhat simpler one, featuring fewer constants.

Exercise 2.3.3-4. Find the *special cases* of (2), obtained by setting to zero some of the 11 constants appearing in this equation, which are invariant in form under the transformation (more general than (3))

$$x_n(t) = a\tilde{x}_n(\tau) + c + vt, \qquad \tau = bt, \tag{4}$$

that features the additional (relative to (3)) arbitrary constant v.

Remark 2.3.3-5. The above correspondence among a linear evolution PDE satisfied by the polynomial of degree *N* in x and the evolution equation of many-body type satisfied by its zeros could be extended to linear PDEs containing x-derivatives of order higher than 2, say up to the order *M* (of course with $M \le N$). The many-body problems would then feature *M*-body forces. Indeed, for instance,

$$\psi_{xxx} = \psi \sum_{n=1}^{N} (x - x_n)^{-1} \sum_{m=1, m\neq n}^{N} (x_n - x_m)^{-1} \sum_{\ell=1, \ell\neq m,n}^{N} \left[(x_n - x_\ell)^{-1} + (x_m - x_\ell)^{-1}\right] . \tag{5}$$

Hereafter we restrict for simplicity our consideration to PDEs of second order, hence to many-body models involving only two-body, and one-body, forces.

Remark 2.3.3-6. Another extension of the approach described above includes in the linear PDE higher powers in x than are featured by (1). It is indeed possible to obtain in this manner more general many-body systems than (2), but one must then also add some restrictions on the initial data.

Exercise 2.3.3-7. Explore this possibility by performing the relevant calculations, and then check with the original literature <C78a>.

Exercise 2.3.3-8. Repeat the entire treatment given above, assuming that $\psi(x,t)$, rather than being a monic polynomial, also feature a time-dependent coefficient multiplying the term of highest (*N*-th) order in x, and then check with the original literature <C78a>.

Exercise 2.3.3-9. Rewrite (2) so that the summand in its right hand side be antisymmetric under exchange of the two indices n and m. *Hint*: use appropriately the identity $x_n = (x_n - x_m) + x_m$.

Let us now discuss the technique to solve the initial-value problem for the Newtonian equations of motion (2).

Given, at the initial time $t = 0$, the initial positions and velocities of the *N* particles, $x_n(0)$ and $\dot{x}_n(0)$, one computes the polynomial (of degree N in x) ψ at $t = 0$, as well as its time-derivative (a polynomial of degree $N-1$ in x):

$$\psi(x,0)=\prod_{n=1}^{N}[x-x_n(0)] \ , \tag{6}$$

$$\psi_t(x,0)=-\psi(x,0)\sum_{n=1}^{N} \dot{x}_n(0)[x-x_n(0)]^{-1} \ , \tag{7}$$

(see (2.3.2-2) and (2.3.2-8), both evaluated at $t=0$).

One then lets ψ evolve according to the *linear* PDE (1), and obtains thereby $\psi(x,t)$. The zeros of this polynomial provide directly the coordinates $x_n(t)$ at time t, see (2.3.2-2).

In some cases the linear PDE (1) can be explicitly solved in closed form, see examples below. Otherwise the natural technique of solution is to follow the (linear!) evolution of the coefficients $c_m(t)$, see (2.3.2-1), which is clearly given by the following equations (implied by (1), see for instance (2.3.2-1, 3, 4, 5, 6)):

$$C\ddot{c}_m+(N+1-m)D_0\dot{c}_{m-1}+[(N-m)D_1+E]\dot{c}_m-mD_2\dot{c}_{m+1}$$

$$+(N+1-m)(N+2-m)A_0c_{m-2}+(N+1-m)[(N-m)A_1+B_0]c_{m-1}$$

$$-m[(2N-m-1)A_2+B_1]c_m+m(m+1)A_3c_{m+1}=0 \ , \ m=1,\ldots,N \ . \tag{8}$$

These N equations must of course be supplemented by the prescriptions $c_{-1}=c_0=c_{N+1}=0$. The consistency of (2.3.2.-1) with (1) is of course entailed by the consistency of (8) with the prescription that c_m vanish if $m<1$ or $m>N$.

The solution of (8) can be reduced to a purely algebraic task in the standard manner, but we do not elaborate on this aspect at this stage, since all the specific examples we treat below allow a simpler treatment than that appropriate to the general case (8).

Let us however note that the system (8) becomes triangular if $A_0=A_1=B_0=D_0=0$, or if $A_3=D_2=0$. Moreover, if $A_3=D_0=D_1=D_2=0$, this system, (8), can be replaced by a much simpler, easily solvable, one, by appropriately modifying the *ansatz* (2.3.2-1): see *Exercise 2.3.3-13* below.

Of course the solution of (8) must be supplemented by the initial data $c_m(0)$, $\dot{c}_m(0)$, which are given, in terms of the initial data $x_n(0)$, $\dot{x}_n(0)$, by the standard formulas (see (2.3.1-2))

$$c_1(0) = -\sum_{n=1}^{N} x_n(0) \quad , \tag{9a}$$

$$c_2(0) = \frac{1}{2} \sum_{n_1,n_2=1;n_1 \neq n_2}^{N} x_{n_1}(0)\, x_{n_2}(0) \quad , \tag{9b}$$

$$c_3(0) = -\frac{1}{6} \sum_{n_1,n_2 n_3=1;n_1 \neq n_2, n_2 \neq n_3, n_3 \neq n_1}^{N} x_{n_1}(0)\, x_{n_2}(0)\, x_{n_3}(0) \quad , \tag{9c}$$

and so on, as well as

$$\dot{c}_1(0) = -\sum_{n=1}^{N} \dot{x}_n(0) \quad , \tag{10a}$$

$$\dot{c}_2(0) = \sum_{n_1,n_2=1;n_1 \neq n_2}^{N} \dot{x}_{n_1}(0)\, x_{n_2}(0) \quad , \tag{10b}$$

$$\dot{c}_3(0) = -\frac{1}{2} \sum_{n_1,n_2 n_3=1;n_1 \neq n_2, n_2 \neq n_3, n_3 \neq n_1}^{N} \dot{x}_{n_1}(0)\, x_{n_2}(0)\, x_{n_3}(0) \quad , \tag{10c}$$

and so on.

Exercise 2.3.3-10. Show that the solution of the following generalization of the Newtonian equations of motion (2),

$$C\,\ddot{x}_n + E\,\dot{x}_n = B_0 + B_1\,x_n - 2(N-1)\,A_3\,x_n^2 + \sum_{m=1,m\neq n}^{N} (x_n - x_m)^{-1} \cdot$$

$$\cdot \left[2C\,\dot{x}_n\,\dot{x}_m - (\dot{x}_n + \dot{x}_m)(D_0 + D_1 x_n) - D_2\,x_n(\dot{x}_n\,x_m + \dot{x}_m\,x_n) \right.$$

$$\left. + 2\,(A_0 + A_1 x_n + A_2 x_n^2 + A_3 x_n^3) \right] - h(x_n, t) \prod_{m=1,m\neq n}^{N} (x_n - x_m)^{-1} \;, \tag{11}$$

where $h(x,t)$ is an *arbitrarily assigned* (possibly time-dependent) polynomial in x of degree *less than* N, is given by the N zeros of the (monic) polynomial $\psi(x,t)$, see (2.3.2-1,2), satisfying the following generalized version of the PDE (1):

$$[\ A_0 + A_1 x + A_2 x^2 + A_3 x^3\ ]\psi_{xx} + [\ B_0 + B_1 x - 2(N-1)A_3 x^2\ ]\psi_x$$

$$+ C\psi_{tt} + [\ E - (N-1)D_2 x\ ]\psi_t + [\ D_0 + D_1 x + D_2 x^2\ ]\psi_{xt}$$

$$-[\ N(N-1)(A_2 - A_3 x) + N B_1\ ]\psi = h(x,t)\ . \qquad (12)$$

Hint: set $x = x_n(t)$ in (12), and use the formulas that obtain from (2.3.2-7$\div$17) by setting there $x = x_n(t)$ and using (2.3.2-2), which of course entails

$$\left\{\psi(x,t) \sum_{m=1}^{N} [x - x_m(t)]^{-1}\right\}_{x=x_n(t)} = \prod_{m=1,m\neq n}^{N} [x_n(t) - x_m(t)]\ . \qquad (13)$$

Exercise 2.3.3-11. Write the generalized version of (8) which, via (2.3.2-2), corresponds to (12). *Hint:* insert the expression of $h(x,t)$ as a (given!) polynomial (of degree less than N) in x, as well as the analogous expression (2.3.2-1) of $\psi(x,t)$ in (12), and equate the coefficients of the powers of x (from 0 to $N-1$).

Exercise 2.3.3-12. Repeat the treatment given in Sect. 2.3.3, but taking as starting point a more general linear evolution equation than (2.3.3-2) or (2.3.3-12), say

$$C\psi_{tt}(x,t) + \sum_{q=1}^{Q'} \sum_{r=1}^{R'} A'_{qr}\,\psi_t(y'_q\,x + z'_r\,,t) + \sum_{q=1}^{Q} \sum_{r=1}^{R} A_{qr}\,\psi(y_q\,x + z_r\,,t) + \sum_{m=1}^{N} h_m\,x^{N-m}$$

$$= x^N \sum_{q=1}^{Q} \sum_{r=1}^{R} A_{qr}\,y_q^N\ , \qquad (14)$$

with $A'_{qr}, A_{qr}, y'_q, z'_r, y_q, z_r, h_m$ arbitrary constants and Q', R', Q, R arbitrary positive integers. *Hint*: insert (2.3-1) in (14). *Solution*: see <C85e>.

Exercise 2.3.3-13. Verify that, if

$$A_3 = D_0 = D_1 = D_2 = 0\ , \qquad (15)$$

the problem of solving the linear PDE (1) can be simplified by replacing the *ansatz* (2.3.2-1) with the following one:

$$\psi(x,t) = \sum_{m=0}^{N} a_m(t)\, P_{N-m}^{(\alpha_+,\alpha_-)}(y) \ , \tag{16a}$$

where $P_n^{(\alpha,\beta)}(y)$ is a Jacobi polynomial in y of degree n <E53>,

$$a_0 = 2^N\, N!\,(N+\alpha_+ +\alpha_-)!\,/\,(2N+\alpha_+ +\alpha_-)!$$

$$= 2^N\, N!\,\Gamma(1+N+\alpha_+ +\alpha_-)/\Gamma(1+2N+\alpha_+ +\alpha_-) \tag{16b}$$

(to guarantee that $\psi(x,t)$ is monic),

$$\alpha_\pm = -1+[B_1/(2A_2)] \pm R\,\{(b_0/A_1)-[B_1/(2A_2)]\}, \tag{16c}$$

$$R = \left(1-4A_0\,A_2\,/\,A_1^2\right)^{1/2}, \tag{16d}$$

$$y = \left(1+2x A_2\,/\,A_1\right) R \tag{16e}$$

(here we are assuming for simplicity that R is real, namely $A_0\,A_2 \le (A_1/2)^2$, and moreover that $\alpha_\pm > -1$, which requires, as necessary but not sufficient condition, $B_1/A_2 > 2$). Indeed insertion of (16) in (1) with (15) yields the *uncoupled* system of linear ODEs

$$C\,\ddot{a}_m + E\,\dot{a}_m = m[(2N-m-1)A_2 + B_1]\,a_m \ , \tag{17a}$$

entailing

$$a_m(t) = \exp[-Et/(2C)]\left\{a_m(0)\cosh(\lambda_m t) + [\dot{a}_m(0) + a_m(0)E/(2C)]\,\lambda_m^{-1}\sinh(\lambda_m t)\right\}, \tag{17b}$$

$$\lambda_m = \left\{[E/(2C)]^2 + m[(2N-m-1)A_2 + B_1]/C\right\}^{1/2} . \tag{17c}$$

Hint: use the second-order linear ODE satisfied by Jacobi polynomials <E53>.

Exercise 2.3.3-14. Repeat the treatment of the preceding *Exercise 2.3.3-13*, but with appropriate, more stringent, restrictions on the constants than (15), so that the role of the Jacobi polynomials, see (16a), is taken over by other classical polynomials (Gegenbauer, Legendre, Laguerre, Hermite). *Hint*: use the relevant second order linear ODEs satisfied by the classical polynomials (see, for instance, <E53>).

2.3.4 Examples

The class of *solvable* many-body models introduced in the preceding Sect. 2.3.3 is rather large, due to the presence of 11 arbitrary constants in (2.3.3-2). In Sect. 2.3.4, and especially in Sects. 2.3.4.1 and 2.3.4.2, we discuss some representative examples, obtained by setting to zero several of the constants appearing in (2.3.3-2). We then introduce certain techniques ("tricks"), and certain findings, associated with such systems; the alert reader will pursue these approaches by applying them in more general cases than those reported below.

In Sect. 2.3.4.1 we treat some systems characterized by evolution equations which are of *first order* in time, in particular the two systems characterized by the evolution equations

$$\dot{x}_n = -a x_n + b \sum_{m=1, m\neq n}^{N} (x_n - x_m)^{-1} , \tag{1}$$

$$\dot{x}_n = -a x_n + c x_n^2 \sum_{m=1, m\neq n}^{N} (x_n - x_m)^{-1} . \tag{2}$$

In Sect. 2.3.4.2 we treat some systems characterized by *second-order* ("Newtonian") equations of motion, in particular the four systems characterized by the equations of motion

$$\ddot{x}_n = -x_n + \sum_{m=1, m\neq n}^{N} (1 + 2\dot{x}_n \dot{x}_m)/(x_n - x_m) , \tag{3}$$

$$\ddot{x}_n = a\dot{x}_n + 2 \sum_{m=1, m\neq n}^{N} \dot{x}_n \dot{x}_m / (x_n - x_m) , \tag{4}$$

$$\ddot{x}_n = \sum_{m=1, m\neq n}^{N} [2\dot{x}_n \dot{x}_m - i\omega (\dot{x}_n + \dot{x}_m) x_n] / (x_n - x_m) , \tag{5}$$

$$\ddot{x}_n = \alpha \dot{x}_n + \beta x_n + \sum_{m=1, m\neq n}^{N} [2\dot{x}_n \dot{x}_m + \lambda (\dot{x}_n + \dot{x}_m) x_n + \mu x_n^2] / (x_n - x_m) . \tag{6}$$

Generalizations, which however generally feature N-body contributions, of these systems are also presented (see the *exercises* in the following two Sects. 2.3.4.1 and 2.3.4.2).

2.3.4.1 First-order systems

In Sect. 2.3.4.1 we consider two models whose time evolution is determined by "equations of motion" of *first order* in time (see (2.3.4-1,2)). Additional models are then introduced via the *exercises*. The alert reader will invent and investigate many others.

The first model obtains by setting in (2.3.3-2) $E=1$, $A_0=b/2$, $B_1=-a$ and all other constants to zero. Hence its equations of motion read

$$\dot{x}_n=-a x_n+b\sum_{m=1,m\neq n}^{N}(x_n-x_m)^{-1} \ . \tag{1}$$

The corresponding PDE, see (2.3.3-1), reads

$$\psi_t+\frac{1}{2}(b\psi_{xx}-2ax\,\psi_x+2aN\psi)=0 \ . \tag{2}$$

By appropriate rescalings of the independent variable t and of the dependent variables $x_n(t)$ one could transform to unity both constants, a and b. We prefer to keep them visible.

The structure of the PDE (2) suggests the introduction of a new representation for the polynomial $\psi(x,t)$ whose zeros $x_n(t)$ are the solutions of the equations of motion (1), namely

$$\psi(x,t)=2^N(b/a)^{N/2}H_N\left[(a/b)^{1/2}x\right]+\sum_{m=1}^{N}b_m(t)H_{N-m}\left[(a/b)^{1/2}x\right] \ , \tag{3}$$

where $H_n(z)$ is the ("Hermite"; see Appendix C) polynomial of degree n in z that satisfies the ODE

$$H_n''(z)-2zH_n'(z)+2nH_n(z)=0 \ , \tag{4}$$

and whose normalization is fixed by the (standard; see Appendix C) condition

$$\lim_{z\to\infty}\left[(2z)^{-n}H_n(z)\right]=1 \ . \tag{5}$$

The definition (3), together with the (standard) normalization (5) of Hermite polynomials, guarantees that the polynomial (3) is monic, consistently with the original definition (2.3.2-1).

The advantage in the present context of the representation (3) over the representation (2.3.2-1) is due to the structure of (2), since (4) entails the formula

$$\frac{1}{2}\left[b(d/dx)^2 - 2ax\,(d/dx) + 2aN\right] H_{N-m}\left[(a/b)^{1/2}x\right] = am H_{N-m}\left[(a/b)^{1/2}x\right] . \quad (6)$$

Hence, see (2), (3) and (6), the coefficients $b_m(t)$ evolve according to the simple (decoupled !) evolution equations

$$\dot{b}_m(t) = -amb_m(t) \ , \quad (7)$$

which can be immediately solved:

$$b_m(t) = b_m(0)\exp(-amt) \ . \quad (8)$$

As for the initial data $b_m(0)$, they must of course be evinced, in terms of the initial data $x_m(0)$, from the polynomial equation

$$2^N\,(b/a)^{N/2} H_N\left[(a/b)^{1/2}x\right] + \sum_{m=1}^{N} b_m(0) H_{N-m}\left[(a/b)^{1/2}x\right] = \prod_{n=1}^{N}\left[x - x_n(0)\right] , \quad (9)$$

which is entailed by (3) and (2.3.2-2).

In conclusion we see that the solution $x_n(t)$ of the evolution equations (1) is provided by the N zeros of the polynomial (3) with (8).

Let us now introduce several *remarks,* which are related to this model, but illustrate techniques more generally applicable.

Remark 2.3.4.1-1. The model (1) admits clearly the equilibrium configuration

$$x_n(t) = r_n, \qquad \dot{x}_n(t) = 0, \quad (10)$$

with (see (1))

$$a r_n = b \sum_{m=1, m \neq n}^{N} (r_n - r_m)^{-1} \quad . \tag{11}$$

Let us set

$$r_n (b/a)^{-1/2} = z_n , \tag{12}$$

so that (11) becomes

$$z_n = \sum_{m=1, m \neq n}^{N} (z_n - z_m)^{-1} \quad . \tag{13}$$

The corresponding solution for the coefficients $b_m(t)$ must of course also be time-independent, hence, see (7), they must all vanish,

$$b_m(t) = b_m(0) = 0 \quad . \tag{14}$$

Via (9), this entails (see (10) and (12)) that the quantities z_n which satisfy (13) are the N zeros of the Hermite polynomial of order N:

$$H_N(z_n) = 0 \quad . \tag{15}$$

Exercise 2.3.4.1-2. Prove this result directly from (4). *Hint*: write for the Hermite polynomial $H_n(z)$ the analogs of (2.3.2-9, 12).

Remark 2.3.4.1-3. Let us consider the behavior of the system (1) in the neighborhood of its equilibrium configuration (10), via the position

$$x_n(t) = r_n + \varepsilon \xi_n(t) \quad , \tag{16}$$

where ε is a small parameter. One thus gets

$$\dot{\xi}_n = -a[\xi_n + \sum_{m=1}^{N} M_{nm} \xi_m] \quad , \tag{17}$$

with

$$M_{nm} = \delta_{nm} \sum_{\ell=1, \ell \neq n}^{N} (z_n - z_\ell)^{-2} - (1 - \delta_{nm})(z_n - z_m)^{-2} \quad . \tag{18}$$

Proof. From (1) and (16)

$$\varepsilon \dot{\xi}_n = -a r_n - a \varepsilon \xi_n + b \sum_{m=1, m \neq n}^{N} \left[r_n - r_m + \varepsilon (\xi_n - \xi_m) \right]^{-1} \ . \tag{19}$$

Expanding the right hand side in ε and using (11) one gets, up to corrections of order ε^2,

$$\varepsilon \dot{\xi}_n = -\varepsilon a \, \xi_n - \varepsilon b \sum_{m=1, m \neq n}^{N} (r_n - r_m)^{-2} (\xi_n - \xi_m) \ , \tag{20}$$

and, via (11), this yields (17) with (18).

From (17) one infers that, at least for small t, the behavior of $\xi_n(t)$ must be of the following type:

$$\xi_n(t) = \sum_{m=1}^{N} c_m \, v_n^{(m)} \exp\left[-a (1 + \mu_m) \, t \right] \ , \tag{21}$$

where the quantities $v_n^{(m)}$ respectively μ_m are the components of the eigenvectors $\underline{v}^{(m)}$ respectively the eigenvalues of the $(N \times N)$-matrix $\underline{M}$,

$$\underline{M} \, \underline{v}^{(m)} = \mu_m \underline{v}^{(m)}, \quad \sum_{m=1}^{N} \left\{ \left[M_{nm} - \mu_n \, \delta_{nm} \right] v_m^{(n)} \right\} = 0 \ . \tag{22}$$

But a comparison of (21) with (8) (via (16) and the property $\psi[x_n(t), t] = 0$, see (3)) entails that the N numbers $1 + \mu_m$ must coincide with the N integers $m = 1, 2, ..., N$. Hence the following

Proposition 2.3.4.1-4. The N eigenvalues μ_m of the $(N \times N)$-matrix $\underline{M}$ (see (18), and recall that the N numbers z_n are the N zeros of the Hermite polynomial of order N, see (15)) coincide with the first N nonnegative integers,

$$\mu_m = m - 1, \quad m = 1, 2, ..., N \ . \tag{23}$$

Exercise 2.3.4.1-5. Prove that, for the N zeros z_n of the Hermite polynomial of order N, see (15), there holds the sum rule

$$\sum_{n,m=1;n\neq m}^{N} (z_n - z_m)^{-2} = N(N-1)/2 \ . \tag{24}$$

Hint: consider the trace of the matrix (18), with the eigenvalues (23).

Remark 2.3.4.1-6. From (1) there obtain the following *second-order* "equations of motions" (with *velocity-independent* forces) for the quantities $x_n(t)$:

$$\ddot{x}_n = a^2 x_n - 2b^2 \sum_{m=1,m\neq n}^{N} (x_n - x_m)^{-3} \ . \tag{25}$$

Proof. Time-differentiation of (1) yields

$$\ddot{x}_n = a\dot{x}_n - b \sum_{m=1,m\neq n}^{N} (x_n - x_m)^{-2} (\dot{x}_n - \dot{x}_m) \ . \tag{26}$$

Using (1) this yields

$$\ddot{x}_n = a^2 x_n + ab \sum_{m=1,m\neq n}^{N} (x_n - x_m)^{-1} - ab \sum_{m=1,m\neq n}^{N} (x_n - x_m)^{-1}$$

$$-b^2 \sum_{m=1,m\neq n}^{N} (x_n - x_m)^{-2} [\sum_{l=1,l\neq n}^{N} (x_n - x_l)^{-1} - \sum_{l=1,l\neq m}^{N} (x_m - x_l)^{-1}] \ , \tag{27a}$$

$$\ddot{x}_n = a^2 x_n - 2b^2 \sum_{m=1,m\neq n}^{N} (x_n - x_m)^{-3} - b^2 Z_n \ , \tag{27b}$$

where

$$Z_n = \sum_{m=1,m\neq n}^{N} (x_n - x_m)^{-2} \sum_{l=1,l\neq m,n}^{N} [(x_n - x_l)^{-1} - (x_m - x_l)^{-1}] \ . \tag{28a}$$

To prove (25) we must show that Z_n vanishes. Indeed, using the *identity* $(x_n - x_l)^{-1} - (x_m - x_l)^{-1} = -(x_n - x_m)(x_n - x_l)^{-1}(x_m - x_l)^{-1}$,

$$Z_n = - \sum_{m,l=1;m\neq n,l\neq n,l\neq m}^{N} (x_n - x_m)^{-1} (x_n - x_l)^{-1} (x_m - x_l)^{-1} \ , \tag{28b}$$

and this entails the vanishing of Z_n, because in the double sum in the right hand side the summand is antisymmetrical under the exchange of the dummy indices l and m.

Note that this result,

$$Z_n = 0 \ , \tag{29}$$

with Z_n defined by (28), is an *identity,* namely it does not require any restriction on the N numbers x_n (other that they be different among themselves, so that the right hand side of (28) is well defined). The diligent reader will check, following the treatment given above and below, that this fact entails that the N zeros z_n of the Hermite polynomial of order N, see (15), besides being characterized by the nonlinear equations (13), also satisfies the nonlinear equations

$$z_n = 2 \sum_{m=1, m \neq n}^{N} (z_n - z_m)^{-3} \ . \tag{30}$$

If we now set

$$a = is\omega \ , \quad b = is'g \ , \quad s^2 = s'^2 = 1 \ , \tag{31}$$

we see that the equations of motion (25) coincide with (2.1.3.3-1). Hence these latter equations of motion, (2.1.3.3-1), can also be solved by the present technique; but only for a set of initial data consistent with (1), namely such that

$$\dot{q}_n(0) = -is\omega \ q_n(0) + is'g \sum_{m=1, m \neq n}^{N} [q_n(0) - q_m(0)]^{-1} \ , \tag{32}$$

with $s^2 = s'^2 = 1$. Note that, to a *real* choice of the initial positions $q_n(0)$, there correspond via (32) *imaginary* values for the initial velocities $\dot{q}_n(0)$; hence the motion in this case becomes necessarily complex. It is, of course, periodic, see (8) and (31).

The diligent reader will, at this point, pause to compare the findings reported so far with those of Sect. 2.1.3.3, including in particular the discussion there of the behavior of the many-body system in the neighborhood of its equilibrium configuration.

Remark 2.3.4.1-7. The evolution equations (1) possess the "similarity solution"

$$x_n(t) = f(t) \ r_n \ , \tag{33}$$

with the (constant) quantities r_n defined by (11) (see also (12), (13) and (15)), and

$$f(t) = g(at) \ , \tag{34a}$$

$$g(\tau) = [1 + c\exp(-2\tau)]^{1/2} \ , \tag{34b}$$

where c is an arbitrary constant.

Proof. Insertion of (33) into (1) yields

$$\dot{f}(t)\, r_n = -a\, f(t) r_n + b[f(t)]^{-1} \sum_{m=1, m\neq n}^{N} (r_n - r_m)^{-1} \ , \tag{35}$$

and, using (11), this yields

$$\dot{f} = a[-f + f^{-1}] \ , \tag{36}$$

which is clearly satisfied by (34).

Remark 2.3.4.1-8. Time-differentiation of (36) yields (using again (36))

$$\ddot{f} = a^2 [f - f^{-3}] \ , \tag{37}$$

consistently with the fact that the similarity solution (33) provides as well a solution to (25) (see (12) and (30)).

Exercise 2.3.4.1-9. Find the most general "similarity solution" of type (33) of the *second-order* "equations of motion" (25) (and of (2.1.3.3-1), see (31)), and discuss its behavior.

The second model we consider in Sect. 2.3.4.1. is characterized by the equations of motion

$$\dot{x}_n = a x_n + c x_n^2 \sum_{m=1, m\neq n}^{N} (x_n - x_m)^{-1} \ . \tag{38}$$

It corresponds to (2.3.3-2) with $E=1$, $B_1=a$ and $A_2=c/2$, hence it yields, in place of (2.3.3-1),

$$\psi_t+\frac{1}{2}cx^2\psi_{xx}+ax\psi_x-N\left[a+(N-1)c/2\right]\psi=0 \; . \tag{39}$$

The corresponding equations of evolution for the coefficients $c_m(t)$, see (2.3.3-8), read simply

$$\dot{c}_m=m\left[c(2N-m-1)/2+a\right]c_m \; , \tag{40}$$

and can therefore be immediately integrated:

$$c_m(t)=c_m(0)\exp\left\{m\left[c(2N-m-1)/2+a\right]t\right\} \; . \tag{41}$$

The evolution equations (38) are not invariant under translation, but they are clearly invariant under the rescaling transformation $x_n(t)\to\tilde{x}_n(t)=c\,x_n(t)$ with c an arbitrary constant. Hence one can obtain from them translation-invariant equations via the following change of (dependent) variables:

$$x_n(t)=\exp\left[b\,q_n(t)\right] \; . \tag{42}$$

Indeed insertion of this position in (38) yields

$$b\dot{q}_n=a+c\sum_{m=1,m\neq n}^{N}\left\{1-\exp\left[b(q_m-q_n)\right]\right\}^{-1} \; . \tag{43}$$

Time-differentiation of (38) yields the following second-order "Newtonian equations of motion":

$$\ddot{x}_n=a^2x_n+c(c+2a)x_n^2\sum_{m=1,m\neq n}^{N}(x_n-x_m)^{-1}-2c^2\sum_{m=1,m\neq n}^{N}(x_nx_m)^2(x_n-x_m)^{-3}$$

$$+c^2x_n^3\sum_{\ell,m=1;\ell\neq m,\ell\neq n,m\neq n}^{N}(x_n-x_\ell)^{-1}(x_n-x_m)^{-1} \; . \tag{44}$$

Proof.

$$\ddot{x}_n = a\dot{x}_n + 2c\,x_n\,\dot{x}_n \sum_{m=1,m\neq n}^{N} (x_n - x_m)^{-1} - c\,x_n^2 \sum_{m=1,m\neq n}^{N} (x_n - x_m)^{-2}(\dot{x}_n - \dot{x}_m) \ , \tag{45a}$$

$$\ddot{x}_n = a^2\,x_n + 2ac\,x_n^2 \sum_{m=1,m\neq n}^{N} (x_n - x_m)^{-1}$$

$$+ 2c^2\,x_n^3 \sum_{m=1,m\neq n}^{N} (x_n - x_m)^{-1} \sum_{\ell=1,\ell\neq n}^{N} (x_n - x_\ell)^{-1}$$

$$- c^2\,x_n^2 \sum_{m=1,m\neq n}^{N} (x_n - x_m)^{-2}\Big[\, x_n^2 \sum_{\ell=1,\ell\neq n}^{N} (x_n - x_\ell)^{-1} - x_m^2 \sum_{\ell=1,\ell\neq m}^{N} (x_m - x_\ell)^{-1} \,\Big] \ , \tag{45b}$$

$$\ddot{x}_n = a^2\,x_n + 2ac\,x_n^2 \sum_{m=1,m\neq n}^{N} (x_n - x_m)^{-1} + 2c^2\,x_n^3 \sum_{m=1,m\neq n}^{N} (x_n - x_m)^{-2}$$

$$- c^2\,x_n^2 \sum_{m=1,m\neq n}^{N} (x_n - x_m)^{-3}\,(x_n^2 + x_m^2) + Y_n \ , \tag{45c}$$

$$Y_n = 2c^2\,x_n^3 \sum_{\ell,m=1;\ell\neq m,\ell\neq n,m\neq n}^{N} (x_n - x_\ell)^{-1}\,(x_n - x_m)^{-1}$$

$$- c^2\,x_n^2 \sum_{\ell,m=1;\ell\neq m,\ell\neq n,m\neq n}^{N} (x_n - x_m)^{-2}\big[\, x_n^2\,(x_n - x_\ell)^{-1} - x_m^2\,(x_m - x_\ell)^{-1} \,\big] \ , \tag{46a}$$

$$\ddot{x}_n = a^2\,x_n + c(2a - c)\,x_n^2 \sum_{m=1,m\neq n}^{N} (x_n - x_m)^{-1} + 4c^2\,x_n^3 \sum_{m=1}^{N} (x_n - x_m)^{-2}$$

$$- 2c^2\,x_n^4 \sum_{m=1}^{N} (x_n - x_m)^{-3} + Y_n \ . \tag{45d}$$

(45a) follows from (38) by t-differentiation. (45b) follows from (45a) via (38). (45d) follows from (45c) using the *identity* $x_m^2 = x_n^2 - 2(x_n - x_m)\,x_n + (x_n - x_m)^2$.

We now use the *identity*

$$x_n^2\,(x_n - x_\ell)^{-1} - x_m^2\,(x_m - x_\ell)^{-1} = (x_n - x_m)(x_n - x_\ell)^{-1}(x_m - x_\ell)^{-1}\big[\,x_n(x_m - x_\ell) - x_\ell\,x_m\,\big]$$

inside the second sum in the right hand side of (46a), getting thereby

$$Y_n = c^2\,x_n^3 \sum_{\ell,m=1;\ell\neq m,\ell\neq n,m\neq n}^{N} (x_n - x_\ell)^{-1}\,(x_n - x_m)^{-1} + Z_n \ , \tag{46b}$$

$$Z_n = c^2 x_n^2 \sum_{\ell,m=1;\ell\neq m,\ell\neq n,m\neq n}^{N} (x_n - x_m)^{-1}(x_n - x_\ell)^{-1}(x_m - x_\ell)^{-1} x_\ell x_m \ . \tag{47}$$

But Z_n clearly vanishes, because the summand in the right hand side of (47) is antisymmetrical under the exchange of the dummy indices ℓ and m. Hence (46b) and (45d) yield

$$\ddot{x}_n = a^2 x_n + c(2a-c)x_n^2 \sum_{m=1,m\neq n}^{N} (x_n - x_m)^{-1}$$

$$+4c^2 x_n^3 \sum_{m=1,m\neq n}^{N} (x_n - x_m)^{-2} - 2c^2 x_n^4 \sum_{m=1,m\neq n}^{N} (x_n - x_m)^{-3}$$

$$+c^2 x_n^3 \sum_{\ell,m=1;\ell\neq m,\ell\neq n,m\neq n}^{N} (x_n - x_\ell)^{-1}(x_n - x_m)^{-1} \ , \tag{48a}$$

or equivalently

$$\ddot{x}_n = a^2 x_n + x_n^2 \sum_{m=1,m\neq n}^{N} \left[c(2a-c)(x_n - x_m)^2 + 4c^2 (x_n - x_m)x_n - 2c^2 x_n^2\right](x_n - x_m)^{-3}$$

$$+c^2 x_n^3 \sum_{\ell,m=1;\ell\neq m,\ell\neq n,m\neq n}^{N} (x_n - x_\ell)^{-1}(x_n - x_m)^{-1} \ . \tag{48b}$$

It is now easily seen that

$$c(2a-c)(x_n - x_m)^2 + 4c^2 (x_n - x_m)x_n - 2c^2 x_n^2 = c(c+2a)(x_n - x_m)^2 - 2c^2 {x_m}^2 \ , \tag{49}$$

and via this *identity* clearly (48b) yields (44), which is thereby proven.

We have therefore seen that the N-body problem characterized by the Newtonian equations of motion (44) (which feature one-, two- and three-body velocity-independent forces) is *partially solvable*: it can be solved for the subset of initial data which satisfy (38) (this entails, for instance, the possibility to assign arbitrarily the initial positions $x_n(0)$ of all N particles, but to forsake any freedom in assigning the initial velocities $\dot{x}_n(0)$, which are then fixed by (38) at $t=0$).

It is likewise seen that the *translation-invariant* "Newtonian equations of motion"

$$\ddot{q}_n = \alpha \sum_{m=1,m\neq n}^{N} \{1+\exp[b(q_m - q_n)]\}\{1-\exp[b(q_m - q_n)]\}^{-3}$$

$$-\alpha \sum_{\ell,m=1;\ell\neq m,\ell\neq n,m\neq n}^{N} \{\exp[b(q_n-q_\ell)]-\exp[b(q_m-q_\ell)]\}\{1-\exp[b(q_m-q_n)]\}^{-2}\cdot$$

$$\cdot\{1-\exp[b(q_n-q_\ell)]-\exp[b(q_m-q_\ell)]+\exp[b(q_n+q_m-2q_\ell)]\}^{-1} \ , \tag{50a}$$

$$\alpha=c^2/b \ , \tag{50b}$$

featuring velocity-independent two- and three-body forces are as well *partially solvable*, since they follow by time-differentiation from (43). Note that the constant a, see (43), is not present in (50).

Exercise 2.3.4.1-10. Verify that (50) follows from (43).

Let us end Sect. 2.3.4.1. devoted to first-order systems with a rather trivial *remark*, and an interesting set of *exercises* (the alert reader is urged to invent other, analogous, ones).

Remark 2.3.4.1-11. The first-order system

$$\dot{x}_n=g_n(\underline{x}) \tag{51}$$

is Hamiltonian for any choice of the N functions $g_n(\underline{x})$, since the equations of motion (51) are just the (first set of) Hamiltonian equations yielded by the Hamiltonian function

$$H(\underline{x},\underline{\xi})=\sum_{n=1}^{N} \xi_n\, g_n(\underline{x}) . \tag{52}$$

Exercise 2.3.4.1-12. Show that the solution of the first-order system

$$\dot{x}_n=h(x_n)\prod_{m=1,m\neq n}^{N}(x_n-x_m)^{-1} \ , \tag{53}$$

where $h(x)$ is an arbitrary (time-independent) polynomial in x of degree less than N, is given by the N roots of the polynomial equation of degree N in x

$$\prod_{m=1}^{N}[x-x_m(0)]=h(x)\,t \ , \tag{54a}$$

namely there holds the polynomial equation

$$\prod_{m=1}^{N} [x - x_m(t)] = \prod_{m=1}^{N} [x - x_m(0)] - h(x)\, t . \tag{54b}$$

Hint: set $h(x,t) = h(x)$, $E = -1$ and all other constants to zero in (2.3.3-11) respectively (2.3.3-12), getting thereby (53) respectively

$$\psi_t(x,t) = -h(x) \ , \tag{55a}$$

which entails

$$\psi(x,t) = \psi(x,0) - h(x)\, t . \tag{55b}$$

Then use (2.3.2-2).

Exercise 2.3.4.1-13. Show that all solutions $x_n(t)$ of the first-order system (53) satisfy the second-order system

$$\ddot{x}_n = 2 \sum_{m=1, m \neq n}^{N} \dot{x}_n \dot{x}_m / (x_n - x_m) \ . \tag{56}$$

Hint: time-differentiate the logarithm of (53), use the *identity* (valid for any polynomial of degree less than N, and for N *arbitrary* (distinct) numbers x_n: see (2.4.1-9))

$$h'(x_n) = b_n(\underline{x}) \sum_{m=1}^{N} D_{nm}(\underline{x}) [b_m(\underline{x})]^{-1} h(x_m) \ , \tag{57}$$

with

$$b_n(\underline{x}) = \prod_{m=1, m \neq n}^{N} (x_n - x_m) \ , \tag{58}$$

$$D_{nm}(\underline{x}) = \delta_{nm} \sum_{l=1, l \neq n}^{N} (x_n - x_l)^{-1} + (1 - \delta_{nm})(x_n - x_m)^{-1} \ , \tag{59}$$

and eliminate $h(x_n)$ and $h(x_m)$ using (53) and (58), namely

$$h(x_n) = b_n(\underline{x}) \dot{x}_n \ . \tag{60}$$

Exercise 2.3.4.1-14. Show, using (53), that the solutions $x_n(t)$ of the *second-order* system (56) are the roots of the following equation in x :

$$\sum_{n=1}^{N} \dot{x}_n(0)/[x-x_n(0)]=1/t \ . \tag{61}$$

Hint: consider (53) at $t=0$,

$$h[x_n(0)]=\dot{x}_n(0) \prod_{m=1,m\neq n}^{N} [x_n(0)-x_m(0)] \ , \tag{62}$$

and note that these relations entail, for all values of x ,

$$h(x)=\sum_{n=1}^{N} \dot{x}_n(0) \prod_{m=1,m\neq n}^{N} [x-x_m(0)] \tag{63}$$

(indeed, two polynomials of degree less than N that take equal values at N distinct points are identical). Then insert this expression of $h(x)$ in (54a).

Exercise 2.3.4.1-15. Compare the findings of the last two *exercises* with those of Sect. 2.1.10. *Hint*: compare (56) respectively (61) with (2.1.10-1) respectively (2.1.10.2-13) (note that this comparison entails that this latter equation, (2.1.10.2-13), is now proven; it is proven again below, see (2.3.4.2-21)).

Exercise 2.3.4.1-16. Show that (53) and (56) entail, via (2.3.2-1),

$$\ddot{c}_n=0 \ , \tag{64}$$

thereby confirming again the findings of Sect. 2.1.10.2, see (2.1.10.2-1,2,3). *Hint*: (55a) entails

$$\psi_{tt}=0 \ . \tag{65}$$

Then use (2.3.2-6).

Exercise 2.3.4.1-17. Discuss the behavior of the solutions of the equations of motion (56), noting the important role played by the (relative) signs of the initial velocities $\dot{x}_n(0)$ in distinguishing whether or not, at any time(s) throughout the motion, two particles collide, thereby causing a singularity in the right hand side of (56) (prove in particular that this does not happen if all the initial velocities have the same sign, and analyze completely the motion in this case). *Hint*: focus on (61), drawing a graph of its left hand side as a function of x.

Exercise 2.3.4.1-18. Show that, for $N > 2$, the Newtonian equations of motion (44) cannot be obtained from a normal Hamiltonian,

$$H(\underline{x},\underline{\xi}) = (1/2)\sum_{n=1}^{N} \xi_n^2 + V(\underline{x}) \quad . \tag{66}$$

Hint: the Newtonian equations of motion yielded by the normal Hamiltonian (66),

$$\ddot{x}_n = f_n(\underline{x}) \quad , \tag{67}$$

$$f_n(\underline{x}) = -\partial V(\underline{x}) / \partial x_n \quad , \tag{68}$$

feature conservative forces $f_n(\underline{x})$ that have the property (entailed by (68))

$$\partial f_n(\underline{x}) / \partial x_m = \partial f_m(\underline{x}) / \partial x_n \quad . \tag{69}$$

Exercise 2.3.4.1-19. Formulate and solve an exercise analogous to that just given, *Exercise 2.3.4.1-18*, with (44) replaced by (50).

2.3.4.2 Second-order systems (Newtonian equations of motion)

In Sect. 2.3.4.2 we consider four models whose time evolution is determined by Newtonian equations of motion (*second-order* in time: see (2.3.4-3,4,5,6)). Other models, which however generally feature N-body forces, are introduced via the *exercises*, see below. The alert reader will easily introduce and investigate many more.

The first model obtains by setting in (2.3.3-2) $C = 1$, $A_0 = 1/2$, $B_1 = -1$ and all other constants equal to zero. Hence its equations of motion read

$$\ddot{x}_n = -x_n + \sum_{m=1, m\neq n}^{N} (1 + 2\dot{x}_n \dot{x}_m)/(x_n - x_m) \quad . \tag{1}$$

The corresponding PDE, see (2.3.3-1), reads

$$\psi_{tt} + \frac{1}{2}(\psi_{xx} - 2x\,\psi_x + 2N\psi) = 0 \quad . \tag{2}$$

By rescaling the independent variable t and the dependent variables $x_n(t)$,

$$\xi_n(\tau) = \beta x_n(t), \quad t = \alpha\tau, \tag{3}$$

the equations of motion (1) can be recast in the form

$$\xi_n'' = -\alpha^2 \xi_n + \sum_{m=1, m\neq n}^{N} (\alpha^2 \beta^2 + 2\xi_n' \xi_m')/(\xi_n - \xi_m) , \tag{4}$$

where of course the primes denote differentiations with respect to τ. In the following we stick for simplicity to the simpler form (1).

In analogy to the treatment of the previous Sect. 2.3.4.1 (see in particular (2.3.4.1-3)) we now set

$$\psi(x,t) = 2^N H_N(x) + \sum_{m=1}^{N} b_m(t)\, H_{N-m}(x) , \tag{5}$$

again with $H_n(x)$ being the Hermite polynomial of order n, see (2.3.4.1-4,5) (and Appendix C).

Insertion of this representation of the monic polynomial $\psi(x,t)$ in (2) yields for the coefficients $b_m(t)$ the simple (decoupled!) evolution equations

$$\ddot{b}_m(t) + m b_m(t) = 0 , \tag{6}$$

whose solution reads

$$b_m(t) = b_m(0)\cos(m^{1/2} t) + \dot{b}_m(0) m^{-1/2} \sin(m^{1/2} t) . \tag{7}$$

Let us recall that the particle positions $x_n(t)$, see (1), are the N zeros of the (time-dependent) polynomial (5) with (7). As for the $2N$ quantities $b_m(0)$ and $\dot{b}_m(0)$ in (5), they are related to the initial positions $x_n(0)$ and velocities $\dot{x}_n(0)$ of the N particles by the polynomial equations

$$2^{-N} H_N(x) + \sum_{m=1}^{N} b_m(0)\, H_{n-m}(x) = \psi(x,0) = \prod_{n=1}^{N} [x - x_n(0)] , \tag{8a}$$

$$\sum_{m=1}^{N} \dot{b}_m(0)\ H_{n-m}(x) = \psi_t(x,0) = -\psi(x,0) \sum_{n=1}^{N} \dot{x}_n(0)/[x - x_n(0)] \qquad (8b)$$

(see (5), (2.3.2-2), and (2.3.2-8)).

Clearly the system (1) admits the equilibrium configuration

$$x_n(t) = z_n, \quad \dot{x}_n(t) = 0, \qquad (9)$$

where the quantities z_n are the N zeros of the Hermite polynomial of order N, see (2.3.4.1-13) and (2.3.4.1-15). This corresponds, for the coefficients $b_m(t)$, see (7), to the trivial solution $b_m(t) = 0$.

Clearly the system behaves generally as a kind of (one-dimensional) crystal, with every particle oscillating around its equilibrium position. If the oscillations are too large, adjacent particles collide, causing a singularity of the equations of motion (1).

After every collision the two zeros of the polynomial (5) that have collided become complex. This suggests that the proper setting to analyze the motion is the complex plane rather than the real line. Cases in which such an extension can be profitably done without loosing contact with "physics" are discussed in Chap. 4.

It is also clear from (7) that all solutions of (1) (namely, the time evolution of the zeros $x_n(t)$ of the polynomial (5) with (7)) are *multiply periodic*: indeed, they are algebraic (nonlinear) functions of the N periodic functions $b_m(t)$, see (7), themselves featuring the N periods

$$T_m = 2\pi\ m^{-1/2} \ . \qquad (10)$$

There exist special solutions of (1) which are periodic; they correspond to the special solutions of (7) with all coefficients $b_m(t)$ vanishing except one (or a few, for instance $b_1(t)$ and $b_4(t)$, see (10)). Indeed if only $b_M(t)$ does not vanish, the particle positions $x_n(t)$ are the N zeros of the polynomial

$$2^{-N}\ H_N(x) + b_M(t)\ H_{N-M}(x) \ , \qquad (11a)$$

which is periodic in t with period $T_M = 2\pi\ M^{-1/2}$, see (10), while if, say, only $b_1(t)$ and $b_4(t)$ do not vanish, the particle positions $x_n(t)$ coincide with the N zeros of the polynomial

$$2^{-N} H_N(x)+b_1(t)\, H_{N-1}(x)+b_4(t) H_{N-4}(x) \ , \tag{11b}$$

which is clearly periodic with period $T_4 = \pi$ (since $T_1 = 2T_4$, see (10)).

An amusing case is that with $b_m(t)=0$ for $m=1,2,...,N-1$. Then (since $H_0(x)=1$, see Appendix C) the particle positions $x_n(t)$ are the N roots of the equation

$$H_N(x) = -2^N b_N(t) = B_N \cos(N^{1/2}\, t+\beta_N) \ , \tag{12}$$

see (7), hence their time-evolution is neatly visualizable via the following construction. Draw first of all the graph of the Hermite polynomial of order N, $H_N(x)$, as a function of x. Consider then a horizontal straight line which oscillates periodically up and down with period $T_N = 2\pi\ N^{-1/2}$. The particle coordinates $x_n(t)$ are then just the abscissas of the points at which the orizontal straight line cuts the graph of $H_N(x)$. It is therefore quite evident, in this case, how the particles oscillates periodically about their equilibrium positions (the zeros of $H_N(x)$) and also the limitations on the amplitude of the oscillations of $b_N(t)$, hence on the initial data (see (7), (8) and (12)), which are required to avoid the occurrence of particle collisions.

Exercise 2.3.4.2-1. Perform the analogous graphical analysis of the behavior of the special solution of (1) corresponding to the case when all coefficients $b_m(t)$ vanish except $b_{N-1}(t)$ (recall that $H_1(x)=2x$, see Appendix C).

The "equations of motion" (6) are clearly Hamiltonian, corresponding to the Hamiltonian function

$$H(\underline{b},\underline{p}) = \frac{1}{2}\sum_{m=1}^{N} (p_m^2 + m b_m^2) \ . \tag{13}$$

On the other hand the transformation among the N "canonical coordinates" $b_m(t)$ and the N "particle coordinates" $x_n(t)$ entailed by the relation

$$2^{-N} H_N(x)+\sum_{m=1}^{N} b_m(t)\, H_{N-m}(x) = \prod_{n=1}^{N} [x-x_n(t)] \ , \tag{14}$$

see (5), can certainly be interpreted as part of a canonical transformation (since it does not involve the canonical momenta). Hence the Newtonian equations of motion (1) are also Hamiltonian (for a discussion of the corresponding Hamiltonian function see <CF97>).

The one-dimensional many-body system defined by the Hamiltonian of normal type

$$H(\underline{q},\underline{p})=\frac{1}{2}\sum_{n=1}^{N}(p_n^2+mq_n^2)+\frac{1}{2}\sum_{m,n=1;m\neq n}^{N}\log(q_n-q_m) \tag{15}$$

yields the Newtonian equations of motion

$$\ddot{q}_n=-q_n+\sum_{m=1;m\neq n}^{N}(q_n-q_m)^{-1}\ . \tag{16}$$

Hence it has the same equilibrium configuration, see (9) and (2.3.4.1-15), as the system (1), and moreover its behavior around equilibrium differs little from that of (1) (only by quadratic terms, see (1) and (16)).

Exercise 2.3.4.2-2. Perform the standard analysis of the behavior of the systems (1) and (16) around their (common) equilibrium configuration, and recover thereby the results reported above as *Remark 2.3.4.1-3.*

The second many-body system we consider is characterized by the equations of motions

$$\ddot{x}_n=a\dot{x}_n+2\sum_{m=1,m\neq n}^{N}\dot{x}_n\dot{x}_m/(x_n-x_m)\ . \tag{17}$$

It is the special case of (2.3.3-2) with $C=1$, $E=a$ and all other constants equal to zero. Hence in this case (2.3.3-1) reads simply

$$\psi_{tt}-a\psi_t=0\ , \tag{18}$$

so that its general solution reads

$$\psi(x,t)=\psi(x,0)+\psi_t(x,0)\left[\exp(at)-1\ \right]/a\ . \tag{19}$$

Hence, using (2.3.3-6) and (2.3.3-7), we conclude that the particle coordinates $x_n(t)$, solutions of the equations of motion (17) with initial data $x_n(0)$ and $\dot{x}_n(0)$, are the N roots of the equation in x

$$\sum_{n=1}^{N}\dot{x}_n(0)/[x-x_n(0)]=a/\left[\exp(at)-1\ \right]\ . \tag{20}$$

In particular, if the constant $a = i\omega$ is imaginary the right hand side of this equation is periodic with period $T = 2\pi/\omega$, hence its roots, namely all solutions of (17), are in this case (in which, however, the motion occurs in the complex plane, see Chap. 4) *completely periodic.*

Remark 2.3.4.2-3. For $a = 0$ the equations of motion (17) coincide (up to a trivial notational change) with (2.1.10-1), while (20) reads

$$\sum_{n=1}^{N} \dot{x}_n(0)/[x - x_n(0)] = 1/t \ , \tag{21}$$

namely (up to the same notational change) it coincides with (2.1.10.2-13). Moreover (18) entails, via (2.3-1),

$$\ddot{c}_m - a\dot{c}_m = 0 \ , \tag{22}$$

hence, for $a = 0$, it implies (2.1.10.2-3).

The statement made at the end of Sect. 2.1.10.2 is thereby proven again (it was previously proven in Sect. 2.3.4.1, see *Exercises 2.3.4.1-13,14*).

Remark 2.3.4.2-4. The complete periodicity of all the solutions of (17) with $a = i\omega$ purely imaginary is merely a special case of the findings discussed in Sects. 2.1.12.3 and 2.1.12.4 (the diligent reader will profitably elaborate on these connections).

The third example we consider obtains by setting $C = 1$, $D_1 = i\omega$ and all other constants equal to zero, in (2.3.3-1) and (2.3.3-2), so that they read

$$\psi_{tt} + i\omega x \psi_{xt} = 0 \ , \tag{23}$$

$$\ddot{x}_n = \sum_{m=1, m\neq n}^{N} [2\dot{x}_n \dot{x}_m - i\omega(\dot{x}_n + \dot{x}_m)x_n]/(x_n - x_m) \ . \tag{24}$$

The corresponding equations for the coefficients $c_m(t)$, see (23) and (2.3-1) (or, more directly, (2.3.3-8)), read

$$\ddot{c}_m + i\omega(N - m)\,\dot{c}_m = 0 \ , \tag{25}$$

so that their solutions read

$$c_m(t) = c_m(0) + i\,\dot{c}_m(0)\{\exp[-i\omega(N-m)t] - 1\}/[\omega(N-m)] \ . \tag{26}$$

Hence, if ω is real, all these coefficients are periodic with period $T=2\pi/\omega$, hence $\psi(x,t)$, as well as all its zeros $x_n(t)$, see (2.3-1), are also periodic. One can therefore conclude that, for ω real and nonvanishing, all solutions of the (complex) many-body problem (24) are periodic.

These equations of motion, (24), as well as the equations of motion (17), feature forces which vanish if the particles do not move. Hence any configuration,

$$x_n(t)=y_n \quad , \tag{27}$$

where the N quantities y_n are *arbitrary* constants, is an equilibrium configuration for the system (24) (as well as (17)). Let us then consider the behavior of the system (24) in the neighborhood of such an equilibrium configuration. To this end we set

$$x_n(t)=y_n+\varepsilon\xi_n(t) \quad , \tag{28}$$

where ε is a small parameter. Insertion of this *ansatz* in (24) yields, up to corrections of order ε,

$$\ddot{\xi}_n+i\omega\sum_{m=1}^{N} M_{nm}\dot{\xi}_m=0 \quad , \tag{29}$$

with

$$M_{nm}=\delta_{nm}\,y_n\sum_{\ell=1,\ell\neq n}^{N}(y_n-y_\ell)^{-1}+(1-\delta_{nm})\,y_n\,(y_n-y_m)^{-1} \quad . \tag{30}$$

A comparison of (29) with (25) suggests that the $(N\times N)$-matrix (30), constructed with the N *arbitrary* quantities y_n, have the first nonnegative integers ($N-m$ with $m=1,2,...,N$) as eigenvalues. This is proved below (see Sect. 2.4.5.1).

Exercise 2.3.4.2-5. Show that the following nonlinear system of N coupled PDEs in S-dimensional space,

$$i\psi_{n,t}+\Delta\psi_n=2\sum_{m=1,m\neq n}^{N}[(\vec{\nabla}\psi_n)\cdot(\vec{\nabla}\psi_m)]/(\psi_n-\psi_m) \quad , \tag{31}$$

where $\psi_n \equiv \psi_n(\vec{r},t)$, $\psi_{n,t} \equiv \partial \psi_n(\vec{r},t)/\partial t$, $\vec{r}$ is a vector in S-dimensional space and $\vec{\nabla}$ respectively $\Delta = \vec{\nabla} \cdot \vec{\nabla}$ are the gradient respectively the Laplacian in S-dimensional space, is linearized via the following prescription:

$$\prod_{n=1}^{N} [\psi - \psi_n(\vec{r},t)] = \psi^N + \sum_{m=1}^{N} \varphi_m(\vec{r},t)\ \psi^{N-m} \ , \tag{32}$$

$$i\varphi_{m,t}(\vec{r},t) + \Delta \varphi_m(\vec{r},t) = 0 \ . \tag{33}$$

Note that (31) is rotation-invariant in S-dimensional space, that the relation between the N functions $\psi_n(\vec{r},t)$ and the N functions $\varphi_m(\vec{r},t)$ is identical to that between the N zeros of a monic polynomial of degree N in the variable ψ and the N coefficients of the same polynomial in ψ, see (32), and that the functions $\varphi_m(\vec{r},t)$ satisfy the *linear* Schroedinger equation in S-dimensional space, see (33). Discuss, on the basis of these results, the solution of the initial-value problem for the nonlinear PDE (31). Write other nonlinear PDEs in S-dimensional space that can be solved by analogous techniques. *Hint*: see (2.3.2-9), (2.3.2-11) and <C94>.

The fourth model we consider in Sect. 2.3.4.2 obtains by setting in (2.3.3-2) $C=1,\ E=-\alpha,\ B_1=\beta,\ D_1=-\lambda,\ A_2=\mu/2$ and all other constants to zero. Hence its equations of motion read

$$\ddot{x}_n = \alpha\,\dot{x}_n + \beta x_n + \sum_{m=1, m\neq n}^{N} \left[2\dot{x}_n\,\dot{x}_m + \lambda(\dot{x}_n + \dot{x}_m)x_n + \mu x_n^2\right]/(x_n - x_m) \ . \tag{34}$$

Note that this model is invariant under the rescaling transformation $x_n \to \tilde{x}_n = c\,x_n$, $\dot{c}=0$.

The corresponding PDE, see (2.3.3-1), reads

$$\psi_{tt} - \lambda x \psi_{xt} - \alpha\psi_t + (1/2)\left\{\mu x^2 \psi_{xx} + 2\beta x \psi_x - N[(N-1)\mu + 2\beta]\psi\right\} = 0 \ , \tag{35}$$

and the equations for the coefficients $c_m(t)$, see (2.3-1) and (2.3.3-8), read

$$\ddot{c}_m = [\alpha + \lambda(N-m)]\dot{c}_m + (m/2)[(2N-m-1)\mu + 2\beta]\,c_m \ . \tag{36}$$

Note that they are decoupled, hence their solutions can be immediately exhibited:

$$c_m(t) = c_m^{(+)} \exp[\ \nu_m^{(+)} t\ ] + c_m^{(-)} \exp[\ \nu_m^{(-)} t\ ] \ , \tag{37a}$$

with $\nu_m^{(\pm)}$ the 2 roots of the following second-order equation in ν_m:

$$\nu_m^2 = [\alpha + \lambda(N-m)]\nu_m + (m/2)\,[(2N-m-1)\,\mu + 2\beta] \quad , \tag{37b}$$

entailing

$$\nu_m^{(\pm)} = [\alpha + \lambda(N-m) \pm \Delta_m]/2 \quad , \tag{37c}$$

$$\Delta_m^2 = [\alpha + \lambda(N-m)]^2 + 2m\,[(2N-m-1)\,\mu + 2\beta] \quad . \tag{37d}$$

Exercise 2.3.4.2-6. Study the behavior of this system, see (34). *Hint:* allow for all quantities to be complex. *Solution*: see Sect. 4.2.

Exercise 2.3.4.2-7. Show that the Newtonian equations of motion (34) are invariant under the following transformation,

$$x_n(t) = \tilde{x}_n(t)\, \exp(at) , \tag{38}$$

where a is an arbitrary constant, in the sense that the "new coordinates" $\tilde{x}_n(t)$ obey analogous equations to (34), except for the replacement of the "coupling constants" $\alpha, \beta, \lambda, \mu$ by the following "new coupling constants":

$$\tilde{\alpha} = \alpha - 2aN, \tilde{\beta} = \beta + a[\alpha - \lambda(N-1)] - a^2(2N-1), \tilde{\lambda} = \lambda + 2a, \tilde{\mu} = \mu + 2\lambda a + 2a^2 . \tag{39}$$

We end Sect. 2.3.4.2 with a set of interesting *exercises* (again, the alert reader is urged to invent additional ones).

Exercise 2.3.4.2-8. Neither the equations of motion (17), nor the equations of motion (24), are invariant under translations ($x_n(t) \to \tilde{x}_n(t) + c, \dot{c} = 0$); but they are both invariant under rescaling ($x_n(t) \to c\,\tilde{x}_n(t), \dot{c} = 0$). Obtain from them equations of motion which are invariant under translations. *Hint*: see (2.3.4.1-42).

Exercise 2.3.4.2-9. Study the *solvable* many-body problem (with many-body forces) characterized by the following Newtonian equations of motion:

$$\ddot{x}_n = -a\,x_n + \sum_{m=1, m\neq n}^{N} (b + 2\,\dot{x}_n\,\dot{x}_m)/(x_n - x_m) + h(x_n) \prod_{m=1, m\neq n}^{N} (x_n - x_m)^{-1} , \tag{40}$$

with a and b arbitrary (positive) constants and $h(x)$ an arbitrarily assigned (time-independent) polynomial of degree less than N; in particular, find conditions on this polynomial, $h(x)$, which are necessary and sufficient in order that this many-body model, see (40), possess some periodic solutions. *Hint*: see *Exercise 2.3.3-10* and (1÷4), and express $h(x)$ as a superposition of (appropriately chosen) Hermite polynomials.

Exercise 2.3.4.2-10. Show that the solutions of the Newtonian equations of motion

$$\ddot{x}_n = a\dot{x}_n + 2\sum_{m=1,m\neq n}^{N} \dot{x}_n\dot{x}_m/(x_n - x_m) + h(x_n)\prod_{m=1,m\neq n}^{N}(x_n - x_m)^{-1}, \tag{41}$$

where a is an arbitrary constant and $h(x)$ is a polynomial in x of degree less than N, are the N roots of the following polynomial equation of degree N in x:

$$\left\{1 + a^{-1}\left[1-\exp(at)\right]\sum_{m=1}^{N}\dot{x}_m(0)\left[x - x_m(0)\right]^{-1}\right\}\prod_{n=1}^{N}\left[x - x_n(0)\right]$$

$$= a^{-2}\,h(x)\left[\exp(at) - 1 - at\right]. \tag{42}$$

Hint: see *Exercise 2.3.3-10.*

Exercise 2.3.4.2-11. Find necessary and sufficient conditions on the *time-dependent* polynomial $h(x,t)$, of degree less than N in x, such that *all* solutions of the (complex) system

$$\ddot{x}_n - i\omega\dot{x}_n = 2\sum_{m=1,m\neq n}^{N}\dot{x}_n\dot{x}_m/(x_n - x_m) + h(x_n,t)\prod_{m=1,m\neq n}^{N}(x_n - x_m)^{-1}, \tag{43}$$

are *completely periodic* (ω being a *real* nonvanishing constant). *Hint*: follow the same procedure used to solve the preceding *Exercise 2.3.4.2-10.*

2.3.5 Trigonometric extension

In Sect. 2.3.5 we consider the extension of the treatment of Sects. 2.3.1, 2.3.2 and 2.3.3 that emerges if, instead of the *ansatz* (2.3-1), we set

$$\psi(x,t)=a^{-N}\prod_{n=1}^{N}\sin\{a[x-x_n(t)]\} \quad . \tag{1}$$

For $a=0$ this expression coincides with (2.3-1), hence the results of Sect. 2.3.5 reduce, for $a=0$, to those obtained and discussed above.

The *ansatz* (1) entails the following formulas:

$$\psi_x(x,t)=\psi(x,t)\ a\sum_{n=1}^{N}\text{cotan}\{a[x-x_n(t)]\} \ , \tag{2}$$

$$\psi_t(x,t)=\psi(x,t)\ a\sum_{n=1}^{N}\text{cotan}\{a[x-x_n(t)]\}[-\dot{x}_n(t)] \ , \tag{3}$$

$$\psi_{xx}(x,t)=\psi(x,t)\cdot$$

$$\cdot\{-N^2\,a^2+a\sum_{n=1}^{N}\text{cotan}\{a\,[x-x_n(t)]\}\,[2a\sum_{m=1,m\neq n}^{N}\text{cotan}\{a[x_n(t)-x_m(t)]\}]\} \tag{4}$$

$$\psi_{xt}(x,t)=\psi(x,t)\{N^2\,a^2\dot{\bar{x}}(t)$$

$$+a\sum_{n=1}^{N}\text{cotan}\{a[x-x_n(t)]\}[-a\sum_{m=1,m\neq n}^{N}[\dot{x}_n(t)-\dot{x}_m(t)]\,\text{cotan}\{a[x_n(t)-x_m(t)]\}]\} \ , \tag{5}$$

$$\psi_{tt}(x,t)=\psi(x,t)\{-N^2\,a^2\,[\dot{\bar{x}}(t)]^2+a\sum_{n=1}^{N}\text{cotan}\{a\,[x-x_n(t)]\}\cdot$$

$$\cdot[-\ddot{x}_n(t)+2a\,\dot{x}_n(t)\sum_{m=1,m\neq n}^{N}\dot{x}_m(t)\,\text{cotan}\{a\,[x_n(t)-x_m(t)]\}]\}, \tag{6}$$

where, in the last two equations (as well as below), $\bar{x}(t)$ is the mean coordinate,

$$\bar{x}(t)=N^{-1}\sum_{n=1}^{N}x_n(t) \quad . \tag{7}$$

Proofs. Logarithmic differentiation of (1) with respect to x respectively t yields directly (2) and (3). Differentiation of (2) with respect to x yields (after using (2))

$$\psi_{xx}=\psi\ \{(a\sum_{n=1}^{N}\text{cotan}[a(x-x_n)])^2-\sum_{n=1}^{N}\ (a\sin[a(x-x_n)])^{-2}\} \ , \tag{8a}$$

$$\psi_{xx} = \psi \left\{-N a^2 + a^2 \sum_{m,n=1;m\neq n}^{N} \text{cotan}[a(x-x_m)]\, \text{cotan}[a(x-x_n)]\right\} . \tag{8b}$$

We now use the trigonometric *identity*

$$\text{cotan}(\alpha)\ \text{cotan}(\beta) = -1 - [\text{cotan}(\alpha) - \text{cotan}(\beta)]\ \text{cotan}(\alpha-\beta) \ , \tag{9}$$

obtaining thereby

$$\psi_{xx} = \psi \left\{-N^2 a^2 + a^2 \sum_{m,n=1;m\neq n}^{N} \{\text{cotan}[a(x-x_n)] - \text{cotan}[a(x-x_m)]\}\ \text{cotan}[a(x_n-x_m)]\right\}, \tag{10}$$

which coincides with (4).

Exercise 2.3.5-1. Prove (5) and (6). *Hint*: see the proofs of (2), (3) and (4), as just given.

We assume now (tentatively, see below) that the function $\psi(x,t)$ satisfy the *linear* partial differential equation

$$A\psi_{xx} + B\psi_x + C\psi_{tt} + D\psi_{xt} + E\psi_t + F\psi = 0 \ , \tag{11}$$

where the quantities A,B,C,D,E and F are independent of x, but might depend on t (see below). Then, via the above formulas (from (1) to (7)) one concludes that the "particle coordinates" $x_n(t)$ evolve according to the equations of motion

$$C\ddot{x}_n + E\dot{x}_n = B + a \sum_{m=1,m\neq n}^{N} [2A - D(\dot{x}_n + \dot{x}_m) + 2C\dot{x}_n\dot{x}_m]\ \text{cotan}[a(x_n-x_m)] \ , \tag{12}$$

with the additional equation

$$F = N^2 a^2 (A - D\dot{\bar{x}} + C\dot{\bar{x}}^2) \ . \tag{13}$$

The compatibility of the two assumptions made above, namely *(i)* that $\psi(x,t)$ be represented by the *ansatz* (1) and *(ii)* that $\psi(x,t)$ satisfy the linear PDE (11) (with A,B,C,D,E and F independent of x) is a crucial, nontrivial, point, on which our entire development hinges. It does follow from the formulas (1)÷(7), since they clearly imply that, given (1), the validity of (12) and (13) is *necessary and sufficient* to guarantee the validity of (11). Such compatibility could not be generally guaranteed if one assumed a different *ansatz*, of type (1) but with a different function $f(z)$ taking the role of $\sin(az)$. Note in this connection the role played by the functional equation (9), that is crucial to yield the formulas (4)÷(7), which are then instrumental (unless $A = C = D = 0$) to imply the compatibility of (1) with (11). See, however, the following Sect. 2.3.6, where a generalization of the kind outlined here is actually introduced.

We now sum (12) over n from 1 to N and thereby get, via (7) (and taking advantage of the vanishing of the double sum in the right-hand side due to the antisymmetry of the summand under exchange of the two dummy indices m and n)

$$C\ddot{\bar{x}} + E\dot{\bar{x}} = B \ , \tag{14}$$

entailing

$$\bar{x}(t) = \bar{x}(0) + (B/E)t + (C/E)\left[\dot{\bar{x}}(0) - (B/E)\right]\left[1 - \exp(-Et/C)\right] \ . \tag{15}$$

Hence from (13) and (15) one gets

$$F(t) = N^2 a^2 \left[\alpha + \beta \exp(-Et/C) + \gamma \exp(-2Et/C)\right] \ , \tag{16a}$$

$$\alpha = A - BD/E + C(B/E)^2 \ , \tag{16b}$$

$$\beta = \left[2(BC/E) - D\right]\left[\dot{\bar{x}}(0) - B/E\right] \ , \tag{16c}$$

$$\gamma = C\left[\dot{\bar{x}}(0) - B/E\right]^2 \ . \tag{16d}$$

One therefore concludes that the solution of the many-body problem characterized by the Newtonian equations of motion (12), where A, B, C, D and E are 5 arbitrary constants (an assumption we hereafter make, for the sake of simplicity; but see *Exercise 2.3.5-9* below), can be reduced to solving the *linear* partial differential equation (11) with (16) and with the initial conditions

$$\psi(x,0) = a^{-N} \prod_{n=1}^{N} \sin\{a[x - x_n(0)]\} \ , \tag{17a}$$

$$\psi_t(x,0) = -\psi(x,0)\, a \sum_{n=1}^{N} \dot{x}_n(0) \operatorname{cotan}\{a[x - x_n(0)]\} \ , \tag{17b}$$

implied by (1) and (3). The particle coordinates $x_n(t)$ are then identified with the zeros of $\psi(x,t)$, see (1).

The most convenient route to solve the linear partial differential equation (11) for the class of functions of interest to us, namely those admitting the representation (1), is via the position

$$\psi(x,t) = \sum_{m=-N}^{N} \gamma_m(t) \exp(i\,a\,m\,x), \tag{18}$$

which entails, via (11), for the coefficients $\gamma_m(t)$ the following set of *de-coupled* ODEs:

$$C\ddot{\gamma}_m + (E + i m a D)\dot{\gamma}_m + (F + i m a B - m^2 a^2 A)\gamma_m = 0 \quad , \qquad m = 0, \pm 1, \ldots, \pm N\,. \tag{19}$$

Hereafter we restrict attention to the case with

$$B = E = 0 \quad , \tag{20}$$

which entails that F is time-independent,

$$F = N^2 a^2 \left\{ A - D\dot{\bar{x}}(0) + C\left[\dot{\bar{x}}(0)\right]^2 \right\}. \tag{21}$$

This formula, (21), is entailed by (13) and (20), which yields, via (14) $\ddot{\bar{x}}(t) = 0$ hence $\dot{\bar{x}}(t) = \dot{\bar{x}}(0)$. Note however that, even when (20) does not hold, for the special initial condition such that $\dot{\bar{x}}(0) = B/E$, F is time-independent, see (16).

In this case, with (20) and (21), the evolution equations (19) are easily solved:

$$\gamma_m(t) = \gamma_m^{(+)} \exp\left[\beta_m^{(+)} t\right] + \gamma_m^{(-)} \exp\left[\beta_m^{(-)} t\right] \quad , \tag{22}$$

where $\beta_m^{(\pm)}$ are the two solutions of the algebraic equation of second degree

$$C\beta_m^2 + i m a D\beta_m + F - m^2 a^2 A = 0 \quad , \tag{23a}$$

$$\beta_m^{(\pm)} = (-i m a D \pm \delta_m)/(2C) \quad , \tag{23b}$$

$$\delta_m = \left[(4AC - D^2)a^2m^2 - 4CF\right]^{1/2} \ . \qquad (23c)$$

The constants $\gamma_m^{(+)}$ and $\gamma_m^{(-)}$, see (22), are easily expressed in terms of the initial values $\gamma_m(0)$ and $\dot{\gamma}_m(0)$, since (22) imply

$$\gamma_m(0) = \gamma_m^{(+)} + \gamma_m^{(-)} \ , \qquad (24a)$$

$$\dot{\gamma}_m(0) = i\left[\beta_m^{(+)}\gamma_m^{(+)} + \beta_m^{(-)}\gamma_m^{(-)}\right] ; \qquad (24b)$$

and the initial values $\gamma_m(0)$ and $\dot{\gamma}_m(0)$ are related to the initial positions $x_n(0)$ and velocities $\dot{x}_n(0)$ of the particles via the relations (see (17) and (18))

$$\sum_{m=-N}^{N} \gamma_m(0) \exp(i\,a\,m\,x) = a^{-N} \prod_{n=1}^{N} \sin\{a[x - x_n(0)]\} = \psi(x,0) \ , \qquad (25a)$$

$$\sum_{m=-N}^{N} \dot{\gamma}_m(0) \exp(i\,a\,m\,x) = -\psi(x,0)\ a \sum_{n=1}^{N} \dot{x}_n(0)\,\mathrm{cotan}\{a[x - x_n(0)]\} = \psi_t(x,0) \ . \qquad (25b)$$

Remark 2.3.5-2. If the particle positions $x_n(t)$ are real, and the constant a is also real (or imaginary, see below), the function $\psi(x,t)$ is real, see (1), hence the coefficients $\gamma_m(t)$ satisfy the conditions

$$[\gamma_m(t)]^* = \gamma_{-m}(t), \qquad m = 0, \pm 1, \ldots, \pm N \ . \qquad (26)$$

Clearly these conditions, (26), are compatible with the time evolution (19): indeed if A, B, C, D, E, F are all real, these equations, (19), are equally affected by the operation of complex conjugation and by the change $m \to -m$.

Remark 2.3.5-3. The treatment applies equally (including the considerations about reality, see *Remark 2.3.5-2* above) if the constant a is replaced everywhere by $i\,a$, entailing the replacement of trigonometric functions by hyperbolic functions. The behavior of the solutions is of course nontrivially affected by such a change.

Exercise 2.3.5-4. Discuss the behavior of the solutions of the many-body problem (12), with particular attention to the case (20). Under which conditions are *all* motions confined ? Or periodic ? Are there *some* periodic trajectories even in the cases when not *all* the motions are periodic ? *Hint:* see (23).

Exercise 2.3.5-5. Show that, and explain in which sense, the many-body systems characterized by the following Newtonian equations of motion,

$$\ddot{x}_n = -8a^3A^2\{ \sum_{m=1,m\neq n}^{N} [\{\sin[a(x_n - x_m)]\}^{-3} \cos[a(x_n - x_m)]$$

$$-(N-2)\{\sin[2a(x_n - x_m)]\}^{-1}$$

$$- \sum_{m,\ell=1;\ell\neq m,\ell\neq n,m\neq n}^{N} \{\sin[2a(x_n - x_m)]\}^{-1} \text{cotan}[a(x_n - x_\ell)]\text{cotan}[2a(x_m - x_\ell)]\}\} , \qquad (27)$$

is *partially solvable.* Note that this model features *velocity-independent* two- and three-body forces. *Hint:* consider the first-order system that obtains by setting $C = D = 0$, $E = 1$ in (12), time-differentiate, use (9).

Exercise 2.3.5-6. Show that the system (12) with $B = 0$ possesses the equilibrium configuration

$$x_n = n\pi/(N\,a) \; . \qquad (28)$$

What can one learn by comparing the behavior of this system near equilibrium with the exact behavior, see (22) ? *Hint:* see <CP78a> and <CP79> as well as Sect. 15.823 of <GRJ94>.

Exercise 2.3.5-7. Obtain the equations of motion (12) from the translation-invariant version of (2.3.3-2) (characterized by $B_1 = D_1 = D_2 = A_1 = A_2 = A_3 = 0$), via the *infinite duplication* technique described in Sect. 2.1.7 and 2.1.13. *Hint:* see (2.1.7-49) and (2.1.13-13).

Exercise 2.3.5-8. Obtain the extension of the model (12) characterized by the presence of two different types of particles, via the technique of Sect. 2.1.7. *Hint*: see (2.1.7-30).

Exercise 2.3.5-9. Repeat the entire treatment forsaking, completely or partially, the assumption that the 5 constants A, B, C, D, E are time-independent.

2.3.6 Further extension

In Sect. 2.3.6 and in its subsections we consider the extension of the treatment of Sects. 2.3.3 and 2.3.5 that obtains by replacing the *ansaetze* (2.3-1) and (2.3.5-1) with the following, more general, one:

$$\psi(x,\tau) = \exp[-\frac{1}{2}\varphi(\tau)] \prod_{n=1}^{N} G[x - \xi_n(\tau)] \qquad (1)$$

Here we keep open the choice of the functions $\varphi(\tau)$ and $G(z)$. We then show, in Sect. 2.3.6.1, that in order to relate a *linear* PDE satisfied by

$\psi(x,\tau)$ to a system of evolution ODEs for the quantities $\xi_n(\tau)$, we must require $G(z)$, or rather its logarithmic derivative $g(z)$,

$$g(z)=G'(z)/G(z) \quad , \tag{2}$$

to satisfy the (new) *functional equation*

$$g(x)g(y)=-f(x-y)[g(x)-g(y)]+h(x-y)+\gamma(x)+\gamma(y) \quad . \tag{3a}$$

Here $f(z)$ and $h(z)$ (as well as $g(z)$) are two *a priori* arbitrary functions, except that the first of them must be odd,

$$f(-z)=-f(z) \quad , \tag{3b}$$

and the second of them must be even,

$$h(-z)=h(z) \quad . \tag{3c}$$

Note the consistency of these parity requirements with (3a). Moreover, in (3a) and below,

$$\gamma(z)\equiv\frac{1}{2}[g'(z)+g^2(z)] \quad . \tag{3d}$$

This functional equation is then investigated in Sect. 2.3.6.2. Its solutions include of course the simple assignment

$$G(z)=z,\ g(z)=f(z)=1/z,\ h(z)=\gamma(z)=0 \quad , \tag{4}$$

which entails that (1) (with $\varphi(\tau)=0$) corresponds to (2.3-1), as well as

$$G(z)=a^{-1}\sin(az),\ g(z)=f(z)=a\,\mathrm{cotan}(az),\ \gamma(z)=-a^2/2,\ h(z)=0 \quad , \tag{5}$$

which entails that (1) (again with $\varphi(\tau)=0$) corresponds to (2.3.5-1).

Exercise 2.3.6-1. Verify that (4) and (5) satisfy (3). *Hint*: for (5), use the trigonometric identity (2.3.5-9).

These two cases, (4) respectively (5), reproduce the treatments of Sect. 2.3.3 respectively 2.3.5. In Sect. 2.3.6.2 we show that the most general solution of the functional equation (3a) reads

$$g(z) = f(z) = a\,\zeta(a z) + \lambda z \quad , \tag{6a}$$

$$h(z) = \frac{1}{2} a^2 \left[\zeta'(a z) + \zeta^2(a z)\right] + \lambda\left[a z \zeta(a z) - 1\right] + \frac{1}{2}\lambda^2 z^2 \quad , \tag{6b}$$

entailing

$$\gamma(z) = h(z) + \frac{3}{2}\lambda \quad , \tag{6c}$$

$$G(z) = a^{-1}\,\sigma(a z)\,\exp(\lambda z^2/2) \quad . \tag{6d}$$

Here $\zeta(y) \equiv \zeta(y|\,\omega,\omega')$ respectively $\sigma(y) \equiv \sigma(y|\,\omega,\omega')$ are the Weierstrass "zeta" respectively "sigma" functions, see Appendix A.

As shown in Sect. 2.3.6.1, this opens the possibility to solve the Newtonian equations of motions

$$\ddot{x}_n = 2 \sum_{m=1, m\neq n}^{N} \dot{x}_n \dot{x}_m f(x_n - x_m) \quad , \tag{7}$$

with $f(x)$ defined by (6a). Here, as usual, $x_n \equiv x_n(t)$ are N particle coordinates, and the superimposed dots denote differentiations with respect to the time t. Note however that in (1) we introduced a different time-like variable, τ, as well as the coordinates $\xi_n \equiv \xi_n(\tau)$ (rather than $x_n \equiv x_n(t)$). The reason for doing so are explained in Sect. 2.3.6.1.

Finally, in Sect. 2.3.6.3, we focus on the Newtonian equations of motion (7) with (6a) and $\lambda = 0$ (the general case with $\lambda \neq 0$ is also treated at the end of that section); namely, we focus mainly on the N-body problem characterized by the Newtonian equations of motion

$$\ddot{x}_n = 2 \sum_{m=1, m\neq n}^{N} \dot{x}_n \dot{x}_m \zeta(x_n - x_m) \quad , \tag{8}$$

and we provide a fairly explicit and straightforward technique to solve the initial-value problem for this N-body system, a technique which is applicable whenever the initial data satisfy the single restriction

$$\sum_{n=1}^{N} \dot{x}_n(0) = 0 \quad . \tag{9a}$$

Note that, due to the odd character of the zeta function, $\zeta(-z) = -\zeta(z)$, the equations of motion (8) entail that the center-of-mass,

$$\bar{x}(t) = N^{-1} \sum_{n=1}^{N} x_n(t) \quad , \tag{10}$$

moves uniformly,

$$\ddot{\bar{x}}(t) = 0, \dot{\bar{x}}(t) = \dot{\bar{x}}(0) \quad . \tag{11}$$

Hence the condition (9a) amounts to the simple requirement that the center of mass not move initially, hence neither throughout the subsequent evolution of the system,

$$\dot{\bar{x}}(t) = \dot{\bar{x}}(0) = 0 \quad , \tag{9b}$$

namely to the property

$$\sum_{n=1}^{N} \dot{x}_n(t) = \sum_{n=1}^{N} \dot{x}_n(0) = 0 \quad . \tag{9c}$$

We already saw in the preceding Sect. 2.3.5 that such a restriction, (9a), entailed a significant simplification of the technique of solution for the "trigonometric" N-body problem considered there.

After treating in detail the problem (8), which corresponds to (7) with (6a) and $\lambda = 0$, we show, at the end of Sect. 2.3.6.3, how the results can be generalized to treat (7) with (6a) and *arbitrary* λ. Indeed we show there quite generally, namely for any problem of type (7) with *arbitrary* (odd!) $f(x)$, that the addition of a term λx to $f(x)$ can always be taken care of by an appropriate change of the independent ("time") variable, provided attention is restricted to initial conditions satisfying the constraint (9a), hence entailing (9c). Note that this finding also applies to the RS many-body models of Sect. 2.1.12. Let us, however, emphasize that the problem treated herein, namely (7) with (6a), does *not* belong to the RS class, see Sect. 2.1.12.

At the end of Sect. 2.3.6.3 we also remind the reader that: *(i)* a simple (complex) deformation of (7) (or, more generally, of (8) with (6a)) has the remarkable property to only feature periodic trajectories; *(ii)* the equations of motion (7) with (3b) are Hamiltonian.

In the following we continue to reserve the notation with superimposed dots to denote differentiations with respect to the time t, while we use appended primes to denote differentiations with respect to τ or indeed, more generally, with respect to the argument of the function the primes are appended to (as we already did in (2), (3d), (6b)).

2.3.6.1 New solvable many-body problems via a new functional equation

Clearly the *ansatz* (2.3.6-1) entails, via logarithmic differentiation, see (2.3.6-2),

$$\psi_x(x,\tau)=\psi(x,\tau)\sum_{n=1}^{N} g[x-\xi_n(\tau)] \quad , \tag{1}$$

$$\psi_\tau(x,\tau)=\psi(x,\tau)\{-\frac{1}{2}\varphi'(\tau)+\sum_{n=1}^{N} g[x-\xi_n(\tau)][-\xi_n'(\tau)]\} \quad . \tag{2}$$

Assume now that $g(z)$ satisfy the *functional equation* (2.3.6-3). There then hold the following (additional) relations:

$$\psi_{xx}(x,\tau)=\psi(x,\tau)\{\sum_{\ell,m=1,\ell\neq m}^{N} h[\xi_m(\tau)-\xi_\ell(\tau)]+2N\sum_{n=1}^{N}\gamma[x-\xi_n(\tau)]$$

$$+\sum_{n=1}^{N} g[x-\xi_n(\tau)]\,2\sum_{m=1,m\neq n}^{N} f[\xi_n(\tau)-\xi_m(\tau)]\} \quad , \tag{3}$$

$$\psi_{x\tau}(x,\tau)=\psi(x,\tau)\{-\frac{1}{2}\sum_{\ell,m=1,\ell\neq m}[\xi_m'(\tau)+\xi_\ell'(\tau)]\,h[\xi_m(\tau)-\xi_\ell(\tau)]$$

$$-\sum_{n=1}^{N}\gamma[x-\xi_n(\tau)]\,[(N-1)\xi_n'(\tau)+N\bar{\xi}'(\tau)]$$

$$-\sum_{n=1}^{N} g[x-\xi_n(\tau)]\{\frac{1}{2}\varphi'(\tau)+\sum_{m=1,m\neq n}^{N}[\xi_n'(\tau)+\xi_m'(\tau)]\,f[\xi_n(\tau)-\xi_m(\tau)]\}\} \quad , \tag{4}$$

$$\psi_{\tau\tau}(x,\tau)=\psi(x,\tau)\{-\frac{1}{2}\varphi''(\tau)+\frac{1}{4}[\varphi'(\tau)]^2+\sum_{\ell,m=1;\ell\neq m}^{N}\{\xi_m'(\tau)\,\xi_\ell'(\tau)\,h[\xi_m(\tau)-\xi_\ell(\tau)]\}$$

$$+2N\bar{\xi}'(\tau)\sum_{n=1}^{N}\gamma[x-\xi_n(\tau)]\,\xi_n'(\tau)$$

$$+\sum_{n=1}^{N} g[x-\xi_n(\tau)]\{-\xi_n''(\tau)+\varphi'(\tau)\,\xi_n'(\tau)+2\sum_{m=1,m\neq n}^{N}\xi_n'(\tau)\xi_m'(\tau)\,f[\xi_n(\tau)-\xi_m(\tau)]\}\} \,. \tag{5}$$

Here and below we denote by $\bar{\xi}(\tau)$ the mean value of the N coordinates $\xi_n(\tau)$:

$$\bar{\xi}(\tau) \equiv N^{-1} \sum_{n=1}^{N} \xi_n(\tau) \ . \tag{6}$$

Proofs. Partial τ-differentiation of (2) yields

$$\psi_{\tau\tau} = \psi\{-\frac{1}{2}\varphi'' - \sum_{n=1}^{N} g(x-\xi_n)\xi_n'' + \sum_{n=1}^{N} [g'(x-\xi_n)\,(\xi_n')^2]$$

$$+[\frac{1}{2}\varphi' + \sum_{n=1}^{N} g(x-\xi_n)\,\xi_n']^2\} \ , \tag{7a}$$

hence, via (2.3.6-3d),

$$\psi_{\tau\tau} = \psi\{-\frac{1}{2}\varphi'' + \frac{1}{4}(\varphi')^2 + 2\sum_{n=1}^{N} [\gamma(x-\xi_n)(\xi_n')^2] + \sum_{n=1}^{N} g(x-\xi_n)[-\xi_n'' + \varphi'\,\xi_n']$$

$$+ \sum_{\ell,m=1;\ell\neq m}^{N} g(x-\xi_\ell)\, g(x-\xi_m)\, \xi_\ell'\, \xi_m'\} \ . \tag{7b}$$

We now use, in the last term in the right hand side, the *functional equation* (2.3.6-3), and we get (5), which is thereby proven.

Exercise 2.3.6.1-1. Prove (3) and (4). *Hint*: see the proof of (5), as given just above.

Exercise 2.3.6.1-2. Obtain the analogs of (1), (3) and (4) (and observe that (2) and (5) are essentially unchanged) if the *ansatz* (2.3.6-1) is replaced by the following, more general, one:

$$\psi(x,\tau) = \exp[-\frac{1}{2}\varphi(x,\tau)] \prod_{n=1}^{N} G[x-\xi_n(\tau)] \ . \tag{8}$$

Let us now assume that $\psi(x,\tau)$ satisfy the *linear* PDE

$$A(\tau)\psi_{xx}(x,\tau) + B(\tau)\psi_x(x,\tau) + C(\tau)\psi_{\tau\tau}(x,\tau) + D(\tau)\psi_{x\tau}(x,\tau) + E(\tau)\psi_\tau(x,\tau)$$

$$+F(\tau)\psi(x,\tau) = 0 \ . \tag{9}$$

Then (1)÷(5) clearly imply that this assumption is compatible with the *ansatz* (2.3.6-1) iff there hold the following equations:

$$2A = D\,\overline{\xi}' \quad , \tag{10a}$$

$$2NC\overline{\xi}' = (N-1)D \quad , \tag{10b}$$

$$F = \frac{C}{4}\left[2\varphi'' - (\varphi')^2\right] + \frac{E}{2}\varphi' + \sum_{\ell,m=1;\ell\neq m}^{N} \left\{\left[-A + \frac{D}{2}(\xi_m' + \xi_\ell') - C\,\xi_m'\,\xi_\ell'\right] h(\xi_m - \xi_\ell)\right\} , \tag{10c}$$

$$C\,\xi_n'' + (E - C\varphi')\,\xi_n'$$

$$= B - \frac{D}{2}\varphi' + \sum_{m=1,m\neq n}^{N} \left\{\left[2A - D(\xi_n' + \xi_m') + 2C\,\xi_n'\,\xi_m'\right] f(\xi_n - \xi_m)\right\} \ . \tag{10d}$$

Here we have omitted, for notational simplicity, to indicate explicitly the τ-dependence of all quantities.

These findings open the prospect to solve the fairly general Newtonian equations of motion (10d), of course with the restrictions (10a,b,c) as well as (2.3.6-3). Hereafter we focus on the simpler, yet quite interesting, model that obtains by setting

$$A = B = D = E = F = 0 \ , \quad C = 1 \ , \tag{11}$$

as well as

$$\overline{\xi}'(\tau) = 0 \ . \tag{12}$$

Clearly, with these restrictions, (10a) and (10b) are identically satisfied, (10c) yields the constraint

$$\varphi'' - \frac{1}{2}(\varphi')^2 = 2 \sum_{\ell,m=1;\ell\neq m}^{N} \xi_m'\,\xi_\ell'\,h(\xi_m - \xi_\ell) \ , \tag{13}$$

while (10d) yields the Newtonian equations of motion

$$\xi_n'' - \varphi'\,\xi_n' = 2 \sum_{m=1,m\neq n}^{N} \xi_n'\,\xi_m'\,f(\xi_n - \xi_m) \ , \tag{14}$$

where the functions $f(z)$ and $h(z)$ are of course always characterized by the requirement to satisfy the functional equation (2.3.6-3). Note that the condition (12), which itself entails

$$\bar{\xi}(\tau)=\bar{\xi}(0) \quad , \tag{15}$$

is consistent with the equations of motion (14), since these clearly entail (see (6) and (2.3.6-3b))

$$\bar{\xi}''-\varphi'\bar{\xi}'=0 \quad . \tag{16}$$

The conditions (11) entail that the PDE (9) take the simple form

$$\psi_{\tau\tau}(x,\tau)=0 \quad , \tag{17a}$$

implying

$$\psi_{\tau}(x,\tau)=\psi_{\tau}(x,0) \quad , \tag{17b}$$

$$\psi(x,\tau)=\psi(x,0)+\tau\;\psi_{\tau}(x,0) \quad . \tag{17c}$$

Hereafter we conveniently set

$$\varphi(0)=0 \quad , \tag{18a}$$

$$\varphi'(0)=0 \quad . \tag{18b}$$

These initial conditions are consistent with (13), indeed they complement this ODE, (13), satisfied by $\varphi(\tau)$, with the additional conditions required to define this function, $\varphi(\tau)$, uniquely. Then (2.3.6-1) yields

$$\psi(x,0)=\prod_{n=1}^{N} G[x-\xi_n(0)] \quad , \tag{19a}$$

and likewise (2) yields

$$\psi_{\tau}(x,0)=-\psi(x,0)\sum_{n=1}^{N} \xi_n'(0)\;g[x-\xi_n(0)] \quad . \tag{19b}$$

Hence from (17c) we get

$$\psi(x,\tau)=\psi(x,0)\left\{1-\tau\sum_{n=1}^{N}\xi_n'(0)\,g\left[x-\xi_n(0)\right]\right\} \ . \tag{20}$$

If the function $G(z)$ vanishes at $z=0$,

$$G(0)=0 \ , \tag{21}$$

and has no other zeros, the *ansatz* (2.3.6-1) entails that the quantities $\xi_n(\tau)$ are the N zeros of $\psi(x,\tau)$:

$$\psi\left[\xi_n(\tau),\tau\right]=0 \ . \tag{22}$$

Then, via (20), we conclude that the initial-value problem for the Newtonian equations of motion (14) is solved by the following neat prescription:

Proposition 2.3.6.1-3. The N coordinates $\xi_n(\tau)$ are the N roots of the following equation in x:

$$\sum_{n=1}^{N}\xi_n'(0)\;g\left[x-\xi_n(0)\right]=1/\tau \ . \tag{23}$$

Note the neat way the initial conditions, $\xi_n(0)$ and $\xi_n'(0)$, enter in this equation. Of course the function $g(z)$ is characterized by the functional equation (2.3.6-3), and it has a pole at $z=0$, consistently with (2.3.6-2) and (21).

The equations of motion (14) are, however, still polluted by the presence of the (*a priori* unknown) function $\varphi(\tau)$, which is determined by the ODE (13) complemented by the initial conditions (18). Since this ODE, (13), contains in its right hand side the quantities $\xi_n(\tau)$ and $\xi_n'(\tau)$, the claim that the equations of motion (14) are *solvable* is moot, or rather, the interest of these equations of motion is somewhat questionable. But, via an appropriate (minor) modification of the *ansatz* (2.3.6-1), we indicate below (see Sect. 2.3.6.3) a convenient way to bypass this difficulty, namely to determine the function $\varphi(\tau)$ *directly from the initial data.*

Once the function $\varphi(\tau)$ is known, the disturbing presence of $\varphi'(\tau)$ in the equations of motion (14) can be easily gotten rid of, via an appropriate change of the time-like variable. Indeed let us set

$$\xi_n(\tau)=x_n(t),\ \ \tau=\tau(t) \ , \tag{24}$$

with the function $\tau(t)$ characterized by the following properties:

$$\tau(0)=0 \quad , \tag{25a}$$

$$\dot{\tau}(0)=1 \quad , \tag{25b}$$

$$\ddot{\tau}(t)+\varphi'(\tau)\left[\dot{\tau}(t)\right]^2=0 \quad . \tag{25c}$$

It is then easily seen that (14), via (24) and (25), become

$$\ddot{x}_n=2\sum_{m=1,m\neq n}^{N}\dot{x}_n\,\dot{x}_m\,f(x_n-x_m) \quad , \tag{26}$$

while (25a) and (25b) entail, via (24),

$$\xi_n(0)=x_n(0) \quad , \tag{27a}$$

$$\xi_n'(0)=\dot{x}_n(0) \quad . \tag{27b}$$

On the other hand integration of (25c) yields, using (25a), (25b) and (18a),

$$t(\tau)=\int_0^\tau d\tau' \exp[\varphi(\tau')] \, , \tag{28}$$

which, up to a quadrature, provides an explicit relation among t and τ. Hence $t(\tau)$, and by functional inversion $\tau(t)$ as well, can be considered known if the function $\varphi(\tau)$ is known.

Proofs. From (24) one gets

$$\dot{x}_n=\xi_n'\,\dot{\tau} \quad , \tag{29a}$$

$$\ddot{x}_n=\xi_n''\,\dot{\tau}^2+\xi_n'\,\ddot{\tau} \quad , \tag{29b}$$

hence, via (14),

$$\ddot{x}_n=\xi_n'\left[\dot{\tau}^2\,\varphi'+\ddot{\tau}\right]+2\sum_{m=1,m\neq n}^{N}\dot{x}_n\,\dot{x}_m\,f(x_n-x_m) \quad , \tag{30}$$

which clearly yields (26) via (25c).

On the other hand (25c) yields

$$\ddot{\tau}(t)/\dot{\tau}(t) = -\varphi'[\tau(t)]\,\dot{\tau}(t) \ , \tag{31a}$$

hence, via (25a,b) and (18a)

$$\log[\dot{\tau}(t)] = -\varphi[\tau(t)] \ , \tag{31b}$$

$$\dot{\tau}(t) = \exp\{-\varphi[\tau(t)]\} \ , \tag{31c}$$

namely

$$dt = d\tau \ \exp[\varphi(\tau)] \ , \tag{31d}$$

whose integration, via (25a), clearly yields (28).

2.3.6.2 General solution of the new functional equation

The developments of the preceding two Sects. 2.3.6 and 2.3.6.1 hinge on the assumption that the *functional equation* (2.3.6-3),

$$g(x)g(y) = -f(x-y)[g(x)-g(y)] + h(x-y) + \gamma(x) + \gamma(y) \ , \tag{1a}$$

$$\gamma(z) \equiv \frac{1}{2}[g'(z) + g^2(z)] \ , \tag{1b}$$

possess nontrivial solutions. In Sect. 2.3.6.2 we confirm this hypothesis, and we find the *most general* solution of (1), consistent with the parity properties (2.3.6-3b,c),

$$f(-z) = -f(z) \ , \tag{1c}$$

$$h(-z) = h(z). \tag{1d}$$

Note that, via (1b), the functional equation (1a) can also be rewritten in the (completely equivalent) form

$$f(x-y)[g(x)-g(y)] = \frac{1}{2}\{[g(x)-g(y)]^2 + g'(x) + g'(y)\} + h(x-y) \ . \tag{1e}$$

This version, (1e), of the functional equation makes particularly evident the validity of the following

Remark 2.3.6.2-1. If $g(z)$, $f(z)$ and $h(z)$ satisfy the functional equation (1), so do

$$\tilde{g}(z)=a\,g(a z+b)+c,\ \tilde{f}(z)=a\,f(a z),\ \ \tilde{h}(z)=a^2\,h(z) \tag{2}$$

with a,b,c three arbitrary constants.

In the following we take advantage of this invariance property to always write the solutions of the *functional equation* (1) in the simplest form, on the understanding that a generalization of type (2) of the solution is always possible.

Exercise 2.3.6.2-2. Trace the effect that addition of a constant, say b, to $g(z)$ (see (2)), has on the *ansatz* (2.3.6-1). *Hint*: see (2.3.6-2).

The functional equation (1) possesses of course the, utterly uninteresting, trivial solution $g(z)=h(z)=0$, with arbitrary $f(z)$. We ignore this solution hereafter.

There is another, somewhat more general, solution, that we also deem *trivial*, of the functional equation (1), which also holds for arbitrary $f(z)$. It reads:

$$g(z)=\lambda z\ , \tag{3a}$$

$$h(z)=\lambda\,[z\,f(z)-1]-\lambda^2\,z^2/2\ , \tag{3b}$$

with λ an arbitrary constant. Hereafter we ignore this uninteresting solution as well.

Exercise 2.3.6.2-3. Verify that (3) satisfies (1).

Exercise 2.3.6.2-4. Understand why the solution (3) is of no help to solve the Newtonian equations of motion (2.3.6.1-26) with arbitrary $f(z)$. *Hint*: note that (3a) yields, via (2.3.6-2),

$$G(z)=\exp(\lambda z^2/2)\ , \tag{3c}$$

and that insertion of this function $G(z)$ in the *ansatz* (2.3.6-1) yields an expression of $\psi(x,\tau)$ that depends on the $\xi_n(\tau)$'s only via the 2 collective coordinates $\overline{\xi}_p(\tau)$,

$$\overline{\xi}_p(\tau)=N^{-1}\sum_{n=1}^{N}\xi_n^p(\tau),\ \ p=1,2, \tag{4}$$

(of course $\overline{\xi}_1(\tau)$, as defined here, coincides with the center of mass $\overline{\xi}(\tau)$, see (2.3.6.1-6)). Hence, even when $\psi(x,\tau)$ is completely known, the individual $\xi_n(\tau)$*'s* cannot, in this case, be recovered from it.

Exercise 2.3.6.2-5. Verify that a nontrivial solution of the functional equation (1) reads as follows:

$$g(z) = f(z) = 1/z \ , \tag{5a}$$

$$h(z) = \gamma(z) = 0 \ . \tag{5b}$$

As noted in Sect. 2.3.6, with this solution, (5), the treatment of Sect. 2.3.6.1 becomes closely analogous to that of Sect. 2.3.3, since (5a), via (2.3.6-2), entails

$$G(z) = z \ , \tag{5c}$$

while (5b), via (2.3.6.1-13) with (2.3.6.1-18), entails

$$\varphi(\tau) = 0 \ . \tag{5d}$$

Exercise 2.3.6.2-6. Verify that a nontrivial solution of the *functional equation* (1) reads as follows:

$$g(z) = f(z) = \mathrm{cotan}(z) \ , \tag{6a}$$

$$h(z) = 0 \ , \tag{6b}$$

$$\gamma(z) = -1/2 \ . \tag{6c}$$

Hint: use the trigonometric identity (2.3.5-9).

Let us again remark that, as already noted in Sect. 2.3.6, with this solution the treatment of Sect. 2.3.6.1 essentially reproduces that of Sect. 2.3.5, since (6a), via (2.3.6-2), entails

$$G(z) = \sin(z) \ , \tag{6d}$$

while (6b), via (2.3.6.1-13) with (2.3.6.1-18), yields again (5d).

Proposition 2.3.6.2-7. The most *general nontrivial solution* of the functional equation (1) reads (up to the transformation (2)):

$$g(z) = f(z) = \zeta(z) + \lambda z \ , \tag{7a}$$

$$h(z) = \frac{1}{2}\left[\zeta'(z) + \zeta^2(z)\right] + \lambda\left[z\,\zeta(z) - 1\right] + \frac{1}{2}\lambda^2 z^2 = \gamma(z) - \frac{3}{2}\lambda \ , \tag{7b}$$

where $\zeta(z) \equiv \zeta(z|\omega,\omega')$ is the Weierstrass zeta function, see (A-39) (so that $\zeta' + \zeta^2 = \sigma''/\sigma$, where $\sigma \equiv \sigma(z|\omega,\omega')$ is the Weierstrass sigma function, see (A-38)), and λ is an arbitrary constant.

Proof. We note first of all that, if $f(0)$ is finite or vanishing (the latter being indeed more in keeping with (2.3.6-3b)), (1e) with $y = x$ yields

$$g'(x) + h(0) = 0 \ , \tag{8}$$

entailing (3a) with $\lambda = -h(0)$. Hence we hereafter exclude the possibility that $f(z)$ not diverge at $z = 0$ (since we stipulated to ignore the trivial solution (3)).

It is on the other hand clear from (1) that, at least as long as we restrict our consideration to *analytic* functions, as we hereafter do, $f(z)$ can have at most a first-order pole at $z = 0$. Hence we set, as $z \to 0$,

$$f(z) = f_{-1}/z + f_1 z + O(z^3) \ , \tag{9}$$

where we have taken account of (1c).

We then set, in (1e),

$$y = x - \varepsilon \ , \tag{10}$$

and let $\varepsilon \to 0$. We thereby get

$$\left[f_{-1}/\varepsilon + f_1 \varepsilon + O(\varepsilon^3)\right]\left[g'(x)\varepsilon - \frac{1}{2}g''(x)\varepsilon^2 + \frac{1}{6}g'''(x)\varepsilon^3 + O(\varepsilon^4)\right]$$

$$= \frac{1}{2}\varepsilon^2\left[g'(x)\right]^2 + g'(x) - \frac{1}{2}g''(x)\varepsilon + \frac{1}{4}g'''(x)\varepsilon^2 + h(0) + \frac{1}{2}h''(0)\varepsilon^2 + O(\varepsilon^3) \ , \tag{11}$$

where we used (1d). Hence, equating the terms of order ε^p with $p = 0,1,2$, we get

$$f_{-1}\, g'(x) = g'(x) + h(0) \ , \tag{12a}$$

$$-\frac{1}{2}f_{-1}\, g''(x) = -\frac{1}{2}g''(x) \ , \tag{12b}$$

$$\frac{1}{6}f_{-1}\, g'''(x) + f_1\, g'(x) = \frac{1}{2}\left[g'(x)\right]^2 + \frac{1}{4}g'''(x) + \frac{1}{2}h''(0) \ . \tag{12c}$$

From (12a) and (12b) we get

$$f_{-1} = 1 \ , \tag{13a}$$

$$h(0)=0 \ , \tag{13b}$$

and using (13a) we then get, from (12c),

$$g'''(x)+6[g'(x)]^2-12 f_1 g'(x)+6 h''(0)=0 \ , \tag{14}$$

hence (see (A-40) and (A-24))

$$g(x)=\zeta(x)+\lambda x \tag{15}$$

with

$$\lambda=-f_1 \ , \tag{16a}$$

$$g_2=12[f_1^2-h''(0)] \ , \tag{16b}$$

where we are using the notation of Appendix A (see in particular (A-24)), and we have also simplified the expression of $g(x)$ by taking advantage of the freedom entailed by the transformation (2), see the *Remark 2.3.6.2-1* and the sentence following it.

Since (14) is a consequence of (1), any solution of (1) must also satisfy (14); hence (15) can now be considered as an *ansatz* that includes all possible solutions of (1), although of course the fact that (15) satisfies (1) remains as yet unproven.

Hence we now insert (15), with λ an *a priori* arbitrary constant, into (1), and we thereby find that, up to the trivial solutions mentioned above, the most general solution of the functional equation (1) is indeed given by (15), or in fact, more specifically, by (7) (and, more generally, via (2), by (2.3.6-6)). Indeed the insertion of (15) in (1a) yields

$$\zeta(x)\zeta(y)+\lambda x\zeta(y)+\lambda y\zeta(x)+\lambda^2 xy=$$

$$-f(x-y)[\zeta(x)-\zeta(y)+\lambda(x-y)]+h(x-y)+\lambda+\frac{1}{2}\lambda^2(x^2+y^2)$$

$$+\frac{1}{2}[\zeta'(x)+\zeta^2(x)+\zeta'(y)+\zeta^2(y)]+\lambda[x\zeta(x)+y\zeta(y)] \ . \tag{17}$$

We now use (A-59c) to eliminate the product $\zeta(x)\zeta(y)$, and we thereby obtain

$$[f(x-y)-\zeta(x-y)-\lambda(x-y)][\ \zeta(x)+\lambda x-\zeta(y)-\lambda y\ ]=H(x-y) \ , \tag{18a}$$

$$H(z)\equiv h(z)-\frac{1}{2}[\zeta'(z)+\zeta^2(z)]+\lambda[1-z\zeta(z)]-\frac{1}{2}\lambda^2 z^2 \ . \tag{18b}$$

Since the first factor in the left hand side, as well as the right hand side, of (18a) only depend on the difference $x-y$, while the second factor in the left hand side of this equation, (18a), does not depend only on this difference, $x-y$, it is necessary and sufficient for the validity of (18) that there hold the two relations (7) (also recall

(15)). It is thereby proven that (7) provides the most general solution of the functional equation (1) (up to the *trivial* solutions mentioned above, and to the transformation (2)).

A crucial role, in the proof we have just given, is played by the key relation (A-59c). Since we have been unable to locate this formula in the literature, we complete our proof, as given above, by reporting a proof of (A-59b,c).

We take as starting point the formula (A-59a), that is well-known (see, for instance, <WW27>, Sect. 20.41, example 1, p. 446):

$$\zeta'(x)+\zeta'(y)+\zeta'(x+y)+\left[\zeta(x)+\zeta(y)-\zeta(x-y)\right]^2=0 \ . \tag{19a}$$

By performing the square, this can be rewritten as follows:

$$\zeta(x)\zeta(y)=\zeta(x+y)\left[\zeta(x)+\zeta(y)\right]-\gamma(x)-\gamma(y)-\gamma(x+y) \ . \tag{19b}$$

Here and below we use the convenient short-hand notation

$$\gamma(z)\equiv\frac{1}{2}\left[\zeta'(z)+\zeta^2(z)\right]=\frac{1}{2}\sigma''(z)/\sigma(z) \ . \tag{19c}$$

(see (1b)). We now replace y with $-y$. Since $\zeta(z)$ is odd, $\zeta(-z)=-\zeta(z)$, see (A-41), while clearly $\gamma(z)$ is even, $\gamma(-z)=\gamma(z)$, we get

$$\zeta(x)\zeta(y)=-\zeta(x-y)\left[\zeta(x)-\zeta(y)\right]+\gamma(x)+\gamma(y)+\gamma(x-y) \ . \tag{19d}$$

which coincides with (A-59c), that is thereby proven. And (A-59b) obtains by subtracting (19b) from (19d).

Remark 2.3.6.2-8. (i) All *nontrivial* solutions of the functional equation (1) have the property (*a priori* far from obvious)

$$g(z)=f(z) \ , \tag{20}$$

(see (7a)). Note however that this relation, (20), is not preserved by the transformation (2); hence it is possible to obtain, via (2), solutions of the functional equation (1) which violate the rule (20). *(ii)* Iff $\lambda=0$, the general solution (7) has the property

$$h(z)=\frac{1}{2}\left[g'(z)+g^2(z)\right]=\gamma(z) \ ; \tag{21}$$

this property is also featured by the solution (5), which indeed corresponds to the general solution (7) with $\lambda=0$ and $\zeta(z)$ completely degenerate, see (A-55b); it is *not* featured by the solution (6), which indeed corresponds to the general solution (7) with $\lambda=-1/3$ and $\zeta(z)$ degenerate (see (A-54c), with $a=i,\ a^2=-1$). *(iii)* The 2 special solutions (5) and (6) both feature a vanishing $h(z)$, see (5b) and (6b); the general solution does not, see (7b), except in the two special cases (5) and (6).

2.3.6.3 A new solvable many-body problem with elliptic-type velocity-dependent forces

In Sect. 2.3.6.3 we focus on the N-body problem

$$\ddot{x}_n = 2 \sum_{m=1,m\neq n}^{N} \dot{x}_n \dot{x}_m \, \zeta(x_n - x_m) \ , \qquad (1)$$

where of course $x_n \equiv x_n(t)$, $\dot{x}_n \equiv d x_n(t)/dt$ and $\zeta(x) \equiv \zeta(x|\omega,\omega')$ is the Weierstrass zeta function, see Appendix A. We show below how to solve the initial-value problem for this system. The initial data consist of the N initial positions, $x_n(0)$, as well as the N initial velocities, $\dot{x}_n(0)$, and we assume that the initial velocities satisfy the single constraint

$$\sum_{n=1}^{N} \dot{x}_n(0) = 0 \ , \qquad (2a)$$

entailing that the center-of-mass of the system (1) is initially at rest. Since, due to the odd character of $\zeta(x)$, $\zeta(-x)=-\zeta(x)$, (1) entails

$$\sum_{n=1}^{N} \ddot{x}_n(t) = 0 \ , \qquad (3a)$$

hence

$$\sum_{n=1}^{N} \dot{x}_n(t) = \sum_{n=1}^{N} \dot{x}_n(0) \ , \qquad (3b)$$

the condition (2a), constraining the center-of-mass of the system to be *initially* at rest, entails that it remains at rest *throughout the motion*,

$$\sum_{n=1}^{N} \dot{x}_n(t) = 0 . \qquad (2b)$$

The motivation for restricting attention to the N-body system (1) with the constraint (2) ensues from the treatment of Sect. 2.3.6.1; and the following developments are of course also closely related to that treatment, although we try to make the presentation below as self-contained as it is compatible with the avoidance of excessive repetition. At the end of Sect. 2.3.6.3, we extend the treatment to the more general equation of motion (2.3.6.1-26) with (2.3.6.2-7a) (which reduce to (1) for $\lambda = 0$).

To solve the problem (1) it is convenient to introduce new coordinates $\xi_n(\tau)$ such that (see (2.3.6.1-24))

$$\xi_n(\tau) = x_n(t)\,, \quad \tau = \tau(t)\,, \quad t = t(\tau)\ . \tag{4}$$

We will choose below the function $\tau(t)$ appropriately, with the properties (see (2.3.6.1-25))

$$\tau(0) = 0\,,\ t(0) = 0\ , \tag{5a}$$

$$\dot{\tau}(0) = 1\,,\ t'(0) = 1\ , \tag{5b}$$

which clearly entail (see (2.3.6.1-27))

$$\xi_n(0) = x_n(0)\ , \tag{6a}$$

$$\xi'_n(0) = \dot{x}_n(0)\ . \tag{6b}$$

Here and below dots denote of course t-differentiations and primes τ-differentiations (or, more generally, differentiations with respect to the argument of the function they are appended to). Note that, via (4), the constraint (2) entails (see (3b))

$$\sum_{n=1}^{N} \xi'_n(\tau) = \sum_{n=1}^{N} \xi'_n(0) = 0\ , \tag{7a}$$

hence also

$$\sum_{n=1}^{N} \xi_n(\tau) = \sum_{n=1}^{N} \xi_n(0) = \sum_{n=1}^{N} x_n(0)\ , \tag{7b}$$

with the first equality entailed by (7a) and the second by (6a).

We introduce now the convenient function

$$\psi(x,\tau)=1-\tau\sum_{n=1}^{N}\dot{x}_n(0)\,\zeta[x-x_n(0)] \quad . \tag{8a}$$

This is a doubly-periodic elliptic function of the variable x; it has N poles at $x=x_n(0)$ with residues $-\tau\dot{x}_n(0)$, see (A-47); its double periodicity is guaranteed by the constraint (2a) (see (A-45)); and evidently

$$\psi(x,0)=1 \quad , \tag{8b}$$

while the (partial) τ-derivative of $\psi(x,\tau)$,

$$\psi_\tau(x,\tau)=\psi_\tau(x,0)=-\sum_{n=1}^{N}\dot{x}_n(0)\,\zeta[x-x_n(0)]=\psi(\xi,1)-1 \quad , \tag{8c}$$

is clearly as well a (τ-independent!) *doubly-periodic elliptic function* of x, with N poles at $x=x_n(0)$ and residues $-\dot{x}_n(0)$; finally, it is obvious that

$$\psi_{\tau\tau}(x,t)=0 \quad . \tag{8d}$$

We now introduce, in addition to the above (standard) representation of the doubly-periodic elliptic function (8a) via its *poles* and *residues*, another (also standard) representation of this *same* doubly-periodic elliptic function, via its *poles* and *zeros:*

$$\psi(x,\tau)=\exp\left[-\frac{1}{2}\varphi(\tau)\right]\prod_{n=1}^{N}\{\sigma[x-\xi_n(\tau)]/\sigma[x-x_n(0)]\} \quad . \tag{9}$$

Here of course $\sigma(x)\equiv\sigma(x|\omega,\omega')$ is the Weierstrass sigma function, see Appendix A, and it is required that the sum of the poles equals the sum of the zeros, a condition that is guaranteed by (7b).

The fact that any doubly periodic elliptic function admit, up to an *additive* respectively *multiplicative* constant, a unique *additive* representation of type (8a) in terms of its *poles* and *residues* (with the condition that the sum of the residues vanish) respectively a unique *multiplicative* representation of type (9) in terms of its *poles* and *zeros* (with the condition that the sum of the poles equals the sum of the zeros) is a well-known result (see for instance Sect. 20.5 of <WW27>). In our case we eliminated the ambiguity associated with the choice of the (additive or multiplicative, as the case may be) constant via (8a), and therefore the (*a priori* unknown) constant φ is introduced in (9) (φ is a constant inasmuch as it does not depend on x; it does, of course, depend on τ, $\varphi\equiv\varphi(\tau)$, see (9)).

Let us emphasize that the representation (9) does not quite coincide with the representation (2.3.6-1) with

$$G(x) = \sigma(x) \ , \tag{10a}$$

which of course corresponds, in the notation of Sect. 2.3.6.1 and 2.3.6.2, to

$$f(x) = g(x) = \zeta(x) \ , \tag{10b}$$

see indeed (2.3.6-2) with (A-39) and compare (1) with (2.3.6.1-26). The difference among (9) and (2.3.6-1) with (10a) is the τ-*independent* multiplicative factor $\prod_{n=1}^{N} \{\sigma[x - x_n(0)]\}^{-1}$, which has little relevance as regards the equation (8d) (or equivalently (2.3.6.1-17a)) which characterizes the time-evolution. But this representation, (9), is quite convenient to solve the initial-value problem for the N-body system (1) with (2), as we now show.

Indeed we now equate (8a) and (9):

$$\exp[-\varphi(\tau)/2] \prod_{n=1}^{N} \{\sigma[x - \xi_n(\tau)]/\sigma[x - x_n(0)]\} = 1 - \tau \sum_{n=1}^{N} \dot{x}_n(0)\, \zeta[x - x_n(0)] \ . \tag{11}$$

It is clear from this equation that the initial data, $x_n(0)$ and $\dot{x}_n(0)$, which characterize its right hand side, determine uniquely the N coordinates $\xi_n(\tau)$ (which are just the N zeros of $\psi(x,\tau)$, see (8a)), as well as the function $\varphi(\tau)$, that feature in its left hand side; and this determination is achieved, via this equation, (11), without having to integrate any differential equation. Note in particular that this equation, (11), or, equivalently, (9) with (8b), entail, via (6a), that $\varphi(\tau)$ vanishes at $\tau = 0$,

$$\varphi(0) = 0 \ . \tag{12a}$$

We now τ-differentiate the logarithm of (9), using (A-39), and equate the result to (8c):

$$\psi_\tau(x,\tau) = \psi(x,\tau)\{-\frac{1}{2}\varphi'(\tau) - \sum_{n=1}^{N} \xi_n'(\tau)\zeta[x - \xi_n(\tau)]\} \ , \tag{13a}$$

$$\psi(x,\tau)\{-\frac{1}{2}\varphi'(\tau) - \sum_{n=1}^{N} \xi_n'(\tau)\zeta[x - \xi_n(\tau)]\} = -\sum_{n=1}^{N} \dot{x}_n(0)\zeta[x - x_n(0)] \ . \tag{13b}$$

Via (8b), (6a) and (6b) this formula, (13b), clearly entails that also the τ-derivative of $\varphi(\tau)$ vanishes at $\tau=0$,

$$\varphi'(0)=0 \quad . \tag{12b}$$

The τ-independence of the left hand side of (13b) (which is less than obvious; but see the right hand side !) should moreover be noted.

The next step is to τ-differentiate (13a), and to use (8d). The relevant computation was detailed in Sect. 2.3.6.1, hence it is not repeated here. The result reads

$$\xi_n''(\tau)=\varphi'(\tau)\,\xi_n'(\tau)+2\sum_{m=1,m\neq n}^{N}\xi_n'(\tau)\,\xi_m'(\tau)\,\zeta[\xi_n(\tau)-\xi_m(\tau)] \quad . \tag{14}$$

This formula coincides with (2.3.6.1-14) via (10b).

Moreover, one gets (2.3.6.1-13) with (2.3.6.2-7b) (with $\lambda=0$):

$$\varphi''-\frac{1}{2}(\varphi')^2=2\sum_{\ell,m=1}^{N}\xi_m'\,\xi_\ell'\,h(\xi_m-\xi_\ell) \quad , \tag{15a}$$

$$h(z)=\frac{1}{2}[\zeta'(z)+\zeta^2(z)]=\frac{1}{2}\sigma''(z)/\sigma(z) \quad . \tag{15b}$$

In (15a) we have omitted the requirement $\ell\neq m$ in the sum, since (15b) implies $h(0)=0$, see (A-47) or (A-46). This ODE, (15a), together with (12), determines in principle the function $\varphi(\tau)$, if the quantities $\xi_n(\tau)$ and $\xi_n'(\tau)$ are known. But, as we have seen above, $\varphi(\tau)$ is also obtainable, without solving any ODE, directly from the initial data $x_n(0)$, $\dot{x}_n(0)$, via (11). Hence, in the context of the initial-value problem for (1), we hereafter consider $\varphi(\tau)$ as a *known* function.

The final step, in the context of the solution of the initial-value problem for (1), is to relate the coordinate $\xi_n(\tau)$ to the coordinate $x_n(t)$. The equations to be compared are (14) and (1), with in addition (6). The link is provided by (4), with (2.3.6.1-25); note in particular the coincidence of (6) with (2.3.6.1-27); as for (2.3.6.1-25c) and (2.3.6.1-28), we refer for their relevance and derivation to the treatment in Sect. 2.3.6.1, without reporting neither their expressions nor their derivations here.

In conclusion: we now see how the initial-value problem for (1) is solved. Given the initial data, $x_n(0)$ and $\dot{x}_n(0)$, via (11) one gets the N coordinates $\xi_n(\tau)$ (the zeros of (11) !), as well as the function $\varphi(\tau)$; then, via (2.3.6.1-28), one gets $t(\tau)$, hence, by functional inversion, $\tau(t)$; and finally from (4) one gets the "particle coordinates" $x_n(t)$ that are entailed by (1) with the assigned initial data $x_n(0)$ and $\dot{x}_n(0)$.

Let us now show, as promised above, how the solution of the initial-value problem for the (more general) N-body system characterized by the equations of motion

$$\ddot{y}_n = 2 \sum_{m=1, m\neq n}^{N} \dot{y}_n \, \dot{y}_m \, [f(y_n - y_m) + \lambda \, (y_n - y_m)] \quad , \tag{16a}$$

with initial conditions satisfying the restriction

$$\sum_{n=1}^{N} \dot{y}_n(0) = 0 \quad , \tag{16b}$$

can be generally reduced to solving the same problem, but with

$\lambda = 0$:

$$\ddot{x}_n = 2 \sum_{m=1, m\neq n}^{N} \dot{x}_n \, \dot{x}_m \, f(x_n - x_m) \quad , \tag{17a}$$

$$\sum_{n=1}^{N} \dot{x}_n(0) = 0 \quad . \tag{17b}$$

Namely, we now show how, from the solution of the initial value-problem for (17), one can, simply via a change of the independent variable, evince a solution of the initial-value problem for (16). Let us emphasize that, in (16a), λ is an *arbitrary* constant, while the function $f(z)$, which appears in both (16a) and (17a), is also, in the present context, *arbitrary*, except for the restriction that it be odd,

$$f(-z) = -f(z) \quad . \tag{18}$$

Note that this restriction, together with the equations of motion (16a) respectively (17a), entail that the restrictions (16b) respectively (17b) on the *initial* conditions hold in fact for *all* time:

$$\sum_{n=1}^{N} \dot{y}_n(t) = 0 \quad , \tag{16c}$$

$$\sum_{n=1}^{N} \dot{x}_n(t) = 0 \quad . \tag{17c}$$

Let us indeed relate the coordinate x_n to y_n via the identification

$$x_n(\tau) = y_n(t) \quad , \tag{19a}$$

with

$$t = \int_0^\tau d\tau' \exp\{(\lambda/2)\sum_{n=1}^{N} \left[x_n^2(\tau') - x_n^2(0)\right]\} \quad . \tag{19b}$$

These relations clearly entail the identities

$$x_n(0) = y_n(0) \quad , \tag{20a}$$

$$\dot{x}_n(0) = \dot{y}_n(0) \quad . \tag{20b}$$

Moreover, if x_n evolves according to (17a), y_n evolves according to (16a); and viceversa. Hence to solve the initial-value problem for (16a) one solves firstly the initial-value problem for (17a), with the same initial data, see (20), and one then uses (19) to get $y_n(t)$; of course to perform the last step a quadrature, and a functional inversion, are generally required, to get $\tau(t)$ from (19b).

Before delving into the proof of the above assertions, let us emphasize that the change of the independent variable discussed here, from t to τ and viceversa, see (19), has nothing to do with the change of independent variable discussed above, see (4), in spite of the similarity of the notations used (also there, from t to τ, and viceversa).

Proofs. From (19b) clearly

$$\tau(0) = 0 \quad , \tag{21a}$$

$$\dot{\tau}(t) = \exp\{-(\lambda/2)\sum_{n=1}^{N} \left[x_n^2(\tau) - x_n^2(0)\right]\} \tag{22a}$$

hence

$$\dot{\tau}(0) = 1 \quad , \tag{21b}$$

and

$$\ddot{\tau}(t)/\left[\dot{\tau}(t)\right]^2 = -\lambda \sum_{n=1}^{N} x_n'(\tau)\, x_n(\tau) \quad . \tag{22b}$$

The prime in the right-hand side of (22b) denotes of course differentiation with respect to τ. Note that (21a) entails (20a) (via (19a)).

Next, from (19a) we get

$$\dot{y}_n(t) = x_n'(\tau)\,\dot{\tau}(t) \tag{23}$$

which of course, via (21), entails (20b).

An additional differentiation yields

$$\ddot{y}_n(t) = x_n''(\tau)\left[\dot{\tau}(t)\right]^2 + x_n'(\tau)\,\ddot{\tau}(t) \ , \tag{24}$$

and, via (17a) (which must now be seen with dots replaced by primes), as well as (23) and (22b), one gets

$$\ddot{y}_n = 2\sum_{m=1,m\neq n}^{N} \dot{y}_n\,\dot{y}_m\,f(y_n - y_m) - 2\,\lambda\,\dot{y}_n \sum_{m=1}^{N} \dot{y}_m\,y_m \ . \tag{25}$$

But, thanks to (16c), the last term in the right-hand side can be rewritten as follows:

$$-2\lambda\,\dot{y}_n \sum_{m=1}^{N} \dot{y}_m\,y_m = 2\lambda \sum_{m=1}^{N} \dot{y}_n\,\dot{y}_m\,(y_n - y_m) \ . \tag{26}$$

Moreover, the term with $m = n$ can be omitted, in the sum in the right-hand side of this last equation, since it vanishes. It is thus seen that (25) coincides with (16a), and this completes the proof of the assertions made above.

Let us end Sect. 2.3.6.3 by emphasizing that the N-body system solved herein, see (1), is Hamiltonian. The corresponding Hamiltonian function reads

$$H(\underline{p},\underline{q}) = \sum_{n=1}^{N} \exp(s\,p_n) \prod_{m=1,m\neq n}^{N} \left[\sigma(q_n - q_m)\right]^{-1} \ , \tag{27}$$

with s an arbitrary constant and $\sigma(q) \equiv \sigma(q|\omega,\omega')$ the Weierstrass sigma function, see Appendix A.

Exercise 2.3.6.3-1. Verify that the Hamiltonian equations entailed by (27) yield the Newtonian equations of motion (1), of course via the identification $q_n(t) = x_n(t)$. *Hint*: see the treatment of Sect. 2.1.12.1, and use (A-39).

Remark 2.3.6.3-2. The *solvable* N-body model (1) does not belong to the RS class treated in Sect. 2.1.12 and the subsections following Sect. 2.1.12.

Remark 2.3.6.3-3. The solvable N-body model (1) admits the (complex) extension characterized by the Hamiltonian (compare with (27))

$$\tilde{H}(\underline{\tilde{p}},\underline{\tilde{q}}) = -i(\Omega/s)\sum_{n=1}^{N} \tilde{q}_n + \sum_{n=1}^{N} \exp(s\,\tilde{p}_n) \prod_{m=1,m\neq n}^{N} [\sigma(\tilde{q}_n - \tilde{q}_m)]^{-1} \quad , \tag{28a}$$

and by the Newtonian equations of motion (compare with (1))

$$\ddot{\tilde{x}}_n - i\Omega\dot{\tilde{x}}_n = 2 \sum_{m=1,m\neq n}^{N} \dot{\tilde{x}}_n \dot{\tilde{x}}_m \zeta(\tilde{x}_n - \tilde{x}_m) \quad , \tag{28b}$$

which obtain from the Hamiltonian equations of motion entailed by (28a) via the identification $\tilde{q}_n(t) = \tilde{x}_n(t)$. Here we denote by Ω an arbitrary *real* (nonvanishing) constant (we use the capital letter Ω to avoid any confusion with the semiperiods, ω and ω', associated with the Weierstrass zeta function $\zeta(x) \equiv \zeta(x|\omega,\omega')$).

Exercise 2.3.6.3-4. Verify that (28b) follows from (28a).

Remark 2.3.6.3-5. Set

$$\tilde{x}_n(t) = x_n[\tilde{t}(t)] \tag{29a}$$

with

$$\tilde{t}(t) = [\exp(i\Omega t) - 1]/(i\Omega) \quad . \tag{29b}$$

Then clearly

$$\tilde{x}_n(0) = x_n(0) \quad , \tag{29c}$$

$$\dot{\tilde{x}}_n(0) = \dot{x}_n(0) \quad , \tag{29c}$$

and moreover, if $x_n(t)$ evolves according to (1), $\tilde{x}_n(t)$ evolves according to (28b), and viceversa. Hence the solutions of the initial-value problem for (28b) can be obtained simply by replacing t with $\tilde{t}(t)$, see (29b), in the solutions of the initial-value problem (with the same initial data) for (1). This clearly entails that *all* solutions of the Newtonian equations of motion (28b), corresponding to the Hamiltonian (28a), are completely periodic, with period $T = 2\pi/\Omega$ (or a multiple of it; see, for instance, Sect. 4.5).

Exercise 2.3.6.3-6. Verify by explicit computation the validity of the above assertions, and review all previous analogous findings. *Hint*: see Sects. 2.1.12.3, 2.1.12.4 and 2.1.13.

Exercise 2.3.6.3-7. Generalize the Hamiltonian (27) so that it yields, rather than (1), the equations of motion (16a) (with $f(z) = \zeta(z)$). *Hint*: see Sect. 2.1.12.1.

Exercise 2.3.6.3-8. Formulate and solve analogous exercises to those written above, but for the more general Hamiltonian introduced in (the immediately preceding) *Exercise 2.3.6.3-7* and, correspondingly, for the equations of motion (16a) (with $f(z) = \zeta(z)$).

2.4 Finite-dimensional representations of differential operators, Lagrangian interpolation, and all that

In previous sections certain *remarkable matrices* emerged (see for instance the discussion preceding (2.1.3.3-46), that leading to (2.3.4.1-23), the statement following (2.3.4.2-30)): they were defined in a neat manner in terms of the zeros of Hermite polynomials or in terms of N arbitrary distinct numbers, and they possessed only integer eigenvalues. Clearly such findings have a (purely mathematical !) interest of their own, independent of the investigation of many-body problems. In Sect. 2.4 we tersely review certain developments on Lagrangian interpolation, finite-dimensional (matrix) representations of differential operators and all that: these classical mathematical results provide an appropriate context to prove and extend the findings about remarkable matrices mentioned above. They also provide tools to manufacture many-body problems amenable to exact treatments; but such developments are mainly postponed to the next Chap. 3, where they are treated on the basis of a somewhat more general formulation of Lagrangian interpolation than that presented here, which is restricted to a *one-dimensional* context and to *polynomial* interpolation.

Before delving, in the next subsections of Sect. 2.4, into the substantive topics indicated by the titles of Sect. 2.4 and of the following subsections, let us again specify the *notation* which is used below. N is a (fixed but arbitrary) positive integer, generally larger than unity; all indices $(n,m,l,j,k,...)$ run, unless otherwise indicated, from 1 to N; $(N \times N)$-matrices are denoted by *underlined* upper-case letters, N-vectors are denoted by *underlined* lower-case letters, and the standard rules are used for matrix-vector algebra: thus the N-vector $\underline{v}$ has the N components v_n, the matrix $\underline{M}$ has the N^2 elements M_{nm}, the N-vectors

$$\underline{w} = \underline{M} \cdot \underline{v} \ , \quad \underline{u} = \underline{v} \cdot \underline{M} \tag{1a}$$

have the N components

$$w_n = \sum_{m=1}^{N} M_{nm} v_m \ , \quad u_n = \sum_{m=1}^{N} M_{mn} v_m \ , \tag{1b}$$

and the scalar product among the two N-vectors $\underline{v}$ and $\underline{u}$ is defined in the standard manner,

$$\underline{u} \cdot \underline{v} = \underline{v} \cdot \underline{u} = \sum_{n=1}^{N} u_n v_n \ . \tag{2}$$

In the following the dot, in equations such as (1a) and (2), will be omitted whenever this is unlikely to cause any misunderstanding.

Finally, in the following, unless otherwise indicated, by the notation x_n (or y_n or ...) we denote N arbitrary distinct numbers (possibly complex); namely, the properties of the $(N \times N)$-matrices which are defined below in terms of the N numbers x_n (or y_n or ...) hold for any arbitrary choice of these numbers, except for the restriction that they be different,

$$x_n \neq x_m \qquad \text{if} \qquad n \neq m \ , \tag{3}$$

which is hereafter assumed to hold (actually most of the formulas written below remain valid, perhaps in modified forms obtained via appropriate limiting processes, even if (3) does not hold; but for simplicity we exclude this possibility hereafter).

2.4.1 Finite-dimensional matrix representations of differential operators

In Sect. 2.4.1 we report the main formulas which provide a convenient finite-dimensional matrix representation of differential operators, exactly applicable in the functional space of the polynomials of degree less than N.

Let the matrices $\underline{X}$ and $\underline{D}$ be defined as follows, in terms of N *arbitrary* (but distinct, see (2.4-3)) numbers x_n:

$$\underline{X} = \text{diag}(x_n), \qquad X_{nm} = \delta_{nm}\, x_n \ , \tag{1}$$

$$D_{nm} \equiv D_{nm}(\underline{x}) = \delta_{nm} \sum_{\ell=1,\ell\neq n}^{N} (x_n - x_\ell)^{-1} + (1-\delta_{nm})(x_n - x_m)^{-1} \ . \tag{2}$$

Let us moreover introduce, for notational convenience, the two N-vectors $\underline{d}$ and $\underline{b}$, and the $(N\times N)$-matrix $\underline{B}$, also defined in terms of the N *arbitrary* numbers x_n, as follows:

$$d_n \equiv d_n(\underline{x}) = \sum_{m=1,m\neq n}^{N} (x_n - x_m)^{-1} \ , \tag{3}$$

$$b_n \equiv b_n(\underline{x}) = \prod_{m=1,m\neq n}^{N} (x_n - x_m) \ , \tag{4a}$$

$$\underline{B} \equiv \underline{B}(\underline{x}) = \text{diag}(b_n)\,, \qquad B_{nm} = \delta_{nm}\, b_n \ . \tag{4b}$$

Note that (2) and (3) entail

$$D_{nn} = d_n \ , \tag{5a}$$

$$\sum_{m=1}^{N} D_{nm} = 2 d_n \ , \tag{5b}$$

$$\sum_{m=1}^{N} D_{mn} = 0 \ , \tag{5c}$$

$$(\underline{D}^2)_{nm} = \delta_{nm}\left[d_n{}^2 - \sum_{\ell=1,\ell\neq n}^{N} (x_n - x_\ell)^{-2}\right] + (1-\delta_{nm})2\left[d_n (x_n - x_m)^{-1} - (x_n - x_m)^{-2}\right], \tag{5d}$$

while (4a) entails

$$b_n = p'_N(x_n)\ , \tag{6a}$$

where $p_N(x)$ is the monic polynomial of degree N in x having the N numbers x_n as its zeros,

$$p_N(x) = \prod_{n=1}^{N} (x - x_n) \tag{6b}$$

The proof of these formulas is plain; the diligent reader might want to work it out in full detail.

The two $(N \times N)$-matrices $\underline{X}$ respectively $\underline{D}$ provide faithful representations of the (multiplicative) operator x respectively of the (differential) operator d/dx, in the N-dimensional functional space of the polynomials in x of degree less than N. A more precise formulation of this statement reads as the following

Lemma 2.4.1-1. Let $f(x)$ be an arbitrary polynomial of degree less than N,

$$f(x) = \sum_{n=1}^{N} \alpha_{n-1} x^{n-1} = \sum_{n=1}^{N} c_n x^{N-n} \quad , \tag{7a}$$

and $f^{(r)}(x)$ its r-th derivative

$$f^{(r)}(x) \equiv (d/dx)^r f(x) = \sum_{n=r+1}^{N} \alpha_{n-1} [(n-1)!/(n-1-r)!] x^{n-1-r} \quad . \tag{7b}$$

Now associate to $f(x)$ respectively $f^{(r)}(x)$ the N-vectors $\underline{f}$ respectively $\underline{f}^{(r)}$ via the following prescriptions: the n-th component of the N-vectors $\underline{f}$ respectively $\underline{f}^{(r)}$ are the values that the functions $f(x)$ respectively $f^{(r)}(x)$ take at the point x_n:

$$f_n \equiv (\underline{f})_n = f(x_n) \quad , \tag{8a}$$

$$f_n^{(r)} \equiv (\underline{f}^{(r)})_n = f^{(r)}(x_n) \quad . \tag{8b}$$

There holds then the following N-vector formula:

$$\underline{f}^{(r)} = \underline{B}\,\underline{D}^r\,\underline{B}^{-1}\,\underline{f} = (\underline{B}\,\underline{D}\,\underline{B}^{-1})^r\,\underline{f} \, , \quad r = 0,1,2,\ldots \, . \tag{9}$$

This important formula demonstrates that, up to the similarity transformation induced by the matrix $\underline{B}$, the matrix $\underline{D}$, given in terms of the N *arbitrary* numbers x_n by the neat definition (2), provides a representation of the differential operator d/dx. In the following two Sects. 2.4.2 and 2.4.3 we will look at this result from other angles, and in so doing we will also prove it (in Sect. 2.4.2). This should not prevent the diligent reader from trying and proving (9) forthwith.

Note that, since clearly (see (7))

$$\underline{f}^{(N)} = 0 \quad , \tag{10}$$

(9) entails that the matrix $\underline{D}$ is *nilpotent*,

$$\underline{D}^N = 0 \ . \tag{11}$$

The diligent reader is also advised to try and prove now this nontrivial property of the matrix $\underline{D}$, see (2).

The *Lemma 2.4.1-1* stated above entails the following important

Proposition 2.4.1-2. Let A be an arbitrary linear differential operator written as follows:

$$\mathsf{A} = \sum_{r=0} a_r(x)(d/dx)^r \ , \tag{12}$$

and let

$$\mathsf{A} f(x) = F(x) \ , \tag{13}$$

$f(x)$ being a polynomial in x of degree less than N, see (7a) (but note: no such condition on $F(x)$). There then holds the N-vector equation

$$\underline{A}\ f(\underline{X})\ \underline{v} = F(\underline{X})\ \underline{v} \tag{14}$$

with

$$\underline{A} = \sum_{r=0} a_r(\underline{X})\ \underline{D}^r \tag{15}$$

and (see (5))

$$\underline{v} = \underline{B}^{-1}\underline{u} \ , \tag{16a}$$

where $\underline{u}$ is the N-vector having all components equal to unity,

$$u_n = 1 \ , \tag{17}$$

so that (see (4))

$$v_n = b_n^{-1} = \left[\prod_{m=1, m\neq n}^{N} (x_n - x_m) \right]^{-1} . \tag{16b}$$

This result provides a simple rule to transform any linear differential equation (including non homogeneous ones), valid for a polynomial $f(x)$ of degree *less* than *N*, into a corresponding *N*-vector equation, via the replacement of the variable x with the diagonal $(N \times N)$-matrix $\underline{X}$, see (1), and of the operator of differentiation d/dx with the $(N \times N)$-matrix $\underline{D}$, see (2).

There clearly moreover hold the following two *Corollaries*.

Corollary 2.4.1-3. If for the differential operator A, see (12), there holds the equation

$$\mathsf{A} f(x) = 0 \ , \tag{18}$$

$f(x)$ being a polynomial in x of degree less than *N*, see (7a), then the $(N \times N)$-matrix $\underline{A}$, see (15), has vanishing determinant

$$\det[\underline{A}] = 0 \ . \tag{19}$$

Corollary 2.4.1-4. If the differential operator A, see (12), has the eigenvalue a,

$$\mathsf{A}\, f_a(x) = a\, f_a(x) \ , \tag{20}$$

and the corresponding eigenfunction, $f_a(x)$, is a polynomial in x of degree less than *N*, see (7a), then the matrix $\underline{A}$, see (15), also has the eigenvalue a,

$$\underline{A}\, \underline{w}^{(a)} = a\, \underline{w}^{(a)} \ , \tag{21}$$

and the corresponding eigenvector $\underline{w}^{(a)}$ is given by the following simple rule:

$$\underline{w}^{(a)} = f_a(\underline{X})\, \underline{v} \ , \tag{22}$$

with $\underline{v}$ defined by (16) (and of course $\underline{X}$ defined by (1)).

Proofs. The *Proposition 2.4.1-2* is an immediate consequence of the *Lemma 2.4.1-1*. Indeed, by setting $x = x_n$ in (13), one gets

$$\sum_{r=0} a_r(x_n)\, f^{(r)}(x_n) = F(x_n) \ , \tag{23a}$$

hence, via (1), (8b), (9) and (17),

$$\underline{B}\,\underline{A}\,\underline{B}^{-1}\underline{f} = F(\underline{X})\,\underline{u} \ , \tag{23b}$$

where moreover, see (17), (8a) and (1),

$$\underline{f} = f(\underline{X})\,\underline{u} \ . \tag{24}$$

Insertion of (24) in (23) yields, after multiplication from the left by $\underline{B}^{-1}$ (and using the commutativity of the diagonal matrices $\underline{X}$ and $\underline{B}$), precisely (14) with (15) and (16). The *Proposition 2.4.1-2* is thereby proven.

The *Corollary 2.4.1-4* is just the special case of the *Proposition 2.4.1-2* with $f(x) = f_a(x)$ and $F(x) = a\, f_a(x)$.

As for the *Corollary 2.4.1-3*, (19) is immediately entailed by the formula

$$\underline{A}\, f(\underline{X})\, \underline{v} = 0 \ , \tag{25}$$

which corresponds to (14) with $F(\underline{X}) = 0$, as indeed (18) corresponds to (13) with $F(x) = 0$.

Exercise 2.4.1-5. Prove the *identities*

$$\sum_{m=1}^{N} (\underline{D}^r)_{nm}\, b_m^{-1} = \delta_{r,0}\, b_n^{-1} \ , \ r = 0,1,,2,... \ , \tag{26}$$

with $\underline{D}$ respectively b_m defined, in terms of the N *arbitrary* (distinct) numbers x_n, by (2) respectively (4a). *Hint*: set $f(x) = 1$ in (9).

Exercise 2.4.1-6. Prove the *identities*

$$\sum_{m=1, m\neq n}^{N} (x_n - x_m)^{-1} (b_n^{-1} + b_m^{-1}) = 0 \ , \tag{27}$$

with b_m defined, in terms of the N arbitrary (distinct) numbers x_n, by (4a). *Hint*: set $r = 1$ in (26) and use (2).

2.4.2 Connection with Lagrangian interpolation

In Sect. 2.4.2 we tersely indicate how the findings reported in the preceding Sect. 2.4.1 fit in the context of the standard theory of Lagrangian (one-dimensional, polynomial) interpolation, and in so doing we also provide a proof of the *Lemma 2.4.1-1*.

Let x_n be N arbitrary distinct numbers, see (2.1-3), and f_n be N assigned values. The problem of Lagrangian (polynomial) interpolation is to construct the (unique !) polynomial $f(x)$, of degree less than N (generally, of degree $N-1$), that takes the assigned values f_n at the N points x_n,

$$f(x_n)=f_n \; . \tag{1}$$

Note the analogy of this formula, (1), with (2.4.1-8a). Also note that, via (2.4.1-7a) (which merely reflects the property of $f(x)$ to be a polynomial of degree less than N), (1) becomes

$$\sum_{m=1}^{N} \alpha_{m-1}(x_n)^{m-1}=f_n \; . \tag{2}$$

This is a system of N linear equations for the N unknowns α_{n-1}. A necessary and sufficient condition to guarantee that this system admit a unique solution is that the determinant of the $(N\times N)$-matrix with (nm)-element $(x_n)^{m-1}$ not vanish,

$$\det\left[(x_n)^{m-1}\right]\neq 0 \; . \tag{3}$$

But the *Vandermonde identity*,

$$\det\left[(x_n)^{m-1}\right]=\prod_{n,m=1;n>m}^{N}(x_n-x_m) \; , \tag{4}$$

guarantees that (3) holds (since the x_n's are, by assumption, distinct, see (2.4-3)).

The standard way to construct explicitly the polynomial $f(x)$ goes as follows. Introduce the N interpolational polynomials, all of them of degree $N-1$,

$$q_{N-1}^{(n)}(x)=\prod_{m=1,m\neq n}^{N}\left[(x-x_m)/(x_n-x_m)\right] \; , \tag{5}$$

which clearly satisfy the property

$$q_{N-1}^{(n)}(x_m) = \delta_{nm} \quad . \tag{6}$$

It is then obvious that

$$f(x) = \sum_{n=1}^{N} f_n \, q_{N-1}^{(n)}(x) \quad . \tag{7}$$

Indeed, clearly $f(x)$, see (7), is a polynomial of degree (at most) $N-1$ (hence, less than N), and, via (6), it satisfies (1).

The diligent reader will verify that the prescription (7) with (5) yields the same result that obtains by solving (2) for the α_n's and inserting the result in (2.4.1-7a).

It is now convenient to introduce the (x-dependent) N-vector $\underline{q}(x)$, whose components are the interpolational polynomials:

$$[\ \underline{q}(x)\ ]_n \equiv q_n(x) = q_{N-1}^{(n)}(x) \quad . \tag{8}$$

Then the right hand side of (7) can be written, via (2.4.1-8a), as a scalar product (see (2.4-2)):

$$f(x) = \underline{f} \cdot \underline{q}(x) = \underline{q}(x) \cdot \underline{f} , \tag{9}$$

and moreover there holds the important N-vector formula

$$(d/dx)\underline{q}(x) = \underline{q}(x) \cdot \underline{B} \cdot \underline{D} \cdot \underline{B}^{-1} = \underline{q}(x) \underline{B}\,\underline{D}\,\underline{B}^{-1} \quad , \tag{10}$$

with the ($N \times N$)-matrices $\underline{B}$ and $\underline{D}$ defined by (2.4.1-4,2).

Proof. Via (8), (5) and (2.4.1-4a) we write the n-th component of $\underline{q}(x)$ as follows:

$$q_n(x) = (b_n)^{-1} \prod_{m=1, m \neq n}^{N} (x - x_m) \quad . \tag{11}$$

Hence

$$(d/dx) q_n(x) \equiv q_n'(x) = (b_n)^{-1} \sum_{\ell=1, \ell \neq n}^{N} \ \prod_{m=1, m \neq n, \ell}^{N} (x - x_m) \quad , \tag{12a}$$

$$q'_n(x) = b_n^{-1}\left[\prod_{m=1}^{N}(x-x_m)\right] \sum_{\ell=1,\ell\neq n}^{N} \left[(x-x_m)(x-x_\ell)\right]^{-1} , \qquad (12b)$$

$$q'_n(x) = \sum_{\ell=1,\ell\neq n}^{N} b_n^{-1}\left[\prod_{m=1}^{N}(x-x_m)\right] \left[(x-x_m)^{-1}-(x-x_\ell)^{-1}\right](x_n-x_\ell)^{-1} , \qquad (12c)$$

$$q'_n(x) = \sum_{\ell=1,\ell\neq n}^{N} b_n^{-1}\left[\prod_{m=1,m\neq n}^{N}(x-x_m) - \prod_{m=1,m\neq\ell}^{N}(x-x_m)\right] (x_n-x_\ell)^{-1} , \qquad (12d)$$

$$q'_n(x) = \sum_{\ell=1,\ell\neq n}^{N} b_n^{-1}(x_n-x_\ell)^{-1}\left[b_n\, q_n(x) - b_\ell\, q_\ell(x)\right] . \qquad (12e)$$

In the last step we used (11); the others were, we trust, clear enough.

It is now clear, via (2.4.1-4b) and (2.4.1-2), that (12e) can be written in the N-vector form

$$\underline{q}'(x) = \underline{B}^{-1}\,\underline{D}^{T}\,\underline{B}\;\underline{q}(x) , \qquad (13)$$

where $\underline{D}^T$ is the transpose of $\underline{D}$ (note that the antisymmetry of the off-diagonal terms in the right hand side of (2.4.1-2) takes thereby care of the minus sign in the right hand side of (12e)).

Since the transpose of $\underline{B}^{-1}\,\underline{D}^{T}\,\underline{B}$ is $\underline{B}\,\underline{D}\,\underline{B}^{-1}$ ($\underline{B}$ being symmetrical, indeed diagonal, see (2.4.1-4b)), clearly (13) coincides with (10), which is therefore proven.

Clearly (10) can be iterated, yielding the more general formula

$$(d/dx)^r\,\underline{q}(x) = \underline{q}(x)(\underline{B}\,\underline{D}\,\underline{B}^{-1})^r = \underline{q}(x)\,\underline{B}\,\underline{D}^r\,\underline{B}^{-1} . \qquad (14)$$

Hence from (9) we get

$$(d/dx)^r f(x) = \underline{q}(x)\cdot\underline{B}\,\underline{D}^r\,\underline{B}^{-1}\,\underline{f} , \qquad (15)$$

which corresponds to the important formula (2.4.1-9), that is therefore now proven.

Indeed, by setting $x = x_n$ in (15) we get, via (2.4.1-8b),

$$f_n^{(r)} = \underline{q}(x_n) \cdot \underline{B}\,\underline{D}^r\,\underline{B}^{-1} \underline{f} \quad , \tag{16}$$

and the scalar product in the right hand side coincides with the n-th component of the N-vector $\underline{B}\,\underline{D}^r\,\underline{B}^{-1}\underline{f}$, since

$$q_m(x_n) = \delta_{mn} \tag{17}$$

(see (8) and (6)).

Let us end Sect. 2.4.2 with some final remarks, which are perhaps more notational than substantive, yet may be quite illuminating in connection with the developments reported in subsequent sections.

The interpolational polynomials (5) are defined in terms of the N numbers x_n, hence they might be conveniently redefined as follows:

$$q_{N-1}^{(n)}(\underline{x}, y) = \prod_{m=1, m \neq n}^{N} \left[(y - x_m)/(x_n - x_m) \right] \ , \tag{18}$$

where of course the N x_n's are the N components of the N-vector $\underline{x}$. Then of course (6) reads

$$q_{N-1}^{(n)}(\underline{x}, x_m) = \delta_{nm} \tag{19}$$

and by setting in (7) $f(x) = x^{m-1}$ one gets

$$\sum_{n=1}^{N} q_{N-1}^{(n)}(\underline{x}, y)(x_n)^{m-1} = y^{m-1} \ , \tag{20}$$

a formula where, for notational convenience, we have replaced x with y, and which of course holds as usual for $m = 1, \ldots, N$.

In particular, if we now introduce the N arbitrary numbers y_n and set $y = y_n$ in (20) we get

$$(y_n)^{m-1} = \sum_{\ell=1}^{N} q_{N-1}^{(\ell)}(\underline{x}, y_n)(x_\ell)^{m-1} \ , \tag{21}$$

which can be then written as the following N N-vector equations:

$$\underline{y}^{(m-1)} = \underline{Q}(\underline{y}, \underline{x})\ \underline{x}^{(m-1)} \ , \tag{22}$$

via the definitions

$$\underline{x}^{(m-1)} = (x_1^{m-1}, x_2^{m-1}, \ldots, x_N^{m-1}) \; , \tag{23a}$$

$$\underline{y}^{(m-1)} = (y_1^{m-1}, y_2^{m-1}, \ldots, y_N^{m-1}) \; , \tag{23b}$$

$$[\; \underline{Q}(\underline{y},\underline{x}) \;]_{nm} = q^{(m)}(\underline{x}, y_n) \; , \tag{24a}$$

$$[\underline{Q}(\underline{y},\underline{x})]_{nm} = \prod_{\ell=1,\ell\neq m}^{N} [(y_n - x_\ell)/(x_m - x_\ell)] \; . \tag{24b}$$

Note that (22) entails

$$\underline{Q}(\underline{x},\underline{x}) = \underline{I} \; , \tag{25}$$

$$[\underline{Q}(\underline{x},\underline{y})]^{-1} = \underline{Q}(\underline{y},\underline{x}) \; , \tag{26}$$

$$\underline{Q}(\underline{x},\underline{y})\, \underline{Q}(\underline{y},\underline{z}) = \underline{Q}(\underline{x},\underline{z}) \; . \tag{27}$$

Finally note that (23) entail

$$\underline{x}^{(0)} = \underline{y}^{(0)} = \underline{u} \tag{28}$$

where $\underline{u}$ is the N-vector with unit components,

$$\underline{u} = (1,1,\ldots,1) \; , \tag{29}$$

as well as

$$\underline{x}^{(m-1)} = \underline{X}^{(m-1)} \underline{u} \; , \tag{30a}$$

$$\underline{y}^{(m-1)} = \underline{Y}^{(m-1)} \underline{u} \; , \tag{30b}$$

with the diagonal $(N \times N)$-matrices $\underline{X}$ and $\underline{Y}$ defined as follows:

$$X_{nm} = \delta_{nm} x_n , \qquad Y_{nm} = \delta_{nm} y_n \; . \tag{31}$$

Finally let us re-emphasize that all these formulas hold for an arbitrary choice of the $2N$ distinct numbers x_n and y_n (and, in (27), as well of the N numbers z_n), and for an arbitrary value of the indices, or exponents, $n,m,...$, provided they are in the range $1,2,...,N$ (this latter restriction is essential !).

Exercise 2.4.2-1. Prove the *identities*

$$\sum_{n=1}^{N} x_n^{l-1} \prod_{m=1,m\neq n}^{N} [(x-x_m)/(x_n-x_m)] = x^{l-1}, \; l=1,2,...,N. . \tag{32}$$

Hint: set $f(x)=x^{l-1}$ in (7); or see (20).

2.4.3 Algebraic approach

In Sect. 2.4.3 we revisit the results of the two preceding Sections, 2.4.1 and 2.4.2, in a more algebraic setting.

The main formula of this approach is the matrix relation

$$[\underline{D},\underline{X}] = \underline{I} - \underline{J} \ . \tag{1}$$

Here of course $\underline{D}$ and $\underline{X}$ are the $(N\times N)$-matrices defined, in terms of the arbitrary N distinct numbers x_n, by (2.4.1-2) and (2.4.1-1), while $\underline{I}$ is the $(N\times N)$-unit matrix,

$$I_{nm} = \delta_{nm} \ , \tag{2}$$

and $\underline{J}$ is the matrix all of whose elements equal unity,

$$J_{nm} = 1 \ . \tag{3}$$

Let us also introduce the matrix

$$\underline{P} = \underline{J}/N \ , \tag{4}$$

the N- vector $\underline{u}$ with unit components (see (2.4.1-17) and (2.4.2-29)),

$$u_n = 1 \ , \tag{5}$$

and N N-vectors $\underline{v}^{(m)}$,

$$\underline{v}^{(m)} = \underline{X}^{m-1}\underline{v} = \underline{X}^{m-1}\underline{B}^{-1}\underline{u}\ ,\ m = 1,2,\ldots,N\ , \tag{6a}$$

with $\underline{v} = \underline{B}^{-1}\underline{u}$ (see (2.4.1-16)), so that their components read as follows:

$$v_n^{(m)} = (x_n)^{m-1}/b_n = (x_n)^{m-1}\Big/\left[\prod_{\ell=1,\ell\neq n}^{N}(x_n - x_\ell)\right],\ m = 1,2,\ldots,N\ , \tag{6b}$$

see (2.4.1-4a) (note that this definition entails $\underline{v}^{(1)} \equiv \underline{v}$); and let us take note of the following formulas:

$$\underline{P}^2 = \underline{P}\ , \tag{7}$$

$$\underline{P}\cdot\underline{D} = \underline{J}\,\underline{D} = 0\ , \tag{8}$$

$$\underline{P}\,\underline{v}^{(n)} = \underline{J}\,\underline{v}^{(n)} = 0 \qquad n = 1,\ldots,N-1\ , \tag{9a}$$

$$\underline{J}\,\underline{v}^{(N)} = \underline{u}\ , \quad \underline{P}\,\underline{v}^{(N)} = \underline{u}/N\ , \tag{9b}$$

$$\underline{P}\,\underline{u} = \underline{u}\ , \tag{9c}$$

$$\underline{D}\,\underline{v}^{(n)} = (n-1)\,\underline{v}^{(n-1)}\ , \tag{10}$$

$$\underline{X}\,\underline{v}^{(n)} = \underline{v}^{(n+1)},\quad n = 1,\ldots,N-1\ . \tag{11}$$

Proofs and comments. The proof of (1) is plain: since $\underline{X}$ is diagonal, see (2.4.1-1), the commutator in the left hand side of (1) has no diagonal part, and this is obviously also true of the right hand side, see (2) and (3). As for the off-diagonal part of (1), it amounts via (2.4.1-1), (2.4.1-2), (2) and (3), to the identity $(x_n - x_m)^{-1}(x_m - x_n) = -1$. The proof of (7), from (4) and (3), is trivial; note that it qualifies $\underline{P}$ as a projector. The proof of (8) is also trivial; indeed, this formula coincides essentially with (2.4.1-5c), via (3) and (4).

Note that (9a) only hold for $n < N$ (see (9b)). It amounts via (3), (6) and (2.4.1-4a), to the sum rule (*identity*)

$$\sum_{n=1}^{N}(x_n)^{s-1}/[\prod_{m=1,m\neq n}^{N}(x_n - x_m)] = 0, \quad s = 1,\ldots,N-1\ . \tag{12}$$

Indeed, consider the polynomial of degree N in z, having the N x_n's as its zeros,

$$p_N(z) = \prod_{n=1}^{N}(z - x_n)\ , \tag{13}$$

and the $N-1$ meromorphic functions

$$\varphi_s(z) = z^{s-1} / p_N(z), \qquad s = 1, \ldots, N-1 \ , \tag{14}$$

whose (only) singularities in the complex z-plane are N simple poles at $z = x_n$, with residues ρ_n,

$$\rho_n = (x_n)^{s-1} / p'_N(x_n) . \tag{15}$$

Clearly (14) and (13) entail that, at large $|z|$,

$$\varphi_s(z) = z^{s-1-N}\left[1 + O\left(|z|^{-1}\right)\right] \quad . \tag{16}$$

Hence all these $N-1$ functions, see (14), vanish at least as $|z|^{-2}$ when the modulus $|z|$ of the complex variable z diverges. This entails

$$(2\pi i)^{-1} \int_C dz \, \varphi_s(z) = 0 \tag{17}$$

if C is a circle of diverging radius in the complex z-plane. But, by the residue theorem, see (15), this integral equals the sum in the left hand side of (12), since (13) entails

$$p'_N(x_n) = \prod_{m=1, m \neq n}^{N} (x_n - x_m) \quad . \tag{18}$$

Hence (17) entails (12) namely (9a), which is thereby proven.

To prove (9b) one repeats the previous argument, but now with $s = N$. Then, at large $|z|$, via (13),

$$\varphi_N(z) \equiv z^{N-1} / p_N(z) = z^{-1}\left[1 + O\left(|z|^{-1}\right)\right] \tag{19}$$

hence

$$(2\pi i)^{-1} \int_C dz \, \varphi_N(z) = 1 \ , \tag{20}$$

and, proceeding as above, one gets now the sum rule (*identity*)

$$\sum_{n=1}^{N} (x_n)^{N-1} / \left[\prod_{m=1, m \neq n}^{N} (x_n - x_m) \right] = 1 \ , \tag{21}$$

which, via (6), (2.1.4-4a) and (5), corresponds to (9b).

Exercise 2.4.3-1. Verify, by direct computation, the validity of (9a) and (9b) for $N=2$ and $N=3$.

The proof of (9c) is trivial, from (3), (4) and (5). Note that this equation, (9c), entails that $\underline{u}$ is the (right) eigenvector of $\underline{P}$ corresponding to its unit eigenvalue ($\underline{P}$ of course also has the eigenvalue 0, with multiplicity $N-1$). Hence the action of the projector $\underline{P}$ on any *N*-vector $\underline{w}$ is to yield the component of $\underline{w}$ in the direction of $\underline{u}$,

$$\underline{P}\,\underline{w} = (\underline{u} \cdot \underline{w})\,\underline{u} \ . \tag{22}$$

Note that (9b) entails that all the vectors $\underline{v}^{(n)}$ with $n < N$, see (6), are orthogonal to $\underline{u}$,

$$\underline{u} \cdot \underline{v}^{(n)} = 0, \qquad n = 1, \ldots, N-1 \ . \tag{23}$$

There remain to prove (10) and (11). The latter is a trivial consequence of the definitions of $\underline{X}$, see (2.4.1-1) and of $\underline{v}^{(n)}$, see (6). But note that this formula, (11), only holds for $n < N$; indeed $\underline{v}^{(N+1)}$ is not defined, since (6) only holds for $m = 1, \ldots, N$. Of course one might define $\underline{v}^{(m)}$ for arbitrary values of m via (6), and then (11) would trivially hold for arbitrary values of n; this would, however, not be the case for (10), see below, and it is therefore preferable to limit the definition of the vectors $\underline{v}^{(n)}$ to the *N* values $n = 1, \ldots, N$, since, in any case, there can only be *N* independent *N*-vectors (indeed, as we see below, the *N N*-vectors $\underline{v}^{(n)}$, $n = 1, \ldots, N$, provide generally a *complete basis* in the finite-dimensional vector space of *N*-vectors).

The simpler way to prove (10) is to note that it is merely a special case of (2.4.1-9) (which has been proven in the preceding Sect. 2.4.2).

Indeed for

$$f(x) = \varphi_n(x) = x^{n-1} \ , \tag{24a}$$

via (2.4.1-8a), (2.4.1-4b) and (6) one gets

$$\underline{B}^{-1}\,\underline{f} = \underline{v}^{(n)} \ , \tag{24b}$$

as well as (using additionally (2.4.1-8b) with $r=1$)

$$\underline{B}^{-1}\,\underline{f}^{(1)} = (n-1)\,\underline{v}^{(n-1)} \ . \tag{24c}$$

But then, multiplication of (2.4.1-9) (with $r=1$) by $\underline{B}^{-1}$ (from the left) yields, via (24b) and (24c), precisely (10), which is thereby proven. Note that the condition $n = 1, \ldots, N$, required for (10) to hold, corresponds precisely to the requirement that $f(x)$, see (24a), be a polynomial in x of degree less than *N*, which is essential for the validity of (2.4.1-9).

The two key formulas (10) and (11) correspond to the two elementary relations

$$(d/dx)\,\varphi_n(x) = (n-1)\,\varphi_{n-1}(x) \quad , \tag{25}$$

$$x\,\varphi_n(x) = \varphi_{n+1}(x) \quad , \tag{26}$$

valid for the set of functions (simple powers)

$$\varphi_n(x) = x^{n-1}, \qquad n = 1,2,\ldots \quad . \tag{27}$$

This underscores the correspondence among, on one side, the $(N \times N)$-matrices $\underline{D}$ respectively $\underline{X}$, and, on the other, the operators d/dx respectively x. Of course (25) and (26) hold for all positive integer values of n, while, as we have seen, in the finite-dimensional framework of the N-vector space spanned by the N N-vectors $\underline{v}^{(n)}$, see (6), (10) indeed holds for all values of n in the range from 1 to N, while (11) only holds for the first $N-1$ N-vectors $\underline{v}^{(n)}$ with $n = 1,\ldots,N-1$, because the vector $\underline{v}^{(N+1)}$ is not defined. Likewise, to the commutation relation

$$[d/dx, x] = 1 \quad , \tag{28}$$

there corresponds, see (1) and (4), the $(N \times N)$-matrix formula (see (1) and (4))

$$[\underline{D}, \underline{X}] = \underline{I} - N\,\underline{P} \quad , \tag{29}$$

with the projector operator $\underline{P}$, see (4), (3) and (7), characterized by the properties (9).

To complement the interpretation of the N N-vectors $\underline{v}^{(n)}$ as a basis in the finite-dimensional N-vector space, let us introduce the N N-vectors $\underline{u}^{(n)}$ orthonormal to the N N-vectors $\underline{v}^{(n)}$:

$$\underline{u}^{(n)} \cdot \underline{v}^{(m)} = \delta_{nm} \quad . \tag{30}$$

There hold then the following formulas:

$$\underline{u}^{(n)} = \sum_{m=n}^{N} \gamma_m\, \underline{X}^{m-n}\, \underline{u} \quad , \tag{31a}$$

$$\underline{u}^{(N)} = \underline{u} \quad , \tag{31b}$$

with $\underline{u}$ defined by (5) and the coefficients γ_m related to the x_n's via the formula (see (2.4.1-6b))

$$p_N(x) = \prod_{n=1}^{N} (x - x_n) = \sum_{m=0}^{N} \gamma_m \, x^m \quad . \tag{31c}$$

Note that (31c) entails

$$\gamma_N = 1 \quad , \tag{31d}$$

and that (31b) is merely the special case of (31a) with $n = N$, see (5).

Let us now introduce the $(N \times N)$-matrices $\underline{V}$ and $\underline{U}$ via the (standard) definitions

$$V_{nm} \equiv v_n^{(m)} = (x_n)^{m-1} / [\prod_{\ell=1, \ell \neq n}^{N} (x_n - x_\ell)] \quad , \tag{32}$$

$$U_{nm} \equiv u_m^{(n)} = \sum_{\ell=n}^{N} \gamma_\ell (x_m)^{\ell - n} \quad , \tag{33}$$

and let us then note the following formulas:

$$\underline{U}\underline{V} = \underline{V}\underline{U} = \underline{I} \quad , \tag{34}$$

$$\det[\underline{V}] = (-)^{N(N-1)/2} \prod_{n,m=1; n>m}^{N} (x_n - x_m)^{-1}, \tag{35}$$

$$\det[\underline{U}] = (-)^{N(N-1)/2} \prod_{n,m=1; n>m}^{N} (x_n - x_m), \tag{36}$$

$$\underline{u}^{(n)} \underline{D} = n \, \underline{u}^{(n+1)}, \qquad n = 1, \ldots, N-1 \quad , \tag{37a}$$

$$\underline{u}^{(N)} \underline{D} = \underline{u} \cdot \underline{D} = 0 \quad , \tag{37b}$$

$$\underline{u}^{(n)} \cdot \underline{X} = \underline{X} \cdot \underline{u}^{(n)} = \underline{u}^{(n-1)} - \gamma_{n-1} \underline{u}, \qquad n = 2, \ldots, N \quad , \tag{38a}$$

$$\underline{u}^{(1)} \cdot \underline{X} = \underline{X} \cdot \underline{u}^{(1)} = -\underline{u} \quad , \tag{38b}$$

$$(\underline{U}\,\underline{D}\,\underline{V})_{nm} = n\,\delta_{n,m-1} \;, \tag{39}$$

$$(\underline{U}\,\underline{X}\,\underline{V})_{nm} = \delta_{n,m+1}, \quad n = 1,\ldots,N, \quad m = 1,\ldots,N-1, \tag{40a}$$

$$(\underline{U}\,\underline{X}\,\underline{V})_{nN} = -\gamma_{n-1} \;. \tag{40b}$$

Proofs. The main result to be proven is the consistency of (31a) with (30), namely validity of the sum rule

$$\sum_{\ell=1}^{N} \sum_{s=n}^{N} \gamma_s (x_\ell)^{s-n+m-1} / [\prod_{j=1, j\neq\ell}^{N} (x_\ell - x_j)] = \delta_{nm} \;, \tag{41}$$

which corresponds to (30) via the definitions (31a) and (6) of $\underline{u}^{(n)}$ and $\underline{v}^{(m)}$. To this end we exchange the order of the two sums in the left hand side of (41), and use as main tool the *identity*

$$\sum_{\ell=1}^{N} x_\ell^p / [\prod_{j=1, j\neq\ell}^{N} (x_\ell - x_j)] = \delta_{p,N-1}, \qquad p = 0,\ldots,N-1 \;, \tag{42}$$

which coincides with (12) (for $p = 0,\ldots,N-2$) and (21) (for $p = N-1$). We note that the exponent of x_ℓ in (41) ranges from $m = 1$ to $N-1+m-n$, and consider firstly the case

$$n \geq m \;. \tag{43}$$

Then clearly (42) is applicable, and (41) becomes

$$\sum_{s=n}^{N} \gamma_s\, \delta_{s,N+n-m} = \delta_{nm} \;, \tag{44}$$

which is clearly satisfied (see (43) and (31d)). There remains to prove (41) for

$$n < m \;. \tag{45}$$

To this end we use the equality

$$\sum_{s=n}^{N} \gamma_s (x_\ell)^s = -\sum_{s=0}^{n-1} \gamma_s (x_\ell)^s \;, \tag{46a}$$

which is implied by the definition (31c), namely by the fact that x_ℓ is a zero of the polynomial $p_N(x)$,

$$p_N(x_\ell)=\sum_{s=0}^{N} \gamma_s (x_\ell)^s = 0 \ . \tag{46b}$$

Via (46a), the formula that remains to be proven reads

$$\sum_{s=0}^{n-1} \gamma_s \sum_{\ell=1}^{N} (x_\ell)^{s-n+m-1} / [\prod_{j=1, j\neq\ell}^{N} (x_\ell - x_j)] = 0 \ . \tag{47}$$

But this is implied by (42), since the exponent of x_ℓ in (47) ranges from the minimum value $p_{\min} = m-n-1$ (which is certainly in the range $0 \le p_{\min} \le N-2$, see (45)) to the maximum value $p_{\max} = m-2$ (which is also in the range $0 \le p_{\max} \le N-2$, see (45)). Hence (31) is proven.

The definitions (32) and (33) are implied by (6) and (31) with (2.4.1-1) and (5); and, via these definitions, (34) coincides with (30) (note that the commutativity of $\underline{V}$ and $\underline{U}$ is entailed by their being each others inverse).

The validity of (35) is implied, via (32), by the *Vandermonde identity* (2.4.2-4), together with the obvious *identity*

$$\prod_{n,m=1, n\neq m}^{N} (x_n - x_m) = (-)^{N(N-1)/2} [\prod_{n,m=1; n>m}^{N} (x_n - x_m)]^2 \ . \tag{48}$$

The validity of (36) follows from (34) and (35).

The validity of (37a) is implied by the *identity*

$$(\underline{u}^{(n)} \cdot \underline{D}) \cdot \underline{v}^{(m)} = \underline{u}^{(n)} \cdot (\underline{D} \cdot \underline{v}^{(m)}) \quad . \tag{49}$$

Indeed, via (10) and (30), its right hand side yields

$$\underline{u}^{(n)} \cdot (\underline{D} \cdot \underline{v}^{(m)}) = (m-1)\, \underline{u}^{(n)} \cdot \underline{v}^{(m-1)} = n\,\delta_{n,m-1} \quad . \tag{50a}$$

For $n = 1,\dots,N-1$ one can, using again (30), rewrite this formula as follows:

$$\underline{u}^{(n)} \cdot (\underline{D} \cdot \underline{v}^{(m)}) = n\, \underline{u}^{(n+1)} \cdot \underline{v}^{(m)} \quad . \tag{50b}$$

Hence, from (49) and (50b),

$$(\underline{u}^{(n)} \cdot \underline{D} - n\, \underline{u}^{(n+1)}) \cdot \underline{v}^{(m)} = 0, \quad n = 1,\dots,N-1 \ , \tag{51}$$

and this equation, which holds for the entire set of values of m, $m = 1,\dots,N$, entails (37a), which is thereby proven.

As for (37b), it coincides, via (31b) and (5), with (2.4.1-5c).

The proof of (38a) is analogous: for $m = 1,\dots,N-1$,

$$(\underline{u}^{(n)} \cdot \underline{X}) \cdot \underline{v}^{(m)} = \underline{u}^{(n)} \cdot (\underline{X} \cdot \underline{v}^{(m)}) = \underline{u}^{(n)} \cdot \underline{v}^{(m+1)} = \delta_{n,m+1} = \underline{u}^{(n-1)} \cdot \underline{v}^{(m)}, \tag{52a}$$

where the last step is only applicable for $n=2,...,N$. Hence

$$(\underline{u}^{(n)}\cdot\underline{X}-\underline{u}^{(n-1)})\cdot\underline{v}^{(m)}=0,\quad m=1,...,N-1\ . \tag{52b}$$

If an N-vector is orthogonal to all the vectors $\underline{v}^{(m)}$ except $\underline{v}^{(N)}$, it must, see (30), be proportional to $\underline{u}^{(N)}=\underline{u}$ (see (31b)). Hence

$$\underline{u}^{(n)}\cdot\underline{X}-\underline{u}^{(n-1)}=\tilde{\gamma}_{n-1}\,\underline{u},\quad n=2,...,N\ . \tag{53}$$

To compute the scalar $\tilde{\gamma}_{n-1}$, we take the scalar product of this equation with $\underline{u}$. This yields (see (5), (31a) and (2.4.1-1))

$$N\tilde{\gamma}_{n-1}=(\underline{u}^{(n)}\cdot\underline{X}-\underline{u}^{(n-1)})\cdot\underline{u}=\sum_{m=1}^{N}\left\{\sum_{s=n}^{N}\gamma_s\,x_m^{s-n+1}-\sum_{s=n-1}^{N}\gamma_s\,x_m^{s-n+1}\right\}=N\gamma_{n-1}. \tag{54}$$

Clearly (53) with (54) imply (38a), which is thereby proven.

The proof of (38b) is analogous, and is left as an *exercise* for the diligent reader.

Finally, (39) and (40) are immediate consequences of the definitions of the $(N\times N)$-matrices $\underline{V}$ and $\underline{U}$ in terms of the N-vectors $\underline{v}^{(n)}$ and $\underline{u}^{(n)}$, see (32) and (33), of (10) and (11) (or, likewise, of (37) and (38)), and of (30).

It is also of interest to display some formulas involving the "number" (or "counting") $(N\times N)$-matrix $\underline{N}$, defined as follows:

$$\underline{N}=\underline{X}\,\underline{D}\ , \tag{55a}$$

$$N_{nm}=\delta_{nm}\,x_n\sum_{\ell=1,\ell\neq n}^{N}(x_n-x_\ell)^{-1}+(1-\delta_{nm})\,x_n\,(x_n-x_m)^{-1}\ . \tag{55b}$$

They read:

$$\underline{N}\,\underline{v}^{(n)}=(n-1)\,\underline{v}^{(n)}\ , \tag{56}$$

$$\underline{u}^{(n)}\,\underline{N}=\underline{N}^T\,\underline{u}^{(n)}=(n-1)\,\underline{u}^{(n)}\ , \tag{57}$$

$$\underline{U}\,\underline{N}\,\underline{U}^{-1}=\underline{V}^{-1}\,\underline{N}\,\underline{V}=\underline{U}\,\underline{N}\,\underline{V}=\text{diag}(n-1;n=1,...,N)\ . \tag{58}$$

Proof. (55b) follows from the definition (55a) and (2.4.1-1,2). (56) and (57) are immediate consequences of the definition (55), and of (10) and (11). Likewise, (58) follows from (56) and (34). *Beware* of a notational awkwardness: $\underline{N}\,\underline{v}^{(n)}=(n-1)\,\underline{v}^{(n)}$, hence $\underline{N}\,\underline{v}^{(N)}=(N-1)\,\underline{v}^{(N)}$ (note: $\underline{N}\,\underline{v}^{(N)}\neq N\underline{v}^{(N)}$).

Note that (56), (57) and (58) entail that the $(N \times N)$-matrix $\underline{N}$ has the first nonnegative integers, $0,1,...,N-1$, as its eigenvalues. This is the counterpart of the property of the operator

$$\mathsf{N} = x(d/dx) \ , \tag{59}$$

to have the nonnegative integers as eigenvalues, and polynomials (indeed, simple powers) as eigenfunctions,

$$\mathsf{N} x^{n-1} = (n-1) x^{n-1}, \qquad n = 1,2,... \ . \tag{60}$$

Let us emphasize that this property of the $(N \times N)$-matrix $\underline{N}$ holds for any arbitrary choice of the N distinct numbers x_n. Hence any variation of these numbers corresponds to an *isospectral deformation* of this matrix, as displayed by the formula

$$\underline{N}(\underline{y}) = \underline{W}(\underline{y},\underline{x}) \ \underline{N}(\underline{x}) \ [\underline{W}(\underline{y},\underline{x})]^{-1} \ . \tag{61}$$

Here we have introduced the two N-vectors $\underline{x}$ respectively $\underline{y}$, of (arbitrary but distinct) components x_n respectively y_n, we have indicated by $\underline{N}(\underline{x})$ respectively $\underline{N}(\underline{y})$ the $(N \times N)$-matrices defined by (55) in terms of the (N components of the) N-vectors $\underline{x}$ respectively $\underline{y}$, and we have introduced the $(N \times N)$-matrix $\underline{W}(\underline{y},\underline{x})$ defined in terms of the two N-vectors $\underline{x}$ and $\underline{y}$ as follows:

$$\underline{W}(\underline{y},\underline{x}) = \underline{V}(\underline{y}) \ \underline{U}(\underline{x}) = \underline{V}(\underline{y}) \ [\underline{V}(\underline{x})]^{-1} = [\underline{U}(\underline{y})]^{-1} \ \underline{U}(\underline{x}) \ . \tag{62}$$

Let us also record the following formulas:

$$[\underline{W}(\underline{y},\underline{x})]_{nm} = [(y_n - x_n)/(y_n - x_m)] \prod_{\ell=1,\ell\neq n}^{N} [(y_n - x_\ell)/(y_n - y_\ell)] \ , \tag{63}$$

$$[\underline{W}(\underline{y},\underline{x})]^{-1} = [\underline{W}(\underline{x},\underline{y})] \ , \tag{64}$$

$$\underline{W}(\underline{x},\underline{x}) = \underline{I} \ , \tag{65}$$

$$\underline{W}(\underline{x},\underline{y}) \ \underline{W}(\underline{y},\underline{z}) = \underline{W}(\underline{x},\underline{z}) \ . \tag{66}$$

Proofs and comments. The eigenvalue equation (60) is a trivial consequence of the definition (59). Likewise, (61) is an immediate consequence of (62) and (58) (which ought to be rewritten with $\underline{U},\underline{N},\underline{V}$ replaced by $\underline{U}(\underline{x}),\underline{N}(\underline{x}),\underline{V}(\underline{x})$ respectively $\underline{U}(\underline{y}),\underline{N}(\underline{y}),\underline{V}(\underline{y})$).

The explicit expression (63) of the $(N\times N)$-matrix $\underline{W}(\underline{x},\underline{y})$ in terms of the (N-components of the) two N-vectors $\underline{x}$ and $\underline{y}$ is less trivial. To obtain it we note that (62), (32) and (33) entail

$$[\underline{W}(\underline{x},\underline{y})]_{nm}=\sum_{j=1}^{N}(y_n)^{j-1}[\prod_{\ell=1,\ell\neq n}^{N}(y_n-y_\ell)]^{-1}\sum_{k=j}^{N}\gamma_k(x_m)^{k-j}\ ,\tag{67a}$$

with the coefficients $\gamma_k\equiv\gamma_k(\underline{x})$ related to the N components of the N-vector $\underline{x}$ via (31c). We now exchange the order of the two sums:

$$[\underline{W}(\underline{x},\underline{y})]_{nm}=[\prod_{\ell=1,\ell\neq n}^{N}(y_n-y_\ell)]^{-1}\sum_{k=1}^{N}\gamma_k(x_m)^{k-1}\sum_{j=1}^{k}(y_n/x_m)^{j-1}\ ,\tag{67b}$$

$$[\underline{W}(\underline{x},\underline{y})]_{nm}=[\prod_{\ell=1,\ell\neq n}^{N}(y_n-y_\ell)]^{-1}\sum_{k=1}^{N}\gamma_k(x_m^k-y_n^k)/(x_m-y_n)\ ,\tag{67c}$$

$$[\underline{W}(\underline{x},\underline{y})]_{nm}=[(y_n-x_m)\prod_{\ell=1,\ell\neq n}^{N}(y_n-y_\ell)]^{-1}\sum_{k=0}^{N}\gamma_k(y_n^k-x_m^k)\ ,\tag{67d}$$

$$[\underline{W}(\underline{x},\underline{y})]_{nm}=[(y_n-x_m)\prod_{\ell=1,\ell\neq n}^{N}(y_n-y_\ell)]^{-1}\prod_{\ell=1}^{N}(y_n-x_l)\ .\tag{67e}$$

In the last step we used (31c) (with $x=y_n$, and with $x=x_m$ which yields a vanishing contribution). It is now clear that (67e) coincides with (63), which is thereby proven.

The other formulas, (64), (65) and (66), are immediate consequences of the definition (62); it is, however, far from trivial that they are satisfied by the $(N\times N)$-matrix whose elements are given by the explicit formula (63) in terms of the $2N$ arbitrary numbers x_n, y_n.

This formula, (63), as written, requires that all the numbers x_n and y_n be different; it also holds, if this requirement is violated, provided appropriate care is taken to treat the ratio of vanishing factors.

Two limiting cases of (61) deserve explicit display:

$$\partial\underline{N}(\underline{x})/\partial x_k=[\underline{N}(\underline{x}),\ \underline{M}^{(k)}(\underline{x})],\tag{68a}$$

with

$$M_{nm}^{(k)}(\underline{x})=(\delta_{nk}-\delta_{nm})(x_m-x_k)^{-1}=(\delta_{nk}-\delta_{nm})D_{mk}; \tag{68b}$$

$$\dot{\underline{N}}=[\underline{N},\underline{M}] , \tag{69a}$$

with

$$\underline{M}=\underline{V}\,\dot{\underline{U}}=-\dot{\underline{V}}\;\underline{U}=\text{diag}[\underline{\dot{x}}\cdot\underline{D}]-\dot{\underline{X}}\,\underline{D} . \tag{69b}$$

In (68), the $(N\times N)$-matrices $\underline{N},\underline{M}$ and $\underline{D}$ are constructed with the N numbers x_n, see (55), (68b) and (2.4.1-2). Likewise, in (69) the $(N\times N)$-matrices $\underline{N},\underline{M},\underline{V},\underline{U},\underline{D}$, are constructed with the N components x_n of the N-vector $\underline{x}$ (see (55), (69b), (32), (33) with (31c) and (2.4.1-2)) which is however now assumed to depend (in some arbitrary manner) on the parameter t, $\underline{x}\equiv\underline{x}(t)$; and the superimposed dot denotes of course differentiation with respect to this parameter. Note the analogy of (69a) with a Lax equation, see (2.1-2); this is elaborated upon in Sect. 2.4.5.3.

Proofs. To obtain (68) we set, in (63),

$$y_j=x_j+\delta_{jk}\,dx_j, \tag{70a}$$

where k is a fixed integer in the range from 1 to N, while j takes all values in that range, namely

$$y_1=x_1,\ y_2=x_2,\dots,\ y_{k-1}=x_{k-1},\ y_k=x_k+dx_k,\ y_{k+1}=x_{k+1},\dots,\ y_N=x_N. \tag{70b}$$

It is then easily seen that

$$W_{nm}(x_j+\delta_{jk}\,dx_k)=\delta_{nm}\left[\delta_{nk}-(1+\delta_{nk})(x_n-x_k)/(x_n-x_k-dx_k)\right]$$

$$+(1+\delta_{nm})\,\delta_{nk}\,dx_k/(x_k-dx_k-x_m) , \tag{71a}$$

hence, in the limit of infinitesimal dx_k,

$$W_{nm}(x_j+\delta_{jk}\,dx_k)=\delta_{nm}\left\{\delta_{nk}+(1+\delta_{nk})\left[1+dx_k/(x_n-x_k)\right]\right\},$$

$$+(1-\delta_{nm})\delta_{nk}\,dx_k/(x_k-x_m)+O[\,(dx_k)^2\,], \tag{71b}$$

$$W_{nm}(x_j+\delta_{jk}\,dx_k)=\delta_{nm}+dx_k(x_n-x_k)^{-1}[\delta_{nm}(1-\delta_{nk})-(1-\delta_{nm})\delta_{nk}]+O[\,(dx_k)^2\,],$$
(71c)

namely

$$\underline{W}(x_j+\delta_{jk}\,dx_k)=\underline{I}-dx_k\,\underline{M}^{(k)}\,(\underline{x})+O[\,(dx_k)^2\,], \tag{72a}$$

as well of course as

$$[\underline{W}\,(x_j+\delta_{jk}\,dx_k)]^{-1}=I+dx_k\underline{M}^{(k)}\,(\underline{x})+O[\,(dx_k)^2\,], \tag{72b}$$

with $\underline{M}^{(k)}\,(\underline{x})$ defined by (68b). Insertion of (72) into (61) yields clearly, in the limit of infinitesimal dx_k, (68), which is thereby proven (the second equality in (68b) is of course entailed by (2.4.1-2); note that only the off-diagonal terms of $\underline{D}$ enter in the right hand side of (68b), since for $m=k$ the factor $\delta_{nm}-\delta_{nk}$ vanishes).

To prove (69), we set in (61)

$$\underline{x}\equiv\underline{x}(t),\quad \underline{y}\equiv\underline{x}(t+d\,t)=\underline{x}(t)+\dot{\underline{x}}(t)\;d\,t\;. \tag{73}$$

We assume of course $d\,t$ to be infinitesimal, hence here and below we neglect contributions of order $(d\,t)^2$ or higher. It is then clear that, via (62), (61) yields (69a) with

$$\underline{M}=-\dot{\underline{V}}\,\underline{U}\;, \tag{74a}$$

or equivalently, see (34)

$$\underline{M}=\underline{V}\,\dot{\underline{U}}\;, \tag{74b}$$

which correspond to the first two equalities in (69b). To get also the last, we *t*-differentiate the definition (32) of $\underline{V}$, getting

$$\dot{V}_{nm}=(m-1)\,\dot{x}_n\,(x_n)^{m-1}/\,b_n-(\dot{b}_n/\,b_n)V_{nm}\;, \tag{75a}$$

$$\dot{V}_{nm}=\dot{x}_n(m-1)\;v_n^{(m-1)}-[\sum_{\ell=1,\ell\neq n}^{N}(\dot{x}_n-\dot{x}_m)(x_n-x_\ell)^{-1}]\,V_{nm}\;, \tag{75b}$$

$$\dot{V}_{nm}=\dot{x}_n[\;\underline{D}\underline{v}^{(m)}\;]_n-[\,\dot{\underline{x}}\cdot\underline{D}\,]_n\,V_{nm}\;, \tag{75c}$$

namely

$$\dot{\underline{V}}=\{\,\dot{\underline{X}}\,\underline{D}-\text{diag}[\,\dot{\underline{x}}\,\underline{D}\,]\}\,\underline{V}\;. \tag{75d}$$

To go from (75a) to (75b) we used (6) and (2.4.1-4a); from (75b) to (75c), we used (10) and (2.4.1-2); from (75c) to (75d), we used (2.4.1-1). Insertion of (75d) in (74a) yields, via (34), the last of the (69b), which is thereby proven.

The reader will ponder on the correspondence among (68) and (69). (*Hint:* set $d\underline{x}=\dot{\underline{x}}dt$).

Let us end Sect. 2.4.3 by noting that, while the results reported herein have been described and proven self-consistently, they can just as well (indeed, perhaps more easily) be obtained from the results reported in the preceding Sect. 2.4.2. A key formula to establish the connection is the relation

$$\underline{W}(\underline{y},\underline{x})=[\underline{B}(\underline{y})]^{-1}\underline{Q}(\underline{y},\underline{x})\underline{B}(\underline{x}) \quad , \tag{76}$$

see (63), (2.4.2-24) and the definition (2.4.1-4) of the $(N\times N)$-matrix $\underline{B}(\underline{x})$. We urge the diligent reader to purse in some detail the connection among the results reported in Sect. 2.4.3 and those reported in Sect. 2.4.2.

2.4.4 The finite-dimensional (matrix) algebra of raising and lowering operators, and its realizations

In Sect. 2.4.4 we revisit tersely the results of previous Sections, and in particular of the preceding Sect. 2.4.3, in the context of the well-known formalism of *raising* and *lowering* operators.

Let $\underline{v}^{(m)}$ and $\underline{u}^{(n)}$ be two orthonormal sets of *N*-vectors,

$$\underline{u}^{(m)}\cdot\underline{v}^{(n)}=\delta_{nm} \quad . \tag{1}$$

Note that, for the moment, we are not committed to any particular realizations of these *N*-vectors, although of course the main realization we have in mind, see below, is that given in the preceding Sect. 2.4.3, see (2.4.3-30).

Let us also introduce, as in the preceding Sect. 2.4.3 (see (2.4.3-32,33), the two $(N\times N)$-matrices $\underline{V}$ respectively $\underline{U}$, associated with these sets $\underline{v}^{(n)}$ respectively $\underline{u}^{(n)}$:

$$V_{nm}=v_n^{(m)} \quad , \tag{2a}$$

$$U_{nm}=u_m^{(n)} \quad , \tag{2b}$$

so that, see (1),

$$\underline{U}\,\underline{V}=\underline{V}\,\underline{U}=\underline{I} \;. \tag{3}$$

We then introduce the "lowering" $(N\times N)$-matrix $\underline{L}$, and the "raising" $(N\times N)$-matrix $\underline{R}$, via the following formulas:

$$\underline{L}\,\underline{v}^{(n)}=(n-1)\,\underline{v}^{(n-1)} \;, \tag{4a}$$

$$\underline{R}\,\underline{v}^{(n)}=(1-\delta_{nN})\,\underline{v}^{(n+1)} \;; \tag{4b}$$

$$\underline{u}^{(n)}\,\underline{L}=\underline{L}^{T}\,\underline{u}=n(1-\delta_{nN})\,\underline{u}^{(n+1)} \;, \tag{5a}$$

$$\underline{u}^{(n)}\,\underline{R}=\underline{R}^{T}\,\underline{u}^{(n)}=n(1-\delta_{n1})\,\underline{u}^{(n-1)} \;. \tag{5b}$$

The formulas (4) respectively (5) specify how the $(N\times N)$-matrices $\underline{L}$ and $\underline{R}$ act on the *complete* sets of vectors $\underline{v}^{(n)}$ respectively $\underline{u}^{(n)}$; hence they provide *complete* definitions of these $(N\times N)$-matrices. Note that $\underline{L}$ acts, *from the left*, as a *lowering* operator on the set of N-vectors $\underline{v}^{(n)}$, see (4a), but it acts, *from the right*, as a *raising* operator on the set of N-vectors $\underline{u}^{(n)}$, see (5a); hence its transpose $\underline{L}^{T}$ acts, *from the left*, as a *raising* operator on this set, see (5a). The converse is true for the raising operator $\underline{R}$ and its transpose $\underline{R}^{T}$, see (4b) and (5b). Of course, in the special case in which the two sets of orthonormal N-vectors coincide, $\underline{u}^{(n)}=\underline{v}^{(n)}$, there hold the (equivalent) relations $\underline{L}=\underline{R}^{T}$, $\underline{R}=\underline{L}^{T}$.

Exercise 2.4.4-1. Prove that (5) is consistent with (4). *Hint*: compute the scalar products $\underline{u}^{(m)}\cdot\underline{L}\cdot\underline{v}^{(n)}$ respectively $\underline{u}^{(m)}\cdot\underline{R}\cdot\underline{v}^{(n)}$ using (1), (4) and (5).

Clearly the $(N\times N)$-matrices $\underline{L}$ and $\underline{R}$ are given, in terms of the N-vectors $\underline{v}^{(n)}$ and $\underline{u}^{(n)}$, by the formulas

$$L_{nm}=\sum_{j=2}^{N}\,(j-1)v_n^{(j-1)}\,u_m^{(j)}=\sum_{j=2}^{N}\,(j-1)V_{n,j-1}U_{jm} \;, \tag{6a}$$

$$R_{nm}=\sum_{j=1}^{N-1}\,v_n^{(j+1)}\,u_m^{(j)}=\sum_{j=1}^{N}\,V_{n,j+1}U_{jm} \;, \tag{6b}$$

and they are nilpotent,

$$\underline{L}^N = \underline{R}^N = 0 \ . \tag{7}$$

It is also convenient to introduce the "counting" $(N \times N)$-matrix $\underline{N}$,

$$\underline{N} = \underline{R} \cdot \underline{L}, \tag{8}$$

as well as the "projector" $(N \times N)$-matrix $\underline{P}^{(N)}$,

$$P_{nm}^{(N)} = v_n^{(N)} u_m^{(N)}, \tag{9}$$

which, acting form the left respectively from the right, projects on the highest vectors, $\underline{v}^{(N)}$ respectively $\underline{u}^{(N)}$ (see (13a,b) below).

There hold then the following relations:

$$\underline{N}\, \underline{v}^{(n)} = (n-1)\, \underline{v}^{(n)}, \tag{10a}$$

$$\underline{u}^{(n)} \underline{N} = \underline{N}^T \underline{u}^{(n)} = (n-1)\, \underline{u}^{(n)}; \tag{10b}$$

$$\underline{N} = \underline{V} \cdot \text{diag}(n-1; n = 1,\ldots,N) \cdot \underline{U}, \tag{11a}$$

$$\underline{U}\, \underline{N}\, \underline{V} = \text{diag}(n-1; n = 1,\ldots,N) \ ; \tag{11b}$$

$$(\underline{U}\, \underline{L}\, \underline{V})_{nm} = n\, \delta_{n,m-1} \ , \tag{12a}$$

$$(\underline{U}\, \underline{R}\, \underline{V})_{nm} = \delta_{n,m+1} \ ; \tag{12b}$$

$$\underline{P}^{(N)} \underline{v}^{(n)} = \delta_{nN} \underline{v}^{(N)}, \tag{13a}$$

$$\underline{u}^{(n)} \underline{P}^{(N)} = \underline{P}^{(N)T} \underline{u}^{(n)} = \delta_{nN} \underline{u}^{(N)}; \tag{13b}$$

$$[\, \underline{P}^{(N)} \,]^2 = \underline{P}^{(N)}; \tag{13c}$$

$$\underline{P}^{(N)} \underline{L} = \underline{R}\, \underline{P}^{(N)} = 0, \tag{13d}$$

$$\underline{P}^{(N)} \underline{N} = \underline{N}\, \underline{P}^{(N)} = (N-1)\, \underline{P}^{(N)}; \tag{13e}$$

$$[\underline{L}, \underline{N}] = \underline{L}, \tag{14a}$$

$$[\underline{R}, \underline{N}] = -\underline{R}, \tag{14b}$$

$$\underline{L}\,\underline{R}=(\underline{N}+\underline{I})(\underline{I}-\underline{P}^{(N)})=\underline{I}+\underline{N}-N\,\underline{P}^{(N)}, \tag{14c}$$

$$[\underline{L},\underline{R}]=\underline{I}-\underline{N}\,\underline{P}^{(N)}\,. \tag{14d}$$

There moreover hold the following *Propositions.*

Proposition 2.4.4-2. The matrix $\underline{M}^{(L)}$,

$$\underline{M}^{(L)}=\underline{N}+\sum_{s=0}^{\infty}\sum_{p=1}^{N-1}c_{sp}\,\underline{L}^{p}\,\underline{N}^{s}, \tag{15a}$$

has the same eigenvalues as the matrix $\underline{N}$, namely the first N nonnegative integers:

$$\underline{M}^{(L)}\,\underline{v}^{(L)(n)}=(n-1)\,\underline{v}^{(L)(n)}, \tag{15b}$$

with

$$\underline{v}^{(L)(n)}=\underline{v}^{(n)}+\sum_{m=1}^{n-1}a_m^{(n)}\,\underline{v}^{(m)}/(m-1)!. \tag{15c}$$

The coefficients c_{sp} in (15a) are *arbitrary* (except for the requirement that the sums $\sum_{s=0}^{\infty}c_{sp}(N-1)^s$ converge), and the coefficients $a_m^{(n)}$ in (15c) are determined recursively by the *triangular* relations

$$(n-m)\,a_m^{(n)}=(n-1)!\sum_{s=0}^{\infty}c_{s,n-m}\,(n-1)^s$$

$$+\sum_{\ell=m+1}^{n-1}a_\ell^{(n)}\sum_{s=0}^{\infty}c_{s,\ell-m}\,(\ell-1)^s,\quad m=n-1,n-2,\ldots,1. \tag{15d}$$

Here, and always below, a sum vanishes if the lower limit exceeds the upper limit.

Proposition 2.4.4-3. The matrix $\underline{M}^{(R)}$,

$$\underline{M}^{(R)}=\underline{N}+\sum_{s=0}^{\infty}\sum_{p=1}^{N-1}c_{sp}\,\underline{R}^{p}\,\underline{N}^{s}, \tag{16a}$$

has the same eigenvalues as the matrix $\underline{N}$, namely the first N nonnegative integers:

$$\underline{M}^{(R)}\,\underline{v}^{(R)(n)}=(n-1)\,\underline{v}^{(R)(n)}, \tag{16b}$$

with

$$\underline{v}^{(R)(n)}=\underline{v}^{(n)}+\sum_{m=n+1}^{N} b_m^{(n)}\,\underline{v}^{(m)}. \tag{16c}$$

The coefficient c_{sp} in (16a) are again *arbitrary* (except for the requirement that the sums $\sum_{s=0}^{\infty}(N-1)^s c_{sp}$ converge), and the coefficients $b_m^{(n)}$ in (16c) are determined recursively by the *triangular* relations

$$(n-m)\,b_m^{(n)}=\sum_{s=0}^{\infty} c_{s,m-n}(n-1)^s+\sum_{\ell=n+1}^{m-1} b_\ell^{(n)}\sum_{s=0}^{\infty} c_{s,m-n}(\ell-1)^s,\quad m=n+1,n+2,\dots,N. \tag{16d}$$

Proposition 2.4.4-4. The eigenvalues μ of the generalized eigenvalue equation

$$\underline{M}^{(1)}(\theta)\,\underline{v}^{(\mu)}(\theta)=\mu\,\underline{M}^{(2)}(\theta)\,\underline{v}^{(\mu)}(\theta) \tag{17a}$$

are independent of θ, provided the two $(N\times N)$-matrices $\underline{M}^{(1)}(\theta)$ and $\underline{M}^{(2)}(\theta)$ are defined as follows:

$$\underline{M}^{(r)}(\theta)=\sum_{s=0}^{\infty}\sum_{p=0}^{N-1}\sum_{q=0}^{N-1} c_{spq}^{(r)}\exp[i(p-q)\theta]\,\underline{L}^p\,\underline{R}^q\,\underline{N}^s,\ r=1,2\ . \tag{17b}$$

Here the coefficients $c_{spq}^{(r)}$ are independent of θ, but otherwise arbitrary, except for the requirement that the sums $\sum_{s=0}^{\infty} c_{spq}^{(r)}(N-1)^s$ converge.

Proofs and comments. (10) follows from (8) and (4), (5). (11) follow from (10) and (2). (12) follows from (4), (5), (2) and (1). (13a, b, c) follow from (9) and (1); (13d) follows from (4a) and (13a), which entail $\underline{P}^{(N)}\,\underline{L}\,\underline{v}^{(n)}=0$ for $n=1,\dots,N$, and likewise (13e) from (5b) and (13a), which entail $\underline{u}^{(n)}\,\underline{R}\,\underline{P}^{(N)}=0$ for $n=1,\dots,N$. (14a,b) follow from (4a,b) and (10a); (14c) from (4b), (4a), (10a) and (13e); (14d) from (8) and (14c).

The *Propositions 2.4.4-2* and *2.4.4-3* are consequences of the "triangular" character of the operators $\underline{M}^{(L)}$ respectively $\underline{M}^{(R)}$, and the diligent reader will easily verify them, namely the validity of (15d) and (16d).

Finally *Proposition 2.4.4-4* is proven by setting in (17a),

$$\underline{v}^{(\mu)}(\theta)=\sum_{n=1}^{N} \beta_n^{(\mu)} \exp(i\,n\,\theta)\ \underline{v}^{(n)} \ , \tag{18}$$

and then noting that, due to the structure (17b) of $\underline{M}^{(1)}(\theta)$ and $\underline{M}^{(2)}(\theta)$, insertion of (18) into (17a) yields for the coefficients $\beta_n^{(\mu)}$ θ-independent (linear) equations. Hence these coefficients are, as our notation indicates, themselves θ-independent, and it is likewise θ-independent the associated "secular equation" (i.e., the condition that the determinant of the coefficients of these linear equations for the $\beta_n^{(\mu)}$ vanish, so that a nonvanishing determination of the coefficient $\beta_n^{(\mu)}$ exist). And it is of course this "secular equation" that determines the *N* eigenvalues μ of (17a), as the *N* roots of a polynomial in μ of degree *N*.

A *trivial* representation of the matrix algebra of lowering and raising operators reads as follows:

$$v_m^{(n)}=u_m^{(n)}=\delta_{nm} , \tag{19a}$$

$$\underline{V}=\underline{U}=\underline{I} \ , \tag{19b}$$

$$L_{nm}=n\,\delta_{n,m-1} , \qquad R_{nm}=\delta_{n,m+1} \ , \tag{19c}$$

$$\underline{N}=\text{diag}(n-1;n=1,\ldots,N) \ . \tag{19d}$$

All other representations can be obtained from this one via similarity transformations (see (11) and (12)).

Less trivial is the representation, featuring N arbitrary parameters x_n, that is yielded by the results of the preceding Sect. 2.4.3 via the identification

$$\underline{L}=\underline{D} \ , \tag{20a}$$

$$\underline{R}=\underline{X}(\underline{I}-\underline{P}^{(N)}) \ , \tag{20b}$$

$$\underline{N}=\underline{R}\ \underline{L}=\underline{X}\,\underline{D} \ , \tag{20c}$$

with $\underline{D}$ and $\underline{X}$ defined by (2.4.1-2,1) and

$$\left[\underline{P}^{(N)}\right]_{nm}=v_n^{(N)}=x_n^{N-1}/\left[\prod_{\ell=1,\ell\neq n}^{N}(x_n-x_\ell)\right] \ , \tag{21}$$

consistently, via (9), with (2.4.3-6) and (2.4.3-31b,5). Let us emphasize that this entails validity of all the formulas written above, in Sect. 2.4.4, of course with these definitions, (20), of the $(N\times N)$-matrices $\underline{L}$, $\underline{R}$ and $\underline{N}$ as well as the definitions (2.4.3-6) respectively (2.4.3-31) of the N-vectors $\underline{v}^{(n)}$ respectively $\underline{u}^{(n)}$.

The diligent reader will ponder and verify, being aware of the difference among the two projectors, $\underline{P}$ (see (2.4.3-4,3)) and $\underline{P}^{(N)}$ (see (21)): although both annihilate (acting from the left) all N-vectors $\underline{v}^{(n)}$ with $n<N$,

$$\underline{P}\,\underline{v}^{(n)}=\underline{P}^{(N)}\,\underline{v}^{(n)}=0,\quad n=1,\dots,N-1 \ , \tag{22}$$

and both satisfy the equation characterizing a projector, see (2.4.3-7) and (13c), they generally act differently on the highest vector $\underline{v}^{(N)}$,

$$\underline{P}\,\underline{v}^{(N)}=N^{-1}\underline{u} \ , \tag{23}$$

$$\underline{P}^{(N)}\,\underline{v}^{(N)}=\underline{v}^{(N)} \tag{24}$$

(see (2.4.3-9b) and (13a)), as well as on the vector $\underline{u}^{(N)}=\underline{u}$ (see (2.4.3-5,31b)),

$$\underline{P}\,\underline{u}=\underline{u} \ , \tag{25a}$$

$$\underline{P}^{(N)}\underline{u}=N\,\underline{v}^{(N)} \tag{26}$$

(see (2.4.3-9c) and (21), (2.4.3-5)). Also note that, while clearly (see (21))

$$\underline{u}^{(n)}\,\underline{P}^{(N)}=\underline{P}^{(N)T}\,\underline{u}^{(n)}=\delta_{nN}\,\underline{u}^{(N)}=\delta_{nN}\,\underline{u} \ , \tag{27}$$

and of course (see (25a), and note that $\underline{P}^T=\underline{P}$, see (2.4.3-4,3))

$$\underline{u}\,\underline{P}=\underline{u} \ , \tag{25b}$$

there is no simple formula for $\underline{P}$ analogous to (27), namely displaying the effect of $\underline{P}$ acting from the right on $\underline{u}^{(n)}$ (rather than only on $\underline{u}^{(N)}=\underline{u}$, see (25b)). There are, however, two simple formulas relating $\underline{P}$ and $\underline{P}^{(N)}$:

$$\underline{P}^{(N)}\,\underline{P}=\underline{P}^{(N)} \ , \qquad \underline{P}\,\underline{P}^{(N)}=\underline{P} \ , \tag{28}$$

as the diligent reader will verify directly from their definitions, (2.4.3-4,3) and (21), using the identity (2.4.3-21). This last relation, (28), is indeed instrumental, as the diligent reader will check, to verify, via (20), the consistency of the commutation relations (14d) and (2.4.3-29).

The diligent reader will also verify the consistency of the two equalities in (20c), as well as the related identity

$$\underline{P}^{(N)}\,\underline{L}=0\ ,\tag{29}$$

which is in fact essentially identical, via (9), (2.4.3-31b,5) and (20a), to (2.4.1-5c).

Exercise 2.4.4-5. Is there a special choice of the N numbers x_n which entails $\underline{P}^{(N)}=\underline{P}$? *Hint:* see (2.4.5.1-2h).

Other convenient instances of $(N\times N)$-matrices $\underline{L}$ and $\underline{R}$ satisfying the algebra of lowering and raising operators can be associated with special choices of the N numbers x_n, for instance according to the following methodology.

Recall the standard relations satisfied by Hermite polynomials (see Appendix C),

$$\left[x(d/dx)-\frac{1}{2}(d/dx)^2\right]H_{n-1}(x)=(n-1)\,H_{n-1}(x)\ ,\tag{30a}$$

$$\frac{1}{2}(d/dx)H_{n-1}(x)=(n-1)\,H_{n-2}(x)\ ,\tag{30b}$$

$$\left[2x-(d/dx)\right]H_{n-1}(x)=H_{n+1}(x)\ ,\tag{30c}$$

and define the set of N N-vectors $\underline{v}^{(H)(n)}$ via the following rule:

$$\underline{v}^{(H)(n)}=H_{n-1}(\underline{X})\,\underline{v}\ ,\tag{31}$$

of course with $\underline{X}$ and $\underline{v}$ defined by (2.4.1-1) and (2.4.1-16).

Now apply *Proposition 2.4.1-2* and thereby get the following N-vector counterparts of (30):

$$(\underline{X}\,\underline{D}-\frac{1}{2}\underline{D}^2)\,\underline{v}^{(H)(n)}=(n-1)\,\underline{v}^{(H)(n)}\ ,\tag{32}$$

$$\frac{1}{2}\underline{D}\,\underline{v}^{(H)(n)}=(n-1)\,\underline{v}^{(H)(n-1)}\ ,\tag{33}$$

$$(2\underline{X}-\underline{D})\,\underline{v}^{(H)(n)}=\underline{v}^{(H)(n+1)}\quad,\quad n=1,...,N-1\ ,\tag{34a}$$

$$(2\underline{X}-\underline{D})\,\underline{v}^{(H)(N)}=H_N(\underline{X})\,\underline{v}\ .\tag{34b}$$

The relations (30) are of course valid for all positive integer values of n, $n=1,2,...$; the definition (31) applies instead only for $n=1,...,N$, and likewise (32) and (33) hold for $n=1,...,N$, as implied by *Proposition 2.4.1-2* and by (31). As for (34a), it only holds, as indicated, for $n=1,...,N-1$, and it is replaced, for $n=N$, by (34b) (consistently with *Proposition 2.4.1-2* and with the validity of the definition (31) for $n=1,...,N$).

As implied by the treatment of Sect. 2.4.1, all these formulas are valid for any arbitrary choice of the N distinct numbers x_n. Let us now assume that these N numbers coincide with the N zeros $x_n^{(H)}$ of the Hermite polynomial of order N,

$$x_n=x_n^{(H)}\ ,\tag{35a}$$

$$H_N(x_n^{(H)})=0\ .\tag{35b}$$

(Note that in Sect. 2.3.4.1 we denoted these numbers simply as z_n, see for instance (2.3.4.1-15)). Then the right hand side of (34b) vanishes, see (2.4.1-1) and (35b), hence (34) can be replaced by

$$(2\underline{X}-\underline{D})\,\underline{v}^{(H)(n)}=(1-\delta_{nN})\,\underline{v}^{(H)(n)}\ .\tag{36}$$

It is then clear, by comparing (33), (36) respectively (32) with (4a), (4b) respectively (10a), that the following identification becomes justified:

$$\underline{v}^{(n)}=\underline{v}^{(H)(n)}\ ,\tag{37a}$$

$$\underline{L}=\frac{1}{2}\underline{D}\ ,\tag{37b}$$

$$\underline{R}=(2\underline{X}-\underline{D})\ ,\tag{37c}$$

$$\underline{N}=\underline{X}\,\underline{D}-\frac{1}{2}\underline{D}^2\ .\tag{37d}$$

Note that, while we got (37d) by comparing (32) with (10a), this formula, (37d), is also consistent with (37b), (37c) and (8).

In conclusion, all the results given above, including in particular *Proposition 2.4.4-2, 2.4.4-3* and *2.4.4-4*, are now applicable with the assignments (37) of $\underline{v}^{(n)}$ (with (31)), of $\underline{L}$, of $\underline{R}$ and of $\underline{N}$, of course with $\underline{X}$ and $\underline{D}$ given by (2.4.1-1) and (2.4.1-2), the N numbers x_n being now uniquely identified (up to permutations) by the condition to coincide with the N zeros $x_n^{(N)}$ of the N-th Hermite polynomial $H_N(x)$, see (35). Recall that these N numbers satisfy the N nonlinear equations (see (2.3.4.1-13))

$$x_n^{(H)} = \sum_{m=1,m\neq n}^{N} \left[x_n^{(H)} - x_m^{(H)}\right]^{-1} , \tag{38a}$$

hence they entail, see (2.4.1-2) and (2.4.1-3),

$$d_n = x_n^{(H)} , \tag{38b}$$

$$D_{nm} = \delta_{nm} x_n^{(H)} + (1-\delta_{nm}) / \left[x_n^{(H)} - x_m^{(H)}\right], \tag{38c}$$

$$b_n = 2^{N-1} N H_{N-1}(x_n^{(H)}) , \tag{38d}$$

$$v_n^{(H)(m)} = 2^{N-1} N^{-1} H_{m-1}(x_n^{(H)}) / H_{N-1}(x_n^{(H)}) , \tag{38e}$$

$$L_{nm} = \frac{1}{2}\left\{\delta_{nm} x_n^{(H)} + (1-\delta_{nm})\left[x_n^{(H)} - x_m^{(H)}\right]^{-1}\right\} , \tag{38f}$$

$$R_{nm} = \delta_{nm} x_n^{(H)} - (1-\delta_{nm})\left[x_n^{(H)} - x_m^{(H)} \right]^{-1} , \tag{38g}$$

$$\underline{N} = (N-1)\, \underline{I} - \underline{M} , \tag{38h}$$

$$M_{nm} = \delta_{nm} \sum_{\ell=1,\ell\neq n}^{N} \left[x_n^{(H)} - x_\ell^{(H)}\right]^{-2} - (1-\delta_{nm}) \left[x_n^{(H)} - x_m^{(H)}\right]^{-2} . \tag{38i}$$

Proofs. (38b) respectively (38c) are merely (2.4.1-3) and (2.4.1-2) with (35a) and (38a). (38d) is implied by the following formulas:

$$H_N(x)=2^N\prod_{n=1}^{N}\left[x-x_n^{(H)}\right]=2^N p_N(x) \ , \tag{39}$$

$$H'_N(x)=2^N\sum_{n=1}^{N}\prod_{m=1,m\neq n}^{N}(x-x_m^{(H)})=2^N p'_N(x) \ , \tag{40a}$$

$$H'_N(x_n^{(H)})=2^N\prod_{m=1,m\neq n}^{N}\left[x_n^{(H)}-x_m^{(H)}\right]=2^N p'_N(x_n^{(H)})=2^N b_n \ , \tag{40b}$$

$$H'_N(x)=2N H_{N-1}(x) \ . \tag{40c}$$

The first of these, (39), is entailed by the normalization of Hermite polynomials and by (35b) (see also (2.4.1-6b)); (40a) obtains by differentiation from (39), and (40b) from (40a) (see also (2.4.1-6a)); (40c) is (30b) with $n=N+1$. And clearly (40b) and (40c) entail (38d), via the definition (2.4.1-4b) of b_n with (35a). (38e), (38f) respectively (38g) are immediate consequences of (31) with (2.4.1-1), (35), (2.4.1-16) and (38d), of (37b) with (38c), respectively of (37c) with (2.4.1-1) and (37c). Finally (38h) follows from (37d) and (2.4.1-5d) with (38b), (35a) and (38c), which entails

$$N_{nm}=\frac{1}{2}\delta_{nm}\left\{\left[x_n^{(H)}\right]^2+\sum_{m=1,m\neq n}^{N}\left[x_n^{(H)}-x_m^{(H)}\right]^{-2}\right\}+(1-\delta_{nm})\left[x_n^{(H)}-x_m^{(H)}\right]^{-2} . \tag{41}$$

This indeed yields (38h) with (38i) via the following additional formula,

$$\sum_{m=1,m\neq n}^{N}\left[x_n^{(H)}-x_m^{(H)}\right]^{-2}=\left\{2(N-1)-\left[x_n^{(H)}\right]^2\right\}/3 \ , \tag{42}$$

valid for the zeros of Hermite polynomials, as proven below (and reported in Appendix C). Clearly (38a) yields

$$\left[x_n^{(H)}\right]^2=\sum_{m,\ell=1;m\neq n,\ell\neq n}^{N}\left[x_n^{(H)}-x_m^{(H)}\right]^{-1}\left[x_n^{(H)}-x_\ell^{(H)}\right]^{-1} \ , \tag{43a}$$

$$\left[x_n^{(H)}\right]^2=\sum_{m=1,m\neq n}^{N}\left[x_n^{(H)}-x_m^{(H)}\right]^{-2}+\sum_{m,\ell=1;m\neq n,\ell\neq n,\ell\neq m}^{N}\left[x_n^{(H)}-x_m^{(H)}\right]^{-1}\left[x_n^{(H)}-x_\ell^{(H)}\right]^{-1} \ , \tag{43b}$$

$$\left[x_n^{(H)}\right]^2=\sum_{m=1,m\neq n}^{N}\left[x_n^{(H)}-x_m^{(H)}\right]^{-2}$$

$$+\sum_{m,\ell=1;m\neq n,\ell\neq n,\ell\neq m}^{N}\left[x_n^{(H)}-x_\ell^{(H)}\right]^{-1}\left\{\left[x_n^{(H)}-x_m^{(H)}\right]^{-1}-\left[x_n^{(H)}-x_\ell^{(H)}\right]^{-1}\right\} \ , \tag{43c}$$

$$\left[x_n^{(H)}\right]^2=3\sum_{m=1,m\neq n}^{N}\left[x_n^{(H)}-x_m^{(H)}\right]^{-2}+2\sum_{m=1,m\neq n}^{N}\left[x_n^{(H)}-x_m^{(H)}\right]^{-1}\sum_{\ell=1,\ell\neq m}^{N}\left[x_n^{(H)}-x_\ell^{(H)}\right]^{-1} \ , \tag{43d}$$

$$\left[x_n^{(H)}\right]^2 = 3\sum_{m=1,m\neq n}^{N}\left[x_n^{(H)}-x_m^{(H)}\right]^{-2} + 2\sum_{m=1,m\neq n}^{N}\left[x_n^{(H)}-x_m^{(H)}\right]^{-1} x_m^{(H)} ,\qquad (43e)$$

$$\left[x_n^{(H)}\right]^2 = 3\sum_{m=1,m\neq n}^{N}\left[x_n^{(H)}-x_m^{(H)}\right]^{-2} - 2\,(N-1) + 2\left[x_n^{(H)}\right]^2 .\qquad (43f)$$

The steps to get (43b) and (43c) are plain; to get (43d), we exchanged the dummy indices m and ℓ in the last term; to get (43e) we used (38a); to get (43f) we replaced, in the last term in (43e), $x_m^{(H)}$ with $(x_m^{(H)}-x_n^{(H)})+x_n^{(H)}$ and used again (38a). Since (43f) coincides with (42), our proof is completed.

Exercise 2.4.4-6. Obtain analogous results involving the zeros of other classical polynomials, for instance Laguerre or Jacobi instead of Hermite, or even some appropriate combinations of such polynomials. *Hint*: see <C81a>.

2.4.5 Remarkable matrices and identities

In Sect. 2.4.5, or rather in Sects. 2.4.5.1, 2.4.5.2, 2.4.5.3 and 2.4.5.4, we obtain and report a number of formulas which follow from the results of the preceding sections. These findings correspond to specific implementations of some of the relations written above, in which the relevant $(N\times N)$-matrices and N-vectors are expressed either in terms of N arbitrary numbers or in terms of N specific numbers (say, the roots of certain polynomials, see above and below).

There is clearly a large number of results of this kind that are implied by the findings detailed above. Our presentation below is rather terse; the results we report are a small subset of those which can be easily obtained by these techniques; the alert reader is welcome to derive others. We have selected for presentation, in the first place, those results which reproduce (and complete) findings that had previously emerged from the study of integrable many-body problems on the line (mainly from their behavior near their equilibrium configurations), as well as those which appeared particularly neat (an aesthetic criterion). These results, in addition to their intrinsic mathematical interest, are likely to have applicative relevance, for instance to test the accuracy of computer programs to invert or diagonalize matrices, as well as didactic uses, for instance to concoct exercises when teaching matrix algebra. This approach also yields many identities, mainly in the nature of explicit evaluation of sums, or their transformation into products. Again, only a representative selection of such results are presented below, and the alert reader will have no difficulty (indeed, some fun) in deriving others.

Some of these findings, as well as others gleaned from the literature, are also presented in user friendly form in Appendices C and D, in the hope that they will be eventually taken over and included in compilations of mathematical formulas such as, for instance, the standard reference text by I. S. Gradshteyn and I. M. Ryzhik, now edited by Alan Jeffrey, which has already made some progress in this direction (see Sect. 15.823 of <GRJ94>).

2.4.5.1 Matrices with known spectrum

In Sect. 2.4.5.1 we exemplify techniques to manufacture remarkable matrices with known spectrum, and we exhibit some such matrices (see also Sect. 2.4.5.4). Analogous results are reported, in user-friendly form (but without detailing their origin) in Appendix D.

Of course a technique to get such matrices is to start from a known diagonal matrix and then undiagonalize it via a similarity transformation. This, however, does not, as a rule, yield a *remarkable* matrix having a neat expression in terms of several parameters, be they arbitrary numbers or well-defined numbers characterized by some definite rule (for instance, to be the N roots of a classical polynomial, or to satisfy some other algebraic equation, see below).

The prototype of *remarkable matrices* having a known spectrum is given by the observation that the $(N\times N)$-matrix $\underline{N}$, explicitly defined in terms of N distinct, but otherwise *arbitrary*, numbers x_n by the neat rule (2.4.3-55b),

$$N_{nm}=\delta_{nm}\,x_n\sum_{\ell=1,\ell\neq n}^{N}(x_n-x_\ell)^{-1}+(1-\delta_{nm})\,x_n\,(x_n-x_m)^{-1}\ , \qquad (1)$$

has the first N nonnegative integers, $0,1,\dots,N-1$, as its eigenvalues, see (2.4.3-56).

A more general result of this kind is entailed by *Propositions 2.4.4-2* and *2.4.4-3* of Sect. 2.4.4, of course associated with the representation (2.4.4-20) of the $(N\times N)$-matrices $\underline{L}$, $\underline{R}$ and $\underline{D}$.

These results hold for any *arbitrary* choice of the N distinct numbers x_n. For *special* choices of these numbers, one obtains additional results. For instance, let

$$x_n=\exp(2i\pi n/N)\ , \qquad (2a)$$

entailing

$$d_n = \frac{1}{2}(N-1)\exp(-2i\pi n/N) \quad , \tag{2b}$$

$$D_{nm} = \exp(-2i\pi n/N)\left[\frac{1}{2}(N-1)\delta_{nm} + (1-\delta_{nm})\{1-\exp[2i\pi(m-n)/N]\}^{-1}\right], \tag{2c}$$

$$b_n = N\exp(-2i\pi n/N) \quad , \tag{2d}$$

$$v_n = N^{-1}\exp(2i\pi n/N) \quad , \tag{2e}$$

$$v_n^{(m)} = N^{-1}\exp(2i\pi m\, n/N) \quad , \tag{2f}$$

$$\underline{u}^{(m)} = \underline{X}^{N-m}\,\underline{u}, \quad u_n^{(m)} = \exp(-2i\pi mn/N) \quad , \tag{2g}$$

$$\underline{P}^{(N)} = \underline{P}, \ P_{nm}^{(N)} = N^{-1} \quad , \tag{2h}$$

$$\underline{L} = \underline{D}, \ L_{nm} = \exp(-2i\pi n/N)\left[\frac{1}{2}(N-1)\delta_{nm} + (1-\delta_{nm})\{1-\exp[2i\pi(m-n)/N]\}^{-1}\right] \tag{2i}$$

$$\underline{R} = \underline{X}\,(\underline{I}-\underline{P}) \quad , \qquad R_{nm} = -(1-\delta_{nm})\exp(2i\pi n/N) \quad , \tag{2l}$$

$$N_{nm} = \frac{1}{2}(N-1)\delta_{nm} + (1-\delta_{nm})\{1-\exp[2i\pi(m-n)/N]\}^{-1} \ . \tag{2m}$$

Proofs. The numbers (2a) are the N-th roots of unity, hence

$$p_N(x) = \prod_{n=1}^{N}(x-x_n) = x^N - 1 \ . \tag{2n}$$

This entails

$$p_N'(x) = N x^{N-1} \quad , \tag{2o}$$

hence, via (2.4.1-6a),

$$b_n = N\exp[2i\pi(N-1)n/N] \ , \tag{2p}$$

which coincides with (2d), that is thereby proven. (2e) is then entailed by (2.4.1-16b), (2f) by (2.4.3-6b) with (2a) and (2d), and (2h) by (2.4.4-21) with (2e) and (2.4.1-4,3).

We then note that (2.4.3-56) with (2.4.3-55), (2a) and (2f) entail the identity

$$\sum_{\ell=1,\ell\neq m}^{N} \left[\exp(2i\pi n/N)-\exp(2i\pi \ell/N)\right]^{-1}\left[\exp(2i\pi nm/N)+\exp(2i\pi n\ell/N)\right]$$

$$=(n-1)\,\exp\left[2i\pi m(n-1)/N\right] \ , \qquad (3)$$

which, for $n=N$, via (2a) and (2.4.1-3), yields (2b), that is thereby proven. Then (2c) is entailed by (2.4.1-2), (2b) and (2a), likewise (2m) is entailed by (2.4.3-55), (2b) and (2a), while (2i) is a simple copy of (2.4.4-20a), and (2l) is entailed by (2.4.4-20b), (2h) and (2a).

There finally remains to prove (2g). This is also easy, since a comparison of (2n) with (2.4.3-31c) entails that all the coefficients γ_m vanish except for $\gamma_0=-1$ and $\gamma_N=1$ (see (2.4.3-31d)), hence (2.4.3-31a) yields (2g), which is thereby proven.

We now set

$$\underline{A}=(N-1)\underline{I}-2\underline{X}^{-1}\,\underline{N}\,\underline{X} \qquad (4a)$$

and, from (2m) and (2a), we get for this $(N\times N)$-matrix the neat expression

$$A_{nm}=(1-\delta_{nm})\left\{1+i\,\mathrm{cotan}\left[(n-m)\pi/N\right]\right\} \ . \qquad (4b)$$

Proof. From (4a), (2m) and (2a) one gets

$$A_{nm}=-2(1-\delta_{nm})\,\exp\left[-2i\pi(n-m)/N\right]\left\{1-\exp\left[2i\pi(n-m)/N\right]\right\}^{-1} \ , \qquad (5a)$$

$$A_{nm}=-2(1-\delta_{nm})\,\exp\left[-i\pi(n-m)/N\right]\left\{\exp\left[i\pi(n-m)/N\right]-\exp\left[-i\pi(n.-m)/N\right]\right\}^{-1} \qquad (5b)$$

$$A_{nm}=i(1-\delta_{nm})\left\{\cos\left[\pi(n-m)/N\right]-i\sin\left[\pi(n-m)/\pi\right]\right\}/\sin\left[\pi(n.-m)/N\right] \ , \qquad (5c)$$

and this clearly entails (4b).

Since the matrix $\underline{N}$ has the N eigenvalues $0,1,\ldots,N-1$, it follows from (4a) that the matrix $\underline{A}$, defined by the neat formula (4b), has the N eigenvalues $-(N-1),-(N-3),\ldots,(N-3),(N-1)$. The diligent reader will write out the corresponding eigenvectors, and will also note the identities entailed by the insertion of the explicit expressions (2) in, say, the formulas from (2.4.4-10a) to (2.4.4-14d).

Two other $(N \times N)$-matrices which both possess the first N nonnegative integers, $n = 0,1,...,N-1$, as their eigenvalues, are defined in terms of the N zeros $x_n^{(H)}$ of the Hermite polynomial $H_N(x)$, see (2.4.4-35), as follows:

$$N_{nm} = \delta_{nm}\left[x_n^{(H)}\right]^2 + (1-\delta_{nm})x_n^{(H)} / \left[x_n^{(H)} - x_m^{(H)}\right] , \tag{6}$$

$$M_{nm} = \delta_{nm} \sum_{\ell=1, l\neq n}^{N} \left[x_n^{(H)} - x_\ell^{(H)}\right]^{-2} - (1-\delta_{nm})x_n^{(H)} / \left[x_n^{(H)} - x_m^{(H)}\right]^{-2} . \tag{7}$$

The matrix (6) coincides, via (2.4.4-38a), with (2.4.3-55); the corresponding eigenvalue equation is (2.4.3-56), with (2.4.3-6) and (2.4.4-35).

The matrix (7) coincides with (2.4.4-38i); its eigenvectors are easily obtainable from (2.4.4-38h), (2.4.4-38e) and (2.4.4-10a).

Both these matrices, (6) and (7), can be written in other, equivalent, forms, by using (2.4.4-38a) and (2.4.4-42).

Exercise 2.4.5.1-1. Obtain analogous results in terms of the zeros of other classical polynomials, for instance Laguerre and Jacobi, instead of Hermite. *Hint:* see <C81a>.

Exercise 2.4.5.1-2. Obtain more general results by taking advantage of the three *Propositions 2.4.4-2,3,4.*

2.4.5.2 Matrices with known inverse

In Sect. 2.4.5.2 we point out that the results reported above provide the possibility to manufacture remarkable matrices whose inverses can be exhibited in explicit form. Such matrices might be useful for didactic purposes or to test the numerical accuracy of computer routines to invert $(N \times N)$-matrices, a task which is far from trivial for large N. In this brief Sect. 2.4.5.2 we merely focus on the prototypical example given by the $(N \times N)$-matrix $\underline{W}(\underline{y},\underline{x})$, see (2.4.3-63). A variation on this same theme is provided in Sect. 2.4.5.4. Analogous results are displayed, in user friendly form (but without detailing their origin) in Appendix D.

The $(N \times N)$-matrix $\underline{W}(\underline{y},\underline{x})$, and its inverse, are defined by the formulas

$$[\ \underline{W}(\underline{y},\underline{x})\ ]_{nm} = [\delta_{nm} + (1-\delta_{nm})(y_n - x_n)/(y_n - x_m)]\, w_n(\underline{y},\underline{x}) \ , \tag{1a}$$

$$\{[\ \underline{W}(\underline{y},\underline{x})\ ]^{-1}\}_{nm} = [\underline{W}(\underline{x},\underline{y})]_{nm} = [\delta_{nm} + (1-\delta_{nm})(y_n - x_n)/(y_m - x_n)]\, w_n(\underline{x},\underline{y}), \tag{1b}$$

where

$$w_n(\underline{y},\underline{x}) = \prod_{m=1,m\neq n}^{N} [(y_n - x_m)/(y_n - y_m)] \ . \tag{1c}$$

Here the $2N$ numbers x_n, y_n are arbitrary.

For instance for $N = 2$

$$\underline{W}(\underline{y},\underline{x}) = \begin{pmatrix} \frac{y_1 - x_2}{y_1 - y_2} & \frac{y_1 - x_1}{y_1 - y_2} \\ \frac{y_2 - x_2}{y_2 - y_1} & \frac{y_2 - x_1}{y_2 - y_1} \end{pmatrix} = (y_1 - y_2)^{-1} \begin{pmatrix} y_1 - x_2 & y_1 - x_1 \\ x_2 - y_2 & x_1 - y_2 \end{pmatrix} \ , \tag{2a}$$

$$[\ \underline{W}(\underline{y},\underline{x})\ ]^{-1} = \begin{pmatrix} \frac{x_1 - y_2}{x_1 - x_2} & \frac{x_1 - y_1}{x_1 - x_2} \\ \frac{x_2 - y_2}{x_2 - x_1} & \frac{x_2 - y_1}{x_2 - x_1} \end{pmatrix} = (x_1 - x_2)^{-1} \begin{pmatrix} x_1 - y_2 & x_1 - y_1 \\ y_2 - x_2 & y_1 - x_2 \end{pmatrix} \ . \tag{2b}$$

Exercise 2.4.5.2-1. Verify that the product of (2a) and (2b) yields the unit matrix.

Exercise 2.4.5.2-2. Write explicitly the matrices $\underline{W}(\underline{x},\underline{y})$ and $\underline{W}(\underline{y},\underline{x})$ for $N = 3$, and verify that their product yields the unit matrix.

2.4.5.3 A remarkable matrix, and some related trigonometric identities

In Sect. 2.4.5.3 we display some formulas associated with a specific $(N \times N)$-matrix, which is merely another avatar of the $(N \times N)$-matrix discussed in the preceding Sect. 2.4.5.2 and features a known spectrum as well as a known inverse, and moreover entails several nontrivial trigonometric identities. These results are displayed, in user friendly form (but without detailing their origin) in Appendix D.

Let the $(N \times N)$-matrix $\underline{R}(\underline{\varphi},\underline{\theta})$ be defined by the neat formula

$$R_{nm}(\underline{\varphi},\underline{\theta}) = \prod_{\ell=1,\ell\neq m}^{N} [\sin(\varphi_n - \theta_\ell)/\sin(\theta_m - \theta_\ell)] \ , \tag{1}$$

the $(N\times N)$-matrix $\underline{B}(\underline{\theta},\alpha)$ be defined by the formulas

$$\underline{B}(\underline{\theta},\alpha)=\underline{R}(\underline{\theta}+\alpha,\underline{\theta}) \quad , \tag{2a}$$

$$B_{nm}(\underline{\theta},\alpha)=\prod_{\ell=1,\ell\neq m}^{N}\left[\sin(\theta_n-\theta_\ell+\alpha)/\sin(\theta_m-\theta_\ell)\right] \quad , \tag{2b}$$

and the N N-vectors $\underline{r}^{(m)}(\underline{\theta})$ be defined by the simple rule

$$r_n^{(m)}(\underline{\theta})=\exp\left[i\,(2m-N-1)\,\theta_n\right] \quad . \tag{3}$$

(*Beware*: do not confuse this matrix $\underline{R}(\underline{\varphi},\underline{\theta})$ with the "raising" matrix introduced in Sect. 2.4.4).

There hold then the following formulas:

$$\underline{R}(\underline{\theta},\underline{\theta})=\underline{I} \quad , \tag{4}$$

$$\underline{R}(\underline{\varphi},\underline{\eta})\,\underline{R}(\underline{\eta},\underline{\theta})=\underline{R}(\underline{\varphi},\underline{\theta}) \quad , \tag{5a}$$

$$\underline{R}(\underline{\varphi},\underline{\eta}^{(1)})\,\underline{R}(\underline{\eta}^{(1)},\underline{\eta}^{(2)})\,\underline{R}(\underline{\eta}^{(2)},\underline{\eta}^{(3)})\cdots\underline{R}(\underline{\eta}^{(p)},\underline{\theta})=\underline{R}(\underline{\varphi},\underline{\theta}),\quad p=1,2,\ldots \quad , \tag{5b}$$

$$\left[\ \underline{R}(\underline{\varphi},\underline{\theta})\ \right]^{-1}=\underline{R}(\underline{\theta},\underline{\varphi}) \quad , \tag{5c}$$

$$\underline{r}^{(m)}(\underline{\varphi})=\underline{R}(\underline{\varphi},\underline{\theta})\,\underline{r}^{(m)}(\underline{\theta}),\quad m=1,\ldots,N \quad ; \tag{6}$$

$$\underline{B}(\underline{\theta},0)=\underline{I} \quad , \tag{7}$$

$$\underline{B}(\underline{\theta},\alpha)\ \underline{B}(\underline{\theta},\beta)=\underline{B}(\underline{\theta},\beta)\ \underline{B}(\underline{\theta},\alpha)=\underline{B}(\underline{\theta},\alpha+\beta) \quad , \tag{8a}$$

$$\underline{B}(\underline{\theta},\alpha_1)\ \underline{B}(\underline{\theta},\alpha_2)\cdots\underline{B}(\underline{\theta},\alpha_p)=\underline{B}(\underline{\theta},\alpha_1+\alpha_2+\cdots+\alpha_p) \quad , \qquad p=1,2,\ldots, \tag{8b}$$

$$\left[\underline{B}(\underline{\theta},\alpha)\right]^{-1}=\underline{B}(\underline{\theta},-\alpha) \quad , \tag{8c}$$

$$\underline{B}(\underline{\theta},\alpha)\ \underline{r}^{(m)}(\underline{\theta})=\lambda_m(\alpha)\ \underline{r}^{(m)}(\underline{\theta}) \quad , \quad m=1,\ldots,N \quad , \tag{9a}$$

$$\lambda_m(\alpha)=\exp\left[i(2m-N-1)\,\alpha\right] \quad , \tag{9b}$$

$$\text{trace}\left[\underline{B}(\underline{\theta},\alpha)\right]=\sin(N\alpha)/\sin(\alpha) \quad , \tag{10a}$$

$$\text{trace}\left[\underline{B}(\underline{\theta},\alpha_1)\ \underline{B}(\underline{\theta},\alpha_2)\cdots\underline{B}(\underline{\theta},\alpha_p)\right]=\sin(N\sum_{s=1}^{p}\alpha_s)/\sin(\sum_{s=1}^{p}\alpha_s)\,,\quad p=1,2,\ldots, \tag{10b}$$

$$\det[\underline{B}(\underline{\theta},\alpha)]=1 \quad . \tag{11}$$

The eigenvalue equation (9) entails that the N eigenvalues $\lambda_n(\alpha)$ of $\underline{B}(\underline{\theta},\alpha)$ are independent of $\underline{\theta}$, see (9b), while its N eigenvectors $\underline{r}^{(n)}(\underline{\theta})$ are independent of α, see (3). Hence any change of the vector $\underline{\theta}$ entails for $\underline{B}(\underline{\theta},\alpha)$ an *isospectral deformation*, as detailed by the following identities:

$$\underline{B}(\underline{\varphi},\alpha)=\underline{R}(\underline{\varphi},\underline{\theta})\underline{B}(\underline{\theta},\alpha)[\underline{R}(\underline{\varphi},\underline{\theta})]^{-1} \quad . \tag{12}$$

By writing out some of the matrix or vector equations written above one gets nontrivial *trigonometric identities*, such as:

$$\sum_{n=1}^{N}\prod_{m=1,m\neq n}^{N}[\sin(\theta_n-\theta_m+\alpha)/\sin(\theta_n-\theta_m)]=\sin(N\alpha)/\sin(\alpha), \tag{13a}$$

$$\prod_{s=1}^{p}\left\{\sum_{n_s=1}^{N}\prod_{m_s=1,m_s\neq n_s}^{N}[\sin(\theta_{n_{s-1}}-\theta_{m_s}+\alpha_s)/\sin(\theta_{n_s}-\theta_{m_s})]\right\}$$

$$=\sin(N\sum_{s=1}^{p}\alpha_s)/\sin(\sum_{s=1}^{p}\alpha_s)\ , \qquad n_0\equiv n_p, \qquad p=1,2,3,\ldots\ ; \tag{13b}$$

$$\sum_{m=1}^{N}\cos[s(\theta_n-\theta_m+\alpha)\prod_{\ell=1,\ell\neq m}^{N}[\sin(\theta_n-\theta_\ell+\alpha)/\sin(\theta_m-\theta_\ell)]=1 \ ,$$

$$s=N-1,\ N-3,\ N-5,\ldots,1 \text{ or } 0;\ n=1,\ldots,N, \tag{14a}$$

$$\sum_{m=1}^{N}\sin[s(\theta_n-\theta_m+\alpha)]\prod_{\ell=1,\ell\neq m}^{N}[\sin(\theta_n-\theta_\ell+\alpha)/\sin(\theta_m-\theta_\ell)]=0 \ ,$$

$$s=N-1,\ N-3,\ N-5,\ldots,1 \text{ or } 0;\ n=1,\ldots,N; \tag{14b}$$

$$\sum_{\ell=1}^{N}\left[\left\{\prod_{j=1,j\neq\ell}^{N}[\sin(\varphi_n-\theta_j)/\sin(\theta_\ell-\theta_j)]\right\}\left\{\prod_{k=1,k\neq m}^{N}[\sin(\theta_\ell-\varphi_k)/\sin(\varphi_m-\varphi_k)]\right\}\right]=\delta_{nm},$$

$$n=1,\ldots,N;\ m=1,\ldots,N \ ; \tag{15}$$

$$\sum_{\ell=1}^{N}\left[\left\{\prod_{j=1,j\neq\ell}^{N}[\sin(\varphi_n-\eta_j)/\sin(\eta_\ell-\eta_j)]\right\}\left\{\prod_{k=1,k\neq m}^{N}[\sin(\eta_\ell-\theta_k)/\sin(\varphi_n-\theta_k)]\right\}\right]=1\ ,$$

$$n=1,\ldots,N;\ m=1,\ldots,N \ . \tag{16}$$

The *identity* (13a) features the N arbitrary numbers θ_n; the (more general) *identities* (13b), of which there is one for each positive integer value of p, feature the N arbitrary numbers θ_n and the p arbitrary numbers α_s; the *identities* (14a) and (14b), of which there are altogether $N(N+1)$ if N is odd and N^2 if N is even, feature the $N+1$ arbitrary numbers θ_n and α; the N^2 *identities* (15) feature the $2N$ arbitrary numbers θ_n, φ_n; the N^2 *identities* (16) feature the $3N$ arbitrary numbers $\theta_n, \varphi_n, \eta_n$. While all these *identities*, as well as the other formulas written above, hold for any choice of the arbitrary numbers they feature, they may require an appropriate interpretation of ambiguous expressions (of type $0/0$) if some of these arbitrary numbers coincide.

Proofs. The $(N\times N)$-matrix $\underline{R}(\underline{\varphi},\underline{\theta})$ and the N-vectors $\underline{r}^{(m)}(\underline{\theta}), \underline{r}^{(m)}(\underline{\varphi})$ are related to the $(N\times N)$-matrix $\underline{Q}(\underline{y},\underline{x})$ and the N-vectors $\underline{x}^{(m-1)}$, $\underline{y}^{(m-1)}$, see (2.4.2-24) and (2.4.2-23), by setting

$$x_n = \exp(2i\theta_n), \quad y_n = \exp(2i\varphi_n) \; , \tag{17a}$$

which entail, as the diligent reader will readily verify, the simple relations

$$R(\underline{\varphi},\underline{\theta}) = \underline{Y}^{-(N-1)/2}\,\underline{Q}(\underline{y},\underline{x})\,\underline{X}^{(N-1)/2} \; , \tag{17b}$$

$$\underline{r}^{(m)}(\underline{\theta}) = \underline{X}^{-(N-1)/2}\,\underline{x}^{(m-1)}, \quad \underline{r}^{(m)}(\underline{\varphi}) = \underline{Y}^{-(N-1)/2}\,\underline{y}^{(m-1)} \; , \tag{17c}$$

with the diagonal $(N\times N)$-matrices $\underline{X}$ and $\underline{Y}$ defined by (2.4.2-31).

It is then immediately seen that (4), (5c) respectively (5a) correspond to (2.4.2-25), (2.4.2-26) respectively (2.4.2-27) (with $z_n = \exp(2i\eta_n)$, see (17a)), and that (6) corresponds to (2.4.2-22).

Then (5b) follows trivially from (5a); (7), (8) and (12) follows, via (2a), from (4) and (5); (9) follows from (6) and the identity

$$\underline{r}^{(m)}(\underline{\theta}+\alpha) = \lambda_m(\alpha)\,\underline{r}^{(m)}(\underline{\theta}) \tag{18}$$

entailed by the definitions (3) and (9b); (10) and (11) follow from (9).

As for the identities, they correspond merely, via (1), (2) and (3), to the explicit expressions of some of the matrix and vector equations written above: (13a,b) correspond to (10a,b), (14a,b) to (9a,b), (15) to (5c), (16) to (5a).

2.4.5.4 Matrices satisfying "fake" Lax equations

In Sect. 2.4.5.4 we mainly call attention to the identity (2.4.3-69), that we rewrite here in the "Lax form"

$$\dot{\underline{L}}=[\underline{L},\underline{M}] \tag{1}$$

with

$$L_{nm}=\delta_{nm}\,x_n\sum_{\ell=1,\ell\neq n}^{N}(x_n-x_\ell)^{-1}+(1-\delta_{nm})\,x_n\,(x_n-x_m)^{-1}\ , \tag{2}$$

$$M_{nm}=\delta_{nm}\sum_{\ell=1,\ell\neq n}^{N}\dot{x}_\ell\,(x_n-x_\ell)^{-1}-(1-\delta_{nm})\,\dot{x}_n\,(x_n-x_m)^{-1}\ . \tag{3}$$

Clearly here we assume the N arbitrary quantities x_n to depend (arbitrarily!) on a parameter t, $x_n\equiv x_n(t)$, and the superimposed dot denotes of course differentiation with respect to this parameter t. (*Beware*: do not confuse this matrix $\underline{L}$ with the "lowering" matrix of Sect. 2.4.4).

Let us recall that the $(N\times N)$-matrix $\underline{L}$, see (2) and (2.4.5.1-1), has the first N nonnegative integers as its eigenvalues; it is therefore automatically isospectral with respect to any change of the parameters x_n that define it. Hence it is not at all surprising that it satisfy a Lax equation, see (1): indeed any matrix satisfying a Lax equation, see (1), is isospectral (its eigenvalues do not depend on the parameter t), and if a matrix $\underline{L}$ is isospectral (namely, if its spectrum does not depend on a parameter t), then one can always associate to it a Lax equation, see (1).

Let us re-emphasize that (1) with (2) and (3) is an *identity*, hence certainly devoid of any "dynamical" content.

It is however easy to manufacture a Lax equation which is only satisfied if the quantities $x_n(t)$ evolve in some definite manner. Indeed consider the Lax equation (1) with (2) but with, instead of (3),

$$M_{nm}=\delta_{nm}\sum_{\ell=1,\ell\neq n}^{N}v_\ell(\underline{x})\,(x_n-x_\ell)^{-1}-(1-\delta_{nm})\,v_n(\underline{x})\,(x_n-x_m)^{-1}\ , \tag{4}$$

where the N-vector $\underline{v}(\underline{x})$ is some assigned N-vector-valued function of the N-vector $\underline{x}$ of components x_n. It is then clear, see (3) and (4), that validity of the Lax equation (1) with (2) and (4) corresponds to validity of the "equation of motion"

$$\underline{\dot{x}} = \underline{v}(\underline{x}) \quad . \tag{5}$$

The Lax equation has thereby acquired a dynamical content! But it would of course be silly to expect that the (completely arbitrary!) equations of motion (5) be *integrable*, because one can associate to them the Lax equation (1) with (2) and (4). Indeed the spectrum of $\underline{L}$ does provide N constant of the motion, but these are merely numbers (the first N nonnegative integers!), rather than nontrivial functions of the quantities x_n.

2.4.5.5 Determinantal representations of polynomials defined by ODEs or by recurrence relations

In Sect. 2.4.5.5 we outline techniques based on the results of previous Sections, whereby determinantal representations are exhibited of polynomials defined by ODEs or by recurrence relations.

The first type of results is exemplified by the following

Proposition 2.4.5.5-1. Let the polynomial $P_p(x)$, of degree $p \leq N$, satisfy the ODE

$$\mathsf{A}\, P_p(x) = 0 \tag{1}$$

with

$$\mathsf{A} = \sum_{r=0}^{p} a_r(x)\,(d/dx)^r \quad . \tag{2}$$

There holds then the formula

$$P_p(x)\; Q_{N-p}(x) = \det\left[(\underline{X} - x)\,\underline{A} + \underline{A}'\right] \quad , \tag{3}$$

with $Q_{N-p}(x)$ a polynomial of degree $N-p$ and the $(N \times N)$-matrices $\underline{A}$ and $\underline{A}'$ defined as follows:

$$\underline{A} = \sum_{r=0}^{p} a_r(\underline{X})\,\underline{D}^r \quad , \tag{4}$$

$$\underline{A}' = \sum_{r=1}^{p} r\, a_r(\underline{X})\,\underline{D}^{r-1} \quad , \tag{5}$$

where $\underline{X}$ and $\underline{D}$ are the $(N \times N)$-matrices defined by (2.4.1-1,2) in terms of the N arbitrary numbers x_n.

Corollary 2.4.5.5-2. Any polynomial $P_N(x)$, of degree N in x, that satisfies the linear ODE (1) with (2), admits the determinantal representation

$$P_N(x) = c_N \det[(\underline{X} - x)\underline{A} + \underline{A}'] \ , \tag{6}$$

with $\underline{X}$, $\underline{A}$ and $\underline{A}'$ defined as above.

Proofs. Let z be a zero of $P_p(x)$,

$$P_p(z) = 0 \ , \tag{7}$$

and define the *polynomial*, of degree $p-1$,

$$P_{p-1}(x) = P_p(x)/(x-z) \ , \tag{8a}$$

so that

$$P_p(x) = (x-z)P_{p-1}(x) \ , \tag{8b}$$

hence

$$(d/dx)^r P_p(x) = [\ (x-z)(d/dr)^r + r(d/dr)^{r-1}\ ]P_{p-1}(x) \ . \tag{8c}$$

To the ODE, (1) with (2), satisfied by $P_p(x)$, there therefore corresponds the following ODE satisfied by $P_{p-1}(x)$:

$$[(x-z)\mathsf{A} + \mathsf{A}']P_{p-1}(x) = 0 \ , \tag{9}$$

with A defined by (2) and

$$\mathsf{A}' = \sum_{r=1}^{p} r\, a_r(x)\, (d/dx)^{r-1} \ . \tag{10}$$

Hence, via *Corollary 2.4.1-3,* one concludes that

$$\det[(\underline{X} - z)\underline{A} + \underline{A}'] = 0 \ . \tag{11}$$

The left hand side of this equation is a polynomial of degree N in z, and we just proved that it has the property to vanish whenever z is a zero of the polynomial $P_p(x)$, see (7). This entails (3). The *Proposition 2.4.5.5-1* is thereby proven.

As for the *Corollary 2.4.5.5-2*, it is merely the special case of *Proposition 2.4.5.5-1* with $p = N$.

Let us now consider a set of polynomials $P_p(x)$ characterized by *recursion relations* of the following type:

$$P_{p+1}(x) = [\alpha(p)x + \beta(p)]\, P_p(x) - p\gamma(p) P_{p-1}(x) \ , \qquad p = 1,2,..., \tag{12a}$$

$$P_0(x) = 1 \ , \tag{12b}$$

$$P_1(x) = \alpha(0)\, x + \beta(0) \ . \tag{12c}$$

We moreover assume, for definiteness (but this is hardly relevant), that the coefficients $\alpha(p)$, $\beta(p)$ and $\gamma(p)$ are polynomials in p.

Our restriction here to *three-term recurrence relations of this type*, see (12), is motivated by the well-known fact that all sets of *orthogonal polynomials* satisfy (indeed, are characterized by) such recurrence relations. It is left for the diligent reader to extend the method outlined below to sets of polynomials characterized by more general recursion relations.

There holds then the following

Proposition 2.4.5.5-3. Let z be one of the N zeros of the polynomial of degree N, $P_N(x)$, determined by the recursion relations (12):

$$P_N(z) = 0 \ . \tag{13}$$

Then z is an eigenvalue of the generalized eigenvalue equation

$$[\underline{M}^{(1)}(\theta) - z\, \underline{M}^{(2)}]\, \underline{w}(\theta) = 0 \ , \tag{14}$$

where

$$\underline{M}^{(1)}(\theta) = \exp(-i\theta)\, \underline{L} - \beta(\underline{N}) + \exp(i\theta)\, \gamma(\underline{N})\, \underline{R} \ , \tag{15a}$$

$$\underline{M}^{(2)} = \alpha(\underline{N}) \ , \tag{15b}$$

and the three $(N \times N)$-matrices $\underline{L}$, $\underline{R}$, $\underline{N}$ satisfy the algebra of "lowering", "raising" and "counting" operators, see Sect. 2.4.4. The eigenvector $\underline{w}(\theta)$, see (14), is given by the formula

$$\underline{w}(\theta) = \sum_{n=1}^{N} \exp(i\,n\theta)\, P_{n-1}(z)\, \underline{v}^{(n)} / (n-1)! \ . \tag{16}$$

Here, above and below θ is an arbitrary parameter, and the N-vectors $\underline{v}^{(n)}$ are those that provide the basis for the action of $\underline{L}$, $\underline{R}$ and $\underline{N}$, see (2.4.4-4) and (2.4.4-1).

Proof. To prove *Proposition 2.4.5.5-3* it is sufficient to verify that the N-vector (16) satisfy the generalized eigenvalue equation (14) with the definitions (15) of $\underline{M}^{(1)}(\theta)$ and $\underline{M}^{(2)}$. Indeed, via (2.4.4-4) and (2.4.4-10a), the left hand side of (14) yields

$$\sum_{n=1}^{N} \exp(i\,n\theta)\, P_{n-1}(z) \left\{ \exp(-i\,\theta)\,(n-1)\,\underline{v}^{(n-1)} \right.$$

$$\left. -\beta(n-1)\underline{v}^{(n)} + \exp(i\theta)\,\gamma(n)\,(1-\delta_{nN})\,\underline{v}^{(n+1)} - z\,\alpha(n-1)\,\underline{v}^{(n)} \right\} / (n-1)!$$

$$= \sum_{n=1}^{N} \left(\left[\exp(i\,n\theta)\,\underline{v}^{(n)} / (n-1)!\right] \left\{ P_n(z) - \left[\beta(n-1) + z\,\alpha(n-1)\right] P_{n-1}(z) \right.\right.$$

$$\left.\left. + (n-1)\,\gamma(n-1)\,P_{n-2}(z) \right\}\right) - \left[\exp(i\,N\theta)\,v^{(N)} / (N-1)!\right] P_N(z) \ , \tag{17}$$

and this clearly vanishes thanks to (12) and (13).

Corollary 2.4.5.5-4. The polynomial $P_N(x)$, characterized by the recursion relations (12) (with $p = 1,\ldots,N-1$), admits the determinantal representation

$$P_N(x) = c_N \det\left[\underline{M}^{(1)}(\theta) - x\,\underline{M}^{(2)}\right] \ , \tag{18}$$

with the $(N \times N)$-matrices $\underline{M}^{(1)}(\theta)$ and $\underline{M}^{(2)}$ defined by (15) in terms of the "lowering", "raising", respectively "counting" $(N \times N)$-matrices $\underline{L}$, $\underline{R}$ respectively $\underline{N}$, and of the arbitrary "angle" θ.

Proof. It is analogous to that of *Proposition 2.4.5.5-1*: the right hand side of (18) is a polynomial of degree N in x, and *Proposition 2.4.5.5-3* (see in particular (14)), together with *Corollary 2.4.1-2*, entail that this polynomial vanishes whenever its argument, x, coincides with a zero, z, of the polynomial $P_N(x)$, see (13). Hence this polynomial, up to a (nonvanishing) multiplicative constant, coincides with $P_N(x)$, see (18).

To implement the *Propositions* and *Corollaries* given above *any* representation can be used for the $(N\times N)$-matrices $\underline{X}$ and $\underline{D}$, as well as $\underline{L}$, $\underline{R}$ and $\underline{N}$. In particular, one can use the representations of these $(N\times N)$-matrices in terms of N *arbitrary* numbers x_n, see (2.4.4-20) with (2.4.1-2,1) as well as the representations that correspond to special choices of these N numbers, see for instance (2.4.4-38).

Exercise 2.4.5.5-5. Obtain explicit representations for the classical polynomials (Hermite, Laguerre, Legendre, Gegenbauer, Jacobi), as implementations of *Corollaries 2.4.5.5-2* and *2.4.5.5-4*. *Hint*: see <C84b>, <C85a> and <C85d>.

Exercise 2.4.5.5-6. Prove the neat formula

$$H_N(x)=2^N\det\left[x-\underline{M}(\varphi)\right] \ , \tag{19a}$$

$$M_{nm}(\varphi)=\delta_{nm}\cos(\varphi)\,x_n^{(H)}+i(1-\delta_{nm})\sin(\varphi)\left[\,x_n^{(H)}-x_m^{(H)}\,\right]^{-1} \ . \tag{19b}$$

Here $x_n^{(H)}$ are the N zeros of the Hermite polynomial of order N, see (2.4.4-35b, 38a). *Hint*: see (18), and use the representation (2.4.4-38) of $\underline{L}$, $\underline{R}$ and $\underline{N}$. Note that this formula, (19), is trivial for $\varphi=0$, but nontrivial for $\varphi\neq 0 \mod(\pi)$: indeed the fact that the eigenvalues of the matrix $\underline{M}(\varphi)$ coincide, for all values of φ, with the N zeros $x_n^{(N)}$ of the Hermite polynomial of order N, is a nontrivial finding originally discovered as a by-product of the study of certain integrable many-body problems on the line, see *Exercise 2.1.3.3-5*.

2.5 Many-body problems on the line solvable via techniques of exact Lagrangian interpolation

The main idea underlying the approach pursued in Sect. 2.3 and in its subsections, in particular in Sects. 2.3.3 and 2.3.4, is based on the nonlinear mapping relating the N coefficients $c_n(t)$ of a (monic, time-dependent) polynomial of degree N in x, to its N zeros $x_n(t)$, see (2.3-1). In Sect. 2.5 we outline an extension of this approach, that follows naturally from the techniques of exact Lagrangian interpolation described in Sect. 2.4 and in its subsections. The new idea is to exploit the (nonlinear) mapping that relates the coefficients of a (monic, time-dependent) polynomial of degree N in x to the values that this polynomial takes at N points $x_n(t)$. The same kind of idea is exploited, in a more general context (nonpolynomial, multidimensional), in the next Chap. 3; hence the presentation given here is mainly an introduction (which indeed corresponds to the chronological unfolding of these developments) to the treatment given in the following Chap. 3.

For definiteness let us consider again a (monic, time-dependent) polynomial of degree N in x,

$$\psi(x,t)=x^N+\sum_{m=1}^{N} c_m(t)x^{N-m} \quad , \tag{1}$$

that satisfies the linear PDE (2.3-2). As noted above, the time evolution of this polynomial, namely the time evolution of its N coefficients $c_m(t)$, is *solvable* via purely algebraic operations; indeed, as we saw in Sects. 2.3.4.1 and 2.3.4.2, in some subcases of the linear PDE (2.3-2), characterized by the vanishing of some of the 11 constants it features, this time evolution can be exhibited in *explicit* form; otherwise to get it one must solve the system of N linear ODEs with constants coefficients (2.3.3-8), which amounts to the purely algebraic task of diagonalizing and inverting an $(N\times N)$-matrix.

Let now $x_n(t)$ indicate N (*a priori arbitrary*) points, and $f_n(t)$ the N values that the polynomial $\psi(x,t)$ takes at these N points $x_n(t)$:

$$f_n(t)=\psi\left[x_n(t),t\right] \quad . \tag{2}$$

It is then convenient to introduce the following polynomial:

$$f(x,t)=\psi(x,t)-\prod_{n=1}^{N}\left[x-x_n(t)\right] \quad . \tag{3}$$

It is clear that this polynomial in x is of degree *less* than N (see (1) and (3)), and that there holds for it the relation (see (2) and (3))

$$f[x_n(t),t]=f_n(t) \quad . \tag{4}$$

The fact that $f(x,t)$ is a polynomial of degree less than N entails the applicability (see (4) and (2.4.1.8a)) of the (exact) Lagrangian interpolation formalism of Sect. 2.4, which shall indeed be exploited below. In that formalism the choice of the N points x_n, as well as that of the N values f_n, is arbitrary. Here these quantities depend on t, $x_n \equiv x_n(t)$, $f_n \equiv f_n(t)$, but this of course is no impediment to the use of the Lagrangian interpolation formalism.

In the following, for notational simplicity, we often omit to indicate explicitly the t-dependence of x_n and f_n.

Let us emphasize the difference of the present approach from that of Sect. 2.3. Clearly the new treatment given herein reduces to the previous one for the special assignment

$$f_n(t)=0 . \tag{5}$$

The requirement that the polynomial (1) satisfy the linear PDE (2.3-2) entails N relations, among the $2N$ quantities x_n and f_n and their time-derivatives, that can be written in explicit, and fairly compact, form, as we show below. The structure of this system of N coupled ODEs looks as follows:

$$c\{\ddot{f}_n+\ddot{x}_n[M_n^{(1)}(\underline{x})+\sum_{m=1}^{N} M_{nm}^{(2)}(\underline{x})\, f_m]\}$$

$$+\sum_{m=1}^{N} \{M_{nm}^{(3)}(\underline{x})\, \dot{f}_m+[M_{nm}^{(4)}(\underline{x})+M_{nm}^{(5)}(\underline{x})\, f_m]\,\dot{x}_m\}$$

$$+\dot{x}_n\sum_{m=1}^{N} [M_{nm}^{(6)}(\underline{x})\,\dot{f}_m+M_{nm}^{(7)}(\underline{x})\,\dot{x}_m+\sum_{\ell=1}^{N} M_{nm\ell}^{(8)}(\underline{x})\, f_\ell\, \dot{x}_m]$$

$$+M_n^{(9)}(\underline{x})+\sum_{m=1}^{N} M_{nm}^{(10)}(\underline{x})\, f_m=0 \quad . \tag{6}$$

The 10 quantities M, variously decorated with superscripted and subscripted indices, are nonlinear functions of the N coordinates x_n, as indicated by our notation, see (6); their explicit form is given below (see

(16)), after we complete this preliminary outline of the new approach used in Sect. 2.5.

This system of N ODEs, (6), is of course insufficient to characterize the time evolution of the $2N$ quantities f_n, x_n; to determine this evolution completely, N additional relations among these $2N$ quantities must be posited. The choice of these relations remains our privilege; it opens the possibility to manufacture a large collection of dynamical systems. Only some of these are discussed below; the alert reader is welcome to consider other possibilities. Of course, as noted above, the particularly simple prescription (5) reproduces the class of models discussed in Sect. 2.3 and in its subsections.

According to the nature of these additional relations, introduced to completely determine the $2N$ quantities $f_n(t)$ and $x_n(t)$, one obtains a dynamical system characterized by equations of motion that are *solvable*, or one that features equations of motion which lack complete solvability, yet are *amenable to a treatment that significantly facilitates their study*. Let us outline two examples, one of each kind.

Consider firstly the assignment

$$f_n(t) = \varphi_n\left[\underline{x}(t)\right] \ , \tag{7}$$

where φ_n are N (arbitrarily) chosen functions of the N "particle coordinates" x_n. Insertion of (7) into (6) yields a system of N second-order ODEs, which can be easily solved for the "accelerations" $\ddot{x}_n$ (being linear in these quantities), so that it take the Newtonian form

$$\ddot{x}_n = F_n\left(\underline{x},\underline{\dot{x}}\right) \ . \tag{8}$$

In this manner one manufactures *solvable* N-body problems (generally featuring many-body forces, see below), the solution of which can indeed be achieved, via purely algebraic operations, by inserting (7) into (2).

A different technique to manufacture Newtonian equations of motion for the "particle coordinates" x_n sets $C = 0$ in (6), so that this become a system of *first-order* time-evolution ODEs for the N particle coordinates x_n. Then one recovers a set of *second-order* time-evolution ODEs, which can again be easily cast in the Newtonian form (8), for the "particle coordinates" x_n, by supplementing (6) with relations, say, of the following kind:

$$f_n(t) = \tilde{\varphi}_n\left[x_n(t),\, \dot{x}_n(t)\right] \ . \tag{9}$$

The (arbitrarily chosen) function $\tilde{\varphi}_n$ depends now on the "particle coordinate" x_n and also on the corresponding "velocity" $\dot{x}_n$. One has thereby manufactured again an N-body problem, see (8), the solution of which, while not being now reducible to a purely algebraic task, is nevertheless much simplified, compared to the task of solving the Newtonian equations (8) (a system of N nonlinear *coupled second-order* ODEs), since it is reduced, via the insertion of (9) into (2), to solving N *decoupled first-order* ODEs (however, these ODEs are generally neither linear nor autonomous, hence generally they are *not* integrable). Below we will also consider certain cases in which the function $\tilde{\varphi}_n$ in the right hand side of (9) is chosen to depend (appropriately!) on *all* the coordinates $x_m(t)$ rather than only on $x_n(t)$, $\tilde{\varphi}_n \equiv \tilde{\varphi}_n(\underline{x}(t), \dot{x}_n(t))$.

To summarize: the idea is again to consider N quantities, for instance the N coefficients $c_n(t)$, see (1) and (2.3-2), the time evolution of which is determined by easily *solvable* (linear) equations, and to then introduce via a *nonlinear mapping*, for instance that induced by (2) and (7) or (9) with the N functions φ_n or $\tilde{\varphi}_n$ assigned according to our choice, N "particle coordinates" $x_n(t)$, whose time-evolution is then *nonlinear* yet *solvable* or at least to some extent *treatable*. It is remarkable that, via this procedure, one obtains time-evolutions, for the N "particle coordinates" $x_n(t)$, which are characterized by a system of nonlinear ODEs that can be explicitly displayed and can be fairly naturally interpreted as the Newtonian equations of motion of an N-body problem.

Let us now implement the scheme we just outlined. To this end we need a more explicit version of (6). This requires a straightforward, if tedious, calculation, whose starting point are the N relations that obtain by evaluating the PDE (2.3-2) at the N points x_n:

$$C\psi_n(x_n,t) + [\, E-(N-1)D_2 x_n \,]\psi_t(x_n,t) + [\, D_0 + D_1 x_n + D_2 x_n^2 \,]\psi_{tt}(x_n,t)$$

$$+[\, A_0 + A_1 x_n + A_2 x_n^2 + A_3 x_n^3 \,]\psi_{xx}(x_n,t) + [\, B_0 + B_1 x_n - 2(N-1)A_3 x_n^2 \,]\psi_x(x_n,t)$$

$$-[\, N(N-1)(A_2 - A_3 x_n) + NB_1 \,]\psi(x_n,t) = 0 \quad . \tag{10}$$

The task is now to express all the quantities that appear in these N equations, whose significance we trust to be self-evident, in terms of the $2N$ quantities f_m, x_m and their time derivatives, taking as starting point the N relations (2), as well as the fact that the function $f(x,t)$, being a polynomial of degree less than N in x, can be expressed, together with its x-derivatives, in terms of the N values, f_n, it takes at the N points

x_n, see (2) and (4), via the (exact) formulas that emerge from the Lagrangian interpolation technique, as reviewed in the preceding Sect. 2.4.

The relevant formulas read as follows:

$$\psi(x_n,t)=f_n \quad , \tag{11a}$$

$$\psi_x(x_n,t)=b_n+f_n^{[1]} \quad , \tag{11b}$$

$$\psi_{xx}(x_n,t)=2b_n \sum_{m=1,m\neq n}^{N} (x_n-x_m)^{-1}+f_n^{[2]} \quad , \tag{11c}$$

$$\psi_t(x_n,t)=-b_n\,\dot{x}_n+\dot{f}_n-\dot{x}_n\,f_n^{[1]} \quad , \tag{11d}$$

$$\psi_{xt}(x_n,t)=-b_n \sum_{m=1,m\neq n}^{N} \left[(\dot{x}_n+\dot{x}_m)/(x_n-x_m)\right]+\dot{f}_n^{[1]}-\dot{x}_n\,f_n^{[2]} \quad , \tag{11e}$$

$$\psi_{tt}(x_n,t)=b_n\left[-\ddot{x}_n+2\sum_{m=1,m\neq n}^{N} \dot{x}_n\,\dot{x}_m/(x_n-x_m)\right]+\ddot{f}_n-\ddot{x}_n\,f_n^{[1]}-2\dot{x}_n\,\dot{f}_n^{[1]}+\dot{x}_n^2\,f_n^{[2]} \; . \tag{11f}$$

Here we have introduced the two quantities $f_n^{[1]}$ and $f_n^{[2]}$ via the convenient definitions

$$f_n^{[1]}=b_n\sum_{m=1}^{N} D_{nm}\,b_m^{-1}\,f_m \quad , \tag{12a}$$

$$f_n^{[2]}=b_n\sum_{m=1}^{N} (\underline{D}^2)_{nm}\,b_m^{-1}\,f_m \quad , \tag{12b}$$

where of course the N-vector $\underline{b}\equiv\underline{b}(\underline{x})$ and the $(N\times N)$-matrix $\underline{D}\equiv\underline{D}(\underline{x})$ are defined by (2.4.1-2,3,4):

$$b_n=\prod_{m=1,m\neq n}^{N} (x_n-x_m), \tag{13a}$$

$$D_{nm}=\delta_{nm}\sum_{\ell=1,\ell\neq n}^{N} (x_n-x_\ell)^{-1}+(1-\delta_{nm})(x_n-x_m)^{-1} \; . \tag{13b}$$

Proofs. (11a) coincides with (2). To prove (11b) we x-differentiate (3), getting

$$\psi_x(x,t)=f_x(x,t)+\sum_{n=1}^{N}\prod_{m=1,m\neq n}^{N}(x-x_m)\quad ; \tag{14a}$$

we then set $x=x_n$, and use (2.4.1-9,8b) (with $r=1$) and (2.4.1-4a) or (13a).

We then note that $\psi_x(x,t)$ is a polynomial of degree less than N, hence, via (2.4.1-9) with $r=1$,

$$\psi_{xx}(x_n,t)=b_n\sum_{m=1}^{N}D_{nm}\,b_m^{-1}\,\psi_x(x_m,t)\quad . \tag{14b}$$

We then insert (11b) in this formula, and use (2.4.1-5b), getting thereby (11c).

We then t-differentiate (11a), getting thereby

$$\psi_t(x_n,t)+\dot{x}_n\,\psi_x(x_n,t)=f_n\quad ; \tag{14c}$$

via (11b) this yields (11d).

Likewise, t-differentiation of (11b) yields

$$\psi_{xt}(x_n,t)+\dot{x}_n\,\psi_{xx}(x_n,t)=\dot{b}_n+\dot{f}_n^{[1]}\quad . \tag{14d}$$

We now use (11c), as well as the formula

$$\dot{b}_n=b_n\sum_{m=1,m\neq n}^{N}(\dot{x}_n-\dot{x}_m)/(x_n-x_m)\quad , \tag{15}$$

which is clearly implied (via logarithmic t-differentiation) by (13a). There thus obtains (11e), which is thereby proven.

Finally, t-differentiation of (11d) yields

$$\psi_{tt}(x_n,t)+\dot{x}_n\,\psi_{xt}(x_n,t)=\ddot{f}_n-\ddot{x}_n\left[b_n+f_n^{[1]}\right]-\dot{x}_n\left[\dot{b}_n+\dot{f}_n^{[1]}\right]\quad . \tag{14e}$$

Via (11e) and (15) this yields (11f), which is thereby proven.

Insertion of (11) into (10) yields the following formula, which provides the more explicit realization of (6) we need:

$$b_n\{C[-\ddot{x}_n+2\sum_{m=1,m\neq n}^{N}\dot{x}_n\,\dot{x}_m/(x_n-x_m)]-E\,\dot{x}_n+B_0+B_1\,x_n-2(N-1)\,A_3\,x_n^2$$

$$+\sum_{m=1,m\neq n}^{N}(x_n-x_m)^{-1}\left[-(\dot{x}_n+\dot{x}_m)(D_0+D_1\,x_n)-D_2\,x_n\,(\dot{x}_n\,x_m+\dot{x}_m\,x_n)\right.$$

$$\left.+2(A_0+A_1\,x_n+A_2\,x_n^2+A_3\,x_n^3)\,\right]\}$$

$$+C\left[\ddot{f}_n+\ddot{x}_n\,f_n^{[1]}-2\dot{x}_n\,\dot{f}_n^{[1]}+\dot{x}_n^2\,f_n^{[2]}\right]+\left[E-(N-1)D_2x_n\right]\left[\dot{f}_n-\dot{x}_n\,f_n^{[1]}\right]$$

$$+\left[D_0+D_1\,x_n+D_2x_n^2\right]\left[\dot{f}_n^{[1]}-\dot{x}_n\,f_n^{[2]}\right]+\left[B_0+B_1\,x_n-2(N-1)\,A_3x_n^2\right]f_n^{[1]}$$

$$+\left[A_0+A_1\,x_n+A_2\,x_n^2+A_3\,x_n^3\right]f_n^{[2]}-\left[N(N-1)\,(A_2-A_3\,x_n)+N\,B_1\right]f_n=0\quad,\qquad(16)$$

where $f_n^{[1]}$ and $f_n^{[2]}$ are of course defined by (12), b_n by (13a), and we omit to indicate explicitly the time-dependence.

As explained above, this system of N coupled ODEs for the $2N$ quantities x_n, f_n must now be supplemented by N additional relations.

Let us consider firstly relations of type (7), that yield *solvable* models. In particular let us set

$$f_n(t)=h\left[x_n(t),\,t\right]\quad,\qquad(17a)$$

where $h(x,t)$ is a (possibly time-dependent) polynomial in x, of degree *less* than N but otherwise *arbitrary*. Note that this entails, via (12) and (2.4.1-9) (which is now applicable),

$$f_n^{[1]}=h_x\,(x_n,t)\quad,\qquad(17b)$$

$$f_n^{[2]}=h_{xx}\,(x_n,t)\quad,\qquad(17c)$$

while (17a) entails (indeed, independently of any restriction on $h(x,t)$, other than its differentiability)

$$\dot{f}_n=h_t\,(x_n,t)+\dot{x}_n\,h_x\,(x_n,t)\quad,\qquad(17d)$$

$$\ddot{f}_n=h_{tt}\,(x_n,t)+2\,\dot{x}_n\,h_{xt}\,(x_n,t)+\ddot{x}_n\,h_x\,(x_n,t)+\dot{x}_n^2\,h_{xx}\,(x_n,t)\quad,\qquad(17e)$$

and (17b) entails

$$\dot{f}_n^{[1]}=h_{xt}\,(x_n,t)+\dot{x}_n\,h_{xx}\,(x_n,t)\quad.\qquad(17f)$$

Insertion of these expressions in (16) yields, after several cancellations and via (13a), the following "equations of motion":

$$C\ddot{x}_n + E\dot{x}_n = B_0 + B_1 x_n - 2(N-1)A_3 x_n^2 + \sum_{m=1,m\neq n}^{N} (x_n - x_m)^{-1} \cdot$$

$$\cdot [\ 2C\, \dot{x}_n \dot{x}_m - (\dot{x}_n + \dot{x}_m)(D_0 + D_1 x_n) - D_2 x_n (\dot{x}_n x_m + \dot{x}_m x_n)$$

$$+2\,(A_0 + A_1 x_n + A_2 x_n^2 + A_3 x_n^3)\ ]$$

$$+\{C h_{tt}(x_n,t) + [\ E - (N-1)D_2 x_n\ ]h_t(x,t) + [\ D_0 + D_1 x_n + D_2 x_n^2\ ]h_{xt}(x_n,t)$$

$$+[\ A_0 + A_1 x_n + A_2 x_n^2 + A_3 x_n^3\ ]h_{xx}(x_n,t) + [\ B_0 + B_1 x_n - 2(N-1)A_3 x_n^2\ ]h_x(x_n,t)$$

$$-[\ N(N-1)(A_2 - A_3 x_n) + N B_1\ ]h(x_n,t)\} \prod_{m=1,m\neq n}^{N} (x_n - x_m)^{-1} \quad . \tag{18}$$

These equations of motion are of Newtonian type, see (8); the product $\prod_{m=1,m\neq n}^{N} (x_n - x_m)^{-1} = b_n^{-1}$ (see (13a)) which multiplies the curly bracket in the right hand side indicates however the presence of "many-body forces". On the other hand the presence of the *arbitrary* polynomial $h(x,t)$, which is only restricted by the requirement to be of degree less than N, entails a significant generality.

Of course for $h(x,t) = 0$ this system, (18), reduces to (2.3.3-2).

The solution $x_n(t)$ of these Newtonian equations of motion are the N roots of the polynomial equation, of degree N in x,

$$\psi(x,t) = h(x,t) \tag{19a}$$

namely

$$\psi[x_n(t),t] = h[x_n(t),t] \tag{19b}$$

where $\psi(x,t)$ is the monic polynomial, of degree N in x, determined by the linear PDE (2.3-1) (which features of course the same 11 constants, $A_0, A_1, A_2, A_3, B_0, B_1, C, D_0, D_1, D_2, E$ that appear in (18)), with the initial conditions, $\psi(x,0)$ and $\psi_t(x,0)$, entailed by (19b) (note that (19a) only holds for $x = x_n(t)$).

Let us indicate how to obtain from (19b) the initial data $\psi(x,0)$ and $\psi_t(x,0)$.

First of all we note that (19b), evaluated at $t=0$, yields

$$\psi[x_n(0),0]=h[x_n(0),0], \tag{20a}$$

and its t-derivative, evaluated at $t=0$, yields

$$\psi_t[x_n(0),0]=-\dot{x}_n(0)\,\psi_x[x_n(0),0]+h_t[x_n(0),0]+\dot{x}_n(0)\,h_x[x_n(0),0]\;. \tag{20b}$$

There are now two (equivalent) routes to determine the (monic, N-th degree) polynomial $\psi(x,0)$. We can focus on the polynomial $\psi(x,0)-x^N$, the degree of which is clearly less than N, and which clearly takes the N values $\psi[x_n(0),0]-[x_n(0)]^N$ at the N nodes $x_n(0)$. Hence, by applying the standard theory of (exact, polynomial) Lagrangian interpolation of Sect. 2.4 to this polynomial we get (via (2.4.2-7,5) and (20a))

$$\psi(x,0)=x^N+\sum_{n=1}^{N}\{h[x_n(0),0]-[x_n(0)]^N\}\prod_{m=1,m\neq n}^{N}\{[x-x_m(0)]/[x_n(0)-x_m(0)]\}. \tag{21a}$$

Alternatively, we can focus on the polynomial $\psi(x,0)-\prod_{n=1}^{N}[x-x_n(0)]$, also of degree less than N. Then via (2.4.2-7,5) and (20a) one gets

$$\psi(x,0)=\prod_{n=1}^{N}[x-x_n(0)]+\sum_{n=1}^{N}h[x_n(0),0]\prod_{m=1,m\neq n}^{N}\{[x-x_m(0)]/[x_n(0)-x_m(0)]\}\;. \tag{21b}$$

The two equivalent formulas (21a) and (21b) provide explicit expressions of the monic polynomial (of degree N) $\psi(x,0)$ in terms of the initial data $x_n(0)$ of the many-body problem (18).

Exercise 2.5-1. Verify the equivalence of (21b) with (21a). *Hint*: note that the Lagrangian interpolation formula (2.4.2-7), applied to the polynomial (of degree less than N) $\prod_{n=1}^{N}(x-x_n)-x^N$, yields the polynomial *identity*

$$\prod_{n=1}^{N}(x-x_n)=x^N-\sum_{n=1}^{N}\left\{x_n^N\prod_{m=1,m\neq n}^{N}[(x-x_m)/(x_n-x_m)]\right\}. \tag{21c}$$

Likewise, since $\psi_t(x,0)$ is a polynomial of degree $N-1$, there holds the formula (see (2.4.2-7))

$$\psi_t(x,0)=\sum_{n=1}^{N} \psi_t\left[x_n(0),0\right] \prod_{m=1,m\neq n}^{N} \{\left[x-x_m(0)\right]/\left[x_n(0)-x_m(0)\right]\} . \tag{22a}$$

Here the N quantities $\psi_t\left[x_n(0),0\right]$ can be obtained from (20b), but to use this formula we need the N quantities $\psi_x\left[x_n(0),0\right]$. To obtain them we use the fact that $\psi(x,0)-x^N$ is a polynomial of degree $N-1$, hence (via (2.4.1-9) with $r=1$, and (20a))

$$\psi_x\left[x_n(0),0\right]=N\left[x_n(0)\right]^{N-1}$$

$$+\sum_{m=1}^{N} b_n\left[\underline{x}(0)\right] D_{nm}\left[\underline{x}(0)\right]\{ b_m\left[\underline{x}(0)\right] \}^{-1}\{h\left[x_m(0),0\right]-\left[x_m(0)\right]^N \}, \tag{22b}$$

with $\underline{D}(\underline{x})$ respectively $\underline{b}(\underline{x})$ defined of course by (13b) respectively (13a). Insertion of (22b) in (20b) yields an explicit expression in terms of the initial data, $\underline{x}(0)$ and $\underline{\dot{x}}(0)$, of $\psi_t\left[x_n(0),0\right]$, which can then be inserted in (22a) to yield finally the expression of the polynomial $\psi_t(x,t)$ at the initial time $t=0$ in terms of the initial data, $\underline{x}(0)$ and $\underline{\dot{x}}(0)$, of the N-body problem (18):

$$\psi_t(x,0)=\sum_{n=1}^{N} \{ h_t\left[x_n(0),0\right]+\dot{x}_n(0)\{ h_x\left[x_n(0),0\right]-N\left[x_n(0)\right]^{N-1}$$

$$-\sum_{m=1}^{N} b_n\left[\underline{x}(0)\right] D_{nm}\left[\underline{x}(0)\right]\{ b_m\left[\underline{x}(0)\right] \}^{-1}\{h\left[x_m(0),0\right]-\left[x_m(0)\right]^N \} \} \}\cdot$$

$$\cdot\prod_{l=1,l\neq n}^{N} \{ \left[x-x_l(0)\right]/\left[x_n(0)-x_l(0)\right] \} . \tag{22c}$$

(Alternatively, as we did above, we could have exploited the fact that $\psi(x,0)-\prod_{n=1}^{N}\left[x-x_n(0)\right]$ is also a polynomial of degree *less* than N (as well as $\psi(x,0)-x^N$), obtaining thereby an equivalent version of (22b) (equivalent in just the same sense as (21b) is equivalent to (21a)), hence an equivalent version of (22c). The diligent reader will write it out explicitly.

This completes our general discussion of the technique to solve (18).The diligent reader, before proceeding to the examples given below (and others (s)he may wish to invent), is urged to ponder on the differences, and similarities, among (18) and (2.3.3-11) (including the developments that led to these equations and the techniques to solve them).

Let us now display a few special cases of (18), including as well some systems of *first-order* ODEs. These are all *solvable* models.

$$\dot{x}_n = b \prod_{m=1,m\neq n}^{N} (x_n - x_m)^{-1} \ . \tag{24}$$

$$\dot{x}_n = -a x_n + b \prod_{m=1,m\neq n}^{N} (x_n - x_m)^{-1} \ . \tag{25}$$

$$\dot{x}_n = -a x_n + [N g_N(x_n) - x_n g'_N(x_n)] \prod_{m=1,m\neq n}^{N} (x_n - x_m)^{-1} \ . \tag{26}$$

$$\ddot{x}_n = 2 \sum_{m=1,m\neq n}^{N} \dot{x}_n \dot{x}_m / (x_n - x_m) + b \prod_{m=1,m\neq n}^{N} (x_n - x_m)^{-1} \ . \tag{27}$$

$$\ddot{x}_n = -a x_n + 2 \sum_{m=1,m\neq n}^{N} \dot{x}_n \dot{x}_m / (x_n - x_m) + [N g_N(x_n) - x_n g'_N(x_n)] \prod_{m=1,m\neq n}^{N} (x_n - x_m)^{-1} \ . \tag{28}$$

In these equations a and b are two arbitrary constants (which could be rescaled away; but we prefer to keep them), and $g_N(x)$ is an *arbitrary* polynomial of degree N (or less; of course for $g_N(x) = b/N$ (26) becomes (25), and (28) with $a = 0$ becomes (27)).

Exercise 2.5-2. Show that the solutions of the initial-value problem for the system of N *first-order* ODEs (24) are the N roots of the following simple polynomial equation in x:

$$\prod_{n=1}^{N} [x - x_n(0)] = b t \ . \tag{29}$$

Hint: set $h(x,t) = bt$, $E = 1$ and all other constants to zero in (18), to check that with these assignments it reduces to (24); then use (2.3-2) with these same assignments, (entailing $\psi(x,t) = \psi(x,0)$), as well as (19) and (21b).

Exercise 2.5-3. Show that the solutions $x_n(t)$ of the initial-value problem for the system of N first-order ODEs (25) are the N roots of the following polynomial equation in x:

$$x^N + \sum_{m=1}^{N} c_m(0) \exp(-m a t)\, x^{N-m} = b/(aN) \ , \tag{30a}$$

where the N constants $c_m(0)$ are of course the solutions of the system of N linear equations for these constants that obtain by setting in (30a) $t=0$ and $x=x_n(0)$, $n=1,...,N$; or equivalently, and more explicitly in terms of the initial data, that the coordinates $x_n(t)$ are the N roots of the following polynomial equation in x:

$$x^N + \exp(-at)\sum_{n=1}^{N} \left\{b/(aN)-[x_n(0)]^N\right\} \prod_{m=1,m\neq n}^{N} \{[x-x_m(0)\exp(-at)]/[x_n(0)-x_m(0)]\}$$

$$\equiv \prod_{n=1}^{N} [x-x_n(0)\exp(-at)]$$

$$+\exp(-at)\sum_{n=1}^{N} [b/(aN)] \prod_{m=1,m\neq n}^{N} \{[x-x_m(0)\exp(-at)]/[x_n(0)-x_m(0)]\} = b/(aN).$$

(30b)

Hint: set $h(x,t)=b/(aN)$, $E=1$, $B_1=-a$ and all other constants to zero in (18), and check that with these assignments it reduces to (25); then insert the same assignments in the PDE (2.3-2) and solve it, either via the *ansatz* (1) or directly by the method of characteristics using the initial value $\psi(x,0)$, see (21a,b); finally use (19a). The identity of the two expressions that are equated to $b/(aN)$ is of course guaranteed by (21c) (verify!).

Exercise 2.5-4. Show that the solutions $x_n(t)$ of the initial-value problem for the system of N first-order ODEs (26) are the N roots of the following polynomial equation in x:

$$x^N + \sum_{m=1}^{N} c_m(0)\, \exp(-mat)\, x^{N-m} = [\, g_N(x) - \gamma x^N \,]/a \quad , \qquad (31a)$$

where the constant γ is defined by the requirement that the polynomial

$$h(x) = [\, g_N(x) - \gamma\, x^N \,]/a \qquad (31b)$$

have degree *less* than N ($g_N(x)$ being a polynomial of degree N, or less), and the N constants $c_m(0)$ are of course the solutions of the system of N linear algebraic equations for them that obtain by setting in (31a) $t=0$ and $x=x_n(0)$, $n=1,...,N$; or equivalently, and more explicitly in terms of the initial data, that the solutions $x_n(t)$ are the N roots of the following polynomial equation in x:

$$x^N + \exp(-at)\sum_{n=1}^{N} \left\{b/(aN)-[x_n(0)]^N\right\} \prod_{m=1,m\neq n}^{N} \{[x-x_m(0)\, \exp(-at)]\,/\,[x_n(0)-x_m(0)]\}$$

$$\equiv \prod_{n=1}^{N} [x - x_n(0)\exp(-at)]$$

$$+\exp(-at)\sum_{n=1}^{N} [b/(aN)] \prod_{m=1, m\neq n}^{N} \{[x - x_m(0)\exp(-at)] / [x_n(0) - x_m(0)]\}$$

$$= [\, g_N(x) - \gamma\, x^N \,]/a \;, \tag{31c}$$

where the constant γ is of course defined as above, see (31b). *Hint*: set $h(x, t) = h(x)$, with $h(x)$ defined by (31b), $E = 1$, $B_1 = -a$ and all other constants to zero in (18), and check that with these assignments it reduces to (26); then proceed as suggested above, see the *hint* after *Exercise 2.5-3*.

Exercise 2.5-5. Show that, for any initial condition, the solutions $x_n(t)$ of the system of N ODEs (26) with $a = -\gamma > 0$ tend, as $t \to \infty$, to the zeros of the polynomial $g_N(x)$ (of degree N, or less):

$$x_n(t) \underset{t\to\infty}{\to} x_n^{(g_N)} \;,\quad g_N(x_n^{(g_N)}) = 0 \;. \tag{32}$$

Hint: see (31a).

Remark 2.5-6. This result, (32), suggests a technique to compute *numerically* the N zeros of any polynomial $g_N(x)$, of degree N: by integrating numerically the first-order system (26), with $a = -\gamma$ (of course with γ identified by the requirement that $h(x)$, see (31b), be a polynomial in x of degree less than N; and with the overall sign of the polynomial adjusted so that $\gamma < 0$ hence $a > 0$).

Exercise 2.5-7. Show that the solution of the initial-value problem for the system of N *second-order* ODEs (27) are the N roots of the following polynomial equation in x:

$$\left[1 - t\sum_{n=1}^{N} \dot{x}_n(0)[x - x_n(0)]^{-1} \left\{N[x_n(0)]^{N-1} - \sum_{m=1}^{N} D_{nm}[\underline{x}(0)]\,\{b_m[\underline{x}(0)]\}^{-1}[x_m(0)]^N\right\}\right]\cdot$$

$$\cdot \prod_{l=1}^{N} [x - x_l(0)] = bt^2/2 \;. \tag{33}$$

Hint: verify that (18) becomes (27) if one sets $h(x,t) = bt^2/2$, $C = 1$ and all other constants to zero; then solve (2.3-2) (with this assignment), using the initial data $\psi(x,0)$ and $\psi_t(x,0)$ as given by (21b) and (22c); finally use (19a).

Exercise 2.5-8. Show that the solution of the system of N *second-order* ODEs (28) are the N roots of the following polynomial equation of degree N in x:

$$x^N + \sum_{m=1}^{N} \{c_m(0) \cos[(m a)^{1/2} t] + (m a)^{-1/2} \dot{c}_m(0) \sin[(m a)^{1/2} t]\} x^{N-m}$$

$$= [\ g_N(x) - \gamma x^N \ ] / a \ , \qquad (34)$$

where the constant γ must be adjusted so that the polynomial in the right hand side of this equation, (34), have degree less than N ($g_N(x)$ being a polynomial of degree N, or less), and the $2N$ constants $c_m(0), \dot{c}_m(0)$ are determined in terms of the initial data, $x_n(0), \dot{x}_n(0)$, by the requirement that the N equations that obtain from (34) by replacing x with $x_n(t)$, as well as their t-derivatives (another N equations), are satisfied at $t=0$ (the first set of N equations is a system of N *linear* algebraic equations for the N constants $c_m(0)$, with known coefficients and known inhomogeneous terms; and, after this system has been solved, the second set yields an analogous system of N *linear* algebraic equations for the N unknown constants $\dot{c}_m(0)$). *Hint*: verify that (18) becomes (28) if one sets $h(x,t) = [\ g_N(x) - \gamma\ x^N \ ] / a$, $B_1 = -a$, $C = 1$ and all other constants to zero; then, via the *ansatz* (1), solve (2.3-2) with this assignment of the constants; finally use (19a).

Exercise 2.5-9. Display the effect on (24)÷(28) of the trivial rescaling transformation

$$x_n(t) = \alpha \xi_n(\tau), \ \tau = \beta t \qquad (35)$$

with α, β arbitrary constants.

Exercise 2.5-10. Show that the Newtonian N-body problem

$$\ddot{x}_n = 2\dot{x}_n \sum_{m=1, m\neq 1}^{N} [\alpha \dot{x}_n + (1+\alpha)\dot{x}_m] / (x_n - x_m) \ , \qquad (36)$$

with α an *arbitrary* constant, is *partially solvable*, since it possesses the solution given by the (α-independent!) recipe (29), where the N constants $x_n(0)$ coincide of course with the initial positions of the N particles and can be assigned arbitrarily, while, for this solution, the N initial velocities $\dot{x}_n(0)$ are given by the prescription

$$\dot{x}_n(0) = b \prod_{m=1, m\neq n}^{N} [x_n(0) - x_m(0)]^{-1} \ , \qquad (37)$$

with b an arbitrary constant. *Hint*: differentiate logarithmically (24) and use the *identity* (2.4.1-27).

Exercise 2.5-11. Noting that, for $\alpha = 0$, the equations of motion (36) coincide with (2.3.4.2-17) (with $a = 0$), show that the recipes (29) and (2.3.4.2-21) yield the same result when (37) holds. *Hint*: insert (37) in (2.3.4.2-21) and use the identity (2.4.2-32); or note that, for $\alpha = 0$, (36) coincides with the completely solvable equations of motion (2.1.10-1).

Exercise 2.5-12. Show that the following two Newtonian N-body problems are *partially solvable*, and exhibit the corresponding (class of) solutions:

$$\ddot{x}_n = 2(\dot{x}_n + a) \sum_{m=1, m \neq n}^{N} [\alpha \dot{x}_n + (1+\alpha)\dot{x}_m + (1+2\alpha)a] / (x_n - x_m) \;, \tag{38}$$

respectively

$$\ddot{x}_n = -2a\dot{x}_n - a^2 x_n$$

$$+ 2(\dot{x}_n + a x_n) \sum_{m=1, m \neq n}^{N} [\alpha(\dot{x}_n + a x_n) + (1+\alpha)(\dot{x}_m + a x_m)] / (x_n - x_m) \;. \tag{39a}$$

Here a and α are 2 arbitrary constants. Note that, for $\alpha = -1/2$, this system, (39a), takes the neater form

$$\ddot{x}_n = -(N+1)a\dot{x}_n - N a^2 x_n - (\dot{x}_n + a x_n) \sum_{m=1, m \neq n}^{N} (\dot{x}_n - \dot{x}_m) / (x_n - x_m) \;. \tag{39b}$$

Hint: set $x_n(t) = \tilde{x}_n(t) + at$ respectively $x_n(t) = \tilde{x}_n(t) \exp(at)$ in (36); then eliminate the tildes!

Remark 2.5-13. For $\alpha = 0$ and arbitrary a these two N-body problems, (38) and (39), are of course *solvable* (indeed, for $\alpha = 0$, (36) is itself *solvable*: it is a special case of (2.3.4-6), (2.3.4.2-34), or see directly (2.1.10-1)).

Exercise 2.5-14. Show that solutions $x_n(t)$ of the following two (quite different !) Newtonian 3-body problems,

$$\ddot{x}_n = -2b^2 (x_n - x_{n+1})^{-2} (x_n - x_{n-1})^{-2} [\, (x_n - x_{n+1})^{-1} + (x_n - x_{n-1})^{-1} \,] \;, \tag{40}$$

$$\ddot{x}_n = -2\dot{x}_n^3 (2x_n - x_{n+1} - x_{n-1}) / b \;, \tag{41}$$

where of course $n=1,2,3 \mod(3)$, are provided by the 3 roots of the following cubic equation in x:

$$[x-x_1(0)][x-x_2(0)][x-x_3(0)]=bt \ . \tag{42}$$

Here the 3 quantities $x_n(0)$ are of course the 3 initial positions of the 3 particles, and they can be assigned arbitrarily, while, for these solutions, the 3 initial velocities $\dot{x}_n(0)$ are given by the prescription

$$\dot{x}_n(0)=b[x_n(0)-x_{n+1}(0)]^{-1}[x_n(0)-x_{n-1}(0)]^{-1} \ . \tag{43}$$

Hint: consider (24) with $N=3$, time-differentiate it (most conveniently logarithmically), and use it again appropriately to get (40) and (41); then use (29) with $N=3$.

Remark 2.5-15. The two *partially-solvable* 3-body problems (40) and (41) feature forces that are *translation-invariant*; (40) features *velocity-independent* 3-body forces; (41) features velocity-dependent 2*-body* forces.

Exercise 2.5-16. Show that solutions $x_n(t)$ of the following three Newtonian 3-body problems,

$$\ddot{x}_n=a^2\,x_n+a\,b\,(x_n-x_{n+1})^{-1}(x_n-x_{n-1})^{-1}$$

$$-2\,b^2(x_n-x_{n+1})^{-2}(x_n-x_{n-1})^{-2}[\,(x_n-x_{n+1})^{-1}+(x_n-x_{n-1})^{-1}\,], \tag{44a}$$

$$\ddot{x}_n=a\dot{x}_n+2\,a^2\,x_n-2\,(\dot{x}_n+a\,x_n)^3(2\,x_n-x_{n+1}-x_{n-1})/b, \tag{44b}$$

$$\ddot{x}_n=a\dot{x}_n+2a^2\,x_n-2\,(\dot{x}_n+a\,x_n)^2[\,(x_n-x_{n+1})^{-1}+(x_n-x_{n-1})^{-1}\,], \tag{44c}$$

where of course $n=1,2,3 \mod(3)$, are provided by the 3 roots of the following cubic equation in x:

$$[x-x_1(0)\exp(-at)][x-x_2(0)\exp(-at)][x-x_3(0)\exp(-at)]=[b/(3a)]\cdot$$

$$\cdot\left\{1-\exp(-at)\sum_{n=1}^{3}\prod_{m=1,m\neq n}^{3}\{[x-x_m(0)\exp(-at)]/[x_n(0)-x_m(0)\exp(-at)]\}\right\}. \tag{44d}$$

Here the 3 quantities $x_n(0)$ are of course the initial positions of the 3 particles, and they can be assigned arbitrarily, while, for these solutions, the 3 initial velocities $\dot{x}_n(0)$ are given by the prescription

$$\dot{x}_n(0) = -a\,x_n(0) + b \prod_{m=1,m\neq n}^{N} \left[x_n(0) - x_m(0)\right]^{-1} \quad . \tag{44e}$$

Hint: as for the preceding *Exercise 2.5-14*, but with (24) respectively (29) replaced by (25) respectively (30b).

Remark 2.5-17. The three *partially solvable* Newtonian 3-body problems (44a,b,c) feature forces that are *not* translation-invariant; (44a) features *velocity-independent* 3-body forces; (44b) and (44c) feature velocity-dependent 2-*body* forces, and the latter (that does not feature the constant b, that can therefore be chosen arbitrarily in (44e)) is remarkably similar to, yet quite different from, (2.3.4-6), or equivalently (2.3.4.2-34) (with $N=3$ and, say, $\alpha = a,\ \beta = \mu = 2a^2, \lambda = 2a$).

Let us now consider a second family of models, those that obtain by positing (9) rather than (7), after having set, in (16),

$$C = 0 \quad , \tag{45}$$

so that these ODEs, (16), become of *first-order*:

$$b_n \{ -E\,\dot{x}_n + B_0 + B_1\,x_n - 2(N-1)\,A_3\,x_n^2$$

$$+2 \sum_{m=1,m\neq n}^{N} (A_0 + A_1\,x_n + A_2\,x_n^2 + A_3\,x_n^3)/(x_n - x_m) \}$$

$$+E[\ \dot{f}_n - \dot{x}_n\,f_n\ ] + [\ B_0 + B_1\,x_n - 2(N-1)\,A_3\,x_n^2\ ]\,f_n^{[1]}$$

$$+[\ A_0 + A_1\,x_n + A_2\,x_n^2 + A_3\,x_n^3\ ]\,f_n^{[2]} - [\ N(N-1)(A_2 - A_3\,x_n) + N\,B_1\ ]\,f_n = 0 \,. \tag{46}$$

To obtain this equation we did also set in (16), for the sake of simplicity,

$$D_0 = D_1 = D_2 = 0 \quad . \tag{47}$$

In (46) of course we are using again the definitions (12) and (13), and we again omit for notational simplicity to indicate explicitly the time dependence.

Let us now set

$$f_n(t) = \mu_n \dot{x}_n(t) + h[x_n(t)] \ , \tag{48a}$$

where μ_n are N arbitrary constants and $h(x)$ is an *arbitrary* polynomial of degree *less* than N.

Then (48a) entails

$$\dot{f}_n = \mu_n \ddot{x}_n + h'(x_n)\, \dot{x}_n \ , \tag{48b}$$

as well as (see (12) and (2.4.1-9) with $r=1,2$)

$$f_n^{[1]} = \mu_n b_n \sum_{m=1}^{N} D_{nm} b_m^{-1} \dot{x}_m + h'(x_n) \ , \tag{48c}$$

$$f_n^{[2]} = \mu_n b_n \sum_{m=1}^{N} (\underline{D}^2)_{nm} b_m^{-1} \dot{x}_m + h''(x_n) \ . \tag{48d}$$

Here of course the N-vector $\underline{b} \equiv \underline{b}(\underline{x})$ and the $(N\times N)$-matrix $\underline{D} \equiv \underline{D}(\underline{x})$ are defined by (13) (see also (2.4.1-2,4a,5d)) and primes denotes derivatives with respect to the argument of the function they are appended to.

Insertion of these relations in (46) yields the following system of Newtonian equations:

$$E\mu_n \ddot{x}_n = [N(N-1)(A_2 - A_3 x_n) + N B_1][\mu_n \dot{x}_n + h(x_n)]$$

$$-[B_0 + B_1 x_n - 2(N-1) A_3 x_n^2]\, h'(x_n) - [A_0 + A_1 x_n + A_2 x_n^2 + A_3 x_n^3]\, h''(x_n)$$

$$+ b_n(\underline{x}) \{ E\dot{x}_n - B_0 - B_1 x_n + 2(N-1) A_3 x_n^2$$

$$-2(A_0 + A_1 x_n + A_2 x_n^2 + A_3 x_n^3) \sum_{m=1, m\neq n}^{N} (x_n - x_m)^{-1}$$

$$+ \mu_n [E\dot{x}_n - B_0 - B_1 x_n + 2(N-1) A_3 x_n^2] \sum_{m=1}^{N} D_{nm} \dot{x}_m / b_m(\underline{x})$$

$$-\mu_n (A_0 + A_1 x_n + A_2 x_n^2 + A_3 x_n^3) \sum_{m=1}^{N} (\underline{D}^2)_{nm} \dot{x}_m / b_m(\underline{x}) \} \ . \tag{49}$$

Exercise 2.5-18. Write out the more general equations that replace (49) if the simplifying condition (47) does not hold, and/or the polynomial h, see (48a), is assumed to depend on time, $h \equiv h(x,t)$.

Let us now consider a few examples; the alert reader will try out many more.

The initial-value problem for the (*translation-invariant*) Newtonian equations of motion

$$\mu_n \ddot{x}_n = \dot{x}_n \cdot$$

$$\cdot [\prod_{\ell=1,\ell \neq n}^{N} (x_n - x_\ell) + \mu_n \sum_{m=1,m\neq n}^{N} (x_n - x_m)^{-1} \{ \dot{x}_n - \dot{x}_m \prod_{\ell=1,\ell\neq n,m}^{N} [(x_n - x_\ell)/(x_m - x_\ell)]\}] \quad (50)$$

is *solved* by the following prescription: $x_n(t)$ is the solution of the (non-differential) equation

$$\prod_{m=1}^{N} \{[x_n(t) - y_m]/[x_n(0) - y_m]\}^{\lambda_m} = \exp(t/\mu_n) \quad , \tag{51a}$$

$$\lambda_m = \prod_{\ell=1,\ell\neq m}^{N} (y_m - y_\ell)^{-1} \quad , \tag{51b}$$

where the N constants y_m are determined, in terms of the initial data $x_n(0)$, $\dot{x}_n(0)$, by the N algebraic equations

$$\prod_{m=1}^{N} [x_n(0) - y_m] = \mu_n \dot{x}_n(0), \quad n = 1, \ldots, N \quad . \tag{51c}$$

Proof. It is easily seen that the system (50) corresponds to (49) with $E = 1$ and all other constants equal to zero. Hence the corresponding PDE, see (2.3-2), reads

$$\psi_t(x,t) = 0 \tag{52a}$$

entailing

$$\psi(x,t) = \psi(x,0) \quad . \tag{52b}$$

Hence, from (48a) and (2) we conclude that

$$\mu_n \dot{x}_n(t) = \prod_{m=1}^{N} [x_n(t) - y_m] \quad , \tag{53a}$$

where we have conveniently introduced, in the right hand side of this equation, a generic monic polynomial of degree N in $x_n(t)$, by displaying explicitly its N zeros y_m (this polynomial is independent of the index n, and of the time t; it coincides with $\psi[x_n(t),0]-h[x_n(t)]$, with $h(x)$ both arbitrary and irrelevant, since this polynomial, $h(x)$, of degree *less* than N, neither features in the equations of motion (50) nor in their solution, see (51)).

These N constants, y_m, are then defined by (53) at $t=0$, namely by (51c).

There then remains to integrate (53). To this end, using the standard "partial-fractions decomposition", namely the *identity*

$$\prod_{m=1}^{N} (x-y_m)^{-1} = \sum_{m=1}^{N} (x-y_m)^{-1} \lambda_m \tag{54}$$

with λ_m defined by (51b), we rewrite (53) in the form

$$\dot{x}_n(t) \sum_{m=1}^{N} [x_n(t) - y_m]^{-1} \lambda_m = \mu_n^{-1} \quad . \tag{53b}$$

Then, via a trivial quadrature, one gets (51a), which is thereby proven.

Let us emphasize that, while in general the approach based on the *ansatz* (9) does *not* yield completely solvable models, in this case we were able to reduce the solution to a purely algebraic task, see (51).

Remark 2.5-19. Clearly the equations of motion (50) entail that, if $\dot{x}_n(0)=0$, then $\dot{x}_n(t)=0$, $x_n(t)=x_n(0)$. Hence any particle that is initially (or at any time) at rest always remains at rest. However, its presence does affect the motion of the other particles.

Exercise 2.5-20. Investigate how the method of solution of the model (50) must be modified if one or more of the N particles do not move. *Hint*: begin by understanding the $N=2$ case.

The next example we report is characterized by the Newtonian equations of motion

$$\mu_n \ddot{x}_n = N a \mu_n \dot{x}_n + (N-1) a c x_n + N a d - b c + (\dot{x}_n - a x_n - b) \cdot$$

$$\cdot [\prod_{\ell=1, \ell\neq n}^{N} (x_n - x_\ell) + \mu_n \sum_{m=1, m\neq n}^{N} (x_n - x_m)^{-1} \{\dot{x}_n - \dot{x}_m \prod_{\ell=1, \ell\neq n,m}^{N} [(x_n - x_\ell)/(x_m - x_\ell)]\}] \tag{55}$$

with a, b, c, d arbitrary constants (for $a=b=c=0$ this model reduces to the previous one, see (50)). We claim that its solution is provided by the solutions of the following *uncoupled first-order ODEs*:

$$\mu_n \dot{x}_n(t)+c\, x_n(t)+d=\prod_{m=1}^{N}\{x_n(t)-x_m(0)\exp(at)+(b/a)[1-\exp(at)]\}$$

$$+\exp(at)\sum_{m=1}^{N}[\mu_m \dot{x}_m(0)+c\, x_m(0)+d]\cdot$$

$$\cdot\prod_{\ell=1,\ell\neq m}^{N}[\{x_n(t)-x_\ell(0)\exp(at)+(b/a)[1-\exp(at)]\}/\{x_m(0)-x_\ell(0)\}]\ . \qquad (56a)$$

Note however that these ODEs are not solvable by quadratures, because they are not autonomous.

Proofs. We begin by noting that (55) obtains from (49) by setting $h(x)=c\,x+d$, $B_0=b$, $B_1=a$, $E=1$ and all other constants to zero. Hence the corresponding PDE, see (2.3-2), reads

$$\psi_t(x,t)+(a\,x+b)\,\psi_x(x,t)-N\,a\,\psi(x,t)=0 \qquad (57a)$$

entailing (as can be easily verified)

$$\psi(x,t)=\exp(N\,at)\,\psi(x\exp(-at)+(b/a)[\exp(-at)-1],\ 0)\ . \qquad (57b)$$

But, from (3),

$$\psi(x,0)=\prod_{n=1}^{N}[x-x_n(0)]+f(x,0) \qquad (58a)$$

hence, via (2.4.2-7,5), (4) and (48a) with the above assignment of $h(x)$,

$$\psi(x,0)=\prod_{n=1}^{N}[x-x_n(0)]$$

$$+\sum_{n=1}^{N}[\mu_n \dot{x}_n(0)+c\,x_n(0)+d]\prod_{\ell=1,\ell\neq n}^{N}\{[x-x_\ell(0)]/[x_n(0)-x_\ell(0)]\}. \qquad (58b)$$

Now insert this expression, (58b), of $\psi(x,0)$ in (57b), set $x=x_n(t)$ in the resulting equation, get thereby an explicit expression for $\psi[x_n(t),t]$, and finally use (2) with (48a). After some trivial steps one obtains (56a), which is thereby proven.

Exercise 2.5-21. Show that (56a) can be rewritten in the following two equivalent forms:

$$\mu_n \dot{x}_n(t) + c\, x_n(t) + d\left[1-\exp(aNt)\right]$$

$$= \prod_{m=1}^{N} \{x_n(t) - x_m(0)\exp(at) + (b/a)[1-\exp(at)]\}$$

$$+\exp(at)\sum_{m=1}^{N} [\mu_m \dot{x}_m(0) + c\, x_m(0)]\cdot$$

$$\cdot \prod_{\ell=1,\ell\neq m}^{N} [\{x_n(t) - x_\ell(0)\exp(at) + (b/a)[1-\exp(at)]\}/\{x_m(0)-x_\ell(0)\}] \;, \tag{56b}$$

$$\mu_n \dot{x}_n(t) + c\,\{\, x_n(t)\{1-\exp[a(N-1)t]\}\,(b/a)\,\exp(aNt)\,[1-\exp(-at)]\}$$

$$+d[1-\exp(aNt)] = \prod_{m=1}^{N} \{x_n(t) - x_m(0)\exp(at) + (b/a)[1-\exp(at)]\}$$

$$+\exp(at)\sum_{m=1}^{N} \mu_n \dot{x}_m(0) \prod_{\ell=1,\ell\neq m}^{N} [\{x_n(t) - x_\ell(0)\exp(at)$$

$$+(b/a)[1-\exp(at)]\}/\{x_m(0)-x_\ell(0)\}] \;. \tag{56c}$$

Hint: use the *identities*

$$\sum_{m=1}^{N} x_m^r \prod_{\ell=1,\ell\neq m}^{N} [(x-x_\ell)/(x_m-x_\ell)] = x^r \,, \quad r=0,1 \;, \tag{59}$$

which correspond to (2.4.2-7,5) with $f(x)=x^r$ (see also (2.4.2-32)).

Exercise 2.5-22. For $a=0$ (55) becomes independent of d, while (56a) seems to still depend on d. How can this be ? *Hint*: see (56b,c).

Let us finally return to (46), but let us now supplement these evolution equations, (46), by positing

$$f_n(t) = b_n[\underline{x}(t)]\,[\,\mu_n \dot{x}_n(t) + \eta_n x_n(t)\,] \;, \tag{60}$$

where of course $b_n[\underline{x}(t)]$ is defined by (13a) and the $2N$ quantities μ_n, η_n are arbitrary constants.

One gets thereby the following Newtonian equations of motion:

$$\mu_n \ddot{x}_n = (1-\eta_n)\dot{x}_n + \left[N(N-1)(A_2 - A_3 x_n) + N B_1\right](\mu_n \dot{x}_n + \eta_n x_n)$$

$$+ \sum_{m=1, m\neq n}^{N} (x_n - x_m)^{-1} \left[(\mu_n + \mu_m)\dot{x}_n \dot{x}_m + \eta_n x_n \dot{x}_m + \eta_m x_m \dot{x}_n\right]$$

$$-\left[B_0 + B_1 x_n - 2(N-1) A_3 x_n^2\right]\left[1 + \sum_{m=1, m\neq n}^{N} (x_n - x_m)^{-1} (\mu_n \dot{x}_n + \mu_m \dot{x}_m + \eta_n x_n + \eta_m x_m)\right]$$

$$-\left[A_0 + A_1 x_n + A_2 x_n^2 + A_3 x_n^3\right] \sum_{m=1, m\neq n}^{N} (x_n - x_m)^{-1} \cdot$$

$$\cdot \left[2 + \sum_{\ell=1; \ell\neq n,m}^{N} (x_n - x_\ell)^{-1} (\mu_n \dot{x}_n + \mu_m \dot{x}_m + \mu_\ell \dot{x}_\ell + \eta_n x_n + \eta_m x_m + \eta_\ell x_\ell)\right] . \quad (61)$$

Proofs. The *ansatz* (60) entails

$$\dot{f}_n = b_n \left[\mu_n \ddot{x}_n + \eta_n \dot{x}_n + (\mu_n \dot{x}_n + \eta_n x_n) \sum_{m=1, m\neq n}^{N} (\dot{x}_n - \dot{x}_m)/(x_n - x_m)\right] , \quad (62a)$$

$$f_n^{[1]} = b_n \sum_{m=1, m\neq n}^{N} (x_n - x_m)^{-1} (\mu_n \dot{x}_n + \mu_m \dot{x}_m + \eta_n x_n + \eta_m x_m) , \quad (62b)$$

$$f_n^{[2]} = b_n \sum_{m=1, m\neq n}^{N} (x_n - x_m)^{-1} \cdot$$

$$\cdot \sum_{\ell=1; \ell\neq n,m}^{N} (x_n - x_\ell)^{-1} (\mu_n \dot{x}_n + \mu_m \dot{x}_m + \mu_\ell \dot{x}_\ell + \eta_n x_n + \eta_m x_m + \eta_\ell x_\ell) . \quad (62c)$$

The first of these formulas, (62a), follows by t-differentiation from (60); the second, (62b), follows from (12a) via (60) and (13b); the third, (62c), follows from (12b) via (60) and (2.4.1-5d,3). Insertion of these formulas in (46) yields (61), which is thereby proven. Note that, for notational simplicity, we did set, in (46),

$$E = 1 . \quad (63)$$

The Newtonian equations of motion (61) feature, in addition to the $2N$ arbitrary constants μ_n and η_n, the 6, also arbitrary, constants

$A_0, A_1, A_2, A_3, B_0, B_1$. These Newtonian equations of motion feature *only one-, two- and three-body forces; the latter are missing if* $A_0 = A_1 = A_2 = A_3 = 0$. They are *invariant under translations* $(x_n \to x_n + x_0,\ \dot{x}_0 = 0)$ iff $\eta_n = A_1 = A_2 = A_3 = B_1 = 0$; they are invariant under rescaling of the dependent variables $(x_n \to c\,x_n,\ \dot{c} = 0)$ iff $A_0 = A_1 = A_3 = B_0 = 0$; they are invariant under rescaling of the independent ("time") variable $(t \to a t,\ \dot{a} = 0)$ iff $A_0 = A_1 = A_2 = A_3 = B_0 = B_1 = 0$, $\eta_n = 1$, in which case (61) becomes simply

$$\mu_n \ddot{x}_n = \sum_{m=1, m \neq n}^{N} (\mu_n + \mu_m) \dot{x}_n \dot{x}_m / (x_n - x_m) \ . \tag{64}$$

The Newtonian equations of motion (61) are *not* solvable; our treatment only guarantees that this set of N coupled *second-order* ODEs can be reduced to the following set of N coupled *first-order* ODEs:

$$\mu_n \dot{x}_n(t) + \eta_n x_n(t) = \psi[x_n(t), t] \prod_{m=1, m \neq n}^{N} [x_n(t) - x_m(t)]^{-1} \tag{65}$$

(see (60), (2) and (13a)), where, as explained above, the monic polynomial $\psi(x,t)$, of degree N in x, can be considered known (see (1), and (2.3-2) with (45) and (47)). However, in the *equal-particle case*,

$$\mu_n = \mu, \ \eta_n = \eta \ , \tag{66}$$

the equations (65) are themselves *solvable*, hence in this case the many-body problem (61), whose equations of motion then read

$$\mu \ddot{x}_n = (1-\eta)\dot{x}_n + [N(N-1)(A_2 - A_3 x_n) + N B_1](\mu \dot{x}_n + \eta x_n)$$

$$+ \sum_{m=1, m \neq n}^{N} (x_n - x_m)^{-1} [2\mu \dot{x}_n \dot{x}_m + \eta(x_n \dot{x}_m + x_m \dot{x}_n)]$$

$$- [B_0 + B_1 x_n - 2(N-1) A_3 x_n^2] \{1 + \sum_{m=1, m \neq n}^{N} (x_n - x_m)^{-1} [\mu(\dot{x}_n + \dot{x}_m) + \eta(x_n + x_m)]\}$$

$$- [A_0 + A_1 x_n + A_2 x_n^2 + A_3 x_n^3] \sum_{m=1, m \neq n}^{N} (x_n - x_m)^{-1} \cdot$$

$$\cdot\left\{2+\sum_{\ell=1;\ell\neq n,m}^{N}(x_n-x_\ell)^{-1}\left[\mu(\dot{x}_n+\dot{x}_m+\dot{x}_\ell)+\eta(x_n+x_m+x_\ell)\right]\right\} \quad , \tag{67}$$

is itself *solvable.*

To prove this assertion, we must show how to solve the system of N ODEs

$$\left[\mu\dot{x}_n(t)+\eta x_n(t)\right]\prod_{m=1,m\neq n}^{N}\left[x_n(t)-x_m(t)\right]=\psi(x_n,t) \ , \tag{68}$$

namely (65) with (66), when $\psi(x,t)$ is a known monic polynomial of degree N in x, see (1). To this end we introduce the monic polynomial, say $\tilde{\psi}(x,t)$, of degree N in x, which has the N coordinates $x_n(t)$ as its N zeros:

$$\tilde{\psi}(x,t)=\prod_{n=1}^{N}\left[x-x_n(t)\right] \ . \tag{69}$$

It is then clear that

$$-\mu\,\tilde{\psi}_t(x,t)+\eta\left[x\,\tilde{\psi}_x(x,t)-N\tilde{\psi}(x,t)\right]=\psi(x,t)-x^N \ . \tag{70}$$

Indeed clearly both sides of this equations are polynomials of degree *less* than N in x, and they clearly coincide, thanks to (68), at the N points $x=x_n$.

Exercise 2.5-23. Prove this result. *Hint*: see (2.3.2-1,8,12).

But the linear PDE (70) can be easily solved, for instance by setting

$$\tilde{\psi}(x,t)=x^N+\sum_{m=1}^{N}\tilde{c}_m(t)\,x^{N-m} \quad , \tag{71}$$

and then by noting that (70) and (1) entail

$$-\mu\,\dot{\tilde{c}}_m-\eta\,m\,\tilde{c}_m=c_m \tag{72a}$$

namely

$$\tilde{c}_m(t)=\tilde{c}_m(0)\exp\left[-\eta mt/\mu\right]-\mu^{-1}\int_o^t dt'c_m(t')\,\exp\left[-\eta m(t-t')/\mu\right] \ . \tag{72b}$$

Here the coefficients $c_m(t)$ can be considered known, being related by (1) to the known polynomial $\psi(x,t)$.

Then, once the polynomial $\tilde{\psi}(x,t)$ has been determined (via (70) with (71)), the coordinates $x_n(t)$ are just its N zeros, see (68), hence their determination reduces to the purely algebraic task of finding the zeros of a given polynomial.

We end Sect. 2.5 by proposing two *exercises*, thereby hinting at further developments.

Exercise 2.5-24. Extend the treatment given in Sect. 2.5 by assuming the polynomial $\psi(x,t)$ not to be monic, namely by assuming that it has the form

$$\psi(x,t) = c_0(t)\,x^N + \sum_{m=1}^{N} c_m(t)\, x^{N-m} \quad , \tag{72}$$

instead of (1). Hint: see <C86a>.

Exercise 2.5-25. Extend the treatment given in Sect. 2.5 by assuming the polynomial $\psi(x,t)$, see (1), to satisfy a more general evolution equation than (2.3-2), albeit one that preserves the property to be *solvable* by algebraic operations (for instance, a linear evolution equation analogous to (2.3-2) but with the space derivatives replaced by finite differences). *Hint*: see <C85e>.

2.N Notes to Chapter 2

The idea of a Lax pair, see (2.1-2), was introduced by P. D. Lax <L68>, to identify *integrable* evolutions in the context of the study of nonlinear (partial differential) evolution equations. The first application of this idea to *integrable dynamical systems* (i.e., ODEs rather than PDEs) was made, independently and more or less simultaneously, by S. V. Manakov <Man74> and H. Flaschka <F74a, F74b>, both of whom applied it to the integrable Hamiltonian one-dimensional many-body problem with exponential “nearest-neighbor” interaction (“Toda lattice”) introduced by M. Toda <T67, T81>, whose integrability was first noted by M. Henon <H74>.

The *ansatz* (2.1.1-2,3) for a Lax pair was introduced in <C75>, as a generalization of the specific Lax pair (2.1.2-6,7) introduced by Juergen Moser <Mo75> to demonstrate the integrability of the one-dimensional problem of N equal particles on the line interacting pairwise with repulsive forces inversely proportional to the cube of their mutual distance, see Sect. 2.1.3. This model with inverse-cube forces had been previously introduced and solved in the *quantal* context (perhaps the first time that a quantum many-body problem has been treated *before* its classical counterpart) in <C71>; for this reason it is often referred to as the "Calogero-Moser" system.

The functional equation (*), see (2.1.1-16), as well as its general solution, see Sect. 2.1.4, were introduced in <C75>. This was the first appearance both of functional equations and of elliptic functions in the context of classical many-body problems integrable via the Lax-pair approach. See also Appendix B.

A proof that the N eigenvalues $\lambda^{(m)}$ of the Lax matrix (2.1.1-1) Poisson-commute if the function $\alpha(q)$ satisfies the functional equation (*), see (2.1.1-16), was first given by A. M. Perelomov <P77> (this proof is also reported in Sect. 3.2 of <P90> and in Chap. 2 of <H92>). S. Wojciechowski, more or less simultaneously, gave an independent proof of the Poisson-commutativity of the N symmetric invariants J_n, see (2.1-10) <W77>.

The result (2.1.3.1-5) was firstly obtained, for arbitrary N, in the quantal context ("no diffraction") in <C71>, and in the classical context by J. Moser <Mo75>. For $N=3$ it had been previously discovered in the quantal context by C. Marchioro <Mar70> (who also solved the problem in the classical case, but did not publish the result). Actually the solvability of the classical one-dimensional problem of 3 particles interacting pairwise with inverse-cube forces had been, much earlier, noted by C. Jacobi <J1866>. (This phenomenology -- namely, the fact that no new asymptotic momenta emerge from the interaction, in spite of its nonlinear nature -- is sometimes characterized by the adjective "solitonic"; to understand the origin of this language see the literature on "solitons", for instance <CD82> and the references quoted there. There is of course more to this than just semantics: see for instance <C78a> and the literature quoted there).

The OP technique of solution (see Sect. 2.1.3.2 and also subsequent sections) was introduced by M. A. Olshanetsky and A. M. Perelomov <OP76a,c>, <OP81>. It is reviewed in several textbooks, see for instance Chap. 3 of <P90> and Chap. 1 of <H92>; for a seminal, more group-theoretical, treatment see <KKS78>. It is sometimes called "the projection method" <P90>. The explicit solvability of the model of Sect. 2.1.3 had been first shown by J. Moser <Mo75> (see also <AMM77>, <Mo80>).

The N-body problem on the line with a harmonic interaction in addition to pair inverse-cube forces, see Sect. 2.1.3.3, was also introduced and solved firstly in the quantal context <C71>. The equispaced character of the corresponding spectrum motivated the conjecture <C71> that all motions of the corresponding classical problem be completely periodic with period T, see (2.1.3.3-17). The first proof of this result (see (2.1.3.3-18)), and of the integrability of this model in the classical context, is due to D. C. Khandekar and S. V. Lawande <KL72> for $N=3$, and to M. Adler <A76, A77> (see also <P76> and <OP76a>) for arbitrary *N*. The transformation (2.1.3.3-22) relating the two classical problems with and without harmonic interactions was discovered by A. M. Perelomov <P78>. The connection among the equilibrium configuration of the classical *N*-body problem of Sect. 2.1.3.3 and the zeros of the Hermite polynomial of order *N*, see (2.1.3.3-37), was pointed out in <C77b>. This finding is particularly intriguing because Hermite polynomials are closely connected to the eigenfunctions of the quantal harmonic oscillator problem. Properties of the zeros of the classical polynomials such as (2.1.3.3-44) were originally discovered in the context of the study of integrable many-body problems on the line via the technique of Sect. 2.3, hence for a more detailed discussion of this type of results see below the notes on Sects. 2.3 and 2.4 (see also the notes on Appendix C and Chap. 3).

The general solution of the functional equation (*), see (2.1.1-16), was firstly exhibited in <C75> and proven in <C76a>. This result was also proven, more or less simultaneously, by A.M. Perelomov <OP76b> and by S. I. Pydkuyko and A. M. Stepin <PS76>.

The integrability of the model treated in Sect. 2.1.5 was firstly noted in <C75> and <CMR75>; its explicit solution was firstly given in <OP76c>. The factorization property, entailing the formula (2.1.5-44) for the asymptotic shifts of the scattering trajectories in the N-body case, is due to P. P. Kulish <K77>. S. Wojciechowski introduced and solved <W84> a generalization of this model, characterized by the additional presence of an external exponential potential (see also Sect. 3.5 of <P90>).

The model treated in Sect. 2.1.6 was introduced and treated in the quantal context by B. Sutherland <S71, S72>, and is therefore generally referred to as "Sutherland model."

For the (more or less explicit) solution of the model, see (2.1.4-32), with elliptic interactions, see <K78>, <K80> and <GP99>.

The trick (2.1.7-27) to generate a model involving particles of two different types was introduced in <C75>; in this same paper the possibility was indicated to generate, in an analogous manner, a model involving different types of particles, starting from the integrable Hamiltonian (2.1.4-32) and taking advantage of the periodicity of the Weierstrass function $\wp(z|\,\omega,\omega')$. The behavior of the many-body system characterized by the Hamiltonian (2.1.7-28) was investigated by M. A. Olshanetsky and V-B. K. Rogov <OR78>.

The possibility described in Sect. 2.1.7 to generate Lax pairs composed of matrices of size *2N, 4N, 8N* and so on was pointed out in

<C76b>; as indicated there, the same trick can be used as well in the context of the more general integrable model of Sect. 2.1.4.

The results of Sect. 2.1.7 based on symmetrical duplications on the (real) line can be given a group-theoretical significance in terms of root systems associated with semisimple Lie algebras <OP76b>, <OP81>, <P90>; this development yielded, over time, a large body of additional findings by many contributors (we list here a few *recent* references: <DHP98>, <BCS98>, <BCS99>, <BS99>, <BST99>, <KST99>, <BMS2000>, <CFS2000>). The idea of duplications involving some kind of complexification was introduced in <CF92> (but see also <C86b>, <C86c>). The treatment of "infinite duplications" given in Sect. 2.1.7 is patterned after those of <C93b> and <C97d>; in particular, the observation that led to (2.1.7-54) was originally made in <C93b>.

The "reduction" trick used in Sect. 2.1.7 to get the model with nearest-neighbor interactions, see (2.1.7-57,60), was taught to me by Simon Ruijsenaars; perhaps it was discovered simultaneously by him and by Bill Sutherland (see the parenthetical remark after eq. (2.1.17) of <R94a>). This exactly treatable model was introduced by M. Toda <T67>, and is therefore generally referred to as the "Toda model" (sometimes as the "Toda lattice": the nearest-neighbor character of the interaction suggest that the more natural context for this model is to investigate nonlinear lattices <T81>).

The *ansatz* (2.1.8-1,2) for a Lax pair (as well as the functional equation (**), see below), were introduced in <BC87>.

The fake Lax pair (2.1.9.1-12, 13) is taken from (Sect. 3.3 of) <C84b>, and the fake Lax pair (2.1.9.1-1,14), as well as the Hamiltonian (2.1.9.1-17), are taken from <CF2000a> (see also Sect. 2.4.5.4). For the notion, and several examples, of fake pairs in the PDE context, see <CN91>.

The solvable Newtonian equations of motion (2.1.10-1) were probably introduced for the first time in <C78a>. The remarkable nature of this system has been underlined by attributing to it the status of "goldfish" <C99b>.

The treatments of Sects. 2.1.10.1, 2.1.10.2 and especially 2.1.10.3 are largely patterned after <CF2000a>.

The functional equation (**), see (2.1.8-19) and (2.1.11-1), as well as the functional equation (2.1.11-23), were introduced, and solved, in <BC87>, whose treatment is closely followed in Sect. 2.1.11; see also <BC90> and <BB97b>, and Appendix B.

The RS model discussed in Sect. 2.1.12 was introduced by Simon Ruijsenaars and Harald Schneider <RS86>; it has been extensively studied by many authors, and especially by S. Ruijsenaars, who has written many original contributions, as well as review papers and lectures notes, on these models and related developments, see, for instance, <R87> (which contains the first proof, for the general elliptic case, that the N constants of motion are in involution), <R88>, <R90>, <R94a>, <R94b>, <R95>, <R97>, and the PhD thesis <vD94> written by Jan Felipe van Diejen under Ruijsenaars' guidance.

The relativistic character of the RS model (see Sect. 2.1.12.2) has been emphasized in the original paper introducing it <RS86>, and in several subsequent publications (see for instance <R87>, <R94a>); it has been questioned by H. Braden and R. Sasaki <BS97>.

The conjecture of Sect. 2.1.12.3 was proposed in <C97c>.

The treatment of Sect. 2.1.12.4 is closely patterned after <CF2000a>.

The treatment of Sect. 2.1.13 owes much to <C97d>. The results based on *symmetrical duplications* (mentioned in the paragraph before (2.1.13-9)) have a group-theoretical underpinning in terms of root systems of Lie groups (see above, re Sect. 2.1.7; also this development yielded, over time, a large body of additional findings). The system (2.1.13-24) was firstly introduced by M. Bruschi and O. Ragnisco <BR89>.

The many-body model treated in Sects. 2.1.14 and 2.1.15 was introduced in <C95b> (eq. (2.20) of this paper is misprinted: the exponents 2 and –2 in the right hand side should not be there, see (2.1.15.1-1b)). The idea to introduce the constants of motion in involution h_M, see (2.1.15-8), is due to S. Rauch (Wojciechowski). For a treatment of quantized versions of this model see <CvD96>. Other treatments, which emphasize the group-theoretical structure underlying this model, have been given by M. Bruschi, O. Ragnisco and S. Rauch (Wojciechowski) (unpublished).

The many-body model treated in Sects. 2.1.16 and 2.1.16.1 was introduced in <CF96>, which we followed rather closely in our presentation; for a treatment of it in terms of the r-matrix (a notion not discussed in this book) see <RB96>.

The integrable many-body model characterized by the Hamiltonian (2.1.16-19) was introduced by R. Camassa, D. D. Holm and J. M. Hyman <CH93, CHH94>; it was this discovery that motivated our investigations of models characterized by Hamiltonians of type (2.1.14-6), see the Sects. from 2.1.14 to 2.1.16.1.

The many-body model treated in Sect. 2.2 was introduced in <C95c>. For a treatment of a quantized version of this model see <CvD95>. Other treatments, which emphasize the group-theoretical structure underlying this model, have been given by M. Bruschi, O. Ragnisco and S. Rauch (Wojciechowski) (unpublished).

The treatment of Sect. 2.3, including Sects. 2.3.1, 2.3.2, 2.3.3, 2.3.4 and 2.3.5, is based on <C78a>, where this approach was introduced. However, not all the findings of <C78a> are reported in this book, and some results reported here are not contained in <C78a>. The generalized treatment of Sect. 2.3.6 and its subsections is new, except for Sect. 2.3.6.3 that follows <CF2000b>.

The treatment of Sect. 2.4, including all its subsections, is mainly based on <C84b>, but some results were previously given, or are more fully elaborated, in <C81a>, as well as <C80a>, <C80b>, <C80c>, <BC81>, <C81b>, <C81c>, <C82a>, <C83a>, <C83b>, <C83c>, <D83>, <C84a>, <C85a>, <C85c>, <C85d>, <CF85>, <D85>, <C86b>, <Ca86>, <C88>, <BCP90>.

The main idea on which the treatment of Sect. 2.5 is based was introduced in <C85e> and <C86a>; the results in Sect. 2.5 are mostly new.

Various portions of the material treated in Chap. 2 have been covered in review papers, lecture notes, conference proceedings and books, see for instance <C78b>, <C80a>, <C81c>, <C82c>, <C85b>, <C86b>, <C92>, <C95a>, <C97d>, <FG76>, <H92>, <Mo80>, <OP81>, <P90>, <T81>.

3 N-BODY PROBLEMS TREATABLE VIA TECHNIQUES OF EXACT LAGRANGIAN INTERPOLATION IN SPACES OF ONE OR MORE DIMENSIONS

In the first part of Chap. 3 we describe a version of the (exact) Lagrangian interpolation technique, which is more general than that outlined above (see Sect. 2.4.2) on two counts: it is not restricted to a one-dimensional environment (namely, it is not limited to considering functions of a single, scalar, variable), and it is not restricted to a polynomial functional space (namely, it is not limited to using polynomials as basic building blocks). Then, in the second part of Chap. 3, we indicate how this generalized technique of interpolation can be utilized to manufacture solvable N-body problems in spaces of one or more dimensions: we discuss a general technique to do so, including a few variations on this theme, and we exhibit several examples.

These techniques of (exact) Lagrangian interpolation can also be exploited to identify remarkable matrices and related identities and to treat certain problems in numerical analysis (for instance to solve numerically eigenvalue problems in S-dimensional space); moreover, certain findings closely connected with these developments are instrumental to uncover and to prove certain theorems in elementary geometry. A (terse) survey of these results is confined to Appendices D, E and F.

3.1 Generalized formulation of Lagrangian interpolation, in spaces of arbitrary dimensions

Notation. We denote by a superimposed arrow vectors in S-dimensional space, say $\vec{r}$: in particular, for $S=2$, $\vec{r}\equiv(x,y)$ is a 2-vector (the environment is the plane), for $S=3$, $\vec{r}\equiv(x,y,z)$ is a 3-vector (the environment is the ordinary space we inhabit). Let N be a positive integer, and $s_n(\vec{r})$, $n=1,2,...,N$, be N functions; we assume they are assigned once and for

F. Calogero: LNPm 66, pp. 311 - 415, 2001

all (although we retain the option to choose them, as well as N, at our convenience), and we refer to them as "seeds". Let $\vec{r}_n$, $n=1,2,...,N$, be N *different* points in S-dimensional space, $\vec{r}_n \neq \vec{r}_m$ if $n \neq m$; in the following we refer to them as "nodes".

As we will soon see, it is moreover convenient to introduce N-vectors respectively $(N\times N)$-matrices; these quantities are denoted by *underlined* lower-case respectively upper-case letters, say $\underline{u}$ for the N-vector whose N components are the N numbers u_n, $n=1,2,...,N$, respectively $\underline{U}$ for the $(N\times N)$-matrix whose N^2 elements are the N^2 numbers U_{nm}, $n,m=1,2,...,N$:

$$\underline{u} \equiv (u_1,u_2,...,u_n) \ , \tag{1a}$$

$$\underline{U} \equiv \begin{pmatrix} U_{11} & U_{12} & \cdots & U_{1N} \\ \vdots & \vdots & \cdots & \vdots \\ U_{N1} & U_{N2} & \cdots & U_{NN} \end{pmatrix} \ . \tag{1b}$$

The N components of an N-vector , as well as the N^2 elements of an $(N\times N)$-matrix, might themselves be S-vectors. For instance we use below the convenient notation

$$\underline{\vec{r}} \equiv (\vec{r}_1,\vec{r}_2,...,\vec{r}_N) \tag{2}$$

to indicate the N-vector whose N components are the N nodes $\vec{r}_n$, $n=1,2,...,N$. Of course $\underline{\vec{r}}$ is both an S-vector-valued N-vector (namely, an N-vector whose N components are S-vectors) and an N-vector-valued S-vector (namely, an S-vector whose S components are N-vectors). Likewise for $(N\times N)$-matrices: for instance in the following it will be convenient to use the diagonal $(N\times N)$-matrix whose N diagonal elements are the N nodes,

$$\underline{\vec{R}} \equiv \text{diag}(\vec{r}_n; n=1,2,...,N), \ \vec{R}_{nm} = \delta_{nm}\vec{r}_n \ . \tag{3}$$

Again, $\underline{\vec{R}}$ is both an S-vector-valued $(N\times N)$-matrix (an $(N\times N)$-matrix whose elements are S-vectors; of course in this particular case, the off-diagonal elements all vanish), and a (diagonal) $(N\times N)$-matrix-valued S-vector (an S-vector whose S components are $(N\times N)$-matrices; in this particular case, diagonal $(N\times N)$-matrices).

Let us emphasize that the choices of the N seeds $s_n(\vec{r})$ and of the N nodes $\vec{r}_n$ remain our privilege -- choices to be done, in the context of this treatment, once and for all; as indeed the choice of the positive integer $N>1$-- and that we assume that these choices are done *independently* (namely, the N seeds $s_n(\vec{r})$ do not depend on the N nodes $\vec{r}_m$). But we hereafter assume that these choices guarantee that the $(N\times N)$-determinant

$$\Delta(\vec{r}_1,\vec{r}_2,...,\vec{r}_N)=\det[s_n(\vec{r}_m)]\equiv\begin{vmatrix} s_1(\vec{r}_1) & s_2(\vec{r}_1) & \cdots & s_N(\vec{r}_1) \\ s_1(\vec{r}_2) & s_2(\vec{r}_2) & \cdots & s_N(\vec{r}_2) \\ \vdots & \vdots & \vdots & \vdots \\ s_1(\vec{r}_N) & s_2(\vec{r}_N) & \cdots & s_N(\vec{r}_N) \end{vmatrix} \tag{4}$$

does not vanish:

$$\Delta(\vec{r}_1,\vec{r}_2,...,\vec{r}_N)\neq 0 \ . \tag{5}$$

Let now f_n, $n=1,2,...,N$, be N given numbers. The problem of (generalized) Lagrangian interpolation is then formulated as follows: *to find a function $f(\vec{r})$ of the S-vector $\vec{r}$ that possesses the following two properties: (i) $f(\vec{r})$ is a linear superposition (with constant, namely $\vec{r}$-independent, coefficients) of the N seeds $s_n(\vec{r})$,*

$$f(\vec{r})=\sum_{m=1}^{N} h_m\, s_m(\vec{r}); \tag{6}$$

(ii) *the N values that $f(\vec{r})$ takes at the N nodes $\vec{r}_n$ coincide with the N assigned values f_n,*

$$f(\vec{r}_n)=f_n\ ,\quad n=1,2,...,N\ . \tag{7}$$

It is clear that this problem always admits one, and only one, solution. Indeed setting $\vec{r}=\vec{r}_n$ in (6) and using (7) one gets

$$\sum_{m=1}^{N} h_m\, s_m(\vec{r}_n)=f_n\ ,\quad n=1,2,...,N\ , \tag{8}$$

and the condition (5) guarantees that this system of N linear equations for the N unknowns h_m admits one and only one solution.

This solution could be easily written out, but we prefer to display directly an expression for the "interpolating function" $f(\vec{r})$ characterized by (6) and (7). To this end we introduce N "interpolational functions" $q^{(n)}(\vec{r}|\underline{\vec{r}})$ defined as follows: *$q^{(n)}(\vec{r}|\underline{\vec{r}})$ is the function that obtains by first replacing, in the n-th line of the determinant $\Delta(\underline{\vec{r}})$, see (4), the node $\vec{r}_n$ with the variable $\vec{r}$, and by then dividing by the determinant $\Delta(\underline{\vec{r}})$ itself:*

$$q^{(n)}(\vec{r}|\underline{\vec{r}})=\Delta(\vec{r}_1,\ldots,\vec{r}_{n-1},\vec{r},\vec{r}_{n+1},\ldots,\vec{r}_N)\,/\,\Delta(\vec{r}_1,\ldots,\vec{r}_m,\ldots,\vec{r}_N)$$

$$\equiv\begin{vmatrix} s_1(\vec{r}_1) & \cdots & s_N(\vec{r}_1) \\ \vdots & \vdots & \vdots \\ s_1(\vec{r}_{n-1}) & \cdots & s_N(\vec{r}_{n-1}) \\ s_1(\vec{r}) & \cdots & s_N(\vec{r}) \\ s_1(\vec{r}_{n+1}) & \cdots & s_N(\vec{r}_{n+1}) \\ \vdots & \vdots & \vdots \\ s_1(\vec{r}_N) & \cdots & s_N(\vec{r}_N) \end{vmatrix} \Bigg/ \begin{vmatrix} s_1(\vec{r}_1) & \cdots & s_N(\vec{r}_1) \\ \vdots & \vdots & \vdots \\ s_1(\vec{r}_{n-1}) & \cdots & s_N(\vec{r}_{n-1}) \\ s_1(\vec{r}_n) & \cdots & s_N(\vec{r}_n) \\ s_1(\vec{r}_{n+1}) & \cdots & s_N(\vec{r}_{n+1}) \\ \vdots & \vdots & \vdots \\ s_1(\vec{r}_N) & \cdots & s_N(\vec{r}_N) \end{vmatrix}. \qquad (9)$$

The notation $q^{(n)}(\vec{r}|\underline{\vec{r}})$ is used to indicate that these interpolational functions of the (S-vector) variable $\vec{r}$ also depend, as implied by their definition, on the choice of the N nodes $\vec{r}_n$, whose N values are encoded in the S-vector-valued N-vector $\underline{\vec{r}}$. Let the diligent reader now pose and ponder on the different significance of the variable $\vec{r}$ and of the N nodes $\vec{r}_n$, whose values are encoded in the S-vector-valued N-vector $\underline{\vec{r}}$ and which enter, as it were parametrically, in the definition of the N interpolational functions $q^{(n)}(\vec{r}|\underline{\vec{r}})$, see (9). As entailed by this definition, these N interpolational functions $q^{(n)}(\vec{r}|\underline{\vec{r}})$ also depend, of course, on the choice of the set of seeds $\{s_n(\vec{r}); n=1,2,\ldots,N\}$. In the following the explicit indication of the dependence on $\underline{\vec{r}}$ is sometimes omitted, namely we sometimes write $q^{(n)}(\vec{r})$ in place of $q^{(n)}(\vec{r}|\underline{\vec{r}})$.

It is now clear that the N functions $q^{(n)}(\vec{r}|\underline{\vec{r}})$ possess the following two properties: *(i)* $q^{(n)}(\vec{r}|\underline{\vec{r}})$ is a linear combination, with constant (i.e., $\vec{r}$-independent) coefficients, of the N seeds $s_m(\vec{r})$,

$$q^{(n)}(\vec{r}|\underline{\vec{r}})=\sum_{m=1}^{N} s_m(\vec{r})\, c_{mn}, \quad n=1,\ldots,N; \qquad (10)$$

(ii) $q^{(n)}(\vec{r}|\underline{\vec{r}})$ vanishes for $\vec{r}=\vec{r}_m$ with $m \neq n$, and takes the value unity at $\vec{r}=\vec{r}_n$,

$$q^{(n)}(\vec{r}_m|\underline{\vec{r}})=\delta_{nm} \quad . \tag{11}$$

Hence an explicit expression of the interpolating function $f(\vec{r})$, characterized by (6) and (7), reads

$$f(\vec{r})=\sum_{n=1}^{N} f_n \, q^{(n)}(\vec{r}|\underline{\vec{r}})=\sum_{n=1}^{N} f(\vec{r}_n) \, q^{(n)}(\vec{r}|\underline{\vec{r}}) \quad . \tag{12}$$

The standard formulas of Lagrangian polynomial interpolation, see Sect. 2.4.2, obtain from those given above for $S=1$ (one-dimensional space) and for the following choice of the N seeds:

$$s_n(x)=x^{n-1}, \quad n=1,2,\ldots,N \quad . \tag{13}$$

In this special case the determinant $\Delta \equiv \Delta(x_1,x_2,\ldots,x_N)$, see (1), becomes the *Vandermonde determinant*,

$$\Delta(x_1,\ldots,x_N)=\det[\ (x_n)^{m-1}\ ] \quad , \tag{14a}$$

and it admits therefore the factorized representation

$$\Delta(x_1,\ldots,x_N)= \prod_{n,m=1;n>m}(x_n-x_m) \quad ; \tag{14b}$$

hence the "interpolational functions" $q^{(n)}(x|\underline{x})$, see (9), become the *polynomials* (2.4.2-5) (of degree $N-1$):

$$q^{(n)}(x|\underline{x})=q_{N-1}^{(n)}(x)= \prod_{m=1,m\neq n}^{N} \left[(x-x_n)/(x_n-x_m)\right] \quad . \tag{15}$$

Exercise 3.1-1. Show that any redefinition of the seeds via a linear (invertible) transformation,

$$\tilde{s}_n(\vec{r})=\sum_{m=1}^{N} a_{nm}\, s_m(\vec{r}) \quad , \tag{16a}$$

where the N^2 coefficients a_{nm} are constant ($\vec{r}$-independent) but otherwise arbitrary except for the condition that the $(N \times N)$-matrix a_{nm} be invertible, namely

$$\det[a_{nm}] \neq 0 \ , \tag{16b}$$

leads to the *same* set of interpolational functions $q^{(n)}(\vec{r}|\underline{r})$.

3.1.1 Finite-dimensional representation of the operator of differentiation

Let us now assume that the set of N seeds $s_n(\vec{r})$ is *closed under the operation of (partial) differentiation*, namely that *the S partial derivatives of every seed can be expressed as a linear superposition (with constant coefficients) of the seeds themselves:*

$$\vec{\nabla} s_n(\vec{r}) = \sum_{m=1}^{N} s_m(\vec{r}) \, \vec{\nabla}_{mn} , \quad n = 1, \dots, N . \tag{1}$$

Of course here, and always below, $\vec{\nabla} \equiv (\partial/\partial x, \partial/\partial y, \dots)$ is the gradient differential operator in S-dimensional space.

Note that, via this formula, we have introduced the (constant, i.e. $\vec{r}$-independent) S-vector-valued $(N \times N)$-matrix $\underline{\vec{\nabla}}$, whose N^2 matrix elements are the N^2 constant S-vectors $\vec{\nabla}_{nm}$. Clearly in order for (1) to hold it is necessary (but not sufficient) that the seeds be constructed out of elementary functions, i.e. (integer) powers and exponentials. For instance, for $S = 2$, the set of (4) seeds

$$\{s_n(\vec{r})\} = \{1, x, y, x^2\} \tag{2}$$

possess the property to be closed under differentiation, as well as the set of 8 seeds

$$\{s_n(\vec{r})\} = \{1, x, y, x^2, \exp(ax + by), \sinh(ax + by),$$

$$(\alpha x + \beta y) \sinh(ax + by), (\alpha x + \beta y) \cosh(ax + by)\} \ , \tag{3}$$

while the set of 5 seeds

$$\{s_n(\vec{r})\} = \{1, x, y, x^2, xy, x^2 y^2\} \tag{4}$$

do not.

Exercise 3.1.1-1. Prove these statements, and compute $\underline{\vec{\nabla}}_{mn}$, see (1), for the sets (2) and (3).

Let us however emphasize that our treatment can be extended to the case when the set of seeds is *not* closed under differentiation, as we show below towards the end of Sect. 3.1.1.

It is now clear that the following formula holds:

$$\vec{\nabla} q^{(n)}(\vec{r}|\underline{\vec{r}}) = \sum_{m=1}^{N} q^{(m)}(\vec{r}|\underline{\vec{r}})\, \vec{D}_{mn}(\underline{\vec{r}}), \quad n = 1, \ldots, N, \tag{5}$$

of course with $q^{(n)}(\vec{r}|\underline{\vec{r}})$ defined by (3.1-9) and with $\vec{D}_{mn}(\underline{\vec{r}})$ the (mn)-th element of the S-vector valued *constant* $(N \times N)$-matrix $\underline{\vec{D}}(\underline{\vec{r}})$. Note that the property of $\underline{\vec{D}}$ to be *constant* refers to its independence from the variable $\vec{r}$; $\underline{\vec{D}}(\underline{\vec{r}})$ depends instead, of course, on the choice of the N seeds $s_n(\vec{r})$ and, as our notation emphasizes (and in contrast to $\underline{\vec{\nabla}}$, see (1)) on the N nodes $\vec{r}_n$, indeed there clearly holds the (important) formula

$$\vec{D}_{nm}(\underline{\vec{r}}) = \vec{\nabla} q^{(m)}(\vec{r}_n|\underline{\vec{r}}) \equiv \vec{\nabla} q^{(m)}(\vec{r}|\underline{\vec{r}})\Big|_{\vec{r}=\vec{r}_n}. \tag{6}$$

Occasionally, in the following, we omit to indicate explicitly the dependence of $\underline{\vec{D}}(\underline{\vec{r}})$ on $\underline{\vec{r}}$, namely we write $\underline{\vec{D}}$ instead of $\underline{\vec{D}}(\underline{\vec{r}})$.

The *proof* of (6) is immediate: set $\vec{r} = \vec{r}_m$ in (5) (but be careful: before doing this you should rename the dummy summation index in the right hand side !) and use (3.1-11). Then (for notational convenience) exchange the two indices n and m.

Exercise 3.1.1-2. Compute the matrices $\underline{\vec{D}}(\underline{\vec{r}})$ for the sets (2) and (3).

It is now clear that the S components of the $(N \times N)$-matrix-valued S-vector $\vec{\underline{D}}(\underline{\vec{r}})$ provide faithful $(N \times N)$-matrix representations of the operators of partial differentiation in S-dimensional space, in the following sense. Let us associate to every function that admits the representation (3.1-6) (to which our consideration is hereafter restricted) the N-vector $\underline{f}$, whose N components are the N values, see (3.1-7), that the function $f(\vec{r})$ takes at the N nodes $\vec{r}_n$,

$$\underline{f} \equiv (f_1, f_2, ..., f_N) = (f(\vec{r}_1), f(\vec{r}_2), ..., f(\vec{r}_N)) \quad . \tag{7}$$

It is then easily seen that there holds the N-vector formula

$$\underline{f_x} = \underline{D}_x(\underline{\vec{r}}) \, \underline{f} \quad , \tag{8a}$$

which features, in the left hand side, the N-vector $\underline{f_x}$ associated to the function $f_x(\vec{r}) \equiv \partial f(\vec{r}) / \partial x$ (i.e., to the partial derivative, with respect to the x-component of the S-vector $\vec{r}$, of the function $f(\vec{r})$),

$$\underline{f_x} \equiv (f_x(\vec{r}_1), f_x(\vec{r}_2), ..., f_x(\vec{r}_N)) \quad , \tag{8b}$$

and, in the right hand side, the $(N \times N)$-matrix $\underline{D}_x(\underline{\vec{r}})$ (i.e., the x-component of the $(N \times N)$-matrix-valued S-vector $\vec{\underline{D}}(\underline{\vec{r}})$, see (6)), acting on the N-vector $\underline{f}$. Here of course x stands for any component of the S-vector $\vec{r}$, indeed a more general version of (8) reads as follows:

$$\underline{\vec{\nabla} f} = \vec{\underline{D}}(\underline{\vec{r}}) \, \underline{f} \quad , \tag{9a}$$

or equivalently (see (3.1-4))

$$\left[\vec{\nabla} f(\vec{r})\right]\Big|_{\vec{r}=\vec{r}_n} = \sum_{m=1}^{N} \vec{D}_{nm}(\underline{\vec{r}}) \, f_m = \sum_{m=1}^{N} \vec{D}_{nm}(\underline{\vec{r}}) \, f(\vec{r}_m), \quad n=1,...,N \; . \tag{9b}$$

The proof of these equations is immediate, since (3.1-12) entails

$$\vec{\nabla} f(\vec{r}) = \sum_{m=1}^{N} f_m \, \vec{\nabla} q^{(m)}(\vec{r}|\underline{\vec{r}}) \quad , \tag{10}$$

hence, via (5),

$$\vec{\nabla} f(\vec{r}) = \sum_{m=1}^{N} f_m \sum_{\ell=1}^{N} q^{(\ell)}(\vec{r}|\underline{\vec{r}})\, \underline{\vec{D}}_{\ell m}(\underline{\vec{r}}) \ . \tag{11}$$

Setting now $\vec{r} = \vec{r}_n$, and using (3.1-11), there obtains (9b), which is thereby proven.

One can then state the following

Proposition 3.1.1-3. Assume that there hold the following partial differential equation:

$$\mathrm{A} f(\vec{r}) = 0 \ , \tag{12}$$

with the linear differential operator A defined, in self-evident notation, by the formula

$$\mathrm{A} = \sum_{\alpha,\beta,\gamma,..} a_{\alpha\beta\gamma..}(\vec{r})\ \partial^{\alpha+\beta+\gamma+\cdots} / \partial x^{\alpha} \partial y^{\beta} \partial z^{\gamma} \cdots \ , \tag{13}$$

where $\alpha, \beta, \gamma, \ldots$ are of course nonnegative integers. There then also holds the N-vector formula

$$\underline{A}\, \underline{f} = 0 \ , \tag{14}$$

with the N-vector $\underline{f}$ defined by (7) and the $(N \times N)$-matrix $\underline{A}$ defined by the formula (see (14))

$$\underline{A} = \sum_{\alpha,\beta,\gamma,..} a_{\alpha\beta\gamma..}(\underline{\vec{R}})\, [\ \underline{D_x}(\underline{\vec{r}})\ ]^{\alpha} [\ \underline{D_y}(\underline{\vec{r}})\ ]^{\beta} [\ \underline{D_z}(\underline{\vec{r}})\ ]^{\gamma} \cdots \tag{15}$$

obtained by applying to the operator A, see (13), the substitution rule

$$\vec{r} \Rightarrow \underline{\vec{R}}, \ \ \partial / \partial x \Rightarrow \underline{D_x}(\underline{\vec{r}}), \ \ \partial / \partial y \Rightarrow \underline{D_y}(\underline{\vec{r}}), \ldots \ . \tag{16}$$

Here we are of course using the definition (3.1-3) of the $(N \times N)$-(diagonal) matrix-valued S-vector $\underline{\vec{R}}$, as well as the definition (6) (see also (9)) of the $(N \times N)$-matrix-valued S-vector $\underline{\vec{D}} = \underline{\vec{D}}(\underline{\vec{r}})$ (and of course $\underline{D_x}$ respectively $\underline{D_y}$ are the x-component respectively the y-component of the S-vector $\underline{\vec{D}}$).

Of course the validity of this *Proposition 3.1.1-3.* is predicated upon the fact that $f(\vec{r})$ be a linear combination (with constant coefficients) of the N seeds $s_n(\vec{r})$ (i.e., that it admit the representation (3.1-3)), and moreover that the set of seeds $s_n(\vec{r})$ be closed under differentiation. Note that this entails that *all* functions obtained from $f(\vec{r})$ by (multiple) differentiation are also expressible as linear combinations (with constant coefficients) of the N seeds, hence they also admit representations of type (3.1-6). This clearly entails that (9) can be iterated, namely that there also holds the more general formula (in self-evident notation)

$$\partial^{\alpha+\beta+\gamma+\cdots} f(\vec{r})/\partial x^{\alpha}\partial y^{\beta}\partial z^{\gamma}\cdots\Big|_{\vec{r}=\vec{r}_n} = \{ [\, \underline{D_x}(\vec{r}) \,]^{\alpha} [\, \underline{D_y}(\vec{r}) \,]^{\beta} [\, \underline{D_z}(\vec{r}) \,]^{\gamma} \cdots \underline{f} \,\}_n , \quad (17)$$

where the notation $\{\underline{u}\}_n$ in the right hand side denotes of course the n-th component of the N-vector $\underline{u}$.

The *proof* of *Proposition 3.1.1-3.* is then immediate: set $\vec{r}=\vec{r}_n$ in (12) with (13) and use (3.1-12) and (3.1-3) to get (14) with (15).

Remark 3.1.1-4. The commutativity of differentiations with respect to different variables which may be expressed, say, as the *operator identity*

$$[\, \partial_x, \partial_y \,] = 0 \ , \quad (18)$$

entails that a corresponding formula, say,

$$[\, \underline{D_x}, \underline{D_y} \,] = 0 \ , \quad (19)$$

hold for the $(N\times N)$-matrices $\underline{D_x}, \underline{D_y}$. Here of course x and y stand for any two components of the S-vector $\vec{r}$.

Exercise 3.1.1-5. Check the validity of this formula using the solutions of *Exercise 3.1.1-2.*

Clearly *Proposition 3.1.1-3* entails that, *to every partial differential equation of the general form (12) with (13) satisfied by a function admitting the representation (3.1-6), there corresponds an* N*-vector equation, immediately obtainable via the substitution rule*

$$\vec{r} \Rightarrow \vec{\underline{R}} \ , \quad \vec{\nabla} \Rightarrow \vec{\underline{D}}(\vec{r}) \ , \quad f(\vec{r}) \Rightarrow \underline{f} \ . \quad (20)$$

This also entails the following

Corollary 3.1.1-6. If the differential operator A, see (13), possesses the eigenvalue a,

$$\mathrm{A}\, f^{(a)}(\vec{r}) = a\, f^{(a)}(\vec{r}) \ , \tag{21}$$

and the corresponding eigenfunction $f^{(a)}(\vec{r})$ belongs to the functional space spanned by the N seeds (namely, it admits the representation (3.1-6)), then the $(N\times N)$-matrix $\underline{A}$, see (15), also possesses the same eigenvalue a,

$$\underline{A}\, \underline{f}^{(a)} = a\, \underline{f}^{(a)} \ , \tag{22}$$

and the corresponding eigenvector is related to the eigenfunction $f^{(a)}(\vec{r})$ by (7),

$$\underline{f}^{(a)} = (f^{(a)}(\vec{r}_1), f^{(a)}(\vec{r}_2), \ldots, f^{(a)}(\vec{r}_n)) \ . \tag{23}$$

It is instructive to consider the relationships of these results with those of Sect. 2.4.

In the one-dimensional case ($S=1$), and for the choice (3.1-13) of seeds, the $(N\times N)$-matrix

$$\underline{\vec{D}}(\vec{r}) \equiv \underline{D_x} \equiv \underline{\tilde{D}} \tag{24}$$

has the explicit representation

$$\underline{\tilde{D}} = \underline{B}\ \underline{D}\ \underline{B}^{-1} \ , \tag{25}$$

with the $(N\times N)$-matrices $\underline{B}$ respectively $\underline{D}$ defined by (2.4.1-4b) respectively (2.4.1-2) in terms of the nodes $\vec{r}_n \equiv x_n$; likewise

$$\underline{\vec{R}} = \underline{X} \ , \tag{26}$$

with the $(N\times N)$-matrix $\underline{X}$ defined by (2.4.1-1).

Proofs. The validity of (26) is an immediate consequence of the definitions (3.1-3) and (2.4.1-1).

As for (25), it follows from (6) and from the expression (3.1-15) of $q_{N-1}^{(n)}(x)$. Indeed this entails

$$d\, q_{N-1}^{(n)}(x)/dx = \sum_{\ell=1,\ell\neq n}^{N} (x_n - x_\ell)^{-1} \prod_{j=1,j\neq n,\ell}^{N} [(x - x_j)/(x_n - x_j)] \;, \tag{27}$$

namely, via (6) and (24),

$$\widetilde{D}_{mn} = \sum_{\ell=1,\ell\neq n}^{N} (x_n - x_\ell)^{-1} \prod_{j=1,j\neq n,\ell}^{N} [(x_m - x_j)/(x_n - x_j)] \;. \tag{28}$$

Hence, for $n = m$,

$$\widetilde{D}_{nn} = \sum_{\ell=1,\ell\neq n}^{N} (x_n - x_\ell)^{-1} = d_n = D_{nn} \tag{29}$$

(see (2.4.1-2,3,5)), while for $n \neq m$

$$\widetilde{D}_{mn} = (x_n - x_m)^{-1} \prod_{j=1,j\neq n,m}^{N} [(x_m - x_j)/(x_n - x_j)] \;, \tag{30a}$$

$$\widetilde{D}_{mn} = b_n^{-1} (x_m - x_n)^{-1}\, b_m \;, \tag{30b}$$

$$\widetilde{D}_{nm} = b_n D_{nm} b_m^{-1} \;. \tag{30c}$$

To get (30a) from (28) we noted that the product in the right hand side of (28) vanishes unless $\ell = m$; to get (30b) we used the definition (2.4.1-4a) of b_n, while to get (30c) we exchanged the indices n and m and we used the definition (2.4.1-2) of D_{nm} (with $n \neq m$).

It is now clear, via the definition (2.4.1-4) of the *diagonal* $(N \times N)$-matrix $\underline{B}$, that (29) and (30c) coincide with (25), which is thereby proven.

Since the $(N \times N)$-matrices $\underline{B}$ and $\underline{X}$ obviously commute (they are both diagonal!), the formulas of *Proposition 3.1.1-3,* see (12), (13), (14) and (15), can now be rewritten, using (24), (25) and (26), as follows:

$$\mathrm{A}\, f(x) = 0 \;, \tag{31}$$

$$\mathrm{A} = \sum_{\alpha=0} a_\alpha(x)\, (d/dx)^\alpha \;, \tag{32}$$

$$\check{\underline{A}}\, \check{\underline{f}} = 0 \;, \tag{33}$$

$$\check{\underline{A}} = \sum_{\alpha=0} a_\alpha(\underline{X})\, \underline{D}^\alpha \;, \tag{34}$$

$$\check{\underline{f}} = \underline{B}^{-1}\,\underline{f} \;. \tag{35}$$

It is thus seen that, in the (one-dimensional) case of standard Lagrangian (polynominal) interpolation, an equivalent transformation rule from linear differential equations to N-vector equations involves the replacement

$$x \Rightarrow \underline{X},\; d/dx \Rightarrow \underline{D},\; f(x) \Rightarrow \underline{B}^{-1}\,\underline{f} \;, \tag{36}$$

with the matrices $\underline{X}$ and $\underline{D}$ defined as in Sect. 2.4, see (2.4.1-1) and (2.4.1-2).

The advantage of the rule (36) over the (one-dimensional version of the) rule (20) resides in the simpler expression of the $(N \times N)$-matrix $\underline{D}$ in terms of the N nodes x_n, see (2.4.1-2), as compared to the analogous expression of the $(N \times N)$-matrix $\underline{\vec{D}} = \underline{\tilde{D}}$, see (24), in terms of the N nodes x_n, see (25) with (2.4.1-4).

As we will see in some of the examples given below, there are also other cases in which a similarity transformation, analogous to (25), generated by a *diagonal* matrix, is instrumental in yielding a convenient simplification.

Before ending Sect. 3.1.1 two important points must be made, and a useful (final) *remark.*

To derive the results reported above (in Sect. 3.1.1) we assumed the functional space spanned by the N seeds $s_n(\vec{r})$ to be *closed under differentiation,* see (1). It is important to note that a (finite-dimensional) $(N \times N)$-matrix representation of the operator of differentiation can be usefully introduced even if this condition does *not* hold, and that even in such a case some formulas remain valid precisely as they have been written above, while others remain valid after appropriate modifications, and others are not valid at all. It is indeed clear that the crucial formula (9) (of course, in both its avatars, (9a) and (9b)) remains valid, together with the fundamental definition (6) of $\underline{\vec{D}}(\underline{\vec{r}})$.

Proof. Differentiation of (3.1-12) yields

$$\vec{\nabla} f(\vec{r}) = \sum_{m=1}^{N} f_m\, \vec{\nabla} q^{(m)}(\vec{r}\,|\,\underline{\vec{r}}) \;, \tag{37}$$

and by setting in this formula $\vec{r}=\vec{r}_n$ one obtains precisely (9) with (6). (Note that we simply reproduced here the proof of (9) given above, which indeed does *not* require the set of seeds to be closed under differentiation).

On the other hand now (11) ceases to hold (except at the nodes, see (9)) and (17) must be modified to read

$$\left[\partial^{\alpha+\beta+\gamma\cdots} f(\vec{r})/\partial x^{\alpha}\,\partial y^{\beta}\,\partial z^{\gamma}\cdots\right]\Big|_{\vec{r}=\vec{r}_n}=\underline{D_{xyz\cdots}^{(\alpha\beta\gamma\cdots)}(\vec{r})}\;\underline{f}\;, \tag{38}$$

with the following definition of the $(N\times N)$-matrix $\underline{D_{xyz\cdots}^{(\alpha\beta\gamma\cdots)}(\vec{r})}$:

$$\left[\underline{D_{xyz\cdots}^{(\alpha\beta\gamma\cdots)}(\vec{r})}\right]_{mn}=\partial^{\alpha+\beta+\gamma\cdots}q^{(n)}(\vec{r}|\underline{\vec{r}})/\partial x^{\alpha}\,\partial y^{\beta}\,\partial z^{\gamma}\cdots\Big|_{\vec{r}=\vec{r}_m}\;. \tag{39}$$

Proof. As above, from (3.1-9) we get, by multiple differentiation,

$$\partial^{\alpha+\beta+\gamma\cdots} f(\vec{r})/\partial x^{\alpha}\,\partial y^{\beta}\,\partial z^{\gamma}\cdots=\sum_{m=1}^{N} f_m\left[\partial^{\alpha+\beta+\gamma\cdots}q^{(m)}(\vec{r}|\underline{\vec{r}})/\partial x^{\alpha}\,\partial y^{\beta}\,\partial z^{\gamma}\cdots\right], \tag{40}$$

and by setting $\vec{r}=\vec{r}_n$ in this formula we get (38) with (39). Of course this formula, (38) with (39), is as well valid in the case treated above (of a seed space closed under differentiation), but in that case there holds the additional $(N\times N)$-matrix formula

$$\underline{D_{xyz\cdots}^{(\alpha\beta\gamma\cdots)}(\vec{r})}=\left[\,\underline{D_x(\vec{r})}\,\right]^{\alpha}\left[\,\underline{D_y(\vec{r})}\,\right]^{\beta}\left[\,\underline{D_z(\vec{r})}\,\right]^{\gamma}\cdots\;, \tag{41}$$

where the (integer) exponents $\alpha,\beta,\gamma,\cdots$ in the right hand side indicate of course arbitrary (*positive integer*) powers. Note that in this formula, (41), the ordering of the $(N\times N)$-matrices in the right hand side is irrelevant, since these matrices commute, see (19). On the other hand (19) does not necessarily hold if the seed space is not closed under differentiation, but it must be replaced by, say,

$$\underline{D_{xy}^{(\alpha\beta)}(\vec{r})}=\underline{D_{yx}^{(\beta\alpha)}(\vec{r})}\;, \tag{42}$$

and by analogous, more general, formulas that are obvious consequences of the definition (39) together with the commutativity of different differential operators.

It is also obvious that *Proposition 3.1.1-3* and *Corollary 3.1.1-6* remain valid even if the seed space is *not* closed under differentiation, provided the definition of the $(N \times N)$-matrix $\underline{A}$ is modified to read (instead of (15))

$$\underline{A} = \sum_{\alpha,\beta,\gamma \cdots} a_{\alpha\beta\gamma}(\vec{R})\, \underline{D_{xyz\cdots}^{(\alpha\beta\gamma\cdots)}(\vec{r})}, \tag{43}$$

of course with $\underline{D_{xyz\cdots}^{(\alpha\beta\gamma\cdots)}(\vec{r})}$ defined by (39). Clearly this formula, (43), reduces to (16) whenever (41) holds. Of course, for the validity of *Proposition 3.1.1-3* (or *Corollary 3.1.1-6*), modified as we just indicated, it remains essential that the function $f(\vec{r})$ that satisfies (12) (or the function $f^{(a)}(\vec{r})$ that satisfies (21)) live in the functional space spanned by the N seeds $s_n(\vec{r})$, namely admit the fundamental representation (3.1-6).

This completes our discussion of the notation, and properties, of finite-dimensional $(N \times N)$-matrix representations of the operators of differentiation in the (more general) context of N-dimensional seed spaces which are *not* closed under differentiation.

The second observation relevant to the introduction of finite-dimensional $(N \times N)$-matrix representations of the differential operator $\vec{\nabla}$ focuses on the following special form of the interpolational functions:

$$q^{(n)}(\vec{r}|\underline{\vec{r}}) = \prod_{\ell=1,\ell\neq n}^{N} \left[\varphi_\ell(\vec{r}-\vec{r}_\ell)/\varphi_\ell(\vec{r}_n-\vec{r}_\ell)\right] , \tag{44a}$$

where the N functions $\varphi_\ell(\vec{r})$ are arbitrary, except for the crucial condition to vanish at the origin,

$$\varphi_n(\vec{0}) \equiv \varphi_n(\vec{r})\Big|_{\vec{r}=\vec{0}} = 0 , \tag{44b}$$

which is clearly sufficient to guarantee the fundamental property (3.1-11) of these interpolational functions $q^{(n)}(\vec{r}|\underline{\vec{r}})$, see (44), hence validity of (3.1-12) with (3.1-7), namely validity of the standard interpolation formula

$$f(\vec{r}) = \sum_{n=1}^{N} f(\vec{r}_n)\, q^{(n)}(\vec{r}|\underline{\vec{r}}) . \tag{45}$$

Before proceeding to investigate the finite-dimensional $(N \times N)$-matrix representations of the differential operator $\vec{\nabla}$ entailed by the interpolation formula (45) with (44), let us pause to interject two remarks. In the first place it must be noted that, except for some very special choices of the seeds $s_n(\vec{r})$ and correspondingly of the functions $\varphi_\ell(\vec{r})$, see (44), the interpolation formula (45) is generally *inconsistent* with the fundamental representation (3.1-6) of $f(\vec{r})$ in terms of the seeds $s_n(\vec{r})$, if one requests that the seeds $s_n(\vec{r})$ be *independent* of the (choice of the) nodes $\vec{r}_n$. Note that, while the assumption that the seeds $s_n(\vec{r})$ are independent of the (choice of the) nodes $\vec{r}_n$ is perhaps implied by our notation, it is in fact *not* required for the validity of the results reported above. But, as we shall see, it plays a crucial role when these findings are used to manufacture many-body problems amenable to exact treatments, see below.

Secondly, let us emphasize that in the prototypical case of polynomial Lagrangian interpolation in one-dimensional space, corresponding to the choice of seeds (3.1-13), the interpolational polynomials $q^{(n)}(x)$ do indeed take the factorized form (44), see (3.1-15).

Let us now compute the $(N \times N)$-matrix representation of the differential operator $\vec{\nabla}$ that corresponds, of course via (6), to the interpolational functions (44). It reads:

$$[\,\underline{\vec{D}}(\vec{r})\,]_{nm} = \delta_{nm} \sum_{\ell=1,\ell\neq n}^{N} \left\{ \left[\vec{\nabla}\varphi_\ell(\vec{r}_n - \vec{r}_\ell)\right] / \varphi_\ell(\vec{r}_n - \vec{r}_\ell) \right\}$$

$$+ (1-\delta_{nm})\,(\beta_n / \beta_m)\,\left\{ \left[\vec{\nabla}\varphi_n(\vec{0})\right] / \varphi_m(\vec{r}_n - \vec{r}_m) \right\} \,, \tag{46a}$$

$$\beta_n \equiv \beta_n(\vec{r}) \equiv \prod_{\ell=1,\ell\neq n}^{N} \varphi_\ell(\vec{r}_n - \vec{r}_\ell) \;. \tag{46b}$$

Here of course

$$\vec{\nabla}\varphi_\ell(\vec{r}_n - \vec{r}_\ell) \equiv \vec{\nabla}\varphi_\ell(\vec{r})\Big|_{\vec{r}\,=\,\vec{r}_n - \vec{r}_\ell} \,, \tag{47a}$$

$$\vec{\nabla}\varphi_\ell(\vec{0}) \equiv \vec{\nabla}\varphi_\ell(\vec{r})\Big|_{\vec{r}\,=\,\vec{0}} \;. \tag{47b}$$

Proof. From (44a), by logarithmic differentiation,

$$\vec{\nabla} q^{(m)}(\vec{r}|\underline{\vec{r}}) \equiv \sum_{\ell=1,\ell\neq m}^{N} \{[\vec{\nabla}\varphi_\ell(\vec{r}-\vec{r}_\ell)]/\varphi_\ell(\vec{r}_m-\vec{r}_\ell)\} \prod_{\ell'=1,\ell'\neq m,\ell}^{N} [\varphi_{\ell'}(\vec{r}-\vec{r}_{\ell'})/\varphi_{\ell'}(\vec{r}_m-\vec{r}_{\ell'})], \quad (48a)$$

hence, for $\vec{r}=\vec{r}_m$,

$$\vec{\nabla} q^{(m)}(\vec{r}_m|\underline{\vec{r}}) = \sum_{\ell=1,\ell\neq m}^{N} \{[\vec{\nabla}\varphi_\ell(\vec{r}_m-\vec{r}_\ell)]/\varphi_\ell(\vec{r}_m-\vec{r}_\ell)\} , \quad (48b)$$

while, for $\vec{r}=\vec{r}_n$ with $n\neq m$,

$$\nabla q^{(m)}(\vec{r}_n|\underline{\vec{r}}) = \{[\vec{\nabla}\varphi_n(\vec{0})]/\varphi_n(\vec{r}_m-\vec{r}_n)\} \prod_{\ell=1,\ell\neq n,m}^{N} [\varphi_\ell(\vec{r}_n-\vec{r}_\ell)/\varphi_\ell(\vec{r}_m-\vec{r}_\ell)] , \quad (48c)$$

since only the term with $\ell=n$ contributes to the sum in the right hand side of (48a), due to (44b). Clearly (48b) and (48c) yield, via (6), precisely (46), which is thereby proven.

Clearly via the definition

$$\underline{B}(\vec{r})=\mathrm{diag}[\beta_n(\underline{\vec{r}});n=1,...,N], \quad B_{nm}(\underline{\vec{r}})=\delta_{nm}\beta_n(\underline{\vec{r}}) , \quad (49)$$

there holds the formula

$$\underline{\vec{D}}(\underline{\vec{r}})=\underline{B}(\underline{\vec{r}})\ \underline{\check{D}}(\underline{\vec{r}})\ [\underline{B}(\underline{\vec{r}})]^{-1} \quad (50a)$$

with

$$\left[\underline{\check{D}}(\underline{\vec{r}})\right]_{nm} = \delta_{nm} \sum_{\ell=1,\ell\neq n}^{N} \{[\vec{\nabla}\varphi_\ell(\vec{r}_n-\vec{r}_\ell)]/\varphi_\ell(\vec{r}_n-\vec{r}_\ell)\}$$

$$+(1-\delta_{nm})\{[\vec{\nabla}\varphi_n(\vec{0})]/\varphi_m(\vec{r}_n-\vec{r}_m)\} . \quad (50b)$$

Note that both the $(N\times N)$-matrices $\underline{\check{D}}(\underline{\vec{r}})$, see (50b), and $\underline{D}(\underline{\vec{r}})$, see (46), become *diagonal* if $\vec{\nabla}\varphi_n(\vec{r})$ (in addition to $\varphi_n(\vec{r})$, see (44b)) vanishes at $\vec{r}=0$,

$$\vec{\nabla}\varphi_n(\vec{0})\equiv\vec{\nabla}\varphi_n(\vec{r})\Big|_{\vec{r}=\vec{0}}=0 \quad . \tag{51}$$

Exercise 3.1.1-7. Repeat the treatment just given, and derive the formula analogous to (46), in the more general case when (44a) is replaced by

$$q^{(n)}(\vec{r}|\underline{\vec{r}})=\prod_{\ell=1,\ell\neq n}^{N}\left[\varphi_{n\ell}(\vec{r}-\vec{r}_\ell)\,/\,\varphi_{n\ell}(\vec{r}_n-\vec{r}_\ell)\right] \;, \tag{52a}$$

which features now N^2 functions $\varphi_{n\ell}(\vec{r})$, arbitrary except for the crucial condition to vanish at the origin,

$$\varphi_{n\ell}(\vec{0})\equiv\varphi_{n\ell}(\vec{r})\Big|_{\vec{r}=\vec{0}}=0 \quad , \tag{52b}$$

which is, again, clearly sufficient to guarantee the fundamental property (3.1-11), hence validity of the representation (3.1-12).

We end Sect. 3.1.1 with the following

Remark 3.1.1-8. Let the $(N\times N)$-matrix $\underline{\vec{D}}(\underline{\vec{r}})$ be the finite-dimensional $(N\times N)$-matrix representation of the differential operator $\vec{\nabla}$ associated with the set of seeds $\{s_n(\vec{r}),\, n=1,2,...,N\}$. Consider then the set of seeds

$$\tilde{s}_n(\vec{r})=w(\vec{r})\; s_n(\vec{r}),\; n=1,2,...,N \quad , \tag{53a}$$

where $w(\vec{r})$ is an arbitrary "weight function", only restricted by the condition not to vanish at any of the nodes,

$$w(\vec{r}_n)\neq 0\,,\quad n=1,2,...,N \quad , \tag{53b}$$

to avoid violating the rule (3.1-5). Then the $(N\times N)$-matrix $\underline{\tilde{\vec{D}}}(\underline{\vec{r}})$ which provides the $(N\times N)$-matrix representation of the differential operator $\vec{\nabla}$ associated with the set of seeds $\{\tilde{s}_n(\vec{r}),\, n=1,...,N\}$ and with the (*same*!) set of nodes $\vec{r}_n$ is given by the following simple rule:

$$\underline{\tilde{\vec{D}}}(\underline{\vec{r}})=\underline{W}(\underline{\vec{r}})\;\left[\underline{\vec{D}}(\underline{\vec{r}})+\underline{\vec{V}}(\underline{\vec{r}})\right]\left[\underline{W}(\underline{\vec{r}})\right]^{-1}=\underline{W}(\underline{\vec{r}})\,\underline{\vec{D}}(\underline{\vec{r}})\left[\underline{W}(\underline{\vec{r}})\right]^{-1}+\underline{\vec{V}}(\underline{\vec{r}}) \quad , \tag{54}$$

with the two *diagonal* (hence commuting) $(N \times N)$-matrices $\underline{W}(\underline{r})$ and $\underline{\vec{V}}(\underline{r})$ defined by the following simple rules:

$$\underline{W}(\underline{r}) = \mathrm{diag}\left[w(\vec{r}_n)\right], \; W_{nm}(\underline{r}) = \delta_{nm}\, w(\vec{r}_n) \quad , \tag{55}$$

$$\underline{\vec{V}}(\underline{r}) = \mathrm{diag}\left\{ \left[\vec{\nabla} w(\vec{r}_n)\right] / w(\vec{r}_n) \right\}, \; \vec{V}_{nm} = \delta_{nm}\left[\ \vec{\nabla} w(\vec{r}_n)\ \right] / w(\vec{r}_n) \quad , \tag{56a}$$

where of course

$$\vec{\nabla} w(\vec{r}_n) \equiv \left[\ \vec{\nabla} w(\vec{r})\ \right]\Big|_{\vec{r}=\vec{r}_n} \quad . \tag{56b}$$

Proof. Let the interpolational functions $q^{(n)}(\vec{r}|\underline{r})$ respectively $\tilde{q}^{(n)}(\vec{r}|\underline{r})$ be associated with the set of seeds $\{s_n(\vec{r}),\, n=1,\ldots,N\}$ respectively $\{\tilde{s}_n(\vec{r}),\, n=1,\ldots,N\}$, see (53). There then follows immediately, from the definition (3.1-6), that they are related as follows:

$$\tilde{q}^{(n)}(\vec{r}|\underline{r}) = \left[w(\vec{r})/w(\vec{r}_n)\right] q^{(n)}(\vec{r}|\underline{r}) \quad . \tag{57}$$

Hence, by logarithmic differentiation,

$$\left[\ \vec{\nabla}\tilde{q}^{(n)}(\vec{r}|\underline{r})\ \right] / \tilde{q}^{(n)}(\vec{r}|\underline{r}) = \left[\ \vec{\nabla} q^{(n)}(\vec{r}|\underline{r})\right] / q^{(n)}(\vec{r}|\underline{r}) + \left[\ \vec{\nabla} w(\vec{r})\ \right] / w(\vec{r}) \quad , \tag{58}$$

entailing, via (57),

$$\vec{\nabla}\tilde{q}^{(n)}(\vec{r}|\underline{r}) = \left[\ \vec{\nabla} q^{(n)}(\vec{r}|\underline{r})\ \right]\left[w(\vec{r})/w(\vec{r}_n)\right] + \left\{\left[\ \vec{\nabla} w(\vec{r})\ \right] / w(\vec{r}_n)\right\} \tilde{q}^{(n)}(\vec{r}|\underline{r}) \quad . \tag{59}$$

It is then clear, via (3.1-11) (which of course holds as well for the interpolational functions $\tilde{q}^{(n)}(\vec{r}|\underline{r})$), that there hold the following relations:

$$\vec{\nabla}\tilde{q}^{(n)}(\vec{r}_n|\underline{r}) = \vec{\nabla} q^{(n)}(\vec{r}_n|\underline{r}) + \left[\ \vec{\nabla} w(\vec{r}_n)\ \right] / w(\vec{r}_n) \quad , \tag{60a}$$

$$\vec{\nabla}\tilde{q}^{(n)}(\vec{r}_m|\underline{r}) = \vec{\nabla} q^{(n)}(\vec{r}_m|\underline{r})\left[w(\vec{r}_m)/w(\vec{r}_n)\right], \; n \neq m \quad . \tag{60b}$$

These relations, via (6), entail (54) (with (55) and (56)), which is thereby proven.

3.1.2 Examples

In the following Sects. 3.1.2.1, 3.1.2.2 respectively 3.1.2.3 we exhibit the expressions of the $(N \times N)$-matrices that provide, according to the treatment of Sect. 3.1.1, finite-dimensional $(N \times N)$-matrix representations of the operators of differentiation. These examples correspond to specific choices for the dimensionality S of ambient space (we limit our consideration to $S=1$, $S=2$ respectively $S=3$), and, in each of these cases, to specific choices for the number N of nodes (which of course coincides with the dimensionality N of the functional seed space), and to specific identifications of the N seeds.

3.1.2.1 One-dimensional space ($S=1$)

In Sect. 3.1.2.1 we focus on *one-dimensional* ambient space ($S=1$), hence denote the nodes by x_n, and the independent variable by x. In the preceding Sect. 3.1.1 we used, for the specific (one-dimensional) case of *polynomial* Lagrangian interpolation, the notation $\underline{\tilde{D}}$, see (3.1.1-30), for the (one-dimensional) version of the $(N \times N)$-matrix $\underline{\vec{D}}(\vec{r})$, to avoid confusion with the $(N \times N)$-matrix $\underline{D}$ introduced in Sect. 2.4.1, see (2.4.1-2) and (3.1.1-25). In Sect. 3.1.2.1 we stick generally to the notation $\underline{D} \equiv \underline{D}(\underline{x})$ for the one-dimensional version of the $(N \times N)$-matrix $\underline{\vec{D}}(\vec{r})$, except when we consider an example (see the next-to-last one treated below) that includes the standard case of (one-dimensional) Lagrangian (polynomial!) interpolation as a subcase.

We start from an elementary example:

$$N=2;\ \ s_1(x)=\exp(x),\ \ s_2(x)=\exp(-x)\ . \tag{1}$$

In this case the seed space is clearly closed under differentiation. The corresponding expression of the (2×2)-matrix $\underline{D}(\underline{x})$ providing a (2×2)-matrix representation of the differential operator d/dx, see (3.1.1-6), reads:

$$\underline{D}(\underline{x})=\begin{pmatrix} \text{cotgh}(x_1-x_2) & -1/\sinh(x_1-x_2) \\ 1/\sinh(x_1-x_2) & -\text{cotgh}(x_1-x_2) \end{pmatrix}\ . \tag{2}$$

Proof.

$$\Delta(x_1,x_2)=\begin{vmatrix}\exp(x_1) & \exp(-x_1)\\ \exp(x_2) & \exp(-x_2)\end{vmatrix}=2\sinh(x_1-x_2)\ ; \tag{3}$$

$$q^{(1)}(x|\underline{x})=\sinh(x-x_2)/\sinh(x_1-x_2)\ , \tag{4a}$$

$$q^{(2)}(x|\underline{x})=-\sinh(x-x_1)/\sinh(x_1-x_2)\ ; \tag{4b}$$

$$q_x^{(1)}(x|\underline{x})=\cosh(x-x_2)/\sinh(x_1-x_2)\ , \tag{5a}$$

$$q_x^{(2)}(x|\underline{x})=-\cosh(x-x_1)/\sinh(x_1-x_2)\ . \tag{5b}$$

Here (3) has been obtained from the definition (3.1-4) with (1); (4) from (3) via the definition (3.1-9); (5) by differentiating (4), and it yields directly (2) via (3.1.1-6).

In the functional space spanned by the seeds (1) the operator $\mathrm{A}=d/dx$ has the eigenvalues ± 1:

$$(d/dx)\exp(\pm x)=\pm\exp(\pm x)\ . \tag{6}$$

Hence *Corollary 3.1.1-6* entails that the matrix $\underline{D}(\underline{x})$, see (2), also have these eigenvalues, $+1$ respectively -1, with eigenvectors $(\exp(x_1),\exp(x_2))$ respectively $(\exp(-x_1),\exp(-x_2))$.

Exercise 3.1.2-1. Check this.

Exercise 3.1.2-2. Check explicitly that the choice of seeds

$$s_1(x)=a\exp(x)+b\exp(-x),\ s_2(x)=c\exp(x)+d\exp(-x)\ , \tag{7}$$

with a,b,c,d arbitrary constants $(ad\neq bc)$, yields the *same* (2×2)-matrix $\underline{D}(\underline{x})$, see (2), indeed the same interpolational functions $q^{(n)}(x|\underline{x})$, see (5) (see *Exercise 3.1-1*).

Next, we consider the following choice of seeds:

$$N=3;\ s_1(x)=1,\ \ s_2(x)=\exp(x),\ \ s_3(x)=\exp(-x)\ . \tag{8}$$

Also in this case the seed space is closed under differentiation. The corresponding (3×3)-matrix $\underline{D}(\underline{x})$ reads:

$$\underline{D}(\underline{x})=(\Delta/2)^{-1}\cdot$$

$$\cdot\begin{pmatrix} \cosh(x_1-x_2)-\cosh(x_1-x_3) & -1+\cosh(x_2-x_3) & 1-\cosh(x_3-x_2) \\ 1-\cosh(x_1-x_3) & \cosh(x_2-x_3)-\cosh(x_1-x_2) & -1+\cosh(x_1-x_3) \\ -1+\cosh(x_1-x_2) & 1-\cosh(x_1-x_2) & \cosh(x_1-x_3)-\cosh(x_2-x_3) \end{pmatrix},$$

(9)

$$(\Delta/2)=\sinh(x_1-x_2)+\sinh(x_2-x_3)+\sinh(x_3-x_1)\,. \tag{10a}$$

Proof.

$$\Delta(x_1,x_2,x_3)=\begin{vmatrix} 1 & 1 & 1 \\ \exp(x_1) & \exp(x_2) & \exp(x_3) \\ \exp(-x_1) & \exp(-x_2) & \exp(-x_3) \end{vmatrix}, \tag{10b}$$

$$\Delta(x_1,x_2,x_3)=2[\sinh(x_1-x_2)+\sinh(x_2-x_3)+\sinh(x_3-x_1)]\,, \tag{10c}$$

$$q^{(1)}(x|\underline{x})=[\sinh(x-x_2)+\sinh(x_2-x_3)+\sinh(x_3-x)]$$

$$/[\ \sinh(x_1-x_2)+\sinh(x_2-x_3)+\sinh(x_3-x_1)\ ]; \tag{11}$$

$$q_x^{(1)}(x|\underline{x})=[\cosh(x-x_2)-\cosh(x-x_3)\,]$$

$$/[\ \sinh(x_1-x_2)+\sinh(x_2-x_3)+\sinh(x_3-x_1)\ ], \tag{12}$$

$$q_x^{(1)}(x_1|\underline{x})=[\cosh(x_1-x_2)-\cosh(x_1-x_3)\,]$$

$$/[\ \sinh(x_1-x_2)+\sinh(x_2-x_3)+\sinh(x_3-x_1)\ ]\,, \tag{13}$$

$$q_x^{(1)}(x_2|\underline{x})=[1-\cosh(x_2-x_3)]/\ [\sinh(x_1-x_2)+\sinh(x_2-x_3)+\sinh(x_3-x_1)\,]. \tag{14}$$

We trust the derivation of these formulas to be self-evident, as well as the derivation of (9) with (10) from these formulas and the analogous ones obtainable from these by appropriate permutations of the relevant indices.

Exercise 3.1.2.1-3. Formulate and solve the analog (with (8) in place of (1)) of *Exercise 3.1.2.1-2.*

The expressions (2) and (9) of $\underline{D}(\underline{x})$ are clearly *translation-invariant*, namely they do not change if all the nodes x_n undergo a common shift, $x_n \to x_n + x_0$.

Exercise 3.1.2.1-4. Prove that this property also holds for the following two choices of seeds, which generalize (1) and (8):

$$N=2M;\ \ s_m(x)=\exp(a_m x),\ s_{m+M}(x)=\exp(-a_m x),\ \ m=1,\dots,M\ ; \tag{15a}$$

$$N=2M+1;\ \ s_1(x)=1,\ s_{m+1}(x)=\exp(a_m x),\ s_{m+M+1}(x)=\exp(-a_m x),\ \ m=1,\dots,M\ . \tag{15b}$$

Here M is an arbitrary positive integer, and the M constants a_m are also arbitrary except for the requirement that they be all different, $a_{m_1} \neq a_{m_2}$ if $m_1 \neq m_2$. *Hint:* note the invariance of $\Delta(x_1,\dots,x_N)$, see (3.1-1), under a common shift of the nodes $x_n, x_n \to x_n + x_0$, $n=1,\dots,N$.

Exercise 3.1.2.1-5. Prove that the condition

$$\sum_{n=1}^{N} b_n = 0 \tag{16}$$

is necessary and sufficient to guarantee translation-invariance (under $x_n \to x_n + x_0$) of the $(N\times N)$-matrix $\underline{D}(\underline{x})$ associated with the following set of seeds (which includes all those treated above):

$$s_n(x)=\exp(b_n\, x),\ \ n=1,\dots,N,\ (b_n \neq b_m\ \text{ if }\ n\neq m)\,. \tag{17}$$

Hint: same as above, see *Exercise 3.1.2.1-4.*

Exercise 3.1.2.1-6. Determine the eigenvalues and eigenvectors of the $(N\times N)$-matrix $\underline{D}(\underline{x})$ associated with the set of seeds (17). *Hint:* see *Corollary 3.1.1-6* and *Exercise 3.1.2-1.*

Let us now consider another simple choice of seeds:

$$N=2;\ \ s_1(x)=x,\ \ s_2(x)=1/x\,. \tag{18}$$

Note that this set is *not* closed under differentiation. As explained in the last part of Sect. 3.1.1 we can nevertheless associate a matrix $\underline{D}(\underline{x})$ to this set. It reads:

$$\underline{D}(\underline{x})=\begin{pmatrix}(x_1^2+x_2^2)/x_1 & -2x_2\\ 2x_1 & -(x_1^2+x_2^2)/x_2\end{pmatrix}/(x_1^2-x_2^2)\,. \tag{19}$$

Proof.

$$\Delta(x_1,x_2)=\begin{vmatrix} x_1 & x_2 \\ 1/x_1 & 1/x_2 \end{vmatrix}=(x_1^2-x_2^2)/(x_1\,x_2)\ , \tag{20}$$

$$q^{(1)}(x|\underline{x})=\left[(x^2-x_2^2)/(x_1^2-x_2^2)\right](x_1/x)\ , \tag{21}$$

$$q_x^{(1)}(x|\underline{x})=\left[1+(x_2^2/x^2)\right]x_1/(x_1^2-x_2^2)\ , \tag{22}$$

$$q_x^{(1)}(x_1|\underline{x})=(x_1^2+x_2^2)/\left[x_1(x_1^2-x_2^2)\right]\ , \tag{23a}$$

$$q_x^{(1)}(x_2|\underline{x})=2x_1/(x_1^2-x_2^2)\ . \tag{23b}$$

Likewise, evaluate $q^{(2)}(x|\underline{x}), q_x^{(2)}(x|\underline{x})$ and $q_x^{(2)}(x_n|\underline{x}), n=1,2$. Then use (3.1.1-6).

Exercise 3.1.2.1-7. The following formulas are clearly true:

$$f'(x)=1,\quad f(x)=x\ , \tag{24}$$

$$x^2 f'(x)=1,\quad f(x)=-1/x\ , \tag{25}$$

$$xf'(x)=f(x),\quad f(x)=x\ , \tag{26a}$$

$$xf'(x)=-f(x),\quad f(x)=1/x\ , \tag{26b}$$

where the primes denote of course differentiations. Note that in all these cases the functions $f(x)$ are in the functional space spanned by the seeds (18), and that the two equations (26) entail that the operator $\mathrm{A}=x\,d/dx$ in this functional space has the eigenvalues $+1$ respectively -1 with eigenfunctions x respectively $1/x$. Using the (2×2)-matrix $\underline{D}(\underline{x})$, see (19), write the 2-vector equations that correspond (recall *Proposition 3.1.1-3*) to these formulas, (24), (25) and (26), and check explicitly their validity.

Exercise 3.1.2.1-8. Repeat the analysis for the set of seeds

$$N=3;\ \ s_1(x)=1,\ \ s_2(x)=x,\ \ s_3(x)=1/x. \tag{27}$$

The next set of seeds we consider in Sect. 3.1.2.1 have arbitrary dimension N. The first of them is a straightforward generalization of the (polynomial) set (3.1-13). It reads

$$s_n(x)=w(x)\,x^{n-1},\ \ n=1,2,\ldots,N\ \ . \tag{28a}$$

This set is closed under differentiation iff the function $w(x)$ is an exponential,

$$w(x)=\exp(ax) \quad . \tag{29}$$

Our treatment below is not restricted to this case, (29); but of course we hereafter assume that the "weight function" $w(x)$ does not vanish at any one of the nodes,

$$w(x_n)\neq 0 \quad , \tag{28b}$$

to avoid violating the condition (3.1-5).

The $(N\times N)$-matrix $\underline{\tilde{D}}$ corresponding to (28) reads as follows:

$$\left[\underline{\tilde{D}}(x)\right]_{nm}=\delta_{nm}\left\{\left[w'(x_n)/w(x_n)\right]+\sum_{\ell=1,\ell\neq n}^{N}(x_n-x_\ell)^{-1}\right\}$$

$$+(1-\delta_{nm})\left[w(x_n)/w(x_m)\right](b_n/b_m)\,(x_n-x_m)^{-1} \quad , \tag{30a}$$

with (see (2.4.1-4a))

$$b_n(\underline{x})=\prod_{\ell=1,\ell\neq n}^{N}(x_n-x_\ell) \quad . \tag{30b}$$

Of course, if $w(x)$ is merely a constant, the present findings reproduce results already discussed above, see (3.1.1-24,25); this fact, incidentally, motivates our use of the tilde-notation, $\underline{\tilde{D}}$ (see (30)), here as well.

The *proof* of (30) is by now standard, and we indicate tersely the relevant steps without any additional comment:

$$\Delta(x_1,\ldots,x_N)=\begin{vmatrix} w(x_1) & w(x_2) & \cdots & w(x_N)\\ \vdots & \vdots & \vdots\vdots\vdots & \vdots\\ w(x_1)\,x_1^{N-1} & w(x_2)\,x_2^{N-1} & \cdots & w(x_N)\,x_N^{N-1}\end{vmatrix}=\left[\prod_{j=1}^{N}w(x_j)\right]\left[\prod_{n,m=1;n>m}^{N}(x_n-x_m)\right], \tag{31}$$

$$q^{(n)}(x|\underline{x})=\left[w(x)/w(x_n)\right]\prod_{\ell=1,\ell\neq n}^{N}\left[(x-x_\ell)/(x_n-x_\ell)\right] \quad , \tag{32}$$

$$q_x^{(n)}(x|\underline{x})=q^{(n)}(x)\left[w'(x)/w(x)+\sum_{\ell=1,\ell\neq n}^{N}(x-x_\ell)^{-1}\right] \quad , \tag{33}$$

$$q_x^{(n)}(x_n|\underline{x}) = w'(x_n)/w(x_n) + \sum_{\ell=1,\ell\neq n}^{N} (x_n - x_\ell)^{-1} \ , \tag{34}$$

$$q_x^{(n)}(x_m|\underline{x}) = \left[q^{(n)}(x)(x - x_m)^{-1}\right]\Big|_{x=x_m} , \ n \neq m \ , \tag{35a}$$

$$q_x^{(n)}(x_m|\underline{x}) = [w(x_m)/w(x_n)]\,(b_m/b_n)\,(x_m - x_n)^{-1}, \quad n \neq m \ . \tag{35b}$$

Clearly (30) can be written in the compact form

$$\underline{\tilde{D}} = (\underline{W}\,\underline{B})\,\underline{D}\,(\underline{W}\,\underline{B})^{-1} + \underline{V} \tag{36}$$

with the 2 *diagonal* (hence commuting) $(N \times N)$-matrices $\underline{W}$ and $\underline{V}$ defined as follows:

$$\underline{W} \equiv \underline{W}(\underline{x}) = \text{diag}[w(x_n)], \quad W_{nm} = \delta_{nm}\, w(x_n) \ , \tag{37}$$

$$\underline{V} \equiv \underline{V}(\underline{x}) = \text{diag}[\ w'(x_n)/w(x_n)\ ], \quad V_{nm} = \delta_{nm}\, w'(x_n)/w(x_n) \ , \tag{38}$$

and the matrices $\underline{B}$ and $\underline{D}$ defined as above, see (2.4.1-4) and (2.4.1-2).

The alert reader should have noticed that the results we just reported could have been obtained by applying the *Remark 3.1.1-8* stated at the end of Sect. 3.1.1 to the treatment of the standard polynomial set given in that section (see after (3.1.1-23)). The reader who did not notice this connection (and who therefore does not deserve to be considered alert!) should pause and ponder over it.

Likewise in the next example we keep an arbitrary weight function in the definition of the set of seeds, even though the effect of its presence is accounted for by *Remark 3.1.1-8* (as the alerted reader will note!).

The next set of seeds we consider in Sect. 3.1.2.1 reads as follows:

$$s_n(x) = w(x)\,\sigma(x - a_n + a_0)/\sigma(x - a_n) \ , \tag{39}$$

where $w(x)$ is an arbitrary function, the $N+1$ quantities $a_k, k = 0,1,...,N$ are arbitrary constants, and the function

$$\sigma(x) \equiv \sigma(x|\omega,\omega') \tag{40}$$

is the "sigma" Weierstrass function, see (A-38). This set of seeds is *not* closed under differentiation, since

$$s_n'(x) = s_n(x) \, \{[w'(x)/w(x)] + \zeta(x-a_n+\alpha) - \zeta(x-a_n)\} \tag{41}$$

cannot generally be written as a linear combination with constant coefficients of the seeds (39). Note the appearance, in the right-hand side of this equation (41), of the Weierstrass zeta functions (via (A-39)). Also note that in this equation, (41), and always below, we omit to indicate the dependence of the Weierstrass functions on the "semiperiods" ω and ω'.

As explained at the end of the preceding Sect. 3.1.1, it is nevertheless possible to introduce, via (3.1.1-6), an $(N\times N)$-matrix $\underline{D}(\underline{x})$ that provides a finite-dimensional representation of the differential operator. It reads

$$[\underline{D}(\underline{x})]_{nm} = \delta_{nm} d_n + (1-\delta_{nm})[w(x_n)/w(x_m)(\beta_n/\beta_m)\cdot$$

$$\cdot[\sigma(\alpha+x_n-x_m)/\sigma(\alpha)][\sigma(x_n-x_m)]^{-1} \; , \tag{42a}$$

$$d_n \equiv d_n(\underline{x}) = [w'(x_n)/w(x_n)] + \zeta(\alpha) + \sum_{\ell=1,\ell\neq n}^{N} \zeta(x_n-x_\ell) - \sum_{j=1}^{N} \zeta(x_n-a_j) \, , \tag{42b}$$

$$\alpha \equiv \alpha(\underline{x}) = a_0 + \sum_{j=1}^{N} (x_j - a_j) \; , \tag{42c}$$

$$\beta_n \equiv \beta_n(\underline{x}) = \left[\prod_{\ell=1,\ell\neq n}^{N} \sigma(x_n-x_\ell)\right] \Big/ \left[\prod_{j=1}^{N} \sigma(x_n-a_j)\right] \; . \tag{42d}$$

Proof. From (3.1-4), (39) and (A-63)

$$\Delta(x_1,\dots,x_n) = \det[w(x_n)\,\sigma(x_m-a_n+a_0)/\sigma(x_m-a_n)]$$

$$= \left[\prod_{n=1}^{N} w(x_n)\right] \sigma[a_0 + \sum_{j=1}^{N}(x_j-a_j)] \, [\sigma(a_0)]^{N-1} \cdot$$

$$\cdot \prod_{n,m=1;n>m}^{N} [\sigma(x_n-x_m)\,\sigma(a_m-a_n)] \Big/ \prod_{n,m=1}^{N} \sigma(x_n-a_m) \; . \tag{43}$$

Hence (see (3.1-9))

$$q^{(n)}(x|\underline{x}) = [w(x)/w(x_n)]\{\sigma[a_0 + x + \sum_{\ell=1,\ell\neq n}^{N} x_\ell - \sum_{j=1}^{N} a_j\ ]/\sigma[a_0 + \sum_{j=1}^{N}(x_j - a_j)]\}\cdot$$

$$\cdot\left\{\prod_{\ell=1,\ell\neq n}^{N}[\sigma(x-x_\ell)/\sigma(x_n-x_\ell)]\right\}\left\{\prod_{j=1}^{N}[\sigma(x_n-a_j)/\sigma(x-a_j)]\right\} \ . \tag{44}$$

Hence, by logarithmic differentiation, and using (A-39),

$$q_x^{(n)}(x|\underline{x}) = q^{(n)}(x)\ \{w'(x)/w(x) + \zeta[a_0 + x + \sum_{\ell=1,\ell\neq n}^{N} x_\ell - \sum_{j=1}^{N} a_j]$$

$$+ \sum_{\ell=1,\ell\neq n}^{N} \zeta(x-x_\ell) - \sum_{j=1}^{N} \zeta(x-a_j)\}. \tag{45}$$

Hence, see (3.1-11),

$$q_x^{(n)}(x_n|\underline{x}) = w'(x_n)/w(x_n) + \zeta[a_0 + \sum_{j=1}^{N}(x_j - a_j)] + \sum_{\ell=1,\ell\neq n}^{N}\zeta(x_n - x_\ell) - \sum_{j=1}^{N}\zeta(x_n - a_j)\ . \tag{46}$$

To compute from (45) the "off-diagonal" element $q_x^{(n)}(x_m|\underline{x})$ with $m \neq n$ one uses again (3.1-11), as well as the property (see (A-46) and (A-47))

$$\lim_{z\to 0}[\sigma(z)\ \zeta(z)] = 1\ . \tag{47}$$

Hence, for $n \neq m$,

$$q_x^{(n)}(x_m|\underline{x}) = [w(x_m)/w(x_n)]\{\sigma[a_0 + x_m - x_n + \sum_{j=1}^{N}(x_n - a_j)]/\sigma[a_0 + \sum_{j=1}^{N}(x_j - a_j)]\}\cdot$$

$$\cdot\prod_{j=1}^{N}[\sigma(x_n - a_j)/\sigma(x_m - a_j)]\left[\prod_{\ell=1,\ell\neq n,m}^{N}\sigma(x_m - x_\ell)\right]/\left[\prod_{l=1,\ell\neq n}^{N}\sigma(x_n - x_\ell)\right]\ . \tag{48}$$

Via (3.1.1-9) these formulas, (46) and (48), yield (42), which is thereby proven.

The diligent reader will ponder on the analogies, and differences, of the present treatment, relative to that presented at the end of the preceding Sect. 3.1.1 (see (3.1.1-44a), and the discussion following it).

Clearly the $(N\times N)$-matrix $\underline{D}$, see (42), admits the convenient representation

$$\underline{D} = (\underline{W}\,\underline{B})\check{\underline{D}}(\underline{W}\,\underline{B})^{-1}\ , \tag{49a}$$

$$\underline{W} = \underline{W}(\underline{x}) = \text{diag}(w(x_n), n = 1,...,N), (\underline{W})_{nm} = \delta_{nm}\, w(x_n) \ , \tag{49b}$$

$$\underline{B} = \underline{B}(\underline{x}) = \text{diag}(\beta_n, n = 1,...,N), (\underline{B})_{nm} = \delta_{nm}\beta_n \ , \tag{49c}$$

$$(\check{\underline{D}})_{nm} = \delta_{nm}\, d_n + (1-\delta_{nm})\left[\sigma(\alpha + x_n - x_m)/\sigma(\alpha)\right]\left[\sigma(x_n - x_m)\right]^{-1} , \tag{49d}$$

with the diagonal elements d_n defined by (42b), α defined by (42c) and the elements β_n of the diagonal $(N \times N)$-matrix $\underline{B}$, see (49c), defined by (42d).

According to *Proposition 3.1.1-3* the differential equation (with j an arbitrary integer in the range $1 \le j \le N$)

$$f'(x) - \{[w'(x)/w(x)] + \zeta(x - a_j + a_0) - \zeta(x - a_j)\} f(x) = 0 \ , \tag{50a}$$

which clearly holds, see (41) and (39), with

$$f(x) \equiv f(x;j) = s_j(x) = w(x)\ \sigma(x - a_j + a_0)/\sigma(x - a_j) \ , \tag{50b}$$

entails an N-vector equation whose n-th component reads

$$\zeta[a_0 + \sum_{k=1}^{N}(x_k - a_k)] - \zeta(x_n - a_j + a_0) + \sum_{\ell=1,\ell\neq n}^{N} \zeta(x_n - x_\ell) - \sum_{k=1,k\neq j}^{N} \zeta(x_n - a_k)$$

$$+ \sum_{m=1,m\neq n}^{N} \{\sigma[a_0 + x_n - x_m + \sum_{k=1}^{N} (x_k - a_k)] / \sigma[a_0 + \sum_{k=1}^{N} (x_k - a_k)]\}[\sigma(x_n - x_m)]^{-1} \cdot$$

$$\cdot [\sigma(a_0 + x_m - a_j)/\sigma(a_0 + x_n - a_j)]\{[\prod_{\ell=1,\ell\neq n}^{N} \sigma(x_n - x_\ell)]/[\prod_{\ell=1,\ell\neq m}^{N} \sigma(x_m - x_\ell)]\}\cdot$$

$$\cdot\{\prod_{k=1}^{N} [\sigma(x_m - a_k)/\sigma(x_n - a_k)]\} = 0 \ . \tag{51}$$

This formula displays N^2 (equivalent!) identities, since the choice of the indices n and j remains arbitrary; it features the $2N+1$ arbitrary constants x_m, $m = 1,2,...,N$ and a_k, $k = 0,1,...,N$. Its *proof* is such a direct consequence of (50), via *Proposition 3.1.1-3* with (42) or (49), not to require any further elaboration here (but try and do for yourself the calculation!).

Exercise 3.1.2.1-9. Prove (A-67). *Hint:* set, in (51), $n=j=1,\ x_1=0,\ a_0-a_1=\tilde{z},\ N=\tilde{N}+1,\ x_\ell=\tilde{x}_{\ell-1},\ a_\ell=-\tilde{y}_{\ell-1},\ \ell=2,\dots,N$, and, after having rewritten it appropriately, eliminate all tildes.

The last set of seeds we consider in Sect. 3.1.2.1 reads as follows:

$$s_1(x)=1,\ s_2(x)=\wp(x),\ s_3(x)=\wp'(x),\ s_4(x)=\wp''(x)\equiv\wp^{(2)}(x),\dots,s_N(x)=\wp^{(N-2)}(x). \tag{52}$$

Note that here we have dispensed with carrying over the multiplicative function $w(x)$, which can of course always be reinstated using *Remark 3.1.1-8.* Here of course $\wp(x)\equiv\wp(x|\omega,\omega')$ is the Weierstrass function, see Appendix A, and the primes appended to it (as well as the parenthetical upper index) denote of course differentiations with respect to the variable x.

This set, (52), is not closed under differentiation, since the x-derivative of the last seed, $s_N(x)$, cannot be expressed as a linear combination of the N seeds.

It is nevertheless useful and easy to obtain an explicit expression of the $(N\times N)$-matrix $\underline{D}(x)$ that corresponds to this choice of seeds, (52):

$$\underline{D}(x)=\underline{\Sigma}(x)\,\hat{\underline{D}}(x)\,[\underline{\Sigma}(x)]^{-1}\ , \tag{53a}$$

$$\underline{\Sigma}(x)=\mathrm{diag}\{[\sigma(x_n)]^{-N}\ [\prod_{m=1,m\neq n}^{N}\sigma(x_n-x_m)],\ n=1,\dots,N\}\ , \tag{53b}$$

$$[\hat{\underline{D}}(x)]_{nm}=\delta_{nm}\,[\zeta(N\bar{x})-N\zeta(x_n)+\sum_{\ell=1,\ell\neq n}^{N}\zeta(x_n-x_\ell)]$$

$$+(1-\delta_{nm})\,\sigma(x_n-x_m+N\bar{x})\,/\,[\sigma(N\bar{x})\,\sigma(x_n-x_m)]\ . \tag{53c}$$

Here $\sigma(x)\equiv\sigma(x|\omega,\omega')$ respectively $\zeta(x)\equiv\zeta(x|\omega,\omega')$ are the sigma respectively zeta functions, see Appendix A, $\bar{x}$ is the "mean coordinate",

$$\bar{x}\equiv N^{-1}\sum_{n=1}^{N}x_n\ , \tag{54}$$

and we assume of course that it does not vanish, $\bar{x}\neq 0$.

Proof. From (3.1-4), (52), (54) and (A-57),

$$\Delta(\underline{x}) = (-)^{(N-1)(N-2)/2} \left[\prod_{n=1}^{N-1} n!\right] \sigma(N\bar{x}) \left[\prod_{n=1}^{N} \sigma(x_n)\right]^{-N} \prod_{\ell,m=1;\ell>m}^{N} \sigma(x_m - x_\ell) \quad . \tag{55}$$

Hence (see (3.1-9))

$$q^{(n)}(x|\underline{x}) = \left[\sigma(x - x_n + N\bar{x})/\sigma(N\bar{x})\right]\left[\sigma(x_n)/\sigma(x)\right]^N \prod_{m=1,m\neq n}^{N} \left[\sigma(x - x_m)/\sigma(x_n - x_m)\right]. \tag{56}$$

From this expression of $q^{(n)}(x|\underline{x})$ one gets rather immediately (53), via (3.1.1-6) and (A-39), (A-46d).

Exercise 3.1.2.1-10. Prove (A-70). *Hint*: note that, for the set (52), there obviously holds the relation

$$s_{k+2}(x) = d\, s_{k+1}(x)/dx, \quad k = 1,2,\ldots,N-2 \ . \tag{57}$$

Hence

$$s_{k+2}(x_n) = \sum_{m=1}^{N} D_{nm}(\underline{x})\, s_{k+1}(x_m), \quad k = 1,2,\ldots,N-2 \ . \tag{58}$$

Now use (52) and (53).

Exercise 3.1.2.1-11. Prove (A-71) (from (A-70), via (A-37b) and (A-55b)), and verify explicitly its validity for $N = 3, k = 1$.

3.1.2.2 Two-dimensional space ($S = 2$)

In Sect. 3.1.2.2 we focus on *two-dimensional* ambient space ($S = 2$), but we use for convenience a "3-dimensional notation for 2-vectors", as follows:

$$\vec{r} \equiv (x,y,0), \quad \hat{k} \equiv (0,0,1), \quad \vec{k} \wedge \vec{r} \equiv (-y,x,0) \quad , \tag{1a}$$

$$\vec{r}_1 \cdot \vec{r}_2 = \vec{r}_2 \cdot \vec{r}_1 = x_1\, x_2 + y_1\, y_2 \quad , \tag{1b}$$

$$\hat{k}\cdot\vec{r}_1\wedge\vec{r}_2=\vec{r}_2\cdot\hat{k}\wedge\vec{r}_1=-\hat{k}\cdot\vec{r}_2\wedge\vec{r}_1=-\vec{r}_1\cdot\hat{k}\wedge\vec{r}_2=x_1\,y_2-y_1\,x_2\ . \tag{1c}$$

Under rotations the scalar respectively pseudoscalar products, $\vec{r}_1\cdot\vec{r}_2$ respectively $\hat{k}\cdot\vec{r}_1\wedge\vec{r}_2$, remain invariant; under inversions (say, $x\to -x, y\to y$), the scalar product $\vec{r}_1\cdot\vec{r}_2$ remains invariant, the pseudoscalar product $\hat{k}\cdot\vec{r}_1\wedge\vec{r}_2$ changes sign.

The first, very simple, choice of seeds we make is

$$N=2;\ \ s_1(\vec{r})=x,\ s_2(\vec{r})=y\ . \tag{2}$$

This set of seeds is *not* closed under differentiation. The corresponding (2×2)-matrix $\underline{\vec{D}}(\underline{\vec{r}})$ representing the differential operator $\vec{\nabla}$ reads as follows:

$$\vec{D}(\underline{\vec{r}})=\begin{pmatrix}-\hat{k}\wedge\vec{r}_2 & \hat{k}\wedge\vec{r}_1\\ -\hat{k}\wedge\vec{r}_2 & \hat{k}\wedge\vec{r}_1\end{pmatrix}/(\hat{k}\cdot\vec{r}_1\wedge\vec{r}_2)\ . \tag{3}$$

Proof: from (3.1-1) and (2)

$$\Delta(\vec{r}_1,\vec{r}_2)=\begin{vmatrix}x_1 & x_2\\ y_1 & y_2\end{vmatrix}=x_1\,y_2-x_2\,y_1=\hat{k}\cdot\vec{r}_1\wedge\vec{r}_2\ . \tag{4}$$

Hence (see (3.1-6)

$$q^{(1)}(\vec{r}\,|\underline{\vec{r}})=(\hat{k}\cdot\vec{r}\wedge\vec{r}_2)/(\hat{k}\cdot\vec{r}_1\wedge\vec{r}_2)\ , \tag{5a}$$

$$q^{(2)}(\vec{r}\,|\underline{\vec{r}})=(\hat{k}\cdot\vec{r}_1\wedge\vec{r})/(\hat{k}\cdot\vec{r}_1\wedge\vec{r}_2)\ . \tag{5b}$$

Hence

$$\vec{\nabla}q^{(1)}(\vec{r}\,|\underline{\vec{r}})=-\hat{k}\wedge\vec{r}_2/(\hat{k}\cdot\vec{r}_1\wedge\vec{r}_2)\ , \tag{6a}$$

$$\vec{\nabla}q^{(2)}(\vec{r}\,|\underline{\vec{r}})=\hat{k}\wedge\vec{r}_1/(\hat{k}\cdot\vec{r}_1\wedge\vec{r}_2)\ . \tag{6b}$$

This, via (3.1.1-9), yields (3), which is thereby proven. Note that the (2×2)-matrix $\underline{\vec{D}}(\underline{\vec{r}})$, see (3), has equal elements in each column, and that all its elements behave as vectors under a plane rotation of the 2-vectors $\vec{r}_1$ and $\vec{r}_2$. There clearly holds moreover the equation (see *Proposition 3.1.1-3*)

$$\underline{\vec{R}}\cdot\underline{\vec{D}}(\underline{\vec{r}})=\underline{I}\ ,\quad [\ \underline{\vec{R}}\cdot\underline{\vec{D}}(\underline{\vec{r}})\ ]_{nm}=\delta_{nm}\ , \tag{7}$$

reflecting the obvious property of the operator $A = \vec{r}\cdot\vec{\nabla}$ to reduce to unity in the functional space spanned by the 2 seeds (2).

The next set of seeds we consider reads as follows:

$$N=3;\ s_1(\vec{r})=1,\ s_2(\vec{r})=x,\ s_3(\vec{r})=y\ . \tag{8}$$

This set is closed under differentiation. The corresponding expression of the (3×3)-matrix $\vec{\underline{D}}$, providing a (3×3)-matrix representation of the 2-vector differential operator $\vec{\nabla}$, can be written in the following compact form:

$$\vec{D}_{nm}(\underline{\vec{r}}) = \left[\hat{k}\wedge(\vec{r}_{m+2}-\vec{r}_{m+1})\right]/\Delta\ ,\qquad m=1,2,3,\ \mathrm{mod}(3), \tag{9}$$

$$\Delta \equiv \Delta(\vec{r}_1,\vec{r}_2,\vec{r}_3) = \hat{k}\cdot(\vec{r}_1\wedge\vec{r}_2+\vec{r}_2\wedge\vec{r}_3+\vec{r}_3\wedge\vec{r}_1)\ , \tag{10a}$$

$$\Delta = \hat{k}\cdot(\vec{r}_1-\vec{r}_2)\wedge(\vec{r}_1-\vec{r}_3)\ . \tag{10b}$$

Proof.

$$\Delta(\vec{r}_1,\vec{r}_2,\vec{r}_3) = \begin{vmatrix} 1 & x_1 & y_1 \\ 1 & x_2 & y_2 \\ 1 & x_3 & y_3 \end{vmatrix} = \begin{vmatrix} 1 & x_1 & y_1 \\ 0 & x_2-x_1 & y_2-y_1 \\ 0 & x_3-x_1 & y_3-y_1 \end{vmatrix}\ , \tag{11a}$$

$$\Delta(\vec{r}_1,\vec{r}_2,\vec{r}_3) = \hat{k}\cdot(\vec{r}_1\wedge\vec{r}_2+\vec{r}_2\wedge\vec{r}_3+\vec{r}_3\wedge\vec{r}_1) = \hat{k}\cdot(\vec{r}_1-\vec{r}_2)\wedge(\vec{r}_1-\vec{r}_3)\ . \tag{11b}$$

The first equality in (11a) corresponds to the definition (3.1-4) with (8), and the second obtains by subtracting the first line from the second and third in the determinant. The two expressions in the right hand side of (11b) obtain by evaluating the two determinants in the right hand side of (11a). Thus (10a) and (10b) are proven. Of course additional, equivalent, expressions of Δ, see (11), obtain by performing cyclic permutations on the indices of the vectors in the right hand side of (10b).

$$q^{(n)}(\vec{r}|\underline{\vec{r}}) = \hat{k}\cdot(\vec{r}\wedge\vec{r}_{n+1}+\vec{r}_{n+1}\wedge\vec{r}_{n+2}+\vec{r}_{n+2}\wedge\vec{r})/\Delta\ ,\quad n=1,2,3\ \mathrm{mod}(3), \tag{12}$$

$$\vec{\nabla}q^{(n)}(\vec{r}|\underline{\vec{r}}) = \hat{k}\wedge(\vec{r}_{n+2}-\vec{r}_{n+1})/\Delta\ ,\quad n=1,2,3\ \mathrm{mod}(3), \tag{13}$$

$$D_{nm}(\underline{\vec{r}}) = \hat{k}\wedge(\vec{r}_{m+2}-\vec{r}_{m+1})/\Delta,\quad n=1,2,3\ \mathrm{mod}(3). \tag{14}$$

The first of these 3 formulas follows from (3.1-9) and (10a); the second from the first, using the 3-vector identities

$$\vec{a} \wedge \vec{b} = -\vec{b} \wedge \vec{a} \ , \tag{15a}$$

$$\vec{a} \cdot \vec{b} \wedge \vec{c} = \vec{b} \cdot \vec{c} \wedge \vec{a} = \vec{c} \cdot \vec{a} \wedge \vec{b} \ ; \tag{15b}$$

the third follows from the second, see (3.1.1-6), and it coincides with (9), which is thereby proven.

Exercise 3.1.2.2-1. The differential operator $\vec{r} \cdot \vec{\nabla}$ has the eigenvalues 0 (with multiplicity 1) respectively 1 (with multiplicity 2), with eigenfunctions 1 respectively x and y. Noting that all these eigenfunctions are expressible as linear combinations of the seeds (8) (actually, they coincide with these 3 seeds !), find eigenvalues and eigenvectors of the (3×3)-matrix $\underline{\vec{R}} \cdot \underline{\vec{D}}(\vec{r})$, see (3.1-3) and (9), with matrix elements

$$\left[\underline{\vec{R}} \cdot \underline{\vec{D}}(\vec{r})\right]_{nm} = \hat{k} \cdot \vec{r}_n \wedge (\vec{r}_{n+1} - \vec{r}_{n+2}) / \Delta \ , \ n,m = 1,2,3 \ \mathrm{mod}(3), \tag{16}$$

see (10). *Hint*: use *Corollary 3.1.1-6.*

Exercise 3.1.2.2-2. Same as *Exercise 3.1.2-1*, but for the operator $\hat{k} \cdot \vec{r} \wedge \vec{\nabla}$, having the eigenvalues $0, +i$ respectively $-i$, with eigenfunctions 1, $x+iy$ respectively $x-iy$.

Exercise 3.1.2.2-3. Clearly, for any function $f(\vec{r})$ living in the 3-dimensional functional space spanned by the seeds (8) (namely, expressible as a linear combination with constant coefficients of the 3 seeds (8)) there hold the equations

$$f_{xx}(\vec{r}) = f_{yy}(\vec{r}) = f_{xy}(\vec{r}) = 0 \ . \tag{17a}$$

Hence, according to *Proposition 3.1.1-3*, the (3×3)-matrix $\underline{\vec{D}}(\vec{r})$, see (9) with (10), must have the property

$$(\underline{D}_x)^2 = (\underline{D}_y)^2 = \underline{D}_x \underline{D}_y = \underline{D}_y \underline{D}_x = 0 \ , \tag{17b}$$

since the functional space (8) is closed under differentiation. Check that this is indeed the case.

Note that the (3×3)-matrix $\underline{\vec{D}}(\vec{r})$, see (9) with (10), is invariant under a common translation of the nodes $\vec{r}_n$ ($\vec{r}_n \to \vec{r}_n + \vec{r}_0$, $n = 1,2,3$), and that it behaves as a vector under a (common) plane rotation of the 3 nodes $\vec{r}_n$. As for the determinant Δ, see (10), it is clearly invariant both under translation and rotation of the nodes, indeed, up to a factor of 2 and possibly a sign, its value coincides with the area of the plane triangle having the 3 nodes $\vec{r}_n$ as its 3 vertices.

Exercise 3.1.2.2-4. Consider the set of seeds

$$N = 4;\ s_1(\vec{r}) = 1,\ s_2(\vec{r}) = x,\ s_3(\vec{r}) = y,\ s_4(\vec{r}) = r^2 = x^2 + y^2\ . \tag{18}$$

This set is closed under differentiation. Calculate the corresponding (4×4)-matrix $\underline{\vec{D}}$. In the process, note that the quantity $\Delta(\vec{r}_1, \vec{r}_2, \vec{r}_3, \vec{r}_4)$, see (3.1-4), is invariant both under (common) rotations and translations of the 4 nodes $\vec{r}_n$. Hence this quantity must have a *geometrical* significance. *(i)* What is it ? *(ii)* What *Theorem* (of elementary plane geometry) is entailed by the possibility to evaluate $\Delta(\vec{r}_1, \vec{r}_2, \vec{r}_3, \vec{r}_4)$ in different manners ? *(iii)* Can you generalize this result to higher-dimensional spaces $(S > 2)$? *Hint* (for question *(iii)*): use the set of seeds

$$N = S + 2;\ s_1(\vec{r}) = 1,\ s_N(\vec{r}) = r^2 = x^2 + y^2 + z^2 ...,\ s_2(\vec{r}) = x,\ s_3(\vec{r}) = y, ...\ . \tag{19}$$

Solutions: see Appendix F.

The next choice we make is characterized by an arbitrary number N of nodes and seeds, and it is clearly closed under differentiation:

$$s_n(\vec{r}) = \exp[(n-1)x + (N-n)y],\ n = 1, 2, ..., N\ . \tag{20}$$

The corresponding expression of the matrix $\underline{\vec{D}}(\vec{r})$ providing an $(N \times N)$-matrix representation of the 2-vector differential operator $\vec{\nabla}$ can be written as follows:

$$[\ \underline{D_x}(\underline{\vec{r}})\ ]_{nm} = \sum_{\ell=1, \ell \neq n}^{N} [1 - \exp(x_\ell - x_n + y_n - y_\ell)]^{-1} \qquad \text{for} \quad n = m, \tag{21a}$$

$$[\ \underline{D_x}(\underline{\vec{r}})\ ]_{nm} = \beta_n [\exp(y_m - y_n) - \exp(x_m - x_n)]^{-1} \beta_m^{-1} \qquad \text{for} \quad n \neq m, \tag{21b}$$

$$\beta_n(\underline{\vec{r}}) = \prod_{\ell=1, \ell \neq n}^{N} [\exp(x_n + y_\ell) - \exp(y_n + x_\ell)]\ , \tag{21c}$$

with an analogous expression for its y-component, $\underline{D_y}(\underline{\vec{r}})$, obtained by performing, in the right hand side of (21), the exchange $x_j \leftrightarrow y_j, j = 1, ..., N$.

Proof. From (3.1-1) and (20)

$$\Delta(\vec{r}_1,...,\vec{r}_N)=\begin{vmatrix}\exp[(N-1)y_1] & \exp[x_1+(N-2)y_1] & \cdots & \exp[(N-1)x_1]\\ \vdots & \vdots & ::: & \vdots \\ \exp[(N-1)y_N] & \exp[x_N+(N-2)y_N] & \cdots & \exp[(N-1)x_N]\end{vmatrix} \tag{22a}$$

$$=\exp\Big[(N-1)\sum_{n=1}^{N}y_n\Big]\begin{vmatrix}1 & \exp(x_1-y_1) & \exp[2(x_1-y_1)] & \cdots & \exp[(N-1)(x_1-y_1)]\\ \vdots & \vdots & \vdots & ::: & \vdots\\ 1 & \exp(x_N-y_N) & \exp[2(x_N-y_N)] & \cdots & \exp[(N-1)(x_N-y_N)]\end{vmatrix} \tag{22b}$$

$$=\exp\Big[(N-1)\sum_{n=1}^{N}y_n\Big]\prod_{n,m=1;m<n}^{N}[\exp(x_n-y_n)-\exp(x_m-y_m)]\ . \tag{22c}$$

The first equation, (22a), corresponds to the definition (3.1-4) with (20). Then one extracts a common factor from each line of the determinant in the right hand side of (22a), obtaining thereby (22b), and finally one notes that this formula features a determinant of Vandermonde type, whose evaluation (via (3.1-14) with x_n replaced by $\exp(x_n-y_n)$) yields (22c).

From (22c) and the definition (3.1-9) one then gets

$$q^{(n)}(\vec{r}|\underline{\vec{r}})=\exp[(N-1)(y-y_n)]\cdot$$

$$\cdot\prod_{\ell=1,\ell\neq n}^{N}\{[\exp(x-y)-\exp(x_\ell-y_\ell)]/[\exp(x_n-y_n)-\exp(x_\ell-y_\ell)]\}\ , \tag{23}$$

and from this, by differentiation,

$$\partial q^{(n)}(\vec{r}|\underline{\vec{r}})/\partial x=\exp[(N-1)(y-y_n)]\sum_{\ell=1,\ell\neq n}^{N}\exp(x-y)[\exp(x_n-y_n)-\exp(x_\ell-y_\ell)]^{-1}\cdot$$

$$\cdot\prod_{j=1,j\neq n,j\neq\ell}^{N}\{[\exp(x-y)-\exp(x_j-y_j)]/[\exp(x_n-y_n)-\exp(x_j-y_j)]\} \tag{24a}$$

hence

$$\partial q^{(n)}(\vec{r}|\underline{\vec{r}})/\partial x\Big|_{\vec{r}=\vec{r}_m}=\sum_{\ell=1,\ell\neq n}^{N}[1-\exp(x_\ell-x_n+y_n-y_\ell)]^{-1}\quad\text{for } n=m\ , \tag{24b}$$

$$\partial q^{(n)}(\vec{r}|\underline{\vec{r}})/\partial x\Big|_{\vec{r}=\vec{r}_m}=-\exp[(N-1)(y_m-y_n)]\,[1-\exp(x_n-x_m+y_m-y_n)]^{-1}\cdot$$

$$\cdot \prod_{j=1, j\neq n, j\neq m}^{N} \{ [\exp(x_m - y_m) - \exp(x_j - y_j)] / [\exp(x_n - y_n) - \exp(x_j - y_j)] \} \text{ for } n \neq m ,$$

(24c)

$$\partial q^{(n)}(\vec{r}|\underline{\vec{r}})/\partial x \Big|_{\vec{r} = \vec{r}_m} = \beta_m(\underline{\vec{r}}) [\exp(y_n - y_m) - \exp(x_n - x_m)]^{-1} [\beta_n(\underline{\vec{r}})]^{-1} \quad \text{for } n \neq m ,$$

(24d)

with $\beta_n(\underline{\vec{r}})$ defined by (21c). Via (3.1.1-6) this yields (21), which is thereby proven.

Clearly the differential operator

$$A = \vec{a} \cdot \vec{\nabla} = a_x \partial / \partial x + a_y \partial / \partial y \tag{25}$$

has eigenvalues

$$\tilde{a}_n = (n-1) a_x + (N-n) a_y , \quad n = 1, ..., N , \tag{26}$$

with the seeds (20) themselves as eigenfunctions:

$$(\vec{a} \cdot \vec{\nabla}) \exp[(n-1)x + (N-n)y] = \tilde{a}_n \exp[(n-1)x + (N-n)y] , \quad n = 1, ..., N . \tag{27}$$

Hence, as a consequence of *Corollary 3.1.1-6*, one may state the following

Proposition 3.1.2.2-5. The $(N \times N)$-matrix

$$(\underline{\tilde{A}})_{nm} = \delta_{nm} \sum_{\ell=1, \ell\neq n}^{N} (a_x \xi_n \eta_\ell - a_y \xi_\ell \eta_n)/(\xi_n \eta_\ell - \xi_\ell \eta_n)$$

$$+ (1 - \delta_{nm})(a_x \xi_n \eta_m - a_y \xi_m \eta_n)/(\xi_n \eta_m - \xi_m \eta_n) \tag{28}$$

has the N eigenvalues $\tilde{a}_n$, see (26), with eigenvectors $\underline{\tilde{v}}^{(n)}$,

$$(\underline{\tilde{v}}^{(n)})_m = \{ \prod_{\ell=1, \ell\neq m}^{N} [\xi_m \eta_\ell - \xi_\ell \eta_m]^{-1} \} \xi_m^{n-1} \eta_m^{N-n} , \tag{29}$$

$$\underline{\tilde{A}} \underline{v}^{(n)} = \tilde{a}_n \underline{\tilde{v}}^{(n)} . \tag{30}$$

This result holds of course for any *arbitrary* assignment of the integer $N \geq 2$ and of the $N+2$ numbers a_x, a_y, ξ_n, η_n $(\xi_n \neq \xi_m, \xi_n \neq 0, \eta_n \neq \eta_m, \eta_n \neq 0)$; note that the eigenvalues, see (26), are independent of the values of the $2N$ numbers ξ_n, η_n, though the $(N \times N)$-matrix $\underline{\tilde{A}}$, see (28), depends non-trivially on these $2N$ parameters; conversely the eigenvectors, see (29), are independent of the 2-vector $\vec{a}$.

Proof. This result corresponds, via *Corollary 3.1.1-6*, to (27), by setting

$$\xi_n = \exp(x_n), \qquad \eta_n = \exp(y_n) \ , \tag{31a}$$

$$\underline{\tilde{A}} = \underline{B}^{-1} \underline{A} \underline{B} \ , \tag{31b}$$

$$\underline{B} = \text{diag}(\beta_n) \ , \tag{31c}$$

$$\beta_n = \prod_{\ell=1, \ell \neq n}^{N} (\xi_n \eta_\ell - \xi_\ell \eta_n) \ . \tag{31d}$$

Exercise 3.1.2.2-6. Check these results by explicit computation for $N=2$ and for $N=3$.

Exercise 3.1.2.2-7. The independence of the eigenvalues $\tilde{a}_n$, see (26), from the parameters ξ_n, η_n, entails that, when these parameters are changed, the $(N \times N)$-matrix $\underline{\tilde{A}}$, see (28), undergoes an *isospectral* deformation. Show that indeed such a deformation corresponds to a similarity transformation, find the matrix that generate it, and, by assuming that the quantities $\xi_n \equiv \xi_n(t)$, $\eta_n \equiv \eta_n(t)$ depend (arbitrarily!) on a parameter t ("time"), write a *Lax equation* for the "time-variation" of the matrix $\underline{\tilde{A}}$. *Hint*: see the analogous treatment given in Sects. 2.4.5.3 and 2.4.5.4.

The last (but perhaps not least interesting?) set of seeds we consider in Sect. 3.1.2.2 reads

$$s_n(\vec{r}) = x^{\alpha+n} y^{\beta-n} = x^{\alpha+1} y^{\beta-1} (x/y)^{n-1}, \quad n = 1, 2, \ldots, N \ , \tag{32}$$

with α and β two arbitrary constants (not necessarily integers).

The corresponding $(N \times N)$-matrix $\underline{\vec{D}}(\vec{r})$ reads

$$\left[\underline{D_x}(\vec{r})\right]_{nm} = \delta_{nm} \left[(\alpha+1)/x_n + \sum_{\ell=1, \ell \neq n}^{N} y_\ell \, (\hat{k} \cdot \vec{r}_n \wedge \vec{r}_\ell)^{-1}\right]$$

$$-(1-\delta_{nm})y_n(x_n/x_m)^{\alpha+1}(y_n/y_m)^{\beta-N}[\sigma_n(\underline{\vec{r}})/\sigma_m(\underline{\vec{r}})](\hat{k}\cdot\vec{r}_n\wedge\vec{r}_m)^{-1} \ , \tag{33a}$$

$$\left[\underline{D_y}(\underline{\vec{r}})\right]_{nm}=\delta_{nm}[(\beta-N)/y_n-\sum_{\ell=1}^{N}x_\ell(\hat{k}\cdot\vec{r}_n\wedge\vec{r}_\ell)^{-1}]$$

$$+(1-\delta_{nm})x_n(x_n/x_m)^{\alpha+1}(y_n/y_m)^{\beta-N}[\sigma_n(\underline{\vec{r}})/\sigma_m(\underline{\vec{r}})](\hat{k}\cdot\vec{r}_n\wedge\vec{r}_m)^{-1} \ , \tag{33b}$$

$$\sigma_n(\underline{\vec{r}})\equiv\prod_{\ell=1,\ell\neq n}^{N}(\hat{k}\cdot\vec{r}_n\wedge\vec{r}_\ell) \ . \tag{34}$$

Proof. From (3.1-4) and (32)

$$\Delta(\vec{r}_1,...,\vec{r}_N)=\prod_{j=1}^{N}[(x_j)^{\alpha+1}(y_j)^{\beta-1}]\prod_{n,m=1;n>m}^{N}[(x_n/y_n)-(x_m/y_m)] \ , \tag{35a}$$

$$\Delta(\vec{r}_1,...,\vec{r}_N)=\prod_{j=1}^{N}[(x_j)^{\alpha+1}y_j{}^{\beta-N}]\prod_{n,m=1,n>m}^{N}(\hat{k}\cdot\vec{r}_n\wedge\vec{r}_m) \ . \tag{35b}$$

To obtain (35a) we used the Vandermonde identity (see (3.1-14), now with x_n replaced by x_n/y_n); to obtain (35b) from (35a) we used the trivial *identity*,

$$\prod_{n,m=1,n>m}^{N}y_n\,y_m=\prod_{j=1}^{N}(y_j)^{N-1} \ , \tag{36}$$

as well as the definition (1c).

Hence, via (3.1-9),

$$q^{(n)}(\vec{r}|\underline{\vec{r}})=(x/x_n)^{\alpha+1}(y/y_n)^{\beta-N}\prod_{\ell=1,\ell\neq n}^{N}[(\hat{k}\cdot\vec{r}\wedge\vec{r}_\ell)/(\hat{k}\cdot\vec{r}_n\wedge\vec{r}_\ell)] \ . \tag{37}$$

Hence, by logarithmic differentiation, and using (1c),

$$q_x^{(n)}(\vec{r}|\underline{\vec{r}})=[(\alpha+1)/x+\sum_{\ell=1,\ell\neq n}^{N}(\hat{k}\cdot\vec{r}\wedge\vec{r}_\ell)^{-1}y_\ell\ ]q^{(n)}(\vec{r}) \ , \tag{38a}$$

$$q_y^{(n)}(\vec{r}|\underline{\vec{r}})=[(\beta-N)/y-\sum_{\ell=1,\ell\neq n}^{N}(\hat{k}\cdot\vec{r}\wedge\vec{r}_\ell)^{-1}x_\ell\ ]q^{(n)}(\vec{r}) \ . \tag{38b}$$

Hence, by using (3.1-11),

$$q_x^{(n)}(\vec{r}_n|\underline{\vec{r}}) = (\alpha+1)/x_n + \sum_{\ell=1,\ell\neq n}^{N} (\hat{k}\cdot\vec{r}\wedge\vec{r}_\ell)^{-1} y_\ell \,, \tag{39a}$$

$$q_y^{(n)}(\vec{r}_n|\underline{\vec{r}}) = (\beta-N)/y_n - \sum_{\ell=1,\ell\neq n}^{N} (\hat{k}\cdot\vec{r}\wedge\vec{r}_\ell)^{-1} x_\ell \,. \tag{39b}$$

To compute the "off-diagonal" terms $q_x^{(n)}(\vec{r}_m|\underline{\vec{r}})$, $q_y^{(n)}(\vec{r}_m|\underline{\vec{r}})$, $n\neq m$, we use again (3.1-11), as well as the formula

$$\lim_{\vec{r}\to\vec{r}_m}\{ \left[\hat{k}\cdot\vec{r}\wedge\vec{r}_m)^{-1}\right] q^{(n)}(\vec{r}|\underline{\vec{r}}) \} = (x_m/x_n)^{\alpha+1}(y_m/y_n)^{\beta-N}\cdot$$

$$\cdot[\prod_{\ell=1,\ell\neq n,m}^{N}(\hat{k}\cdot\vec{r}_m\wedge\vec{r}_\ell)]/[\prod_{\ell=1,\ell\neq n}^{N}(\hat{k}\cdot\vec{r}_n\wedge\vec{r}_\ell)] \,, \tag{40}$$

which is clearly entailed by (37). Hence, for $n\neq m$,

$$q_x^{(n)}(\vec{r}_m|\underline{\vec{r}}) = y_m (x_m/x_n)^{\alpha+1}(y_m/y_n)^{\beta-N}\left[\sigma_m(\underline{\vec{r}})/\sigma_n(\underline{\vec{r}})\right](\hat{k}\cdot\vec{r}_n\wedge\vec{r}_m)^{-1} \,, \tag{41a}$$

$$q_y^{(n)}(\vec{r}_m|\underline{\vec{r}}) = -x_m (x_m/x_n)^{\alpha+1}(y_m/y_n)^{\beta-N}\left[\sigma_m(\underline{\vec{r}})/\sigma_n(\underline{\vec{r}})\right](\hat{k}\cdot\vec{r}_n\wedge\vec{r}_m)^{-1} \,, \tag{41b}$$

with $\sigma_n(\underline{\vec{r}})$ defined by (34). These formulas, (39) and (41), yield, via (3.1.1-6), the expressions (33) with (34), which are thereby proven.

Exercise 3.1.2.2-8. Do the two matrices $\underline{D_x}(\underline{\vec{r}})$ and $\underline{D_y}(\underline{\vec{r}})$, see (33), commute? If not, why not (recall the *Remark 3.1.1-4*)? *Hint*: is the set (32) closed under differentiation?

Let the differential operator A be defined as follows:

$$\mathrm{A} = a x\,\partial/\partial x + b y\,\partial/\partial y \,, \tag{42}$$

with a and b two arbitrary constants. It is then clear, see (32), that

$$(\mathrm{A}-\gamma_n)s_n(\vec{r}) = 0 \quad , \quad n=1,\ldots,N \,, \tag{43}$$

$$\gamma_n = a(\alpha+n)+b(\beta-n) = a\alpha+b\beta+(a-b)n \,. \tag{44}$$

This shows that the operator A has the N eigenvalues γ_n, see (44), with eigenvectors $s_n(\vec{r})$, in the N-dimensional space spanned by the seeds (32). There holds therefore, according to *Proposition 3.1.1-3* and *Corollary 3.1.1-6,* the following

Proposition 3.1.2.2-9. There holds the following N-vector eigenvalue equation:

$$(\underline{A}-\gamma_j)\,\underline{v}^{(j)}=0 \quad , \qquad j=1,2,...,N \ , \tag{45}$$

$$\underline{A}=a\,\underline{X}\,\underline{D}_x+b\,\underline{Y}\,\underline{D}_y \ , \tag{46}$$

$$v_n^{(j)}=s_j(\vec{r}_n) \ , \tag{47}$$

with $\underline{D}_x, \underline{D}_y$ defined by (33) with (34), γ_j defined by (44), and of course

$$\underline{X}=\text{diag}(x_n), \qquad X_{nm}=\delta_{nm}\,x_n \ , \tag{48a}$$

$$\underline{Y}=\text{diag}(y_n), \qquad Y_{nm}=\delta_{nm}\,y_n \ . \tag{48b}$$

Exercise 3.1.2.2-10. Compare this *Proposition 3.1.2.2-9* with *Proposition 3.1.2.2-5*.

For

$$\alpha=-1, \qquad \beta=N \ , \tag{49}$$

the matrix (33) can be conveniently written in the neat form

$$\underline{\vec{D}}(\vec{r})=\underline{\Sigma}(\vec{r})\,\underline{\overset{\vec{\vee}}{D}}(\vec{r})\,[\underline{\Sigma}(\vec{r}))]^{-1} \ , \tag{50a}$$

$$\left[\underline{\overset{\vec{\vee}}{D}}(\vec{r})\right]_{nm}=-\delta_{nm}\sum_{\ell=1,\ell\neq n}^{N}(\hat{k}\wedge\vec{r}_\ell)\,/\,(\hat{k}\cdot\vec{r}_n\wedge\vec{r}_\ell)+(1-\delta_{nm})\,(\hat{k}\wedge\vec{r}_n)\,/\,(\hat{k}\cdot\vec{r}_n\wedge\vec{r}_m) \ , \tag{50b}$$

$$\underline{\Sigma}(\vec{r})=\text{diag}[\sigma_n(\vec{r}),\ n=1,...,N)] \ , \tag{50c}$$

with $\sigma_n(\vec{r})$ defined of course by (34). Note that $\underline{\Sigma}(\vec{r})$ behaves as a scalar under a (common) plane rotation of the nodes $\vec{r}_n$ (actually as a pseudo-scalar, if N is even: see (34)), and the $(N\times N)$-matrix $\underline{\overset{\vec{\vee}}{D}}(\vec{r})$ (as well, in this case (49), as $\underline{\vec{D}}(\vec{r})$ itself), as a 2-vector.

3.1.2.3 Three-dimensional space ($S = 3$)

In Sect. 3.1.2.3 we focus on *three-dimensional* ambient space $(S = 3)$, for which we use the standard 3-vector notation:

$$\vec{r} \equiv (x, y, z) \ , \tag{1a}$$

$$\vec{r}_1 \wedge \vec{r}_2 = -\vec{r}_2 \wedge \vec{r}_1 \equiv (y_1\, z_2 - z_2\, y_1, z_1\, x_2 - x_1\, z_2, x_1\, y_2 - y_1 x_2) \ , \tag{1b}$$

$$\vec{r}_1 \cdot \vec{r}_2 = \vec{r}_2 \cdot \vec{r}_1 = x_1\, x_2 + y_1 y_2 + z_1 z_2 \ , \tag{1c}$$

$$\vec{r}_1 \cdot \vec{r}_2 \wedge \vec{r}_3 = \vec{r}_2 \cdot \vec{r}_3 \wedge \vec{r}_1 = \vec{r}_3 \cdot \vec{r}_1 \wedge \vec{r}_2 = -\vec{r}_1 \cdot \vec{r}_3 \wedge \vec{r}_2 = -\vec{r}_2 \cdot \vec{r}_1 \wedge \vec{r}_3 = -\vec{r}_3 \cdot \vec{r}_2 \wedge \vec{r}_1$$

$$= x_1\, y_2\, z_3 + x_2\, y_3\, z_1 + x_3\, y_1\, z_2 - x_1\, y_3\, z_2 - x_2\, y_1\, z_3 - x_3\, y_2\, z_1$$

$$= x_1\, y_2\, z_3 + y_1\, z_2\, x_3 + z_1\, x_3\, y_1 - x_1\, z_2\, y_3 - z_1\, y_2\, x_3 - y_1\, x_2\, z_3 \ , \tag{1d}$$

$$\vec{r}_1 \wedge (\vec{r}_2 \wedge \vec{r}_3) = (\vec{r}_1 \wedge \vec{r}_2) \wedge \vec{r}_3 = (\vec{r}_1 \cdot \vec{r}_3)\vec{r}_2 - (\vec{r}_1 \cdot \vec{r}_2)\vec{r}_3 \ . \tag{1e}$$

Note that the triple product (1d) is a pseudoscalar: it remains invariant under (collective) rotations of the three 3-vectors $\vec{r}_1, \vec{r}_2, \vec{r}_3$, and it changes sign under (collective) inversions $(x \to -x, y \to -y, z \to -z)$. It has a simple geometrical meaning: it is, possibly up to a sign, 6 times the volume of the tetrahedron having the origin of coordinates, and the 3 nodes $\vec{r}_n$, has its 4 vertices.

The first choice of seeds we consider reads as follows:

$$N = 3; \quad s_1(\vec{r}) = x, \quad s_2(\vec{r}) = y, \quad s_3(\vec{r}) = z \ . \tag{2}$$

This set of seeds is *not* closed under differentiation.

The corresponding expression of the (3×3)-matrix $\underline{\vec{D}}(\vec{r})$ reads

$$[\ \underline{\vec{D}}(\vec{r})\ ]_{nm} = (\vec{r}_{m+1} \wedge \vec{r}_{m+2})/\Delta, \quad m = 1,2,3, \ \text{mod}(3) \ , \tag{3}$$

$$\Delta \equiv \Delta(\vec{r}_1, \vec{r}_2, \vec{r}_3) = \vec{r}_1 \cdot \vec{r}_2 \wedge \vec{r}_3 \ . \tag{4}$$

Proof. From (3.1-4) and (2)

$$\Delta(\vec{r}_1,\vec{r}_2,\vec{r}_3)=\begin{vmatrix} x_1 & y_1 & z_1 \\ x_2 & y_2 & z_2 \\ x_3 & y_3 & z_3 \end{vmatrix}=\vec{r}_1\cdot\vec{r}_2\wedge\vec{r}_3=\vec{r}_n\cdot\vec{r}_{n+1}\wedge\vec{r}_{n+2} \quad . \tag{5}$$

The index n in the right-hand-side of (5) is defined $\mathrm{mod}(3)$, and it can take any value (1, 2 or 3; see (1c)). In the remaining part of this proof we consider the index n to be always defined $\mathrm{mod}(3)$. Hence, from (5) and (3.1-6),

$$q^{(n)}(\vec{r}\,|\underline{\vec{r}})=(\vec{r}\cdot\vec{r}_{n+1}\wedge\vec{r}_{n+2})/\Delta \quad , \tag{6}$$

entailing

$$\vec{\nabla}q^{(n)}(\vec{r}\,|\underline{\vec{r}})=(\vec{r}_{n+1}\wedge\vec{r}_{n+2})/\Delta \quad . \tag{7}$$

This formula, via (3.1.1-6), yields immediately (3) with (4), which is thereby proven.

Exercise 3.1.2.3-1. Verify the property

$$\underline{\vec{R}}\cdot\underline{\vec{D}}(\underline{\vec{r}})=\underline{1}, \qquad (\underline{1})_{nm}=\delta_{nm} \quad . \tag{8}$$

Can you explain this remarkable fact ? *Hint*: consider the effect of the linear differential operator $\mathrm{A}=\vec{r}\cdot\vec{\nabla}$ in the 3-dimensional functional space spanned by the seeds (2); and recall *Proposition 3.1.1-3*.

The next, again quite simple, example we consider is characterized by the following choice of seeds:

$$N=4; \quad s_1(\vec{r})=1, \quad s_2(\vec{r})=x, \quad s_3(\vec{r})=y, \quad s_4(\vec{r})=z \quad . \tag{9}$$

In this case the set of seeds is closed under differentiation, and a faithful representation of the operator of differentiation $\vec{\nabla}$ is provided by the (4×4)-matrix

$$[\ \underline{\vec{D}}(\underline{\vec{r}})\ ]_{nm}=[(\vec{r}_{m+1}-\vec{r}_{m+3})]\wedge[(\vec{r}_{m+2}-\vec{r}_{m+3})]/\Delta \quad , \tag{10}$$

$$\Delta\equiv\Delta(\vec{r}_1,\vec{r}_2,\vec{r}_3,\vec{r}_4)=(\vec{r}_2-\vec{r}_1)\cdot[(\vec{r}_3-\vec{r}_1)\wedge(\vec{r}_4-\vec{r}_1)] \quad . \tag{11}$$

Here and below the indices are defined $\mathrm{mod}(4)$. Note that Δ is a pseudo-scalar (invariant under a collective rotation of the nodes), and it is moreover invariant under a translation of the nodes $(\vec{r}_n \to \vec{r}_n + \vec{r}_0,\ n=1,...,4)$, as well of course as under any cyclic permutation of the nodes appearing in the right hand side of (11). Indeed, up to a sign, the value of Δ coincides with 6 times the volume of the tetrahedron whose 4 vertices coincides with the 4 nodes $\vec{r}_n,\ n=1,...,4$. Likewise, the (4×4)-matrix $\underline{\vec{D}}(\underline{\vec{r}})$ behaves as a vector under rotations, and it is invariant under a translation of the nodes $(\vec{r}_n \to \vec{r}_n + \vec{r}_0,\ n=1,...,4)$; and it features 4 equal lines.

Proof. From (3.1-4) and (9)

$$\Delta(\vec{r}_1,\vec{r}_2,\vec{r}_3) = \begin{vmatrix} 1 & x_1 & y_1 & z_1 \\ 1 & x_2 & y_2 & z_2 \\ 1 & x_3 & y_3 & z_3 \\ 1 & x_4 & y_4 & z_4 \end{vmatrix} = \begin{vmatrix} 1 & x_1 & y_1 & z_1 \\ 0 & x_2-x_1 & y_2-y_1 & z_2-z_1 \\ 0 & x_3-x_1 & y_3-y_1 & z_3-z_1 \\ 0 & x_4-x_1 & y_4-y_1 & z_4-z_1 \end{vmatrix} =$$

$$= \begin{vmatrix} x_2-x_1 & y_2-y_1 & z_2-z_1 \\ x_3-x_1 & y_3-y_1 & z_3-z_1 \\ x_4-x_1 & y_4-y_1 & z_4-z_1 \end{vmatrix} = (\vec{r}_2-\vec{r}_1)\cdot\left[(\vec{r}_3-\vec{r}_1)\wedge(\vec{r}_4-\vec{r}_1)\right]$$

$$= (\vec{r}_n-\vec{r}_{n+3})\cdot\left[(\vec{r}_{n+1}-\vec{r}_{n+3})\wedge(\vec{r}_{n+2}-\vec{r}_{n+3})\right]\ . \tag{12}$$

From this formula and (3.1-6)

$$q^{(n)}(\vec{r}\,|\underline{\vec{r}}) = (\vec{r}-\vec{r}_{n+3})\cdot\left[(\vec{r}_{n+1}-\vec{r}_{n+3})\wedge(\vec{r}_{n+2}-\vec{r}_{n+3})\right]/\Delta\ , \tag{13}$$

hence

$$\vec{\nabla} q^{(n)}(\vec{r}\,|\underline{\vec{r}}) = \left[(\vec{r}_{n+1}-\vec{r}_{n+3})\wedge(\vec{r}_{n+2}-\vec{r}_{n+3})\right]/\Delta\ , \tag{14}$$

and this formula, via (3.1.1-6), yields (10) with (11), which is thereby proven.

3.2 N-body problems in spaces of one or more dimensions

In the 3 subsections of Sect. 3.2 we discuss N-body problems in one-, two-, respectively three-dimensional space, obtained by using the exact (generalized) Lagrangian interpolation technique described in the first part of Chap. 3. The examples exhibited below are meant to *illustrate* this approach to manufacture N-body problems amenable to exact treatment, not to provide a systematic survey of this methodology, which lends itself to several variations and modifications; much less do we try an exhaustive display of all the models that can be manufactured in this manner or even of all those that have already been investigated (references to these are provided below, see Sect. 3.N).

But before presenting specific examples, in the rest of Sect. 3.2 we outline the general methodology to manufacture, in S-dimensional space, N-body models which are amenable, as we explain below, to exact treatment.

The starting point of our treatment is an S-vector-valued function $\vec{f}(\vec{r},t)$ whose S components admit, for some specific choice of the set of N seeds $\{s_n(\vec{r}),\ n=1,...,N\}$, the (exact!) interpolational representation (3.1-6) hence also (3.1-12), which are now written as follows:

$$\vec{f}(\vec{r},t)=\sum_{m=1}^{N}\vec{h}_m(t)\,s_m(\vec{r}) \quad , \tag{1}$$

$$\vec{f}(\vec{r},t)=\sum_{m=1}^{N}\vec{f}_n(t)\,q^{(n)}\left[\vec{r}\,|\,\underline{\vec{r}}(t)\right] \quad , \tag{2}$$

entailing (see (3.1-11))

$$\vec{f}_n(t)=\vec{f}\left[\vec{r}_n(t),t\right] \quad , \tag{3}$$

$$\sum_{m=1}^{N}\vec{h}_m(t)\,s_m\left[\vec{r}_n(t)\right]=\vec{f}_n(t) \quad . \tag{4}$$

The diligent reader should now pause for a moment, to digest the novelties of these formulas, relative to the analogous ones given above, see (3.1-6), (3.1-12), (3.1-7) and (3.1-8). First of all here not only the space variable $\vec{r}$ is an S-vector, but also the function $\vec{f}(\vec{r},t)$, as well of course as the N values, $\vec{f}_n(t)$, that $\vec{f}(\vec{r},t)$ takes at the N nodes $\vec{r}_n(t)$, see (3). Secondly, and most importantly, we have now introduced an additional variable t ("time"), and we have assumed (see below) that the S-

vector-valued function $\vec{f}(\vec{r},t)$ depend on it, as well, most importantly, as the N nodes $\vec{r}_n(t)$ (see (2), (3) and (4)). The dependence of the N nodes,

$$\vec{r}_n \equiv \vec{r}_n(t) \ , \tag{5}$$

on the time t entails that the N interpolational functions $q^{(n)}\left[\vec{r}\middle|\underline{\vec{r}}(t)\right]$ depend as well on the time t, via their dependence, see (3.1.9), on the N nodes $\vec{r}_n(t)$. We assume on the other hand that the N seeds $s_n(\vec{r})$, see (1), are time-independent; as we shall soon see, this entails an important simplification.

Let us now assume that the S-vector-valued function $\vec{f}(\vec{r},t)$ satisfy the following *linear* partial differential equation, characterizing its time-evolution:

$$\vec{f}_t(\vec{r},t) = \overleftrightarrow{A}\ \vec{f}(\vec{r},t) \ . \tag{6}$$

The subscripted variable t of course denotes partial differentiation

$$\vec{f}_t(\vec{r},t) \equiv \partial \vec{f}(\vec{r},t)/\partial t \ . \tag{7}$$

Note that the evolution PDE (6) is of *first* order in time (the extension to linear PDEs of *second* order in time is an avenue of generalization we will not pursue here). The linear differential operator in the right hand side of (6) is assumed to be of the type considered above, see (3.1.1-14),

$$\overleftrightarrow{A} = \sum_{\alpha,\beta,\gamma,\cdot\cdot} \overleftrightarrow{a}_{\alpha\beta\gamma\cdot\cdot}(\vec{r})\partial^{\alpha+\beta+\gamma+\cdot\cdot}/\partial x^{\alpha}\,\partial y^{\beta}\,\partial z^{\gamma}\cdots \ , \tag{8}$$

where of course $x,y,z,\ldots$ are the Cartesian coordinates of the S-vector $\vec{r} \equiv (x,y,z,\ldots)$ and $\alpha,\beta,\gamma,\ldots$ are nonnegative integers. Note that, in writing (8), we assumed the functions $\overleftrightarrow{a}_{\alpha\beta\gamma\ldots}(\vec{r})$ to be time-independent; this is again for the sake of simplicity (indeed, it entails a significant simplification, see below). But we do not forsake here the possibility that the operator $\overleftrightarrow{A}$, hence the coefficients $\overleftrightarrow{a}_{\alpha\beta\gamma\ldots}(\vec{r})$ (see (8)), act as *tensors* on the S-vector $\vec{f}(\vec{r},t)$, and the double-headed arrow on $\overleftrightarrow{A}$ and $\overleftrightarrow{a}_{\alpha\beta\gamma\ldots}(\vec{r})$ is a reminder of this possibility.

Let us re-emphasize that we posit the time-evolution (6) to be compatible with the interpolational representation, see (1) and (2), of $\vec{f}(\vec{r},t)$, namely we assume that the time-evolution (6) of $\vec{f}(\vec{r},t)$ maintains for all time every component of the S-vector $\vec{f}(\vec{r},t)$ inside the class of functions representable as a linear superposition (with coefficients independent of the space variable $\vec{r}$) of the N seeds $s_n(\vec{r})$ (which are here assumed to have been chosen once and for all, as starting point for the treatment): see (1). This entails that the time evolution (6) can be generally mapped, via this representation (1) of $\vec{f}(\vec{r},t)$ as a superposition of the seeds, into a set of N *linear* coupled first-order evolution ODEs for the N S- vectors $\vec{h}_n(t)$, namely into a set of SN *linear* evolution equations, with constants coefficients (thanks to the assumption made above, that the operator $\tilde{A}$ is time-independent, see (8), and that the seeds $s_n(\vec{r})$ are also time-independent), for SN functions of time (the SN components of the N S- vectors $\vec{h}_n(t)$). As it is well known, the solution of such a system is a matter of *linear algebra*, the "most difficult" task entailed by it consisting in finding the eigenvalues and eigenvectors of a matrix of rank (at most) SN.

Let us now proceed and manufacture the equations of motion of our N-body problem. To this end we time-differentiate (3):

$$\dot{\vec{f}}_n(t)=\left\{\vec{f}_t(\vec{r},t)+\left[\dot{\vec{r}}_n(t)\cdot\vec{\nabla}\right]\vec{f}(\vec{r},t)\right\}\Big|_{\vec{r}=\vec{r}_n(t)} \quad . \tag{9}$$

Here and below we denote by superimposed dots time-differentiations (for functions depending only on the time t).

We now use the results, and notation, introduced in the first part of this chapter, to rewrite this equation as follows:

$$\dot{\underline{\vec{f}}}(t)=\underline{\vec{f}}_t(t)+\left\{\dot{\underline{\vec{R}}}(t)\cdot\underline{\vec{D}}\left[\underline{\vec{r}}(t)\right]\right\}\underline{\vec{f}}(t) \quad . \tag{10a}$$

On the left hand side this equation features the time-derivative of the (S-vector valued) N-vector $\underline{\vec{f}}(t)$ (which is, of course, as well an N-vector-valued S-vector), whose n-th component $\vec{f}_n(t)$ is given by (3). The first term in the right hand side is the (S-vector-valued) N-vector $\underline{\vec{f}}_t(t)$, whose n-th component is the value taken, at the n-th node, namely for $\vec{r}=\vec{r}_n$, by the partial derivative with respect to time, $\vec{f}_t(\vec{r},t)$, of the S-vector $\vec{f}(\vec{r},t)$. As for the second term on the right hand side, the (S-vector-valued) $(N\times N)$-matrices $\underline{\vec{R}}(t)$ and $\underline{\vec{D}}[\underline{\vec{r}}(t)]$ are those defined in the first part of this Chapter, see (3.1-3) and (3.1.1-6), (3.1.1-9); their time-dependence is inherited via the time-dependence of the nodes $\vec{r}_n(t)$, and of course $\dot{\underline{\vec{R}}}(t)$ denotes the time derivative of $\underline{\vec{R}}(t)$. The dot between $\dot{\underline{\vec{R}}}(t)$ and $\underline{\vec{D}}(t)$ denotes the scalar product in S-space, while the usual rules for matrix-matrix and matrix-vector products are moreover operational for (S-vector-valued) $(N\times N)$-matrices, such as $\dot{\underline{\vec{R}}}(t)$ and $\underline{\vec{D}}[\underline{\vec{r}}(t)]$, and for ($S$-vector-valued) N-vectors, such as $\dot{\underline{\vec{f}}}(t)$, $\underline{\vec{f}}_t(t)$ and $\underline{\vec{f}}(t)$. Hence, componentwise (and after using (3.1-3)) the N-vector equation (10a) reads

$$\dot{\vec{f}}_n(t)=\vec{f}_t[\vec{r}_n(t),t]+\sum_{m=1}^{N}\vec{f}_m(t)\{\dot{\vec{r}}_n(t)\cdot\vec{D}_{nm}[\underline{\vec{r}}(t)]\} \ , \tag{10b}$$

with the dot interposed between $\dot{\vec{r}}_n(t)$ and $\vec{D}_{nm}[\underline{\vec{r}}(t)]$ denoting of course the scalar product among S-vectors.

We now set $\vec{r}=\vec{r}_n(t)$ in (6), obtaining thereby, via *Proposition 3.1.1-3*,

$$\underline{\vec{f}}_t(t)=\underline{\vec{A}}(t)\,\underline{\vec{f}}(t) \ . \tag{11}$$

Here we have written again this formula in (S-vector-valued) N-vector form; the $(N\times N)$-matrix $\underline{\vec{A}}(t)$,

$$\underline{\vec{A}}(t)=\sum_{\alpha,\beta,\gamma,\dots}\vec{a}_{\alpha\beta\gamma\dots}(\vec{R})\{\underline{D_x}[\underline{\vec{r}}(t)]\}^{\alpha}\{\underline{D_y}[\underline{\vec{r}}(t)]\}^{\beta}\{\underline{D_z}[\underline{\vec{r}}(t)]\}^{\gamma}\cdots \tag{12}$$

or

$$\underline{\vec{A}}(t)=\sum_{\alpha,\beta,\gamma}\vec{a}_{\alpha\beta\gamma}(\vec{R})\ \underline{D^{(\alpha\beta\gamma\cdots)}_{xyz\cdots}}[\underline{\vec{r}}(t)] \ , \tag{13}$$

is obtained of course from the (possibly tensorial in S-space) linear operator $\vec{\mathsf{A}}$, see (8), and the validity and significance of (12) and (13) have been explained in Sect. 3.1.1 (see (3.1.1-16) or (3.1.1-43): in particular (12) applies if the set of seeds is closed under differentiation, otherwise the more general formula (13) must be used). The $(N\times N)$-matrix $\underline{\vec{A}}(t)$ acts of course according to the standard rules of matrix-vector multiplication in N-space on the (S-vector-valued) N-vector $\underline{\vec{f}}(t)$, and according to the standard (possibly tensorial, as entailed by the structure of $\vec{\mathsf{A}}$) rules in S-space on the (N-vector-valued) S-vector $\underline{\vec{f}}(t)$.

From (10) and (11) we get

$$\dot{\underline{\vec{f}}}(t)=\{\underline{\vec{A}}[\underline{\vec{r}}(t)]+\dot{\underline{\vec{R}}}(t)\cdot\underline{\vec{D}}[\underline{\vec{r}}(t)]\}\,\underline{\vec{f}}(t) \tag{14a}$$

or equivalently (componentwise; after using (3.1-3))

$$\dot{\vec{f}}_n(t)=\sum_{m=1}^{N}\left\{\vec{A}_{nm}\left[\underline{\vec{r}}(t)\right]+\dot{\vec{r}}_n(t)\cdot\vec{D}_{nm}\left[\underline{\vec{r}}(t)\right]\right\}\vec{f}_m(t) \quad . \tag{14b}$$

Since $\underline{\vec{A}}[\underline{\vec{r}}(t)]$ and $\underline{\vec{D}}[\underline{\vec{r}}(t)]$ are explicitly known in terms of the N nodes $\vec{r}_n(t)$, see (12) or (13) and (3.1.1-6) or (3.1.1-39), this equation provides an explicit set of N relations among the N S-vectors $\vec{f}_n(t)$, their time-derivatives $\dot{\vec{f}}_n(t)$, the N S-vectors $\vec{r}_n(t)$ and their time-derivatives $\dot{\vec{r}}_n(t)$ (see (3.1-3): a relation which is *linear*, and structurally independent from the choice of the set of N seeds $\{s_n(\vec{r}), n=1,\ldots,N\}$, for the $3N$ S-vectors $\vec{f}_n(t)$, $\dot{\vec{f}}_n(t)$ and $\dot{\vec{r}}_n(t)$, but is instead generally highly *nonlinear*, and directly dependent on the choice of the set of N seeds $\{s_n(\vec{r}), n=1,\ldots,N\}$, for the N nodes $\vec{r}_n(t)$.

We are now at liberty to posit another set of N relations among these quantities, generating thereby a dynamical system, which shall look (see below) like a Newtonian N-body problem in S-dimensional space, with the nodes $\vec{r}_n(t)$ being interpreted as particle positions; a model which is (partially or completely; see below) solvable. Here we limit our choice of such a relation to the following special form:

$$\vec{f}_n(t)=\rho_n\left[\underline{\vec{r}}(t)\right]\dot{\vec{r}}_n(t)+\vec{\gamma}_n\left[\underline{\vec{r}}(t)\right] \tag{15a}$$

or equivalently, in N-vector form,

$$\underline{\vec{f}}(t)=\underline{\mathrm{P}}\left[\underline{\vec{r}}(t)\right]\underline{\dot{\vec{r}}}(t)+\underline{\vec{\gamma}}\left[\underline{\vec{r}}(t)\right] \quad . \tag{15b}$$

Here we have of course introduced the *diagonal* $(N\times N)$-matrix $\underline{\mathrm{P}}(\underline{\vec{r}})$, and the ($S$-vector-valued) N-vector $\underline{\vec{\gamma}}(\underline{\vec{r}})$,

$$\underline{\mathrm{P}}(\underline{\vec{r}})=\mathrm{diag}\left[\rho_n(\underline{\vec{r}});n=1,2,\ldots,N\right],\quad \mathrm{P}_{nm}(\underline{\vec{r}})=\delta_{nm}\rho_n(\underline{\vec{r}}), \tag{16}$$

$$\underline{\vec{\gamma}}(\underline{\vec{r}})=\left[\vec{\gamma}_n(\underline{\vec{r}});n=1,2,\ldots,N\right] \quad , \tag{17}$$

as functions of the N-vector $\underline{\vec{r}}$, see (3.1-2). Note that we assume both $\underline{\mathrm{P}}$ and $\underline{\vec{\gamma}}$ to depend on the time t only via the t-dependence of the "particle coordinates" $\vec{r}_n(t)$.

To get our evolution equations we must now combine (14) with (15). To this end we time-differentiate (15a):

$$\dot{\vec{f}}_n = \rho_n \ddot{\vec{r}}_n + \sum_{m=1}^{N} \{\dot{\vec{r}}_n [(\partial\rho_n / \partial\vec{r}_m) \cdot \dot{\vec{r}}_m] + [(\partial\vec{\gamma}_n / \partial\vec{r}_m) \cdot \dot{\vec{r}}_m]\} \ . \tag{18}$$

The reader is advised to scrutinize this formula attentively (in particular, the significance of the scalar products in the right hand side!); if there remains any uncertainty about the precise significance of any term, re-obtaining this formula by t-differentiation of (15a) shall eliminate any doubt. Note that, for notational convenience, we have omitted to indicate explicitly the functional dependence of each quantity: be it directly on the time t, as is the case for $\dot{\vec{f}}_n$, $\ddot{\vec{r}}_n$ and $\dot{\vec{r}}_m$, or be it, as it is the case for all the other quantities, on the S-vector-valued N-vector $\underline{\vec{r}}$, and of course, through it, $\underline{\vec{r}} = \underline{\vec{r}}(t)$, on the time t as well.

We now insert (15a) in the right hand side of (14b) and equate the expression obtained in this manner to the right hand side of (18), getting thereby, after some trivial rearrangements, an evolution equation which can be conveniently written as follows:

$$\rho_n(\underline{\vec{r}})\, \ddot{\vec{r}}_n = \vec{F}_n(\dot{\underline{\vec{r}}}, \underline{\vec{r}}), \quad n = 1, 2, \ldots, N \ , \tag{19a}$$

$$\vec{F}_n(\dot{\underline{\vec{r}}}, \underline{\vec{r}}) = \sum_{m=1}^{N} \{ -\dot{\vec{r}}_n [\ (\partial\rho_n(\underline{\vec{r}}) / \partial\vec{r}_m) \cdot \dot{\vec{r}}_m\ ] + \rho_m(\underline{\vec{r}}) \dot{\vec{r}}_m [\ \dot{\vec{r}}_n \cdot \vec{D}_{nm}(\underline{\vec{r}})\ ]$$

$$+ \vec{\gamma}_m(\underline{\vec{r}}) [\ \dot{\vec{r}}_n \cdot \vec{D}_{nm}(\underline{\vec{r}})\ ] - [\ (\partial\vec{\gamma}_n(\underline{\vec{r}}) / \partial\vec{r}_m) \cdot \dot{\vec{r}}_m\ ] + \rho_m(\underline{\vec{r}}) [\ \vec{A}_{nm} \cdot \dot{\vec{r}}_m\ ]$$

$$+ [\ \vec{A}_{nm} \cdot \vec{\gamma}_m\ ] \} \ . \tag{19b}$$

Clearly (19a) is interpretable as the Newtonian equation of motion ("mass times acceleration equals force") for N particles of mass ρ_n, moving in S-dimensional space. The "force" $\vec{F}_n(\dot{\underline{\vec{r}}}, \underline{\vec{r}})$ acting on the n-th particle depends on the position $\vec{r}_m$ and the velocities $\dot{\vec{r}}_m$ of all particles, as detailed by (18b), whose right hand side has been organized to highlight three different types of velocity-dependence, respectively of degree 2 (*quadratic*: first line), of degree 1 (*linear*: second line), of degree 0 (*no dependence*: third line). In contrast to this relatively simple dependence on the N velocities $\dot{\vec{r}}_m$, the dependence on the N particle coordinates $\vec{r}_m$ is generally *highly nonlinear* , originating from the dependence upon the set of nodes $\{\vec{r}_m; m = 1, 2, \ldots, N\}$ of $\rho_n \equiv \rho_n(\underline{\vec{r}})$ (see (15); of course only for constant $\rho_n(\underline{\vec{r}}) = \mu_n$ with μ_n independent of $\underline{\vec{r}}$, can (19a), as written, be directly interpreted as Newtonian equations of motion), of $\underline{D} \equiv \underline{D}(\underline{r})$ (see (3.1.1-6) with (3.1-9)), of $\vec{A} \equiv \vec{A}(\underline{r})$ (see (12) or (13), as the case may be), and of $\vec{\gamma}_n \equiv \vec{\gamma}_n(\underline{r})$ (see (15)). (To avoid any possible ambiguity let us emphasize that the middle term in the second line of (19b) is an S-vector whose j-th component reads $-\sum_{k=1}^{S} (\partial \gamma_{nj} / \partial r_{mk})\, \dot{r}_{mk}$, where of course γ_{nj} is the j-th component of the S-vector $\vec{\gamma}_n$, r_{mk} is the k-th component of the S-vector $\vec{r}_m$ and likewise $\dot{r}_{mk}$ is the k-th component of the S-vector $\dot{\vec{r}}_m$).

In the following three Sects. 3.2.1, 3.2.2 respectively 3.2.3 we discuss representative examples of such Newtonian N-body problems in spaces of one, two respectively three dimensions. But before doing that we must still discuss, in Sect. 3.2, to what extent, and how, the class of N-body problems in S-dimensional space we just manufactured, see (19), is amenable to exact treatment. Indeed, let us now indicate how to deal with the *initial-value problem* for (19), namely how to determine, for $t > 0$, the set $\underline{r}(t) \equiv \{\vec{r}_n(t); n = 1,...,N\}$ from the (assumedly given) initial data $\underline{r}(0) \equiv \{\vec{r}_n(0); n = 1,...,N\}$ and $\dot{\underline{r}}(0) \equiv \{\dot{\vec{r}}_n(0); n = 1,...,N\}$.

The *first* step is to evaluate, from the initial data, the corresponding initial values of $\vec{f}_n(t)$,

$$\vec{f}_n(0) = \rho_n[\underline{r}(0)]\, \dot{\vec{r}}_n(0) + \vec{\gamma}_n[\underline{r}(0)] , \quad n = 1,...,N . \tag{20}$$

The *second* step is to evaluate the initial value of the S-vector function $\vec{f}(\vec{r},t)$,

$$\vec{f}(\vec{r},0) = \sum_{n=1}^{N} \vec{f}_n(0)\; q^{(n)}\left[\vec{r} \,|\, \underline{r}(0)\right] , \tag{21a}$$

(see (2)). Here of course the interpolational functions $q^{(n)}\left[\vec{r} \,|\, \underline{r}(0)\right]$ are constructed, see (3.1-9), with N nodes which coincide with the *initial* positions of the particles, $\vec{r}_n = \vec{r}_n(0)$. In fact, in view of the next (third) step, what is actually needed are the initial values $\vec{h}_m(0)$ of the quantities $\vec{h}_m(t)$, see (1), which can be obtained by solving the *linear algebraic* equations (see (4) and (20))

$$\sum_{m=1}^{N} \vec{h}_m(0)\; s_m[\vec{r}_n(0)] = \rho_n[\underline{r}(0)]\, \dot{\vec{r}}_n(0) + \vec{\gamma}_n[\underline{r}(0)], \quad n = 1,2,...,N . \tag{21b}$$

The *third* step is to evaluate the S-vector $\vec{f}(\vec{r},t)$ from its initial value $\vec{f}(\vec{r},0)$. This is achieved by solving (the initial-value problem for) the *linear* evolution PDE (6). As already noted above, under our assumptions this task is essentially algebraic, as it can be reduced to solving the corresponding evolution equations for the N S-vectors $\vec{h}_m(t)$, see (1), a set of *first-order linear evolution equations with constant coefficients* whose solution is generally reducible to the diagonalization and inversion of matrices of maximal rank NS. (In fact in most of the following applications this step will be quite trivial, as we will often limit consideration to time-independent functions $\vec{f}(\vec{r},t)=f(\vec{r},0)$, entailing that the corresponding quantities $\vec{h}_m(t)=\vec{h}_m(0)$, see (1), are constants of the motion; see below).

The *fourth* step is to insert the quantities

$$\vec{f}_n(t)=\vec{f}[\vec{r}_n(t),t] \tag{22}$$

in (15), obtaining thereby the equations

$$\rho_n[\underline{\vec{r}}(t)]\,\dot{\vec{r}}_n(t)=\vec{f}[\vec{r}_n(t),t]-\vec{\gamma}_n[\underline{\vec{r}}(t)] \ , \tag{23}$$

which are the final formulas of our technique of solution. Note that the function $\vec{f}(\vec{r},t)$, as well of course as the given functions $\rho_n(\underline{\vec{r}})$ and $\vec{\gamma}_n(\underline{\vec{r}})$, the choice of which remains our privilege, are now known. Hence this equation, (23), is now an explicitly known system of *first-order* evolution ODEs for the "particle coordinates" $\vec{r}_n(t)$, which must of course be complemented with the initial data $\vec{r}_n(0)$ (the initial data $\dot{\vec{r}}_n(0)$ are no more needed; they have already been used to determine $\vec{f}(\vec{r},t)$, and this guarantees the consistency of (22) at $t=0$). The "degree of solvability" of the many-body problem characterized by the Newtonian equations of motion (19) is therefore generally tantamount to the possibility to reduce the *second-order* system (19) to the *first-order* system (23). However, the system (23), in contrast to (19), is generally *nonautonomous*, due to the explicit time-dependence of $\vec{f}(\vec{r},t)$; although there are simple, yet interesting cases (see below), in which such an explicit time-dependence does not emerge. Two additional simplifications emerge moreover in special cases: *(i)* the evolution equations (23) *decouple*, *(ii)* the evolution equations (23) *linearize*. The simplification *(i)* occurs if the N scalar functions $\rho_n(\underline{\vec{r}})$, as well as the N S-vector-valued functions $\vec{\gamma}_n(\underline{\vec{r}})$, *decouple*, namely both ρ_n and $\vec{\gamma}_n$ only depend on $\vec{r}_n$ (rather than on the entire set $\underline{\vec{r}}=\{\vec{r}_m;m=1,\dots,N\}$):

$$\rho_n(\underline{\vec{r}}) = \rho_n(\vec{r}_n) \quad , \tag{24a}$$

$$\vec{\gamma}_n(\underline{\vec{r}}) = \vec{\gamma}_n(\vec{r}_n) \quad . \tag{24b}$$

The simplification *(ii)* occurs in particularly simple cases, which may nevertheless correspond to nontrivial N-body problems in S-dimensional space (see below).

A class of N-body problems we consider in the following corresponds to the special case of (6) with

$$\vec{\mathrm{A}} = a \quad , \tag{25a}$$

entailing of course

$$\vec{A}_{nm} = a\,\delta_{nm} \quad , \tag{25b}$$

where a is a (possibly vanishing; see below) scalar constant. Note that this choice is clearly always compatible with the essential requirement that the time-evolution equation (6) be consistent with the *ansatz* (1). We moreover now set

$$\rho_n(\underline{\vec{r}}) = \mu_n \quad , \tag{26}$$

and

$$\vec{\gamma}_n(\vec{r}) = \sum_{m=1}^{N} \eta_{nm}\,\vec{r}_m \tag{27a}$$

entailing

$$(\partial\vec{\gamma}_n(\underline{\vec{r}})/\partial\vec{r}_m)\cdot\vec{v} = \eta_{nm}\,\vec{v} \tag{27b}$$

for any S-vector $\vec{v}$. Here of course the N quantities μ_n and the N^2 quantities η_{nm} are (arbitrary) constants. Then the many-body problem (19) takes the form

$$\mu_n\,\ddot{\vec{r}}_n = \vec{F}_n^{(1)}(\dot{\vec{r}}_n,\vec{r}_n) + \vec{F}_n^{(*)}(\dot{\underline{\vec{r}}},\underline{\vec{r}}) \quad , \tag{28a}$$

where, in the right hand side, we have separated, in the force $\vec{F}_n(\dot{\underline{\vec{r}}},\underline{\vec{r}})$ acting on the n-th particle, see (19b), the one-body contribution depending only on the velocity and the coordinate of the n-th particle,

$$\vec{F}_n^{(1)}(\dot{\vec{r}},\vec{r}) = (a\,\mu_n - \eta_{nn})\,\dot{\vec{r}} + a\,\eta_{nn}\,\vec{r} \quad , \tag{28b}$$

from the many-body contribution,

$$\vec{F}_n^{(*)}(\underline{\dot{r}},\underline{r}) = \left[\dot{\vec{r}}_n \cdot \vec{D}_{nn}(\underline{r})\right]\left[\mu_n \dot{\vec{r}}_n + \sum_{j=1}^{N} \eta_{nj}\vec{r}_j\right]$$

$$+ \sum_{m=1,m\neq n}^{N} \left\{ \left[\, \dot{\vec{r}}_n \cdot \vec{D}_{nm}(\underline{r}) \,\right]\left[\, \mu_m \dot{\vec{r}}_m + \sum_{j=1}^{N} \eta_{mj}\vec{r}_j \,\right] - \eta_{nm}\dot{\vec{r}}_m + a\,\eta_{nm}\vec{r}_m \right\} . \tag{28c}$$

The corresponding version of (23) reads

$$\mu_n \dot{\vec{r}}_n(t) = \vec{f}\left[\vec{r}_n(t),0\right] \exp(a\,t) - \sum_{j=1}^{N} \eta_{nj}\,\vec{r}_j(t) \tag{29a}$$

with

$$\vec{f}\left[\vec{r},0\right] = \sum_{m=1}^{N} \vec{h}_m(0)\, s_m(\vec{r}) \tag{29b}$$

(see (1)), where the values of the N constant (i.e, time-independent) S-vectors $\vec{h}_m(0)$ can be determined, from the initial data, via (21b) with (20) and (27a).

To obtain this equation, (29), we used the relation

$$\vec{f}(\vec{r},t) = \vec{f}(\vec{r},0)\exp(a\,t) \quad , \tag{30}$$

which is an immediate consequence of (6) with (25a).

Exercise 3.2-1. Write the more general equations that replace (28) and (29) if (25a) is replaced by

$$\vec{\vec{A}} = \vec{\vec{a}} \quad , \tag{31}$$

with $\vec{\vec{a}}$ a constant tensor in S-dimensional space.

If

$$a = 0 \tag{32a}$$

the system of first-order evolution equations (29a) (or, for that matter, (23); see below) becomes *autonomous*. In this case of course (see (25a))

$$\vec{\mathrm{A}} = 0 \ , \tag{32b}$$

entailing (see (25b))

$$\underline{\vec{A}} = 0 \ , \tag{32c}$$

as well as (see (6))

$$\vec{f}_t(\vec{r},t) = 0 \tag{32d}$$

hence

$$\vec{f}(\vec{r},t) = \vec{f}(\vec{r},0) \ . \tag{32e}$$

Therefore in this case the N S-vectors $\vec{h}_m$, see (1) and (32e), provide N constants of motion. Their explicit expressions in terms of the particle coordinates $\vec{r}_n(t)$ and their velocities $\dot{\vec{r}}_n(t)$, $\vec{h}_m(\underline{\vec{r}},\underline{\dot{\vec{r}}})$, can be obtained from (4), with $\vec{f}_n(t)$ given by (15), namely from the relations

$$\sum_{m=1}^{N} \vec{h}_m \, s^{(m)}\left[\vec{r}_n(t)\right] = \rho_n(\underline{\vec{r}})\dot{\vec{r}}_n(t) + \vec{\gamma}_n(\underline{\vec{r}}) \ , \quad n = 1,2,\ldots,N \ , \tag{33a}$$

and more particularly, for the special choices (26), (27) of $\rho_n(\underline{\vec{r}})$ and $\vec{\gamma}_n(\underline{\vec{r}})$,

$$\sum_{m=1}^{N} \vec{h}_m \, s^{(m)}\left[\vec{r}_n(t)\right] = \mu_n \dot{\vec{r}}_n(t) + \sum_{m=1}^{N} \eta_{nm} \vec{r}_m(t) \ , \quad n = 1,2,\ldots,N \ . \tag{34a}$$

The corresponding class of many-body problems is particularly interesting: indeed, the explicit availability of the N constants of motion $\vec{h}_m\left[\underline{\vec{r}}(t),\underline{\dot{\vec{r}}}(t)\right]$, which, as we just mentioned, can be obtained by solving for these quantities this set of N *linear algebraic equations*, (33a) respectively (34a), justifies considering these models as *integrable*. Their Newtonian equations of motion read (from (19) with (32))

$$\rho_n(\underline{\vec{r}})\ddot{\vec{r}}_n = \sum_{m=1}^{N} \left\{ -\dot{\vec{r}}_n\left[(\partial\rho_n(\underline{\vec{r}})/\partial\vec{r}_m)\cdot\dot{\vec{r}}_m\right] + \rho_m(\underline{\vec{r}})\dot{\vec{r}}_m\left[\dot{\vec{r}}_n\cdot\vec{D}_{nm}(\underline{\vec{r}})\right]\right.$$

$$+\vec{\gamma}_m(\underline{\vec{r}})[\ \dot{\vec{r}}_n\cdot\vec{D}_{nm}(\underline{\vec{r}})\ ]-[\ (\partial\vec{\gamma}_n(\underline{\vec{r}})/\partial\vec{r}_m)\cdot\dot{\vec{r}}_m\ ],\ n=1,2,...,N, \qquad (33b)$$

respectively (for the special choices (26), (27) of $\rho_n(\underline{\vec{r}})$ and $\vec{\gamma}(\underline{\vec{r}})$)

$$\mu_n\ddot{\vec{r}}_n=\sum_{m=1}^{N}\{-\eta_{nm}\dot{\vec{r}}_m+[\dot{\vec{r}}_n\cdot\vec{D}_{nm}(\underline{\vec{r}})][\mu_m\dot{\vec{r}}_m+\sum_{\ell=1}^{N}\eta_{m\ell}\vec{r}_\ell]\},\ n=1,2,...,N. \qquad (34b)$$

Let us emphasize what this finding entails: for any arbitrary choice of the number N of particles and of the set of seeds $s_n(\vec{r})$ (which of course yield, correspondingly, a specific definition of the S-vector-valued $(N\times N)$-matrix $\underline{\vec{D}}(\underline{\vec{r}})$, as explained in detail in the first part of this chapter, see in particular (3.1:.1-6) and (3.1-9)), (33b) respectively (34b) provide the Newtonian equations of motion of an *integrable* N-body problem in S-dimensional space, whose initial-value problem can be solved by integrating the set of *first-order*, autonomous, evolution equations (33a) respectively (34a), which can be rewritten here as follows:

$$\rho_n(\underline{\vec{r}})\,\dot{\vec{r}}_n(t)=-\vec{\gamma}_n(\underline{\vec{r}})+\sum_{m=1}^{N}\vec{h}_m\,s^{(m)}[\vec{r}_n(t)]\,,\quad n=1,2,...,N\ , \qquad (33c)$$

respectively

$$\mu_n\dot{\vec{r}}_n=\sum_{m=1}^{N}[-\eta_{nm}\vec{r}_m+\vec{h}_m\,s^{(m)}(\vec{r}_n)],\quad n=1,2,...,N. \qquad (34c)$$

Of course the N constant S-vectors $\vec{h}_m\equiv\vec{h}_m(\underline{\vec{r}},\dot{\underline{\vec{r}}})$ which appear in these equations can themselves be obtained, in terms of the initial data $\underline{\vec{r}}(0)$ and $\dot{\underline{\vec{r}}}(0)$, from these same equations (perhaps in their completely equivalent avatars (33a) respectively (34a)) at $t=0$, which, as indicated above, should then be considered a system of *linear algebraic equations* for the N unknowns $\vec{h}_m$.

The special case of (34) with

$$\eta_{nm}=\delta_{nm}\,\eta_n \qquad (35a)$$

is sufficiently important to deserve explicit mention. Then the Newtonian equations of motion (34b) become

$$\mu_n\ddot{\vec{r}}_n=-\eta_n\dot{\vec{r}}_n+\sum_{m=1}^{N}(\mu_m\dot{\vec{r}}_m+\eta_m\vec{r}_m)[\ \dot{\vec{r}}_n\cdot\vec{D}_{nm}(\underline{\vec{r}})\ ]\ ,\ n=1,2,...,N\ , \qquad (35b)$$

and (34a) become

$$\sum_{m=1}^{N} \vec{h}_m \; s^{(m)}(\vec{r}_n) = \mu_n \dot{\vec{r}}_n + \eta_n \vec{r}_n, \quad n = 1,2,\ldots,N \quad , \tag{35c}$$

or, completely equivalently,

$$\mu_n \dot{\vec{r}}_n = -\eta_n \; \vec{r}_n + \sum_{m=1}^{N} \vec{h}_m \; s^{(m)}(\vec{r}_n), \quad n = 1,2,\ldots,N \quad . \tag{35d}$$

Let us repeat: the first version of this equation, (35c), is meant as a set of linear algebraic equations, to be solved, at $t = 0$, to get the N (constant) S-vectors $\vec{h}_m$ in terms of the initial data; the second version, (35d), is meant as a set of ODEs to be integrated in order to evaluate the time-evolution of the coordinates $\vec{r}_n(t)$. Note that in this special case this set of ODEs consists of N copies of just one (decoupled!) *first-order autonomous* ODE for the S-vector $\vec{r}_n(t)$, whose evolution, as given by (the n-th one of) these equations (35d), appears decoupled from all other coordinates (the coupling entailed by the Newtonian equations (35b) is now completely encoded in the values of the *constant* S-vectors $\vec{h}_m$). And of course, in the one-dimensional case, (35d) can be solved *by quadratures*, as we indeed discuss in the following Sect. 3.2.1.

Remark 3.2-2. Clearly (35b) (but not (34b)) imply that any particle that is initially at rest, say the k-th one,

$$\dot{\vec{r}}_k(0) = 0 \quad , \tag{36a}$$

remains at rest for all time,

$$\vec{r}_k(t) = \vec{r}_k(0), \quad \dot{\vec{r}}(t) = 0 \quad . \tag{36b}$$

However the presence of one, or more, such fixed particles affects non-trivially the movement of the other particles. Hence each of the N-body problems in S-dimensional space characterized by the equations of motion (35b) contains in itself a family of $(N-M)$-body problems, $M = 1,2,\ldots,N-1$, characterized by the presence of M constant vectors.

Exercise 3.2-3. How should the equations of motion (35b) be modified (making them complex in the process), so as to guarantee that they possess a (large) set of *completely periodic* solutions? *Hint: see Proposition 2.1.13-1,* as well as Sect. 4.5.

3.2.1 One-dimensional examples

The results reported at the end of the preceding Sect. 3.2 identify a class of one-dimensional N-body models whose initial-value problem can be solved by *quadratures*. We start Sect. 3.2.1 by reviewing the relevant equations, which are written below in the notation appropriate to the one-dimensional context to which consideration is restricted in Sect. 3.2.1. We then apply these results to the examples treated in Sect. 3.1.2.1. The alert reader will have no difficulty in experimenting with additional examples.

Then, in the latter part of Sect. 3.2.1, we consider a different approach, which also emerges as natural follow-up to the treatment of Sect. 3.2. An interesting feature of the findings obtained in this manner is to yield many-body problems with *one- and two-body forces only* (see below), rather than the many-body forces that are instead characteristic of the models treated in the first part of Sect. 3.2.1. Another interesting aspect of these many-body problems is their close relationship with models treated previously, see Sect. 2.3.3.

The *solvable* Newtonian equations of motion treated in the first part of Sect. 3.2.1 read simply

$$\mu_n \ddot{x}_n = \dot{x}_n \{-\eta_n + \sum_{m=1}^{N} D_{nm}(\underline{x}) [\mu_m \dot{x}_m + \eta_m x_m]\} \ , \tag{1}$$

with μ_n and η_n $2N$ arbitrary constants, and the $(N \times N)$-matrix $\underline{D}(\underline{x})$ a representation of the differential operator, see Sects. 3.1.1 and 3.1.2.1. As usual, here and below the indices n,m,l run from 1 to N, unless otherwise indicated.

The solution of the corresponding initial-value problem is provided, in implicit form, by the quadrature formula

$$\mu_n \int_{x_n(0)}^{x_n(t)} dx \{\sum_{m=1}^{N} [h_m s_m(x)] - \eta_n x\}^{-1} = t \ , \tag{2}$$

where the N functions $s_n(x)$, $n=1,...,N$, are the N seeds which we can choose arbitrarily, and which determine the $(N \times N)$-matrix $\underline{D}(\underline{x})$, see (1), as explained in Sects. 3.1.1 (and see Sect. 3.1.2.1 for several examples). The N constants h_m, $m=1,2,...,N$, in (2) are determined by the initial data via the following set of linear algebraic equations:

$$\sum_{m=1}^{N} h_m s_m[x_n(0)] = \mu_n \dot{x}_n(0) + \eta_n x_n(0) \ . \tag{3}$$

Proof. The equations of motion (1) are merely the one-dimensional version of (3.2-35). The quadrature formula (2) is obtained from the following one-dimensional version of (3.2-36):

$$\mu_n \dot{x}_n = -\eta_n x_n + \sum_{m=1}^{N} [h_m s_m(x_n)] \ . \tag{4}$$

The formula (3) is of course merely (4) evaluated at $t = 0$; and (4) also provides the system of N linear algebraic equations which define the N *constants of the motion* h_m, $m = 1, 2, ..., N$, in terms of the N particle coordinates $x_n \equiv x_n(t)$ and of their velocities $\dot{x}_n \equiv \dot{x}_n(t)$.

The following solvable N-body problems correspond to the examples of Sect. 3.1.2.1.

$$\mu_1 \ddot{x}_1 = \dot{x}_1 \{-\eta_1 + [(\mu_1 \dot{x}_1 + \eta_1 x_1)\operatorname{cotgh}(x_{12}) - (\mu_2 \dot{x}_2 + \eta_2 x_2)]/\sinh(x_{12})\} \ , \tag{5a}$$

$$\mu_2 \ddot{x}_2 = \dot{x}_2 \{-\eta_2 - [(\mu_2 \dot{x}_2 + \eta_2 x_2)\operatorname{cotgh}(x_{12}) - (\mu_1 \dot{x}_1 + \eta_1 x_1)]/\sinh(x_{12})\} \ . \tag{5b}$$

Here and below we occasionally use the shorthand notation

$$x_{nm} \equiv x_{n,m} \equiv x_n - x_m \ . \tag{6}$$

$$\mu_n \ddot{x}_n = \dot{x}_n \{-\eta_n + [(\mu_n \dot{x}_n + \eta_n x_n)\,[\cosh(x_{n,n+1}) - \cosh(x_{n,n+2})]$$

$$+ (\mu_{n+2} \dot{x}_{n+2} - \mu_{n+1} \dot{x}_{n+1} + \eta_{n+2} x_{n+2} - \eta_{n+1} x_{n+1})[1 - \cosh(x_{n+1,n+2})]] \cdot$$

$$\cdot [\sinh(x_{12}) + \sinh(x_{23}) + \sinh(x_{31})]^{-1}\}, \quad n = 1, 2, 3, \bmod(3) \ . \tag{7}$$

$$\mu_1 \ddot{x}_1 = \dot{x}_1 \{\ -\eta_1 + [\ (\mu_1 \dot{x}_1 + \eta_1 x_1)(x_1^2 + x_2^2) - 2(\mu_2 \dot{x}_2 + \eta_2 x_2)x_1^2\ ] \cdot$$

$$\cdot [\ x_1(x_1^2 - x_2^2)\ ]^{-1}\ \}, \tag{8a}$$

$$\mu_2 \ddot{x}_2 = \dot{x}_2 \{\ -\eta_2 - [\ (\mu_2 \dot{x}_2 + \eta_2 x_2)(x_1^2 + x_2^2) - 2(\mu_2 \dot{x}_2 + \eta_2 x_2)x_2^2\ ] \cdot$$

$$\cdot [\ x_2(x_1^2 - x_2^2)\ ]^{-1}\ \}. \tag{8b}$$

$$\mu_n \ddot{x}_n = \dot{x}_n \{ -\eta_n + (\mu_n \dot{x}_n + \eta_n x_n)[w'(x_n)/w(x_n)]$$

$$+ \sum_{m=1, m\neq n}^{N} \{ \mu_n \dot{x}_n + \eta_n x_n + (\mu_m \dot{x}_m + \eta_m x_m)[w(x_n)/w(x_m)][b_n(\underset{\sim}{x})/b_m(\underset{\sim}{x})] \} \cdot$$

$$\cdot (x_n - x_m)^{-1} \}. \tag{9a}$$

Here $w(x)$ is an arbitrary function, while (see (3.1.2.1-30b))

$$b_n(\underset{\sim}{x}) \equiv \prod_{\ell=1, \ell\neq n}^{N} (x_n - x_\ell) \ . \tag{9b}$$

$$\mu_n \ddot{x}_n = \dot{x}_n \{ -\eta_n + (\mu_n \dot{x}_n + \eta_n x_n)\{ [w'(x_n)/w(x_n)] + \zeta(\alpha) - \zeta(x_n - a_n)$$

$$+ \sum_{\ell=1, \ell\neq n}^{N} [\zeta(x_n - x_\ell) - \zeta(x_n - a_\ell)]\}$$

$$+ \sum_{m=1, m\neq n}^{N} \{ (\mu_m \dot{x}_m + \eta_m x_m)[w(x_n)/w(x_m)][\beta_n(\underset{\sim}{x})/\beta_m(\underset{\sim}{x})] \cdot$$

$$\cdot \sigma(x_n - x_m + \alpha)/[\ \sigma(\alpha)\ \sigma(x_n - x_m)]^{-1} \}\}. \tag{10a}$$

Here $w(x)$ is an arbitrary function, $\zeta(z) \equiv \zeta(z|\omega,\omega')$ respectively $\sigma(z) \equiv \sigma(z|\omega,\omega')$ are the Weierstrass zeta respectively sigma functions (see Appendix A), the $N+1$ quantities a_k, $k = 0,1,\ldots,N$, are arbitrary constants,

$$\alpha \equiv \alpha(\underset{\sim}{x}) \equiv a_0 + \sum_{j=1}^{N} (x_j - a_j) \ , \tag{10b}$$

$$\beta_n(\underset{\sim}{x}) \equiv \left[\prod_{\ell=1, \ell\neq n}^{N} \sigma(x_n - x_\ell) \right] / \left[\prod_{j=1}^{N} \sigma(x_n - a_j) \right] \ . \tag{10c}$$

$$\mu_n \ddot{x}_n = \dot{x}_n \{ -\eta_n + (\mu_n \dot{x}_n + \eta_n x_n)\ [\zeta(N\overline{x}) - N\zeta(x_n) + \sum_{\ell=1, \ell\neq n}^{N} \zeta(x_n - x_\ell)]$$

$$+ \sum_{m=1, m\neq n}^{N} \{(\mu_m \dot{x}_m + \eta_m x_m)\{[\sigma(x_m)/\sigma(x_n)]^N [\ \prod_{\ell=1, \ell\neq n}^{N} \sigma(x_n - x_\ell)] / [\ \prod_{\ell=1, \ell\neq m}^{N} \sigma(x_m - x_\ell)]\} \cdot$$

$$\cdot \sigma(x_n - x_m + N\overline{x})/[\sigma(N\overline{x})\ \sigma(x_n - x_m)]\}\}\ . \tag{11a}$$

Here $\zeta(x) \equiv \zeta(x|\omega,\omega')$ respectively $\sigma(x) \equiv \sigma(x|\omega,\omega')$ are again the Weierstrass zeta respectively sigma functions (see Appendix A), and $\bar{x} \equiv \bar{x}(t)$ denotes the "mean coordinate",

$$\bar{x}(t) \equiv N^{-1} \sum_{n=1}^{N} x_n(t) \quad , \tag{11b}$$

which is of course assumed not to vanish.

These Newtonian equations of motion all correspond to (1), with the various choices for N and for the seeds $s_n(x)$, hence for $\underline{D}(\underline{x})$, of Sect. 3.1.2.1: specifically, (5) corresponds to (3.1.2.1-2), (7) to (3.1.2.1-9), (8) to (3.1.2.1-19), (9) to (3.1.2.1-30), (10) to (3.1.2.1-42), and (11) to (3.1.2.1-53). Clearly the last model, (11), appears as the special case of the preceding one, (10), with all the constants a_k, $k = 0,1,\ldots,N$, vanishing, $a_k = 0$, and $w(x) = 1$.

Exercise 3.2.1-1. Solve the one-dimensional few-body problems (5), (7) and (8), and discuss the corresponding motions. *Hint*: see (2).

Let us now discuss a bit the three many-body problems (9), (10) and (11).

We begin from (9), assuming for the sake of simplicity that

$$w(x) = \exp(a\,x) \quad , \tag{12}$$

with a an arbitrary (possibly vanishing) constant so that (9) become

$$\mu_n \ddot{x}_n = \dot{x}_n \{ -\eta_n + a(\mu_n \dot{x}_n + \eta_n x_n)$$

$$+ \sum_{m=1, m\neq n}^{N} \{ \mu_n \dot{x}_n + \eta_n x_n + (\mu_m \dot{x}_m + \eta_m x_m) \exp[a(x_n - x_m)]\, [b_n(\underline{x})/b_m(\underline{x})] \} \cdot$$

$$\cdot (x_n - x_m)^{-1} \} , \tag{13a}$$

of course always with (9b). The corresponding version of (2) reads then (see (3.1.2.1-28a) with (12))

$$\mu_n \int_{x_n(0)}^{x_n(t)} dx \{ \sum_{m=1}^{N} h_m \exp(a\,x) x^{m-1} - \eta_n\, x \}^{-1} = t \ . \tag{13b}$$

Note that, iff

$$\eta_n = 0 \quad , \tag{14}$$

the Newtonian equations (13a) become invariant under (space) translations $(x_n \rightarrow x_n + x_0 \, , \, \dot{x}_0 = 0)$, as they then read

$$\mu_n \ddot{x}_n = \dot{x}_n \{ a \mu_n \dot{x}_n$$

$$+ \sum_{m=1, m \neq n}^{N} \{ \mu_n \dot{x}_n + \mu_m \dot{x}_m \exp[a(x_n - x_m)] [b_n(\underline{x}) / b_m(\underline{x})] \} / (x_n - x_m) \} \; . \tag{15}$$

Exercise 3.2.1-2. Modify these equation of motion, (15), so that they possess a (large) set of *completely periodic* solutions. *Hint*: see *Exercise 3.2-3.*

There is another manner to manufacture *translation-invariant* equations from (13a), without requiring (14) but imposing instead the condition

$$a = 0 \tag{16}$$

(see (12); hence $w(x)$ becomes a constant and drops completely out from consideration). It is based on the remark that (13a) with (16) are invariant under rescaling of the particle coordinates $(x_n \rightarrow c \, x_n \, , \, \dot{c} = 0)$. Hence, via the position,

$$x_n(t) = \exp[b \, \xi_n(t)] \quad , \tag{17}$$

with b an arbitrary (nonvanishing) constant, one obtains for the new "particle coordinates" $\xi_n(t)$ the *translation-invariant* equations of motion

$$b \, \mu_n \ddot{\xi}_n = \dot{\xi}_n \{ -\eta_n + \sum_{m=1, m \neq n}^{N} \{ \mu_n b \dot{\xi}_n + \eta_n + (\mu_m b \dot{\xi}_m + \eta_m) \exp[(N-2) b (\xi_m - \xi_n)]$$

$$[\, \tilde{b}_n(\underline{\xi}) / \tilde{b}_m(\underline{\xi}) \,] \} \{ 1 - \exp[b(\xi_m - \xi_n)] \, \}^{-1} \, \} \, , \tag{18a}$$

$$\tilde{b}_n(\underline{\xi}) = \prod_{\ell=1, \ell \neq n}^{N} \{ \, 1 - \exp[b(\xi_\ell - \xi_n)] \, \} \; . \tag{18b}$$

Next, some considerations on the many-body system (10).

Perhaps the main reason that recommends this system, (10), to our attention is the fact that it features the Weierstrass functions ζ and σ. It is however of some interest to also display the more special models, featuring *hyperbolic* (or, equivalently, *trigonometric*) respectively *rational* functions in place of elliptic functions, that obtain from (10) in the degenerate cases when one of, respectively both, the semiperiods of the *elliptic* functions diverge (see (A-54) respectively (A-55)). They read, respectively, as follows:

$$\mu_n \ddot{x}_n = \dot{x}_n \{ -\eta_n + (\mu_n \dot{x}_n + \eta_n x_n) \{ [w'(x_n)/w(x_n)]$$

$$-(a^2/3) a_0 + a \text{ cotanh}(a\alpha) - a \text{ cotanh}[a(x_n - a_n)]$$

$$+ \sum_{\ell=1, \ell \neq n}^{N} [a \text{ cotanh}[a(x_n - x_\ell)] - a \text{ cotanh}[a(x_n - a_\ell)]] \}$$

$$+ \sum_{m=1, m\neq n}^{N} [(\mu_m \dot{x}_m + \eta_m x_m) [w(x_n)/w(x_m)] [\hat{\beta}_m(\underline{x}) / \hat{\beta}_m(\underline{x})] \cdot$$

$$\cdot \{ \sinh[a(\alpha + x_n - x_m)] / \sinh(a\alpha) \} \exp\{ (a^2/6)(x_n - x_m)[(N-1)(x_n + x_m) - 2a_0] \}] \cdot$$

$$\cdot \{ a^{-1} \sinh[a(x_n - x_m)] \}^{-1} \} , \tag{19a}$$

with (10b) and

$$\hat{\beta}_n(\underline{x}) = a \left\{ \prod_{\ell=1, \ell\neq n}^{N} \sinh[a(x_n - x_\ell)] \right\} / \left\{ \prod_{j=1}^{N} \sinh[a(x_n - a_j)] \right\} ; \tag{19b}$$

$$\mu_n \ddot{x}_n = \dot{x}_n \{ -\eta_n + (\mu_n \dot{x}_n + \eta_n x_n) \{ [w'(x_n)/w(x_n)] + \alpha^{-1} - (x_n - a_n)^{-1}$$

$$+ \sum_{m=1, m\neq n}^{N} [(x_n - x_m)^{-1} - (x_n - a_m)^{-1}] \}$$

$$+ \sum_{m=1, m\neq n}^{N} [(\mu_m \dot{x}_m + \eta_m x_m) [w(x_n)/w(x_m)] [\check{\beta}_n(\underline{x}) / \check{\beta}_m(\underline{x})] \cdot$$

$$\cdot [(\alpha + x_n - x_m)/\alpha] / (x_n - x_m)] \} , \tag{20a}$$

again with (10b) and

$$\check{\beta}_n(\underline{x}) \equiv \left[\prod_{\ell=1,\ell\neq n}^{N} (x_n - x_\ell) \right] / \left[\prod_{j=1}^{N} (x_n - a_j) \right] . \qquad (20b)$$

Exercise 3.2.1-3. Derive these equations of motion, (19) and (20), as well as the corresponding quadrature formulas, see (2).

Exercise 3.2.1-4. Verify that, for $a = 0$, (19) yields (20).

Exercise 3.2.1-5. Investigate the model that obtains from (19) in the limit $a_n \to \infty$, $n = 1, 2, ..., N$.

Exercise 3.2.1-6. In the limit $a_k \to \infty$, $k = 0, 1, ..., N$ (and with $w(x)$ given by (12)) (20) become (13a). Verify this fact, and understand why it happens.

Likewise, let us display the form that the Newtonian evolution equations (11) take in the degenerate cases when one of, respectively both, the semiperiods of the elliptic functions diverge (see (A-54) respectively (A-55)):

$$\mu_n \ddot{x}_n = \dot{x}_n \{ -\eta_n$$

$$+ (\mu_n \dot{x}_n + \eta_n x_n) \{ a \operatorname{cotanh}(a N \overline{x}) - N a \operatorname{cotanh}(a x_n) + \sum_{\ell=1,\ell\neq n}^{N} a \operatorname{cotanh}[a(x_n - x_\ell)] \}$$

$$+ \sum_{m=1,m\neq n}^{N} (\mu_m \dot{x}_m + \eta_m x_m) [\sinh(a x_m) / \sinh(a x_n)]^N [s_n(\underline{x}) / s_m(\underline{x})] \cdot$$

$$\cdot a \sinh[a(x_n - x_m + N\overline{x})] / \{ \sinh(a N \overline{x}) \sinh[a(x_n - x_m)] \} \} , \qquad (21a)$$

$$s_n(\underline{x}) \equiv \prod_{\ell=1,\ell\neq n}^{N} \sinh[a(x_n - x_\ell)] \ ; \qquad (21b)$$

$$\mu_n \ddot{x}_n = \dot{x}_n \{ -\eta_n + (\mu_n \dot{x}_n + \eta_n x_n) \ [(N\overline{x})^{-1} - N x_n^{-1} + \sum_{\ell=1,\ell\neq n}^{N} (x_n - x_\ell)^{-1}]$$

$$+ \sum_{m=1,m\neq n}^{N} (\mu_m \dot{x}_m + \eta_m x_m)(x_m / x_n)^N [b_n(\underline{x}) / b_m(\underline{x})] [(N\overline{x})^{-1} + (x_n - x_m)^{-1}] , \qquad (22a)$$

$$b_n(\underline{x}) \equiv \prod_{\ell=1,\ell\neq n}^{N} (x_n - x_\ell) \quad . \tag{22b}$$

Of course in both these equations, (21a) and (22a), $\bar{x}$ is the "mean coordinate", see (11b).

Exercise 3.2.1-7. Derive these equations of motion, (21) and (22), as well as the corresponding quadrature formulas. *Hint*: see (11) with (A-54,55), and (2) with (3.1.2.1-52) and again (A-54,55).

Exercise 3.2.1-8. How should all these equations of motion, (10), (19), (20), (21), (22), be modified, to guarantee that they possess a (large) set (or perhaps only) *completely periodic* solutions? *Hint*: as for *Exercise 3.2-3*.

Exercise 3.2.1-9. Investigate in as much explicit detail as you can the solutions of the many-body problems (9), (10), (11), (13), (15), (18), (19), (20), (21), (22) for $N=2$.

In the last part of Sect. 3.2.1 we return to the treatment of Sect. 3.2, and consider the results that emerge, again for the choices of seeds (3.1.2.1-28), (3.1.2.1-39) respectively (3.1.2.1-52), but now coupled with an appropriate choice, different from (3.2-26) (indeed, different from (3.2-24a)), for the N functions $\rho_n(\underline{x})$, and a correspondingly appropriate choice for $\gamma_n(\underline{x})$ (see (3.2-15)). The equations of motion then read

$$\rho_n(\underline{x})\,\ddot{x}_n = \sum_{m=1}^{N} \{\dot{x}_n \dot{x}_m \left[D_{nm}(\underline{x})\,\rho_m(\underline{x}) - \partial\,\rho_n(\underline{x})/\partial x_m\right]$$

$$+\dot{x}_n\, D_{nm}(\underline{x})\gamma_m(\underline{x}) - \dot{x}_m\,\partial\gamma_n(\underline{x})/\partial x_m\} \;, \tag{23}$$

of course with $D_{nm}(\underline{x})$ given by (3.1.2.1-30), (3.1.2.1-42) respectively (3.1.2.1-53). (*Warning*: in the case of (3.1.2.1-30), the reader should be aware that the $(N\times N)$-matrix denoted here as $\underline{D}(\underline{x})$ coincides with the $(N\times N)$-matrix $\underline{\tilde{D}}(\underline{x})$ of (3.1.2.1-30)).

These equations of motion are merely the one-dimensional version of (3.2-19), with in addition the assignment

$$\underline{\tilde{A}} = 0 \;, \tag{24a}$$

which corresponds of course to

$$\ddot{A}=0 \quad , \tag{24b}$$

and entails, see (3.2-6),

$$f_t(x,t)=0 \quad . \tag{24c}$$

Of this fact we will take advantage below when we shall discuss how to solve the equations of motion we now derive. Note that, as a consequence of (24), it is justified to consider all these models as *integrable* ones, since they possess N constants of motion h_m (see the discussion of this important point in Sect. 3.2). The diligent reader will consider also the more general case without the assumption (24a,b), hence with (3.2-6) (and what follows from it, see Sect. 3.2) instead of (24c) (but beware: the time-evolution (3.2-6) must be compatible with the *ansatz* (3.2-1)).

Let us discuss firstly the model corresponding to the choice of seeds (3.1.2.1-28), hence to the expression (3.1.2.1-30) of $D_{nm}(\underline{x})$. (Once more, beware! : the $(N\times N)$-matrix denoted here $D_{nm}(\underline{x})$ (see for instance (21)), has been instead denoted $\tilde{D}_{nm}(\underline{x})$ in (3.1.2.1-30)).

It is then natural to set (see (3.1.2.1-30b))

$$\rho_n(\underline{x})=\mu_n\, w(x_n)\, b_n(\underline{x})=\mu_n\, w(x_n)\prod_{\ell=1,\ell\neq n}^{N}(x_n-x_\ell) \quad , \tag{25a}$$

$$\gamma_n(\underline{x})=g_n(x_n)\, w(x_n)\, b_n(\underline{x})=g_n(x_n)\, w(x_n)\prod_{\ell=1,\ell\neq n}^{N}(x_n-x_\ell) \quad , \tag{25b}$$

where the μ_n are N arbitrary constants, $g_n(x)$ are N arbitrary functions and of course $b_n(\underline{x})$ is defined by (9b). Thereby (23) become simply

$$\mu_n\ddot{x}_n=-\dot{x}_n\, g_n'(x_n)+\sum_{m=1,m\neq n}^{N}\{[(\mu_n+\mu_m)\dot{x}_n\dot{x}_m+g_m(x_m)\dot{x}_n+g_n(x_n)\dot{x}_m\,]\cdot$$

$$\cdot(x_n-x_m)^{-1}\} \quad . \tag{26}$$

Let us emphasize that these equations of motion, (26), only feature *one- and two-body* forces, in contrast to the many-body problems considered previously in Sect. 3.2.1 (see (9) and (10), and note the presence there of the quantities $b_n(\underline{x})$ and $b_m(\underline{x})$ which depend on *all* the coordinates $x_1,x_2,...,x_N$, see (9b)).

Proof. We like to consider here a slightly more general *ansatz* than (25), since it is of some interest to report the equations of motion that correspond to this more general case. Hence we write, in place of (25a),

$$\rho_n(\underline{x}) = \mu_n(x_n)\, w(x_n)\, b_n(\underline{x}) = \mu_n(x_n)\, w(x_n) \prod_{\ell=1,\ell\neq n}^{N} (x_n - x_\ell) \ , \tag{27}$$

without modifying (25b). These equations, (27) and (25b), entail

$$\partial \rho_n(\underline{x}) / \partial x_m = \rho_n(\underline{x}) \big[\delta_{nm} \{ [\mu_n'(x_n)/\mu_n(x_n)] +$$

$$[w'(x_n)/w(x_n)] + \sum_{\ell=1,\ell\neq n}^{N} (x_n - x_\ell)^{-1} \} - (1-\delta_{nm})(x_n - x_m)^{-1} \big] \ , \tag{27c}$$

$$\partial \gamma_n(\underline{x}) / \partial x_m = \gamma_n(\underline{x}) \{ \ \delta_{nm} \{ [g_n'(x_n)/g_n(x_n)] + [w'(x_n)/w(x_n)] + \sum_{\ell=1,\ell\neq n}^{N} (x_n - x_\ell)^{-1} \}$$

$$-(1-\delta_{nm})(x_n - x_m)^{-1} \ \} \ . \tag{27d}$$

Insertion of these expressions, and of (3.1.2.1-30), in (23) yield

$$\mu_n(x_n)\ddot{x}_n = -\dot{x}_n^2 \mu_n'(x_n) - \dot{x}_n\, g_n'(x_n)$$

$$+ \sum_{m=1,m\neq n}^{N} \big[\{ \dot{x}_n \dot{x}_m [\mu_n(x_n) + \mu_m(x_m)] + \dot{x}_n\, g_m(x_m) + \dot{x}_m\, g_n(x_n) \} / (x_n - x_m) \big] \ . \tag{27e}$$

For $\mu_n(x) = \mu_n$ (independent of x: see (27) and (25a)), this equation, (27e), yields (26), which is thereby proven.

If

$$g_n(x) = \eta_n \ , \qquad g_n'(x) = 0 \ , \tag{28a}$$

these equations of motion, (26), obviously entail that the center-of-mass

$$\bar{x}(t) = \left[\sum_{n=1}^{N} \mu_n\, x_n(t) \right] / \left[\sum_{n=1}^{N} \mu_n \right] \tag{29}$$

moves uniformly,

$$\ddot{\bar{x}}(t) = 0 \ , \tag{28b}$$

$$\bar{x}(t) = \bar{x}(0) + t\,\dot{\bar{x}}(0) \quad . \tag{28c}$$

Let us now review the solution technique for these equations of motion, (26). The fundamental formula, see (3.2-23) and (3.2-1), now reads

$$\rho_n(\underline{x})\dot{x}_n + \gamma_n(\underline{x}) = \sum_{m=1}^{N} h_m\, s_m(x_n) \quad , \tag{30}$$

with the N seeds $s_m(x)$ given by (3.1.2.1-28), $\rho_n(\underline{x})$ and $\gamma_n(\underline{x})$ given by (25), and with the time-independence of the N constants of motion h_m guaranteed by (24c) with (3.2-1). We can therefore rewrite these equations as follows:

$$[\mu_n \dot{x}_n + g_n(x_n)]\, b_n(\underline{x}) = \sum_{m=1}^{N} h_m\, (x_n)^{m-1} \quad , \tag{31}$$

of course with $b_n(\underline{x})$ given by (9b).

Note that the weight function $w(x)$ has completely dropped out from this equation, (31), as indeed from the equations of motion (24). To simplify the rest of this discussion we therefore set hereafter

$$w(x) = 1 \quad . \tag{32}$$

As explained in Sect. 3.2, the degree of solvability of the many-body problem (26) amounts to the availability of the N constants of motion h_m, entailing the possibility of reducing this system, (26), of N coupled *second-order* nonlinear ODEs, to the system (31) of N coupled *first-order* nonlinear ODEs. But in special cases one can go much beyond this. Indeed, let us consider the special case

$$\mu_n = 1 \quad , \tag{33a}$$

$$g_n(x) = g(x) = -(D_0 + D_1\, x) \quad , \tag{33b}$$

so that (26), respectively (31), read

$$\ddot{x}_n = N D_1\, \dot{x}_n + \sum_{m=1, m\neq n}^{N} \{[2\dot{x}_n\, \dot{x}_m - (\dot{x}_n + \dot{x}_m)(D_0 + D_1\, x_n)]/(x_n - x_m)\} \quad , \tag{34}$$

respectively

$$\left[-\dot{x}_n + (D_0 + D_1\, x_n)\right] b_n(\underline{x}) = -\sum_{m=1}^{N} h_m(x_n)^{m-1} \quad . \tag{35}$$

To obtain (34), in addition to inserting (33) in (26) one must use, in the numerator in the right hand side, the simple trick to replace x_m with $x_n + (x_m - x_n)$ and then note that the second term cancels with the denominator. As for (35), it follows directly from (33) via (32).

It is now convenient to introduce the (monic, time-dependent) polynomial of degree N in x that has the N coordinates $x_n(t)$ as its zeros:

$$\psi(x,t) = \prod_{n=1}^{N} \left[x - x_n(t)\right] \quad , \tag{36a}$$

$$\psi(x,t) = x^N + \sum_{m=1}^{N} c_m(t)\, x^{N-m} \quad , \tag{36b}$$

and to set

$$\psi_t(x,t) + (D_0 + D_1\, x)\psi_x(x,t) - N\, D_1\, \psi(x,t) = -f(x,t) \quad . \tag{37}$$

Then this function $f(x,t)$ coincides (as our notation suggests) with the function $f(x,t)$ introduced above, hence it satisfies the extremely simple evolution equation (24c).

Proof. The strategy is to show firstly that $f(x,t)$, as now defined, see (37), lies within the functional space spanned by the seeds (3.1.2.1-28) with (32), then secondly (and sufficiently) that this function $f(x,t)$, evaluated at the N nodes, $x = x_n(t)$, yields precisely the N quantities $f_n(t)$ defined above. Indeed, the first statement coincides with the requirement that $f(x,t)$, see (37), be a polynomial in x of degree at most $N-1$ (see (3.1.2.1-28) with (32)); and this is clear from (37) and (36b). To prove the second statement one must show that

$$f\left[x_n(t),t\right] = f_n(t) = \rho_n\left[\underline{x}(t)\right] \dot{x}_n(t) + \gamma_n\left[\underline{x}(t)\right] \tag{38}$$

(see (3.2-15)), with $\rho_n(x)$ and $\gamma_n(\underline{x})$ given by (25) with (32) and (33). Hence (see (36) and (25) with (32) and (33)) the relation we must prove reads

$$-\psi_t[x_n(t),t]-[D_0+D_1x_n(t)]\,\psi_x[x_n(t),t]+N\,D_1\,\psi[x_n(t),t]$$

$$=\{\dot{x}_n(t)-[D_0+D_1x_n(t)]\}\,b_n[\underline{x}(t)] \tag{39}$$

with $b_n(\underline{x})$ defined by (9b). Indeed the definition (36a) of $\psi(x,t)$ entails

$$\psi_x(x,t)=\sum_{n=1}^{N}\;\prod_{\ell=1,\ell\neq n}^{N}\,[x-x_\ell(t)]\ , \tag{40a}$$

$$\psi_t(x,t)=-\sum_{n=1}^{N}\,\dot{x}_n(t)\prod_{\ell=1,\ell\neq n}^{N}\,[x-x_\ell(t)]\ , \tag{40b}$$

hence (see (9b))

$$\psi_x[x_n(t),t]=b_n[\underline{x}(t)]\ , \tag{41a}$$

$$\psi_t[x_n(t),t]=-\dot{x}_n(t)\,b_n[\underline{x}(t)]\ , \tag{41b}$$

while of course (see (36a))

$$\psi[x_n(t),t]=0\ . \tag{41c}$$

The validity of (39) is now obvious.

We therefore conclude that the coordinates $x_n(t)$, evolving according to the Newtonian equations of motion (34), are just the N zeros of the (time-dependent, monic) polynomial of degree N in x, see (36), that satisfies the linear PDE

$$\psi_{tt}(x,t)+(D_0+D_1\,x)\,\psi_{xt}(x,t)-N\,D_1\,\psi_t(x,t)=0\ , \tag{42}$$

which is clearly implied by (37) with (24c). Now compare (42) respectively (34) to (2.3.3-1) respectively (2.3.3-2) with (in both cases, (2.3.3-1) and (2.3.3-2)) $C=1$, $E=-N\,D_1$, $A_0=A_1=A_2=A_3=B_0=B_1=D_2=0$: clearly we have recovered (a special case of) the results of Sect. 2.3.3!

Exercise 3.2.1-10. Recover, in an analogous manner, the full result of Sect. 2.3.3, see (2.3.3-1) and (2.3.3-2). *Hint*: replace appropriately the assignment (24a,b) with a more general position. (*Warning*: the solution $f(x,t)$ of (3.2-6) must remain in the functional space spanned by the seeds (3.1.2.1-28)).

Let us now return to (23), but now with the seed choice (3.1.2.1-39) rather than (3.1.2.1-28) (and let us immediately also use the simplification (32)). Natural choices for $\rho_n(\underline{x})$ and $\gamma_n(\underline{x})$ in this case then read (see (3.1.2.1-42))

$$\rho_n(\underline{x}) = \beta_n(\underline{x}) \quad , \tag{43a}$$

$$\gamma_n(\underline{x}) = \eta_n \, \beta_n(\underline{x}) \quad , \tag{43b}$$

with $\beta_n(\underline{x})$ defined by (10c). There thus obtain the following Newtonian equations of motion:

$$\ddot{x}_n = \dot{x}_n \, (\dot{x}_n + \eta_n) \, \zeta(\alpha)$$

$$+ \sum_{m=1, m\neq n}^{N} \{ \, \dot{x}_n \, (\dot{x}_m + \eta_m) \frac{\sigma(x_n - x_m + \alpha)}{\sigma(\alpha) \, \sigma(x_n - x_m)} + \dot{x}_m \, (\dot{x}_n + \eta_n) \, \zeta(x_n - x_m) \, \} \quad , \tag{44a}$$

of course with α defined again by (10b). Note that, except for the dependence on the (collective) mean coordinate $\bar{x}(t)$, now defined as follows,

$$\bar{x}(t) = N^{-1} \sum_{n=1}^{N} x_n(t) \quad , \tag{44b}$$

so that (see (10b))

$$\alpha \equiv \alpha(\underline{x}) = N\left[\bar{x}(t) - \bar{a}\right] \quad , \tag{44c}$$

$$\bar{a} \equiv N^{-1}\left[-a_0 + \sum_{j=1}^{N} a_j\right] \quad , \tag{44d}$$

this many-body model only features *one- and two-body* forces (in contrast to the model (9)). Moreover the $N+1$ constants a_k, $k = 0,1,\ldots,N$, only enter via the single constant $\bar{a}$, see (44d).

Proof. Note that, by logarithmic differentiation, (43a) entails, via (A-39),

$$\partial \rho_n(\underline{x}) / \partial x_m = \beta_n(\underline{x}) \{ \, \delta_{nm} \left[\sum_{\ell=1, \ell\neq n}^{N} \zeta(x_n - x_\ell) - \sum_{j=1}^{N} \zeta(x_n - a_j) \right]$$

$$- (1 - \delta_{nm}) \zeta(x_n - x_m) \, \} \, , \tag{43c}$$

and likewise, from (43b),

$$\partial \gamma_n(\underline{x})/\partial x_m = \eta_n \beta_n(\underline{x})\{\ \delta_{nm}\left[\sum_{\ell=1,\ell\neq n}^{N} \zeta(x_n - x_\ell) - \sum_{j=1}^{N} \zeta(x_n - a_j)\right]$$

$$-(1-\delta_{nm})\zeta(x_n - x_m)\}\ . \tag{43d}$$

Insertion of these expressions, (43c,d) as well as (43a,b) and (3.1.2.1-42) with (32), in (23) yields (after a bit of trivial algebra) (44), which is thereby proven.

Exercise 3.2.1-11. Show that the solutions of the *first-order* system

$$\dot{x}_n + \eta_n = \frac{\prod_{j=1}^{N} \sigma(x_n - a_j)}{\prod_{\ell=1,\ell\neq n}^{N} \sigma(x_n - x_l)} \sum_{m=1}^{N} h_m \frac{\sigma(x_n - a_m + a_0)}{\sigma(x_n - a_m)}\ , \tag{45}$$

where the h_m are N constants (whose values, in the context of the initial value problem, are determined by this very equation, (44a), at $t=0$), yield the solutions of the *second-order* system (44a). *Hint*: insert (43a,b) with (10c) and (3.1.2.1-39) with (32) in (30).

Exercise 3.2.1-12. (i) Verify that, in the (degenerate) case when $\sigma(x) = x$, $\zeta(x) = x^{-1}$, see (A-55), the model (44a) takes the simple form

$$\ddot{x}_n = u\,\dot{x}_n + \sum_{m=1,m\neq n}^{N} (2\dot{x}_n \dot{x}_m + \eta_n \dot{x}_m + \eta_m \dot{x}_n)/(x_n - x_m)\ , \quad n = 1,\ldots,N\ , \tag{46a}$$

with

$$u \equiv u(t) = (\dot{\overline{x}} + \overline{\eta})/(\overline{x} - \overline{a})\ , \tag{46b}$$

with $\overline{x} = \overline{x}(t)$ respectively $\overline{a}$ defined by (44c) respectively (44d), and

$$\overline{\eta} = N^{-1} \sum_{n=1}^{N} \eta_n\ . \tag{46c}$$

(ii) Verify that in this case

$$\overline{x}(t) = \overline{x}(0)\ \exp[u(0)t]\{\ 1 + \overline{\eta}[\overline{x}(0) - \overline{a}]^{-1}\{\exp[-u(0)t] - 1\}\ \}\ , \tag{46d}$$

$$u(t)=\left\{\dot{\bar{x}}(0)+\bar{\eta}\,\exp[-u(0)\,t]\right\}/\left[\bar{x}(0)-\bar{a}\right] \quad , \tag{46e}$$

where of course $u(0)$ can be obtained from (46b) (or equivalently from (46e)) at $t=0$. *Hint:* sum (46a) over n from 1 to N, and then integrate the resulting nonlinear ODE for $\bar{x}(t)$ (after dividing it by $\dot{\bar{x}}-\bar{\eta}$).

(iii) Write the *first-order* system of evolution equations equivalent to the *second-order* system (46a). *Solution*: see (45) and (A-55).

Exercise 3.2.1-13. In the equal particle case, entailing

$$\eta_n=\eta \quad , \tag{47}$$

(46) is similar, albeit not identical, to (34). Try and repeat, for (46) with (47), the discussion given after (34). Can this treatment be extended to (46) without (47) ?

Finally, let us return once again to (23), but now with the seed choice (3.1.2.1-52). A natural choice for $\rho_n(\underline{x})$ and $\gamma_n(\underline{x})$ in this case then read (see (3.1.2.1-53))

$$\rho_n(\underline{x})=[\sigma(x_n)]^{-N}\prod_{m=1,m\neq n}^{N}[\sigma(x_n-x_m)] \quad , \tag{48a}$$

$$\gamma_n(\underline{x})=\eta_n\,\rho_n(\underline{x}) \quad . \tag{48b}$$

There thus obtain the following equations of motion:

$$\ddot{x}_n=\dot{x}_n\,(\dot{x}_n+\eta_n)\,\zeta(N\bar{x})$$

$$+\sum_{m=1,m\neq n}^{N}\{\dot{x}_n\,(\dot{x}_m+\eta_m)\sigma(x_n-x_m+N\bar{x})/[\sigma(N\bar{x})\,\sigma(x_n-x_m)]$$

$$+\dot{x}_m\,(\dot{x}_n+\eta_n)\,\zeta(x_n-x_m)\} \quad , \tag{49}$$

where of course $\bar{x}\equiv\bar{x}(t)$ is defined by (44b), and $\zeta(x)\equiv\zeta(x|\omega,\omega')$ respectively $\sigma(x)\equiv\sigma(x|\omega,\omega')$ are the zeta respectively sigma Weierstrass functions, see Appendix A.

Proof. Logarithmic differentiation of (48a, b) yields

$$\partial \rho_n(\underline{x})/\partial x_m = \delta_{nm}\, \rho_n(\underline{x})[-N\zeta(x_n) + \sum_{\ell=1,\ell\neq n}^{N} \zeta(x_n - x_\ell)]$$

$$-(1-\delta_{nm})\, \rho_n(\underline{x})\, \zeta(x_n - x_m) \ , \qquad (48c)$$

as well as well as

$$\partial \gamma_n(\underline{x})/\partial x_m = \delta_{nm}\, \eta_n\, \rho_n(\underline{x})\, [-N\zeta(x_n) + \sum_{\ell=1,\ell\neq n}^{N} \zeta(x_n - x_\ell)]$$

$$-(1-\delta_{nm})\, \eta_n\, \rho_n(\underline{x})\, \zeta(x_n - x_m) \ . \qquad (48d)$$

Insertion of these expressions, (48c,d), as well as (48a,b) and (3.1.2.1-53), in (23) yields (after a bit of trivial algebra) (49), which is thereby proven.

Remark 3.2.1-14. The Newtonian equations of motion (49) are the special case of the Newtonian equations of motion (44) with $a_0 = \sum_{j=1}^{N} a_j$, hence $\bar{a} = 0$ (see (44d) and (44c)).

Exercise 3.2.1-15. Show that the solutions of these (Newtonian, *second-order*) many-body equations of motion, (49), are provided by the solutions of the following system of *first-order* ODEs:

$$\dot{x}_n + \eta_n = [\sigma(x_n)]^N \{ \prod_{m=1,m\neq n}^{N} [\sigma(x_n - x_m)] \}^{-1} [h_1 + \sum_{m=2}^{N} h_m\, \wp^{(m-2)}(x_n)] \ , \qquad (50a)$$

where the h_n's are N constants of motion, $\sigma(x) \equiv \sigma(x|\omega,\omega')$ respectively $\wp(x) \equiv \wp(x|\omega,\omega')$ are the "sigma" respectively the doubly-periodic "pee" Weierstrass functions (see Appendix A), and of course

$$\wp^{(0)}(x) \equiv \wp(x),\ \wp^{(1)}(x) \equiv \wp'(x) = d\,\wp(x)/dx,\ \wp^{(2)}(x) \equiv \wp''(x) = d^2\,\wp(x)/dx^2, \qquad (50b)$$

and so on. *Hint*: insert (48a,b) and (3.1.2.1-52) in (30).

Exercise 3.2.1-16. Ponder on the similarities and differences of (44) respectively (45) with (49) respectively (50). *Hint*: consider to begin with the $N = 2$ case; and note that generally, if $\bar{a}$ vanishes, $\bar{a} = 0$, see (44d), the Newtonian equations of motion (44a) and (49) coincide.

Exercise 3.2.1-17. (i) Verify that, in the (degenerate) case when $\sigma(x)=x$, $\zeta(x)=x^{-1}$, see (A-55), the Newtonian equations of motion (49) take the simple form

$$\ddot{x}_n=\nu\dot{x}_n+\sum_{m=1,m\neq n}^{N}(2\dot{x}_n\dot{x}_m+\eta_n\dot{x}_m+\eta_m\dot{x}_n)/(x_n-x_m) \ , \tag{51a}$$

$$\nu\equiv\nu(t)=\left[\dot{\bar{x}}(t)+\bar{\eta}\right]/\bar{x}(t) \ , \tag{51b}$$

with $\bar{x}(t)$ defined by (44b) and $\bar{\eta}$ by (46c). *(ii)* Verify that in this case

$$\nu(t)=\left\{\dot{\bar{x}}(0)+\bar{\eta}\exp\left[-\nu(0)t\right]\right\}/\bar{x}(0) \ , \tag{51c}$$

$$\bar{x}(t)=\left\{\bar{\eta}\,\bar{x}(0)+\dot{\bar{x}}(0)\bar{x}(0)\exp\left[\nu(0)t\right]\right\}/\left[\dot{\bar{x}}(0)+\bar{\eta}\right] , \tag{51d}$$

where of course $\nu(0)$ can be obtained from (51b) or (51c). *(iii)* Verify that the system of *first-order* ODEs

$$\dot{x}_n+\eta_n=\Big[\prod_{m=1,m\neq n}^{N}(x_n-x_m)\Big]^{-1}\Big[h_1\,x_n^N+\sum_{m=2}^{N}(-)^m(m-1)!\,h_m\,x^{N-m}\Big] \tag{52}$$

is equivalent to the system of *second-order* ODEs (51). *(iv)* When is this system of ODEs, (52), *solvable*? *Hints*: for *(i)* and *(ii)*, see *Exercises 3.2.1-11* and *3.2.1-12*; for *(iii)*, insert (A-55b) and (A-37b) in (50); for *(iv)*, see (2.5-26).

Exercise 3.2.1-18. Find the flaw in the following general treatment, and identify the exceptional cases when it is correct (and it reproduces results given above). Consider the set of interpolational functions

$$q^{(n)}(x|\underline{x})=\prod_{\ell=1,\ell\neq n}^{N}\left[\varphi_{n\ell}(x-x_\ell)/\varphi_{n\ell}(x_n-x_\ell)\right] \ , \tag{53a}$$

where we maintain the freedom to choose at our convenience the N^2 functions $\varphi_{n\ell}(x)$, see below, except that we require them to satisfy the conditions

$$\varphi_{n\ell}(0)=0 \ , \tag{53b}$$

which is sufficient to guarantee the fundamental property of the interpolational functions,

$$q^{(n)}(x_m|\underline{x})=\delta_{nm} \ , \tag{54}$$

so that, if we define the function $f(x,t)$ via the interpolational formula

$$f(x,t)=\sum_{n=1}^{N} f_n(t)\, q^{(n)}[x|\underline{x}(t)] \ , \tag{55a}$$

there holds the relation

$$f_n(t)=f[x_n(t),t] \ . \tag{55b}$$

The $(N\times N)$-matrix $\underline{D}(\underline{x})$ that corresponds to the set of interpolational functions (53),

$$D_{nm}(\underline{x})=q_x^{(m)}(x_n|\underline{x}) \tag{56}$$

(see (3.1.1-6)), takes the explicit form (see *Exercise 3.1.1-7*, or verify now by explicit computation)

$$D_{nm}(\underline{x})=\delta_{nm}\sum_{\ell=1,\ell\neq n}^{N}\left[\varphi'_{n\ell}(x_n-x_\ell)/\varphi_{n\ell}(x_n-x_\ell)\right]$$

$$+(1-\delta_{nm})\left[\varphi'_{mn}(0)/\varphi_{mm}(x_n-x_m)\right]\left[\beta_{mn}(\underline{x})/\beta_{mm}(\underline{x})\right] \ , \tag{57a}$$

$$\beta_{nm}(\underline{x})=\prod_{\ell=1,\ell\neq n}^{N}\varphi_{m\ell}(x_n-x_\ell) \ . \tag{57b}$$

(How can this formula be correct ? Evidently the definition (53a) does not depend on the assignment of the "diagonal" functions $\varphi_{mm}(x)$, yet $D_{nm}(\underline{x})$ seems to feature a presence of these functions in its off-diagonal component, see (57a). Yet (57) is correct: verify it, and understand the way out of this (fictitious) paradox).

The property of $\underline{D}(\underline{x})$, to provide a representation of the differential operator d/dx in the functional space spanned by the interpolational functions (53), corresponds to the formula

$$f_x[x_n(t),t]\equiv \partial f(x,t)/\partial x\Big|_{x=x_n(t)}=\sum_{m=1}^{N} D_{nm}[\underline{x}(t)]f_m(t) \ . \tag{58}$$

Therefore time-differentiation of (55b) yields the relations

$$\dot f_n(t)=f_t[x_n(t),t]+\dot x_n(t)\sum_{m=1}^{N} D_{nm}[\underline{x}(t)]\, f_m(t) \ , \tag{59}$$

where of course (here and below)

$$f_t(x,t)\equiv\partial f(x,t)/\partial t \tag{60}$$

Let us now assume that $f(x,t)$ satisfy the very simple PDE

$$f_t(x,t) = 0 \quad , \tag{61a}$$

entailing

$$f(x,t) = f(x,0) \quad , \tag{61b}$$

and let us moreover set

$$f_n(t) = \mu_n \, \dot{x}_n(t) \quad , \tag{62}$$

where we reserve the privilege to choose later the N constants μ_n. This equation, via (55b), entails

$$\mu_n \, \dot{x}_n(t) = f[x_n(t), t] \, , \tag{63}$$

hence, via (56b),

$$\mu_n \, \dot{x}_n(t) = f[x_n(t), 0] \, . \tag{64}$$

But (55a) entails

$$f(x,0) = \sum_{n=1}^{N} f_n(0) \; q^{(n)}[x | \underline{x}(0)] \quad , \tag{65}$$

hence

$$f[x_n(t),0] = \sum_{m=1}^{N} f_m(0) \; q^{(m)}[x_n(t) | \underline{x}(0)] \quad , \tag{66}$$

hence (via (62) and (53a))

$$\mu_n \, \dot{x}_n(t) = \sum_{m=1}^{N} \mu_m \, \dot{x}_m(0) \prod_{\ell=1, \ell \neq m}^{N} \{\varphi[x_n(t) - x_\ell(0)] / \varphi[x_m(0) - x_\ell(0)]\} \; . \tag{67}$$

This is a *first-order autonomous* ODE for $x_n(t)$ which can clearly be solved *by quadratures*.

On the other hand, insertion of (62) in (59) yields, via (61a), the Newtonian evolution equations

$$\mu_n \, \ddot{x}_n = \dot{x}_n \sum_{m=1}^{N} D_{nm}(\underline{x}) \, \mu_m \, \dot{x}_m \quad , \tag{68}$$

with $D_{nm}(\underline{x})$ defined by (57). In this equation, and sometimes again below, merely for notational simplicity we omit to indicate explicitly the time-dependence.

We are still free to choose the N^2 functions $\varphi_{n\ell}(x)$, except for the constraint (53b). Let us choose them so that they satisfy the additional limitation

$$\varphi'_{n\ell}(0) \equiv d\varphi_{n\ell}(x)/dx\Big|_{x=0} = 0 \ , \tag{69}$$

which entails, see (57a), that the $(N \times N)$-matrix $\underline{D}(\underline{x})$ become *diagonal*,

$$D_{nm}(\underline{x}) = \delta_{nm} \sum_{\ell=1,\ell\neq n}^{N} g_{n\ell}(x_n - x_\ell)/\mu_n \ . \tag{70}$$

Here we have introduced the function $g_{n\ell}(x)$ via the position

$$g_{n\ell}(x) = \mu_n \, \varphi'_{n\ell}(x)/\varphi_{n\ell}(x) \ , \tag{71a}$$

which entails

$$\varphi_{n\ell}(x) = \exp\{\mu_n^{-1} \int^x dx\, g_{n\ell}(x')\} \ . \tag{71b}$$

In this formula, (71b), the lower limit of integration is arbitrary (but different from zero), since any arbitrary constant multiplying $\varphi_{n\ell}(x)$ has no relevance whatsoever, see (53a).

Insertion of (70) into (68) yields the Newtonian evolution equations

$$\mu_n \ddot{x}_n = \dot{x}_n^2 \sum_{m=1,m\neq n}^{N} g_{nm}(x_n - x_m) \ . \tag{72}$$

We have therefore concluded that the initial-value problem for the many-body model characterized by these equations of motion, (72), can be solved *by quadratures*, see (67). Note that we have a large latitude in the choice of the functions $g_{n\ell}(x)$, which can still be assigned arbitrarily, except for the requirement that the functions $\varphi_{n\ell}(x)$ yielded via (71b) satisfy the conditions (53b) and (69): this of course entails that $g_{n\ell}(x)$ diverge at $x = 0$. Alternatively, one can choose $\varphi_{n\ell}(x)$ to satisfy (53b) and (69), and compute the corresponding $g_{n\ell}(x)$, for instance

$$\varphi_{n\ell}(x) = x^{\lambda_{n\ell}/\mu_n} \tag{73a}$$

with $\lambda_{n\ell}/\mu_n > 1$ (sufficient to guarantee (48b) and (60)) yields

$$g_{n\ell}(x) = \lambda_{n\ell}/x \ ; \tag{73b}$$

$$\varphi_{n\ell}(x) = \exp[\ (-\lambda_{n\ell}/\mu_n) x^{-2p}\ ] \tag{74a}$$

with $\lambda_{n\ell} / \mu_n > 0$ and $p = 1,2,3,...$ (also sufficient to guarantee (53b) and (69)), yields

$$g_{n\ell}(x) = 2 p \lambda_{n\ell} \, x^{-(2p+1)} \quad ; \tag{74b}$$

and so on.

Given the large freedom in the choice of $g_{n\ell}(x)$, one can moreover set (see, for instance, (73b))

$$g_{n\ell}(x) = \varepsilon^2 f_{n\ell}(x) \quad , \tag{75a}$$

as well as

$$x_n(t) = t / \varepsilon + \xi_n(t) \quad , \tag{75b}$$

and then take the limit $\varepsilon \to 0$, transforming thereby (72) into

$$\ddot{\xi}_n(t) = \sum_{m=1, m \neq n}^{N} f_{nm}(\xi_n - \xi_m) \quad , \tag{76}$$

which are the Newtonian equations of motion of the standard many-body problem with velocity-independent pair forces!

Note finally that, while we have discussed here this (flawed!) approach in a one-dimensional context, the treatment we have just reported could be repeated just as well (nay, just as badly -- since there is a flaw!) in the S-dimensional case, see indeed the discussion given at the end of Sect. 3.1.1.

Hint: is the simple time-evolution (61) compatible with the representation (55, 53)?

3.2.2 Two-dimensional examples (in the plane)

In Sect. 3.2.2 we use the notation for 2-vectors introduced at the beginning of Sect. 3.1.2; the reader is advised to review it. And we also review here, in 2-vector notation, the two formulas on which our treatment will be based in Sect. 3.2.2:

$$\rho_n \ddot{\vec{r}}_n = \sum_{m=1}^{N} \{ -\dot{\vec{r}}_n [(\partial \rho_n / \partial \vec{r}_m) \cdot \dot{\vec{r}}_m] + \rho_m \dot{\vec{r}}_m [\dot{\vec{r}}_n \cdot \vec{D}_{nm}]$$

$$+ \vec{\gamma}_m [\dot{\vec{r}}_n \cdot \vec{D}_{nm}] - [(\partial \vec{\gamma}_n / \partial \vec{r}_m) \cdot \dot{\vec{r}}_m] \} \quad , \tag{1}$$

$$\rho_n \, \dot{\vec{r}}_n + \vec{\gamma}_n = \sum_{m=1}^{N} \vec{h}_m \, s_m(\vec{r}_n) \quad . \tag{2}$$

In both these equations ρ_n and $\vec{\gamma}_n$ are functions of the N particle coordinates $\vec{r}_m(t)$, $\rho_n \equiv \rho_n[\underline{\vec{r}}(t)]$, $\vec{\gamma}_n \equiv \vec{\gamma}_n[\underline{\vec{r}}(t)]$, and the choice of these functions remains our privilege; the 2-vector-valued $(N \times N)$-matrix $\underline{\vec{D}} \equiv \underline{\vec{D}}[\underline{\vec{r}}(t)]$ in (1) is determined, as detailed in Sects. 3.1.1 and 3.1.2.2, by the choice we make for the N seeds $s_n(\vec{r})$; the N "constants of motion" $\vec{h}_m$ in (2) can be determined, in terms of the initial data, from (2) at $t = 0$, namely from the (system of linear algebraic) relations

$$\sum_{m=1}^{N} \vec{h}_m \, s_m[\vec{r}_n(0)] = \rho_n[\underline{\vec{r}}(0)] \, \dot{\vec{r}}_n(0) + \vec{\gamma}_n[\underline{\vec{r}}(0)] \quad ; \tag{3}$$

and of course in (1) $\ddot{\vec{r}}_n \equiv \ddot{\vec{r}}_n(t)$, $\dot{\vec{r}}_m \equiv \dot{\vec{r}}_m(t)$, and likewise in (2) $\dot{\vec{r}}_n \equiv \dot{\vec{r}}_n(t)$, $\vec{r}_n \equiv \vec{r}_n(t)$. Needless to say, in all these equations, (1), (2) and (3), as well as in those written below, the index n takes all integer values from 1 to N.

These formulas follow directly from those of Sect. 3.2; but note that we are now restricting attention to the simple case with

$$\vec{\mathsf{A}} = 0 \quad , \tag{4a}$$

entailing

$$\underline{\vec{A}} = 0 \tag{4b}$$

hence (see (3.2-6))

$$\vec{f}(\vec{r},t) = \vec{f}(\vec{r},0) \quad . \tag{4c}$$

In particular (1) corresponds to (3.2-19) with (4b); (2) corresponds to (3.2-23) via (4c) and (3.2-1), of course with $\vec{h}_m = \vec{h}_m(0)$ (indeed, see (3.2-1) and (4)), which justifies considering the N 2-vectors $\vec{h}_m$ as constants of motion (and therefore the Newtonian equations of motion (1) as *integrable*); and likewise (3) corresponds to (3.2-21b) (as well as to (2) at $t = 0$). Of course, for any given choice of N, of $\rho_n(\underline{\vec{r}})$, of $\gamma_n(\underline{\vec{r}})$, and of $s_n(\vec{r})$ hence $\underline{\vec{D}}(\underline{\vec{r}})$, the specific form taken by (1) is interpreted as Newtonian equations of motion of our N-body problem in the plane, and the corresponding form taken by (2) provides substantial progress towards solving that N-body problem.

Before going over to the discussion of specific examples, let us interject the following.

Remark 3.2.2-1. The evolution equations (1) admit the (trivial) solution

$$\vec{r}_n(t)=\vec{r}_n(0),\ \dot{\vec{r}}_n(t)=0,\ , \tag{5}$$

for any (arbitrary) choice of the initial positions $\vec{r}_n(0)$.

Now, to the examples. Firstly we restrict attention to the following simple choice for $\rho_n(\underline{\vec{r}})$ and $\vec{\gamma}_n(\underline{\vec{r}})$:

$$\rho_n(\underline{\vec{r}})=\mu_n\ , \tag{6a}$$

$$\vec{\gamma}_n(\underline{\vec{r}})=\sum_{m=1}^{N}\eta_{nm}\vec{r}_n\ , \tag{6b}$$

with μ_n and η_{nm} arbitrary (scalar) constants. Then (1) yields the Newtonian equations of motion

$$\mu_n\ddot{\vec{r}}_n=\sum_{m=1}^{N}\{\ -\eta_{nm}\dot{\vec{r}}_m+[\ \dot{\vec{r}}_n\cdot\vec{D}_{nm}(\underline{\vec{r}})\ ][\ \mu_m\dot{\vec{r}}_m+\sum_{\ell=1}^{N}\eta_{m\ell}\vec{r}_\ell\ ]\}\ , \tag{7}$$

while (2) becomes

$$\mu_n\dot{\vec{r}}_n=\sum_{m=1}^{N}\vec{h}_m s_m(\vec{r}_n)-\sum_{m=1}^{N}\eta_{nm}\vec{r}_m\ . \tag{8}$$

This latter equation, (8), is a system of N coupled, generally nonlinear (up to exceptions, see below), *first-order ODEs with constant coefficients* for N 2-vectors, namely a system of $2N$ (first-order, constant-coefficient) ODEs for $2N$ (scalar) unknowns. Moreover, if the $(N\times N)$-matrix with elements η_{nm} is diagonal, namely if

$$\eta_{nm}=0\quad\text{for}\quad n\neq m\ , \tag{9}$$

then this system, (8), decouples into N separate 2-vector (first-order, constant-coefficient) ODEs, each of them involving 2 (scalar) unknowns.

The alert reader should have noted that (5) respectively (6) are merely copies of (3.2-26) respectively (3.2-27a), as (4a), (4b) respectively (4c) are copies of (3.2-32b), (3.2-32c) respectively (3.2-32e), and (7) respectively (8) are copies of (3.2-34) respectively (3.2-33).

Exercise 3.2.2-2. Suggest a (possibly complex) modification of these equations of motion, (7), adequate to guarantee that they then possess a (large) set of *completely periodic* solutions. *Hint:* see *Exercise 3.2-3.*

The simplest example we consider corresponds to the choice of seeds (3.1.2.2-2). This yields the following *solvable* 2-body problem in the plane:

$$\mu_1 \ddot{\vec{r}}_1 = -\eta_{11} \dot{\vec{r}}_1 - \eta_{12} \dot{\vec{r}}_2 + (\hat{k} \cdot \vec{r}_1 \wedge \vec{r}_2)^{-1} \cdot$$

$$\cdot [(\mu_1 \dot{\vec{r}}_1 + \eta_{11} \vec{r}_1 + \eta_{12} \vec{r}_2)(\hat{k} \cdot \dot{\vec{r}}_1 \wedge \vec{r}_2) - (\mu_2 \dot{\vec{r}}_2 + \eta_{21} \vec{r}_1 + \eta_{22} \vec{r}_2)(\hat{k} \cdot \dot{\vec{r}}_1 \wedge \vec{r}_1)] \quad , \qquad (10a)$$

$$\mu_2 \ddot{\vec{r}}_2 = -\eta_{21} \dot{\vec{r}}_1 - \eta_{22} \dot{\vec{r}}_2 + (\hat{k} \cdot \vec{r}_1 \wedge \vec{r}_2)^{-1} \cdot$$

$$\cdot [(\mu_1 \dot{\vec{r}}_1 + \eta_{11} \vec{r}_1 + \eta_{12} \vec{r}_2)(\hat{k} \cdot \dot{\vec{r}}_2 \wedge \vec{r}_2) - (\mu_2 \dot{\vec{r}}_2 + \eta_{21} \vec{r}_1 + \eta_{22} \vec{r}_2)(\hat{k} \cdot \dot{\vec{r}}_2 \wedge \vec{r}_1)] \quad . \qquad (10b)$$

The corresponding equations that demonstrate the *solvable* nature of this problem read as follows:

$$\mu_1 \dot{x}_1 = h_{1x} x_1 + h_{2x} y_1 - \eta_{11} x_1 - \eta_{12} x_2 \quad , \qquad (11a)$$

$$\mu_1 \dot{y}_1 = h_{1y} x_1 + h_{2y} y_1 - \eta_{11} y_1 - \eta_{12} y_2 \quad , \qquad (11b)$$

$$\mu_2 \dot{x}_2 = h_{1x} x_2 + h_{2x} y_2 - \eta_{21} x_1 - \eta_{22} x_2 \quad , \qquad (11c)$$

$$\mu_2 \dot{y}_2 = h_{1y} x_2 + h_{2y} y_2 - \eta_{21} y_1 - \eta_{22} y_2 \quad . \qquad (11d)$$

The equations of motion (10) follow straightforwardly by inserting (3.1.2.2-3) in (7). Note their rotation-invariant character. They are also invariant under inversions, and under rescaling of the particle coordinates ($\vec{r}_n \to c\vec{r}_n$, $\dot{c} = 0$).

Likewise, (11) follow straightforwardly by inserting (3.1.2.2-2) in (8). Note that the solution of (the initial-value for) these set of 4 coupled *linear* ODEs is a purely algebraic task. The first step is to determine the values of the 4 constants $h_{1x}, h_{2x}, h_{1y}, h_{2y}$ in terms of the initial data, $\vec{r}_1(0), \vec{r}_2(0), \dot{\vec{r}}_1(0), \dot{\vec{r}}_2(0)$, by solving

(10) at $t=0$. Note that the condition on the initial data sufficient to guarantee that this system of 4 linear equations for the 4 unknowns $h_{1x}, h_{2x}, h_{1y}, h_{2y}$ have a unique solution, namely the requirement that the relevant "determinant of coefficients" not vanish, reads (after an elementary computation)

$$\hat{k}\cdot\vec{r}_1(0)\wedge\vec{r}_2(0)\neq 0 \ , \tag{12}$$

and its significance for the equations of motion (10) is clear. After the 4 constants $h_{1x}, h_{2x}, h_{1y}, h_{2y}$ have been computed, the solution of (11), now to be considered a system of 4 coupled *first-order linear ODEs with constant coefficients* for the 4 unknown functions $x_1(t), y_1(t), x_2(t), y_2(t)$, generally requires the diagonalization and inversion of (4×4)-matrices. However, if (9) holds, this task reduces merely to the diagonalization and inversion of (2×2)-matrices and can therefore be easily accomplished in completely explicit form.

Exercise 3.2.2-3. Solve explicitly this case and discuss the character of the motion. Are there periodic trajectories ? *Solution*: see <CJX93b>.

Exercise 3.2.2-4. Investigate the nature of the motion in the more general case when (9) does not hold: what about periodic trajectories ? *Hint*: see *Exercise 3.2.2-5* below.

The second model we consider corresponds to the choice of seeds (3.1.2.2-8). Hence this *solvable* 3-body problem in the plane reads as follows:

$$\mu_n\ddot{\vec{r}}_n=\sum_{m=1,2,3 \bmod(3)}\{\ -\eta_{nm}\dot{\vec{r}}_m+[\ \hat{k}\cdot(\vec{r}_1-\vec{r}_2)\wedge(\vec{r}_1-\vec{r}_3)\ ]^{-1}$$
$$[\ \mu_m\dot{\vec{r}}_m+\sum_{\ell=1}^{3}\eta_{m\ell}\vec{r}_\ell\ ][\ \hat{k}\cdot\dot{\vec{r}}_n\wedge(\vec{r}_{m+1}-\vec{r}_{m+2})\ ]\ \},\quad n=1,2,3 \bmod(3)\ , \tag{13}$$

while the equations from which its solution can be obtained in almost explicit form are given by (8) with (3.1.2.2-8).

These equations of motion, (13), are clearly invariant under (plane) rotations, as well as under rescaling of the particle coordinates ($\vec{r}_n\to c\vec{r}_n,\ n=1,2,3,\ \dot{c}=0$). They are moreover invariant under translations ($\vec{r}_n\to\vec{r}_n+\vec{r}_0,\ n=1,2,3,\ \dot{\vec{r}}_0=0$), if the 9 constants η_{nm} satisfy the 3 constraints

$$\sum_{m=1}^{3}\eta_{nm}=0,\quad n=1,2,3\ . \tag{14}$$

And let us recall the geometrical significance of the denominator in the right hand side of (13): up to its sign, it is twice the area of the (plane) triangle whose vertices coincide with the positions of the 3 particles.

Exercise 3.2.2-5. Show that the solution of (13) has the explicit form

$$\vec r_n(t) = \vec r_{n0} + \sum_{p=1}^{6} \vec r_{np} \exp(\lambda_p t), \quad n = 1,2,3, \tag{15}$$

obtain the algebraic equations that determine, in terms of the initial data $\underline{\vec r}(0)$ and $\dot{\underline{\vec r}}(0)$, the 6 scalars λ_p, $p = 1,\ldots,6$, and the 21 2-vectors $\vec r_{np}$, $n = 1,2,3$; $p = 0,\ldots 6$, and discuss the solution (15), focussing on the eventual presence of (possibly multiply) periodic motions. *Hint*: write out (8) with (3.1.2.2-8) and take advantage of its character: linear, constant coefficients. *Solution*: see <CJX93>, <C93a>.

The third model we consider corresponds to the choice of seeds (3.1.2.2-32) with (3.1.2.2-49); we moreover restrict consideration to the case (9) (for simplicity; the diligent reader will also explore the general case, without (9)).

We thus obtain the following *solvable* N-body problem in the plane:

$$\mu_n \ddot{\vec r}_n = -\eta_n \dot{\vec r}_n + \sum_{m=1, m\neq n}^{N} \{(\hat k \cdot \vec r_n \wedge \vec r_m)^{-1} \cdot$$

$$\cdot \left[(\mu_n \dot{\vec r}_n + \eta \vec r_n)(\hat k \cdot \dot{\vec r}_n \wedge \vec r_m) + [\sigma_n(\underline{\vec r})/\sigma_m(\underline{\vec r})]\,(\mu_m \dot{\vec r}_m + \eta_m \vec r_m)(\hat k \cdot \dot{\vec r}_n \wedge \vec r_n)\right]\}, \tag{16a}$$

$$\sigma_n(\underline{\vec r}) \equiv \prod_{\ell=1, \ell\neq n}^{N} (\hat k \cdot \vec r_n \wedge \vec r_\ell) \ . \tag{16b}$$

The corresponding equations, providing the clue to the solvability of this model, read

$$\mu_n \dot{\vec r}_n = -\eta_n \vec r_n + \sum_{m=1}^{N} \vec h_m \, (x_n)^{m-1} \, (y_n)^{N-m} \ , \qquad n = 1,2,\ldots,N \ , \tag{17a}$$

or equivalently

$$\mu_n \dot x_n = -\eta_n x_n + y_n^{N-1} \sum_{m=1}^{N} h_{mx} \, (x_n / y_n)^{m-1} \ , \tag{17b}$$

$$\mu_n \dot{y}_n = -\eta_n y_n + y_n^{N-1} \sum_{m=1}^{N} h_{my} (x_n / y_n)^{m-1} \quad . \tag{17c}$$

The Newtonian equations of motion (16) are obtained by inserting (9) and (3.1.2.2-50) in (7); likewise (17) are obtained by inserting (9) and (3.1.2.2-32) with (3.1.2.2-49) in (8).

We called this model, (16), *solvable*. Indeed the evolution equations (17), with the N (time-independent!) 2-vectors $\vec{h}_m$ determined in terms of the initial data $\vec{\underline{r}}(0)$ and $\dot{\vec{\underline{r}}}(0)$ by the same equations (17) at $t=0$, can be *solved by quadratures*. Let us tersely indicate how.

For $N=2$ these equations, (17), are linear hence easily solvable; indeed they reduce tó (a special case of) (10). Hereafter we assume $N>2$.

We rewrite these equations, (17b,c), as follows:

$$\mu \, \dot{x} = -\eta x + y^{N-1} H^{(x)}(x/y) \quad , \tag{18a}$$

$$\mu \, \dot{y} = -\eta y + y^{N-1} H^{(y)}(x/y) \quad , \tag{18b}$$

with

$$H^{(x)}(u) = \sum_{m=1}^{N} h_{mx} \, u^{m-1} \quad , \tag{19a}$$

$$H^{(y)}(u) = \sum_{m=1}^{N} h_{my} \, u^{m-1} \quad . \tag{19b}$$

Note that, merely for notational convenience, we omit here to indicate explicitly the index n.

We now set

$$u = x/y \, , \qquad v = y^{N-1}/x \quad , \tag{20a}$$

$$x = (u^{N-1} v)^{1/(N-2)} \quad , \qquad y = (u\, v)^{1/(N-2)} \quad , \tag{20b}$$

and obtain for u and v the ODEs

$$\mu \dot{u} = v \, F(u) \quad , \tag{21a}$$

$$\mu \dot{v} = (2-N) \, \eta \, v + v^2 \, G(u) \quad , \tag{21b}$$

with

$$F(u) = u \left[\, H^{(x)}(u) - u \, H^{(y)}(u) \, \right] \quad , \tag{22a}$$

$$G(u) = H^{(x)}(u) - (N-1)\, u\, H^{(y)}(u) \quad . \tag{22b}$$

We now set

$$v(t) = V[u(t)] \tag{23a}$$

so that (in self-evident notation)

$$\dot{v} = V'(u)\, \dot{u} \quad . \tag{23b}$$

Hence, from (21) and (23),

$$V'(u) = [F(u)]^{-1}\,[(2-N)\,\eta + V(u)\, G(u)] \quad . \tag{24}$$

This linear ODE for $V(u)$ can be solved by quadratures:

$$V(u) = V[u(0)] \exp\{ \int_{u(0)}^{u} du' \, [G(u')/F(u')] \}$$

$$+ (2-N)\eta \int_{u(0)}^{u} du' \, [F(u')]^{-1} \exp\{ \int_{u'}^{u} du'' \, [G(u'')/F(u'')] \} \; . \tag{25}$$

Then (21a), which now reads

$$\mu \dot{u}(t) = V[u(t)]\, F[u(t)] \quad , \tag{26}$$

can also be integrated by a quadrature, and the solution of the problem is thereby completed (up to functional inversions), since once $u(t)$ is known, $v(t)$ is easily obtained from (21a), (22a) and (19), and then x and y by (20b).

Indeed we can go a bit further, by noticing that (19) and (22) entail

$$F(u) = c \prod_{j=0}^{N} (u - u_j) \quad , \tag{27}$$

$$G(u)/F(u) = c \sum_{j=0}^{N} \rho_j/(u-u_j) \quad , \tag{28}$$

with (see (22a) and (19b)) $u_0 = 0$ and $c = -h_y^{(N)}$ (we assume this latter constant not to vanish). The computation of the other $2N+1$ constants (namely: u_m, $m = 1, 2, \ldots, N$; ρ_j, $j = 0, 1, \ldots, N$) is a purely algebraic task. Then the integrations in the right hand side of (25) can be performed to yield:

$$V(u) = V[u(0)] \prod_{j=0}^{N} \{(u-u_j)/[u(0)-u_j]\}^{\rho_j}$$

$$+[(2-N)\eta/c][\prod_{j=0}^{N}(u-u_j)^{\rho_j}]\int_{u(o)}^{u}du'\prod_{k=0}^{N}[(u'-u_k)^{-(1+\rho_k)}] \quad . \tag{29}$$

The many-body model (16) features many-body forces, due to the factors $\sigma_\ell(\vec{r})$ in the right hand side. It is clearly invariant both under a (common) rotation and (time-independent) rescaling of all particle coordinates $\vec{r}_n$, $n=1,2,...,N$. It is, instead, *not* translation-invariant, and it features forces which become singular whenever 2 particles are aligned with the origin of coordinates. It may be of some interest to write the Newtonian equations of motion of this model, (16), in the polar coordinates r_n, θ_n, defined in the standard manner:

$$\vec{r}_n = r_n(\cos\theta_n, \sin\theta_n) \quad , \tag{30a}$$

entailing of course (see (3.1.2.2-1))

$$\vec{r}_n\cdot\vec{r}_m = r_n r_m\cos(\theta_n-\theta_m) \quad , \tag{30b}$$

$$\hat{k}\cdot\vec{r}_n\wedge\vec{r}_m = -r_n r_m\sin(\theta_n-\theta_m) \quad . \tag{30c}$$

They read

$$\mu_n\ddot{r}_n = r_n\dot{\theta}_n^2 - \eta_n\dot{r}_n + \sum_{m=1,m\neq n}^{N}\{\,\mu_n\dot{r}_n+\eta_n r_n)[(\dot{r}_n/r_n)+\dot{\theta}_n\,\mathrm{cotan}(\theta_n-\theta_m)]$$

$$+\dot{\theta}_n[\,(\mu_m\dot{r}_m+\eta_m r_m)\,\mathrm{cotan}(\,\theta_n-\theta_m)+\mu_m r_m\dot{\theta}_m\,][\,\gamma_n(\underline{\theta})/\gamma_m(\underline{\theta})]\} \,, \tag{31a}$$

$$\mu_n r_n\ddot{\theta}_n = \dot{\theta}_n[\,-2\,\dot{r}_n-\eta_n r_n+\sum_{m=1,m\neq n}^{N}\{\,\mu_n[(\dot{r}_n+r_n\dot{\theta}_n\,\mathrm{cotan}(\theta_n-\theta_m)]$$

$$+[\mu_m r_m\dot{\theta}_m\,\mathrm{cotan}(\,\theta_n-\theta_m)-\mu_m\dot{r}_m-\eta_m r_m][s_n(\underline{\theta})/s_m(\underline{\theta})]\,\}\,] \;, \tag{31b}$$

$$\gamma_n(\underline{\theta}) \equiv \prod_{\ell=1,\ell\neq n}^{N}[\sin(\theta_n-\theta_\ell)] \;. \tag{31c}$$

Exercise 3.2.2-6. Verify these equations.

Exercise 3.2.2-7. Modify the equations of motion (16) and (31) so that they possess a (large) set of *completely periodic* solutions. *Hint*: see *Exercise 3.2-3*, and do not hesitate to go complex if need be.

To manufacture an example of many-body problem in the plane that only features one- and two-body forces we now maintain the choice of seeds (3.1.2.2-32) with (3.1.2.2-49), but we abandon the special assignment (6) and we replace it, as indeed suggested by (3.1.2.2-50), as follows:

$$\rho_n(\underline{\vec{r}}) = \mu_n \, \sigma_n(\underline{\vec{r}}) \ , \tag{32a}$$

$$\vec{\gamma}_n(\underline{\vec{r}}) = \eta_n \, \vec{r}_n \, \sigma_n(\underline{\vec{r}}) \ , \tag{32b}$$

where μ_n and η_n are again $2N$ arbitrary constants and of course $\sigma_n(\underline{\vec{r}})$ is defined by (16b) (or (3.1.2.2-34)). We thus get the following Newtonian equations of motion:

$$\mu_n \ddot{\vec{r}}_n = -\eta_n \dot{\vec{r}}_n + \sum_{m=1, m\neq n}^{N} \{ (\hat{k}\cdot\vec{r}_n \wedge \vec{r}_m)^{-1} \cdot$$

$$\cdot [(\mu_n \dot{\vec{r}}_n + \eta_n \vec{r}_n)(\hat{k}\cdot\dot{\vec{r}}_m \wedge \vec{r}_n) + (\mu_m \dot{\vec{r}}_m + \eta_m \vec{r}_m)(\hat{k}\cdot\dot{\vec{r}}_n \wedge \vec{r}_n)]\} \ . \tag{33}$$

Indeed, logarithmic differentiation of (32) with (16b) yields

$$\partial \rho_n(\underline{\vec{r}}) / \partial \vec{r}_m = \mu_n \, \sigma_n(\underline{\vec{r}}) \cdot$$

$$\cdot \{ -\delta_{nm} \sum_{\ell=1, \ell\neq n}^{N} [(\hat{k}\wedge\vec{r}_\ell)/(\hat{k}\cdot\vec{r}_n \wedge \vec{r}_\ell)] + (1-\delta_{nm})(\hat{k}\wedge\vec{r}_n)/(\hat{k}\cdot\vec{r}_n\wedge\vec{r}_m), \tag{34a}$$

$$[\partial \vec{\gamma}_n(\underline{\vec{r}}) / \partial \vec{r}_m] \cdot \vec{a} = \eta_n \, \sigma_n(\underline{\vec{r}}) [\, \delta_{nm} \{ \, \vec{a} + \vec{r}_n \sum_{\ell=1, \ell\neq n}^{N} [(\hat{k}\cdot\vec{a}\wedge\vec{r}_\ell)/(\hat{k}\cdot\vec{r}_n\wedge\vec{r}_\ell)] \, \}$$

$$-(1-\delta_{nm}) \, \vec{r}_n \, (\hat{k}\cdot\vec{a}\wedge\vec{r}_n)/(\hat{k}\cdot\vec{r}_n\wedge\vec{r}_m) \,] \ , \tag{34b}$$

with $\vec{a}$ an arbitrary 2-vector. And insertion in (1) of these expressions, as well as of the expression (3.1.2.2-50) of $\underline{\vec{D}}(\underline{\vec{r}})$, yields, after some nice cancellations, precisely (33).

The Newtonian equations of motion (33) only feature *one- and two-body forces*; the curious property of these forces, to become singular whenever a pair of particles is aligned to the origin, remains. Again, the model (33) is *not* invariant under translations; it is instead clearly invariant both under rotation and rescaling.

Note that the equations of motion (33) differ from (16) not only because of the presence in (16) of the (multi-body) term $[\sigma_n(\underline{x})/\sigma_m(\underline{x})]$. Indeed the equations of motion (16) entail that, if a single particle, say the k-th one, has at any one time a vanishing velocity, $\dot{\vec{r}}_k(0)=0$, then $\dot{\vec{r}}_k(t)=0$, $\vec{r}_k(t)=\vec{r}_k(0)$, for all time; namely this particle never moves, although its presence does affect the motion of the other particles. Such a property is not featured by the equations of motion (33), which are inconsistent with any one particle remaining still while all the others move; although these equations of motion, (33), admit as equilibrium configuration any configuration with *all* particle standing still, namely $\vec{r}_n(t)=\vec{r}_n(0)$, $n=1,2,\ldots,N$, is a solution of (33) for any arbitrary assignment of the N 2-vectors $\vec{r}_n(0)$.

Exercise 3.2.2-8. Write the equations in polar coordinates that correspond to (33).

Exercise 3.2.2-9. Modify the equations of motion (33) so that they possess a (large) set of *completely periodic* solutions. *Hint*: see *Exercise 3.2-3*.

The "degree of solvability" of the Newtonian equations of motion (33) is lower than for the equations of motion (16). It amounts to the availability of N 2-vector integrals of motion $\vec{h}_m$ defined (implicitly), in terms of the N 2-vector coordinates $\vec{r}_n$ and the N velocities $\dot{\vec{r}}_n$ of the N particles, by the N *linear* 2-vector equations

$$\sum_{m=1}^{N} \vec{h}_m\,(x_n)^{m-1}\,(y_n)^{N-m}=\sigma_n(\underline{\vec{r}})\,\left[\mu_n\,\dot{\vec{r}}_n+\eta_n\,\vec{r}_n\right] \quad , \tag{35}$$

with $\sigma_n(\underline{\vec{r}})$ defined by (16b).

These equations, (35), are merely (2) with (32) and (3.1.2.2-32) with (3.1.2.2-49). They have, as usual, a double role. At $t=0$, they can be considered a set of N 2-vector (i.e., $2N$ scalar) *linear algebraic* equations for the N 2-vector (i.e., $2N$ scalar) unknowns $\vec{h}_m$: their solution, which is a standard algebraic task, determines these quantities in terms of the initial data (for the N-body problem) $\underline{\vec{r}}(0)$ and $\dot{\underline{\vec{r}}}(0)$. (This task is facilitated by noticing that the determinant of the coefficients in the left hand side of (35) -- an algebraic set of linear equations that should of course be solved componentwise, i.e. firstly for the x-component of the 2-vectors $\vec{h}_m$, then for the y-component of these 2-vectors -- is of Vandermonde type). Once the N (time-independent !) 2-vectors $\vec{h}_m$ have been determined, the equations (35) characterize the time-evolution of the coordinates $\vec{r}_n(t)$, hence they can be solved instead of the equations of motion (33). To obtain a unique solution from them one must of course assign the initial data $\underline{\vec{r}}(0)$, and the theory detailed above then guarantees that the solution of this initial-value problem for (35) also provides the (unique) solution to the initial-value problem, of course with the same initial data, for the Newtonian equations of motion (33). The advantage, in terms of solvability, is that (35) is a system of N 2-vector, generally coupled and nonlinear, albeit autonomous, *first-order* ODEs, while (33) is a system of *second-order* ODEs.

3.2.3 Few-body problems in ordinary (3-dimensional) space

In Sect. 3.2.3 we use the notation of Sect. 3.1.2.3, which the diligent reader should immediately review. Our presentation here is limited to *solvable* few-body (specially: 3-body and 4-body) problems, of course in ordinary (3-dimensional) space: those that correspond to the two cases treated in Sect. 3.1.2.3. Indeed the results of Sect. 3.2.3 emerge directly from the insertion of the relevant expressions of $\underline{\vec{D}}(\underline{\vec{r}})$ and of the seeds $s_n(\vec{r})$, as given in Sect. 3.1.2.3, into the two fundamental formulas (see (3.2-34) and (3.2-33))

$$\mu_n \ddot{\vec{r}}_n = \sum_{m=1}^{N} \{ -\eta_{nm} \dot{\vec{r}}_m + [\dot{\vec{r}}_n \cdot \vec{D}_{nm}(\underline{\vec{r}})][\mu_m \dot{\vec{r}}_m + \sum_{\ell=1}^{N} \eta_{m\ell} \vec{r}_\ell] \} , \tag{1}$$

$$\mu_n \dot{\vec{r}}_n = \sum_{m=1}^{N} [-\eta_{nm} \vec{r}_m + \vec{h}_m s^{(m)}(\vec{r}_n)] . \tag{2}$$

Here, of course, all vectors are 3-vectors.

The *solvable* 3-body problem that obtains by inserting (3.1.2.3-2) in (1) reads

$$\mu_n \ddot{\vec{r}}_n = \sum_{m=1}^{3} \{ -\eta_{nm} \dot{\vec{r}}_m + \dot{\vec{r}}_n \cdot (\vec{r}_{m+1} \wedge \vec{r}_{m+2}) [\mu_m \dot{\vec{r}}_m + \sum_{\ell=1}^{3} \eta_{m\ell} \vec{r}_\ell]/\Delta \} , \tag{3a}$$

with (see (3.1.2.3-3))

$$\Delta \equiv \Delta(\vec{r}_1, \vec{r}_2, \vec{r}_3) = \vec{r}_1 \cdot \vec{r}_2 \wedge \vec{r}_3 \ . \tag{3b}$$

Here, and always below when discussing this 3-body model, all indices take the values $1,2,3 \bmod(3)$.

These equations of motion, (3), are invariant under rotation and rescaling, but not under translation. And let us recall the geometrical significance of the denominator Δ in the right hand side of (3a) ; it is, up to its sign, six times the volume of the tetrahedron having the origin of coordinates, and the positions $\vec{r}_n$ of the 3 particles, as its 4 vertices (see (3b)).

The solvability of this model follows from the form that (2), via (3.1.2.3-2), take in this case:

$$\mu_n \dot{\vec{r}}_n = -\sum_{m=1}^{3} [\eta_{nm} \vec{r}_m] + \vec{h}_1 x_n + \vec{h}_2 y_n + \vec{h}_3 z_n \ , \quad n = 1,2,3 \ , \tag{4a}$$

or equivalently,

$$\mu_n \dot{u}_{nj} = \sum_{m=1}^{3} (-\eta_{nm} u_{mj} + h_{mj} u_{nm}) \ , \quad n = 1,2,3, \qquad j = 1,2,3 \ , \tag{4b}$$

where we have set, for notation convenience,

$$x_n \equiv u_{n1} \ , \ y_n \equiv u_{n2} \ , \ z_n \equiv u_{n3} \ , \ n = 1,2,3 \ , \tag{4c}$$

$$h_{mx} = h_{m1} \ , \ h_{my} \equiv h_{m2} \ , \ h_{mz} \equiv h_{m3} \ , \ m = 1,2,3 \ , \tag{4d}$$

of course with $\vec{h}_m \equiv (h_{mx}, h_{my}, h_{mz})$. By a technique which should by now be familiar the 9 scalar equations, (4b), at $t = 0$, determine the values of the 9 scalar constants of motion h_{mj} in terms of the 18 scalar initial data, $u_{nm}(0)$ and $\dot{u}_{nm}(0)$, see (4c), and then determine the time evolution of the 9 quantities $u_{nm}(t)$, namely of the 3 3-vectors $\vec{r}_n(t)$, see (4c). Both steps require only algebraic operations: the first one obviously so, see (4b); and the second one as well, since the required task is to solve a system of *linear first-order ODEs with constant coefficients*, see again (4b).

Exercise 3.2.3.1-1. Solve as explicitly as possible the 3-body model (3), and discuss the possibility of (completely, or multiply) periodic motions. Consider the simplification that occur if (i) $\eta_{nm}=\delta_{nm}\eta_n$, (ii) $\eta_{nm}=0$. Modify case (ii) so that *all* its motions are completely periodic. Is there a modification of case (i) (with $\eta_n \neq 0$) that yields the same effect (i.e., only completely periodic motions) ? *Hint*: see <CJX94>, and note that only the simplified case (i) is fully treated in this reference. However, the case considered in this reference is marginally more general: it also features, in its equations of motion, 3 constant 3-vectors $\vec{a}_k$, which however break the rotation-invariance of the model. Does such a generalization correspond just to a shifting of the 3 particle coordinates ? If not, trace the appropriate place in the treatment of Sect. 3.2. where such a generalization should be introduced.

Exercise 3.2.3-2. Discuss the 2-body respectively 1-body problems that obtain from the 3-body problem (3) if one respectively two particles are fixed at some arbitrarily assigned positions. *Hint*: see the *Remark 3.2.1.*

Next, let us consider the *solvable* 4-body problem that is obtained by inserting (3.1.2.3-10) in (1):

$$\mu_n \vec{r}_n = \sum_{m=1}^{4} \{-\eta_{nm}\dot{\vec{r}}_m + \dot{\vec{r}}_n \cdot [\ (\vec{r}_{m+1}-\vec{r}_{m+3}) \wedge (\vec{r}_{m+2}-\vec{r}_{m+3})\ ][\mu_m \dot{\vec{r}}_m + \sum_{\ell=1}^{4} \eta_{m\ell}\vec{r}_\ell]/\Delta\ \}, \quad (5a)$$

$$\Delta \equiv \Delta(\vec{r}_1,\vec{r}_2,\vec{r}_3,\vec{r}_4) = (\vec{r}_2-\vec{r}_1)\cdot[(\vec{r}_3-\vec{r}_1)\wedge(\vec{r}_4-\vec{r}_1)] \ . \quad (5b)$$

Here, and always below while discussing this 4-body problem, the indices n,m,ℓ run from 1 to 4 and are defined $\text{mod}(4)$, while the indices j,k run from 1 to 3 (see below: corresponding to the 3-dimensional nature of ambient space).

These equations of motion, (5), are invariant under any common (time-independent) *rotation*, and under any common constant *rescaling* ($\vec{r}_n \to c\vec{r}_n$, $\dot{c}=0$), of the 4 particle coordinates (and as well, of course, of the 4 velocities $\dot{\vec{r}}_n$ and the 4 accelerations $\ddot{\vec{r}}_n$); they are moreover invariant under a common, time-independent, *translation* ($\vec{r}_n \to \vec{r}_n+\vec{r}_0$, $\dot{\vec{r}}_0=0$), iff the 16 parameters η_{nm} satisfy the following 4 constraints:

$$\sum_{m=1}^{4} \eta_{nm}=0\ , \quad n=1,2,3,4\ . \quad (6)$$

And let us recall the geometrical significance of the denominator Δ in the right hand side of (5): up to its sign, it is 6 times the volume of the tetrahedron that has the 4 particle coordinates $\vec{r}_n$ has its 4 vertices (see (5b)).

In close analogy to the 3-body model discussed above, the formula that displays the solvability of this 4-body problem obtains by inserting (3.1.2.3-9) in (2). Hence it reads:

$$\mu_n \dot{\vec{r}}_n = -\sum_{m=1}^{4} [\eta_{nm} \vec{r}_m] + \vec{h}_1 + \vec{h}_2 x_n + \vec{h}_3 y_n + \vec{h}_4 z_n , \quad n = 1,2,3,4 \ , \tag{7a}$$

or equivalently

$$\mu_n \dot{u}_{n,j+1} = \sum_{m=1}^{4} (-\eta_{nm} u_{m,j+1} + h_{mj} u_{n,m}) , \quad n = 1,2,3,4 \qquad j = 1,2,3 \ , \tag{7b}$$

with

$$u_{n,1} \equiv 1 , \ u_{n,2} \equiv x_n , \ u_{n,3} \equiv y_n , \ u_{n,4} \equiv z_n \ , \ n = 1,2,3,4 \ , \tag{7c}$$

$$h_{m1} = h_{mx} , \ h_{m2} = h_{my} , \ h_{m3} = h_{mz} , \quad m = 1,2,3,4 \quad . \tag{7d}$$

The discussion of how to evince from this formula, (7), the solution of the initial-value problem for the 4-body model (5) is so closely analogous to that given above (in Sect. 3.2.3, as well indeed as in other preceding sections), that we forsake any additional elaboration here, except to note that, while (4b) are 9 scalar equations and serve to determine firstly 9 constants of motion h_{jk}, see (4d), and then the time-evolution of 9 quantities $u_{jk}(t)$, see (4c), now (7b) are 12 scalar equations, to evaluate firstly the 12 constants h_{nj}, see (7d), and then the time evolution of the 12 quantities $u_{n,j+1}(t)$, see (7c).

Exercise 3.2.3-3. Formulate and solve exercises analogous to *Exercises 3.2.3-1* and *3.2.3-2*, but for the solvable 4-body model (5) (with or without (6)). *Hint:* see <CJX94>.

3.2.4 M-body problems in M-dimensional space, or M^2-body problems in one-dimensional space

Any $\tilde{N}$-body problem in $\tilde{S}$-dimensional space can also be interpreted as an N-body problem in S-dimensional space provided $\tilde{N}\tilde{S} = NS$, since both models refer to the time-evolution of the same number, $NS = \tilde{N}\tilde{S}$, of (scalar) quantities. However, by such a trick, one generally obtains N-body models in S-dimensional space whose physical interpretation is moot, and in particular which do not live up to the requirement we consider mandatory for any set of evolution equations to be interpretable as those characterizing an N-body problem in S-dimensional space, namely that they be expressible in terms of S-vectors so that their property to be rotation-invariant in S-dimensional space be immediately apparent.

Such a criterion has however no relevance for one-dimensional space $(S = 1)$, since such an environment has no room for rotations. In such a case one might, but one need not, consider as a requirement for "physical interpretability" that the Newtonian equations of motion be invariant under *translations* $(x_n \to x_n + x_0,\ n = 1,\dots,N,\ \dot{x}_0 = 0)$.

In Sect. 3.2.4 we outline how certain *translation-invariant* N-body problems in one-dimensional space, with

$$N = M^2 \ , \tag{1}$$

can be obtained by (appropriately) reinterpreting certain M-body problems in M-dimensional space. As we show below, these models are *solvable* by purely algebraic operations.

Our starting point is, in M-dimensional space, the set of M "linear" seeds

$$s_1(\vec{r}) = x,\ s_2(\vec{r}) = y,\ s_3(\vec{r}) = z,\dots \tag{2a}$$

We evidently need here a change of notation, which we perform by using for the M components of the M-vector $\vec{r}$ the notation ξ_j,

$$\vec{r} \equiv (\xi_1, \xi_2, \dots, \xi_M) \ , \tag{3a}$$

which also entails, in connection with the M nodes (or particle coordinates) $\vec{r}_n$, the corresponding assignment

$$\vec{r}_j \equiv (x_j, y_j, z_j, \dots) = (\xi_{j1}, \xi_{j2}, \xi_{j3}, \dots, \xi_{jM}) \ . \tag{3}$$

Hereafter the indices $j,k,\ell,\ldots$ run from 1 to M, and we focus on the evolutions of $N=M^2$ "particle coordinates in one-dimensional space" $\xi_{jk}\equiv\xi_{jk}(t)$.

The second-order evolution equations of motion of our solvable N-body problem then read

$$\mu_{jk}\,\ddot{\xi}_{jk}=\sum_{j',k'=1}^{M}\left\{-\eta_{jkj'k'}\,\dot{\xi}_{j'k'}+\dot{\xi}_{jk'}\left[\mu_{j'k}\,\dot{\xi}_{j'k}+\sum_{\ell,\ell'=1}^{M}(\eta_{j'k\ell\ell'}\,\xi_{\ell\ell'})\right](\underline{\Xi}^{-1})_{k'j'}\right\}\,, \tag{4a}$$

with the M^2 elements of the $(M\times M)$-matrix $\underline{\Xi}$ given by the simple rule

$$(\underline{\Xi})_{jk}=\xi_{jk}\ . \tag{4b}$$

In the right hand side of (4a) $\underline{\Xi}^{-1}$ denotes the $(M\times M)$-matrix which is the inverse of $\underline{\Xi}$, and $(\underline{\Xi}^{-1})_{jk}$ indicates of course the jk-th element of this inverse matrix.

The corresponding first-order ODEs read

$$\mu_{jk}\,\dot{\xi}_{jk}=-\sum_{j'k'=1}^{M}(\eta_{jkj'k'}\,\xi_{j'k'})+\sum_{\ell=1}^{M}(\xi_{j\ell}\,h_{k\ell})\ , \tag{5}$$

where the quantities h_{jk} are N^2 constants of motion.

Let us show directly that (4) follows from (5). Indeed time-differentiation of (5) yields

$$\mu_{jk}\,\ddot{\xi}_{jk}=-\sum_{j'k'=1}^{M}(\eta_{jkj'k'}\,\dot{\xi}_{j'k'})+\sum_{\ell=1}^{M}(\dot{\xi}_{j\ell}\,h_{k\ell})\ . \tag{6}$$

On the other hand (5) can be written, in M-vector notation,

$$\underline{\Xi}\,\underline{h}^{(k)}=\underline{p}^{(k)}\ ,\quad k=1,2,\ldots,M\ , \tag{7a}$$

by using (4b) and by introducing the M-vectors

$$\underline{h}^{(k)}\equiv(h_{k1},h_{k2},\ldots,h_{kM})\ ,\quad k=1,2,\ldots,M\ , \tag{7b}$$

$$\underline{p}^{(k)}\equiv(p_{1k},p_{2k},\ldots,p_{Mk})\ ,\quad k=1,2,\ldots,M\ , \tag{7c}$$

$$p_{jk} \equiv \mu_{jk} \dot{\xi}_{jk} + \sum_{j',k'=1}^{M} (\eta_{jkj'k'} \xi_{j'k'}) \quad , \tag{7d}$$

as well, of course, as the $(M \times M)$-matrix (4b). Hence

$$\underline{h}^{(k)} = \underline{\underline{\Xi}}^{-1} \underline{p}^{(k)}, \quad k = 1,2,\dots,M \quad , \tag{8a}$$

or equivalently (see (7b), (7c) and (7d))

$$h_{k\ell} = \sum_{j'=1}^{M} \left\{ \left[\mu_{j'k} \dot{\xi}_{j'k} + \sum_{j'',k''=1}^{M} (\eta_{j'kj''k''} \xi_{j''k''}) \right] (\Xi^{-1})_{\ell j'} \right\} . \tag{8b}$$

Insertion of this expression of $h_{k\ell}$ in (5) yields (4), whose consistency with (5) is thereby proven. (*Advise* to the diligent reader: do the detailed calculation, and *beware* of the different role the index k plays in the right hand sides of (7b) and (7c)).

Let us emphasize that in (5), as well as in (4), the $N = M^2$ quantities μ_{jk}, as well as the $N^2 = M^4$ quantities $\eta_{jkj'k'}$, are arbitrary constants. The $N = M^2$ quantities $h_{k\ell}$ in (5) are also constant but their role is quite different, and indeed they do not appear in the equations of motion (4). They must rather be considered as $N = M^2$ functions of the $N = M^2$ "particle coordinates" $\xi_{jk} = \xi_{jk}(t)$ and velocities $\dot{\xi}_{jk} = \dot{\xi}_{jk}(t)$, as indeed given explicitly by (8); N functions which remain constant over time ("constants of motion") when the $\xi_{jk}(t)$ evolve according to the equations of motion (4).

As usual, the initial-value problem for the Newtonian evolution equations (4) can be solved by focussing instead on (5). Firstly one evaluates the $N = M^2$ constants of motion $h_{k\ell}$ from (5) at $t = 0$, in terms of the initial data, $\xi_{jk}(0)$ and $\dot{\xi}_{jk}(0)$: this requires the solution of M disjoint systems of M linear algebraic equations in M variables. Next, one integrates the *system of linear ODEs with constant coefficients (5)* to get the $N = M^2$ coordinates $\xi_{jk}(t)$. This is also an algebraic task, and it entails, in the general case, diagonalizing an $(N \times N)$-matrix, unless there holds the constraint

$$\eta_{jkj'\ell} = \delta_{jj'} \eta_{jk\ell} \quad , \tag{9}$$

in which case it reduces to M disjoint systems, each entailing the diagonalization of an $(M \times M)$-matrix.

The Newtonian equations of motion (4) are not translation-invariant, but they are clearly invariant under rescaling $(\xi_{jk}(t) \to c\,\xi_{jk}(t),\ \dot{c}=0)$. Hence via the following position,

$$\xi_{jk}(t) = \exp[\ x_{jk}(t)\ ] \ , \tag{10}$$

one obtains for the "new particle coordinates" $x_{jk}(t)$ a system of Newtonian equations of motion which are *translation-invariant*. It reads:

$$\mu_{jk}\,\ddot{x}_{jk} = -\mu_{jk}\,\dot{x}_{jk}^2 + \sum_{j',k'=1}^{M} \{-\eta_{jkj'k'}\,\dot{x}_{j'k'} \exp(x_{j'k'} - x_{jk})$$

$$+\dot{x}_{jk'}[\ \mu_{j'k}\,\dot{x}_{j'k} \exp(x_{jk'} + x_{j'k} - x_{jk}) + \sum_{\ell,\ell'=1}^{M} \eta_{j'k\ell\ell'} \exp(x_{jk'} + x_{\ell\ell'} - x_{jk})\ ](\underline{E}^{-1})_{k'j'}\ \},$$

(11a)

$$(\underline{E})_{jk} = \exp(x_{jk}) \ . \tag{11b}$$

Exercise 3.2.4-1. Verify that the equations of motion (11) are *translation-invariant.*

Exercise 3.2.4-2. Prove that the equations of motion (11) are solved by the formula

$$x_{jk}(t) = \log\left[\sum_{n=1}^{N} c_{jkn} \exp(\lambda_n t)\right] \ , \tag{12a}$$

and obtain the formulas that determine the $N = M^2$ constants λ_n and the $N^2 = M^4$ constants c_{jkn}. *Hint*: see (10) and (5).

Exercise 3.2.4-3. Show that, if (9) holds, (10) can be replaced by the simpler formula

$$x_{jk}(t) = \log\left[\sum_{\ell=1}^{M} c_{jk\ell} \exp(\lambda_\ell t)\right] \ , \tag{12b}$$

which contains only M constants λ_ℓ and M^3 constants $c_{jk\ell}$.

Exercise 3.2.4-4. Correct the misprints that mar the presentation of these results in <CJX94>. *Hint*: focus on the equations of motion, rather than their solutions.

Exercise 3.2.4-5. Write explicitly the equations of motion (4) and (11) in the special cases $M = p$, and compare your findings with appropriate equations of Sect. 3.2. p, for $p = 2,3$. *Hint*: compute explicitly $\underline{\Xi}^{-1}$ and $\underline{E}^{-1}$.

Exercise 3.2.4-6. Discuss the behavior of the N-body system (11), with particular attention to the presence of: *(i)* singularities; *(ii)* confined (multiply or completely periodic) solutions. If neither *(i)* nor *(ii)* apply, discuss the behavior of the system as $t \to \pm\infty$. Perform the analysis more completely in the cases $M = 2$ and $M = 3$, if need be with (9).

Exercise 3.2.4-7. Modify the equations of motion (11) so that they feature lots of (perhaps only?) *periodic* solutions. *Hint*: see *Exercise 3.2-3* (and do not hesitate to go complex if need be).

3.3 First-order evolution equations and partially solvable N-body problems with velocity-independent forces

In Sect. 3.2 we described a general technique to manufacture exactly treatable N-body problems characterized by Newtonian equations of motion, see (3.2-19). Examples of such models were then discussed in Sects. 3.2.1, 3.2.2 respectively 3.2.3 in 1-, 2- respectively 3-dimensional space, and in Sect. 3.2.4 again in 1-dimensional space. For some of these models the initial-value problem is solvable via (a sequence of) purely algebraic operations; for others the "exact treatment" available amounts merely to the possibility of reducing the original Newtonian *second-order* evolution equations to an equivalent set of *first-order* ODEs ("equivalent" means in this context that the solution of the initial-value problem for the original, *second-order*, Newtonian equations can as well be achieved by solving the "equivalent" set of *first-order* ODEs). All these N-body problems feature Newtonian equations of motions with *velocity-dependent* forces; and in all cases we considered problems with *unrestricted initial data.*

In Sect. 3.3 we tersely describe a modification of the approach of Sect. 3.2 that yields Newtonian equations of motion featuring *velocity-independent* forces. However, these N-body problems are only *partially solvable*, namely they are solvable only for a restricted set of initial data: generally the initial positions can be assigned arbitrarily, while the initial velocities are determined by the initial positions (in order for the solution technique to be applicable). Moreover, these models generally feature many-body forces.

The main idea to manufacture such models is fairly trivial: one considers to begin with *solvable* evolution equations of *first-order*, and then obtains from these, by time-differentiation, *second-order* equations of Newtonian type.

This trick has already been used in preceding sections. The diligent reader is advised to retrieve and review these previous developments before proceeding further.

In this Section we outline a treatment of this approach in the general context of multidimensional ambient space. We confine however our presentation to such an outline, and to one, fairly trivial, one-dimensional example: the alert reader will have no difficulty in inventing and treating additional examples.

We take as starting point of our treatment the equations that obtain by setting $\rho_n[\underline{\vec{r}}]$ identically to zero, $\rho_n[\underline{\vec{r}}]=0$, in (3.2-15) and (3.2-19). Then the latter, (3.2-19), yield the evolution equations

$$\underline{\vec{\Gamma}}\cdot\underline{\dot{\vec{r}}}=\underline{\vec{A}}\cdot\underline{\vec{\gamma}} \tag{1a}$$

or equivalently, componentwise,

$$\sum_{m=1}^{N}\sum_{k=1}^{S}\Gamma_{nm,jk}\,\dot{r}_{m,k}=\sum_{m=1}^{N}\sum_{k=1}^{S}A_{nm,jk}\,\gamma_{m,k}\ , \tag{1b}$$

where

$$\Gamma_{nm,jk}=\partial\gamma_{n,j}/\partial r_{m,k}-\delta_{nm}\sum_{\ell=1}^{N}\gamma_{\ell,j}\,D_{n\ell,k}\ . \tag{2}$$

Let us recall that here $\underline{\vec{\gamma}}\equiv\underline{\vec{\gamma}}(\underline{\vec{r}})$ can be chosen arbitrarily, while $\underline{\vec{D}}\equiv\vec{D}(\underline{\vec{r}})$ is determined by the choice of the N seeds $s_m(\vec{r})$, see Sect. 3.1 and its subsections. As for $\underline{\vec{A}}$, we limit hereafter our consideration to the simple case (3.2-25), so that (1) in fact read

$$\underline{\vec{\Gamma}}\cdot\underline{\dot{\vec{r}}}=a\underline{\vec{\gamma}}\ , \tag{3a}$$

or equivalently

$$\sum_{m=1}^{N} \sum_{k=1}^{S} \Gamma_{nm,jk} \dot{r}_{m,k} = a\gamma_{n,j} \quad , \tag{3b}$$

with a an arbitrary *scalar* constant.

The corresponding expression of (3.2-15),

$$\vec{f}_n(t) = \vec{\gamma}_n[\underline{\vec{r}}(t)] \quad , \tag{4}$$

then yields

$$\exp(at)\sum_{m=1}^{N} \vec{h}_m s_m[\vec{r}_n(t)] = \vec{\gamma}_n[\underline{\vec{r}}(t)] \quad . \tag{5}$$

At $t=0$ this (nondifferential) system of N S-vector equations determines the N *constant* S-vectors $\vec{h}_m$; for $t>0$, it determines the N S-vectors $\vec{r}_n(t)$, which are the solutions of (3).

Exercise 3.3-1. Prove (5). *Hint*: see (3.2-1,4,6,25) and (4), and note that notational *consistency* would have suggested to write $\vec{h}_m(0)$ in place of $\vec{h}_m$ in (5), while we heeded instead the call for maximal notational simplicity, thereby introducing in (5) the *constant* S-vectors $\vec{h}_m$.

The evolution equations (3) can be conveniently rewritten in the following form:

$$\dot{\underline{\vec{r}}} = a\,\underline{\vec{\vec{G}}}\cdot\underline{\vec{\gamma}} \quad , \tag{6a}$$

or equivalently

$$\dot{\vec{r}}_n = a\sum_{m=1}^{N} \vec{\vec{G}}_{nm}\cdot\vec{\gamma}_m \tag{6b}$$

or, still equivalently,

$$\dot{r}_{n,j} = a\sum_{m=1}^{N} \sum_{k=1}^{N} G_{nm,jk}\,\gamma_{mk} \tag{6c}$$

Note that the equivalence of (6) to (3) defines $\underline{\vec{\vec{G}}}$, which is clearly, in an appropriate sense, the inverse of $\underline{\Gamma}$. Here of course both $\underline{\vec{\gamma}}$ and $\underline{\vec{\vec{G}}}$ (may) depend on all the coordinates $\vec{r}_n$, $\underline{\vec{\gamma}} \equiv \underline{\vec{\gamma}}(\underline{\vec{r}})$ and $\underline{\vec{G}} \equiv \underline{\vec{G}}(\underline{\vec{r}})$; the dependence of $\underline{\vec{\gamma}}$ on $\underline{\vec{r}}$ is a matter of direct choice (as well as the choices of the dimension S of the ambient space, and of the number N of particles); the dependence of $\underline{\vec{G}}$ on $\underline{\vec{r}}$ is also a matter of choice, but less directly so as it emerges from the choice of the set of seeds $s_m(\vec{r})$, and also from the choice of $\underline{\vec{\gamma}}(\underline{\vec{r}})$, as entailed by (2), (3) and (6).

We now time-differentiate (6):

$$\ddot{\vec{r}}_n = a \sum_{m=1}^{N} \sum_{\ell=1}^{N} \{\partial[\vec{G}_{nm} \cdot \vec{\gamma}_m] / \partial \vec{r}_\ell \} \cdot \dot{\vec{r}}_\ell \quad , \tag{7}$$

and then use (6) again, to eliminate $\dot{\vec{r}}_\ell$. The resulting evolution equations read

$$\ddot{\vec{r}}_n = a^2 \sum_{\ell,m_1,m_2=1}^{N} \{ \partial[\vec{G}_{nm_1} \cdot \vec{\gamma}_{m_1}] / \partial \vec{r}_\ell \} \cdot [\vec{G}_{\ell m_2} \cdot \vec{\gamma}_{m_2}] \ , \tag{8a}$$

$$\ddot{r}_{n,j} = a^2 \sum_{\ell,m_1,m_2=1}^{N} \sum_{k,k_1,k_2=1}^{S} \left\{ \frac{\partial\left[G_{nm_1,jk_1}\, \gamma_{m_1,k_1}\right]}{\partial r_{\ell,k}} \right\} G_{\ell m_2,kk_2}\, \gamma_{m_2,k_2} \quad . \tag{8b}$$

These are then our *partially solvable* Newtonian equations of motion (read again the statement after (6c)!). Their solutions are provided by the solutions of the (nondifferential!) equation (5), but only for the set of initial data that satisfy (3) and (6) (at $t=0$; actually these solutions satisfy (3) and (6) for all time).

Of course a necessary condition for this approach to work is that (3) be actually solvable for $\dot{\underline{\vec{r}}}$, to yield (6). That this may fail to happen is illustrated by the example given below.

Let us begin by considering the *one-dimensional* setting, in which case (5), (6) and (8) take the following, somewhat simpler, forms:

$$\exp(at) \sum_{m=1}^{N} h_m\, s_m[x_n(t)] = \gamma_n[\underline{x}(t)] \quad , \tag{9}$$

$$\dot{x}_n = a \sum_{m=1}^{N} G_{nm}(\underline{x})\, \gamma_m(\underline{x}) \quad , \tag{10}$$

$$\ddot{x}_n = a^2 \sum_{\ell,m_1,m_2=1}^{N} \left\{ \frac{\partial\left[G_{nm_1}(\underline{x})\, \gamma_{m_1}(\underline{x})\right]}{\partial x_\ell} \right\} G_{\ell m_2}(\underline{x})\, \gamma_{m_2}(\underline{x}) \quad , \tag{11}$$

where of course now the $(N \times N)$-matrix $\underline{G}(\underline{x})$ is just the inverse (see (3) and (6)) of the $(N \times N)$-matrix $\underline{\Gamma}(\underline{x})$,

$$\Gamma_{nm}(\underline{x}) = \partial \gamma_n(\underline{x}) / \partial x_m - \delta_{nm} \sum_{\ell=1}^{N} D_{n\ell}(\underline{x}) \gamma_\ell(\underline{x}) \tag{12}$$

(see (2)).

Let us then limit our consideration to the simple case with

$$\gamma_n(\underline{x}) = \gamma_n(x_n) \quad , \tag{13}$$

entailing that $\underline{\Gamma}(\underline{x})$ becomes diagonal,

$$\Gamma_{nm}(\underline{x}) = \delta_{nm} \left[\gamma_n'(x_n) - u_n(\underline{x}) \right] \quad , \tag{14}$$

$$u_n(\underline{x}) = \sum_{m=1}^{N} D_{nm}(\underline{x}) \gamma_m(x_m) \quad . \tag{15}$$

Hence in this case

$$G_{nm}(x) = \delta_{nm} \left[\gamma_n'(x_n) - u_n(\underline{x}) \right]^{-1} \quad , \tag{16}$$

and the two evolution equation (10) respectively (11) read

$$\dot{x}_n = a \left[\gamma_n'(x_n) - u_n(\underline{x}) \right]^{-1} \gamma_n(x_n) \quad , \tag{17}$$

respectively

$$\ddot{x}_n = a \left[\gamma_n'(x_n) - u_n(\underline{x}) \right]^{-2} \gamma_n(x_n) \left\{ \gamma_n'(x_n) - \gamma_n''(x_n) \gamma_n(x_n) \left[\gamma_n'(x_n) - u_n(\underline{x}) \right]^{-1} \right.$$

$$\left. + \sum_{m=1}^{N} \frac{\partial u_n(\underline{x})}{\partial x_m} \gamma_m(x_m) \left[\gamma_m'(x_m) - u_m(\underline{x}) \right]^{-1} \right\} \quad , \tag{18}$$

of course with $u_n(\underline{x})$ defined by (15).

The derivation of these two equations is plain: (17) follows from (10) and (16), while (18) follows from (11), (10) and (16), or directly from (17) by t-differentiation and by using again (17).

Note that, if

$$\gamma_n(x) = f(x) \tag{19a}$$

(note the independence of the right hand side from the index n), and if the function $f(x)$ is representable as a linear superposition of the seeds $s_m(x)$, then (15) entails

$$u_n(\underline{x}) = f'(x_n) = \gamma_n'(x_n) \ , \tag{19b}$$

hence the diagonal matrix $\underline{\Gamma}$ vanishes, see (14), and the matrix $\underline{G}$ does not exist.

Let us simplify still further the evolution equations (17) and (18), by positing

$$\gamma_n(x) = \eta_n \, x \tag{20}$$

Then (17) respectively (18) become

$$\dot{x}_n = a x_n \left[1 - v_n(\underline{x})\right]^{-1} \ , \tag{21}$$

respectively

$$\ddot{x}_n = a^2 x_n \left[1 - v_n(\underline{x})\right]^{-2} \{1 - \sum_{m=1}^{N} x_m \left[1 - v_m(x)\right]^{-1} \left[\partial v_n(\underline{x}) / \partial x_m\right]\} \ , \tag{22}$$

with

$$v_n(\underline{x}) = \sum_{m=1}^{N} D_{nm}(\underline{x}) \, (\eta_m / \eta_n) \, x_m \ . \tag{23}$$

The solutions of these equations, (21) and (22), are then given by the roots $x_n \equiv x_n(t)$ of the (uncoupled !) equations

$$\exp(at) \sum_{m=1}^{N} h_m \, s_m(x_n) = \eta_n \, x_n \ . \tag{24}$$

Note that the evolution equations (22), as indeed, more generally, the evolution equations (8), (11) and (18), can be interpreted as the (Newtonian !) equations of motion of an N-body problem with *velocity-independent* forces.

Note that, if the constants η_n were all equal,

$$\eta_n = \eta \ , \tag{25a}$$

then (20) would correspond to (19a) with

$$f(x) = \eta\, x \ . \tag{25b}$$

Exercise 3.3-2. (i) Verify that, for the choice of seeds (3.1.2.1-18), the 3 equations (22), (21), respectively (24) yield

$$\ddot{x}_n = (a/2)^2\, \eta_n\, (\eta_1 - \eta_2)^{-2}\, (x_n / x_{n+1}^4)(x_n^2 - x_{n+1}^2)\left[3\eta_n\, x_n^2 - (\eta_n + 2\eta_{n+1})\, x_{n+1}^2\right] , \tag{26a}$$

$$\dot{x}_n = -(a/2)\eta_n\, (\eta_1 - \eta_2)^{-1}\, (x_n / x_{n+1}^2)(x_1^2 - x_2^2) \ , \tag{26b}$$

respectively

$$[x_n(t)]^2 = (\eta_2 - \eta_1)\left[x_1(0)\ x_2(0)\right]^2 \cdot$$

$$\cdot \left\{\eta_n\left[x_1^2(0) - x_2^2(0)\right]\exp(-at) + \left[\eta_2\, x_2^2(0) - \eta_1\, x_1^2(0)\right]\right\}^{-1} , \tag{26c}$$

where of course $n = 1,2 \bmod(2)$ and $\eta = \eta_2 / \eta_1$ is an arbitrary constant ($\eta \neq 1$: note that in all these 3 equations η_1 and η_2 only enter via this ratio). *(ii)* Verify that (26a) follows from (26b) by t-differentiation. *(iii)* Verify that (26c) satisfies (26b) and (26a). *(iv)* Verify that, by setting $x_n^2(t) = y_n(t)$, the 3 equations (26) take the following, simpler, form:

$$\ddot{y}_n = a^2\eta_n\, (\eta_1 - \eta_2)^{-2}\, (y_n / y_{n+1}^2)\, (y_n - y_{n+1})\left[2\eta_n\, y_n - (\eta_1 + \eta_2)\, y_{n+1}\right] , \tag{27a}$$

$$\dot{y}_n = -a\eta_n\, (\eta_1 - \eta_2)^{-1}\, (y_n / y_{n+1})(y_1 - y_2) \ , \tag{27b}$$

$$y_n(t) = (\eta_2 - \eta_1)\, y_1(0)\, y_2(0) / \{\ \eta_n[\ y_1(0) - y_2(0)\ ]\exp(-at) + [\ \eta_2\, y_2(0) - \eta_1\, y_1(0)\ ]\ \}. \tag{27c}$$

(v) Solve (27b) with $n = 1$ for $y_2(t)$, insert the expression of $y_2(t)$ so obtained in (27b) with $n = 2$, and obtain thereby the following *single second-order* ODE for $y_1(t)$:

$$\ddot{y}_1 = -a\,\dot{y}_1 + 2\,\dot{y}_1^2 / y_1 \ . \tag{28}$$

Note that there is now no restriction on the initial data, namely, for this evolution equation, one can assign arbitrarily both $y_1(0)$ and $\dot{y}_1(0)$ (the latter assignment corresponds to assigning $y_2(0)$). Also note that (28) does not contain the constants η_1 and η_2, and that (of course !) the *same* equation, (28), can be derived for $y_2(t)$ (from the 2 equations (27b)). *(vi)* Set

$$y_1(t) = 1/z(t) \quad , \tag{29a}$$

and obtain thereby, from (28), the following (linear!) equation for $z(t)$:

$$\ddot{z} = -a\,\dot{z} \quad . \tag{29b}$$

Verify, via (29a), the consistency of this equation with (27c).

3.N Notes to Chapter 3

The main idea underpinning the results presented in Chap. 3 was introduced in <C93a>; it is unlikely that the generalization of standard (one-dimensional, polynomial) Lagrangian interpolation presented in this paper, <C93a>, is entirely new, but its application to manufacture many-body problems amenable to exact treatments, as described in Sect. 3.2, is, to the best of my knowledge, new, as well, perhaps, as its exploitation to obtain *identities*, such as, for instance, (3.1.2.1-51), (A-62), (A-70).

Most of the many-body models discussed in Sects. 3.2.1, 3.2.2, 3.2.3 and 3.2.4 are gleaned from the following papers: <CJX93>, <CJX94>, <CJX95>, <C96a> (the last of these papers is the only one which features *elliptic* functions). The diligent reader will find in these papers more material than has been reported in Chapter 3. On the other hand some findings reported in Chapter 3 are new, in particular the *integrable* many-body problems with only (or, at least, mainly) *one- and two-body forces* treated at the end of Sects. 3.2.1 and 3.2.2, see in particular the one-dimensional models (featuring elliptic functions) (3.2.1-44) and (3.2.1-49) (actually the second is a special case of the first) as well as the two-dimensional model (3.2.2-33) (actually also the *solvable* many-body model in the plane (3.2.2-16), featuring *many-body* forces, is new). Also new is the material of Sect. 3.3.

4 SOLVABLE AND/OR INTEGRABLE MANY-BODY PROBLEMS IN THE PLANE, OBTAINED BY COMPLEXIFICATION

The findings reported and outlined in Chap. 4 are all based on the idea to obtain *two-dimensional* models, i.e. models describing motions *in the plane*, from *one-dimensional* models, i.e. from models describing motions *on the line*, via a very simple trick: *complexification*. How that works is explained in Sect. 4.1 The method is then illustrated by the discussion of a *solvable* model in Sect. 4.2 and its subsections, of some other *solvable* models in Sect. 4.3 and its subsections, and by a survey of *solvable* and/or *integrable* many-body problems *in the plane* obtainable by such an approach in Sect. 4.4 and its subsections. In Sect. 4.5 we investigate a many-rotator problem in the plane, which is rather closely related to the *solvable* model treated in Sect. 4.2.5, that only features *completely periodic* motions. A remarkable novelty is the possibility to treat variants of this *solvable* model which are instead, presumably, *nonintegrable*, yet exhibit sets of *completely periodic* motions which correspond to sets of initial data having nonvanishing measure in phase space. The mechanism which underlies this phenomenology, as analyzed in Sect. 4.5, brings to light an interesting connection among *analyticity* properties in the time variable, and *integrable* features of these motions, as manifested by their *complete periodicity*. Finally, Sect. 4.6 provides an outlook on future developments; the enterprising reader might like to browse through it immediately.

Let us end these introductory words with a remark, which we consider sufficiently important to proffer it here and to repeat it in Sect. 4.6.

An important message entailed by the approach introduced and developed in Chap. 4 is, that it is in many cases worthwhile to investigate many-body problems amenable to exact treatments in the *complex*, rather than only in the *real*, domain. Indeed, as we will see, such an extension yields a much richer gamut of behaviors: for instance in any one-dimensional many-body problem with forces which are singular at zero separation, if the motion is constrained to the *real* axis, the ordering of the particles cannot change throughout the motion without the system going through a singularity, and moreover (as well as because of this), in

F. Calogero: LNPm 66, pp. 417 - 510, 2001

the case of unconfined motions, the scattering behavior is not too interesting. The collection of allowed behaviors, without the system hitting any singularity, is instead much richer if the particles are allowed to roam throughout the complex plane, rather than being forced to move only on the real axis; and the scattering phenomenology in the plane is clearly richer, more interesting, than on the line.

However, physics deals with the real world, and many-body problems are meant to describe motions taking place here. The main observation on which the developments reported in Chap. 4 are based, is that one can, in certain, appropriate, cases, identify the *complex* plane with the *physical* (real) plane, so that not only motions in the complex plane become motions in the physical plane, but the complexified one-dimensional many-body problem becomes a (two-dimensional) many-body problem *in the physical (real) plane*, whose (Newtonian) equations of motion are *invariant under rotations* (in the plane). Indeed, we consider this latter requirement -- *invariance of the Newtonian equations of motion under rotations* -- a *sine qua non* condition for interpreting an (appropriate) set of second-order ODEs as a many-body problem in multidimensional space.

4.1 How to obtain by complexification rotation-invariant many-body models in the plane from certain many-body problems on the line

Let us consider a one-dimensional N-body problem characterized by Newtonian equations of motion,

$$\ddot{z}_n = f_n^{(1)}(z_n;\dot{z}_n) + \sum_{m=1,m\neq n}^{N} f_{nm}^{(2)}(z_n,z_m;\dot{z}_n,\dot{z}_m) \quad . \tag{1}$$

Here of course $z_n \equiv z_n(t)$ is the coordinate of the n-th particle, and as usual, throughout Chap. 4, superimposed dots denote time-differentiations, and the particle indices $(n,m,...)$ range from 1 to N unless otherwise indicated.

The N-body model (1) only features one- and two-body time-independent forces. These restrictions are introduced here merely for notational simplicity (and because such models are generally more interesting); the diligent reader will have no difficulty in extending the treatment to more general cases.

We now assume that the functions $f^{(1)}$ and $f^{(2)}$ depend *analytically* on their arguments. Then the equations of motion (1) also hold for *complex* values of the coordinates z_n; hence the one-dimensional problem (1) describing motions on the real line can be extended to a two-dimensional model describing motions which roam throughout the complex plane. Trivial as this extension might appear, it generally entails, as we already emphasized, a substantial qualitative enrichment of the permitted motions. It is on the other hand clear that, if the original model, see (1), is *solvable and/or integrable*, the motion in the complex plane obtained by this *complexification* procedure, is generally as well *solvable and/or integrable.*

By identifying the *complex plane* with the *real physical plane* (see below), one can identify in this manner solvable and/or integral two-dimensional models which, in our opinion, do qualify as genuine N*-body problems in the plane* iff the corresponding equations of motions are *invariant under rotations in the plane.* How to identify or manufacture models that possess this property is described below, after an interlude devoted to notational material.

Notation. The notation we employ for two-vectors has already been introduced above (see Sect. 3.1.2.2), yet for the convenience of the reader we report here the main formulas, as well as some new ones -- but with minimal commentary.

$$\vec{r} \equiv (x,y,0)\,,\ \ \hat{k}\wedge\vec{r} \equiv (-y,x,0),\ \ \hat{k} \equiv (0,0,1)\ \ , \tag{2a}$$

$$\hat{k}\wedge(\hat{k}\wedge\vec{r}) = -\vec{r}\ \ ; \tag{2b}$$

$$\vec{r} = r\,(\cos\theta,\ \sin\theta,0)\ \ , \tag{3a}$$

$$\hat{k}\wedge\vec{r} = r(-\sin\theta,\ \cos\theta,0) = r(\cos(\theta+\pi/2),\ \sin(\theta+\pi/2),0)\ \ , \tag{3b}$$

$$x = r\,\cos\theta,\ \ y = r\,\sin\theta\ \ , \tag{4a}$$

$$r^2 = x^2+y^2\,,\quad \tan(\theta) = y/x\ \ ; \tag{4b}$$

$$\vec{r}_n\cdot\vec{r}_m = \vec{r}_m\cdot\vec{r}_n \equiv c_{nm} = c_{mn} = x_n\,x_m + y_n\,y_m = r_n\,r_m\cos(\theta_n-\theta_m)\ \ ; \tag{5}$$

$$\vec{r}_m\cdot\hat{k}\wedge\vec{r}_n = -\vec{r}_n\cdot\hat{k}\wedge\vec{r}_m = \hat{k}\cdot\vec{r}_n\wedge\vec{r}_m = -\hat{k}\cdot\vec{r}_m\wedge\vec{r}_n = s_{nm} = -s_{mn}$$

$$= x_n\,y_m - x_m y_n = r_n\,r_m\sin(\theta_m-\theta_n)\,. \tag{6}$$

Note that both the *scalar product* (5), and the *pseudoscalar product* (6), are invariant under rotations in the plane,

$$x \to \tilde{x} = x \cos\varphi - y \sin\varphi, \quad y \to \tilde{y} = x \sin\varphi + y \cos\varphi, \tag{7a}$$

$$\vec{r} \to \tilde{\vec{r}} = r\,(\cos(\theta+\varphi), \sin(\theta+\varphi), 0) \ . \tag{7b}$$

The above definitions entail the following identities (and many others!):

$$c_{11} = r_1^2 \ , \quad s_{11} = 0 \ ; \tag{8}$$

$$\tan(\theta_1 - \theta_2) = -s_{12} / c_{12} \ ; \tag{9}$$

$$\vec{r}_1 = (c_{12}\,\vec{r}_2 - s_{12}\,\hat{k} \wedge \vec{r}_2)/(r_2)^2 \ ; \tag{10}$$

$$(r_1 r_2)^2 = c_{11}\,c_{22} = (c_{12})^2 + (s_{12})^2 \ ; \tag{11}$$

$$s_{12}\,\vec{r}_3 = s_{23}\,\vec{r}_1 - s_{13}\,\vec{r}_2 \ , \tag{12a}$$

$$s_{12}\,\vec{r}_3 = c_{23}\,\hat{k} \wedge \vec{r}_1 - c_{13}\hat{k} \wedge \vec{r}_2 \ , \tag{12b}$$

$$s_{12}\,s_{23} = c_{12}\,c_{23} - c_{13}\,c_{22} \ ; \tag{12c}$$

$$s_{12}\,s_{34} = c_{13}\,c_{24} - c_{14}\,c_{23} = r_1 r_2 r_3 r_4 \sin(\theta_1 - \theta_2)\sin(\theta_3 - \theta_4) \ ; \tag{13}$$

$$c_{12}\,c_{34} + s_{12}\,s_{34} = c_{13}\,c_{24} + s_{13}\,s_{24} = r_1 r_2 r_3 r_4 \cos(\theta_1 - \theta_2 - \theta_3 + \theta_4) \ , \tag{14a}$$

$$c_{12}\,s_{34} - s_{12}\,c_{34} = c_{13}\,s_{24} - s_{13}\,c_{24} -= r_1 r_2 r_3 r_4 \sin(\theta_1 - \theta_2 - \theta_3 + \theta_4) \ . \tag{14b}$$

In the following we often use the convenient short-hand notation

$$\vec{r}_{nm} \equiv \vec{r}_n - \vec{r}_m \ , \tag{15a}$$

entailing

$$\vec{r}_{nm} \cdot \vec{r}_{nm} \equiv (r_{nm})^2 = r_n^2 + r_m^2 - 2\vec{r}_n \cdot \vec{r}_m = c_{nn} + c_{mm} - 2c_{nm} \ . \tag{15b}$$

The one-to-one correspondence, denoted hereafter by the convenient symbol $\doteq$, among the physical plane spanned by the real 2-vector $\vec{r}$, see (2a), and the complex plane spanned by the complex number

$$z=x+iy \ , \tag{16}$$

is given by the following formulas:

$$z\doteq\vec{r}, \quad iz\doteq\hat{k}\wedge\vec{r} \ , \tag{17a}$$

$$z=r\exp(i\theta), \quad |z|^2=r^2, \quad \theta=\arg(z) \ , \tag{17b}$$

which clearly entail the following identities (and many others!):

$$z_1/z_2=(c_{12}-i s_{12})/(r_2)^2=(r_1/r_2)\exp[i(\theta_1-\theta_2)] \ , \tag{18}$$

$$z_1 z_2/z_3\doteq(\vec{r}_1 c_{23}+\vec{r}_2 c_{13}-\vec{r}_3 c_{12})/(r_3)^2 \ , \tag{19}$$

$$z_1 z_2/(z_3 z_4)=[c_{13} c_{24}-s_{13} s_{24}-i(c_{13} s_{24}+s_{13} c_{24})]/(r_3 r_4)^2 \ , \tag{20a}$$

$$z_1 z_2/(z_3 z_4)=[r_1 r_2/(r_3 r_4)]\,[\cos(\theta_1+\theta_2-\theta_3-\theta_4)+i\sin(\theta_1+\theta_2-\theta_3-\theta_4)] \ . \tag{20b}$$

Note the mixed notation used here, in particular the fact that (19) features the "correspondence" symbol, $\doteq$, while (18) and (20) feature the standard "equality" symbol, $=$.

Exercise 4.1-1. Write analogous formulas for $z_1 z_2 z_3/(z_4 z_5)$ and $z_1 z_2 z_3/(z_4 z_5 z_6)$.

It is now obvious that, if the (complexified) equations of motion (1) are invariant under the rescaling transformation $z_n \to \tilde{z}_n = c z_n$, with c an arbitrary (complex) constant ($\dot{c}=0$), then the real 2-vector equations of motion in the plane, obtained from (1) via the correspondence introduced above among real 2-vectors $\vec{r}$ and complex numbers z, are *invariant under rotations in the plane*, since for $c=\exp(i\varphi)$, with φ a real arbitrary "angle", the rescaling transformation

$$z_n \to \tilde{z}_n = z_n \exp(i\varphi) \ , \tag{21}$$

corresponds to the rotation (7) in the plane. Hence a general prescription to obtain, via the complexification technique described above, a *rotation-invariant N-body problem in the plane*, is to start from a *scale-invariant model on the line.* Several such models, amenable to exact treatments, have been obtained in the preceding chapters; some of the two-dimensional models that obtain from them are listed in Sects. 4.3 and 4.4.

There are moreover one-dimensional models whose equations of motion are *not* invariant under rescaling of the particle coordinates, but become invariant under rescaling after an appropriate change of (dependent) variables. Suppose for instance that the equations of motion (1), which via a convenient notational change we now write as follows,

$$\ddot{u}_n = \sum_{m=1}^{N} f_{nm}(u_n - u_m;\, \dot{u}_n, \dot{u}_m) \ , \tag{22}$$

are invariant under *translations* ($u_n \to u_n + u_0$, $\dot{u}_0 = 0$; (22) is written so that this invariance property is immediately apparent; this equation is of course a less general evolution equation than (1), but good enough to illustrate the point we wish to make). It is then evident that via the change of dependent variables

$$u_n = \log(z_n) \, , \quad z_n = \exp(u_n) \ , \tag{23a}$$

which of course entails

$$\dot{u}_n = \dot{z}_n / z_n \, , \ \ddot{u}_n = (\ddot{z}_n - \dot{z}_n^2 / z_n)/ z_n \ , \tag{23b}$$

one obtains new equations of motion for the new "particle coordinates" z_n which are *scaling-invariant*:

$$\ddot{z}_n = \dot{z}_n^2 / z_n + z_n \sum_{m=1}^{N} f_{nm}\left[\log(z_n / z_m); \dot{z}_n / z_n, \dot{z}_m / z_m\right] \ . \tag{24}$$

The invariance of (24) under rescaling ($z_n \to \tilde{z}_n = c\, z_n$, $\dot{c} = 0$) is apparent; it corresponds of course to the translation-invariance of (22), since clearly *translation* of the coordinates u_n, $u_n \to u_n + u_0$, corresponds, via (23a), to *rescaling* of the coordinates z_n, $z_n \to c\, z_n$, with $c = \exp(u_0)$.

Hence it is clearly possible to obtain via this trick, from any one-dimensional N-body problem that features Newtonian equations of motion which are analytic and *translation-invariant*, a two-dimensional N-body problem that features *rotation-invariant* Newtonian equations of motion, and therefore qualifies as a *bona fide* N-body problem in the plane. Note however that this latter model generally turns out not to be invariant under translations. Examples will be given in Sects. 4.3 and 4.4.

We have now seen how *rotation-invariant* N*-body problems in the plane* can be obtained, by (appropriate) *complexification*, from *one-dimensional* N-body problems featuring Newtonian equations of motion which are analytic and either *scaling-invariant* or *translation-invariant.* Of course, if the original one-dimensional N-body problem is *solvable and/or integrable*, the N-body problems in the plane obtained in this manner are as well *solvable and/or integrable*; a survey of such examples is provided in Sect. 4.4. But let us also mention that there is a way to obtain, essentially from *any* one-dimensional many-body problem that features Newtonian equations of motion which depend analytically on the particle coordinates and their velocities, a corresponding many-body problem in the plane whose Newtonian equations of motion are *rotation-invariant*; although the method to achieve this goal is deemed by us too artificial to warrant further elaboration beyond the description we now provide.

The trick is to set, instead of (16),

$$z = \hat{r}_0 \cdot \vec{r} + i\hat{k} \cdot \hat{r}_0 \wedge \vec{r} \quad , \tag{25a}$$

namely, for the particle coordinates,

$$z_n = \hat{r}_0 \cdot \vec{r}_n + i\hat{k} \cdot \hat{r}_0 \wedge \vec{r}_n , \quad n = 1, 2, \ldots, N \quad , \tag{25b}$$

with $\hat{r}_0$ the unit 2-vector in the director of $\vec{r}_0$,

$$\hat{r}_0 = \vec{r}_0 / r_0 \quad , \tag{26}$$

and to then supplement the Newtonian equations of motion for $\vec{r}_n(t)$, $n = 1, 2, \ldots, N$, obtained via this position, (25), from those (complexified) for $z_n(t)$, see for instance (1), with the additional *rotation-invariant* (and trivial !) equation

$$\dot{\vec{r}}_0 = 0 \quad . \tag{27}$$

In Sect. 4.1, as indeed often in this book, we have focussed on many-body problems characterized by equations of motion of Newtonian type, and we have discussed the complexification issue on the basis of its impact on these equations of motion. Let us end Sect. 4.1 by discussing the question of complexification in the context of many-body problems susceptible to a Hamiltonian, or Lagragian, formulation. In particular we now show how any one-dimensional N-body problem which can be formulated in Hamiltonian, or Lagrangian, form, with Hamiltonian and Lagrangian functions depending *analytically* on the particle variables (particle coordinates and canonical momenta in the Hamiltonian case, particle coordinates and their time-derivatives in the Lagragian case), yields by complexification an N-body problem in the (physical, real) plane, whose 2-vector equations of motion can be formulated in Hamiltonian or Lagragian form (in fact, generally via two alternative but equivalent prescriptions).

To this end, let us first make the following observation. Assume $f(z)$ to be an *analytic* function of the complex variable z,

$$z = x + i s y \quad , \tag{28a}$$

with $s = +1$ or $s = -1$, so that

$$d f(z) / d z = \partial f(z) / \partial x = -i s \partial f(z) / \partial y \quad . \tag{28b}$$

Then set

$$f(z) \equiv f(x + i s y) = F(\vec{r}) + i \hat{F}(\vec{r}) \quad , \tag{29}$$

with $F(\vec{r})$ and $\hat{F}(\vec{r})$ *real* functions of the real 2-vector (see (2))

$$\vec{r} \equiv (x, y, 0) \tag{30}$$

(note that, according to the notation introduced above, see (17), for $s = 1$, $z \doteq \vec{r}$, while for $s = -1$, $z^* \doteq \vec{r}$). It is then clear that (28b) and (29) entail (see (2a)) the 2-vector relations

$$(\partial / \partial \vec{r}) F(\vec{r}) = s (\hat{k} \wedge \partial / \partial \vec{r}) \hat{F}(\vec{r}) \quad , \tag{31a}$$

$$(\partial / \partial \vec{r}) \hat{F}(\vec{r}) = -s (\hat{k} \wedge \partial / \partial \vec{r}) F(\vec{r}) \quad . \tag{31b}$$

Consider now a system characterized by the Hamiltonian equations of motion

$$\dot{z}_n = \partial h(\underline{z},\underline{\zeta})/\partial \zeta_n \quad , \tag{32a}$$

$$\dot{\zeta}_n = -\partial h(\underline{z},\underline{\zeta})/\partial z_n \quad , \tag{32b}$$

where the Hamiltonian $h(\underline{z},\underline{\zeta})$ is an analytic function of the N canonical coordinates z_n and of the N canonical momenta ζ_n. We now set

$$z_n = x_n + i\,y_n, \quad \zeta_n = \xi_n - i\eta_n \tag{33a}$$

namely

$$z_n \doteq \vec{r}_n, \quad \zeta_n^* = \vec{\rho}_n \tag{33b}$$

and

$$h(\underline{z},\underline{\zeta}) = H(\underline{r},\underline{\rho}) + i\,\hat{H}(\underline{r},\underline{\rho}) \quad . \tag{34}$$

Note that this formula, (34), defines 2 real functions, $H(\underline{r},\underline{\rho})$ and $\hat{H}(\underline{r},\underline{\rho})$, of the $2N$ real 2-vectors

$$\vec{r}_n \equiv (x_n, y_n, 0) \tag{35a}$$

and

$$\vec{\rho}_n \equiv (\xi_n, \eta_n, 0) \quad . \tag{35b}$$

Then (32), via (31), yield

$$\dot{\vec{r}}_n = (\partial/\partial\vec{\rho}_n) H(\underline{\vec{r}},\underline{\vec{\rho}}) \quad , \tag{36a}$$

$$\dot{\vec{\rho}}_n = -(\partial/\partial\vec{r}_n) H(\underline{\vec{r}},\underline{\vec{\rho}}) \quad , \tag{36b}$$

as well as

$$\dot{\vec{r}}_n = [\ \hat{k}\wedge(\partial/\partial\vec{\rho}_n)\ ]\hat{H}(\underline{\vec{r}},\underline{\vec{\rho}}) \quad , \tag{37a}$$

$$\dot{\vec{\rho}}_n = [\ \hat{k}\wedge(\partial/\partial\vec{r}_n)\ ]\hat{H}(\underline{\vec{r}},\underline{\vec{\rho}}) \quad . \tag{37b}$$

The equations (36) are *real* Hamiltonian equations for the (real) 2-vector canonical coordinates $\vec{r}_n$ and the corresponding (real) 2-vector canonical momenta $\vec{\rho}_n$, see (35). It is thereby seen that, under the sole assumption of analyticity, the one dimensional Hamiltonian equations (32) yield, by appropriate complexification (indeed, note the difference in sign among the 2 equations (33a)), real two-dimensional Hamiltonian equations, see (36).

The equations (37) are instead not Hamiltonian. But they take the standard Hamiltonian form via the following redefinition of the canonical coordinates and momenta:

$$\tilde{\vec{r}}_n = \vec{r}_n \,, \quad \tilde{\vec{\rho}}_n = \hat{k} \wedge \vec{\rho}_n \,; \quad \tilde{H}(\underline{\tilde{\vec{r}}}, \underline{\tilde{\vec{\rho}}}) = \hat{H}(\underline{\vec{r}}, \underline{\vec{\rho}}) \quad , \tag{38}$$

which indeed, see (2a,b), transform (37) into

$$\dot{\tilde{\vec{r}}}_n = (\partial / \partial \tilde{\vec{\rho}}_n) \tilde{H}(\underline{\tilde{\vec{r}}}, \underline{\tilde{\vec{\rho}}}) \quad , \tag{39a}$$

$$\dot{\tilde{\vec{\rho}}}_n = -(\partial / \partial \tilde{\vec{r}}_n) \tilde{H}(\underline{\tilde{\vec{r}}}, \underline{\tilde{\vec{\rho}}}) \quad . \tag{39b}$$

It is thereby seen that the *complex* Hamiltonian system (32), with *analytic* Hamiltonian function $h(\underline{z}, \underline{\zeta})$, yields two distinct *real* Hamiltonian structures, see (36) and (39); and let us emphasize that these three Hamiltonian evolutions, (32), (36) and (39), are completely equivalent.

Likewise, consider a system characterized by the (complex) Lagrangian evolution equations

$$(d/dt) \left[\partial \ell(\underline{z}, \underline{\dot{z}}) / \partial \dot{z}_n \right] = \partial \ell(\underline{z}, \underline{\dot{z}}) / \partial z_n \quad , \tag{40}$$

with the Lagrangian function $\ell(\underline{z}, \underline{\dot{z}})$ depending *analytically* on all its $2N$ arguments, z_n and $\dot{z}_n$, $n = 1, 2, \ldots, N$, and set

$$\ell(\underline{z}, \underline{\dot{z}}) = L(\underline{\vec{r}}, \underline{\dot{\vec{r}}}) + i \tilde{L}(\underline{\vec{r}}, \underline{\dot{\vec{r}}}) \quad , \tag{41}$$

with $L(\underline{\vec{r}}, \underline{\dot{\vec{r}}})$ and $\tilde{L}(\underline{\vec{r}}, \underline{\dot{\vec{r}}})$ *real* functions of the *real* 2-vectors $\vec{r}_n \doteq z_n$, $\dot{\vec{r}}_n \doteq \dot{z}_n$. It is then easily seen that the two real 2-vector Lagrangian evolution equations

$$(d/dt) \left[(\partial / \partial \dot{\vec{r}}_n) \, L(\underline{\vec{r}}, \underline{\dot{\vec{r}}}) \right] = (\partial / \partial \vec{r}_n) L(\underline{\vec{r}}, \underline{\dot{\vec{r}}}) \quad , \tag{42a}$$

$$(d/dt)\,[(\partial/\partial\dot{\vec{r}}_n)\,\tilde{L}(\underline{\vec{r}},\underline{\dot{\vec{r}}})\,]=(\partial/\partial\vec{r}_n)\tilde{L}(\underline{\vec{r}},\underline{\dot{\vec{r}}})\quad, \tag{42b}$$

are both equivalent to (40).

Let us finally recall, for completeness, that equivalence among the Hamiltonian and Lagrangian "complex plane" evolutions, (32) and (40), is entailed by the relations

$$h(\underline{z},\underline{\zeta})=\sum_{n=1}^{N}\zeta_n\,\dot{z}_n-\ell(\underline{z},\underline{\dot{z}})\quad, \tag{43a}$$

$$\zeta_n=\partial\ell(\underline{z},\underline{\dot{z}})/\partial\dot{z}_n\quad. \tag{43b}$$

The corresponding relations in the "real plane" cases read:

$$H(\underline{\vec{r}},\underline{\vec{p}})=\sum_{n=1}^{N}(\vec{p}_n\cdot\dot{\vec{r}}_n)-L(\underline{\vec{r}},\underline{\dot{\vec{r}}})\quad, \tag{44a}$$

$$\vec{p}_n=(\partial/\partial\dot{\vec{r}}_n)\,L(\underline{\vec{r}},\underline{\dot{\vec{r}}})\quad; \tag{44b}$$

$$\tilde{H}(\underline{\vec{r}},\underline{\vec{p}})=\sum_{n=1}^{N}(\hat{k}\cdot\vec{p}_n\wedge\dot{\vec{r}}_n)-\tilde{L}(\underline{\vec{r}},\underline{\dot{\vec{r}}})\quad, \tag{45a}$$

$$\vec{p}_n=[\,\hat{k}\wedge(\partial/\partial\dot{\vec{r}}_n)\,]\tilde{L}(\underline{\vec{r}},\underline{\dot{\vec{r}}})\quad. \tag{45b}$$

Exercise 4.1-1. Let $h(z_n,\zeta_n;n=1,...,N)$ be a Hamiltonian function that depends *analytically* on the N canonical coordinates z_n and on the N canonical momenta ζ_n, but is otherwise quite arbitrary. Construct a *corresponding* Hamiltonian entailing motions in the *real physical (two-dimensional) plane*, characterized by *rotation-invariant* (Hamiltonian) equations of motion. *Hint:* see (25, 26), and note that the "equation of motion" (27) is produced by *any* Hamiltonian that depends on the canonical coordinate $\vec{r}_0$ but not on the corresponding canonical momentum $\vec{p}_0$.

4.2 Example: a family of solvable many-body problems in the plane

In Sect. 4.2, which is conveniently broken down into a few subsections, see below, we discuss the (family of) *solvable* many-body problems in the plane that obtain by applying the technique of *complexification* described in the preceding Sect. 4.1 to the subclass of many-body problems characterized by the Newtonian equations of motion (2.3.3-2) which has the property to be *scaling-invariant* (see (2.3.4.2-34)). As explained below, the Newtonian equations of motion of this family of solvable many-body problems read as follows:

$$\ddot{\vec{r}}_n = (\alpha + \alpha' \, \hat{k} \wedge) \dot{\vec{r}}_n + (\beta + \beta' \, \hat{k} \wedge) \vec{r}_n$$

$$+ \sum_{m=1, m \neq n}^{N} (r_{nm})^{-2} \Big(\, 2 \left[\dot{\vec{r}}_n \, (\dot{\vec{r}}_m \cdot \vec{r}_{nm}) + \dot{\vec{r}}_m \, (\dot{\vec{r}}_n \cdot \vec{r}_{nm}) - \vec{r}_{nm} \, (\dot{\vec{r}}_n \cdot \dot{\vec{r}}_m) \right]$$

$$+ (\lambda + \lambda' \, \hat{k} \wedge) \{ \, (\dot{\vec{r}}_n + \dot{\vec{r}}_m) \left[r_n^2 - (\vec{r}_n \cdot \vec{r}_m) \right] - \vec{r}_n \left[\vec{r}_m \cdot (\dot{\vec{r}}_n + \dot{\vec{r}}_m) \right] + \vec{r}_m \left[\vec{r}_n \cdot (\dot{\vec{r}}_n + \dot{\vec{r}}_m) \right] \, \}$$

$$+ (\mu + \mu' \, \hat{k} \wedge) \{ \, \vec{r}_n \left[r_n^2 - 2 (\vec{r}_n \cdot \vec{r}_m) \right] + \vec{r}_m \, r_n^2 \, \} \Big) \quad . \qquad (1)$$

These equations of motion have been written using the short-hand notation (4.1-15), $\vec{r}_{nm} \equiv \vec{r}_n - \vec{r}_m$; they feature one- and two-body velocity-dependent forces, and they contain 8 arbitrary (real) "coupling constants" $\alpha, \alpha', \beta, \beta', \lambda, \lambda', \mu, \mu'$. Accordingly we refer to a "family" of solvable many-body problems, different members of this family being characterized by different choices for these coupling constants: for instance the "simplest" member of the family (see Sect. 4.2.4) is characterized by the vanishing of all the coupling constants. Depending on such choices, the many-body problems feature different behaviors, and these are surveyed in the following subsections.

Let us re-emphasize that, as it is evident from their structure, these Newtonian equations of motion in the plane are *rotation-invariant*; they are moreover *translation-invariant* iff $\beta = \beta' = \lambda = \lambda' = \mu = \mu' = 0$.

Exercise 4.2-1. Rewrite (1) so that the summand in the right hand side is antisymmetric in the two indices n, m. *Hint*: use the identities

$$r_n^2 - \vec{r}_n \cdot \vec{r}_m = (r_n^2 - r_m^2 + r_{nm}^2)/2 \; , \qquad (2a)$$

$$\vec{r}_n \left[\, r_n^2 - 2(\vec{r}_n \cdot \vec{r}_m) \, \right] + \vec{r}_m \, r_n^2 = \vec{r}_m \, r_n^2 - \vec{r}_n \, r_m^2 + \vec{r}_n \, r_{nm}^2 \; , \qquad (2b)$$

where of course $\vec{r}_{nm} \equiv \vec{r}_n - \vec{r}_m$, see (4.1-15).

Exercise 4.2-2. Investigate the behavior entailed by (1) for the "center-of-mass" coordinate

$$\bar{\vec{r}} = N^{-1} \sum_{n=1}^{N} \vec{r}_n \ . \tag{3}$$

Hint: see the preceding *Exercise 4.2-1.*

4.2.1 Origin of the model and technique of solution

Let us consider the one-dimensional many-body problem characterized by the Newtonian equations of motion

$$\ddot{z}_n = (\alpha + i\alpha')\dot{z}_n + (\beta + i\beta')z_n$$

$$+ \sum_{m=1, m\neq n}^{N} (z_n - z_m)^{-1} \left[2\dot{z}_n \dot{z}_m + (\lambda + i\lambda')(\dot{z}_n + \dot{z}_m) z_n + (\mu + i\mu') z_n^2 \right] \ , \tag{1}$$

which features the 8 *real* coupling constants $\alpha, \alpha', \beta, \beta', \lambda, \lambda', \mu, \mu'$. Clearly this model is invariant under the rescaling transformation $z_n \to \tilde{z}_n = c\, z_n$, with c an arbitrary constant $(\dot{c} = 0)$. Hence, as explained in the preceding Sect. 4.1, via complexification and the correspondence (4.1-17) it gets transformed into a *rotation-invariant* N-body problem in the plane. It is a matter of trivial algebra to check, using if need be appropriate formulas from Sect. 4.1, that the corresponding model is precisely (4.2-1). Hence the technique of solution, and the behavior, of the many-body problem in the plane (4.2-1), coincide with the technique of solution, and the behavior, of (1) *in the complex plane* (identified, via (4.1-17), with the *physical real* plane).

On the other hand we know that the equations of motion (1) are *solvable.* The technique of solution has been described in Sect. 2.3, and in particular in Sect. 2.3.4.2 (indeed (1) coincides, up to trivial notational changes, with the complexified version of (2.3.4.2-34)). Let us tersely review the relevant results here.

The solution $z_n(t)$ of the equations of motion (1) are the zeros of a monic polynomial of degree N in z,

$$\psi(z,t)=\prod_{n=1}^{N}\left[z-z_n(t)\right] \quad , \tag{2a}$$

$$\psi(z,t)=z^N+\sum_{m=1}^{N} c_m(t)\,z^{N-m} \quad , \tag{2b}$$

whose N coefficients evolve in time as follows:

$$c_m(t)=c_m^{(+)}\exp\left[\,\nu_m^{(+)}t\,\right]+c_m^{(-)}\exp\left[\,\nu_m^{(-)}t\,\right] \quad , \tag{3a}$$

$$\nu_m^{(\pm)}=\left\{\,\alpha+\lambda(N-m)+i\left[\alpha'+\lambda'(N-m)\right]\pm\Delta_m\,\right\}/2 \quad , \tag{3b}$$

$$\Delta_m^2=\left[\alpha+\lambda\,(N-m)\right]^2-\left[\alpha'+\lambda'(N-m)\right]^2+2m\left[2\beta+\mu(2N-m-1)\right]$$

$$+i\left\{\,2\left[\alpha+\lambda\,(N-m)\right]\left[\alpha'+\lambda'(N-m)\right]+2m\left[2\beta'+\mu'(2N-m-1)\right]\,\right\} \quad . \tag{3c}$$

Note the completely explicit character of these expressions, whose validity is only predicated upon the conditions

$$\Delta_m\neq 0\ , \quad m=1,\dots,N \quad . \tag{4}$$

Exercise 4.2.1-1. Prove (3), and obtain the formulas that replace (3) if the condition (4) fails to hold for some value of m. *Hint*: see (2.3.4.2-37) and (2.3.4.2-36).

The $2N$ constants $c_m^{(+)}$ in the right-hand-side of (3a) are of course related to the initial values of $c_m(t)$ and $\dot{c}_m(t)$ by the formulas

$$c_m^{(\pm)}=\pm\left[\dot{c}_m(0)-\nu_m^{(\mp)}c_m(0)\right]/\Delta_m \quad . \tag{5}$$

As for the initial values $c_m(0)$ and $\dot{c}_m(0)$, they can be obtained, in terms of the original initial data, via the formulas

$$\sum_{m=1}^{N} c_m(0)\,z^{N-m}=-z^N+\prod_{n=1}^{N}\left[z-z_n(0)\right] \quad , \tag{6a}$$

$$\sum_{m=1}^{N} \dot{c}_m(0)\,z^{N-m}=-\sum_{n=1}^{N}\dot{z}_n(0)\prod_{m=1,m\neq n}^{N}\left[z-z_m(0)\right] \quad , \tag{6b}$$

which are implied by (2), see (2.3.3-6,7), and which of course entail, see (2.3.3-9,10),

$$c_1(0) = -\sum_{n=1}^{N} z_n(0) \quad , \tag{7a}$$

$$c_2(0) = \frac{1}{2} \sum_{n,m=1;m\neq n}^{N} z_n(0)\, z_m(0) \quad , \tag{7b}$$

$$c_3(0) = -\frac{1}{6} \sum_{n,m,\ell=1;m\neq n,\ell\neq n,\ell\neq m}^{N} z_n(0)\, z_m(0)\, z_\ell(0) \quad , \tag{7c}$$

and so on, as well as

$$\dot{c}_1(0) = -\sum_{n=1}^{N} \dot{z}_n(0) \quad , \tag{8a}$$

$$\dot{c}_2(0) = \sum_{n,m=1;m\neq n}^{N} \dot{z}_n(0)\, z_m(0) \quad , \tag{8b}$$

$$\dot{c}_3(0) = -\frac{1}{2} \sum_{n,m,\ell=1;m\neq n,\ell\neq n,\ell\neq m}^{N} \dot{z}_n(0)\, z_m(0)\, z_\ell(0) \quad , \tag{8c}$$

and so on.

In conclusion we see that the motions yielded by the *many-body problem in the plane* (4.2-1) coincide, via the identification, see (4.1-17), of the *physical plane* with the *complex plane*, with the motions, as the time t evolves, of the N zeros of the monic polynomial (2b) in the complex plane. Note how natural it is to interpret (4.2-1) as a many-body problem in the plane, inasmuch as the natural environment to investigate the behavior of the zeros of a polynomial is the complex plane, rather than the real line. But also note that this choice is not merely suggested by mathematical consistency; it acquires a legitimacy of its own, and a sound physical interpretation, from the *rotation-invariance* of the equations of motion in the plane (4.2-1).

Finally, since the positions $\vec{r}_n(t)$ of the particles that move in the plane according to (4.2-1), or equivalently (4.2.1-1), coincide with the zeros $z_n(t)$ in the complex plane of the monic polynomial of degree N (2b), let us end Sect. 4.2.1 by reporting a standard result on the location of the zeros of such a polynomial (see, for instance, Sect. 5.1 of <DM73>):

Proposition 4.2.1-2. All the zeros of the polynomial (2b) are located inside the annulus, centered in the origin in the complex z-plane, whose inner respectively outer radii, $r \equiv r(t)$ respectively $R \equiv R(t)$, are given by the following expressions:

$$r = 1/(1 + B/|c_N|) \ , \tag{9a}$$

$$R = 1 + C \ , \tag{10a}$$

with

$$B = \underset{n=1,2,\ldots,N-1}{\mathrm{Max}} [1, |c_n|] \ , \tag{9b}$$

$$C = \underset{n=1,2,\ldots,N}{\mathrm{Max}} [|c_n|] \ . \tag{10b}$$

Note that in (9b) the index n ranges from 1 to $N-1$, in (10b) from 1 to N.

4.2.2 The generic model; behavior in the remote past and future

The generic model is characterized by the equations of motion (4.2-1), with generic values of the 8 coupling constants $\alpha, \alpha', \beta, \beta', \lambda, \lambda', \mu, \mu'$, namely values which do not satisfy any of the restrictions that characterize the "less generic" models discussed below. In Sect. 4.2.2 we discuss the behavior of the solutions of this model, in particular in the remote past and future.

Clearly the parameters that play a key role in determining the behavior at large, positive and negative, time of the polynomial $\psi(z,t)$, see (4.2.1-2b), hence of its zeros $z_n(t)$, see (4.2.1-2a), are the exponents $\nu_m^{(\pm)}$, see (4.2.1-3), or rather their real parts. Hence we set

$$\nu_m^{(\pm)} = \rho_m^{(\pm)} + i\gamma_m^{(\pm)} \tag{1a}$$

and we rewrite (4.2.1-3) as follows:

$$c_m(t) = c_m^{(+)} \exp\{ [\rho_m^{(+)} + i\gamma_m^{(+)}] t \} + c_m^{(-)} \exp\{ [\rho_m^{(-)} + i\gamma_m^{(-)}] t \} \quad . \tag{1b}$$

Here $\rho_m^{(\pm)}$ and $\gamma_m^{(\pm)}$ are of course *real*, and they are given, see (4.2.1-3), by the following explicit expressions:

$$\rho_m^{(\pm)} = [\alpha + \lambda(N-m) \pm \delta_m^{(+)}]/2 \ , \tag{1c}$$

$$\gamma_m^{(\pm)} = \left[\alpha' + \lambda'(N-m) \pm \delta_m^{(-)}\right]/2 \quad , \tag{1d}$$

where (see (4.2.1-3c))

$$\Delta_m = \delta_m^{(+)} + i\,\delta_m^{(-)} \quad , \tag{1e}$$

of course again with $\delta_m^{(\pm)}$ *real* indeed *nonnegative*:

$$\delta_m^{(\pm)} = \left\{ \left[(a_m^2 + b_m^2)^{1/2} \pm a_m\right]/2 \right\}^{1/2} \geq 0 \quad , \tag{1f}$$

$$a_m = [\alpha + \lambda(N-m)]^2 - [\alpha' + \lambda'(N-m)]^2 + 2m[2\beta + \mu(2N-m-1)] \quad , \tag{1g}$$

$$b_m = 2\left\{ [\alpha + \lambda(N-m)]\,[\alpha' + \lambda'(N-m)] + m[2\beta' + \mu'(2N-m-1)] \right\} \quad . \tag{1h}$$

Note that these expressions, (1c$\div$h), only depend on the parameters characterizing the model -- namely, the number N of particles and the 8 coupling constants α, α', β, β', λ, λ', μ, μ' -- they do not depend on the initial values of the particle positions and velocities, $z_n(0)$ and $\dot{z}_n(0)$, which only affect the values of the coefficients $c_m^{(\pm)}$, see (1b) and (4.2.1-5$\div$8), and thereby characterize the particular trajectories in the complex z-plane of the N zeros $z_n(t)$, hence the particle trajectories, $\vec{r}_n(t) \doteq z_n(t)$, of the N particles in the physical plane which correspond to specific initial data, $\vec{r}_m(0) \doteq z_m(0)$ and $\dot{\vec{r}}_m(0) \doteq \dot{z}_m(0)$ (see (4.1-17)).

Let us now consider the values of the $2N$ real numbers $\rho_m^{(\pm)}$, $m = 1, 2, \ldots, N$; since we are looking here at the *generic* case, we assume they are *all different*, and we call ρ_+ the largest of all of them and m_+ the corresponding value of m (of course $1 \leq m_+ \leq N$); likewise we call ρ_- the smallest of all of them, and m_- the corresponding value of m:

$$\rho_{m_+}^{(+)} = \rho_+; \; \rho_m^{(+)} < \rho_+, \quad m \neq m_+, \tag{2a}$$

$$\rho_{m_-}^{(-)} = \rho_-; \; \rho_m^{(-)} > \rho_-, \quad m \neq m_- \quad . \tag{2b}$$

Genericity entails

$$\delta_m^{(\pm)} > 0 \quad , \tag{3a}$$

see (1f); hence

$$\rho_m^{(+)} > \rho_m^{(-)} \quad , \tag{3b}$$

see (1c). Hence (2) entails

$$\rho_m^{(-)} < \rho_+, \; \rho_m^{(+)} > \rho_-, \quad m = 1, 2, \ldots, N \quad . \tag{3c}$$

The motion of the many-body problem in the remote past and future is determined by the behavior in the complex plane of the zeros of the polynomial (4.2.1-2b) with (1), as $t \to \mp\infty$. An analysis of this mathematical problem is provided in Appendix G, whose findings we hereafter assume the reader to have mastered.

The motion in the remote future is mainly determined by the value of ρ_+, and of m_+ if ρ_+ is positive. Indeed if ρ_+ is negative,

$$\rho_+ < 0 \quad , \tag{4a}$$

as $t \to \infty$ *all* N particles converge to the origin; and this happens for *all* initial conditions. If instead ρ_+ is positive,

$$\rho_+ > 0 \quad , \tag{4b}$$

then as $t \to \infty$ generally m_+ particles escape to infinity, and $N - m_+$ converge to the origin (except for the special initial conditions that cause $c_{m_+}^{(+)}$ to vanish).

The behavior as $t \to -\infty$ is analogous, with an obvious exchange of roles: if ρ_- is positive,

$$\rho_- > 0 \quad , \tag{5a}$$

all particles tend to (or rather, in the remote past, came from) the origin (for arbitrary initial data); if instead ρ_- is negative,

$$\rho_- < 0 \quad , \tag{5b}$$

generally m_- particles tend to (or rather, in the remote past, came from) infinity, and $N - m_-$ tend to (i.e., came from) the origin (except for the special initial conditions that cause $c_{m_-}^{(-)}$ to vanish). Hence in particular, if

$$\rho_- < 0 < \rho_+ \quad , \tag{6}$$

then the generic solution of this generic model is characterized, in the remote past, by m_- particles incoming from large distance and $N-m_-$ coming from the origin, and in the remote future, by m_+ particles outgoing towards infinity while $N-m_+$ converge towards the origin.

Exercise 4.2.2-1. What happens for the special initial conditions that cause either $c_{m_+}^{(+)}$, or $c_{m_-}^{(-)}$, or both these quantities, to vanish? *Hint*: see Appendix G.

Exercise 4.2.2-2. Verify that the many-body model (4.2.1-1) possesses the ring-like *similarity solution*

$$z_n(t) = \varphi(t) \exp(2\pi i n / N) \ , \tag{7}$$

and find $\varphi(t)$. *Hint*: insert the *ansatz* (7) in (4.2.1-1), use the appropriate identities to check its consistency, and solve the resulting ODE satisfied by $\varphi(t)$. *Solution*: see *Remark 4.2.3-14* below.

4.2.3 Some special cases: models with a limit cycle, models with confined and periodic motions, Hamiltonian models, translation-invariant models, models featuring equilibrium and spiraling configurations, models featuring only completely periodic motions

We now survey several solvable many-body problems in the plane belonging to the class (4.2-1), but with some restrictions on the 8 coupling constants $\alpha, \alpha', \beta, \beta', \lambda, \lambda', \mu, \mu'$ which cause their solutions to exhibit the gamut of behaviors indicated in the (long) title of Sect. 4.2.3.

The first model we consider is the borderline case which falls between the two instances (4.2.2-4a,b) considered above, namely the case in which the quantity ρ_+, see (4.2.2-2a), vanishes:

$$\rho_+ = 0 \ . \tag{1}$$

Note that this entails a single (algebraic) constraint on the 8 coupling constants, and that we are otherwise assuming the system to be generic, in particular that one only of the N quantities ρ_m attains the maximal value $\rho_+ = \rho_{m_+} = 0$ (see (4.2.2-2a) and (1)). It is then clear (see Appendix G) that, in the remote future $(t \to \infty)$, $N - m_+$ particles tend to the origin, and m_+ approach (exponentially in time) the circular (limit-cycle) trajectories

$$\tilde{z}_m(t) = \exp(2\pi i m / m_+) \, (-c_{m_+})^{1/m_+} \exp(i \gamma_{m_+} t / m_+) \quad , \qquad m = 1, \ldots, m_+ \quad . \tag{2}$$

Two comments about this formula are now in order.

First of all let us re-emphasize that here (and below) we identify a (real) 2-vector in the plane with the complex number that corresponds to it via (4.1-17). Clearly, via this identification, (2) describes a circular ring of m_+ particles, equispaced on a circle, of constant radius $\left| c_{m_+} \right|^{1/m_+}$, centered at the origin and rotating uniformly with angular velocity γ_{m_+} / m_+.

Secondly, the notation $\tilde{z}_m(t)$ emphasizes two points: *(i)* the "particle coordinates" $z_n(t)$ do not coincide with the quantities $\tilde{z}_m(t)$; *(ii)* as $t \to \infty$, m_+ of the N "particle coordinates" $z_n(t)$ approach asymptotically (exponentially fast, see Appendix G) the m_+ quantities $\tilde{z}_m(t)$, see (2), but to ascertain whether, say, $z_1(t)$ tends to the origin or to one of the quantities $\tilde{z}_m(t)$, and in such a case to which one, a more detailed analysis of the motion is required than that given here (indeed, the choice between these different outcomes depends nontrivially upon the initial data).

This behavior of the system in the remote future emerges out of any initial data, except of course (see (2)) for the special set such that $c_{m_+}^{(+)}$ vanishes.

Exercise 4.2.3-1. What happens as $t \to \infty$ if the initial data entail $c_{m_+}^{(+)} = 0$? *Hint*: see Appendix G.

Exercise 4.2.3-2. What happens in this case (see (1)) as $t \to -\infty$? *Hint*: see Sect. 4.2.2.

There clearly exist a plethora of other, not-completely-generic, models; for instance it might happen that, for some value of m in the range $1, 2, \ldots, N$, there hold the equality $v_m^{(+)} = v_m^{(-)}$, in which case the formula (4.2.1-3a) would have to be modified.

Exercise 4.2.3-3. How? *Hint*: see (2.3.4.2-36,37) and (4.2.1-3).

We forsake the investigation of such possibilities, and proceed to analyze some other, more special, cases, to continue with the illustration of the phenomenology outlined in the title of Sect. 4.2.3.

Let us consider next the subclass of models of type (4.2-1) with

$$\alpha = 0, \; \lambda = 0, \; \beta' = 0, \; \mu' = 0 \quad , \tag{3a}$$

and moreover with the remaining 4 nonvanishing coupling constants $\alpha', \lambda', \beta, \mu$ satisfying the following 3 inequalities:

$$(\alpha')^2 \geq 2N[2\beta+(N-1)\mu] \quad , \tag{3b}$$

$$[\alpha'+(N-1)\lambda']^2 \geq 4\,[\beta+(N-1)\mu] \quad , \tag{3c}$$

$$(\alpha'+N\lambda')^2 \geq [\lambda'(\alpha'+N\lambda')+2\beta+(2N-1)\mu]^2/(\lambda'^2+2\mu) \quad . \tag{3d}$$

It is then clear that the equations of motion (4.2-1), for *any* initial condition, yield *confined motions*, namely trajectories which remain confined, for all times, to a finite region of the plane.

Proof. The above conclusion is clearly warranted by our treatment, see above, if all the (real) exponents $\rho_m^{(\pm)}$ vanish, so that the exponents $\nu_m^{(\pm)}$, see (4.2.1-3) and (4.2.2-1), are *imaginary*, hence the coefficients $c_m(t)$ of the polynomial (4.2.1-2b) remain limited, see (4.2.1-3a), for *all* time (*including* the limits $t \to \pm\infty$). But clearly this condition is entailed by (3), since the equalities (3a) imply that $\rho_m^{(\pm)}$ is proportional to $\delta_m^{(+)}$, see (4.2.2-1c), and that b_m, see (4.2.2-1h), vanishes for all values of m; and the vanishing of b_m clearly entails, see (4.2.2-1f), that $\delta_m^{(+)}$, hence $\rho_m^{(\pm)}$, also vanishes for all the values of m such that a_m is negative (or zero). The 3 inequalities (3b,c,d) guarantee precisely that a_m is negative (or zero), $a_m \leq 0$, for all values of m in the range $1 \leq m \leq N$. (Verify!).

As implied by this analysis, the inequalities (3b,c,d) are best-possible, namely they provide, together with (3a), *sufficient* and *necessary* conditions to guarantee that *all* the motions entailed by (4.2-1) remain confined for *all* time.

Note, however, that this statement is not quite correct.

Exercise 4.2.3-4. Provide a counterexample. *Hint*: the condition $a_m \leq 0$ need not hold for all *real* values of m in the range $1 \leq m \leq N$: it is enough that it hold for $m = 1,2,\ldots,N$. Moreover: what about the possibility that, for some value of m, the quantity a_m (see (4.2.2-1g)) vanish, entailing (together with the vanishing of b_m) the vanishing of both $\delta_m^{(+)}$ and $\delta_m^{(-)}$, $\delta_m^{(\pm)} = 0$, hence the coincidence of $\nu_m^{(+)}$ and $\nu_m^{(-)}$, $\nu_m^{(+)} = \nu_m^{(-)}$ (see (4.2.2-1a,c,d))?

Exercise 4.2.3-5. Does it make sense, in a physics, mathematics, or mathematical-physics context, to assert that a statement is *not quite correct*? Discuss the matter with a colleague, taking turns at arguing one way or the other.

Clearly the simple requirements

$$\beta \le 0, \; \mu < -\lambda'^2/2 \tag{4}$$

are sufficient to guarantee validity of (3b,c,d), hence they are *sufficient* (albeit *not necessary*), together with (3a), to also guarantee that *all* motions entailed by (4.2-1) remain confined for all time.

Exercise 4.2.3-6. Try and suggest a more "physical", if less "mathematically rigorous", interpretation of this conclusion, than the proof given above on the basis of the exact solution technique. *Hint*: look at the form the equations of motion (4.2-1) take for a particle that tries to escape to infinity.

Let us next look at a subclass, of the many-body problems (4.2-1), which is certainly *Hamiltonian*. We already saw in Sect. 4.1 that any model in the plane obtained via the *complexification* trick from a one-dimensional Hamiltonian model is itself amenable to a Hamiltonian formulation. We now remark that, in the context of the models we are now considering, see (4.2-1) and (4.2.1-1), a condition sufficient to guarantee that the evolution equations (4.2.1-1) satisfied by the N quantities $z_n(t)$ be *Hamiltonian* is that the corresponding evolution equations satisfied by the quantities $c_m(t)$ be themselves *Hamiltonian*, since the relations linking the N "canonical coordinates" $z_n(t)$ to the N "canonical coordinates" $c_m(t)$ can certainly be embedded in a canonical transformation: indeed, a "point transformation", which relates old and new canonical coordinates without involving the canonical momenta. But the evolution equations satisfied by the N quantities $c_m(t)$ are indeed Hamiltonian if

$$\alpha = \alpha' = \lambda = \lambda' = 0 \; , \tag{5}$$

since they then read (see (2.3.4.2-36))

$$\ddot{c}_m = m\left[\beta + i\,\beta' + \frac{1}{2}(\mu + i\,\mu')(2N - m - 1)\right] c_m \; , \tag{6}$$

namely they are the ("Newtonian") equations of motion yielded by the Hamiltonian

$$h(\underline{c}, \underline{\gamma}) = \frac{1}{2}\sum_{m=1}^{N} \left\{\gamma_m^2 - m\left[\beta + i\,\beta' + \frac{1}{2}(\mu + i\,\mu')(2N - m - 1)\right] c_m^2 \right\} \; . \tag{7}$$

Proof. The Hamiltonian equations yielded by (7) read

$$\dot{c}_m = \partial h / \partial \gamma_m = \gamma_m \quad , \tag{8a}$$

$$\dot{\gamma}_m = -\partial h / \partial c_m = m\left[\beta + i\,\beta' + \frac{1}{2}(\mu + i\,\mu')(2N - m - 1)\right] c_m \;; \tag{8b}$$

and time-differentiation of (8a) yields, via (8b), precisely (6).

Exercise 4.2.3-7. What about the Hamiltonian $H(\underline{z},\underline{\zeta})$ that yields, when (5) holds, directly (4.2.1-1) ? *Hint*: see <CF97>.

Exercise 4.2.3-8. Prove that all the motions of the (Hamiltonian) N-body problem (4.2-1) with (5) are *completely periodic*, with the same period $T = 2\pi(2/\mu)^{1/2}$ (or an integer multiple of it, no larger than $N\,!$), if there hold the following restrictions (additional to (5)) on the coupling constants:

$$\beta' = \mu' = 0, \quad 2\beta + \mu(2N - 1) = 0\,, \quad \mu > 0 \;. \tag{9}$$

Let us emphasize that the conditions (5) are sufficient, but not necessary, to imply that the system (4.2.1-1) be *Hamiltonian*; indeed examples are given below (see, for instance, Sect. 4.2.5) which violate (5) yet are amenable to a Hamiltonian formulation.

The next subclass of models of type (4.2-1) we single out for additional consideration are characterized by the following *translation-invariant* equations of motion:

$$\ddot{\vec{r}}_n = (\alpha + \alpha' \hat{k} \wedge) \dot{\vec{r}}_n$$

$$+ 2 \sum_{m=1, m\neq n}^{N} (r_{nm})^{-2} \left[\dot{\vec{r}}_n (\dot{\vec{r}}_m \cdot \vec{r}_{nm}) + \dot{\vec{r}}_m (\dot{\vec{r}}_n \cdot \vec{r}_{nm}) - \vec{r}_{nm} (\dot{\vec{r}}_n \cdot \dot{\vec{r}}_m)\right], \qquad (\vec{r}_{nm} \equiv \vec{r}_n - \vec{r}_m) \;, \tag{10}$$

which of course correspond to (4.2-1), hence as well to (4.2.1-1), with $\beta = \beta' = \lambda = \lambda' = \mu = \mu' = 0$.

These equations of motion clearly imply that a particle is acted upon by a nonvanishing forces, and contributes by its presence to the force acting on other particles, *only if it moves* (with nonvanishing velocity; except possibly at the instant of a two-body collision, see (10)). Hence, in the context of the initial-value problem, only particles whose initial velocities do not vanish need be taken into account; those whose initial velocities vanish can simply be ignored, since they will never move nor will they influence the motion of the other particles.

In this case, see (10), the solution of the initial-value problem is of course still given by (4.2.1-2) (with (4.1-17)), but now with (4.2.1-3) replaced by

$$c_m(t) = c_m(0) + \dot{c}_m(0)(\alpha + i\alpha')^{-1}\{\exp[(\alpha + i\alpha')^{-1} t] - 1\} \quad . \tag{11}$$

The initial values $c_m(0)$ and $\dot{c}_m(0)$ are still related to the initial particle positions $\vec{r}_n(0)$ and to the initial particle velocities $\dot{\vec{r}}_n(0)$ by (4.2.1-6,7,8) (via (4.1-17)), and it is easily seen that these formulas, together with (11), yield the following compact prescription for the solution of (the initial-value problem for) (10): the complex coordinates $z_n(t)$ (related to the real 2-vector particle positions $\vec{r}_n(t)$ by (4.1-17)) are the N roots of the following equation in z:

$$\sum_{m=1}^{N} \dot{z}_m(0)/[z - z_m(0)] = (\alpha + i\alpha')/\{\exp[(\alpha + i\alpha')t] - 1\} \quad . \tag{12}$$

Exercise 4.2.3-9. Prove this statement. *Hint*: see (2.3.4.2-20).

Let us now discuss the motion of the N particles, with coordinates $\vec{r}_n(t) \doteq z_n(t)$, as entailed by this finding. But before doing this, let us note that the equations of motion (10) entail, for the center-of-mass coordinate,

$$\bar{\vec{r}}(t) = N^{-1} \sum_{n=1}^{N} \vec{r}_n(t) \quad , \tag{13a}$$

or equivalently, via $\bar{\vec{r}}(t) \doteq \bar{z}(t)$, for the complex coordinate $\bar{z}(t)$,

$$\bar{z}(t) = N^{-1} \sum_{n=1}^{N} z_n(t) \quad , \tag{13b}$$

the evolution equation

$$\ddot{\bar{\vec{r}}}(t) = (\alpha + \alpha' \hat{k} \wedge)\, \dot{\bar{\vec{r}}}(t) \quad , \tag{14a}$$

or equivalently,

$$\ddot{\bar{z}}(t) = (\alpha + i\alpha')\dot{\bar{z}}(t) \quad , \tag{14b}$$

entailing

$$\bar{z}(t)=\bar{z}(0)+\dot{\bar{z}}(0)(\alpha+i\alpha')^{-1}\left\{\exp[(\alpha+i\alpha')t]-1\right\} \quad . \tag{15}$$

The evolution of $\bar{z}(t)$, see (15), all but coincides with the evolution of the quantities $c_m(t)$, which in the case under present consideration all evolve in the same way, see (11). This coincidence is not surprising: compare (2.3.1-2a) with (13b).

Clearly the main element determining the qualitative character of the motion is the value of the coupling constant α, and in particular its sign: see (12).

Let us consider firstly the case of positive α,

$$\alpha>0 \quad . \tag{16}$$

Then, as $t\to\infty$, one of the particles escapes to infinity, and $N-1$ do not, approaching asymptotically, up to corrections of order $\exp(-\alpha t)$, $N-1$ fixed positions, whose configuration depends on the initial conditions: they are the $N-1$ zeros of the function of z appearing in the left hand side of (12) (which, up to a common factor, is indeed a polynomial in z of degree $N-1$). As for the particle that escapes to infinity, its coordinate, say $\vec{r}_j(t)\doteq z_j(t)$, coincides asymptotically, up to finite corrections, with that of the center-of-mass multiplied by N, see (15):

$$z_j(t)=N\bar{z}(t)\left\{1+O[\exp(-\alpha t)]\right\} \quad , \tag{17a}$$

$$z_j(t)=N(\alpha+i\alpha')^{-1}\dot{\bar{z}}\,(0)\exp[(\alpha+i\alpha')t]+O(1) \quad . \tag{17b}$$

Note that we have attached the label j to the particle that escapes to infinity; which one of the N particles does so depends nontrivially on the initial conditions. It is clear, see (17b), what the character of the asymptotic motion of this particle is.

The outcome we just described obtains for a *generic* set of initial data. There are, however, *special* initial data that entail other, different, outcomes. Indeed a necessary and sufficient condition to yield the asymptotic outcome we just described is the requirement that the center-of-mass of the system not be initially (hence, throughtout the motion: see (15)) at rest,

$$\sum_{n=1}^{N} \dot{z}_n(0) \neq 0 \quad , \tag{18a}$$

or equivalently (see (2.3.3-9a) and (13b)), and more significantly (see below), that the initial value, $\dot{c}_1(0)$, of the time-derivative of the coefficient $c_1(t)$ (see (4.2.1-2b)) not vanish,

$$\dot{c}_1(0) \neq 0 \quad . \tag{18b}$$

Exercise 4.2.3-10. Prove this statement. *Hint*: consider (12) for large t and large z and remember (2.3.3-9a).

If instead (18) does not hold, indeed if, for some positive integer $M \leq N$,

$$\dot{c}_m(0) = 0 \,, \quad m = 1,2,\ldots,M-1; \quad \dot{c}_M(0) \neq 0 \quad , \tag{19}$$

then, as $t \to \infty$, $N-M$ particles remain confined and M escape to infinity. Specifically, the $N-M$ particles that remain confined tend asymptotically, as $t \to \infty$, to the $N-M$ zeros of the polynomial in z,

$$\sum_{m=M}^{N} \dot{c}_m(0)\, z^{N-m} = 0 \quad , \tag{20}$$

while those which escape to infinity are characterized by the asymptotic formulas

$$z_j(t) = \tilde{z}_j(t) + o(1) \quad , \tag{21a}$$

$$\tilde{z}_j(t) = \exp(2\pi i\, j/M)\, \left[-\dot{c}_M(0)\right]^{1/M} \exp\left[(\alpha + i\alpha')\,t/M\right], \qquad j = 1,\ldots,M \quad . \tag{21b}$$

Exercise 4.2.3-11. Prove these statements, see (20) and (21). *Hint*: as for the preceding *Exercise 4.2.3-10*, or see Appendix G.

This concludes our analysis of the behavior of the system (10), with α positive, see (16), as $t \to \infty$. As for the behavior in the remote past ($t \to -\infty$), clearly in this case *all* particles approach asymptotically the configuration corresponding via (4.1-17) to the N (complex, finite) roots of the polynomial equation in z

$$z^N + \sum_{m=1}^{N} c_m(-\infty)\, z^{N-m} = 0 \tag{22}$$

with (see (11) and (16))

$$c_m(-\infty) = c_m(0) - \dot{c}_m(0)/(\alpha + i\alpha') \quad . \tag{23}$$

Let us summarize our findings re: the qualitative behavior in the remote past and future of the translation-invariant system (10) with (16).

In the remote past, the N particles are *almost at rest* at some positions (which could of course be arbitrarily assigned); it is of course essential that none of them be *completely* at rest (in which case, they would remain so throughout time and they could simply be ignored). Then the particles begin to move, and $N-1$ of them always remain in a finite region of the plane, approaching asymptotically, in the remote future, fixed positions; while one of them shoots eventually off to infinity, around a straight direction if $\alpha' = 0$ or a spiraling one if $\alpha' \neq 0$. This outcome is the generic one; in special cases (corresponding to special initial, at $t = 0$, or asymptotic, as $t \to -\infty$, conditions), only $N - M$ (with $2 \leq M \leq N$) particles always remain in a finite region of the plane, approaching asymptotically, as $t \to \infty$, a configuration determined by the initial data, while M shoot eventually off to infinity along outgoing stellar straight (if $\alpha' = 0$) or spiraling (if $\alpha' \neq 0$) lines.

This ends our discussion of the case with α positive, see (16). The opposite case, with negative α, $\alpha < 0$, need not be discussed, since the analysis is essentially identical to that just given (for the $\alpha > 0$ case), except for an exchange of the behaviors as $t \to +\infty$ with that as $t \to -\infty$, and viceversa.

Let us finally consider the case in which α vanishes,

$$\alpha = 0 \tag{24}$$

(with $\alpha' \neq 0$; the case $\alpha = \alpha' = 0$ is treated in the following Sect. 4.2.4; the case with $\alpha = 0, \alpha' \neq 0$ we consider now will be treated in more detail in Sect. 4.2.5).

In this case the right hand side of (12) is periodic in t, with period

$$T = 2\pi / \alpha' \quad . \tag{25}$$

Hence the roots of this equation are also periodic, with the same period. One therefore concludes that in this case, see (24), *all* solutions of the system (10) are completely periodic, with (at most) the (same) period $T' = T \cdot N!$ (the factor $N!$ accounts for the possibility that the correspondence between individual particles and roots of (12) get permuted through the motion).

Let us now return to the general model (4.2-1), to investigate the cases when there exist nontrivial equilibrium configurations, namely time-independent solutions of (4.2-1),

$$\vec{r}_n(t)=\bar{\vec{r}}_n\,,\ \dot{\vec{r}}_n(t)=0\,,\quad n=1,\ldots,N\quad , \tag{26a}$$

or equivalently (see (4.2.1-1)),

$$z_n(t)=\bar{z}_n\,,\ \dot{z}_n(t)=0\,,\quad n=1,\ldots,N\quad . \tag{26b}$$

Hereafter we exclude from consideration the *unphysical* solution

$$z_n(t)=0\,,\quad n=1,\ldots,N\ , \tag{27a}$$

corresponding, see (4.2.1-2,3), to

$$c_m(t)=0\,,\quad m=1,\ldots,N\quad . \tag{27b}$$

It is clear (see (4.2.1-2,3)) that a necessary and sufficient condition for the existence of such nontrivial equilibrium solutions is that, for some value $m=M$ (with M a positive integer not exceeding N; but we shall find below that only $M=N$ or $M=N-1$ are actually acceptable possibilities), either $\nu_m^{(+)}$ or $\nu_m^{(-)}$ vanish:

$$\nu_M^{(+)}=0\quad\text{or}\quad\nu_M^{(-)}=0\quad . \tag{28}$$

Via (4.2.1-3b,3c), this requirement corresponds to the following two conditions:

$$(2N-M-1)\,\mu+2\beta=0\quad , \tag{29a}$$

$$(2N-M-1)\,\mu'+2\beta'=0\quad . \tag{29b}$$

Indeed insertion of (29) in (4.2.1-3c) yields

$$\Delta_M^2=\{\alpha+\lambda(N-M)+i[\alpha'+\lambda'(N-M)]\}^2\,, \tag{30}$$

and this, via (4.2.1-3b), entails (28).

Note that the conditions (29) only constrain the 4 coupling constants β,β',μ,μ'; indeed the other 4 coupling constants, $\alpha,\alpha',\lambda,\lambda'$, play no role at equilibrium, since the corresponding forces vanish when the particles are at rest, see (4.2-1).

Let us then assume that the 2 conditions (29) hold, for some positive integer value of M not exceeding N. Then clearly the equilibrium configuration is characterized by the condition that all the coefficients $c_m(t)$ with $m \neq M$ vanish,

$$c_m(t) = 0, \quad m = 1,...,N, \quad m \neq M \tag{31}$$

(corresponding to $c_m^{(+)} = c_m^{(-)} = 0$ for $m \neq M$, see (4.2.1-3a)). Hence the equilibrium configuration, see (26b), is identified by the condition (see (4.2.1-2b))

$$\prod_{n=1}^{N} (z - \bar{z}_n) = z^N + c_M \, z^{N-M} = z^{N-M}(z^M + c_M) \quad , \tag{32}$$

where c_M is an arbitrary complex constant.

Clearly, if $\nu_m^{(+)}$ vanishes, see (28), the equilibrium configuration (corresponding to a nonvanishing time-independent c_M) obtains for vanishing $c_M^{(-)}$ and nonvanishing $c_M^{(+)}$, see (4.2.1-3a):

$$\nu_m^{(+)} = 0 \, , \quad c_M^{(-)} = 0 \, , \, c_M^{(+)} = c_M \, ; \tag{33a}$$

likewise, if $\nu_M^{(-)}$ vanishes,

$$\nu_m^{(-)} = 0 \, , \quad c_M^{(+)} = 0 \, , \, c_M^{(-)} = c_M \, . \tag{33b}$$

For $M = N$ (32) has (up to permutations) the N distinct roots

$$\hat{z}_n = \exp(2\pi i n / N) \, (-c_N)^{1/N} \quad , \quad n = 1,...,N \; ; \tag{34}$$

for $M = N-1$, (32) has (again, up to permutations) the N distinct roots

$$\hat{z}_n = \exp[2\pi i n/(N-1)] \, (-c_{N-1})^{1/(N-1)} \quad , \quad n = 1,...,N-1; \;\; \hat{z}_N = 0 \, , \tag{35}$$

while for $M \leq N-2$ (32) has at least two vanishing roots, yielding therefore a solution, with two or more particles sitting at the origin, that we rule out as “unphysical”, see (4.2-1).

We see, in conclusion, that the many-body problem (4.2-1) admits a nontrivial equilibrium configuration only in two cases. If the 4 coupling constants β, β', μ, μ' satisfy the 2 constraints (see (29), with $M=N$)

$$(N-1)\mu+2\beta=(N-1)\mu'+2\beta'=0 \quad , \tag{36}$$

then the system admits the equilibrium configuration (34): the N particles sit equispaced on a circle, of arbitrary radius, centered at the origin. If instead the 4 coupling constants β, β', μ, μ' satisfy the 2 constraints (see (29), with $M=N-1$)

$$N\mu+2\beta=N\mu'+2\beta'=0 \quad , \tag{37}$$

then the system admits the equilibrium configuration (35): one particle sits at the origin, and $N-1$ sit equispaced on a circle, of arbitrary radius, centered at the origin. Of course, in each case, there exist $N!$ such configurations, corresponding to all possible permutations of the N particles on their equilibrium positions.

Exercise 4.2.3-12. Check directly that, at such configurations, the forces in the right hand side of (4.2-1) do balance off. *Hint*: use (4.2.1-1) (rather than (4.2-1)), and use the appropriate trigonometric identity.

Exercise 4.2.3-13. Investigate, by the standard approach, the behaviors of these systems around their equilibrium configurations and, by comparison with their exact behaviors, discover some "remarkable matrices". *Hint*: see the treatment that led to (2.1.3.3-46).

These equilibrium configurations cannot be completely stable, due to the invariance of the equations of motion under rotations and dilations (rescalings). Indeed it is clear that the equilibrium configurations (34) respectively (35) are special cases of the following circularly symmetrical solutions of the equations of motion (4.2.1-1):

$$z_n(t)=\hat{z}_n\left[-c_M(t)\right]^{1/M} \quad , \tag{38a}$$

with $\hat{z}_n$ given by (34) respectively (35) and

$$c_M(t)=c_M(0)+\dot{c}_m(0)\,\nu_M^{-1}\left[\exp(\nu_M t)-1\right] \quad , \tag{38b}$$

$$\nu_M = \alpha + \lambda(N-M) + i[\alpha' + \lambda'(N-M)] \ , \tag{38c}$$

with $M = N$ respectively $M = N-1$. And of course an additional element of instability of the equilibrium configuration (34) and (35), and as well, more generally, of the spiraling configuration (38), arises from the possibility that a perturbation excite other "nonlinear modes", namely that it induce other coefficients $c_m(t)$, with $m \neq M$, to become different from zero. Whether such perturbations would grow or decay depends of course on the sign of the corresponding real exponents $\rho_m^{(\pm)}$, see (4.2.2-1).

Remark 4.2.3-14. Clearly these circularly symmetrical configurations of type (38a) are featured even by the most general system (4.2-1), without any restriction on the 8 coupling constants: in such a case, of course, the quantity $c_M(t)$ would be given by (4.2.1-3) (rather than by (38b,c)).

Exercise 4.2.3-15. Consider the system (4.2.1-1), with the following circularly symmetrical initial conditions:

$$z_n(0) = \dot{z}_n(0) = 0 \ , \quad n = 1,\ldots,p \ ; \tag{39a}$$

$$z_n(0) = x_j(0) \exp\{2\pi i [n - p - (j-1)\, q]/q\} \ ,$$

$$\dot{z}_n(0) = \dot{x}_j(0) \exp\{2\pi i [n - p - (j-1)\, q]/q\} \ ,$$

$$n = p+1+(j-1)q, \ p+2+(j-1)q, \ldots, p+jq; \ j = 1,2,\ldots,k \ , \tag{39b}$$

where $x_j(0)$ and $\dot{x}_j(0)$ are arbitrary *real* numbers and p, q, k are arbitrary positive *integers* ($N = p + qk; p$ could also vanish, and it should be restricted to be less than 2 to avoid the "unphysical" piling up of particles at the origin, although this is hardly relevant, since the particles there will not move). *(i)* Draw this initial configuration in the plane, to make sure you have understood its layout. *(ii)* Clearly, for reasons of symmetry, this configuration is preserved throughout the motion, hence the solution will read

$$z_n(t) = x_j(t) \ \exp\{2\pi i [n - p - (j-1)\, q]\}, \ n = p+1+(j-1)q, \ldots, \ p+jq; \ j = 1,\ldots,k . \tag{39c}$$

Find the equations of motion of the (real) quantities $x_j(t)$, $j = 1,\ldots,k$. Since the system (4.2.1-1) is *solvable*, the equations of motion satisfied by the quantities $x_j(t)$ must also be *solvable*. Is this a new system? If not, what is its relation with the technique of solution of (4.2.1-1) ?

Remark 4.2.3-16. There are other models, more general than that considered above (namely, than (10) with $\alpha = 0$, see also Sect. 4.2.5), which generally lack translation-invariance, but also possess the remarkable property to only feature *completely periodic* motions. They are characterized by the restrictions

$$\alpha = \beta = \lambda = \mu = \beta' = \mu' = 0, \ \alpha' = p\,\omega, \ \lambda' = q\,\omega \tag{40}$$

with $\omega > 0$ and p, q two *arbitrary* integers (not both vanishing), so that the corresponding Newtonian equations of motion read

$$\ddot{\vec{r}}_n = p\,\omega\,\hat{k} \wedge \dot{\vec{r}}_n + \sum_{m=1, m\neq n}^{N} (r_{nm})^{-2} \left(2\left[\dot{\vec{r}}_n (\dot{\vec{r}}_m \cdot \vec{r}_{nm}) + \dot{\vec{r}}_m (\dot{\vec{r}}_n \cdot \vec{r}_{nm}) - \vec{r}_{nm} (\dot{\vec{r}}_n \cdot \dot{\vec{r}}_m)\right]\right.$$

$$\left. + q\,\omega\,\hat{k} \wedge \left\{(\dot{\vec{r}}_n + \dot{\vec{r}}_m)\left[r_n^2 - (\vec{r}_n \cdot \vec{r}_m)\right] - \vec{r}_n \left[\vec{r}_m \cdot (\dot{\vec{r}}_n + \dot{\vec{r}}_m)\right] + \vec{r}_m \left[\vec{r}_n \cdot (\dot{\vec{r}}_n + \dot{\vec{r}}_m)\right]\right\}\right) . \tag{41}$$

Exercise 4.2.3-17. Verify that the conditions (40) are consistent with the conditions (3) that guarantee *confined* motions.

Exercise 4.2.3-18. Prove that *all* the motions entailed by (41) are *completely periodic*. *Hint*: insert (40) in (4.2.1-3).

4.2.4 The simplest model: explicit solution (the game of musical chairs), Hamiltonian structure

In Sect. 4.2.4 we discuss the simplest of the models (4.2-1), characterized by the vanishing of all 8 coupling constants:

$$\alpha = \alpha' = \beta = \beta' = \lambda = \lambda' = \mu = \mu' = 0 \ . \tag{1}$$

Its equations of motion read

$$\ddot{\vec{r}}_n = 2 \sum_{m=1, m\neq n}^{N} \left[\, \dot{\vec{r}}_n (\dot{\vec{r}}_n \cdot \vec{r}_{nm}) + \dot{\vec{r}}_m (\dot{\vec{r}}_n \cdot \vec{r}_{nm}) - \vec{r}_{nm} (\dot{\vec{r}}_n \cdot \dot{\vec{r}}_m) \,\right] / \, r_{nm}^2 \quad , \quad (\vec{r}_{nm} \equiv \vec{r}_n - \vec{r}_m), \tag{2a}$$

or equivalently, via (4.1-17),

$$\ddot{z}_n = 2 \sum_{m=1, m\neq n}^{N} \dot{z}_n \dot{z}_m / (z_n - z_m) \ . \tag{2b}$$

The solution of the initial-value problem for these equations of motion, (2b), is given by the following neat recipe: *the* N *(complex) coordinates* $z_n(t)$ *are the* N *roots of the following algebraic equations in* z*:*

$$\sum_{m=1}^{N} \dot{z}_m(0)/[z - z_m(0)] = 1/t \ . \tag{3}$$

Exercise 4.2.4-1. Prove this statement. *Hints*: see *Exercise 4.2.3-9*; or show directly that the N roots of (3) coincide with the N zeros of the polynomial

$$\psi(z,t) = z^N + \sum_{m=1}^{N} c_m(t)\ z^{N-m} \tag{4}$$

with

$$\ddot{c}_m(t) = 0 \ , \tag{5a}$$

$$c_m(t) = c_m(0) + \dot{c}_m(0)\ t \ , \tag{5b}$$

$c_m(0)$, $\dot{c}_m(0)$ being related to $z_n(0)$, $\dot{z}_n(0)$ by the standard relations between the coefficients and the zeros of a monic polynomial such as (4), see (4.2.1-7,8) (as for the validity of (5a), see for instance (2.3.4.2-36) with (1)); or see the *Remark 2.3.4.2-3*.

Note that the equations of motion (2) are invariant under translations, and that they entail that the center-of-mass,

$$\bar{z}(t) = N^{-1} \sum_{n=1}^{N} z_n(t) \ , \tag{6}$$

moves uniformly:

$$\ddot{\bar{z}} = 0 \ , \tag{7a}$$

$$\bar{z}(t) = \bar{z}(0) + \dot{\bar{z}}(0)\ t \ . \tag{7b}$$

Let us now discuss the main qualitative features of the time-evolution of the system (2), focussing firstly on the (generic!) case in which the center-of-mass is *not* at rest,

$$\dot{\bar{z}}(t) = \dot{\bar{z}}(0) \neq 0 \ . \tag{8}$$

It is then easily seen that, as $t \to \pm\infty$, $N-1$ particles approach asymptotically (up to corrections $O(t^{-1})$) $N-1$ fixed locations, whose configuration depends on the initial conditions, consisting of the $N-1$ values of z for which the left hand side of (3) vanishes; while one of them goes to infinity, approaching asymptotically (again, up to corrections $O(t^{-1})$) the free trajectory

$$\tilde{z}(t) = \sum_{n=1}^{N} \dot{z}_n(0) \{ t + [z_n(0) / \sum_{m=1}^{N} \dot{z}_m(0)] \} \quad , \tag{9a}$$

$$\tilde{z}(t) = N \dot{\bar{z}}(0)\, t + \sum_{n=1}^{N} \dot{z}_n(0)\, z_n(0) / [N \dot{\bar{z}}(0)] \quad . \tag{9b}$$

Exercise 4.2.4-2. Prove this statement. *Hint*: (3) possesses, at all (finite) times, N solutions; as $t \to \pm\infty$, its right hand side vanishes, ... Alternatively, see (4) with (5b).

Hence the system looks *overall* exactly the same in the remote future as in the remote past ("solitonic behavior": for this terminology see, for instance, <CD82>). Note however that the particle that escapes to infinity in the remote future need not be the same one that came in from infinity in the remote past, and moreover that, through the motion, some particles may change location ("game of musical chairs": the locations of the $N-1$ "chairs" remain fixed through the motion, the identity of their $N-1$ occupants -- who are "seated" only at the beginning and at the end -- may change, from the time $t=-\infty$ when they begin to wander off in the plane and an extra player comes in from afar, to the time $t=+\infty$ when $N-1$ of them sit down again and one player -- the same or a different one -- goes off, along the same straight line, and with the same speed, as the player who came in from afar in the remote past).

This concludes our discussion of the behavior of the system (2) in the (generic) case when the center-of-mass moves, see (8). If instead the center-of-mass is at rest, namely (8) does *not* hold, then the generic behavior sees only $N-2$ particles tend to finite locations as $t=\pm\infty$, and 2 go to infinity (but they move no more as free particles; see below). And if the initial data are further specialized by requiring them to satisfy appropriate additional conditions, the number of particles that go to infinity increases. This phenomenology is analogous (albeit with a difference: the symmetrical behavior now as $t \to -\infty$ and $t \to +\infty$) to that already discussed for the system (4.2.3-10): if the initial conditions entail (see (4.2.3-19))

$$\dot{c}_m(0)=0\,,\quad m=1,2,\ldots,M-1;\;\dot{c}_M(0)\neq 0\ , \tag{10}$$

then, as $t=\pm\infty$, $N-M$ particles converge asymptotically to fixed locations, whose configuration consists of the $N-M$ zeros of the following polynomial of degree $N-M$ in z,

$$\sum_{m=M}^{N}\dot{c}_m(0)\;z^{N-M}=0\ , \tag{11}$$

while M particles go to infinity approaching (up to multiplicative corrections, generally $O(t^{-1})$) the asymptotic stellar trajectories

$$\tilde{z}_j(t)=\exp(2\pi\,i\,j/M)\,\left[-\dot{c}_M(0)\;t\right]^{1/M},\quad j=1,\ldots,M\ . \tag{12}$$

Exercise 4.2.4-3. Prove these results. *Hint*: see *Exercise 4.2.3-11.*

But let us again emphasize that, while these analyses allow to predict that the N particles are asymptotically distributed so that $N-M$ approach the $N-M$ roots of (11) and M approach the M trajectories (12), a more detailed investigation is needed of the solution entailed by (3), or equivalently by (4) with (5), to ascertain the individual fate of each specific particle, and/or to correlate the individual behavior of the particles in the remote past and future.

Exercise 4.2.4-4. Can you imagine some special initial conditions for which the outcome, in the remote past and future, is easily predicted for each individual particle? *Hint*: think of the real subcase, or of analogous one-dimensional configurations.

The system (2) is clearly a special case of the Hamiltonian systems considered in Sect. 4.2.3: indeed (5a) is the special case of (4.2.3-6) entailed by (1) (which is itself a special case of (4.2.3-5)). But it is also, up to complexification, a special case of the RS model (see (2b) and (2.1.10-1), as well as (2.1.12-1,7a)). Hence the results of Sect. 2.1.12.1 entail that the Newtonian equations of motion (2b) follow from the Hamiltonian equations

$$\dot{z}_n=a\,\exp(a\zeta_n)\left[\prod_{m=1,m\neq n}^{N}(z_n-z_m)\right]^{-1}\ , \tag{13a}$$

$$\dot{\zeta}_n = a^{-1} \sum_{m=1, m\neq n}^{N} (\dot{z}_n + \dot{z}_m)/(z_n - z_m) \quad , \tag{13b}$$

which are themselves yielded by the Hamiltonian function

$$H(\underline{z}, \underline{\zeta}) = \sum_{n=1}^{N} \exp(a\zeta_n) \left[\prod_{m=1, m\neq n}^{N} (z_n - z_m) \right]^{-1} \quad , \tag{14}$$

with a an arbitrary (possibly complex) constant.

Exercise 4.2.4-5. Prove these two statements, see (2b), (13) and (14), noting that the second set of Hamiltonian equations, (13b), has been written in more compact form by taking advantage of (13a). *Hint*: to derive (2b), differentiate the logarithm of (13a) and use (13b).

The procedure described in the last part of Sect. 4.1 can now be used to obtain (two, real) Hamiltonians, each of which yields the real, two-dimensional, equations of motion (2a) (rather than the complex, one-dimensional, equations of motion (2b)). For instance, for $N = 2$ the two real Hamiltonians

$$H(\vec{r}_1, \vec{r}_2; \vec{p}_1, \vec{p}_2) = |\vec{r}_1 - \vec{r}_2|^{-2} \cdot$$

$$\cdot \{ (x_1 - x_2) [\exp(\vec{a}\cdot\vec{p}_1) \cos(\hat{k}\cdot\vec{a}\wedge\vec{p}_1) - \exp(\vec{a}\cdot\vec{p}_2) \cos(\hat{k}\cdot\vec{a}\wedge\vec{p}_2)]$$

$$+ (y_1 - y_2) [-\exp(\vec{a}\cdot\vec{p}_1) \sin(\hat{k}\cdot\vec{a}\wedge\vec{p}_1) + \exp(\vec{a}\cdot\vec{p}_2) \sin(\hat{k}\cdot\vec{a}\wedge\vec{p}_2)] \} \quad , \tag{15}$$

$$\tilde{H}(\vec{r}_1, \vec{r}_2; \vec{\tilde{p}}_1, \vec{\tilde{p}}_2) = |\vec{r}_1 - \vec{r}_2|^{-2} \cdot$$

$$\cdot \{ (x_1 - x_2) [\exp(-\hat{k}\cdot\vec{a}\wedge\vec{\tilde{p}}_1) \sin(\vec{a}\cdot\vec{\tilde{p}}_1) - \exp(-\hat{k}\cdot\vec{a}\wedge\vec{\tilde{p}}_2) \sin(\vec{a}\cdot\vec{\tilde{p}}_2)]$$

$$+ (y_1 - y_2) [-\exp(-\hat{k}\cdot\vec{a}\wedge\vec{\tilde{p}}_1) \cos(\vec{a}\cdot\vec{\tilde{p}}_1) + \exp(-\hat{k}\cdot\vec{a}\wedge\vec{\tilde{p}}_2) \cos(\vec{a}\cdot\vec{\tilde{p}}_2)] \} \quad , \tag{16}$$

both yield the equations of motion (2a) (with $N = 2$). Here $\vec{a}$ is of course an arbitrary (constant) 2-vector. Note that neither one of these two Hamiltonians, nor indeed the corresponding Hamiltonian equations, are *rotation-invariant*; while the Newtonian equations of motion (2a) are themselves, of course, *rotation-invariant*.

Exercise 4.2.4-6. Derive (15) and (16). *Hint*: see the last part of Sect. 4.1.

Exercise 4.2.4-7. Verify by explicit computation that both these Hamiltonians, (15) and (16), yield the equations of motion (2b).

4.2.5 The simplest model featuring only completely periodic motions

In Sect. 4.2.5 we discuss another, very simple, special case of the family of models (4.2-1), which is characterized by the remarkable property to feature *only completely periodic trajectories*. It obtains from (4.2-1) by setting all the coupling constants to zero except α',

$$\alpha' = \omega \neq 0, \ \alpha = \beta = \beta' = \lambda = \lambda' = \mu = \mu' = 0 \ , \tag{1}$$

so that its equations of motion read (see (4.2-1))

$$\ddot{\vec{r}}_n = \omega \hat{k} \wedge \dot{\vec{r}}_n + 2 \sum_{m=1, m \neq n}^{N} \left[\ \dot{\vec{r}}_n (\dot{\vec{r}}_m \cdot \vec{r}_{nm}) + \dot{\vec{r}}_m (\dot{\vec{r}}_n \cdot \vec{r}_{nm}) - \vec{r}_{nm} (\dot{\vec{r}}_n \cdot \dot{\vec{r}}_m) \ \right] / \ r_{nm}^2 , \ (\vec{r}_{nm} \equiv \vec{r}_n - \vec{r}_m) , \tag{2a}$$

or equivalently (see (4.2.1-1)),

$$\ddot{z}_n = i \omega \dot{z}_n + 2 \sum_{m=1, m \neq n}^{N} \dot{z}_n \dot{z}_m / (z_n - z_m) \ . \tag{2b}$$

The solution of the initial-value problem for these equations of motion, (2b), is given by the following neat recipe: *the* N *(complex) coordinates* $z_n(t)$ *are the* N *roots of the following algebraic equation in* z *:*

$$\sum_{m=1}^{N} \dot{z}_m(0) / [z - z_m(0)] = i \omega / [\exp(i \omega t) - 1] \ . \tag{3}$$

Exercise 4.2.5-1. Prove this result in more than one way. *Hints*: see (2.3.4.2-20); apply the transformation (2.1.13-2,4,5) to (4.2.4-2, 3).

Since the right hand side of (3) is periodic in t with period

$$T = 2\pi / \omega \quad , \tag{4}$$

clearly the set of roots of (3) is also periodic with the same period. Hence all solutions of (2) are completely periodic, with period (at most) $\tilde{T} = T \cdot N!$, where the factor $N!$ accounts for the possibility that the one-to-one correspondence between the N positions $z_n(t)$ of the particles and the N roots of (3) get permuted through the motion.

Exercise 4.2.5-2. Try and understand this phenomenon, namely: granted that all motions are completely periodic with period (at most!) $\tilde{T} = T \cdot N!$, are there motions periodic with periods $\tilde{T} = T \cdot M$, where the positive integer M is larger than unity (but of course smaller than $N!$, $M < N!$, and such that $N!$ is a multiple of M); and if there are such motions, how are they separated, and how are they identified in terms of their initial data? *Hint*: try some numerical experiments (and see below *Exercises 4.2.5-3* and *Exercises 4.2.5-4*, as well as Sect. 4.5).

Exercise 4.2.5-3. Obtain (and discuss) the general solution of (2a) with $N = 2$. *Hint*: solve (3) (which, for $N = 2$, is a second-degree algebraic equation), and use (4.1-17).

Exercise 4.2.5-4. Obtain (and discuss) the general solution of (2a) with $N = 3$. *Hint*: as for the preceding *Exercise 4.2.5-3*.

The equations of motion (2b) can also be obtained (as (4.2.4-2b)) from a Hamiltonian. Indeed the Hamiltonian

$$H(\underline{z}, \underline{\zeta}) = i(\omega / a) \sum_{n=1}^{N} z_n + \sum_{n=1}^{N} \exp(a\zeta_n) \left[\prod_{m=1, m \neq n}^{N} (z_n - z_m) \right]^{-1} \quad , \tag{5}$$

yields the Hamiltonian equations of motion

$$\dot{z}_n = a \exp(a\zeta_n) \left[\prod_{m=1, m \neq n}^{N} (z_n - z_m) \right]^{-1} \quad , \tag{6a}$$

$$\dot{\zeta}_n = i\,\omega / a + a^{-1} \sum_{m=1, m \neq n}^{N} (\dot{z}_n + \dot{z}_m) / (z_n - z_m) \quad , \tag{6b}$$

and logarithmic differentiation of (6a) yields, via (6b), precisely (2b).

Exercise 4.2.5-5. Verify !

As in the preceding Sect. 4.2.4, two real Hamiltonians can now be obtained from (5) by using the procedure described in the last part of Sect. 4.1, each of which yields the real, two-dimensional, equations of motion (2a) (rather than the complex, one-dimensional, equations of motion (2b)). For $N=2$ clearly they coincide with (4.2.4-15) respectively (4.2.4-16), except for the addition of a term $\hat{k}\cdot\vec{a}\wedge(\vec{r}_1+\vec{r}_2)/a^2$ respectively $-\vec{a}\cdot(\vec{r}_1+\vec{r}_2)/a^2$.

Actually in this case it is also of some interest to display the two real Hamiltonians that yield (2a) for $N=1$. They read

$$H(\vec{r},\vec{p})=\exp(\vec{a}\cdot\vec{p})\cos(\hat{k}\cdot\vec{a}\wedge\vec{p})+(\omega/a^2)(\hat{k}\cdot\vec{a}\wedge\vec{r}) \ , \tag{7a}$$

respectively

$$\tilde{H}(\vec{r},\vec{\tilde{p}})=\exp(\hat{k}\cdot\vec{a}\wedge\vec{\tilde{p}})\sin(\vec{a}\cdot\vec{\tilde{p}})-(\omega/a^2)(\vec{a}\cdot\vec{r}) \ , \tag{8a}$$

and they yield the Hamiltonian equations

$$\dot{\vec{r}}=\exp(\vec{a}\cdot\vec{p})\left[\vec{a}\cos(\hat{k}\cdot\vec{a}\wedge\vec{p})-\hat{k}\wedge\vec{a}\sin(\hat{k}\cdot\vec{a}\wedge\vec{p})\right] \ , \tag{7b}$$

$$\dot{\vec{p}}=-(\omega/a^2)\,\hat{k}\wedge\vec{a} \ , \tag{7c}$$

respectively

$$\dot{\vec{r}}=\exp(\hat{k}\cdot\vec{a}\wedge\vec{\tilde{p}})\left[\vec{a}\cos(\vec{a}\cdot\vec{\tilde{p}})+\hat{k}\wedge\vec{a}\sin(\vec{a}\cdot\vec{\tilde{p}})\right] \ , \tag{8b}$$

$$\dot{\vec{\tilde{p}}}=(\omega/a^2)\,\vec{a} \ . \tag{8c}$$

Both these Hamiltonian equations of motion, (7b,c) and (8b,c), yield the same Newtonian equation,

$$\ddot{\vec{r}}=\omega\,\hat{k}\wedge\dot{\vec{r}} \ , \tag{9}$$

whose general solution reads

$$\vec{r}(t)=\vec{r}(0)+\dot{\vec{r}}(0)\,\omega^{-1}\sin(\omega t)+\hat{k}\wedge\dot{\vec{r}}(0)\,\omega^{-1}\left[1-\cos(\omega t)\right] \ . \tag{10}$$

Note that both Hamiltonians, (7a) and (8a), and both sets of Hamiltonian equations, (7b,c) and (8b,c), depend on the 2-vector $\vec{a}$ hence lack rotation-invariance ($\vec{a}$ provides a preferred direction), while the Newtonian equation of motion (9) is independent of $\vec{a}$ and *rotation-invariant.* Moreover, (9) is *invariant under translations* ($\vec{r} \rightarrow \vec{r} + \vec{r}_0$, $\dot{\vec{r}}_0 = 0$), again in contrast to (7a) and (8a) (but not to (7b,c) and (8b,c)). And (9) has a clear physical interpretation: it is the equation of motion of a charged point particle moving in the plane, under the influence of a constant magnetic field orthogonal to the plane ("cyclotron"): indeed the right hand side of (9) corresponds then to the Lorentz force, and ω to the "Larmor" circular frequency.

Exercise 4.2.4-6. Verify these formulas, from (7) to (10).

Since the many-body system in the plane characterized by the Newtonian equations of motion (2a) is Hamiltonian, and all its trajectories are *completely periodic,* it is natural to conjecture that there exist corresponding quantum systems whose spectrum is *equispaced* (perhaps up to a continuous quantum number, corresponding to the translation-invariant character of the equations of motion (2a)).

As already hinted at above (under *Exercise 4.2.5-2*), additional insight about the behavior of this *integrable* indeed *solvable* system, (2), is provided in Sect. 4.5, where a *nonintegrable* generalization of it is discussed.

4.2.6 First-order evolution equations, and a partially solvable many-body problem with velocity-independent forces, in the plane

The results discussed in the preceding subsections of Sect. 4.2 are all based on the findings of Sect. 2.3.4.2. But the transition from one-dimensional to two-dimensional space via complexification, see Sect. 4.1, can be applied as well to certain results of Sect. 2.3.4.1. In particular it is easily seen that, by complexification and via (4.1-17), the *first-order* evolution equations (2.3.4.1-38) read

$$\dot{\vec{r}}_n = (\alpha + \alpha' \hat{k}\wedge)\vec{r}_n + (\gamma + \gamma' \hat{k}\wedge) \sum_{m=1, m\neq n}^{N} \left\{ \vec{r}_n \left[r_n^2 - 2(\vec{r}_n \cdot \vec{r}_m) \right] + \vec{r}_m r_n^2 \right\} / r_{nm}^2 \ . \qquad (1)$$

These evolution equations follow readily from (2.3.4.1-38), after we set in it

$$a = \alpha + i\alpha' , \quad c = \gamma + i\gamma' \ . \qquad (2)$$

Hence the solutions of (1) are given, via (4.1-17), by the N zeros of the time-dependent polynomial, of degree N in z,

$$\psi(z,t) = z^N + \sum_{m=1}^{N} c_m(t)\, z^{N-m} \quad , \tag{3}$$

where

$$c_m(t) = c_m(0)\exp\{m\,[(\gamma + i\gamma')(2N-m-1)/2 + (\alpha + i\alpha')]\,t\} \quad , \tag{4}$$

(see (2.3.4.1-41)). Here of course the quantities $c_m(0)$ are given, in terms of the initial data $z_n(0)$, by the polynomial equality

$$z^N + \sum_{m=1}^{N} c_m(0)\, z^{N-m} = \prod_{n=1}^{N} [z - z_n(0)] \quad , \tag{5}$$

entailing (4.2.1-7). Note that the motion in the plane of all the coordinates $\vec{r}_n(t)$ is always (even as $t \to \pm\infty$) confined to a finite region of the plane if

$$\alpha = \gamma = 0 \quad , \tag{6a}$$

and it is moreover completely periodic, with a period independent of the initial data, if there holds the additional condition

$$2\,\alpha'/\gamma' = p/q \tag{6b}$$

with p and q integers.

Likewise, from (2.3.4.1-44) we obtain the N-body problem in the plane characterized by the Newtonian equations of motion

$$\ddot{\vec{r}}_n = (\lambda + \lambda'\hat{k}\wedge)\vec{r}_n$$

$$+(\nu + \nu'\hat{k}\wedge) \sum_{m=1, m\neq n}^{N} \{\, \vec{r}_n[\, r_n^2 - 2(\vec{r}_n \cdot \vec{r}_m)\,] + \vec{r}_m r_n^2 \,\}/r_{nm}^2$$

$$-2(\mu + \mu'\hat{k}\wedge) \sum_{m=1, m\neq n}^{N} \{\, \vec{r}_n r_m^2 [\, 3r_n^2 r_m^2 - 2r_m^2(\vec{r}_n \cdot \vec{r}_m) - r_n^4 \,]$$

$$-\vec{r}_m r_n^2 [\, 3r_n^2 r_m^2 - 2r_n^2(\vec{r}_n \cdot \vec{r}_m) - r_m^2 \,]\}/\, r_{nm}^6$$

$$+(\mu+\mu'\hat{k}\wedge)\sum_{\ell,m=1;\ell\neq m,\ell\neq n,m\neq n}^{N}\{\ \vec{r}_n[\ 4(\vec{r}_n\cdot\vec{r}_\ell)(\vec{r}_n\cdot\vec{r}_m)$$

$$+r_n^2[\ r_n^2-(\vec{r}_\ell\cdot\vec{r}_m)-2(\vec{r}_n\cdot\vec{r}_\ell)-2(\vec{r}_n\cdot\vec{r}_m)\ ]$$

$$+\vec{r}_\ell\, r_n^2[\ r_n^2-(\vec{r}_n\cdot\vec{r}_m)\ ]+\vec{r}_m\, r_n^2[\ r_n^2-(\vec{r}_n\cdot\vec{r}_\ell)\ ]\ \}/(r_{n\ell}\, r_{nm})^2\ , \tag{7}$$

where the 6 constants $\lambda,\lambda',\mu,\mu',\nu,\nu'$ are given, in terms of the 4 (*a priori* arbitrary) constants $\alpha,\alpha',\gamma,\gamma'$, by the following rules:

$$\lambda=\alpha^2-\alpha'^2,\ \ \lambda'=2\,\alpha\,\alpha'\ , \tag{8a}$$

$$\mu=\gamma^2-\gamma'^2,\ \ \mu'=2\,\gamma\gamma'\ , \tag{8b}$$

$$\nu=-\gamma^2+\gamma'^2-2(\alpha\gamma-\alpha'\gamma'),\ \ \nu'=2(\alpha\gamma'+\alpha'\gamma-\gamma\gamma')\ . \tag{8c}$$

Note that (the right hand side of) (7) features one-, two-, and three-body *velocity-independent* forces. This N-body problem, (7) with (8), is *partially solvable*; indeed any solution of (1) is also a solution of (7) (since (7) is obtained, by t-differentiation, from (1): see (2.3.4.1-44) and (2.3.4.1-38)). Hence the partially solvable character of this N-body problem, (7) with (8), entails, in the context of the initial-value problem, the freedom to assign the initial positions, $\vec{r}_n(0)$, of the N particles in the plane, but the restriction to then choose the N initial velocities $\dot{\vec{r}}_n(0)$ so that (1) hold (at $t=0$). These restrictions (as well as (8)) must be enforced in order that the technique of solution described above, see $(3\div 5)$, apply: then the solutions not only satisfy (7) with (8), they satisfy as well (1), for all time. Of course the N-body problem (7) is well-defined more generally, namely even when these restrictions do not hold.

Note that the conditions (6a) entail, via (8),

$$\lambda'=\mu'=\nu'=0\ , \tag{9a}$$

$$\lambda=-\alpha'^2\leq 0\ ,\ \ \mu=-\gamma'^2\leq 0\ ,\ \ \nu=\gamma'^2+2\alpha'\gamma'\ . \tag{9b}$$

The derivations of (1) from (2.3.4.1-38), and specially of (7) from (2.3.4.1-44), is facilitated by the following identities (see (4.1-17)):

$$z_n^2\,(z_n^*-z_m^*)\doteq\vec{r}_n[\ r_n^2-2(\vec{r}_n\cdot\vec{r}_m)\ ]+\vec{r}_m\, r_n^2\ , \tag{10a}$$

$$z_n^2\, z_m^2\, (z_n^* - z_m^*)^3 \doteq \vec{r}_n\, r_m^2 [\, 3 r_n^2\, r_m^2 - 2 r_m^2 (\vec{r}_n \cdot \vec{r}_m) - r_n^4 \,]$$

$$- \vec{r}_m\, r_n^2 [\, 3 r_n^2\, r_m^2 - 2 r_n^2 (\vec{r}_n \cdot \vec{r}_m) - r_m^2 \,] \;, \qquad (10b)$$

$$z_n^3\, (z_n^* - z_\ell^*)\, (z_n^* - z_m^*)$$

$$\doteq \vec{r}_n \{\, 4 (\vec{r}_n \cdot \vec{r}_\ell)(\vec{r}_n \cdot \vec{r}_m) + r_n^2 [\, r_n^2 - (\vec{r}_\ell \cdot \vec{r}_m) - 2\, (\vec{r}_n \cdot \vec{r}_\ell) - 2\, (\vec{r}_n \cdot \vec{r}_m) \,] \}$$

$$+ \vec{r}_\ell\, r_n^2 [\, r_n^2 - (\vec{r}_n \cdot \vec{r}_m) \,] + \vec{r}_m\, r_n^2 [\, r_n^2 - (\vec{r}_n \cdot \vec{r}_\ell) \,] \;. \qquad (10c)$$

Exercise 4.2.6-1. Prove these identities. *Hint*: see (4.1-16,17).

4.3 Examples: other families of solvable many-body problems in the plane

In Sect. 4.3, and in its subsections, we manufacture three families of *solvable* many-body problems in the plane, and we indicate how they can be viewed as the first specimens of an infinite hierarchy of such models. The methodology we follow is to manufacture firstly appropriate *solvable* (one-dimensional) many-body models *on the line*, and then to reinterpret them, via the complexification trick (see Sect. 4.1), as *solvable* (two-dimensional) rotation-invariant many-body problems *in the plane*. These models, of which the Newtonian equations of motion are displayed below (in Sect. 4.3) and are derived in the subsequent subsections, are presented here mainly to illustrate the kind of methods and tricks that allow to identify them as *solvable* models.

The first model obtained in this manner is characterized by the following Newtonian equations of motion:

$$\ddot{\vec{r}}_n = [\, 2\, \dot{\vec{r}}_n (\dot{\vec{r}}_n \cdot \vec{r}_n) - \vec{r}_n (\dot{\vec{r}}_n \cdot \dot{\vec{r}}_n) \,] / r_n^2 + (\alpha_n + \alpha'_n\, \hat{k} \wedge) \vec{r}_n$$

$$+ \sum_{m=1}^{N} \{ (\beta_{nm} + \beta'_{nm}\, \hat{k} \wedge) [\, \dot{\vec{r}}_m (\vec{r}_n \cdot \vec{r}_m) + \vec{r}_n (\dot{\vec{r}}_m \cdot \vec{r}_m) - \vec{r}_m (\dot{\vec{r}}_m \cdot \vec{r}_n) \,] / r_m^2 \} \;, \qquad (1)$$

which features the $2N(N+1)$ arbitrary (real) “coupling” constants $\alpha_n, \alpha'_n, \beta_{nm}, \beta'_{nm}$. This abundance of arbitrary coupling constants justify our mention of a *family* of models. Note that these Newtonian equations of motion (“acceleration equal force”) are of course invariant under plane rotations; they are also invariant under rescaling of the particle coordinates ($\vec{r}_n(t) \to \tilde{\vec{r}}_n(t) = c\,\vec{r}_n(t)$, $\dot{c}=0$), while they are *not* translation-invariant. They feature only one- and two-body forces. The former are of two (or rather three) kinds: a contribution that is quadratic in the velocity, and depends nonlinearly on the coordinate, of the moving particle, and an additional velocity-independent term linear in the particle coordinate (actually a one-body term linear in the velocity of the moving particle is also entailed by the term in the sum with $m=n$). The latter depend linearly on the particle velocities and nonlinearly on the particle coordinates.

A *translation-invariant* generalization, involving two types of particles, reads

$$\ddot{\vec{r}}_n^{(\pm)} = \pm\left[2\,\dot{\vec{r}}_n(\dot{\vec{r}}_n\cdot\vec{r}_n) - \vec{r}_n(\dot{\vec{r}}_n\cdot\dot{\vec{r}}_n)\right]/r_n^2 \pm (\alpha_n + \alpha'_n\,\hat{k}\wedge)\vec{r}_n$$

$$\pm\sum_{m=1}^{N}(\beta_{nm} + \beta'_{nm}\,\hat{k}\wedge)\left[\dot{\vec{r}}_m(\vec{r}_n\cdot\vec{r}_m) + \vec{r}_n(\dot{\vec{r}}_m\cdot\vec{r}_m) - \vec{r}_m(\dot{\vec{r}}_m\cdot\vec{r}_n)\right]/r_m^2$$

$$+\sum_{m=1}^{N}(\gamma_{nm} + \gamma'_{nm}\,\hat{k}\wedge)\dot{\vec{R}}_m \quad , \qquad \text{(2a)}$$

with, in the right-hand-side,

$$\vec{r}_n \equiv \left[\vec{r}_n^{(+)} - \vec{r}_n^{(-)}\right]/2 \quad , \qquad \text{(2b)}$$

$$\vec{R}_n \equiv \left[\vec{r}_n^{(+)} + \vec{r}_n^{(-)}\right]/2 \quad . \qquad \text{(2c)}$$

These $2N$ equations of motion for the $2N$ particle positions $\vec{r}_n^{(+)}(t)$, $\vec{r}_n^{(-)}(t)$ feature the $2N(2N+1)$ arbitrary (real) coupling constants $\alpha_n, \alpha'_n, \beta_{nm}, \beta'_{nm}, \gamma_{nm}, \gamma'_{nm}$. Note however the presence of 4-body forces, as demonstrated by the presence in the right hand side of (2a) of terms that depend on the coordinates of 4 different particles, say $\vec{r}_n^{(+)}, \vec{r}_n^{(-)}, \vec{r}_m^{(+)}, \vec{r}_m^{(-)}$.

The equations of motion of the third model read

$$\ddot{\vec{r}}_n^{(\pm)} = \vec{u}(\dot{\vec{r}}_n^{(\pm)}, \dot{\vec{r}}_n^{(\pm)}; \vec{r}_n^{(\pm)})$$

$$\pm\left[\mu(\vec{r}_n^{(+)}, \vec{r}_n^{(-)}) - \mu'(\vec{r}_n^{(+)}, \vec{r}_n^{(-)})\,\hat{k}\wedge\right]$$

$$\cdot\left[\vec{u}(\dot{\vec{r}}_n^{(\pm)}, \dot{\vec{r}}_n^{(\pm)}; \vec{r}_n^{(\pm)}) + \vec{v}(\dot{\vec{r}}_n^{(\pm)}; \dot{\vec{r}}_n^{(\mp)}; \vec{r}_n^{(\mp)}) - 2\vec{u}(\dot{\vec{r}}_n^{(+)}, \dot{\vec{r}}_n^{(-)}; \vec{r}_n^{(\mp)})\right]$$

$$\pm\{\alpha_n \lambda(\vec{r}_n^{(+)}, \vec{r}_n^{(-)}) - \alpha'_n \lambda'(\vec{r}_n^{(+)}, \vec{r}_n^{(-)}) + [\alpha_n \lambda'(\vec{r}_n^{(+)}, \vec{r}_n^{(-)}) + \alpha'_n \lambda(\vec{r}_n^{(+)}, \vec{r}_n^{(-)})]\hat{k}\wedge\}\vec{r}_n^{(\pm)}$$

$$\pm\sum_{m=1}^{N}\{\beta_{nm}\nu(\vec{r}_n^{(+)}; \vec{r}_n^{(-)}; \vec{r}_m^{(+)}; \vec{r}_m^{(-)}) - \beta'_{nm}\nu'(\vec{r}_n^{(+)}; \vec{r}_n^{(-)}; \vec{r}_m^{(+)}; \vec{r}_m^{(-)})$$

$$+[\beta_{nm}\nu'(\vec{r}_n^{(+)}; \vec{r}_n^{(-)}; \vec{r}_m^{(+)}; \vec{r}_m^{(-)}) + \beta'_{nm}\nu(\vec{r}_n^{(+)}; \vec{r}_n^{(-)}; \vec{r}_m^{(+)}; \vec{r}_m^{(-)})]\hat{k}\wedge\}\cdot$$

$$\cdot[\vec{u}(\dot{\vec{r}}_n^{(\pm)}, \vec{r}_m^{(+)}; \vec{r}_m^{(+)}) - \vec{u}(\vec{r}_n^{(\pm)}; \dot{\vec{r}}_m^{(-)}; \vec{r}_m^{(-)})]$$

$$+\sum_{m=1}^{N}(\gamma_{nm} + \gamma'_{nm}\hat{k}\wedge)[\vec{u}(\vec{r}_n^{(\pm)}, \dot{\vec{r}}_m^{(+)}; \vec{r}_m^{(+)}) + \vec{u}(\vec{r}_n^{(\pm)}, \dot{\vec{r}}_m^{(-)}; \vec{r}_m^{(-)})] \quad , \tag{3a}$$

$$\vec{u}(\vec{r}_1, \vec{r}_2; \vec{r}_3) \equiv [\vec{r}_1(\vec{r}_2\cdot\vec{r}_3) + \vec{r}_2(\vec{r}_1\cdot\vec{r}_3) - \vec{r}_3(\vec{r}_1\cdot\vec{r}_2)]/r_3^2 \quad , \tag{3b}$$

$$\vec{v}(\vec{r}_1, \vec{r}_2; \vec{r}_3) \equiv \{\vec{r}_1[2(\vec{r}_2\cdot\vec{r}_3)^2 - r_2^2 r_3^2] + 2\vec{r}_2(\vec{r}_1\cdot\vec{r}_3)(\vec{r}_2\cdot\vec{r}_3) - 2\vec{r}_3(\vec{r}_1\cdot\vec{r}_2)(\vec{r}_2\cdot\vec{r}_3)\}/r_3^4 \; , \tag{3c}$$

$$\lambda(\vec{r}_1, \vec{r}_2) = -\lambda(\vec{r}_2, \vec{r}_1) \equiv \log(r_1/r_2) \quad , \tag{3d}$$

$$\lambda'(\vec{r}_1, \vec{r}_2) = -\lambda'(\vec{r}_2, \vec{r}_1) \equiv \theta_1 - \theta_2 \quad , \tag{3e}$$

$$\mu(\vec{r}_1, \vec{r}_2) = -\mu(\vec{r}_2, \vec{r}_1) \equiv \log(r_1/r_2)/[\log^2(r_1/r_2) + (\theta_1 - \theta_2)^2] \; , \tag{3f}$$

$$\mu'(\vec{r}_1, \vec{r}_2) = -\mu'(\vec{r}_2, \vec{r}_1) \equiv (\theta_1 - \theta_2)/[\log^2(r_1/r_2) + (\theta_1 - \theta_2)^2] \; , \tag{3g}$$

$$\nu(\vec{r}_1; \vec{r}_2; \vec{r}_3; \vec{r}_4) = -\nu(\vec{r}_2; \vec{r}_1; \vec{r}_3; \vec{r}_4) = -\nu(\vec{r}_1; \vec{r}_2; \vec{r}_4; \vec{r}_3) = \nu(\vec{r}_2; \vec{r}_1; \vec{r}_4; \vec{r}_3)$$

$$\equiv [\log(r_1/r_2)\log(r_3/r_4) + (\theta_1 - \theta_2)(\theta_3 - \theta_4)]/[\log^2(r_3/r_4) + (\theta_3 - \theta_4)^2] \; , \tag{3h}$$

$$\nu'(\vec{r}_1; \vec{r}_2; \vec{r}_3; \vec{r}_4) = -\nu'(\vec{r}_2; \vec{r}_1; \vec{r}_3; \vec{r}_4) = -\nu'(\vec{r}_1; \vec{r}_2; \vec{r}_4; \vec{r}_3) = \nu'(\vec{r}_2; \vec{r}_1; \vec{r}_4; \vec{r}_3)$$

$$\equiv [\log(r_1/r_2)(\theta_3 - \theta_4) - \log(r_3/r_4)(\theta_1 - \theta_2)]/[\log^2(r_3/r_4) + (\theta_3 - \theta_4)^2] \; . \tag{3i}$$

Note that, in (3d÷i), we found convenient to use the circular coordinates r, θ, see (4.1-4).

4.3.1 A rescaling-invariant solvable one-dimensional many-body problem

In Sect. 4.3.1 we manufacture the scale-invariant solvable one-dimensional many-body problem that corresponds to (4.3-1) via the complexification trick of Sect. 4.1.

We take as starting point the system of N linear ODEs

$$\dot{g}_n = a_n + \sum_{m=1}^{N} b_{nm}\, g_m \quad , \tag{1}$$

which feature in their right-hand-sides the N arbitrary constants a_n and the N^2 arbitrary constants b_{nm}.

We then set

$$g_n(t) = \dot{z}_n(t) / z_n(t) \quad , \tag{2a}$$

entailing

$$\dot{g}_n(t) = \ddot{z}_n(t)/z_n(t) - [\dot{z}_n(t)/z_n(t)]^2 \quad , \tag{2b}$$

and we thereby get from (1) the system of second-order ODEs

$$\ddot{z}_n = \dot{z}_n^2 / z_n + a_n\, z_n + z_n \sum_{m=1}^{N} [b_{nm}\, \dot{z}_m / z_m] \ . \tag{3}$$

These equations are clearly scale-invariant, and it is easily seen that, via the complexification trick, see (4.1-17), they yield (4.3-1), with the identification

$$a_n = \alpha_n + i\alpha'_n , \quad b_{nm} = \beta_{nm} + i\beta'_{nm} \ . \tag{4}$$

Exercise 4.3.1-1. Verify !

On the other hand these equations, (3), are easily solved by algebraic operations. Indeed (1) entails

$$g_n(t) = \sum_{m=1}^{N} [\{c_m(0)\exp(\lambda^{(m)} t) + (\underline{u}^{(m)}, \underline{a})[\exp(\lambda^{(m)} t) - 1]/\lambda^{(m)}\} v_n^{(m)}] \ , \tag{5}$$

where the N numbers $\lambda^{(m)}$ are the N eigenvalues of the $(N \times N)$-matrix $\underline{B}$ having matrix elements b_{nm},

$$\underline{B}\,\underline{v}^{(m)} = \lambda^{(m)}\,\underline{v}^{(m)} \quad , \tag{6}$$

the quantities $v_n^{(m)}$, $n = 1,\ldots,N$, are the N components of the eigenvectors of $\underline{B}$, see (6), the N N-vectors $\underline{u}^{(m)}$ are orthonormal to the N N-vectors $\underline{v}^{(m)}$,

$$(\underline{u}^{(n)}, \underline{v}^{(m)}) \equiv \sum_{\ell=1}^{N} u_\ell^{(n)}\, v_\ell^{(m)} = \delta_{nm} \quad , \tag{7}$$

of course

$$(\underline{u}^{(m)}, \underline{a}) \equiv \sum_{n=1}^{N} u_n^{(m)}\, a_n \quad , \tag{8}$$

and the N constants $c_m(0)$ are defined by (5) at $t = 0$:

$$g_n(0) = \sum_{m=1}^{N} c_m(0)\, v_n^{(m)} \quad , \tag{9a}$$

hence (see (7))

$$c_m(0) = \sum_{n=0}^{N} u_n^{(m)}\, g_n(0) \quad . \tag{9b}$$

Proofs. Let us rewrite (1) in N-vector form:

$$\dot{\underline{g}}(t) = \underline{a} + \underline{B}\,\underline{g} \quad . \tag{10}$$

Then set

$$\underline{g}(t) = \sum_{n=1}^{N} c_n(t)\, \underline{v}^{(n)} \tag{11}$$

so that, via (6),

$$\sum_{n=1}^{N} \dot{c}_n(t)\, \underline{v}^{(n)} = \underline{a} + \sum_{n=1}^{N} \lambda^{(n)}\, c_n(t)\, \underline{v}^{(n)} \quad , \tag{12}$$

hence, via (7)

$$\dot{c}_n(t) = (\underline{u}^{(n)}, \underline{a}) + \lambda^{(n)} c_n(t) \quad , \tag{13}$$

hence

$$c_n(t) = c_n(0) \exp(\lambda^{(n)} t) + (\underline{u}^{(n)}, \underline{a}) \left[\exp(\lambda^{(n)} t) - 1\right] / \lambda^{(n)} \tag{14}$$

Insertion of this formula, (14), into (11), yields, componentwise, (5), which is thereby proven.

As for the derivation of (9b) from (9a), it is plain, via (7).

Note however that we have implicitly assumed that the (constant) $(N \times N)$-matrix $\underline{B}$ is diagonalizable,

$$(\underline{u}^{(m)}, \underline{B}\, \underline{v}^{(n)}) = \delta_{nm} \lambda^{(n)} \quad ,$$

and that it possesses N distinct (possibly complex) eigenvalues ($\lambda^{(n)} \neq \lambda^{(m)}$ if $n \neq m$).

Exercise 4.3.1-2. Discuss how the solution formulas written above get modified when these assumptions cease to hold.

The next step to solve (3) is via the integration of (2a) with (5). This clearly yields

$$z_n(t) = z_n(0) \exp\{\sum_{m=1}^{N} [\{c_m(0)[\exp(\lambda^{(m)} t) - 1] / \lambda^{(m)}$$

$$+ (\underline{u}^{(m)}, \underline{a}) [\exp(\lambda^{(m)} t) - 1 - \lambda^{(m)} t] / [\lambda^{(m)}]^2 \} v_n^{(m)}]\} \quad , \tag{15a}$$

with

$$c_m(0) = \sum_{n=0}^{N} u_n^{(m)} \dot{z}_n(0) / z_n(0) \tag{15b}$$

(see (9b) and (2a)).

The time-evolution of these coordinates $z_n(t)$ in the complex plane corresponds directly, via (4.1-17), to the motion of the particles $\vec{r}_n(t)$ that satisfy the Newtonian equations of motion (4.3-1) in the physical plane. While a detailed analysis of these motions is left as an instructive exercise for the diligent reader, we tersely outline here some of its most interesting features, focussing on the case treated above, characterized by a diagonalizable (generally complex) $(N \times N)$-matrix $\underline{B}$.

Clearly the most important elements that characterize the behavior of the system (4.3-1) are (see (15a)) the eigenvalues $\lambda^{(m)}$ of the $(N \times N)$-matrix $\underline{B}$, whose elements are the two-body coupling constants,

$$(\underline{B})_{nm} = \beta_{nm} + i\,\beta'_{nm} \quad , \tag{16}$$

as well as the values of the one-body coupling constants $a_n = \alpha_n + i\alpha'_n$. For instance, if all the eigenvalues $\lambda^{(m)}$ are imaginary,

$$\lambda^{(m)*} = -\lambda^{(m)} \neq 0, \quad m = 1, \ldots, N \quad , \tag{17}$$

and moreover the N constants

$$\mu_n = \sum_{m=1}^{N} (\underline{u}^{(m)}, \underline{a})\, v_n^{(m)} / \lambda^{(m)} \tag{18}$$

are also all imaginary (or vanishing)

$$\mu_n^* = -\mu_n \quad , \tag{19}$$

then clearly (see (15)) *all* solutions of the many-body system (4.3-1) remain confined to a finite region of the plane for *all* time, including the asymptotic limits $t \to \pm\infty$. The behavior of such a system depends quite sensitively on the initial data (which enter exponentially in the solution, see (15)). Note that sufficient (but not necessary) conditions to guarantee (17) are the following (symmetry and antisymmetry) properties of the two-body coupling constants:

$$\beta_{nm} = -\beta_{mn}, \ \beta'_{nm} = \beta'_{mn} \quad , \tag{20a}$$

as well as

$$\det[\underline{B}] \neq 0 \tag{20b}$$

(indeed (20a) guarantees that the $(N \times N)$-matrix $\underline{B}$ be antihermitian, namely that $i\underline{B}$ be hermitian, while (20b) excludes that any one of the eigenvalues $\lambda^{(m)}$ of $\underline{B}$ vanish), while the simpler way to satisfy (19) is to assume, see (18), that all the one-body coupling constants vanish,

$$\alpha_n = \alpha_n' = 0 \ . \tag{21}$$

Likewise, if any one of the eigenvalues $\lambda^{(m)}$ is imaginary, say

$$\lambda^{(k)*} = -\lambda^{(k)} \neq 0 \tag{22a}$$

(without any restriction on the other eigenvalues, $\lambda^{(m)}$ with $m \neq k$) and the corresponding value, $\mu^{(k)}$, of the constants $\mu^{(m)}$, see (18), is congruent to $\lambda^{(k)}$,

$$\mu^{(k)} = (p/q)\ \lambda^{(k)} \tag{22b}$$

with p and q both (arbitrary!) integers ($q \neq 0; p$ might also vanish), then clearly (see (15)) the many-body system (4.3-1) possesses a periodic solution (with period $T = 2\pi q / \lambda^{(k)}$ if $p \neq 0$, $T = 2\pi / \lambda^{(k)}$ if $p = 0$), characterized by initial data, see (15b), such that $c_m(0) = 0$ for $m \neq k$.

Exercise 4.3.1-3. Show that *all* solutions of the many-body problem in the plane (4.3-1), with all one-body coupling constants vanishing, see (21), and with the following values of the two-body coupling constants,

$$\beta_{nm} = c(1-\delta_{nm}) \sin[\,2\pi(m-n)/N\,] / \{1-\cos[2\pi(m-n)/N\,]\} \ , \tag{23a}$$

$$\beta'_{nm} = c(1+N\delta_{nm}) \ , \tag{23b}$$

where c is an arbitrary (real, nonvanishing) constant, are *completely periodic*, with period $T = \pi / c$; and construct other many-body problems of type (4.3-1) that also feature *only completely periodic motions* of a given period T. *Hint*: see Sect. 2.4.5.1.

Another interesting instance of the system (4.3-1) is characterized by the vanishing of the one-body coupling constants, see (21), and by appropriate restrictions on the two-body coupling constants, such as to guarantee that all the eigenvalues of the matrix $\underline{B}$, see (16), have negative real parts,

$$\mathrm{Re}[\ \lambda^{(m)}\ ] < 0, \ \ m = 1,\ldots,N \ . \tag{24}$$

Then clearly (see (15)) all solutions of the many-problem (4.3-1) remain confined as $t \to \infty$ and tend to stand-still configurations,

$$z_n(\infty) = z_n(0)\,\exp\{-\sum_{m=1}^{N}\,[\,c_m(0)\,v_n^{(m)}\,/\,\lambda^{(m)}\,]\}\ , \tag{25}$$

while the velocities of all particles vanish asymptotically (exponentially fast) as $t \to \infty$.

If instead one or more of the eigenvalues $\lambda^{(m)}$ of $\underline{B}$ have positive real parts, the behavior of the system (4.3-1) in the remote future depends dramatically on the initial conditions: as $t \to \infty$, one or more particles may shoot off to infinity doubly-exponentially fast, while others may converge to the origin (doubly exponentially fast: this entails no divergence of the forces, since the corresponding velocities also vanish, even faster, as $t \to \infty$: see (3) and (15)).

4.3.2 A rescaling- and translation-invariant solvable one-dimensional many-body problem

In the preceding Sect. 4.3.1 we have manufactured, and tersely analyzed, the solvable one-dimensional many-body problem, see (4.3.1-3), which by complexification yields the first many-body problem in the plane reported in Sect. 4.3, see (4.3-1). As noted above, these models, (4.3.1-3) and (4.3-1), are not invariant under translations. In Sect. 4.3.2, via a simple trick which involves a doubling of the number of particles, we manufacture a translation-invariant extension of (4.3.3-3), which involves the introduction of two different types of particles, and whose two-dimensional version, obtained by complexification, coincides with the second model reported in Sect. 4.3, see (4.3-2).

The trick is to introduce, in parallel to (4.3.1-3), the set of ODEs

$$\ddot{Z}_n = \sum_{m=1}^{N} c_{nm}\,\dot{Z}_m\ \ . \tag{1}$$

Here the N^2 constants c_{nm} are generally complex,

$$c_{nm} = \gamma_{nm} + i\gamma'_{nm}\ \ . \tag{2}$$

Note that these equations of motion are translation-invariant, and that they are, rather trivially, solvable:

$$Z_n(t) = Z_n(0) + \sum_{m=1}^{N}\,[\,\dot{f}_m(0)\{\,\exp[\,\eta^{(m)}\,t\,] - 1\,\}/\eta^{(m)}\,]\,v_n^{(m)}\ , \tag{3a}$$

$$\dot{f}_m(0)=\sum_{n=1}^{N} u_n^{(m)}\,\dot{Z}_n(0) \quad , \tag{3b}$$

where $\underline{u}^{(m)}$ respectively $\underline{v}^{(m)}$ are the left- respectively right-eigenvectors of the $(N\times N)$-matrix $\underline{C}$, with elements c_{nm}, see (1), and $\eta^{(m)}$ are the corresponding eigenvalues:

$$\underline{C}\,\underline{v}^{(m)}=\eta^{(m)}\,\underline{v}^{(m)} \quad , \tag{4a}$$

$$\underline{u}^{(m)}\,\underline{C}=\eta^{(m)}\,\underline{u}^{(m)} \quad , \tag{4b}$$

$$(\underline{u}^{(m)},\underline{v}^{(n)})=\delta_{nm} \quad , \tag{4c}$$

$$(\underline{u}^{(m)},\underline{C}\,\underline{v}^{(n)})=(\underline{u}^{(m)}\,\underline{C},\underline{v}^{(n)})=\delta_{nm}\,\eta^{(n)} \quad . \tag{4d}$$

Exercise 4.3.2-1. Prove the solution formula (3), whose validity is conditional on the two assumptions *(i)* that the $(N\times N)$-matrix $\underline{C}$ be diagonalizable and *(ii)* that its N eigenvalues $\eta^{(m)}$ be all distinct ($\eta^{(n)}\neq\eta^{(m)}$ if $n\neq m$); find how it must be modified when these assumptions do not hold. *Hint:* see Sect. 4.3.1.

We now introduce two kinds of particles, respectively characterized by the coordinates $z_n^{(+)}(t)$ and $z_n^{(-)}(t)$, by setting

$$z_n^{(\pm)}(t)=Z_n(t)\pm z_n(t) \quad , \tag{5}$$

entailing

$$z_n(t)=[\,z_n^{(+)}(t)-z_n^{(-)}(t)\,]/2 \quad , \tag{6a}$$

$$Z_n(t)=[\,z_n^{(+)}(t)+z_n^{(-)}(t)\,]/2 \quad , \tag{6b}$$

and we assume that $Z_n(t)$ respectively $z_n(t)$ evolve according to (1) respectively (4.3.1.3). These implies that these coordinates, $z_n^{(\pm)}(t)$, evolve according to the following equations of motion:

$$\ddot{z}_n^{(\pm)}=\pm\{\,\dot{z}_n^2/z_n+a_n\,z_n+z_n\sum_{m=1}^{N}\,[\,b_{nm}\,\dot{z}_m/z_m\,]\}+\sum_{m=1}^{N}\,c_{nm}\,\dot{Z}_m \quad , \tag{7}$$

where, in the right-hand-side, z_n, z_m and Z_m must be expressed in terms of $z_n^{(+)}$ and $z_n^{(-)}$ (or $z_m^{(+)}$ and $z_m^{(-)}$) via (6).

These equations are invariant under translations ($z_n^{(\pm)}(t) \to \tilde{z}_n^{(\pm)}(t) = z_n^{(\pm)}(t) + z_0$, $\dot{z}_0 = 0$); and they are clearly solvable via (5), (6), (3) and (4.3.1-15). And it is easily seen that, by complexification, they correspond to (4.3-2) (via (2), as well as (4.3.1-4) and of course (4.1-17)).

Exercise 4.3.2-2. Verify !

Exercise 4.3.2-3. Discuss the behavior of the solutions of the many-body problem in the plane (4.3-2), making appropriate hypotheses on the coupling constants featured by this model. *Hint*: see (3) and (4.3.1-15).

Exercise 4.3.2-4. Generalize the treatment of Sect. 4.3.2 by adding N arbitrary (complex) constants to the right-hand side of (1) (note that this does not spoil the translation invariance of (1)).

4.3.3 Another rescaling-invariant solvable one-dimensional many-body problem

In the preceding Sect. 4.3.2 we manufactured a solvable one-dimensional many-body problem, see (4.3.2-7,6), that is invariant under (common, time-independent) translations of all particle coordinates. This model is also invariant under (time-independent) rescaling of the particle coordinates, hence, by complexification, it yields the *rotation-invariant* many-body problem in the plane (4.3-2).

But, as explained in Sect. 4.1, the property of translation-invariance can be turned into rescaling-invariance by the change of variables

$$z_n = \log(\zeta_n) , \quad \zeta_n = \exp(z_n) \quad , \tag{1a}$$

which of course entails

$$\dot{z}_n = \dot{\zeta}_n / \zeta_n , \ddot{z}_n = [\ddot{\zeta}_n - \dot{\zeta}_n^2 / \zeta_n] / \zeta_n . \tag{1b}$$

We now perform this change of dependent variables on the model (4.3.2-7,6), namely we set

$$z_n^{(\pm)} = \log[\ \zeta_n^{(\pm)}\ ] , \ \zeta_n^{(\pm)} = \exp[\ z_n^{(\pm)}\ ] \ , \tag{2}$$

and we thereby obtain a new many-body problem, which is now no more translation-invariant, but it is instead invariant under rescaling of the particle coordinates $\zeta_n^{(\pm)}(t)$. It reads:

$$\ddot{\zeta}_n^{(\pm)} = [\ \dot{\zeta}_n^{(\pm)}\ ]^2 / \zeta_n^{(\pm)}$$

$$\pm \{ [\ \dot{\zeta}_n^{(\pm)}\ ]^2 / \zeta_n^{(\pm)} + \zeta_n^{(\pm)} [\ \dot{\zeta}_n^{(\mp)} / \zeta_n^{(\mp)}\ ]^2 - 2 \dot{\zeta}_n^{(+)} \dot{\zeta}_n^{(-)} / \zeta_n^{(\mp)} \} / \log[\ \zeta_n^{(+)} / \zeta_n^{(-)}\ ]$$

$$\pm a_n \zeta_n^{(\pm)} \log[\ \zeta_n^{(+)} / \zeta_n^{(-)}\ ]$$

$$\pm \zeta_n^{(\pm)} \sum_{m=1}^{N} \{ b_{nm} [\ \dot{\zeta}_m^{(+)} / \zeta_m^{(+)} - \dot{\zeta}_m^{(-)} / \zeta_m^{(-)}\ ] \log[\ \zeta_n^{(+)} / \zeta_n^{(-)}\ ] / \log[\ \zeta_m^{(+)} / \zeta_m^{(-)}\ ] \} \}$$

$$+ \zeta_n^{(\pm)} \sum_{m=1}^{N} \{ c_{nm} [\ \dot{\zeta}_m^{(+)} / \zeta_m^{(+)} + \dot{\zeta}_m^{(-)} / \zeta_m^{(-)}\ ] \} \ . \qquad (3)$$

This model is obviously solvable, and it is easily seen that, via the complexification technique of Sect. 4.1, it yields the third of the many-body problems of Sect. 4.3, see (4.3-3).

Exercise 4.3.3-1. Verify ! *Hint*: prove firstly the identities

$$z_1 z_2 / z_3 \doteq \vec{u}(\vec{r}_1, \vec{r}_2; \vec{r}_3) \ , \qquad (4a)$$

$$z_1 (z_2 / z_3)^2 \doteq \vec{v}(\vec{r}_1; \vec{r}_2; \vec{r}_3) \ , \qquad (4b)$$

$$\log(z_1 / z_2) = \lambda\ (\vec{r}_1, \vec{r}_2) + i \lambda'(\vec{r}_1, \vec{r}_2) \ , \qquad (4c)$$

$$[\log(z_1 / z_2)]^{-1} = \mu\ (\vec{r}_1, \vec{r}_2) + i \mu'(\vec{r}_1, \vec{r}_2) \ , \qquad (4d)$$

$$[\log(z_1 / z_2)] / [\log(z_3 / z_4)] = \nu\ (\vec{r}_1; \vec{r}_2; \vec{r}_3; \vec{r}_4) + i \nu'(\vec{r}_1; \vec{r}_2; \vec{r}_3; \vec{r}_4) \ , \qquad (4e)$$

with the quantities in the right hand side of these equations defined by (4.3-3b÷3i) (see (4.1-17,18), and note that (4a) coincides with (4.1-19,5) ...).

Exercise 4.3.3-2. Discuss the behavior of the solutions of the many-body problem in the plane (4.3-3), making appropriate hypotheses on the coupling constants featured by this model. *Hint*: see *Exercise 4.3.2-3*, and use (2).

The model (3) is not translation-invariant. But a translation-invariant model can be manufactured from it by using the same trick, involving a doubling in the number of particles, that was used in Sect. 4.3.2 to manufacture the translation-invariant model (4.3.2-7,6). Then, having obtained in this manner a translation-invariant model, a rescaling-invariant model can be manufactured using the trick (1).

In this manner a hierarchy of translation-invariant, and rescaling-invariant, models can be manufactured, and each of the rescaling-invariant models yields, by complexification, a rotation-invariant model in the plane. And all these models are of course *solvable*. They do however look more and more artificial (the equations of motion feature logarithms of logarithms of logarithms ...), and their solutions become more and more complicated (they feature exponentials of exponentials of exponentials ...).

Exercise 4.3.3-3. Manufacture the next many-body problem in the plane yielded by this procedure, and discuss the behavior of its solutions, making appropriate hypotheses on the coupling constants featured by this model.

4.4 Survey of solvable and/or integrable many-body problems in the plane obtained by complexification

In Sect. 4.4 and in its subsections we list several many-body problems in the plane which have been obtained by the complexification technique described in previous sections of Chap. 4, see in particular Sect. 4.1. Our presentation follows closely (sometimes *verbatim*) <C98c>, and it is limited to exhibiting the relevant equations of motion, all of them (except those of Sect. 4.4.10) of Newtonian type and satisfying the essential requirement to be invariant under rotations in the plane. We do not elaborate on the techniques to solve them, much less do we discuss, except for some occasional remark, the behavior of their solutions. We do however mention, for each model, the extent to which it can be treated (in particular whether it is *solvable* or *integrable*). Bibliographical clues that shall suffice for the diligent reader who wishes to pursue the matter in more detail than is provided herein -- an incentive to do so is the fact that, for several of the models exhibited below, a detailed analysis of the behavior of the solution has not yet been done-- are provided in Sect. 4.N.

Throughout the following subsections

$$\vec{r}_{nm} \equiv \vec{r}_n - \vec{r}_m \,, \; r_{nm}^2 = r_n^2 + r_m^2 - 2\,\vec{r}_n \cdot \vec{r}_m \quad . \tag{1}$$

In those examples, see Sects. 4.4.5÷9, which feature nearest-neighbor interactions, we deliberately leave vague the "end-point" conditions, namely how the relevant equations of motion are to be interpreted for $n=1$ and for $n=N$, and, in the expressions of the Hamiltonians and Lagrangians, the range of the sums.

4.4.1 Example one

The Newtonian equations of motion of this *solvable* many-body problem in the plane read as follows:

$$\ddot{\vec{r}}_n = [\; 2\dot{\vec{r}}_n (\dot{\vec{r}}_n \cdot \vec{r}_n) - \vec{r}_n \left|\dot{\vec{r}}_n\right|^2 \;] / r_n^2 - (\alpha + \alpha' \hat{k} \wedge) \sum_{m=1, m\neq n}^{N} r_{nm}^{-6} \cdot$$

$$\cdot \{\; \vec{r}_n r_m^2 [\; r_{nm}^4 - r_{nm}^2 (r_n^2 - 3 r_m^2) + 2 (r_n^2 - r_m^2)^2 \;] + \vec{r}_m r_n^2 [\; r_{nm}^2 (r_n^2 + r_m^2) - 2 (r_n^2 - r_m^2)^2 \;] \} \,. \tag{1}$$

Here α and α' are two arbitrary (real) coupling constants.

This model features one-body velocity-dependent forces (quadratic in the velocities), and two-body velocity-independent forces. It is obviously rotation-invariant; it is *not* translation-invariant; it is invariant under rescaling of the particle coordinates ($\vec{r}_n \to \tilde{\vec{r}}_n = c\,\vec{r}_n$, $\dot{c} = 0$).

The complex version of the equations of motion (1) reads as follows:

$$\ddot{z}_n = \dot{z}_n^2 / z_n + (\alpha + i\alpha') \sum_{m=1, m\neq n}^{N} z_n^2 \, z_m (z_n + z_m) / (z_n - z_m)^3 \quad , \tag{2}$$

or equivalently, via

$$z_n = \exp(2 q_n) \quad , \tag{3}$$

$$\ddot{q}_n = \frac{1}{8} (\alpha + i\alpha') \sum_{m=1, m\neq n}^{N} \cosh(q_n - q_m) \, [\sinh(q_n - q_m)]^{-3} \quad . \tag{4}$$

Exercise 4.4.1-1. Verify that, via (3), the equations of motion (2) and (4) are equivalent.

Note the coincidence of (4) with (2.1.5-5) (with $a=1$ and $g^2=(\alpha+i\alpha')/16$). This justifies the claim made above about the *solvability* of this model. It also entails that this model is Hamiltonian, see (2.1.5-3).

Exercise 4.4.1-2. Write Hamiltonians and Lagrangians that yield (1). *Hint*: see Sects. 2.1.5 and 4.1 *Solution*: see <C98c>.

Exercise 4.4.1-3. Discuss the motions in the plane entailed by (1), in particular the behaviors as $t \to \pm\infty$. *Hint*: use (3), and see Sect. 2.1.5.

Exercise 4.4.1-4. Verify that the equation of motion (2) admit the $N!$ similarity solutions

$$z_n(t) = A\exp(b+ct+2\pi i m/N)\,,\ \mathrm{n}=1,\ldots,\mathrm{N},\ \mathrm{m}=1,\ldots,\mathrm{N}, \tag{5}$$

where n and m are matched by an arbitrary permutation and A, b and c are 3 arbitrary (complex) constants, and analyze the corresponding configuration and its time-evolution in the plane. *Hint*: use the identity

$$\sum_{m=1,m\neq n}^{N} \exp[2\pi i(m+n)/N]\,[\exp(2\pi i n/N)+\exp(2\pi i m/N)]$$

$$/[\exp(2\pi i n/N)-\exp(2\pi i m/N)]^3 = 0\quad, \tag{6}$$

which can be proven by first showing that its left hand side is in fact independent of n (dividing by $\exp(6\pi i n/N)$ and setting $m=j+m' \bmod(N)$), and by then summing over n from 1 to N, which yields a vanishing result due to the antisymmetry of the summand under the exchange of the two dummy indices m and n.

Exercise 4.4-5. Show that the system (1) possesses the following two constants of motion:

$$C=\sum_{n=1}^{N} \dot{\vec{r}}_n\cdot\vec{r}_n/r_n^2\ , \tag{7a}$$

$$C'=\sum_{n=1}^{N} \hat{k}\cdot\dot{\vec{r}}_n\wedge\vec{r}_n/r_n^2\ . \tag{7b}$$

Hint: note that the equations of motion (2) entail

$$(d/dt)\left[\sum_{n=1}^{N}(\dot{z}_n/z_n)\right]=0\ . \tag{8}$$

4.4.2 Example two

The Newtonian equations of motion of this many-body problem, written in complex form, read as follows:

$$\ddot{z}_n = \dot{z}_n^2 / z_n + a\dot{z}_n + 4 \sum_{m=1, m\neq n}^{N} \dot{z}_n \dot{z}_m z_n (z_n + z_m) / \{(z_n - z_m)[4 z_n z_m + b(z_n - z_m)^2]\} \, , \tag{1}$$

with a and b two arbitrary *complex* constants. These equations of motion, via the position

$$z_n(t) = \exp[2 u_n(t)] \, , \tag{2}$$

take the form

$$\ddot{u}_n = a\dot{u}_n + 2 \sum_{m=1, m\neq n}^{N} \dot{u}_n \dot{u}_m \operatorname{cotanh}(u_n - u_m) / [1 + b \sinh^2(u_n - u_m)]^2 \, , \tag{3}$$

hence they are integrable, indeed *solvable*, see Sect. 2.1.12.4.

On the other hand these equations of motion, (1), are clearly invariant under rescaling of the particle coordinates, hence, by complexification, they yield a *rotation-invariant* many-body problem in the plane.

Exercise 4.4.2-1. Write the Newtonian equations of motion of the many-body problem which obtains by complexification from (1). *Hint*: see Sect. 4.1.

Exercise 4.4.2-2. Discuss the motions in the plane of the many-body problem of the previous *Exercise 4.4.2-1*, and note the special (completely periodic!) behavior if $a = i\omega$ with ω real and nonvanishing. *Hint*: see Sect. 2.1.12.4.

Exercise 4.4.2-3. Verify that, if one sets

$$w(t) = \sum_{n=1}^{N} [\dot{z}_n(t) / z_n(t)] \, , \tag{4}$$

then (1) entails

$$w(t) = w(0) \exp(a t) \, . \tag{5}$$

Exercise 4.4.2-4. Verify that the system (1) possesses the $N!$ similarity solutions

$$z_n(t) = \varphi(t) \exp(2\pi\, i m / N) \ , \tag{6}$$

with the indices m and n matched to each other according to any permutation of the N numbers $1, 2, \ldots, N$, and with

$$\varphi(t) = \lambda \exp\{\mu\,[\exp(at) - 1]/a\} \ , \tag{7}$$

where λ and μ are two arbitrary complex constants. *Hint*: first show that (6), with some appropriate $\varphi(t)$, indeed provides a solution to (1) (*hint*: see *Exercise 4.4.1-4*); then note that (6), (4) and (5) entail the ODE

$$N\,\dot{\varphi}(t)/\varphi(t) = w(0)\exp(at) \ . \tag{8}$$

Exercise 4.4.2-5. Verify that the Hamiltonian

$$H(\underline{z},\underline{\zeta}) = \sum_{n=1}^{N} \{-(a/s)\log(z_n)$$

$$+\exp(s\, z_n\, \zeta_n) \prod_{m=1, m\neq n}^{N} \{\, [\, b(z_n^2 + z_m^2) - 2(2+b)\, z_n\, z_m \,]^{1/2} / (z_n - z_m) \,\}\} \tag{9}$$

yields the equations of motion (1) (s being an arbitrary constant that does not show up in (1)).

Exercise 4.4.2-6. Verify that the Lagrangian

$$L(\underline{z},\underline{\dot{z}}) = \sum_{n=1}^{N} \{(a/s)\log(z_n) - [\dot{z}_n/(s\, z_n)]\cdot$$

$$\cdot\{1 + \log[\,\dot{z}_n/(s\, z_n)\,] - \sum_{m=1, m\neq n}^{N} \log\{\,[\, b(z_n^2 + z_m^2) - 2(2+b)\, z_n\, z_m \,]^{1/2} / (z_n - z_m)\,\}\}\} \tag{10}$$

yields the equations of motion (1) (s being again an arbitrary constant that does not show up in (1)).

4.4.3 Example three

The Newtonian equations of motion of this *integrable* (indeed, *solvable*) many-body problem in the plane read as follows:

$$\ddot{\vec{r}}_n = [\ 2\dot{\vec{r}}_n (\dot{\vec{r}}_n \cdot \vec{r}_n) - \vec{r}_n \left|\dot{\vec{r}}_n\right|^2\ ]/r_n^2 + [\ g_n(\underline{\vec{r}}) + \hat{g}_n(\underline{\vec{r}})\hat{k}\wedge\ ]\vec{r}_n\ , \tag{1}$$

or equivalently, in circular coordinates

$$\ddot{r}_n = (\dot{r}_n)^2 / r_n + r_n\, g_n(\underline{\vec{r}})\ , \tag{2a}$$

$$\ddot{\theta}_n = \hat{g}_n(\underline{\vec{r}})\ , \tag{2b}$$

with

$$g_n(\underline{\vec{r}}) \equiv \sum_{m=1,m\neq n}^{N} \{\alpha \log(r_n / r_m) - \alpha' (\theta_n - \theta_m)$$

$$+\{\beta \log(r_n / r_m)\{[\log(r_n / r_m)]^2 - 3(\theta_n - \theta_m)^2\}$$

$$-\beta' (\theta_n - \theta_m)\{-3[\log(r_n / r_m)]^2 + (\theta_n - \theta_m)^2\}\}\{\ [\log(r_n / r_m)]^2 + (\theta_n - \theta_m)^2\ \}^{-3}\}\ , \tag{3a}$$

$$\hat{g}_n(\underline{\vec{r}}) \equiv \sum_{m=1,m\neq n}^{N} \{\alpha' \log(r_n / r_m) - \alpha (\theta_n - \theta_m)$$

$$+\{\beta' \log(r_n / r_m)\{[\log(r_n / r_m)]^2 - 3(\theta_n - \theta_m)^2\}$$

$$+\beta (\theta_n - \theta_m)\{-3[\log(r_n / r_m)]^2 + (\theta_n - \theta_m)^2\}\}\{\ [\log(r_n / r_m)]^2 + (\theta_n - \theta_m)^2\ \}^{-3}\ \}\ . \tag{3b}$$

Note that we use here preferentially circular coordinates (see Sect. 4.1).

This model is Hamiltonian (see below); it features one-body velocity-dependent forces (quadratic in the velocities), and two-body velocity-independent forces. It is not translation-invariant, but it is of course invariant under both rotations and rescaling, indeed it is easily seen that, if $r_n(t)$, $\theta_n(t)$ is, in circular coordinates, a solution of (1), then

$$\tilde{r}_n(t) = r_0 \exp(u\,t)\, r_n(t)\ , \tag{4a}$$

$$\tilde{\theta}_n(t)=\theta_0+vt+\theta_n(t) \tag{4b}$$

is also a solution, with r_0, θ_0, u, v arbitrary (real) constants.

The Newtonian equations of motion (1) feature the 4 arbitrary (real) coupling constants α, α', β, β'. If $\alpha<0$ and $\alpha'=0$, the generic solution is, up to a transformation of type (4), completely periodic with period $T=2\pi(-\alpha)^{1/2}$ (see below for a justification of this statement).

If the quantity b/a is *real* and *negative*, with a and b defined as follows

$$a=\alpha+i\alpha' \quad , \qquad b=\beta+i\beta' \quad , \tag{5}$$

this model, (1), possesses the following $N!$ similarity solutions:

$$r_n(t)=r_0\exp(u\,t) \quad , \tag{6a}$$

$$\theta_n(t)=\theta_0+vt+\left[-b/(2aN)\right]^{1/4}\xi_m \tag{6b}$$

which are conveniently written here in circular coordinates. These formulas, (6), feature 4 arbitrary (real) constants r_0,θ_0,u,v; the indices n and m in (6b) are related by an arbitrary permutation; and the N real numbers ξ_m are the N zeros of the Hermite polynomial of order N,

$$H_N(\xi_m)=0 \quad , \tag{7a}$$

hence (see Appendix C)

$$\sum_{n=1}^{N}\xi_n=0 \quad , \tag{7b}$$

$$\xi_n=2\sum_{m=1}^{N}\,(\xi_n-\xi_m)^{-3} \quad . \tag{7c}$$

Exercise 4.4.3-1. Verify that (6) satisfies (1). *Hint*: first try directly, then see below.

The "complex-plane" avatar of (1) reads

$$\ddot{z}_n=\dot{z}_n^2/z_n+z_n\sum_{m=1,m\neq n}^{N}\{\,a\,\log(z_n/z_m)+b\,[\log(z_n/z_m)]^{-3}\,\} \quad , \tag{8}$$

with the 2 complex coupling constants a, b related to the 4 real coupling constants $\alpha, \alpha', \beta, \beta'$ by (5).

Exercise 4.4.3-2. Verify !

These (complex) equations of motion are produced by the Hamiltonian

$$h(\underline{z},\underline{\zeta}) = \frac{1}{2}\sum_{n=1}^{N} z_n^2 \zeta_n^2 + V(\underline{z}) \quad , \tag{9a}$$

as well as by the Lagrangian

$$\ell(\underline{z},\underline{\dot{z}}) = \frac{1}{2}\sum_{n=1}^{N} z_n^{-2} \dot{z}_n^2 - V(\underline{z}) \quad , \tag{9b}$$

with

$$V(\underline{z}) = \frac{1}{4}\sum_{m,n=1;m\neq n}^{N} \{ -a[\ \log(z_n / z_m)\ ]^2 + b[\ \log(z_n / z_m)\ ]^{-2}\ \} \quad . \tag{9c}$$

Exercise 4.4.3-3. Verify !

Exercise 4.4.3-4. Write 2 real Hamiltonians, and 2 real Lagrangians, that yield the Newtonian equations of motion in the plane (1). *Hint*: see Sect. 4.1.

The *integrability* (in fact, *solvability*) of this model, as well as quite straightforward proofs of the results stated above, follow from the findings of Sect. 2.1.3.3, since, by setting

$$z_n(t) = \exp[u_n(t)] \quad , \tag{10}$$

the equations of motion (8) become

$$\ddot{u}_n = \sum_{m=1,m\neq n}^{N} [a(u_n - u_m) + b(u_n - u_m)^{-3}\ ] \quad . \tag{11}$$

Exercise 4.4.3-5. Verify !

4.4.4 Example four

The Newtonian equations of motion of this *solvable* many-body problem in the plane read

$$\ddot{\vec{r}}_n = [2\dot{\vec{r}}_n(\dot{\vec{r}}_n \cdot \vec{r}_n) - \vec{r}_n(\dot{\vec{r}}_n \cdot \dot{\vec{r}}_n)]/r_n^2 + (\alpha + \alpha' \hat{k}\wedge)\,\dot{\vec{r}}_n$$

$$+2\, r_n^{-2} \sum_{m=1, m\neq n}^{N} [\gamma(\vec{r}_n, \vec{r}_m) - \hat{\gamma}(\vec{r}_n, \vec{r}_m)\,\hat{k}\wedge][\,\dot{\vec{r}}_n(\dot{\vec{r}}_m \cdot \vec{r}_n) + \dot{\vec{r}}_m(\dot{\vec{r}}_n \cdot \vec{r}_m) - \vec{r}_n(\dot{\vec{r}}_n \cdot \dot{\vec{r}}_m)]\;, \qquad (1)$$

$$\gamma(\vec{r}_1, \vec{r}_2) = -\gamma(\vec{r}_2, \vec{r}_1) \equiv \log(r_1/r_2)/\{[\log(r_1/r_2)]^2 + (\theta_1 - \theta_2)^2\}\;, \qquad (2a)$$

$$\hat{\gamma}(\vec{r}_1, \vec{r}_2) = -\hat{\gamma}(\vec{r}_2, \vec{r}_1) \equiv (\theta_1 - \theta_2)/\{[\log(r_1/r_2)]^2 + (\theta_1 - \theta_2)^2\}\;. \qquad (2b)$$

Hence they feature one- and two-body velocity-dependent forces (to write the latter we conveniently employed circular coordinates, see (2)).

This many-body problem is Hamiltonian (see below). Its equations of motion are of course invariant under rotations in the plane (in circular coordinates: $\theta_n \to \tilde{\theta}_n = \theta_n + \theta_0$, $\dot{\theta}_0 = 0$), and under coordinate rescaling ($\vec{r}_n \to \vec{\tilde{r}}_n = c\,\vec{r}_n$, $\dot{c} = 0$). They contain 2 arbitrary (real) coupling constants, α and α'; if α vanishes and α' does not, $\alpha = 0, \alpha' \neq 0$, its generic motions are completely periodic, with period (at most) $\tilde{T} = T \cdot N!$, $T = 2\pi/|\alpha'|$.

Exercise 4.4.4-1. Prove the last statement made above. *Hint*: see *Proposition 2.1.13-1*, and (8, 4) below together with Sect. 4.2.5.

The "complex plane" avatar of the Newtonian equations of motion (1) reads

$$\ddot{z}_n = \dot{z}_n^2/z_n + a\,\dot{z}_n + 2 \sum_{m=1, m\neq n}^{N} \dot{z}_n\,\dot{z}_m / [z_n \log(z_n/z_m)]\;, \qquad (3)$$

with

$$a = \alpha + i\alpha'\;. \qquad (4)$$

Exercise 4.4.4-2. Verify !

Exercise 4.4.4-3. Verify that the complex equations of motion (3) possess the $N!$ similarity solutions

$$z_n(t) = \exp\{ c[b + \exp(at)]^{1/N} \exp(2\pi i m/N) \} \ , \tag{5}$$

with b and c arbitrary complex constants, and the indices m and n related by an arbitrary permutation.

Exercise 4.4.4-4. Verify that the equations of motion (3) are yielded by the Hamiltonian

$$h(\underline{z},\underline{\zeta}) = \sum_{n=1}^{N} \{ -(a/s)\log(z_n) + \exp(s\, z_n\, \zeta_n) \prod_{m=1, m\neq n}^{N} [\log(z_n/z_m)]^{-1} \} \ , \tag{6}$$

as well as by the Lagrangian

$$\ell(\underline{z},\underline{\dot{z}}) = \sum_{n=1}^{N} \{ (a/s)\log(z_n)$$

$$+ [\dot{z}_n/(s\, z_n)] \{ -1 + \log[\dot{z}_n/(s\, z_n)] + \sum_{m=1, m\neq n}^{N} \log[\log(z_n/z_m)] \} \} \ , \tag{7}$$

where s is an arbitrary complex constant (that does not show up in (3)).

Exercise 4.4.4-5. Write out two (real) Hamiltonians and two (real) Lagrangians, that yield the (real) Newtonian equations of motion in the plane (1).

The equations of motion (3), hence (1) as well, are *integrable*, indeed *solvable*, since, via (4.4.3-10), they take the form (see (4.2.3-10), or equivalently (4.2.1-1) with $a = \alpha + i\alpha'$, $\beta = \beta' = \lambda = \lambda' = \mu = \mu' = 0$)

$$\ddot{u}_n = a\, \dot{u}_n + 2 \sum_{m=1, m\neq n}^{N} \dot{u}_n \dot{u}_m / (u_n - u_m) \ . \tag{8}$$

Exercise 4.4.4-5. Verify !

Exercise 4.4.4-6. Discuss the behavior of the many-body problem (1). *Hint*: see the discussion after (4.2.3-10).

4.4.5 Example five

The Newtonian equations of motion in the plane of this *solvable* model read as follows:

$$\ddot{\vec{r}}_n = [\ 2\dot{\vec{r}}_n(\dot{\vec{r}}_n \cdot \vec{r}_n) - \vec{r}_n|\dot{\vec{r}}_n|^2\ ]/r_n^2$$

$$+(\alpha+\alpha'\,\hat{k}\wedge)\,\{\vec{r}_{n+1} - [\ 2\,\vec{r}_n(\vec{r}_n \cdot \vec{r}_{n-1}) + \vec{r}_{n-1}\,r_n^2\ ]/r_{n-1}^2\ \}\,. \quad (1)$$

Hence they feature velocity-dependent one-body forces and velocity-independent "nearest-neighbor" two-body forces.

These equations of motion, (1), are obviously invariant under rotations and rescaling, but not under translations. They feature (rather trivially, see below) the 2 arbitrary real coupling constants α and α'.

These equations of motion, (1), are yielded by either one of the following two Hamiltonians, $H(\underline{\vec{r}},\underline{\vec{p}})$ and $\tilde{H}(\underline{\vec{r}},\underline{\tilde{\vec{p}}})$, as well as by either one of the following two Lagrangians, $L(\underline{\vec{r}},\underline{\dot{\vec{r}}})$ and $\tilde{L}(\underline{\vec{r}},\underline{\dot{\vec{r}}})$:

$$H(\underline{\vec{r}},\underline{\vec{p}}) = \frac{1}{2}\sum_n\ \left[(\vec{p}_n \cdot \vec{r}_n)^2 - (\hat{k}\cdot\vec{p}_n \wedge \vec{r}_n)^2\right] + V(\underline{\vec{r}})\ , \quad (2a)$$

$$\tilde{H}(\underline{\vec{r}},\underline{\tilde{\vec{p}}}) = \sum_n\ \left[(\tilde{\vec{p}}_n \cdot \vec{r}_n)(\hat{k}\cdot\tilde{\vec{p}}_n \wedge \vec{r}_n)^2\right] + \hat{V}(\underline{\vec{r}})\ , \quad (2b)$$

$$L(\underline{\vec{r}},\underline{\dot{\vec{r}}}) = \frac{1}{2}\sum_n\ \left[(\dot{\vec{r}}_n \cdot \vec{r}_n)^2 - (\hat{k}\cdot\dot{\vec{r}}_n \wedge \vec{r}_n)^2\right]/r_n^4 - V(\underline{\vec{r}})\ , \quad (3a)$$

$$\tilde{L}(\underline{\vec{r}},\underline{\dot{\vec{r}}}) = -\sum_n\ (\dot{\vec{r}}_n \cdot \vec{r}_n)(\hat{k}\cdot\dot{\vec{r}}_n \wedge \vec{r}_n)/r_n^4 - \hat{V}(\underline{\vec{r}})\ , \quad (3b)$$

with

$$V(\underline{\vec{r}}) = (\alpha^2+\alpha'^2)^{-1}\sum_n\ \left[-\alpha(\vec{r}_n \cdot \vec{r}_{n-1}) + \alpha'(\hat{k}\cdot\vec{r}_n \wedge \vec{r}_{n-1})\ \right]/r_{n-1}^2\ , \quad (4a)$$

$$\hat{V}(\underline{\vec{r}}) = (\alpha^2+\alpha'^2)^{-1}\sum_n\ \left[\alpha'(\vec{r}_n \cdot \vec{r}_{n-1}) + \alpha\ (\hat{k}\cdot\vec{r}_n \wedge \vec{r}_{n-1})\ \right]/r_{n-1}^2\ . \quad (4b)$$

Exercise 4.4.5-1. Verify that (2) or (3) with (4) yield (1).

The "complex plane" avatar of (1) reads as follows:

$$\ddot{z}_n = \dot{z}_n^2 / z_n + a(z_{n+1} - z_n^2 / \dot{z}_{n+1}) \tag{5}$$

with

$$a = \alpha + i\alpha' \ . \tag{6}$$

By setting

$$z_n = a^n \exp(u_n) \tag{7}$$

it goes into the *integrable* (indeed, *solvable*: see *Exercise 2.1.7-5*) "Toda" equations

$$\ddot{u}_n = \exp(u_{n+1} - u_n) - \exp(u_n - u_{n-1}) \ . \tag{8}$$

Note that the complex constant a has dropped out of these equations.

Exercise 4.4.5-2. Verify !

Exercise 4.4.5-3. Verify that the complex equations of motion (5) possess the similarity solutions

$$z_n(t) = \exp\left[(b + nc)\, t \right] \ , \tag{9}$$

with b and c arbitrary complex constant, and analyze the behavior of the corresponding solutions of (1).

4.4.6 Example six

The *solvable* Newtonian equations of motion in the plane of the first example exhibited in Sect. 4.4.6 read

$$\ddot{\vec{r}}_n = (\alpha + \alpha' \wedge \hat{k})\, \dot{\vec{r}}_n - \sum_{m=n\pm1} \left[\dot{\vec{r}}_n (\dot{\vec{r}}_m \cdot \vec{r}_{nm}) + \dot{\vec{r}}_m (\dot{\vec{r}}_n \cdot \vec{r}_{nm}) - \vec{r}_{nm} (\dot{\vec{r}}_n \cdot \dot{\vec{r}}_m) \right] / r_{nm}^2 \ , \tag{1}$$

of course with (4.4-1). Hence they feature velocity-dependent one-body and nearest-neighbor forces. The similarity, and especially the *difference*, should be noted, among these equations of motion, (1), and, say, (4.2-1) (the latter, of course, with $\beta = \beta' = \lambda = \lambda' = \mu = \mu' = 0$): not only, of course, in the range of the sum, but as well in the factor in front of it, which is -1 here, $+2$ there.

These Newtonian equations of motion are obviously invariant under rotations in the plane, under rescaling of the particle coordinates, and also under translations. They feature the two arbitrary (real) coupling constants α and α', and it is clear (see *Proposition 2.1.13-1*, and below) that their generic solution is *completely periodic* if $\alpha = 0$ and $\alpha' \neq 0$.

The complex-plane avatar of (1) reads as follows:

$$\ddot{z}_n = a\dot{z}_n - \sum_{m=n\pm 1} \dot{z}_n \dot{z}_m / (z_n - z_m) \quad , \tag{2}$$

with

$$a = \alpha + i\alpha' \quad . \tag{3}$$

Exercise 4.4.6-1. Verify !

Exercise 4.4.6-2. Verify that the (complex) equations of motion (2) possess the similarity solutions

$$z_n(t) = A^n \exp\{ \, b + c[\, \exp(at) - 1 \,] / a \, \} \quad , \tag{4}$$

with A, b and c arbitrary complex constants, and discuss the behavior in the plane of the corresponding solutions of (1).

The *integrability* (indeed, *solvability*) of the many-body problem in the plane (1) is entailed by the coincidence, up to trivial notational changes and to the insertion of the term proportional to the constant a (for which, see *Proposition 2.1.13-1*), of (2) with (2.1.13-24).

As suggested by the treatment of Sect. 4.1, we now set

$$z_n(t) = \log[u_n(t)\,], \quad u_n(t) = \exp[z_n(t)\,] \quad , \tag{5}$$

and thereby transform the translation-invariant system (3) into the following rescaling-invariant system:

$$\ddot{u}_n = \dot{u}_n^2 / u_n + a\,\dot{u}_n - \sum_{m=n\pm 1} \dot{u}_n \dot{u}_m / [\, u_m \log(u_n / u_m) \,] \quad . \tag{6}$$

Then we complexify this system, and we thereby get the following *solvable* many-body problem in the plane:

$$\ddot{\vec{r}}_n = [\, 2\,\dot{\vec{r}}_n (\dot{\vec{r}}_n \cdot \vec{r}_n) - \vec{r}_n \left| \dot{\vec{r}}_n \right|^2 \,] / r_n^2 + (\alpha + \alpha' \, \hat{k} \wedge) \, \dot{\vec{r}}_n$$

$$- \sum_{m=n\pm 1} \{ [\log(r_n / r_m) - (\theta_n - \theta_m) \hat{k} \wedge] \cdot$$

$$\cdot [\, \dot{\vec{r}}_n (\dot{\vec{r}}_m \cdot \vec{r}_m) + \dot{\vec{r}}_m (\dot{\vec{r}}_n \cdot r_m) - \vec{r}_m (\dot{\vec{r}}_n \cdot \dot{\vec{r}}_m) \,] / \{ [\, \log(r_n / r_m) \,]^2 + (\theta_n - \theta_m)^2 \, \} \} , \qquad (7)$$

which we have conveniently written using, in part, circular coordinates.

These Newtonian equations of motion, (7), are obviously rotation- and rescaling-invariant, but they are not translation-invariant. They feature again the two arbitrary (real) coupling constants α and α', and, if $\alpha = 0$ and $\alpha' \neq 0$, their generic solution is *completely periodic* with period $T = 2\pi / |\alpha'|$.

Exercise 4.4.6-3. Verify that the (complex) equations of motion (6) possess the similarity solution

$$u_n(t) = n A + b + c[\, \exp(at) - 1 \,] / a \quad , \qquad (8)$$

with A, b and c arbitrary complex constants, and discuss the behavior in the plane of the corresponding solutions of (7).

Exercise 4.4.6-4. Verify that the following Hamiltonian $h(\underline{z}, \underline{\zeta})$ and Lagrangian $\ell(\underline{z}, \underline{\dot{z}})$ yield the (complex) equations of motion (3):

$$h(\underline{z}, \underline{\zeta}) = \sum_n \{ -(a/s) z_n + \exp(s \zeta_n) \, (z_{n+1} - z_n) \} \quad , \qquad (9)$$

$$\ell(\underline{z}, \underline{\dot{z}}) = \sum_n \{ a z_n + \dot{z}_n \log[\, \dot{z}_n (z_{n+1} - z_n) \,] \} \quad , \qquad (10)$$

where s is an arbitrary (nonvanishing) complex constant that does not show up in (3).

Exercise 4.4.6-5. Write out two (real) Hamiltonians, and two (real) Lagrangians, that yield the equations of motion (1). *Hint*: see Sect. 4.1.

Exercise 4.4.6-6. Write out Hamiltonians and Lagrangians that yield (6) and (7). *Hint*: see (5).

Exercise 4.4.6-7. Write the rotation-invariant (real) equations of motion in the plane that correspond by complexification to

$$\ddot{z}_n = \dot{z}_n^2 / z_n + a\dot{z}_n - \dot{z}_n \{ \dot{z}_{n-1} / (z_n - z_{n-1}) + \dot{z}_{n+1} z_n / [z_{n+1} (z_n - z_{n+1})] \} \,, \qquad (11)$$

and discuss their integrability/solvability. *Hint*: see (2.1.13-23) and *Proposition 2.1.13-1*.

Exercise 4.4.6-8. Verify that the following two Hamiltonians, as well as the following two Lagrangians, yield the equations of motion (11):

$$h^{(\sigma)}(\underline{z},\underline{\zeta}) = \sum_n \{ -(a/s)\log(z_n) + \exp(s\, z_n\, \zeta_n)[1 - (z_{n+(1+\sigma)/2} / z_{n+(1-\sigma)/2})] \} \,, \qquad (12)$$

$$\ell^{(\sigma)}(\underline{z},\underline{\dot{z}}) = \sum_n \{ (a/s)\log(z_n) + (\dot{z}_n / z_n) \log\{ \dot{z}_n\, z_{n+(1+\sigma)/2} / [(z_n - z_{n+\sigma})] \} \}. \qquad (13)$$

Here $\sigma = +1$ or $\sigma = -1$, while s is an arbitrary (complex) constant, that does not show up in (11).

Exercise 4.4.6-9. Write out 4 (real) Hamiltonians, as well as 4 (real) Lagrangians, that yield directly the real equations of motion obtained in *Exercise 4.4.6-7*. *Hint*: see *Exercise 4.4.6-8* and Sect. 4.1.

4.4.7 Example seven

The Newtonian equations of motion of this *integrable* many-body problem in the plane read as follows:

$$\ddot{\vec{r}}_n = (1+\gamma+\gamma'\hat{k}\wedge)\,[\, 2\dot{\vec{r}}_n(\dot{\vec{r}}_n \cdot \vec{r}_n) - \vec{r}_n |\dot{\vec{r}}_n|^2 \,]/r_n^2 + (\alpha+\alpha'\hat{k}\wedge)\,\dot{\vec{r}}_n + (\beta+\beta'\hat{k}\wedge)\,\vec{r}_n$$

$$- \sum_{m=n\pm 1} \{ (\gamma+\gamma'\hat{k}\wedge)\,[2\dot{\vec{r}}_n(\dot{\vec{r}}_n \cdot \vec{r}_{nm}) - \vec{r}_{nm}|\dot{\vec{r}}_n|^2\,]$$

$$+ (\alpha+\alpha'\hat{k}\wedge)\{\, \dot{\vec{r}}_n[\, -(\vec{r}_n \cdot \vec{r}_m) + r_n^2\,] - \vec{r}_n(\dot{\vec{r}}_n \cdot \vec{r}_m) + \vec{r}_m(\dot{\vec{r}}_n \cdot \vec{r}_n)\,\}$$

$$+ (\beta+\beta'\hat{k}\wedge)\{\vec{r}_n[-2\,(\vec{r}_n \cdot \vec{r}_m) + r_n^2\,] + \vec{r}_m r_n^2\,\}\}/r_{nm}^2 \,, \qquad (1)$$

where of course $\vec{r}_{nm} = \vec{r}_n - \vec{r}_m$, see (4.1-1).

Hence this N-body problem features velocity-dependent one-body and "nearest-neighbor" forces. The 6 (real) coupling constants $\alpha, \alpha', \beta, \beta', \gamma, \gamma'$ are arbitrary.

The equations of motion (1) are obviously rotation- and rescaling-invariant. They are, generally, not translation-invariant (see, however, below).

The complex-plane avatar of this system reads as follows:

$$\ddot{z}_n = (1+c)\dot{z}_n^{\,2}/z_n + a\dot{z}_n + b z_n - (c\dot{z}_n^2 + a\dot{z}_n z_n + b z_n^2)\left[\, (z_n - z_{n+1})^{-1} + (z_n - z_{n-1})^{-1} \,\right] \tag{2}$$

with

$$a = \alpha + i\alpha', \;\; b = \beta + i\beta', \;\; c = \gamma + i\gamma' \;\; . \tag{3}$$

Exercise 4.4.7-1. Verify !

These equations of motion, (2), are yielded by the following Hamiltonian $h(\underline{z},\underline{\zeta})$ and Lagrangian $\ell(\underline{z},\underline{\dot{z}})$:

$$h(\underline{z},\underline{\zeta}) = \sum_n \;\{\lambda\, z_n \zeta_n + \nu\{\,\mu \log[\sinh(z_n \zeta_n/\nu)] + \rho \log[1-(z_{n+1}/z_n)\,]\}\} \;, \tag{4}$$

$$\ell(\underline{z},\underline{\dot{z}}) = \nu \sum_n \;\{\frac{1}{2}\mu[(w_n+1)\log(w_n+1) - (w_n-1)\log(w_n-1)] - \rho\log[1-(z_{n+1}/z_n)\,]\}, \tag{5}$$

with

$$w_n \equiv [(\dot{z}_n/z_n) - \lambda\,]/\mu \;\; , \tag{6}$$

and with the 4 (complex) constants λ, μ, ν, ρ related to the 3 (complex) constants a, b, c (see (3)) as follows:

$$a = 2\lambda\rho/(\mu\nu), \;\; b = \rho(\mu^2 - \lambda^2)/(\mu\nu), \;\; c = -\rho/(\mu\nu) \;. \tag{7}$$

Exercise 4.4.7-2. Verify that the (complex) Hamiltonian (4) and Lagrangian (5) yield the (complex) equations of motion (2), and write out real Hamiltonians and Lagrangians that yield the real Newtonian equations of motion in the plane (1). *Hint*: see Sect. 4.1.

The *integrable* nature of the equations of motion (2) (hence as well of (1)) is demonstrated by setting

$$z_n(t) = \exp[-c\,u_n(t)\,] \;. \tag{8}$$

Thereby the equations of motion (2) become the *integrable* equations

$$\ddot{u}_n = R(\dot{u}_n)\,[\; g(u_{n+1}-u_n) - g(u_n - u_{n-1}) \;] \quad , \tag{9}$$

$$R(v) \equiv E v^2 + A v + B \quad , \tag{10}$$

$$g(u) = [\; g_1 - g_2 \exp(-cu) \;] / [\; 1 - \exp(-cu) \;] \; , \tag{11a}$$

entailing

$$g'(u) = E[g(u) - g_1][g(u) - g_2] \quad , \tag{11b}$$

with

$$c = E(g_2 - g_1), \;\; a = A(g_2 - g_1), \;\; b = B(g_2 - g_1) \quad . \tag{12}$$

Exercise 4.4.7-3. Verify that (9) with (10), (11) and (12) correspond to (2) via (8).

Exercise 4.4.7-4. Verify that (2) admits the similarity solution

$$z_n(t) = \eta^n \exp(\varepsilon t) \quad , \tag{13}$$

with η and ε two arbitrary (complex) constants, and analyze the behavior in the plane of the corresponding solution of (1).

An interesting variation of the many-body problem in the plane (1) reads

$$\ddot{\vec{r}}_n = (1+\gamma+\gamma'\hat{k}\wedge)\,[\; 2\dot{\vec{r}}_n(\dot{\vec{r}}_n\cdot\vec{r}_n) - \vec{r}_n\left|\dot{\vec{r}}_n\right|^2 \;]/r_n^2 + (\alpha+\alpha'\hat{k}\wedge)\dot{\vec{r}}_n$$

$$-(\gamma+\gamma'\hat{k}\wedge)\sum_{m=n\pm1}\{\,[\; 2\dot{\vec{r}}_n(\dot{\vec{r}}_n\cdot\vec{r}_{nm}) - \vec{r}_{nm}\left|\dot{\vec{r}}_n\right|^2 \;]/r_{nm}^2\,\} \;. \tag{14}$$

In can be obtained from (1) by firstly setting $\alpha=\alpha'=\beta=\beta'=0$, and then by applying the change of dependent variable (2.1.13-5a) with $a=\alpha+i\alpha'$. Hence if $\alpha=0$ and $\alpha'\neq 0$, one can assert (via *Proposition 2.1.13-1*) that the generic solution of this many-body problem in the plane, (14), is *completely periodic*.

Exercise 4.4.7-5. Verify that the equations of motion (14) possess the similarity solution (written in circular coordinates)

$$r_n(t) = r_0\,\rho^n \exp\{\ \exp(\alpha t)\,[\ \lambda\cos(\alpha' t) - \mu\sin(\alpha' t)\ ]\ \}\ , \tag{15a}$$

$$\theta_n(t) = \theta_0 + \eta\, n + \exp(\alpha t)[\ \mu\cos(\alpha' t) + \lambda\ \sin(\alpha' t)\ ]\ , \tag{15b}$$

featuring the 6 arbitrary real constants $r_0, \rho, \lambda, \mu, \theta_0, \eta$ ($r_0 > 0, \rho > 0$; the presence of r_0 and θ_0 reflects merely the invariance under rescaling), and discuss the corresponding motion in the plane (in particular, consider the 3 cases: $\alpha > 0$, $\alpha < 0$, $\alpha = 0$).

Let us end Sect. 4.4.7 by highlighting certain special cases of (2) and of (14) that deserve to be singled out.

If $c = a = 0$, and by setting in (2)

$$z_n(t) = \exp\,[u_n(t)]\ , \tag{16}$$

one gets equations of motion which are *translation-invariant* and which feature only *velocity-independent* forces:

$$\ddot{u}_n = b\{1 - [1 - \exp(u_{n+1} - u_n)]^{-1} - [1 - \exp(u_{n-1} - u_n)]^{-1}\}\ , \tag{17a}$$

$$\ddot{u}_n = b[\ \exp(u_{n+1} + u_{n-1} - 2u_n) - 1]\cdot$$

$$\cdot\{[1 - \exp(u_{n+1} - u_n)] - \exp(u_{n-1} - u_n) + \exp(u_{n+1} + u_{n-1} - 2u_n)\ ]\}^{-1}\ . \tag{17b}$$

On the other hand, if $\gamma = -1$ and $\gamma' = 0$, the Newtonian equations of motion (14) become *translation-invariant* and read

$$\ddot{\vec{r}}_n = (\alpha + \alpha' \hat{k} \wedge)\, \dot{\vec{r}}_n + \sum_{m=n\pm1} \{\ [2\,\dot{\vec{r}}_n\,(\dot{\vec{r}}_n \cdot \vec{r}_{nm}) - \vec{r}_{nm}\left|\dot{\vec{r}}_n\right|^2\ ]/r_{nm}^2\ \}\ . \tag{18}$$

Note the similarity (and difference!) between these equations of motion, (18), and (4.4.6-1) as well as (4.2-1) (the latter, of course, with $\beta = \beta' = \lambda = \lambda' = \mu = \mu' = 0$).

Of course the complex-plane avatar of these equations of motion, (18), is also translation-invariant and it reads

$$\ddot{z}_n = a\,\dot{z}_n + \dot{z}_n^2 \sum_{m=n\pm1} \left[(z_n - z_{n+1})^{-1} + (z_n - z_{n-1})^{-1} \right] \; . \tag{19}$$

Exercise 4.4.7-6. Use the transformation (4.4.6-5) on this equation, (19), and thereby obtain an equation similar to, yet different from, (4.4.6-6), which provides via complexification another instance of *integrable* many-body problem in the plane.

4.4.8 Example eight

The Newtonian equations of motion of this *integrable* many-body problem in the plane read as follows:

$$\ddot{\vec{r}}_n = \left[\, 2\dot{\vec{r}}_n\,(\dot{\vec{r}}_n\cdot\vec{r}_n) - \vec{r}_n\left|\dot{\vec{r}}_n\right|^2 \,\right]/\,r_n^2 + (\beta+\beta'\hat{k}\wedge)\,\{\, r_{n-1}^{-4}\,\cdot$$

$$\cdot\{\, 2\dot{\vec{r}}_n\,(\dot{\vec{r}}_n\cdot\vec{r}_{n-1})(\dot{\vec{r}}_{n-1}\cdot\vec{r}_{n-1}) + \dot{\vec{r}}_{n-1}\left[\, 2(\dot{\vec{r}}_n\cdot\vec{r}_{n-1})^2 - \left|\dot{\vec{r}}_n\right|^2 r_{n-1}^2 \,\right] - 2\vec{r}_{n-1}\,(\dot{\vec{r}}_n\cdot\dot{\vec{r}}_{n-1})(\dot{\vec{r}}_n\cdot\vec{r}_{n-1}) \,\}$$
$$- r_n^{-4}\,\{\, 2\dot{\vec{r}}_n\,(\dot{\vec{r}}_n\cdot\vec{r}_n)(\dot{\vec{r}}_{n+1}\cdot\vec{r}_n) - \dot{\vec{r}}_{n+1}\left[2\,(\dot{\vec{r}}_n\cdot\vec{r}_n)^2 - \left|\dot{\vec{r}}_n\right|^2 r_n^2 \,\right] + 2\vec{r}_n\,(\dot{\vec{r}}_n\cdot\dot{\vec{r}}_{n+1})(\dot{\vec{r}}_n\cdot\vec{r}_n) \,\}\,\}. \tag{1}$$

Hence they feature one-body forces quadratic in the velocities and nearest-neighbor forces cubic in the velocities, and, rather trivially (see below), the two coupling constants β and β'.

These equations of motion are obviously rotation- and rescaling-invariant. They are *not* translation-invariant.

The complex-plane avatar of these equations of motion reads

$$\ddot{z}_n = \dot{z}_n^{\;2}/\,z_n + b\,(\dot{z}_n)^2\left[\,(\dot{z}_{n-1}/\,z_{n-1}^2) - (\dot{z}_{n+1}/\,z_n^2)\,\right] \;\; , \tag{2}$$

with

$$b = \beta + i\,\beta' \quad . \tag{3}$$

Exercise 4.4.8-1. Verify ! *Hint*: prove first the identity

$$z_1\,(z_2/\,z_3)^2 \doteq \{\, \vec{r}_1\left[\,2\,(\vec{r}_2\cdot\vec{r}_3)^2 - r_2^{\,2}\, r_3^{\,2}\,\right] + 2\,\vec{r}_2\,(\vec{r}_1\cdot\vec{r}_3)(\vec{r}_2\cdot\vec{r}_3) - 2\vec{r}_3\,(\vec{r}_1\cdot\vec{r}_2)(\vec{r}_2\cdot\vec{r}_3)\,\}/\,r_3^{\,4} . \tag{4}$$

Remark 4.4.8-2. If $z_n(t)$ satisfies (2), so does

$$\tilde{z}_n(t) = c^n\, z_n(ct) \;\; , \tag{5}$$

with c any arbitrary (complex) constant.

Exercise 4.4.8-3. Verify that the equations of motion (2) admit the similarity solution (4.4.7-13).

Exercise 4.4.8-4. Verify that the (complex) equations of motion (2) are yielded by the Hamiltonian

$$h(\underline{z},\underline{\zeta}) = -\sum_n \log\{z_n[\zeta_n - b(z_n/z_{n-1})]\} \quad , \tag{6}$$

as well as by the Lagrangian

$$\ell(\underline{z},\underline{\dot{z}}) = \sum_n [b(\dot{z}_n/z_{n-1}) - \log(\dot{z}_n/z_n)] \quad . \tag{7}$$

Exercise 4.4.8-5. Write out two real Hamiltonians, and two real Lagrangians, that yield the (real) equations of motion in the plane (1). *Hint*: see the preceding *Exercise 4.4.8-4*, and Sect. 4.1.

The *integrability* of this many-body problem in the plane, (1), is demonstrated by setting

$$z_n = b^{-n}\exp(u_n) \quad , \tag{8}$$

whereby (2) become the *integrable* equations

$$\ddot{u}_n = (\dot{u}_n)^2[\dot{u}_{n-1}\exp(u_n - u_{n-1}) - \dot{u}_{n+1}\exp(u_{n+1} - u_n)] \quad . \tag{9}$$

Exercise 4.4.8-6. Verify that the equations of motion (2) are invariant under the transformation $z_n(t) \to [z_{N+1-n}(t_0 - t)]^{-1}$, namely that if one sets

$$\tilde{z}_n(t) = [z_{N+1-n}(t_0 - t)]^{-1}, \tag{10}$$

with t_0 an arbitrary constant, then (2) entail

$$\ddot{\tilde{z}}_n = \dot{\tilde{z}}_n^{\,2}/\tilde{z}_n + b(\dot{\tilde{z}}_n)^2[(\dot{\tilde{z}}_{n-1}/\tilde{z}_{n-1}^2) - (\dot{\tilde{z}}_{n+1}/\tilde{z}_n^2)] \quad . \tag{11}$$

Exercise 4.4.8-7. Set

$$\hat{z}_n(t) = \exp(\lambda t)\, z_n(t), \quad b = a/\lambda^3 \quad , \tag{12}$$

and write out the (autonomous) equations of motion entailed by (2) for $\hat{z}_n(t)$. Then take the limit $\lambda \to \infty$, and show that, up to trivial notational changes $(\hat{z}_n(t) \to z_n(t))$, these equations of motion reduce to (4.4.5-5).

4.4.9 Example nine

The Newtonian equations of motion of this *integrable* many-body problem in the plane read as follows:

$$\ddot{\vec{r}}_n = [2\,\dot{\vec{r}}_n\,(\dot{\vec{r}}_n \cdot \vec{r}_n) - \vec{r}_n \left|\dot{\vec{r}}_n\right|^2\,]/r_n^2 + (\beta + \beta' \hat{k} \wedge)\,\{\,\dot{\vec{r}}_{n+1} - r_{n-1}^{-4} \cdot$$

$$\cdot\{\,\dot{\vec{r}}_{n-1}\,[2\,(\vec{r}_n \cdot \vec{r}_{n-1})^2 - r_n^2\, r_{n-1}^2\,] + 2\,\vec{r}_n\,(\dot{\vec{r}}_{n-1} \cdot \vec{r}_{n-1})(\vec{r}_n \cdot \vec{r}_{n-1})\ - 2\vec{r}_{n-1}(\dot{\vec{r}}_{n-1} \cdot \vec{r}_n)(\vec{r}_n \cdot \vec{r}_{n-1})\ \}\,\}$$

$$+(\beta^2 - \beta'^2 + 2\,\beta\,\beta'\,\hat{k} \wedge)\,\{\,r_n^{-2}[\ -2\,\vec{r}_{n+1}\,(\vec{r}_n \cdot \vec{r}_{n+1}) + \vec{r}_n\, r_{n+1}^2\ ]$$

$$+ r_{n-1}^{-4}\,\{\,\vec{r}_n\,[\ 4(\vec{r}_n \cdot \vec{r}_{n-1})^2 - r_n^2\, r_{n-1}^2\ ] - 2\vec{r}_{n-1}\, r_n^2\,(\vec{r}_n \cdot \vec{r}_{n-1})\ \}\,\}\ . \qquad (1)$$

Hence they feature a one-body term quadratic in the velocities, a nearest-neighbor term linear in the velocities and another nearest-neighbor velocity-independent term. They contain, linearly and quadratically, the two arbitrary real coupling constants β and β'. They obviously are both rotation- and rescaling-invariant; they are not translation-invariant.

The complex-plane avatar of these equations of motion reads

$$\ddot{z}_n = \dot{z}_n^{\,2}/z_n + b\,[\ \dot{z}_{n+1} - \dot{z}_{n-1}\,(z_n/z_{n-1})^2\ ] + b^2[\ -(z_{n+1}^2/z_n) + z_n^3/z_{n-1}^2\ ]\ , \qquad (2)$$

with

$$b = \beta + i\,\beta'\ . \qquad (3)$$

Exercise 4.4.9-1. Verify ! *Hint*: see (4.4.8-4).

Exercise 4.4.9-2. Verify that the *Remark 4.4.8-2* applies also to these equations of motion, (2).

Exercise 4.4.9-3. Verify that the equations of motion (2) admit the similarity solution (4.4.7-13).

Exercise 4.4.9-4. Verify that the (complex) equations of motion (2) are yielded by the Hamiltonian

$$h(\underline{z},\underline{\zeta})=\sum_n \left(s z_n^2 \zeta_n^2 + b z_{n+1} \zeta_n\right) \;, \tag{4}$$

where s is an arbitrary (complex) constant that does not show up in (2), as well as by the Lagrangian

$$\ell(\underline{z},\underline{\dot{z}})=\sum_n \left[(\dot{z}_n / z_n)^2 - 2b\dot{z}_n z_{n+1} / z_n^2 + b^2 (z_{n+1}/z_n)^2 \right] . \tag{5}$$

Exercise 4.4.9-5. Write out two real Hamiltonians, and two real Lagrangians, that yield the equations of motion in the plane (1). *Hint*: see the preceding *Exercise 4.4.9-4*, and Sect. 4.1.

The *integrability* of this many-body problem in the plane, (1), is demonstrated using (4.4.8-8), since (2) is thereby transformed into the *integrable* equations

$$\ddot{u}_n = \dot{u}_{n+1} \exp(u_{n+1} - u_n) - \dot{u}_{n-1} \exp(u_n - u_{n-1})$$

$$- \exp\left[2(u_{n+1} - u_n) \right] + \exp\left[2(u_n - u_{n-1}) \right] . \tag{6}$$

Remark 4.4.9-6. The equations of motion (2) are invariant under the transformation $z_n(t) \to \left[z_{N+1-n}(t_0 - t)\right]^{-1}$, namely if one sets

$$\tilde{z}_n(t) = \left[z_{N+1-n}(t_0 - t)\right]^{-1}, \tag{7}$$

with t_0 an arbitrary constant, then (2) entail

$$\ddot{\tilde{z}}_n = \dot{\tilde{z}}_n^{\,2} / \tilde{z}_n + b\left[\dot{\tilde{z}}_{n+1} - \dot{\tilde{z}}_{n-1} (\tilde{z}_n / \tilde{z}_{n-1})^2 \right] + b^2 \left[-(\tilde{z}_{n+1}^{\,2} / \tilde{z}_n) + (\tilde{z}_n^{\,3} / \tilde{z}_{n-1}^{\,2}) \right] . \tag{8}$$

Exercise 4.4.9-7. Set

$$\hat{z}_n(t) = \exp(\lambda t)\, z_n(t), \quad b = a / \lambda \;, \tag{9}$$

and write out the (autonomous) equations of motion entailed by (2) for $\hat{z}_n(t)$. Then take the limit $\lambda \to \infty$, and show that, up to trivial notational changes ($\hat{z}_n(t) \to z_n(t)$), these equation of motion reduce to (4.4.5-5).

4.4.10 A Hamiltonian example

In the preceding subsections of Sect. 4.4 we generally focussed on Newtonian equations of motion in the plane, although we often also provided the corresponding Hamiltonians and Lagrangians. In this Section we exhibit an example in which one must deal directly with the Hamiltonian equations of motion, since they cannot be easily written in Newtonian form.

Let us start from the Hamiltonian

$$H(\underline{z},\underline{\zeta})=\frac{1}{2}\sum_{n,m=1}^{N}\zeta_n\zeta_m(z_n-z_m)^2 \quad , \tag{1}$$

whose *solvability* is guaranteed by its coincidence, up to trivial notational changes, with (2.1.15.2-6,4) (with $a=1$, $\lambda=1$, $\mu=-1$; we stick hereafter, for simplicity, to this simple choice of these 3 constants, the diligent reader is welcome to consider the more general, albeit essentially equivalent, case when they have generic *complex* values).

The corresponding Hamiltonian equations of motion read

$$\dot{z}_n=\sum_{m=1}^{N}\zeta_m(z_n-z_m)^2 \quad , \tag{2a}$$

$$\dot{\zeta}_n=-2\,\zeta_n\sum_{m=1}^{N}\zeta_m(z_n-z_m) \quad . \tag{2b}$$

Hence, via (4.1-33,15) they can be rewritten as the following Hamiltonian equations in the plane:

$$\dot{\vec{r}}_n=\sum_{m=1}^{N}\left[2\,\vec{r}_{nm}(\vec{\rho}_m\cdot\vec{r}_{nm})-\vec{\rho}_m\vec{r}_{nm}^{\,2}\right] \quad , \tag{3a}$$

$$\dot{\vec{\rho}}_n=2\sum_{m=1}^{N}\left[\vec{r}_{nm}(\vec{\rho}_n\cdot\vec{\rho}_m)-\vec{\rho}_n(\vec{\rho}_m\cdot\vec{r}_{nm})-\vec{\rho}_m(\vec{\rho}_n\cdot\vec{r}_{nm})\right] \quad . \tag{3b}$$

Note the obvious invariance under rotations, and also under translations ($\vec{r}_n\to\tilde{\vec{r}}_n=\vec{r}_n+\vec{r}_0$, $\dot{\vec{r}}_0=0$; $\vec{\rho}_n\to\tilde{\vec{\rho}}_n=\vec{\rho}_n$), of these equations of motion, (3).

Exercise 4.4.10-1. Verify that (3) corresponds to (2). *Hint*: see (4.1-33,15).

An equivalent version of these equations of motion, (3), reads

$$\dot{\vec{r}}_n = \sum_{m=1}^{N} \left[-2\,\vec{r}_{nm}\,(\hat{k}\cdot\tilde{\vec{\rho}}_m \wedge \vec{r}_{nm}) + \hat{k}\wedge\tilde{\vec{\rho}}_m\, r_{nm}^2 \right] \tag{4a}$$

$$\dot{\tilde{\vec{\rho}}}_n = 2\sum_{m=1}^{N} \left[\hat{k}\wedge\vec{r}_{nm}\,(\tilde{\vec{\rho}}_n\cdot\tilde{\vec{\rho}}_m) + \tilde{\vec{\rho}}_n(\hat{k}\cdot\tilde{\vec{\rho}}_m\wedge\vec{r}_{nm}) + \tilde{\vec{\rho}}_m\,(\hat{k}\cdot\tilde{\vec{\rho}}_n\wedge\vec{r}_{nm}) \right] \tag{4b}$$

with

$$\tilde{\vec{\rho}}_n = \hat{k}\wedge\vec{\rho}_n,\ \vec{\rho}_n = -\hat{k}\wedge\tilde{\vec{\rho}}_n\ . \tag{5}$$

Exercise 4.4.10-2. Verify that the following two real Hamiltonians, obtained from (1),

$$H(\underline{\vec{r}},\underline{\vec{\rho}}) = \frac{1}{2}\sum_{m,n=1}^{N} \left[2(\vec{\rho}_n\cdot\vec{r}_{nm})(\vec{\rho}_m\cdot\vec{r}_{nm}) - (\vec{\rho}_n\cdot\vec{\rho}_m)r_{nm}^2 \right]\ , \tag{6}$$

respectively

$$\tilde{H}(\underline{\vec{r}},\underline{\tilde{\vec{\rho}}}) = \frac{1}{2}\sum_{m,n=1}^{N} \left[(\tilde{\vec{\rho}}_n\cdot\vec{r}_{nm})(\hat{k}\cdot\tilde{\vec{\rho}}_m\wedge\vec{r}_{nm}) + (\tilde{\vec{\rho}}_m\cdot\vec{r}_{nm})(\hat{k}\cdot\tilde{\vec{\rho}}_n\wedge\vec{r}_{nm}) \right]\ , \tag{7}$$

yield the Hamiltonian equations (3) respectively (4) (equivalent via (5)).

4.5 A many-rotator, possibly nonintegrable, problem in the plane, and its periodic motions

In Sect. 4.5 we discuss the many-body problem in the plane characterized by the Newtonian equations of motion

$$\ddot{\vec{r}}_n = \omega\,\hat{k}\wedge\dot{\vec{r}}_n$$

$$+2\sum_{m=1,\,m\neq n}^{N} (r_{nm})^{-2}(\alpha_{nm} + \alpha'_{nm}\,\hat{k}\wedge)\,[\dot{\vec{r}}_n(\dot{\vec{r}}_m\cdot\vec{r}_{nm}) + \dot{\vec{r}}_m(\dot{\vec{r}}_n\cdot\vec{r}_{nm}) - \vec{r}_{nm}(\dot{\vec{r}}_n\cdot\dot{\vec{r}}_m)]\,, \tag{1a}$$

where of course (see (4.1-15))

$$\vec{r}_{nm} \equiv \vec{r}_n - \vec{r}_m,\quad r_{nm}^2 = r_n^2 + r_m^2 - 2\,\vec{r}_n\cdot\vec{r}_m\ . \tag{1b}$$

These equations characterize the motion of N, generally different, particles, all having the same mass (set conveniently to unity), all interacting with the *same* one-body force (first term in the right hand side of (1a)), and also interacting pairwise via forces whose strengths may vary for every particle pair. The dynamics is clearly invariant under both rotations and translations (in the plane), but not under Galileian transformations ($\vec{r}_n(t) \rightarrow \tilde{\vec{r}}_n(t) = \vec{r}_n(t) + \vec{v}\,t$).

The one-body force is of Lorentz type: it suggests interpreting the N particles as point-like massive charges moving in a plane in the presence of a constant magnetic field orthogonal to that plane ("cyclotron" configuration). The (real) "coupling constant" ω is then the "Larmor circular frequency". We assume, without loss of generality, that this quantity is positive, $\omega > 0$, and we denote by T the corresponding period,

$$T = 2\pi / \omega . \tag{2}$$

If only this one-body force is present, namely if the two-body interactions are switched off by setting to zero all the corresponding coupling constants ($\alpha_{nm} = \alpha'_{nm} = 0$), then the n-th particle rotates (of course, independently of all the others) on a circular trajectory whose center $\vec{c}_n$ and radius ρ_n are determined by its initial position $\vec{r}_n(0)$ and velocity $\dot{\vec{r}}_n(0)$:

$$\vec{r}_n(t) = \vec{c}_n + \vec{\rho}_n \sin(\omega t) - \hat{k} \wedge \vec{\rho}_n \cos(\omega t) , \tag{3a}$$

$$\vec{c}_n = \vec{r}_n(0) + \hat{k} \wedge \vec{\rho}_n , \tag{3b}$$

$$\vec{\rho}_n = \omega^{-1} \dot{\vec{r}}_n(0) . \tag{3c}$$

Of course in this case the motion is *completely periodic*, with period T.

Exercise 4.5-1. Verify that (3) satisfies (1) with $\alpha_{nm} = \alpha'_{nm} = 0$.

If the two-body interactions are not altogether switched off, namely if not all the $2N(N-1)$ (real) "coupling constants" α_{nm}, α'_{nm} in (1a) vanish, then the motions are much more complicated, since every particle interacts pairwise with every other particle, via two-body forces proportional to the speeds of the two interacting particles and depending nonlinearly on their (relative) positions. Previous results entail the following information on this model.

If the coupling constants depend symmetrically on the two particles of the relevant interacting pair,

$$\alpha_{nm} = \alpha_{mn} , \qquad \alpha'_{nm} = \alpha'_{mn} , \tag{4}$$

the Newtonian equations of motion (1) are Hamiltonian (see below). In this case the center-of-mass of the system,

$$\vec{\bar{r}}(t) = N^{-1} \sum_{n=1}^{N} \vec{r}_n(t) , \tag{5}$$

moves itself as a single rotator:

$$\vec{\bar{r}}(t) = \vec{\bar{c}} + \vec{\bar{\rho}} \sin(\omega t) - \hat{k} \wedge \vec{\bar{\rho}} \cos(\omega t) , \tag{6a}$$

$$\vec{\bar{c}} = \vec{\bar{r}}(0) + \hat{k} \wedge \vec{\bar{\rho}} , \tag{6b}$$

$$\vec{\bar{\rho}} = \omega^{-1} \dot{\vec{\bar{r}}}(0) . \tag{6c}$$

Exercise 4.5-2. Prove this result. *Hint*: note the antisymmetry of the summand in the right hand side of (1a), under the interchange of the two indices n and m, when (4) holds.

If moreover *all* the constants α'_{nm} vanish and *all* the coupling constants α_{nm} equal unity,

$$\alpha_{nm} = 1 , \qquad a'_{nm} = 0 , \tag{7}$$

then the system of interacting particles (1) is *integrable* indeed *solvable*, and all its motions are *completely periodic*, with period (at most) $\tilde{T} = T \cdot N!$; note that this does not exclude the presence of completely periodic motions with period $T_M = T \cdot M$ where $1 \leq M < N!$, provided $N!$ is an integer multiple of M (so that periodicity with period T_M automatically entail periodicity with period $\tilde{T}$).

Exercise 4.5-3. Trace these findings! *Hint*: see Sect. 4.2.5, and below.

The mechanism whereby the initial data for the *solvable* model (1) with (7) divide into sets of finite measures which yield motions with *different* periods T_M, see above, will be evidenced below via the analysis of the periodic character of the motions of the system (1) (in the generic case, without (7)), which constitutes the core of Sect. 4.5.

The equations of motion (1) are clearly invariant under (time-independent) rescaling of the dependent variables $\vec{r}_n(t)$. Indeed, via the standard correspondence (4.1-17), they can be conveniently recast in the complex one-dimensional format:

$$\ddot{z}_n = i\omega z_n + 2 \sum_{m=1, m\neq n}^{N} a_{nm}\, \dot{z}_n\, \dot{z}_m / (z_n - z_m), \tag{8a}$$

with

$$a_{nm} = \alpha_{nm} + i\alpha'_{nm}. \tag{8b}$$

Hereafter, we use this avatar of the equations of motion (1), and we moreover exploit its connection, via the change of variable

$$z_n(t) = \zeta_n(\tau), \qquad \tau = [\exp(i\omega t) - 1]/(i\omega), \tag{9a}$$

with the system

$$\zeta''_n = 2 \sum_{m=1, m\neq n}^{N} a_{nm}\, \zeta'_n\, \zeta'_m / (\zeta_n - \zeta_m). \tag{10}$$

Here and below, the primes denote of course differentiations with respect to the independent variable τ. Note that the constant ω has completely disappeared from (10); nor does it feature in the relations among the initial data for (8) and (10), which read simply

$$z_n(0) = \zeta_n(0), \qquad \dot{z}_n(0) = \zeta'_n(0). \tag{9b}$$

All this of course entails that, to obtain the solution $z_n(t)$ of (8) corresponding to a given set of initial data, one can instead solve (10) with the *same* set of initial data, thereby determine $\zeta_n(\tau)$, and then use (9a) to obtain $z_n(t)$ (hence, as well, the solution of the initial value problem for (1)).

Exercise 4.5-4. Verify the connection, via (9a), among (8a) and (10), as well as the validity of (9b).

Exercise 4.5-5. Verify that the (complex) equations of motion (8a) are yielded by the (complex) Hamiltonian

$$H(\underline{z},\underline{\rho}) = \sum_{n=1}^{N} [i(\omega / c) z_n + \exp(c\rho_n) \prod_{m=1, m \neq n}^{N} (z_n - z_m)^{-a_{nm}}], \tag{11a}$$

with c an arbitrary (non vanishing) constant, provided the constraints (4) hold, namely

$$a_{nm} = a_{mn}; \tag{11b}$$

and note that this fact entails (see Sect. 4.1) the Hamiltonian character of (1) with (4).

Exercise 4.5-6. Review all other spots in this book where the change of independent variable (9a) has been used, and ponder on the fact that it is now being applied in the context of (possibly) *nonintegrable* many-body problems (namely, (1) or (8) possibly *without* (7)).

Let us now proceed and discuss the motions in the plane entailed by the (presumably *nonintegrable*) Newtonian equations of motions (1) describing N pairwise-interacting rotators, or equivalently the motions in the complex plane entailed by (8a), *without* any restriction on the coupling constants a_{nm}, see (8b). The cornerstone of our findings is the connection among (8) and (10) via (9), and the following property entailed by the change of independent variable (9):

Lemma 4.5-7. If $\zeta_n(\tau)$, considered as a function of the *complex* variable τ, is *analytic* in the closed disk C, with radius $1/\omega$, centered, in the complex τ-plane, at $\tau = \tau_0 = i/\omega$, then $z_n(t)$, considered as a function of the *real* variable t, is periodic in t with the period T, see (2),

$$z_n(t+T) = z_n(t). \tag{12}$$

This Lemma is merely a special case of *Proposition 2.1.13.1*; ist validity is implied by the observation that $\tau(t)$, see (7a), is itself periodic in t with period T,

$$\tau(t+T) = \tau(t). \tag{13}$$

The diligent reader is in any case advised to draw the disk C in the complex τ-plane.

But on the other hand the standard existence/uniqueness/analyticity theorem, applied to the initial-value problem for the system (10), entails that the solution $\zeta_n(\tau)$, corresponding to given initial data $\underline{\zeta}(0)$, $\underline{\zeta}'(0)$ (arbitrarily assigned at $\tau = 0$, but of course nonsingular, namely such that

$\zeta_j(0) \neq \zeta_k(0)$ if $j \neq k$, see (10)), is an *analytic* function of the complex variable τ in a disk D centered at $\tau = 0$, whose (nonvanishing!) radius d depends on the given initial data, and on the coupling constants a_{nm}, see (10) (note that d cannot depend on ω, since this quantity does not appear in the evolution equation (10); also note that the dependence of d on the initial data $\underline{\zeta}(0)$ and $\underline{\zeta}'(0)$ entails, via (9b), a completely analogous dependence on the initial data, $\underline{z}(0)$ and $\underline{\dot{z}}(0)$, of (8)). It is moreover clear that, if the coupling constant a_{nm}, or the initial data $\underline{z}(0)$ and $\underline{\dot{z}}(0)$, are changed by setting

$$a_{nm} = a\, \bar{a}_{nm}, \qquad z_m(0) = \lambda\, \tilde{z}_m(0), \qquad \dot{z}_m(0) = \mu\, \dot{\tilde{z}}_m(0), \tag{14}$$

and by then letting the (scaling, real) parameters a, λ, μ vary while keeping the barred and tilded quantities in (13) fixed, then d diverges, $d \to \infty$, as $a \to 0$ or $\lambda \to \infty$ or $\mu \to 0$.

These assertions follow clearly from the structure of the right hand side of (10): the radius d of the disk D is determined by the closest singularity to the origin, in the complex τ-plane, that is developed by the solution $\zeta_n(\tau)$ of (10), due to the nonlinear character of the evolution equations (10); clearly this singularity gets pushed farther away from the origin by "making smaller" the right hand side of (10), and this is precisely what the changes detailed above achieve, by decreasing the scale of the coupling constant a_{nm} or of the speeds $\dot{z}_m$, or by increasing the scale of the coordinates z_m.

Exercise 4.5-8. Write out a formal proof of the statements made above, about the behavior of the radius d when the coupling constants a_{nm}, or the initial data $z_m(0)$ and $\dot{z}_m(0)$, are varied as indicated above. *Hint:* see the proof of the standard existence/uniqueness/analyticity theorem for (systems of) ODEs.

But clearly, as soon as the radius d of the disk D exceeds the diameter, $2/\omega$, of the disk C, it encloses C and this entails, via *Lemma 4.5-7*, that $z_n(t)$ is periodic in t with period T , see (2) and (12). Hence we can state the following

Proposition 4.5-9. Let $\vec{r}_n(t)$ be the solution of (1) with

$$\alpha_{nm} = a\, \bar{\alpha}_{nm}, \qquad \alpha'_{nm} = a\, \bar{\alpha}'_{nm}, \qquad \omega = b\, \bar{\omega}, \tag{15a}$$

and with assigned initial data

$$\vec{r}_n(0) = \lambda\,\vec{u}_n, \qquad \dot{\vec{r}}_n(0) = \mu\,\vec{v}_n, \qquad [\vec{u}_n \neq \vec{u}_m \text{ if } n \neq m], \tag{15b}$$

where the positive numbers a, b, λ, μ play the role of scaling constants (as we shall see immediately). Then the solution $\vec{r}_n(t)$ is *completely periodic* with period T, see (2),

$$\vec{r}_n(t+T) = \vec{r}_n(t), \tag{16}$$

provided one of the following conditions hold:

(i) for given α_{nm}, α'_{nm}, ω, $\vec{r}_n(0)$ and $\vec{v}_n$, the scaling number μ, hence as well the initial velocities $\dot{\vec{r}}_n(0)$, are sufficiently small: $0 \leq \mu \leq \mu_c$, where μ_c is a *positive* number, $\mu_c > 0$, whose value depends on the given quantities;

(ii) for given α_{nm}, α'_{nm}, ω, $\dot{\vec{r}}_n(0)$ and $\vec{u}_n$, the scaling number λ is sufficiently large, $\lambda > \lambda_c$ (hence the initial positions of the N particles in the plane are sufficiently well separated), with λ_c a *positive* number, $\lambda_c > 0$, whose value depends on the given quantities;

(iii) for given $\bar{\alpha}_{nm}$, $\bar{\alpha}'_{nm}$, ω, $\vec{r}_n(0)$ and $\dot{\vec{r}}_n(0)$, the scaling number a, hence as well the magnitude of the coupling constants α_{nm}, α'_{nm}, are sufficiently small, $0 \leq a \leq a_c$, where a_c is a *positive* number, $a_c > 0$, whose value depends on the given quantities;

(iv) for given α_{nm}, α'_{nm}, $\bar{\omega}$, $\vec{r}_n(0)$ and $\dot{\vec{r}}_n(0)$, the scaling number b, hence as well the Larmor circular frequency ω, is sufficiently large, $b > b_c$, where b_c is a *positive* number, $b_c > 0$, whose value depends on the given quantities.

Note that we have chosen to formulate this *Proposition 4.5-9* in terms of the physical system (1) rather than the (equivalent) *complex* system (8). As for its validity, clearly in its first three formulations (see items *(i)*, *(ii)*, *(iii)* above) it is implied by the discussion given above, hence it requires no additional elaboration here. As for the last formulation (see item *(iv)* above), its proof follows even more directly from our treatment: since the (*positive!*) radius d of the disk D is independent of ω, for sufficiently large ω (indeed, precisely for $\omega > \omega_c = 2/d$) the disk C gets completely inside the disk D (draw diagram!), hence the corresponding solution of $z_n(t)$ of (8) (or, equivalently, the solution $\vec{r}_n(t)$ of (1)) is completely periodic with period T, see (2).

The first two aspects of this *Proposition 4.5-9* (see items *(i)* and *(ii)*) refer to a given system of equations of motion, (1), and to different choices of initial data: we have formulated this *Proposition 4.5-9* imagining to vary, and in a very special manner (by a common rescaling), either only the initial velocities, or only the initial positions; clearly the essence of the results holds much more generally, and it amounts to the statement that, for any given system of type (1), there exist sets of initial data, having nonvanishing measure in phase space, which yield *completely periodic* motions, with period T, see (2). Qualitatively, these initial data are characterized by the requirement not to entail too much interaction among the (interacting!) rotators, which should be sufficiently far apart from each other, and not move too fast. On the other hand it is clear (see below) that, if the initial data entail a sufficiently strong interaction among the rotators, in the *nonintegrable* case (when (7) does not hold) the corresponding motions cease to be periodic.

And even in the *integrable* case (namely when the restriction (7) on the coupling constants does hold), if the initial data entail a sufficiently strong interaction among the rotators, the corresponding motion ceases to be periodic with period T, although in that *integrable* case it remains *completely periodic*, but with a larger period which is a multiple of T, at most $\tilde{T} = T \cdot N!$. We will revisit this issue below, after we have gained a better understanding of the mechanism that underlies this phenomenon via the discussion of a simple example.

The other two aspects of *Proposition 4.5-9* (see items *(iii)* and *(iv)*) refer instead to the (somewhat less "physical") consideration of a variation in the parameters, be they the "coupling constants" α_{nm}, α'_{nm} or the "Larmor circular frequency" ω, that characterize the model (1); a variation performed while keeping fixed the initial data that determine the motion.

Remark 4.5-10. Of course *Proposition 4.5-9* does not exclude that there might also be some values of the parameters ω, α_{nm}, α'_{nm} of the model (1) (or equivalently ω, a_{nm} in (8)), or of the initial conditions $\underline{\vec{r}}(0)$, $\dot{\underline{\vec{r}}}(0)$ (or $\underline{z}(0)$, $\dot{\underline{z}}(0)$) which do *not* lie within the bounds dictated by *Proposition 4.5-9*, yet do yield solutions *completely periodic* with period T, see (2).

Let us now pause to consider a simple, but quite illuminating, example.

Exercise 4.5-11. Show that the solution of the equations of motion (8) with $N=2$, in the (Hamiltonian) case with

$$a_{12} = a_{21} = a \equiv \alpha + i\alpha', \tag{17a}$$

and with the restriction that the center of mass, $\bar{z}(t) = (1/2)[z_1(t) + z_2(t)]$, *not* move (indeed, without loss of generality, let it sit just at the origin of coordinates in the complex z-plane, $\bar{z}(t) = 0$, so that

$$z_1(t) = -z_2(t) = z(t)\,; \tag{17b}$$

and check that the condition that the center-of-mass not move, $\dot{\bar{z}}(t)=0$, is in this case, see (17), consistent with the equations of motion (8); *hint*: see (6)), is given by the following formula: if $a \neq -1$,

$$z(t) = z(0)\ \{[\rho - \exp(i\omega t)]/(\rho - 1)\}^{1/(1+a)}, \tag{18a}$$

with

$$\rho = 1 - i\omega\ z(0)/[(1+a)\dot{z}(0)]; \tag{18b}$$

if $a = -1$,

$$z(t) = z(0)\ \exp(\{i\dot{z}(0)\ /\ [\omega\, z(0)]\}\,[1 - \exp(i\omega t)]). \tag{18c}$$

Hint: rather than looking at the ODE satisfied by *z(t)*,

$$\ddot{z} = i\omega z - a\dot{z}^2 / z, \tag{19a}$$

focus on the ODE satisfied by $\zeta(\tau)$ (see (9)),

$$\zeta'' = -a\zeta'^2 / \zeta; \tag{19b}$$

then divide this ODE by ζ' and integrate (twice, sequentially, using the initial conditions, see (9b)); finally use (9a) to recover $z(t)$.

Let us now analyze this solution, (18). Clearly, if the initial conditions entail $|\rho| > 1$ (see (18b)), $z(t)$ is periodic in t with period T, see (2) and (12) or (16) (if this conclusion is not evident see below). Note the consistency of this result with *Proposition 4.5-9*.

Exercise 4.5-12. Compute the (minimal respectively maximal) values of the quantities μ_c and a_c respectively λ_c and b_c, see *Proposition 4.5-9*, for this case, see (18a,b).

Exercise 4.5-13. Use this example, see (18), to illustrate a case in which *Remark 4.5-10* becomes applicable. *Hint*: see (18c).

If instead the parameters of the model (namely, the real number ω and the, possibly complex, number a), and the initial conditions (namely, the generally complex numbers $z(0)$ and $\dot{z}(0)$) yield a value of ρ , see

(18b), such that $|\rho|<1$, then $z(t)$ is *not* periodic in t with period T. Let us look in detail at the mechanism whereby this happens.

But firstly let us dispose of the special case $a=-1$. In this case, see (18c), the solution $z(t)$ is *always* periodic in t with period T, see (2). Indeed, this case ((1) with $N=2$ and $a=-1$) corresponds to the *integrable* (indeed *solvable*) model (4.4.6-1), of course with $N=2$ and $\vec{r}_0(t)=\vec{r}_3(t)=0$ (see (2.1.13-24)), and this periodicity property had been already predicted, see the second paragraph in Sect. 4.4.6 (beware of the notational differences: the parameters α respectively α' of Sect. 4.4.6 should now be replaced by *zero* respectively ω).

Hereafter we assume $a\neq-1$, so that the solution $z(t)$ is given by (18a) with (18b), which, for the purpose of the following discussion, we rewrite as follows:

$$z(t)=z(0)\,(\rho-1)^{-1/(1+a)}\,f[w(t)], \tag{20a}$$

$$f(w)\equiv w^{1/(1+a)}, \tag{20b}$$

$$w(t)\equiv[\rho-\exp(i\omega t)]. \tag{20c}$$

Let us now discuss how $z(t)$ behaves as a function of (real) t. For a generic value of a, the function $f(w)$, see (20b), of the complex variable w, has a branch point at $w=0$ (and another one at $w=\infty$); while $w(t)$ is a periodic function of (real) t, which, as t varies over one period, say from 0 to T, travels in the complex w-plane over a circle c of radius 1, centered at $w=\rho$. If $|\rho|>1$ the branch point at $w=0$ lies *outside* this circle c; hence f (see (20b)), considered as a function of t, is periodic in t with period T, see (2), and the same conclusion applies to $z(t)$, see (20a), as already noted above. If instead $|\rho|<1$, the branch point at $w=0$ lies *inside* the circle c, hence, when w completes its travel over the circle c and comes back to its original value (say, it goes from $\rho-1$ to $\rho-1$, as t varies from 0 to T, see (2) and (20c)), the function f does *not* come back to its original value, because now the contour c over which w traveled crosses necessarily the branch cut of the function $f(w)$, that starts inside the circle c (at $w=0$) and ends outside it (at $w=\infty$). However, if the exponent $1/(1+a)$ is a rational number, say

$$\frac{1}{1+a}=\frac{p}{q}, \tag{21}$$

with p and q integers (and relatively prime, namely such that their decomposition into primes contains no common factor), then clearly f, considered as a function of t, is periodic with period $T q$ (since in this case the Riemann surface associated with the branch cut has a finite number of sheets, q; hence by traversing the branch cut q times in the same direction one gets back to the original sheet).

The *integrable* case of (1) corresponds, in the example under present consideration, to $a=1$; hence in this case $p=1,\ q=2,$ see (21). We thus conclude that, in the *integrable* case of the example under consideration, the solution $z(t)$ is periodic in t, with period T if $|\rho|>1$, with period $2T$ if $|\rho|<1$. Note the consistency of this finding with the general result, based on the exact technique (see Sect. 4.2.5) of solution of (1) in the *integrable* (indeed *solvable*) case (see (7)), according to which *all* solutions of (1), in this *integrable* case (see (7)), are *completely periodic* in t, with period $\tilde{T}=T\cdot N!$ (for $N=2$, $\tilde{T}=2\,T$; of course a solution periodic with period T is also periodic with period $2\,T$).

And what about the boundary case with $|\rho|=1$? The subcase with $\rho=1$ is of course forbidden, since it corresponds to *singular* initial data (either $z(0)=0$ or $\dot{z}(0)=\infty$, see (18b) and (19a)). If $|\rho|=1,\ \rho\neq 1$, let us uniquely define the "critical time" t_c by setting (see (2))

$$\rho=\exp(i\,\omega\,t_c), \qquad 0<t_c<T. \tag{22}$$

Then clearly, at $t=t_c$, $w=0$ (see (20c)), hence, as $t\to t_c$, either $z(t)\to 0$ or $z(t)\to\infty$ (see (20)); hence we conclude that, in this case ($|\rho|=1,\ \rho\neq 1$), at $t=t_c$ the evolution equation (19a) becomes *singular*. There is, however, at least in the *integrable* $a=1$ case, a "natural" continuation of the solution, as a function of t, beyond the singularity; the resulting solution is then periodic (with period T or $2T$, depending on the sign convention for the square root adopted to perform the continuation), and it displays of course a recurring singularity at $t=t_c \bmod(T)$.

In conclusion we see that, for the special example we are now discussing (see *Exercise 4.5-11*), in the *integrable* case ($a=1$) the initial data get divided into two sets, both clearly of nonvanishing measure in the (restricted) phase space under consideration: the first set, identified by the condition $|\rho|>1$, see (18b), yields solutions periodic with period T, see (2); the second set, identified by the condition $|\rho|<1$, see (18b), yields solutions periodic with period $2T$; and these two sets of initial data are separated by a (lower dimensional) set of initial data, characterized by the condition $|\rho|=1,\ \rho\neq 1$, that yields *singular* solutions (which may be made periodic, with periods T or $2T$, by providing an appropriate prescription to continue them beyond the singularity). The physical significance of the singularity is a collision of the two particles, at the finite time t_c: indeed as $t\to t_c$, $z(t)\sim(t-t_c)^{1/2}$, $\dot{z}(t)\sim(t-t_c)^{-1/2}$, $\ddot{z}(t)\sim(t-t_c)^{-3/2}$; note the consistency of this behavior, entailed by the solution (20) with $a=1$, with the equation of motion (19a).

An analogous phenomenology occurs in the general case, with arbitrary a ($\neq\pm 1$); but of course with a difference. Again the initial data are divided into two sets, characterized respectively by the relations $|\rho|>1$ and $|\rho|<1$, see (18b). In the first case the solution is again periodic with period T, see (2); in the second, it is instead, generally, not periodic, except in the special cases (21), when it is periodic with period $T\,q$. And these two sets are again separated by the (lower dimensional) set of initial data characterized by the condition $|\rho|=1,\ \rho\neq 1$, see (18b), which defines via (22) the value t_c of the critical time, $t=t_c$, when the solution develops a singularity, whose physical significance is again a collision of the 2 interacting rotators.

We have discussed at considerable length this very special example, see *Exercise 4.5-11*, because it provides a neat illustration of the mechanism – the interplay of *analyticity* in the complex variable τ and *periodicity* in the real time variable t – that underlies the validity of *Proposition 4.5-9*, and indeed evidences a phenomenology whose validity may well extend beyond the particular many-body model treated in Sect. 4.5, see (1).

In this connection let us emphasize that, in the particular example we just discussed (see *Exercise 4.5-11*), the case with a generic, possibly complex, value of $a=\alpha+i\alpha'$ (see (17) and (19)) may be considered to mimic the generic, hence presumably *nonintegrable*, case of the model (1); in spite of the fact that the specific example of *Exercise 4.5-11*, mainly due to the $N=2$ restriction, is of course *integrable* for generic a (at least in the subcase with fixed center-of-mass), as evidenced by the fact that we did indeed integrate it, see (18).

Let us end Sect. 4.5 by pointing out that the same kind of results expressed by *Proposition 4.5-9* for the system (1) or (8) can be extended (in some cases with minimal modifications; but not in other cases) to more general systems than (1).

Exercise 4.5-14. Formulate and prove results analogous to *Proposition 4.5-9* for the system of ODEs (more general than (8))

$$\ddot{z}_n = i\omega \dot{z}_n + \sum_{l,m=1}^{N} \dot{z}_m \dot{z}_l \varphi_{nml}(\underline{z}), \quad (23)$$

Hint: see <CF2000c>.

Exercise 4.5-15. Verify that the change of dependent and independent variables

$$z_n(t) = \exp(-i\omega t)\, \zeta_n(\tau), \qquad \tau = [\exp(2i\omega t) - 1]/(2i\omega), \quad (24)$$

transforms the system

$$\ddot{z}_n + \omega^2 z_n = 2 \sum_{m=1, m\neq n}^{N} g_{nm}^2 (z_n - z_m)^{-3}, \quad (25)$$

(which, for

$$g_{nm}^2 = g^2, \quad (26)$$

is *integrable* indeed *solvable*, and in this solvable case only possesses, for real ω, completely periodic solutions with period T, see (2); see Sect. 2.1.3.3), into the system

$$\zeta_n'' = 2 \sum_{m=1, m\neq n}^{N} g_{nm}^2 (\zeta_n - \zeta_m)^{-3}, \quad (27)$$

which does not feature at all the parameter ω.

Exercise 4.5-16. The possibility to transform (25) into (27) via (24) suggests that an analogous result to *Proposition 4.5-9* (which has been shown above to be applicable to the system (1) or (8) and to (23), see *Exercise 4.5-14*) be also applicable to (25), entailing the existence of a set, having nonvanishing measure in phase space, of initial values which entail that the corresponding solutions of (25) are *completely periodic* with period T, see (2), even in the, presumably *nonintegrable*, case with different coupling constants g_{nm}^2 for different particle pairs, namely when (26) does *not* hold. Why is this hunch (presumably) false (at least for the "physical" model (25), with *real* particle coordinates z_n)? *Hint*: the transformation (24) entails

$$\zeta_n(0) = z_n(0), \qquad \zeta_n'(0) = \dot{z}_n(0) + i\omega\, z_n(0) \quad (28)$$

(and note that this is *different* from (9b)).

Exercise 4.5-17. Show that the main results about the *many-rotator* model in the plane (1) (in particular the existence of *completely periodic* motions, both in the *integrable* and in the *nonintegrable* cases) are as well valid, up to obvious notational changes, for the (of course rotation-invariant, but *not* translation-invariant) *many-oscillator* model in the plane characterized by the following Newtonian equations of motion (see (1b)):

$$\ddot{\vec{r}}_n + \Omega^2 \vec{r} = 2 \sum_{m=1, m \neq n}^{N} (r_{nm})^{-2} \cdot$$

$$\cdot \{ (\alpha_{nm} + \alpha'_{nm} \hat{k} \wedge) [\dot{\vec{r}}_n (\dot{\vec{r}}_m \cdot \vec{r}_{nm}) + \dot{\vec{r}}_m (\dot{\vec{r}}_n \cdot \vec{r}_{nm}) - \vec{r}_{nm} (\dot{\vec{r}}_n \cdot \dot{\vec{r}}_m) + \Omega^2 (\vec{r}_n r_m^2 - \vec{r}_m r_n^2)]$$

$$- (\alpha'_{nm} - \alpha_{nm} \hat{k} \wedge) \Omega [\dot{\vec{r}}_n (\vec{r}_m \cdot \vec{r}_{nm}) + \dot{\vec{r}}_m (\vec{r}_n \cdot \vec{r}_{nm}) - \vec{r}_n ((\dot{\vec{r}}_n + \dot{\vec{r}}_m) \cdot \vec{r}_m) + \vec{r}_m ((\dot{\vec{r}}_n + \dot{\vec{r}}_m) \cdot \vec{r}_n)] \}. \tag{29}$$

Hint: note that via the position

$$z_n(t) = w_n(t) \exp(i \Omega t), \tag{30a}$$

with

$$\Omega = \omega / 2 , \tag{30b}$$

the equations of motion (8a) become

$$\ddot{w}_n + \Omega^2 w_n = 2 \sum_{m=1, m \neq n}^{N} a_{nm} [\dot{w}_n \dot{w}_m + i \Omega (\dot{w}_n w_m + \dot{w}_m w_n) - \Omega^2 w_n w_m] / (w_n - w_m), \tag{31}$$

and that via the relation (see (4.1-17))

$$w_n \doteq \vec{r}_n \tag{32}$$

these equations of motion, (31), coincide with (29).

Exercise 4.5-18. Verify that the 3-body problem in the plane characterized by the following (rotation-invariant!) Newtonian equations of motion is *solvable*:

$$\ddot{\vec{r}}_n = 2(\alpha + \alpha' \hat{k}\wedge)\dot{\vec{r}}_n + (\beta + \beta' \hat{k}\wedge)\vec{r}_n + \left[2(\gamma_1 + \gamma_1' \hat{k}\wedge)\dot{\vec{r}}_n + (\gamma_2 + \gamma_2' \hat{k}\wedge)\vec{r}_n \right] / D , \tag{33a}$$

$$\gamma_1 \equiv \gamma_1(\vec{r}_1,\vec{r}_2,\vec{r}_3;\dot{\vec{r}}_1,\dot{\vec{r}}_2,\dot{\vec{r}}_3) = \gamma\, \mu_1(\vec{r}_1,\vec{r}_2,\vec{r}_3;\dot{\vec{r}}_1,\dot{\vec{r}}_2,\dot{\vec{r}}_3) + \gamma'\, \mu_2(\vec{r}_1,\vec{r}_2,\vec{r}_3;\dot{\vec{r}}_1,\dot{\vec{r}}_2,\dot{\vec{r}}_3) , \tag{33b}$$

$$\gamma_1' \equiv \gamma_1'(\vec{r}_1,\vec{r}_2,\vec{r}_3;\dot{\vec{r}}_1,\dot{\vec{r}}_2,\dot{\vec{r}}_3) = \gamma'\, \mu_1(\vec{r}_1,\vec{r}_2,\vec{r}_3;\dot{\vec{r}}_1,\dot{\vec{r}}_2,\dot{\vec{r}}_3) - \gamma\, \mu_2(\vec{r}_1,\vec{r}_2,\vec{r}_3;\dot{\vec{r}}_1,\dot{\vec{r}}_2,\dot{\vec{r}}_3) , \tag{33c}$$

$$\gamma_2 \equiv \gamma_2(\vec{r}_1,\vec{r}_2,\vec{r}_3;\dot{\vec{r}}_1,\dot{\vec{r}}_2,\dot{\vec{r}}_3) = -\gamma\, \nu_1(\vec{r}_1,\vec{r}_2,\vec{r}_3;\dot{\vec{r}}_1,\dot{\vec{r}}_2,\dot{\vec{r}}_3) - \gamma'\, \nu_2(\vec{r}_1,\vec{r}_2,\vec{r}_3;\dot{\vec{r}}_1,\dot{\vec{r}}_2,\dot{\vec{r}}_3) , \tag{33d}$$

$$\gamma_2' \equiv \gamma_2'(\vec{r}_1,\vec{r}_2,\vec{r}_3;\dot{\vec{r}}_1,\dot{\vec{r}}_2,\dot{\vec{r}}_3) = -\gamma'\, \nu_1(\vec{r}_1,\vec{r}_2,\vec{r}_3;\dot{\vec{r}}_1,\dot{\vec{r}}_2,\dot{\vec{r}}_3) + \gamma\, \nu_2(\vec{r}_1,\vec{r}_2,\vec{r}_3;\dot{\vec{r}}_1,\dot{\vec{r}}_2,\dot{\vec{r}}_3) , \tag{33e}$$

$$\mu_1 \equiv \mu_1(\vec{r}_1,\vec{r}_2,\vec{r}_3;\dot{\vec{r}}_1,\dot{\vec{r}}_2,\dot{\vec{r}}_3) \equiv \sum_{l,m=1}^{3} \left[(\dot{\vec{r}}_m \cdot \vec{r}_l)(\vec{r}_m \cdot \vec{r}_l) - (\hat{k} \cdot \dot{\vec{r}}_m \wedge \vec{r}_l)(\hat{k} \cdot \vec{r}_m \wedge \vec{r}_l) \right] , \tag{33f}$$

$$\mu_2 \equiv \mu_2(\vec{r}_1,\vec{r}_2,\vec{r}_3;\dot{\vec{r}}_1,\dot{\vec{r}}_2,\dot{\vec{r}}_3) \equiv \sum_{l,m=1}^{3} \left[(\dot{\vec{r}}_m \cdot \vec{r}_l)(\hat{k} \cdot \vec{r}_m \wedge \vec{r}_l) + (\vec{r}_m \cdot \vec{r}_l)(\hat{k} \cdot \dot{\vec{r}}_m \wedge \vec{r}_l) \right] , \tag{33g}$$

$$\nu_1 \equiv \nu_1(\vec{r}_1,\vec{r}_2,\vec{r}_3;\dot{\vec{r}}_1,\dot{\vec{r}}_2,\dot{\vec{r}}_3) \equiv \sum_{l,m=1}^{3} \left[(\dot{\vec{r}}_m \cdot \vec{r}_l)^2 - (\hat{k} \cdot \dot{\vec{r}}_m \wedge \vec{r}_l)^2 \right] , \tag{33h}$$

$$\nu_2 \equiv \nu_2(\vec{r}_1,\vec{r}_2,\vec{r}_3;\dot{\vec{r}}_1,\dot{\vec{r}}_2,\dot{\vec{r}}_3) \equiv \sum_{l,m=1}^{3} 2(\dot{\vec{r}}_m \cdot \vec{r}_l)(\hat{k} \cdot \dot{\vec{r}}_m \wedge \vec{r}_l) , \tag{33i}$$

$$D \equiv D(\vec{r}_1,\vec{r}_2,\vec{r}_3;\dot{\vec{r}}_1,\dot{\vec{r}}_2,\dot{\vec{r}}_3) \equiv \sum_{l,m=1}^{3} \left[r_m^2 r_l^2 - 2(\hat{k} \cdot \vec{r}_m \wedge \vec{r}_l)^2 \right] ; \tag{33l}$$

(ii) find conditions on the 6 "coupling constants" $\alpha,\alpha',\beta,\beta',\gamma,\gamma'$ sufficient to guarantee that *all* nonsingular solutions of these equations of motion, (33), are completely periodic, and conditions on the initial data, $\vec{r}_n(0),\dot{\vec{r}}_n(0)$, sufficient to guarantee that the corresponding solution of (33) is *not* singular. *Hints*: use the results of Sect. 5.1 below, in particular *(i)* identify (33) with the *complexified* version of the *solvable* 3-vector equation of motion (5.1-29), by considering each component of the 3-vector satisfying this *solvable* equation of motion, (5.1-29), as a separate "particle variable", say $\rho_n \equiv \rho_n(t), n = 1,2,3$, and by then identifying the complex numbers ρ_n with the real 2-vectors $\vec{r}_n$, see (4.1-17) (with z_n replaced by ρ_n), and of course by also setting $a = \alpha + i\alpha'$, $b = \beta + i\beta'$, $c = \gamma + i\gamma'$ in (5.1-29); *(ii)* see *Exercise 5.1-14*.

4.6 Outlook

In Chap. 4 it has been pointed out that the technique of complexification – amounting to the consideration of evolving systems in the *complex*, rather than the *real*, field – can in appropriate instances be interpreted as giving rise to genuine, rotation-invariant, many-body problems in the *physical* two-dimensional space: *real motions in the plane.* Several such systems amenable to exact treatments have been exhibited, but clearly many more can be introduced by this approach. Few of these systems have been studied in any detail: much therefore remains to be done in this direction as well, both for the systems described in Chap. 4, and for new ones which can be manufactured by analogous techniques (for instance, by complexifying the *partially solvable* 3-body problem (2.5-44c), and/or the *solvable translation-invariant* N-body problem (2.5-50) (for instance for $N=2$ and $N=3$), and/or the *solvable* N-body problem (2.5-67) with $A_0 = A_1 = A_3 = B_0 = 0, \ldots$).

Several of the systems exhibited above are characterized by remarkable behaviors, for instance by the presence of many, in some case of only, *completely periodic* motions. Especially such systems are interesting candidates for quantum mechanical treatment.

Finally, let us once more emphasize that the gamut of possible motions is much richer in the plane than along the line; hence, even when the technique on which these results are based corresponds merely to an extension from the *real* to the *complex*, it entails a *physically* very interesting step.

And let us end Chap. 4 by drawing attention to the results of Sect. 4.5, both because they, remarkably, are also applicable to systems which are *not* integrable (nor linearizable), and because they uncover and display certain relations among *analyticity* in the time variable and the emergence of *integrable* behaviors (in particular, in this specific case, *completely periodic* motions), which might have a much wider relevance, namely provide illuminating insights in other contexts, well beyond the specific model treated herein.

4.N Notes to Chapter 4

The notion that certain (one-dimensional) many-body problems on the line yield, by complexification, *rotation-invariant* many-body problems in the plane was introduced in <C96b>, and further elaborated in <C98c> and <C97d>. As already mentioned, the idea to complexify comes naturally in the context of models whose treatment invokes the study of the motion of the zeros of a polynomial, since the natural setting for any such investigation is the *complex plane* rather than the *real line*; and this hunch was reinforced by the discovery that certain *complex* deformations of certain exactly treatable many-body problems feature the remarkable property to possess *only completely periodic motions* <C97c>, <C97d>, <CF2000a>.

The treatment of Sect. 4.1 is based on the three papers <C96b>, <C98c> and <C97d>.

The material of Sect. 4.2 and its subsections (except the last one, Sect. 4.2.6) is gleaned from the first of these three papers, <C96b> (for a review of results associated with the "simplest" models treated in Sects. 4.2.4 and 4.2.5 see also <C99b>).

An extension (of possible interest in fluid dynamic) of the results reported in Sect. 4.2, corresponding to the limit in which the number of particles, N, diverges giving rise to a continuum distribution of particles in the plane is treated in <C97b>.

The material of Sect. 4.3 and its subsections is largely based on <C98d>, but it also contains new results.

The material of Sect. 4.4, as indeed indicated there, is mainly reported from <C98c> (with several corrections!), as well as from <C98a>. The integrability of (4.4.7-9,10,11) has been reported by R.I. Yamilov (on the basis of previous work with A.B. Shabat) <Y92>. The integrability of (4.4.8-9) and (4.4.9-6) has been discovered by Yu. B. Suris <S97> (see also <AS97>).

The treatment of Sect. 4.5 follows closely <CF2000c> (except for the last four *exercises*).

For some follow-up to the remarks proffered in the last paragraph of Sect. 4.6 see <CF2001>.

5 MANY-BODY SYSTEMS IN ORDINARY (THREE-DIMENSIONAL) SPACE: SOLVABLE, INTEGRABLE, LINEARIZABLE PROBLEMS

In Chap. 5 we outline a technique to manufacture many-body problems in ordinary (three-dimensional) space amenable to exact treatments, and we exhibit a fairly large collection of such examples, generally characterized by *rotation-invariant* equations of motion, mostly of Newtonian type (see (1.1-18)).

Of course throughout Chap. 5, a superimposed arrow identifies three-vectors, namely vectors in ordinary (three-dimensional) space, say $\vec{r} \equiv (x,y,z)$; the exceptions are in Sects. 5.3, 5.5 and (mainly) 5.6.5, where we also use (with appropriate warning!) this notation to denote S-vectors, with S an *arbitrary* positive integer.

The main idea from which the results of Chap. 5 flow is *(i)* to identify treatable evolution equations for *matrices*, *(ii)* to parametrize the evolving matrices in terms of *three-vectors* (and perhaps also of scalars), and *(iii)* to obtain thereby *evolution equations for three-vectors*. This simple recipe yields equations of motion of Newtonian type for three-vectors (see (1.1-18)), provided the evolution equations for the matrices are suitable (for instance, they should be of second-order in the time-derivative), and provided moreover the parametrization of these matrices in terms of three-vectors is compatible with their time-evolution, in the sense of transforming it into a *covariant* (hence obviously *rotation-invariant*) time-evolution for three-vectors. Several examples are given below.

Most of these examples feature equations of motion of Newtonian type ("acceleration equal force"); these equations are generally covariant (*rotation-invariant*); in some cases they are *translation-invariant* as well. In most cases the forces are *velocity-dependent*, although some cases with *velocity-independent* forces are also presented (see in particular Sect. 5.6.5). Some models only feature *one-* and *two-body* forces. Some models are *Hamiltonian*.

The plan of the presentation is clear enough from the titles of the following sections (see Contents), not to require additional elaboration here. Let us emphasize that our presentation below is aimed at explaining how *solvable* and/or *integrable* and/or *linearizable* many-body problems in

F. Calogero: LNPm 66, pp. 511 - 662, 2001

three-dimensional space can be manufactured, by detailing the methodology, by providing relevant tools, and by exhibiting several examples (the worth of the pudding is in the eating). But we do not delve into any detailed analysis of the solutions of the many-body problems we exhibit (so, we hardly eat the pudding: we only cook and serve it!), except for occasional observations, for instance about those models which feature only *confined, multiply periodic* or *completely periodic* motions.

Let us finally mention that the approach described in Chap. 5 could clearly be as well applied to manufacture treatable N-body problems in spaces of $2, 4$ or more dimensions; but we do not delve into these developments in this book (except, as mentioned above, for reporting some results applicable in S-dimensional space, with S an *arbitrary* positive integer).

5.1 A simple example: a solvable matrix problem, and the corresponding one-body problem in three-dimensional space

In Sect. 5.1 we illustrate the methodology outlined above on a simple example.

Firstly, we exhibit an explicitly *solvable* nonlinear matrix evolution equation. It reads

$$\ddot{\underline{M}} = 2a\,\dot{\underline{M}} + b\,\underline{M} + c\,\dot{\underline{M}}\,\underline{M}^{-1}\,\dot{\underline{M}} \quad . \tag{1}$$

Here $\underline{M} \equiv \underline{M}(t)$ is an invertible square matrix of arbitrary rank, and a, b, c are 3 arbitrary (possibly complex) scalar constants (the factor 2 in front of a is introduced for notational convenience, see below).

The solution of (1) reads as follows:

$$\underline{M}(t) = \exp\left[\, a\,t/(1-c)\,\right] \cdot$$

$$\cdot \left\{\cosh(\Delta t) + \Delta^{-1}\sinh(\Delta t)\left[(1-c)\,\dot{\underline{M}}(0)\left[\underline{M}(0)\right]^{-1} - a\,\right]\right\}^{1/(1-c)}\,\underline{M}(0) \quad , \tag{2a}$$

$$\Delta = \left[\; a^2 + b(1-c)\;\right]^{1/2} \quad , \tag{2b}$$

where of course $\underline{M}(0)$ and $\dot{\underline{M}}(0)$ are the (arbitrary; but of course $\underline{M}(0)$ must be invertible) input data of the initial-value problem for (1).

This solution, as written, is applicable provided $c \neq 1$. The $c=1$ case can be obtained by a limiting process, but the corresponding solutions deserve to be displayed explicitly: if $c=1$ and $a \neq 0$,

$$\underline{M}(t) = \exp\{[b/(2a)][-t+\exp(at)\, a^{-1}\sinh(at)]\}\cdot$$

$$\cdot\exp\{\exp(at)\, a^{-1}\sinh(at)\,\dot{\underline{M}}(0)\,[\underline{M}(0)]^{-1}\}\,\underline{M}(0) \quad ; \tag{3}$$

if $c=1$ and $a=0$,

$$\underline{M}(t) = \exp(bt^2/2)\,\exp\{\,t\,\dot{\underline{M}}(0)[\underline{M}(0)]^{-1}\}\,\underline{M}(0) \quad . \tag{4}$$

Exercise 5.1-1. Verify that (2), and, in the respective special cases, (3) and (4), solve the initial-value problem for (1).

To discuss the behavior of the solution (2) of the matrix evolution equation (1) it is convenient to introduce the matrix

$$\underline{A} = \Delta^{-1}\{\,(1-c)\dot{\underline{M}}(0)[\underline{M}0(\;)]^{-1} - a\,\} \tag{5}$$

and, assuming for simplicity it is diagonalizable, to denote by α_n its eigenvalues and by $\underline{W}$ the matrix that diagonalizes it:

$$\underline{A} = \underline{W}\,\mathrm{diag}(\alpha_n)\,\underline{W}^{-1} \;. \tag{6}$$

Then (2) can be conveniently rewritten as follows:

$$\underline{M}(t) = \underline{W}\;\mathrm{diag}[\mu_n(t)]\,\underline{W}^{-1} \;, \tag{7}$$

so that the time-dependence is all carried by the diagonal elements $\mu_n(t)$, defined by one of the following three equivalent expressions:

$$\mu_n(t) = \exp[at/(1-c)][\;\cosh(\Delta t)+\alpha_n\sinh(\Delta t)\;]^{1/(1-c)} \;, \tag{8a}$$

$$\mu_n(t) = \exp[(a-\Delta)t/(1-c)][\;(1+\alpha_n)/2\;]^{1/(1-c)}[\;\exp(2\Delta t)-\beta_n\;]^{1/(1-c)} \;, \tag{8b}$$

$$\mu_n(t) = \exp[(a+\Delta)t/(1-c)][\;(1-\alpha_n)/2\;]^{1/(1-c)}[\;\exp(-2\Delta t)-\beta_n^{-1}\;]^{1/(1-c)} \;, \tag{8c}$$

where

$$\beta_n = (\alpha_n - 1)/(\alpha_n + 1) \ . \tag{9}$$

Exercise 5.1-2. Verify!

Exercise 5.1-3. Show that, if the three "coupling constants" a, b, c are all *real* (which entails that Δ, see (2b), is either *real* or *imaginary*), a condition on the initial data *sufficient* to guarantee that the solution (2) be *nonsingular* for all (real) values of the time t is the requirement that the matrix $\underline{A}$, see (5), possess no *real* eigenvalues. *Hint*: see (8a).

The expressions (8b,c) are particularly convenient to discuss the possibility that (1) possess *periodic* solutions,

$$\underline{M}(t + \tilde{T}) = \underline{M}(t) \ . \tag{10}$$

Let us to this end consider values of the three coupling constants a, b, c such that Δ, see (2b), be *imaginary*,

$$\Delta = i\omega/2 \ , \tag{11}$$

assuming hereafter, without loss of generality, that the quantity ω is *positive*, $\omega > 0$, and denoting by T the corresponding period,

$$T = 2\pi/\omega \ . \tag{12}$$

But before proceeding with this discussion, let us pause to state two elementary results:

Lemma 5.1-4. The function

$$f(t) = [\exp(i\omega t) - \beta]^{\gamma} \tag{13}$$

is periodic in the (real) variable t, with period T, see (12),

$$f(t + T) = f(t) \, , \tag{14}$$

if the modulus of the complex number β exceeds unity,

$$|\beta| > 1 \ . \tag{15}$$

Note that no assumption is made on γ, which might be an arbitrary (complex) number.

Lemma 5.1-5. If

$$\beta = (\alpha - 1)/(\alpha + 1) \tag{16}$$

(see (9)), the conditions $|\beta| > 1, |\beta| = 1,$ respectively $|\beta| < 1$ correspond to $\mathrm{Re}(\alpha) < 0, \mathrm{Re}(\alpha) = 0,$ respectively $\mathrm{Re}(\alpha) > 0$.

Proof of *Lemma 5.1-4*. Clearly the complex number

$$z(t) = \exp(i\,\omega t) - \beta \tag{17}$$

is periodic in the real variable t with period T, see (12), indeed as t varies over a period it travels full circle on a circular contour in the complex z-plane, centered at β and of unit radius. Hence, if (15) holds, this contour does *not* include the origin, $z = 0$ (draw diagram!). Hence

$$f(t) = [z(t)]^{\gamma} \tag{18}$$

(see (13) and (17)) is periodic as well, with period T. This completes the proof of *Lemma 5.1-4*. But note that this conclusion would *not* have been warranted if (15) did not hold, because in such a case the circle traveled by $z(t)$ would traverse the branch cut, from $z = 0$ to $z = \infty$, of the function z^{γ}, see (18). It is indeed clear that, if $|\beta| = 1$, hence

$$\beta = \exp(i\,\varphi) \tag{19a}$$

with φ *real*, the function $f(t)$ would hit the branch point at $z = 0$ whenever $t = t_c$,

$$t_c = \varphi / \omega \quad \mathrm{mod}(T)\ ; \tag{19b}$$

while if $|\beta| < 1$, the function $f(t)$ would be periodic only if γ were rational, $\gamma = p/q$, but then with period $T' = qT$ (because by traversing the branch cut $z(t)$ would get on a different Riemann sheet of the function z^{γ}; only if γ were rational, $\gamma = p/q$, the number q of different sheets would be finite, hence by traversing the cut q times $z(t)$ would return to the origin sheet).

Exercise 5.1-6. Prove *Lemma 5.1-5. Hint:* compute $|\beta|^2$ from (16).

Let us return to our discussion of the solution (2), or rather (7) with (8), in the case (11). The following results are now plain.

Proposition 5.1-7. If (11) holds and moreover

$$a=\Delta+i(1-c)(p/q)\omega=i[1+2(1-c)(p/q)](\omega/2) \ , \tag{20a}$$

then all solutions (2) of the matrix evolution equation (1) which ensue from initial data $\underline{M}(0)$, $\underline{\dot{M}}(0)$ such that *all* the eigenvalues α_n of the matrix $\underline{A}$, see (5), have (strictly) *negative* real parts,

$$\text{Re}(\alpha_n)<0 \ , \tag{20b}$$

are *periodic*, see (10), with period

$$\tilde{T}=qT \ ; \tag{21}$$

if (11) holds and moreover

$$a=-\Delta+i(1-c)(p/q)\omega=i[-1+2(1-c)(p/q)](\omega/2) \ , \tag{22a}$$

then all solutions (2) of the matrix evolution equation (1) which ensue from initial data $\underline{M}(0)$, $\underline{\dot{M}}(0)$ such that *all* the eigenvalues α_n of the matrix $\underline{A}$, see (5), have (strictly) *positive* real parts,

$$\text{Re}(\alpha_n)>0 \ , \tag{22b}$$

are *periodic* with period $\tilde{T}$, see (21). Here p and q are the two integers which define the (arbitrary) rational number p/q, hence they are arbitrary, except for the requirement that q be *positive*, $q>0$, and that they be relatively prime (namely, such that their decompositions into products of primes possess no common factor).

Exercise 5.1-8. Prove *Proposition 5.1-7. Hint*: insert (11) and (20a) respectively (20b) in (8b) respectively (8c), and use the *Lemmata 5.1-4* and *5.1-5*.

Remark 5.1-9. If (11) and (20a) or (22a) hold (namely, two relations, or rather one and a half -- since ω, see (11), is constrained to be *real*, but is otherwise *arbitrary* -- for the three, *a priori* arbitrary, and possibly complex, "coupling constants" a,b,c), while, say, c is generic (possibly complex; certainly not rational), then the matrix evolution equation (1) possesses a lot of *nonperiodic* solutions; but there also are sets of initial data, clearly having nonvanishing measure in phase space, that yield *periodic* solutions, as detailed by *Proposition 5.1-7*.

Proposition 5.1-10. If the three coupling constants a,b,c in (1) can be expressed as follows:

$$a=i\left[(q_1\,p_2+q_2\,p_1)/(q_1\,p_2-q_2\,p_1)\right](\omega/2)\ , \tag{23a}$$

$$b=\left[p_1\,p_2/(q_1\,p_2-q_2\,p_1)\right]\omega^2\ , \tag{23b}$$

$$c=1-q_1\,q_2/(q_1\,p_2-q_2\,p_1)\ , \tag{23c}$$

where ω is *positive*, $\omega>0$, and $p_1,p_2\,q_1,q_2$ are 4 integers, arbitrary except for the following restrictions: p_1 and q_1, and likewise p_2 and q_2, are relatively prime, q_1 and q_2 are *positive*, and $q_1\,p_2-q_2\,p_1\neq 0$ (namely, p_1/q_1 and p_2/q_2 are two arbitrary *different* rational numbers; note that a,b,c, as given by (23), only depend on these two rational numbers, in addition to the positive real number ω); then *all nonsingular* solutions of (1) are *periodic*, see (10), with the period $\tilde{T}$ given by the following rules: if the initial data $\underline{M}(0)$, $\underline{\dot{M}}(0)$ are such that *all* the eigenvalues α_n of the matrix $\underline{A}$, see (5), have (strictly) *negative* real parts, $\mathrm{Re}(\alpha_n)<0$, then $\tilde{T}=q_1\,T$, see (12); if the initial data $\underline{M}(0)$, $\underline{\dot{M}}(0)$ are such that *all* the eigenvalues α_n of the matrix $\underline{A}$, see (5), have (strictly) *positive* real parts, $\mathrm{Re}(\alpha_n)>0$, then $\tilde{T}=q_2\,T$, see (12); if the initial data $\underline{M}(0)$, $\underline{\dot{M}}(0)$ are such that *all* the eigenvalues α_n of the matrix $\underline{A}$, see (5), have *nonvanishing* real parts, $\mathrm{Re}(\alpha_n)\neq 0$, which however do not all have the same sign, then $\tilde{T}=q\,T$, with q the minimum common multiple of q_1 and q_2. Note that the cases we just enumerated exhaust *all* the initial data $\underline{M}(0)$, $\underline{\dot{M}}(0)$ which yield *nonsingular* solutions of the matrix evolution equation (1): indeed, if the initial data, $\underline{M}(0)$, $\underline{M}(0)$, entail that one or more of the eigenvalues α_n of the matrix $\underline{A}$, see (5), are *imaginary*, say $\mathrm{Re}(\alpha_k)=0$, then there exist one or more (finite, real) values t_c, defined $\mathrm{mod}(T)$, see (19), at which times, $t=t_c$, the solution $\underline{M}(t)$ becomes singular (divergent and/or noninvertible).

Exercise 5.1-11. Prove *Proposition 5.1-10*. *Hint*: note firstly that (23) entail (11) (via (2b)), as well as the simultaneous validity of both (20a) (with $q = q_1, p = p_1$) and of (22a) (with $q = q_2, p = p_2$); then use *Proposition 5.1-7.*

Remark 5.1-12. The expressions (23b) respectively (23c) entail that b is *real*, respectively that c is (*real* and) *rational*, while (23c) entails that a is *imaginary*, unless it vanishes, which is indeed the case if the 4 integers $p_1, p_2\, q_1, q_2$ satisfy the single restriction $q_1\, p_2 + q_2\, p_1 = 0$, namely if the two rational numbers $p_1 / q_1, p_2 / q_2$ are equal in modulus and opposite in sign, entailing, say, $q_1 = q_2 = q > 0, p_1 = -p_2 = p$, hence

$$a = 0 \ , \tag{24a}$$

$$b = \frac{1}{2}(p/q)\omega^2 \ , \tag{24b}$$

$$c = 1 + \frac{1}{2}(q/p) \tag{24c}$$

(note that in this case the 3 periods $\tilde{T} = q_1\, T, \tilde{T} = q_2\, T, \tilde{T} = q\, T$ coincide, hence *all nonsingular* solutions (2) of (1) are, in this case (24), *periodic* with the *same* period $\tilde{T}$, while of course the matrix evolution equation (1) is *real*).

Exercise 5.1-13. Discuss the consistency of these findings (see *Proposition 5.1-7*, *Remark 5.1-9* , *Proposition 5.1-10* and *Remark 5.1-12*) with *Proposition 2.1.13-1.*

The next step is to introduce a convenient parametrization of the matrix $\underline{M}(t)$ in terms of (one or more) 3-vectors, as well as, possibly, of (one or more) scalars. An analysis of various such possibilities is given below (see Sect. 5.5). Here we assume $\underline{M}(t)$ to be a (2×2)-matrix, and we set

$$\underline{M}(t) = \rho(t)\underline{1} + i\, \vec{r}(t) \cdot \vec{\underline{\sigma}} \ . \tag{25}$$

Here and below $\underline{1}$ is the unit (2×2)-matrix, and the 3 components of the 3-vector $\vec{\underline{\sigma}}$ are the standard Pauli matrices:

$$\underline{1} = \begin{pmatrix} 1 & 0 \\ 0 & 1 \end{pmatrix}, \ \underline{\sigma}_x = \begin{pmatrix} 0 & 1 \\ 1 & 0 \end{pmatrix}, \ \underline{\sigma}_y = \begin{pmatrix} 0 & i \\ -i & 0 \end{pmatrix}, \ \underline{\sigma}_z = \begin{pmatrix} 1 & 0 \\ 0 & -1 \end{pmatrix} \ . \tag{26}$$

Generally, in the following, the unit matrix is omitted.

This parametrization, (25), entails the following relation (see Appendix H):

$$\underline{\dot{M}}\,[\underline{M}]^{-1}\,\underline{\dot{M}} = \{\dot{\rho}\,[\dot{\rho}\,\rho + 2(\dot{\vec{r}}\cdot\vec{r})\,] - \rho(\dot{\vec{r}}\cdot\dot{\vec{r}})$$

$$+i\,(2\,\dot{\vec{r}}\,[\,\dot{\rho}\,\rho + (\dot{\vec{r}}\cdot\vec{r})\,] - \vec{r}\,[\,\dot{\rho}^2 + (\dot{\vec{r}}\cdot\dot{\vec{r}})\,])\cdot\underline{\vec{\sigma}}\,\}/(\rho^2 + r^2)\;. \tag{27}$$

Hence (1) gets transcribed, via (25), into the following evolution equations:

$$\ddot{\rho} = 2a\dot{\rho} + b\rho + c\{\dot{\rho}\,[\,\dot{\rho}\rho + 2(\dot{\vec{r}}\cdot\vec{r})\,] - \rho(\dot{\vec{r}}\cdot\dot{\vec{r}})\,\}/(\rho^2 + r^2)\;, \tag{28a}$$

$$\ddot{\vec{r}} = 2a\,\dot{\vec{r}} + b\,\vec{r} + c\{2\,\dot{\vec{r}}\,[\,\dot{\rho}\rho + (\dot{\vec{r}}\cdot\vec{r})\,] - \vec{r}\,[\dot{\rho}^2 + (\dot{\vec{r}}\cdot\dot{\vec{r}})]\}/(\rho^2 + r^2)\;. \tag{28b}$$

These Newtonian equations of motion describe the (coupled, nonlinear) time-evolutions of the scalar $\rho(t)$ and of the 3-vector $\vec{r}(t)$. They are clearly rotation-invariant. They are *not* translation-invariant. Their *solvable* character is entailed, via (25), by the solvable character of (1).

These equations of motion, (28), are clearly compatible with the reduction $\rho(t) = 0$ (namely, if initially $\rho(0) = \dot{\rho}(0) = 0$, then $\rho(t) = 0$ for all time), in which case they take the simpler, 3-vector form

$$\ddot{\vec{r}} = 2a\,\dot{\vec{r}} + b\,\vec{r} + c\left[2\,\dot{\vec{r}}\,(\dot{\vec{r}}\cdot\vec{r}) - \vec{r}\,(\dot{\vec{r}}\cdot\dot{\vec{r}})\right]/r^2\;. \tag{29}$$

Exercise 5.1-14. Verify that (27) corresponds to (1) via (25). *Hint*: verify firstly that the following relations are entailed by (25) with (26):

$$\underline{M}^{-1} = (\rho - i\,\vec{r}\cdot\vec{\sigma})/(\rho^2 + r^2)\;, \tag{30}$$

$$\underline{\dot{M}}\,\underline{M}^{-1} = \{\dot{\rho}\rho + (\dot{\vec{r}}\cdot\vec{r}) + i[\rho\,\dot{\vec{r}} - \dot{\rho}\,\vec{r} + \dot{\vec{r}}\wedge\vec{r}\,]\cdot\vec{\sigma}\,\}/(\rho^2 + r^2)\;, \tag{31a}$$

$$\underline{M}^{-1}\,\underline{\dot{M}} = \{\dot{\rho}\rho + (\dot{\vec{r}}\cdot\vec{r}) + i[\rho\,\dot{\vec{r}} - \dot{\rho}\,\vec{r} - \dot{\vec{r}}\wedge\vec{r}\,]\cdot\vec{\sigma}\,\}/(\rho^2 + r^2)\;. \tag{31b}$$

Exercise 5.1-15. Show that, if a, b, c are (*real*, as) given by (24) (with $q > 0$, p relatively prime to q, and of course both *integers*), then *(i) all* nonsingular solutions of the Newtonian many-body problem (29) are *completely periodic* with period $\tilde{T} = qT$, see (12), and *(ii)* the necessary and sufficient condition for the motion to be *nonsingular* is that the initial position and velocity *not* be aligned,

$$\left|\vec{r}(0)\wedge\dot{\vec{r}}(0)\right| \neq 0\;. \tag{32}$$

If this condition, (32), does *not* hold, compute the time t_c (defined $\bmod(T)$, see (19)) at which the solution $\vec{r}(t)$ of the (*real*) equation of motion (29) vanishes (if $p<0$) or diverges (if $p>0$). *Hint*: see *Proposition 5.1-10* and (25) (with $\rho=0$).

Exercise 5.1-16. Verify that the Newtonian equation of motion (29), in the $a=0$ case, yielded by the Hamiltonian

$$H(\vec{r},\vec{p})=p^2\,f(r^2)+g(r^2)(\vec{r}\cdot\vec{p})+h(r^2)\ , \tag{33a}$$

with

$$f(s)=k\,s^c, \tag{33b}$$

$$(1-c)\,g^2(s)+2s\,g'(s)\,g(s)-4k\,s^c\,h'(s)=b\,. \tag{33c}$$

Here k is an arbitrary (nonvanishing) constant; and note the ample choice permitted, by the *single* constraint (33c), for the *two* arbitrary functions $g(s),h(s)$ (for instance a very simple choice consistent with (33c) is $g(s)=[b/(1-c)]^{1/2}$, $h(s)=0$).

Exercise 5.1-17. Formulate and solve the analogs of the preceding *Exercises 5.1-15* and *5.1-16*, but for the scalar/vector equations of motion (28) (rather than for the three-vector equation of motion (29)).

5.2 Another simple example: a linearizable matrix problem, and the corresponding one-body problem in three-dimensional space

In Sect. 5.2 we illustrate again the main idea on which the results of Chap. 5 are based, via another simple example.

The matrix evolution equation that serves as point of departure for our treatment reads now

$$\ddot{\underline{U}}=G(\dot{\underline{U}},\underline{U})+[\,\dot{\underline{U}},F(\underline{U})\,]\ . \tag{1}$$

Here $\underline{U}=\underline{U}(t)$, the dependent variable, is a square matrix of arbitrary rank; dots denote of course differentiations with respect to the independent variable t ("time"); and $G(\dot{\underline{U}},\underline{U})$, $F(\underline{U})$ are scalar/matrix functions, namely they depend on scalar arguments (including possibly various "coupling constants", as well as the time t, although for notational convenience these dependencies are not indicated explicitly) as well as on their indicated matrix arguments (in the case of $G(\dot{\underline{U}}, \underline{U})$ their order is of course important, since $\dot{\underline{U}} \equiv \dot{\underline{U}}(t)$ and $\underline{U} \equiv \underline{U}(t)$ need not commute), but they do *not* depend on any other matrix, so that there holds the essential property

$$\underline{W}\, G(\dot{\underline{U}},\underline{U})\, \underline{W}^{-1} = G(\underline{W}\,\dot{\underline{U}}\,\underline{W}^{-1}, \underline{W}\,\underline{U}\,\underline{W}^{-1}) \ , \tag{2a}$$

$$\underline{W}\, F(\underline{U})\, \underline{W}^{-1} = F(\underline{W}\,\underline{U}\,\underline{W}^{-1}) \ , \tag{2b}$$

for any matrices $\dot{\underline{U}}, \underline{U}$ and $\underline{W}$ (the latter being, of course, invertible). Of course both $G(\dot{\underline{U}}, \underline{U})$ and $F(\underline{U})$ are matrices (of the same rank as $\dot{\underline{U}}$ and $\underline{U}$), while $G(a, b)$, $F(c)$ are scalars when a, b, c are scalars.

We moreover assume that the matrix evolution equation (1) with $F = 0$, say

$$\ddot{\tilde{\underline{U}}} = G(\dot{\tilde{\underline{U}}}, \tilde{\underline{U}}), \tag{3}$$

is *solvable* and/or *integrable*; this is, for instance, the case if

$$G(\dot{\underline{U}}, \underline{U}) = 2a\dot{\underline{U}} + b\underline{U} + c\dot{\underline{U}}\,\underline{U}^{-1}\underline{U} \ , \tag{4}$$

with a, b, c three arbitrary constants, see the preceding Sect. 5.1.

The first point we now make is that (1) is then *linearizable*, for any arbitrary choice of the scalar/matrix function $F(\underline{U})$; in some special cases, see below, it might itself be *solvable* and/or *integrable*.

Indeed, let us introduce the matrix $\tilde{\underline{U}} \equiv \tilde{\underline{U}}(t)$ related to $\underline{U}(t)$ by the similarity transformation

$$\tilde{\underline{U}} = \underline{W}\,\underline{U}\,\underline{W}^{-1} \ , \tag{5a}$$

$$\underline{U} = \underline{W}^{-1}\,\tilde{\underline{U}}\,\underline{W} \ , \tag{5b}$$

with the matrix $\underline{W} \equiv \underline{W}(t)$ satisfying the first-order matrix differential equation

$$\dot{\underline{W}} = \underline{W}\, F(\underline{U}) \ , \tag{6a}$$

$$\dot{\underline{W}} = F(\tilde{\underline{U}})\, \underline{W} \ . \tag{6b}$$

The equivalence of (6a) and (6b) is of course entailed by (5) (see (2)).

This linear differential equation, (6a) respectively (6b), defines uniquely the matrix $\underline{W}(t)$ in terms of the matrix $\underline{U}(t)$ respectively $\underline{\tilde{U}}(t)$, provided it is supplemented by an initial datum $\underline{W}(0)$. Hereafter, for simplicity, we assume this to be given by the simple rule

$$\underline{W}(0)=\underline{1}. \tag{7}$$

It is now easily seen that (5) and (6) entail the relations

$$\dot{\underline{\tilde{U}}}=\underline{W}\,\dot{\underline{U}}\,\underline{W}^{-1} \quad , \tag{8a}$$

$$\dot{\underline{U}}=\underline{W}^{-1}\,\dot{\underline{\tilde{U}}}\,\underline{W} \quad , \tag{8b}$$

as well as

$$\ddot{\underline{\tilde{U}}}=\underline{W}\{\ddot{\underline{U}}-[\,\dot{\underline{U}},F(\underline{U})\,]\}\underline{W}^{-1} \quad , \tag{9a}$$

$$\ddot{\underline{U}}=\underline{W}^{-1}\{\ddot{\underline{\tilde{U}}}+[\,\dot{\underline{\tilde{U}}},F(\underline{\tilde{U}})\,]\}\underline{W} \quad . \tag{9b}$$

Here and always below $[\underline{A},\underline{B}]$ is the commutator of the two matrices $\underline{A}$, $\underline{B}$:

$$[\underline{A},\underline{B}]\equiv\underline{A}\,\underline{B}-\underline{B}\,\underline{A} \quad . \tag{10}$$

Proofs. Time-differentiation of (5a) yields

$$\dot{\underline{\tilde{U}}}=\underline{W}\,\dot{\underline{U}}\,\underline{W}^{-1}+\dot{\underline{W}}\,\underline{U}\,\underline{W}^{-1}-\underline{W}\,\underline{U}\,\underline{W}^{-1}\dot{\underline{W}}\,\underline{W}^{-1} \quad , \tag{11a}$$

$$\dot{\underline{\tilde{U}}}=\underline{W}\{\,\dot{\underline{U}}+[\,F(\underline{U}),\underline{U}\,]\,\}\underline{W}^{-1} \quad . \tag{11b}$$

Here the second step is a consequence of (6a), which clearly entails

$$F(\underline{U})=\underline{W}^{-1}\dot{\underline{W}} \quad . \tag{6c}$$

But $F(\underline{U})$ clearly commutes with $\underline{U}$ (since it is a function of no other matrix but $\underline{U}$),

$$[\,F(\underline{U}),\underline{U}\,]=0 \ , \tag{12}$$

hence (11b) entails (8a), and (8a) entails of course (8b).

Likewise, time-differentiation of (8a) yields, via (6a),

$$\ddot{\underline{\tilde{U}}}=\underline{W}\,\underline{\ddot{U}}\,\underline{W}^{-1}+\underline{\dot{W}}\,\underline{\dot{U}}\,\underline{W}^{-1}-\underline{W}\,\underline{\dot{U}}\,\underline{W}^{-1}\underline{\dot{W}}\,\underline{W}^{-1}\ , \tag{13a}$$

$$\ddot{\underline{\tilde{U}}}=\underline{W}\{\,\underline{\ddot{U}}+F(\underline{U})\,\underline{\dot{U}}-\underline{\dot{U}}\,F(\underline{U})\,\}\underline{W}^{-1}\ , \tag{13b}$$

which coincides with (9a), that is thereby proven. As for (9b), it follows from (9a) via (2), (5) and (8).

Note moreover that (7) entails, via (5) and (8), that the initial data for the two matrices $\underline{U}(t)$ and $\underline{\tilde{U}}(t)$ coincide:

$$\underline{\tilde{U}}(0)=\underline{U}(0),\ \ \dot{\underline{\tilde{U}}}(0)=\underline{\dot{U}}(0)\ \ . \tag{14}$$

It is now clear, from (8) and (9), that if $\underline{U}(t)$ satisfies (1), then $\underline{\tilde{U}}(t)$, related to $\underline{U}(t)$ by (5) with (6), satisfies the *solvable* and/or *integrable* evolution equation (3),

$$\ddot{\underline{\tilde{U}}}=G(\dot{\underline{\tilde{U}}},\underline{\tilde{U}})\ \ . \tag{15}$$

Hence the matrix evolution equation (1) is *linearizable.*

To substantiate this claim, let us indicate how to solve the initial-value problem for (1). The *first* step is to obtain, via (14), the initial data for (15). The *second* step is to obtain $\underline{\tilde{U}}(t)$ from these initial data, see (14), via the evolution equation (15), which is by assumption *solvable* and/or *integrable*, see (3). The *third* step is to compute $\underline{W}(t)$, by solving, with the initial condition (7), the first-order ODE (6b); note that this is generally a *nonautonomous* matrix differential equation, since $\underline{\tilde{U}}(t)$ is generally time-dependent; it is, however, a *linear* evolution equation, and this fact justifies the claim made above: since indeed, once $\underline{\tilde{U}}(t)$ and $\underline{W}(t)$ are known, $\underline{U}(t)$ is given by the explicit formula (5b), whose utilization constitutes the *fourth*, and last, step to solve the initial-value problem for (1).

As we indicate below, in some cases in which (15), or equivalently (3), is *solvable*, the linear matrix ODE (6b) can also be explicitly solved in terms of known special functions. In those cases the matrix evolution equation (1) is, more appropriately, called *solvable.*

To proceed with the illustration of our method we must now implement its second step, namely introduce an appropriate parametrization of the matrices in terms of three-vectors. But firstly we must make a more specific choice for the matrix equation under consideration. So we make the specific choice (4), namely we focus on the matrix evolution equation

$$\ddot{\underline{U}}=2a\dot{\underline{U}}+b\underline{U}+c\dot{\underline{U}}\underline{U}^{-1}\dot{\underline{U}}+[\dot{\underline{U}},F(\underline{U})]\ ,\tag{16}$$

so that $\tilde{U}$ satisfy the matrix evolution equation

$$\ddot{\tilde{\underline{U}}}=2a\dot{\tilde{\underline{U}}}+b\tilde{\underline{U}}+c\dot{\tilde{\underline{U}}}\underline{U}^{-1}\dot{\tilde{\underline{U}}}\ ,\tag{17}$$

whose solvable character has been demonstrated in the preceding Sect. 5.1. We moreover assume, as we did in the second part of the preceding Sect. 5.1, that the matrix $\underline{U}(t)$ (as well as $\tilde{\underline{U}}(t)$) have rank 2. And we also use for $\underline{U}(t)$ the following parametrization in terms of (only!) a three-vector:

$$\underline{U}(t)=i\,\vec{r}(t)\cdot\vec{\underline{\sigma}}\ ,\tag{18}$$

with $\vec{\underline{\sigma}}$ defined as in Sect. 5.1, see (5.1-26). It is easily seen that this parametrization (which is analogous to (5.1-25), but with $\rho(t)=0$, entailing that the matrix $\underline{U}(t)$ is *traceless*) is compatible with the matrix evolution (16), and that it transforms it into the 3-vector evolution equation

$$\ddot{\vec{r}}=2a\,\dot{\vec{r}}+b\,\vec{r}+c\left[2\dot{\vec{r}}\,(\dot{\vec{r}}\cdot\vec{r})-\vec{r}\,(\dot{\vec{r}}\cdot\dot{\vec{r}})\right]/r^2+f(r)\,\vec{r}\wedge\dot{\vec{r}}\ ,\tag{19a}$$

where $f(r)$ is related to the function $F(u)$, see (16), by the formula

$$f(r)=[F(ir)-F(-ir)]/(ir)\ .\tag{19b}$$

Hence $f(r)$ is generally an (arbitrary) *even* function (but, as we will see below, the restriction to even functions can be by-passed).

Exercise 5.2-1. Verify that (19) corresponds to (16) via (18). *Hint*: prove first the identity

$$F(\vec{u}\cdot\vec{\underline{\sigma}})=F_e(u)+F_o(u)(\vec{u}\cdot\vec{\underline{\sigma}})/u=F_e(u)+\frac{1}{2}f(u)\,(\vec{u}\cdot\vec{\underline{\sigma}})\ ,\tag{20a}$$

where $F_e(u)$ respectively $F_o(u)$ are the even respectively odd parts of $F(u)$,

$$F(u)=F_e(u)+F_o(u),\ F_e(u)=\frac{1}{2}[F(u)+F(-u)],\ F_o(u)=\frac{1}{2}[F(u)-F(-u)],\tag{20b}$$

and of course $f(u)$ is defined by (19b). Note that $F_e(u)$, $F_o(u)$ and $f(u)$, as defined here, are all *scalar* (namely, neither matrix- nor vector-valued) functions of the scalar $u \equiv |\vec{u}| \equiv \left(u_x^2 + u_y^2 + u_z^2\right)^{½}$.

Exercise 5.2-2. Prove that, if the parametrization (18) is replaced by the more general parametrization (5.1.25), then in place of the 3-vector equation (19) one gets from (16) the (more general) scalar/three-vector equations of motion

$$\ddot{\rho} = 2a\rho + b\rho + c\{\,\dot{\rho}[\,\dot{\rho}\rho + 2(\dot{\vec{r}}\cdot\vec{r})\,] - \rho(\dot{\vec{r}}\cdot\dot{\vec{r}})\,\}/(\rho^2 + r^2) \;, \tag{21a}$$

$$\ddot{\vec{r}} = 2a\vec{r} + b\,\dot{\vec{r}} + c\{2\dot{\vec{r}}[\dot{\rho}\,\rho + (\dot{\vec{r}}\cdot\vec{r})] - \vec{r}[\dot{\rho}^2 + (\dot{\vec{r}}\cdot\dot{\vec{r}})]\}/(\rho^2 + r^2) + \varphi(r)\,\vec{r}\wedge\dot{\vec{r}} \;, \tag{21b}$$

$$\varphi(r) = [\,F(\rho + ir) - F(\rho - ir)\,]/(ir) \;, \tag{21c}$$

which are of course compatible with $\rho(t) = 0$, whereby they reduce back to (19). *Hint*: prove firstly the *identity*

$$F(\rho + i\vec{r}\cdot\underline{\vec{\sigma}}) = \frac{1}{2}[\,F(\rho + ir) + F(\rho - ir)\,] + \frac{1}{2}\,[\,F(\rho + ir) - F(\rho - ir)\,](ir)^{-1}\,\vec{r}\cdot\underline{\vec{\sigma}} \;. \tag{22}$$

We have thus seen that the Newtonian equation of motion (19) (as well, indeed, as (21)) is *linearizable*. Let us now pursue in more explicit detail the technique of solution of this equation of motion, (19). But, in the interest of maximal simplicity, let us restrict attention to the case with $a = b = c = 0$, so that (16) take the simple form

$$\underline{\ddot{U}} = [\,\underline{\dot{U}},\, F(\underline{U})\,] \;, \tag{23}$$

and correspondingly (19) take the simple (yet quite interesting, see below) form

$$\ddot{\vec{r}} = f(r)\,\vec{r}\wedge\dot{\vec{r}} \;, \tag{24}$$

with $f(r)$ again related to $F(u)$ by (19b). Then the evolution equation of the matrix $\underline{\tilde{U}}(t)$ takes the very simple form

$$\ddot{\underline{\tilde{U}}} = 0 \tag{25}$$

entailing

$$\underline{\tilde{U}}(t) = \underline{U}(0) + \underline{\dot{U}}(0)\, t \quad , \tag{26}$$

where we also used (14). Hence (6b) now reads

$$\underline{\dot{W}}(t) = F(\underline{A} + \underline{B}\, t)\, \underline{W}(t) \quad , \tag{27}$$

with the two *constant* matrices $\underline{A}$ and $\underline{B}$ defined, via (18), in terms of the initial data as follows:

$$\underline{A} = \underline{U}(0) = i\, \vec{r}\,(0) \cdot \underline{\vec{\sigma}} \quad , \tag{28a}$$

$$\underline{B} = \underline{\dot{U}}(0) = i\, \dot{\vec{r}}\,(0) \cdot \underline{\vec{\sigma}} \quad . \tag{28b}$$

Remark 5.2-3. Via (20) and (28) the evolution equation (23) can be recast in the following form:

$$\underline{\dot{W}}(t) = \{ F_e(i |\, \vec{r}(0) + \dot{\vec{r}}(0)\, t\, |)$$

$$+ \frac{1}{2} f(i |\, \vec{r}(0) + \dot{\vec{r}}(0)\, t\, |)\, (i [\vec{r}(0) + \dot{\vec{r}}(0)\, t\,] \cdot \underline{\vec{\sigma}})\, \}\, \underline{W}(t) \quad , \tag{29}$$

of course with $f(u)$ defined by (19b).

Remark 5.2-4. For our purposes (namely, to eventually insert $\underline{W}(t)$ and $[\underline{W}(t)]^{-1}$ in (5)), the evolution equation (29) can be equivalently replaced by the equation

$$\underline{\dot{W}}(t) = \frac{1}{2} f(|\, \vec{r}(0) + \dot{\vec{r}}(0)\, t\, |)\, (i [\vec{r}(0) + \dot{\vec{r}}(0)\, t\,] \cdot \underline{\sigma})\, \underline{W}(t) \quad , \tag{30}$$

with $f(r)$ defined by (19b), and with the same initial condition (7).

Proof. Set

$$\underline{\hat{W}}(t) = \exp[- \int_0^t dt'\, F_e(i |\, \vec{r}(0) + \dot{\vec{r}}(0)\, t'\, |)\,]\, \underline{W}(t) \quad . \tag{31}$$

Then, also using (19b) and (29),

$$\underline{\dot{\hat{W}}}(t) = \frac{1}{2} f(|\, \vec{r}(0) + \dot{\vec{r}}(0)\, t\, |)([\vec{r}(0) + \dot{\vec{r}}(0)\, t\,] \cdot \underline{\vec{\sigma}})\, \underline{\hat{W}}(t) \quad , \tag{32a}$$

while (7) and (31) entail

$$\underline{\hat{W}}(0)=\underline{1} \quad . \tag{32b}$$

On the other hand the insertion of $\underline{\hat{W}}(t)$, see (31), in place of $\underline{W}(t)$ in (5) entails no difference. *Nota bene*: throughout this argument the fact that F_e is a scalar, not a matrix, plays an essential role, since it entails that it commutes with every matrix.

Exercise 5.2-5. If the matrices $\underline{A}$ and $\underline{B}$ commute, the matrix evolution equation (27) can be immediately solved by a quadrature; hence in this case the equation of motion (24) is *solvable*. But this is a trivial finding. Why ? *Hint*: a necessary and sufficient condition for the commutativity of $\underline{A}$ and $\underline{B}$, see (28), is the following relation among the initial data for (24):

$$\dot{\vec{r}}(0)=\eta\,\vec{r}(0) \quad , \tag{33}$$

with η an arbitrary (scalar !) constant (possibly vanishing); but then the right hand side of (24) vanishes; initially, and not only initially.

Let us now focus on the 3-vector equation of motion (24), with $f(r)$ an *arbitrary* function (not necessarily *even*, see below). It is easily seen that this equation of motion entails

$$\dot{\vec{r}}(t)\cdot\dot{\vec{r}}(t)=v^2 \quad , \tag{34a}$$

$$r^2(t)=s^2+2\beta t+v^2t^2 \quad , \tag{34b}$$

with the 3 constants s, β, v defined as follows in terms of the initial data:

$$s^2=r^2(0),\ \beta=\dot{\vec{r}}(0)\cdot\vec{r}(0),\ v^2=\dot{\vec{r}}(0)\cdot\dot{\vec{r}}(0) \quad . \tag{34c}$$

Proof. The scalar product of (24) with $\dot{\vec{r}}(t)$ respectively $\vec{r}(t)$ yields

$$\ddot{\vec{r}}(t)\cdot\dot{\vec{r}}(t)=0 \quad , \tag{35a}$$

respectively

$$\ddot{\vec{r}}(t)\cdot\vec{r}(t)=0 \quad . \tag{35b}$$

The first of these equations, (35a), entails (34a), which is thereby proven. Then (35b) with (34a) entails

$$d^2\left[r^2(t)\right]/dt^2 = 2v^2 \tag{36}$$

hence

$$r^2(t) = s^2 + 2\beta t + v^2 t^2 \tag{37}$$

with s^2 and β appropriate constants. The identification (34c) is then confirmed by setting $t = 0$ in (37) and in its t-derivative.

Remark 5.2-6. The finding, (34), we just proved, not only opens the possibility to treat the 3-vector evolution equation (24) even when the function $f(r)$ is not even (indeed, no such hypothesis was required to derive (34) from (24)); it also entails that, from the point of view of the initial-value problem, solving (24) is equivalent to solving the (nonautonomous 3-vector) ODE

$$\ddot{\vec{r}}(t) = h(t)\, \vec{r} \wedge \dot{\vec{r}}(t) \quad , \tag{38}$$

or equivalently the (2×2)-matrix equation that corresponds to this 3-vector ODE via (18),

$$\ddot{\underline{U}}(t) = \frac{1}{2} h(t) [\, \dot{\underline{U}}(t), \underline{U}(t)\,] \, , \tag{39}$$

with (in both cases; see (34b))

$$h(t) = f\left[(s^2 + 2\beta t + v^2 t^2)^{1/2}\right] . \tag{40}$$

The rest of Sect. 5.2 is devoted to the study of the (2×2)-matrix evolution equation (39) (or equivalently of the 3-vector evolution equation (38), which corresponds to it via (18)). The developments reported above entail that the solution of this matrix evolution equation, (39), is given by the prescription (5b), with $\underline{\tilde{U}}(t)$ given by (26) and $\underline{W}(t)$ given by (30), which now takes the form (via (40))

$$\dot{\underline{W}}(t) = \frac{1}{2}\, h(t)\, (\underline{A} + \underline{B}\, t)\, \underline{W}(t) \quad , \tag{41}$$

with (28) and (7). Hence the solvability of (39) (hence of (38), hence as well of (24)) hinges on the solvability of this first-order linear (2×2)-matrix evolution equation, (41).

An additional simplification flows from an appropriate choice of the reference frame (or equivalently by performing an appropriate, time-independent, similarity transformation of the matrix $\underline{U}$). Let us assume that the z-axis of the 3-dimensional reference frame for 3-vectors is parallel to the initial velocity $\dot{\vec{r}}(0)$, and that the y-axis is orthogonal to the initial coordinate $\vec{r}(0)$:

$$\vec{r}(0)=(x(0),0,z(0)),\quad \dot{\vec{r}}(0)=(0,0,v) \tag{42}$$

entailing (via (28) and (5.1-9))

$$\underline{A}=i\begin{pmatrix} b & c \\ c & -b \end{pmatrix},\quad \underline{B}=i\begin{pmatrix} v & 0 \\ 0 & -v \end{pmatrix}\quad , \tag{43a}$$

with (see (34c))

$$b=z(0)=\vec{r}\,(0)\cdot\dot{\vec{r}}(0)/v=\beta/v\quad , \tag{43b}$$

$$c=x(0)=\left|\vec{r}(0)\wedge\dot{\vec{r}}(0)\right|/v=s\left\{1-\left[\beta/(s\,v)\right]^2\right\}^{1/2}\quad , \tag{43c}$$

where of course (see (34c))

$$v=\left|\dot{\vec{r}}(0)\right|=\left|\dot{\vec{r}}(t)\right|\ . \tag{43d}$$

Note that the matrices $\underline{A}$ and $\underline{B}$ are traceless,

$$\mathrm{trace}\left[\underline{A}\right]=\mathrm{trace}\left[\underline{B}\right]=0\quad , \tag{44}$$

as it is generally entailed by the representations (18) and (28). Hence the evolution equation (41) with (7) entails

$$\det\left[W(t)\right]=\det\left[W(0)\right]=1\ . \tag{45}$$

Exercise 5.2-7. Prove this result. *Hint*: use the *identity*

$$(d/dt)\log\left\{\det\left[\underline{W}(t)\right]\right\}=\mathrm{trace}\left\{\dot{\underline{W}}(t)\ \left[W(t)\right]^{-1}\right\}\quad , \tag{46}$$

valid for any invertible matrix, and (41), (44), (7).

Let us now discuss the matrix evolution equation (41). To this end we introduce the following explicit parametrization of the (2×2)-matrix $\underline{W}(t)$:

$$\underline{W}(t)=\begin{pmatrix} w_+(t) & -\tilde{w}_-(t) \\ w_-(t) & \tilde{w}_+(t) \end{pmatrix} \quad . \tag{47}$$

Thereby (41) with (43) become

$$\dot{w}_\pm(t)=\frac{i}{2}\,h(t)\,\left[\pm(b+vt)\,w_\pm(t)+c\,w_\mp(t)\,\right] \quad , \tag{48a}$$

$$\dot{\tilde{w}}_\pm(t)=-\frac{i}{2}\,h(t)\,\left[\pm(b+vt)\,\tilde{w}_\pm(t)+c\,\tilde{w}_\mp(t)\,\right] \quad , \tag{48b}$$

while (7) yields

$$w_+(0)=\tilde{w}_+(0)=1 \quad , \; w_-(0)=\tilde{w}_-(0)=0 \quad . \tag{49}$$

Exercise 5.2-8. Verify !

Exercise 5.2-9. Prove the relation

$$w_+(t)\,\tilde{w}_+(t)+w_-(t)\,\tilde{w}_-(t)=1 \quad . \tag{50}$$

Hint: see (47) and (45).

It is clear from (48) and (49) that if, as we hereafter assume, the function $h(t)$, see (40), and the 3 constants b,c and v, see (43), are *real*, then there holds the relations

$$\tilde{w}_+(t)=w_+^*(t) \quad , \qquad \tilde{w}_-(t)=\tilde{w}_-^*(t) \quad , \tag{51}$$

which incidentally entail, via (50), the formula

$$|w_+(t)|^2+|w_-(t)|^2=1 \quad . \tag{52}$$

The relations (51) entail that, to obtain the matrix $\underline{W}(t)$, see (47), it is sufficient to determine the 2 functions $w_+(t)$ and $w_-(t)$. Moreover from (the first of the) (48a) we get

$$w_-(t) = -c^{-1}\{(b+vt)\,w_+(t) + 2i[\dot{w}_+(t)/h(t)]\}\ , \qquad (53)$$

which provides an explicit expression of $w_-(t)$ in terms of $w_+(t)$ and its time-derivative $\dot{w}_+(t)$. We thus conclude that, to obtain an explicit expression of the matrix $\underline{W}(t)$, see (47), it is sufficient to know the function $w_+(t)$, which we hereafter denote, for notational simplicity, as $w(t)$:

$$w_+(t) \equiv w(t)\ . \qquad (54)$$

On the other hand, by inserting (53) in (the second of the) (48a) we get for $w \equiv w(t)$ the following *second-order linear nonautonomous* ODE:

$$\ddot{w} = (\dot{h}/h)\,\dot{w} + [\,-(s^2 + 2\beta t + v^2 t^2)(h/2)^2 + i v h/2\,]w\ , \qquad (55a)$$

with s, β and v defined by (34c), and with the initial conditions (see (49) and (48a))

$$w(0) = 1,\ \ \dot{w}(0) = i b\, h(0)/2\ . \qquad (55b)$$

Exercise 5.2-10. Verify !

By setting

$$w(t) = [h(t)]^{1/2}\,\psi(t) \qquad (56)$$

the ODE (55) takes the "Schroedinger-like" form

$$\ddot{\psi}(t) = \psi(t)\{\,i v h(t)/2 - (s^2 + 2\beta t + v^2 t^2)[h(t)/2]^2 + \frac{3}{4}[\dot{h}(t)/h(t)]^2 - \frac{1}{2}\ddot{h}(t)/h(t)\,\}\ , \qquad (57a)$$

with initial conditions

$$\psi(0) = [h(0)]^{1/2}\ ,\quad \dot{\psi}(0) = [h(0)]^{-3/2}\,[\,i b\, h(0) - \dot{h}(0)\,]/2\ . \qquad (57b)$$

Exercise 5.2-11. Verify !

Let us now survey some cases in which these second-order linear nonautonomous ODEs, (55) or (57), are *solvable* in terms of well-known special functions; the corresponding (2×2)-matrix evolution equation (39) (or equivalently (23), via (19) and (40)), as well as the corresponding 3-vector Newtonian evolution equation (38) (or equivalently (24), via (40)), can then be considered *solvable* as well.

The first case we consider corresponds to the (simplest!) choice

$$h(t) = f(r) = k \quad , \tag{58a}$$

namely to the (2×2)-matrix evolution equation

$$\ddot{\underline{U}} = \frac{k}{2}[\dot{\underline{U}},\underline{U}] \quad , \tag{58b}$$

and to the 3-vector Newtonian equation of motion

$$\ddot{\vec{r}} = k\,\vec{r} \wedge \dot{\vec{r}} \quad . \tag{58c}$$

Here of course k is an arbitrary constant. Then (55) reads

$$\ddot{w} = \left[\,-(s^2 + 2\beta t + v^2 t^2)(k/2)^2 + i v k/2\,\right] w \tag{58d}$$

and can therefore be solved in terms of the *parabolic cylinder* function $D_p(\tau)$ (see f.i. eq. 9.255 of <GRJ94>):

$$w(t) = c_+ D_p(\tau) + c_- D_{-p}(\tau) \quad , \tag{58e}$$

$$p = -1 - i\,k c^2/v \quad , \tag{58f}$$

$$\tau \equiv \tau(t) = (i k v)^{1/2}\,(t - \beta/v^2) \quad . \tag{58g}$$

Exercise 5.2-12. Verify, obtain the expressions of the 2 constants $c_\pm$, see (58e), in terms of the initial data, and discuss the behavior of the solution $\vec{r}(t)$ of the Newtonian equation of motion (58c) which describes the motion in ordinary (three-dimensional) space of a particle acted upon by a force proportional to its angular momentum. *Hint*: see (55b), (54), (53), (51), (47), (43c), (34c), and (26) with (18) (at $t = 0$), and finally (5b) with (18) and (5.1-26).

The next case we consider is characterized by the assignment

$$f(r)=k/r^2 \tag{59a}$$

entailing (see (40))

$$h(t)=k/(s^2+2\beta t+v^2t^2) \quad . \tag{59b}$$

Hence in this case the (2×2)-matrix evolution equation (23) reads

$$\ddot{\underline{U}}=\frac{k}{2}[\underline{U}^{-1},\dot{\underline{U}}] \quad , \tag{59c}$$

and the 3-vector Newtonian equation of motion (24) reads

$$\ddot{\vec{r}}=k\,r^{-2}\vec{r}\wedge\dot{\vec{r}} \quad . \tag{59d}$$

Here k is again an arbitrary constant. The corresponding version of (55), via the change of independent variable

$$w(t)=\varphi(\tau) \quad , \tag{59e}$$

$$t=i(c/v)(1-2\tau)\,,\quad \tau=\frac{1}{2}[\,1+i(v/c)(t+\beta v^2)\,] \quad , \tag{59f}$$

becomes then the *hypergeometric* differential equation (see item 9.151 of <GRJ94>)

$$\tau(1-\tau)\,\varphi''+(1-2\tau)\,\varphi'-\lambda\,\varphi=0 \quad , \tag{59g}$$

where the primes indicate of course differentiations with respect to the "dimensionless time" τ and

$$\lambda=k(k-2iv)/(2v)^2 \quad , \tag{59h}$$

whose general solution is provided by eq. 9.153.2 of <GRJ94> (with $z=\tau$, $\alpha=[\,1+(1-\lambda)^{1/2}\,]/2$, $\beta=[1-(1-\lambda)^{1/2}\,]/2$).

Exercise 5.2-13. Verify !

There is a third case in which the Newtonian equation of motion (24) is *solvable*: if $f(r)=k/r^3$. This is particular interesting because it turns out to be solvable in terms of *elementary* functions, and moreover because it corresponds to a model having a direct physical interpretation (motion of a magnetic monopole in the electric field of a point charge). But for these reasons we prefer to devote to this case the entire Sect. 5.2.2.

We end Sect. 5.2 by pointing out one more case in which (55) is *solvable*, entailing the corresponding solvability of (38) and (39). It obtains for the assignment

$$h(t)=2k/(1+2\omega t) \quad , \tag{60a}$$

where k and ω are 2 arbitrary constants. Thereby the (2×2)-matrix evolution equation (39) reads

$$\ddot{\underline{U}}(t)=k(1+2\omega t)^{-1}[\ \dot{\underline{U}}(t),\underline{U}(t)\ ] \quad . \tag{60b}$$

Note however that this assignment, (60a), does not have in any natural sense the form (40). The motivation for treating this *solvable* case here will become apparent in Sect. 5.2.3.

The solvability of (55) with this assignment, (60a), is manifested via the change of dependent variable

$$w(t)=\chi(\tau) \quad , \tag{60c}$$

$$t=i\omega\tau/(vk)-1/(2\omega), \quad \tau=-\frac{i}{2}(vk/\omega^2)(1+2\omega t) \quad , \tag{60d}$$

whereby (55) becomes the confluent hypergeometric equation (in its *Whittaker* version, see eq. 9.220.1 of <GRJ94>),

$$\chi''+[(\frac{1}{4}-\mu^2)\tau^{-2}+\lambda\tau^{-1}-\frac{1}{4}]\chi=0 \quad . \tag{60e}$$

Here of course the primes denote differentiations with respect to the dimensionless time τ, see (60d), and the 2 dimensionless constants λ and μ are defined as follows:

$$\lambda=\frac{1}{2}-ik(v^2-2\beta\omega)/(4v\omega^2) \quad , \tag{60f}$$

$$\mu^2 = k^2 (4\beta\omega - v^2 - 4s^2\omega^2)/(4\omega^4) \quad . \tag{60g}$$

Exercise 5.2-14. Verify !

Exercise 5.2-15. Obtain the nonlinear (quadratic) relations entailed by (52) with (53) and (54) for the special (parabolic cylinder, hypergeometric, respectively confluent hypergeometric) functions that solve (58d) (see (58e)), (59g) respectively (60e).

Remark 5.2-16. The Newtonian equation of motion (24) is clearly invariant under space *rotations*, but *not* under space *inversions*, namely it describes a *parity-non-invariant* evolution.

5.2.1 Motion of a magnetic monopole in a central electric field

In Sect. 5.2.1 we investigate the (nonrelativistic) motion that a magnetic monopole (if it existed as a separate pointlike particle with mass) would perform in a *central* electric field, or equivalently the motion that a charged massive point-like particle would perform in a *central* magnetic field (if the latter could be realized). The corresponding Newtonian-Lorentzian equation of motion reads of course

$$\ddot{\vec{r}} = f(r)\, \vec{r} \wedge \dot{\vec{r}} \quad , \tag{1}$$

hence it coincides with (5.2-24). We will pay particular attention to the case of a power law of force,

$$f(r) = k\, r^p \quad ; \tag{2}$$

here and below k is a constant, having clearly (see (1) and (2)) the dimensions $(\text{length})^{-(p+1)} \cdot (\text{time})^{-1}$. In particular we show that the problem (1) with (2) is *solvable* if $p=-2$ or $p=-3$. The first of these two cases, $p=-2$ (as well as the case $p=0$) were already identified as solvable in the preceding Sect. 5.2. The second case, $p=-3$, is the most interesting one from the physical point of view, because it corresponds to the Coulomb electric field produced by a charged particle fixed at the origin of coordinates; remarkably, it is also the most interesting case mathematically, as it turns out to be solvable in terms of *elementary* functions. A more detailed elaboration of this special case is provided in the following Sect. 5.2.2.

First of all, let us report some results that are directly implied by (1). We have seen that (1) is equivalent to the 3-vector equation

$$\ddot{\vec{r}}(t) = h(t)\, \vec{r}(t) \wedge \dot{\vec{r}}(t) \quad , \tag{3}$$

with

$$h(t) = f[\ (s^2 + 2\beta t + v^2 t^2)^{1/2}\ ] \quad , \tag{4}$$

since it entails

$$r^2(t) = s^2 + 2\beta t + v^2 t^2 \quad , \tag{5a}$$

with

$$s^2 = r^2(0)\,, \quad \beta = \vec{r}(0) \cdot \dot{\vec{r}}(0)\,, \quad v = \left|\dot{\vec{r}}(0)\right| = \left|\dot{\vec{r}}(t)\right| \quad . \tag{5b}$$

Exercise 5.2.1-1. Prove these formulas. *Hint*: see (5.2-34) and their proof. Be aware of the difference among the modulus $\left|\dot{\vec{r}}(t)\right| = v$ of the 3-vector $\dot{\vec{r}}(t)$ ("velocity"), and the time-derivative,

$$\dot{r}(t) = \dot{\vec{r}}(t) \cdot \vec{r}(t) / r(t) = (\beta + vt) / r(t) \quad , \tag{5c}$$

of the modulus of the 3-vector $\vec{r}(t)$ ("position"), $r(t) \equiv \left|\vec{r}(t)\right|$.

Remark 5.2.1-2. The formula (5a) displays neatly a general feature of the motion entailed by the 3-vector evolution equation (4): the motion is never confined; the moving particle comes in, in the remote past, from afar, reaches a minimum distance a from the origin, and then returns far away in the remote future. Note that this result holds independently of the form of the function $f(r)$.

Exercise 5.2.1-3. Calculate, in terms of the initial data $\vec{r}(0)$ and $\dot{\vec{r}}(0)$, the distance a of closest approach to the origin, of the trajectory associated to those data. *Hint*: use (5a). *Solution*: see below.

Exercise 5.2.1-4. The distance of closest approach a is of course (strictly) *positive*, except in the special case in which the vectors $\vec{r}(0)$ and $\dot{\vec{r}}(0)$ are parallel; then they remain parallel throughout the motion, the trajectory is a straight line going through the origin, and the moving particle travels on it freely

$$\vec{r}(t) = \vec{r}(0) + \dot{\vec{r}}(0)\, t \quad , \tag{6}$$

because the parallelism of $\vec{r}(t)$ and $\dot{\vec{r}}(t)$ entails that the Lorentz-type force appearing in the right hand side of (1) vanishes. Hereafter we exclude from consideration this trivial situation (note that in this case the argument of the square root in the right hand side of (5a) becomes a perfect square, consistently with (6)).

It is now easily seen that (each component of) the 3-vector $\vec{r}(t)$ satisfies a *third-order linear nonautonomous* ODE. Indeed time-differentiation of (3) yields

$$\dddot{\vec{r}}(t) = \left[\dot{h}(t)/h(t)\right]\ddot{\vec{r}}(t) + h^2(t)\left\{ -r^2(t)\dot{\vec{r}}(t) + \left[\vec{r}(t)\cdot\dot{\vec{r}}(t)\right]\vec{r}(t) \right\} , \tag{7a}$$

and this can be rewritten as the following third-order *linear* nonautonomous ODE:

$$\dddot{\vec{r}}(t) = \left[\ \dot{h}(t)/h(t)\ \right]\ddot{\vec{r}}(t) + h^2(t)\left[\ -(s^2 + 2\beta t + v^2 t^2)\dot{\vec{r}}(t) + (\beta + vt)\vec{r}(t)\ \right] . \tag{7b}$$

Exercise 5.2.1-5. Prove (7). *Hint*: to prove (7a), time-differentiate (3) and use again (3), as well as the standard *identity* for a double vector product,

$$\vec{r}_1 \wedge (\vec{r}_2 \wedge \vec{r}_3) \equiv -(\vec{r}_1\cdot\vec{r}_2)\vec{r}_3 + (\vec{r}_1\cdot\vec{r}_3)\vec{r}_2 \quad ; \tag{8}$$

to obtain (7b) from (7a) use (5a) and its time-derivative.

We therefore see that every component of the 3-vector $\vec{r}(t)$ satisfies the *third-order linear nonautonomous* ODE (now written for the dependent variable $\rho(t)$)

$$\dddot{\rho}(t) = \left[\ \dot{h}(t)/h(t)\ \right]\ddot{\rho}(t) + h^2(t)\left[\ -(s^2 + 2\beta t + v^2 t^2)\,\dot{\rho}(t) + (\beta + vt)\,\rho(t)\ \right] . \tag{9}$$

Let us identify 3 independent solutions, $\rho_s(t)$, of this ODE, via the following initial conditions

$$\rho_0(0) = 1 , \quad \dot{\rho}_0(0) = 0 , \quad \ddot{\rho}_0(0) = 0 \quad , \tag{10a}$$

$$\rho_1(0)=0\ ,\quad \dot\rho_1(0)=1\ ,\quad \ddot\rho_1(0)=0\quad , \tag{10b}$$

$$\rho_2(0)=0\ ,\quad \dot\rho_2(0)=0\ ,\quad \ddot\rho_2(0)=1\quad , \tag{10c}$$

namely

$$d^\ell \rho_s(t)/dt^\ell \Big|_{t=0} = \delta_{s\ell}\ ,\quad s,\ell=0,1,2\ . \tag{10d}$$

It is then plain that the solution of the initial-value problem for the 3-vector evolution equation (7) is provided by the formula

$$\vec r(t)=\vec r(0)\,\rho_0(t)+\dot{\vec r}(0)\,\rho_1(t)+\ddot{\vec r}(0)\,\rho_2(t)\quad , \tag{11}$$

hence the solution of the initial-value problem for the original 3-vector evolution equations (3) or (1) reads

$$\vec r(t)=\vec r(0)\,\rho_0(t)+\dot{\vec r}(0)\,\rho_1(t)+h(0)\ \vec r(0)\wedge\dot{\vec r}(0)\,\rho_2(t)\quad . \tag{12}$$

Exercise 5.2.1-6. Verify !

Let us now restrict attention to the case (2), namely to the Newtonian 3-vector evolution equation (see (1))

$$\ddot{\vec r}(t)=k[r(t)]^p\,\vec r(t)\wedge\dot{\vec r}(t)\quad . \tag{13}$$

It is then convenient to define dimensionless variables, by first introducing 2 quantities, a and T, that set the space and time scales,

$$a^2=s^2-(\beta/v)^2=[r(0)]^2-[\vec r(0)\cdot\dot{\vec r}(0)]^2/|\dot{\vec r}(0)|^2=|\vec r(0)\wedge\dot{\vec r}(0)|^2/|\dot{\vec r}(0)|^2\ , \tag{14a}$$

$$T=a/v\quad , \tag{14b}$$

and by then positing

$$t=t_0+T\tau\quad , \tag{15a}$$

$$t_0=-\beta/v^2=-\vec r(0)\cdot\dot{\vec r}(0)/|\dot{\vec r}(0)|^2\quad , \tag{15b}$$

so that (see (5a))

$$r(t) = a(1+\tau^2)^{1/2} \quad . \tag{16}$$

Note that, as it is clear from these formulas, the quantities a and t_0 have a clear physical meaning (in fact, irrespective of the restriction to (2)): a is the distance of closest approach to the origin (see (16): hence (14a) provides the solution to *Exercise 5.2.1-3*!), and t_0 is (see (15a) and (16)) the time at which the moving particle is closest to the origin.

We now also set

$$\rho(t) = a\,\varphi(t) \quad , \tag{17}$$

and, using (3), we rewrite (9) as follows:

$$\varphi''' = p\tau(1+\tau^2)^{-1}\varphi'' + \lambda^2(1+\tau^2)^p\left[-(1+\tau^2)\varphi' + \tau\varphi\right] \quad , \tag{18a}$$

or equivalently

$$(1+\tau^2)^{-p}\varphi''' = p\tau(1+\tau^2)^{-p-1}\varphi'' - \lambda^2(1+\tau^2)\varphi' + \lambda^2\tau\varphi \quad , \tag{18b}$$

where of course the primes denote differentiations with respect to the "dimensionless time" τ, and the dimensionless constant λ is defined as follows:

$$\lambda = k\,a^{p+2}/v \quad . \tag{19}$$

It is now natural to introduce yet another change of variables, by setting

$$w = 1+\tau^2,\ \ \varphi(\tau) = g(w) \quad , \tag{20}$$

whereby one gets, in place of (18),

$$8(w-1)\,w^3\,g'''(w) + 4\left[(3-p)\,w + p\right]w^2\,g''(w)$$

$$+2(-p\,w + \lambda^2\,w^{3+p})\,w\,g'(w) - \lambda^2\,w^{3+p}\,g(w) = 0 \quad . \tag{21}$$

Here of course the primes denote differentiations with respect to w.

Exercise 5.2.1-7. Verify !

It is now plain, from (21), that for $p=-2$ and for $p=-3$ this third-order differential equation, (21), can be explicitly solved by representing $g(w)$ as a power series, say

$$g(w)=w^{\mu}\sum_{m=0}^{\infty} g_m\, w^{-m} \quad , \tag{22}$$

since the insertion of this *ansatz* in (21) yields for the coefficients g_n the 3-term recursion relation

$$2(m-\mu+1)\left[4(m-\mu)^2+2(7+p)(m-\mu)+3(4+p)\right]c_{m+1}$$

$$-4(m-\mu)(m-\mu+1)\left[2(m-\mu)+4+p\right]c_m+\lambda^2\left[2(m-\mu)+7+2p\right]c_{m+3+p}=0, \tag{23}$$

which reduces to a 2-term (hence *solvable*!) recursion relation for $p=-2$ as well as for $p=-3$. Indeed for

$$p=-2 \tag{24a}$$

(23) becomes

$$\left[2(m-\mu)+3\right]\left[\,4(m-\mu+1)^2+\lambda^2\,\right]c_{m+1}$$

$$=8(m-\mu)(m-\mu+1)^2\,c_m\,, \quad m=0,1,2,..., \tag{24b}$$

with μ taking one of the following 3 values (which correspond to the 3 independent solutions of (21) with (22) and (24a)):

$$\mu=1/2,\ \ \mu=\pm i\lambda/2\,; \tag{24c}$$

while for

$$p=-3 \tag{25a}$$

(23) becomes

$$2(m-\mu+1)[2(m-\mu)+3]\,c_{m+1}=[\,4(m-\mu)(m-\mu+1)-\lambda^2\,]\,c_m\,,\quad m=0,1,2,...,\tag{25b}$$

with μ taking one of the following 3 values (which correspond to the 3 independent solutions of (21) with (22) and (25a):

$$\mu=0,\ \mu=\pm 1/2\ .\tag{25c}$$

Exercise 5.2.1-8. Verify !

Exercise 5.2.1-9. Use these results (see in particular (24)) to solve (21) with $p=-2$, and verify the consistency of what you find with the corresponding results of Sect. 5.2 (see in particular (5.2-59g)).

Exercise 5.2.1-10. Use the results of the preceding *Exercise 5.2.1-9* to analyze (1) with (2) and $p=-2$. *Hint*: see (11), (10), (17), (14a), (19), (20).

Exercise 5.2.1-11. Use the above results (see in particular (25)) to solve (21) with $p=-3$. *Solution*: see Sect. 5.2.2.

Before ending Sect. 5.2.1 let us return to the general case of a central, but otherwise arbitrary, field of force, see (1), to associate a physical interpretation to the two constants of motion

$$\left|\dot{\vec{r}}\right|=v\tag{26}$$

and

$$(\vec{r}\cdot\vec{r})(\dot{\vec{r}}\cdot\dot{\vec{r}})-(\vec{r}\cdot\dot{\vec{r}})^2=L^2\ .\tag{27}$$

The time-independence of the first of these two constants of motion, v, see (26) and (5.2-34a), corresponds simply to the transverse character of the Lorentzian force, see (1), that changes the direction of the velocity but not its modulus. The time-independence of the second one of these two constants of motion, L^2, see (27), corresponds to the conservation of the modulus of angular momentum, since it is easily seen that (27) coincides with the expression

$$L^2=\left|\vec{r}\wedge\dot{\vec{r}}\right|^2\ .\tag{28}$$

The time-independence of this quantity (which is easily proven by time-differentiating (27) and by using (5.2-35a,b)), can also be proven in the following, less direct but no less interesting, fashion. Introduce the scalar quantity

$$\Delta(t)=\left[\vec{r}(t)\wedge\dot{\vec{r}}(t)\right]\cdot\ddot{\vec{r}}(t) \quad , \tag{29}$$

which has a clear geometrical meaning: it is (up to a factor $1/6$ and possibly to a sign) the volume of the (3-dimensional) tetrahedron defined by the 3 three-vectors $\vec{r}(t)$, $\dot{\vec{r}}(t)$ and $\ddot{\vec{r}}(t)$, all 3 of them drawn from the origin. Clearly this definition entails

$$\dot{\Delta}(t)=\left[\vec{r}(t)\wedge\dot{\vec{r}}(t)\right]\cdot\dddot{\vec{r}}(t) \quad , \tag{30}$$

hence, via (1)

$$\dot{\Delta}(t)/\Delta(t)=\left\{(d/dt)f[r(t)]\right\}/f[r(t)] \quad . \tag{31}$$

Proof: time-differentiation of (1) yields

$$\dddot{\vec{r}}=[(d/dt)f(r)]\,\vec{r}\wedge\dot{\vec{r}}+f(r)\vec{r}\wedge\ddot{\vec{r}} \tag{32a}$$

hence, via (1) and (8),

$$\dddot{\vec{r}}=\left\{[(d/dt)f(r)]/f(r)\right\}\ddot{\vec{r}}+[f(r)]^2\left[-r^2\dot{\vec{r}}+(\vec{r}\cdot\dot{\vec{r}})\vec{r}\right] \quad . \tag{32b}$$

Insertion of this expression of $\dddot{\vec{r}}$ in the right hand side of (30) yields

$$\dot{\Delta}=\left\{[(d/dt)f(r)]/f(r)\right\}\Delta \quad , \tag{33}$$

and this coincides with (31), which is thereby proven.

By integrating (31) one then gets

$$\Delta(t)=C\,f[r(t)] \quad , \tag{34}$$

with C constant. One the other hand the definition (28) of L, which can be re-written as follows,

$$L^2=(\vec{r}\wedge\dot{\vec{r}})\cdot(\vec{r}\wedge\dot{\vec{r}}) \tag{35}$$

clearly entails, via (1) and (29),

$$L^2 = \Delta(t) / f[r(t)] \quad . \qquad (36)$$

It is now plain that (34) and (36) entail that L^2 is constant.

Exercise 5.2.1-12. Show that

$$L^2 = a^2 v^2 \qquad (37)$$

and·discuss the physical significance of this formula in terms of the physical significance of L, v respectively a. *Hint*: see (28), (14a) and the sentence after (16).

5.2.2 Motion of a magnetic monopole in a central Coulomb field

In Sect. 5.2.2 we investigate the (nonrelativistic) motion of a point-like (massive) monopole in the Coulomb electric field produced by a fixed electric charge (which we locate at the origin of the coordinate system); or equivalently, the motion of a point-like (massive) electrically charged particle moving in the magnetic field produced by a fixed magnetic monopole. The Newtonian-Lorentz equation of motion for this problem reads

$$\ddot{\vec{r}} = k \, r^{-3} \, \vec{r} \wedge \dot{\vec{r}} \quad , \qquad (1)$$

with k a constant (proportional to the electric and magnetic charges, and inversely proportional to the mass of the moving particle), clearly having the dimensions $(\text{length})^2 \, (\text{time})^{-1}$.

As we saw in the preceding Sect. 5.2.1, the following two properties,

$$\left| \dot{\vec{r}}(t) \right| = \left| \dot{\vec{r}}(0) \right| = v \quad , \qquad (2)$$

$$r^2(t) = s^2 + 2\beta \, t + v^2 t^2 \quad , \qquad (3)$$

are entailed by (5.2-1), hence a *fortiori* by (1). Here the 3 constants v, s^2 and β are defined by (2) and (5.2.1-5). But hereafter we assume (without significant loss of generality, but with a significant notational improvement) that the origin of time (namely, the value $t = 0$) is set at the moment when the moving particle is closer to the center of force, which is also the time when its velocity is orthogonal to its position (relative to the origin). Hence in place of (3) we write (see (5.2.1-5,14a))

$$r^2(t) = a^2 + v^2 t^2 \quad , \tag{4}$$

with the advantage that both constants, a respectively v, entering this formula have a clear physical meaning: v is the constant (modulus of the) velocity with which the moving particle travels along its trajectory, a is the distance of closest approach of that trajectory to the origin (as in the case discussed in the preceding Sect. 5.2.1, we exclude from consideration the trivial, free motion that goes through the origin, namely we assume $a > 0$; and we exclude as well the trivial case of a particle that does not move at all, namely we assume $v > 0$).

Given these (essentially notational) assumptions it is convenient to pursue the notational simplification, by choosing the distance a as the scale for length, and the time (see (5.2.1.14b))

$$T = a / v \tag{5}$$

as the scale for time. Hence we set

$$t = T\tilde{t} \,, \quad \vec{r}(t) = a\,\vec{\tilde{r}}(\tilde{t}) \,, \quad k = (a^2/T)\,\tilde{k} = (av)\,\tilde{k} \quad , \tag{6}$$

introducing thereby the dimensionless quantities $\tilde{t}, \vec{\tilde{r}}$ and $\tilde{k}$. Note that the definition of these quantities is specifically adjusted to treat the (family of) trajectories characterized by the same velocity v as well as by the same distance of closest approach to the origin a.

Hereafter, however, for typographical convenience we drop the tildes from the quantities introduced in (6). Hence our (now dimensionless) equation of motion still reads just like (1), while (2) respectively (4) are now replaced by

$$\left|\dot{\vec{r}}(t)\right| = 1 \tag{7}$$

respectively

$$r^2(t) = 1 + t^2 \quad . \tag{8}$$

Note the analogy, as well as the (notational) differences, of this introduction of dimensional quantities relative to that performed in the preceding Sect. 5.2.1, see after (5.2.1-13).

We saw in the preceding Sect. 5.2.1 that, as general consequence of the central character of the electric field, the modulus of the angular momentum of the moving monopole,

$$L^2 = \vec{L}\cdot\vec{L} = (\vec{r}\wedge\dot{\vec{r}})^2 = (\vec{r}\cdot\vec{r})(\dot{\vec{r}}\cdot\dot{\vec{r}}) - (\vec{r}\cdot\dot{\vec{r}})^2 \quad , \tag{9a}$$

is a constant of motion, a relation that in our dimensionless units now reads

$$L^2 = 1 \quad . \tag{9b}$$

Note that the angular momentum itself,

$$\vec{L} = \vec{r}\wedge\dot{\vec{r}} \quad , \tag{10}$$

is *not* a constant of motion, contrary to what happens when a particle moves under the influence of a central force (the Lorentz force in the right hand side of (1), as well as (5.2.1-1), is not central!). However, in the special (Coulomb) case of equation (1), there is a three-vector which is conserved over time:

$$\vec{J} = \vec{L} + k\hat{r} \quad , \tag{11a}$$

$$\dot{\vec{J}} = 0 \; . \tag{11b}$$

Here and below

$$\hat{r} \equiv \vec{r}/r \tag{12}$$

is the unit-vector in the direction of $\vec{r}$.

Proof.

$$\dot{\vec{J}} = \dot{\vec{L}} + k[\, r^{-1}\dot{\vec{r}} - r^{-2}\dot{r}\,\vec{r}\,] , \tag{13}$$

$$\dot{\vec{L}} = \vec{r} \wedge \ddot{\vec{r}} \ , \qquad (14a)$$

$$\dot{\vec{L}} = k r^{-3} \vec{r} \wedge (\vec{r} \wedge \dot{\vec{r}}) \ , \qquad (14b)$$

$$\dot{\vec{L}} = -k r^{-1} \dot{\vec{r}} + k r^{-3} (\vec{r} \cdot \dot{\vec{r}}) \vec{r} \ , \qquad (14c)$$

$$\dot{\vec{L}} = -k [\, r^{-1} \dot{\vec{r}} - r^{-2} \dot{r} \, \vec{r} \,] \ . \qquad (14d)$$

(14a) follows by t-differentiation from (10), (14b) from (14a) via the equation of motion (1); (14c) follows from (14b) via the three-vector identity (5.2.1-8); (14d) follows from (14c) via the relation $r\dot{r} = \vec{r} \cdot \dot{\vec{r}}$ (which of course obtains by time-differentiating $r^2 = \vec{r} \cdot \vec{r}$). Clearly (14d) and (13) entail (11b), which is thereby proven.

The fact that the three-vector $\vec{J}$, see (11), is a constant of motion for (1) was pointed out by H. Poincaré over a century ago <P1896>; the vector $\vec{J}$ is therefore generally called the Poincaré vector.

Since $\vec{L}$ and $\hat{r}$ are orthogonal, see (10) and (12), the definition (11a) of the Poincaré vector entails

$$J^2 = L^2 + k^2 \ , \qquad (15)$$

as well as

$$\hat{r} \cdot \vec{J} = k \ . \qquad (16)$$

This entails that the angle, call it θ, among the unit vector $\hat{r} \equiv \hat{r}(t)$ (or, equivalently, the three-vector $\vec{r} \equiv \vec{r}(t)$) and the constant vector $\vec{J}$ is also constant, and it takes the value

$$\cos\theta = k / J \, . \qquad (17)$$

Hence the motion takes place on a *fixed circular half-cone*, whose vertex is at the origin, whose axis coincides with the (constant) three-vector $\vec{J}$, and whose (constant) half-angle θ is given by (17). If we denote as $r \equiv r(t)$, θ and $\varphi \equiv \varphi(t)$ the spherical coordinates of the moving particle in a coordinate system whose origin coincides with the (fixed) electrical charge and whose azimuthal axis coincides with the direction of $\vec{J}$, so that

$$\vec{r}(t) \equiv r(t) \ (\sin\theta \cos\varphi(t), \sin\theta \sin\varphi(t), \cos\theta) \ , \qquad (18)$$

then there remains to know only $\varphi(t)$ in order to acquire complete information on the particle motion, namely on $\vec{r}(t)$, since $r(t)$ is given by (8) and θ remains constant throughout the motion, see (17) with (15) and (9b), which entail

$$\cos\theta = k(1+k^2)^{-1/2} \tag{19a}$$

hence

$$\sin\theta = (1+k^2)^{-1/2} \ , \tag{19b}$$

$$\operatorname{cotan}\theta = k \ . \tag{19c}$$

The computation of $\varphi(t)$ is now easy. Indeed time-differentiation of (18) yields, via (7) and (8),

$$\dot{\varphi}(t) = (1+k^2)^{1/2}/(1+t^2) \quad . \tag{20}$$

Proof. From (18)

$$\dot{\vec{r}} = \dot{r}(\sin\theta\ \cos\varphi, \sin\theta\ \sin\varphi, \cos\theta)$$

$$+ r\dot{\varphi}\ (-\sin\theta\ \sin\varphi, \sin\theta\ \cos\varphi, 0) \ . \tag{21}$$

Hence

$$\dot{\vec{r}}\cdot\dot{\vec{r}} = (\dot{r}^2 + r^2\,\dot{\varphi}^2)\sin^2\theta + \dot{r}^2\cos^2\theta \ , \tag{22}$$

hence, via (7),

$$1 = \dot{r}^2 + r^2\,\dot{\varphi}^2\sin^2\theta \ . \tag{23}$$

From this formula one easily gets, via (8) and (19b),

$$1 = t^2/(1+t^2) + \dot{\varphi}^2\,(1+t^2)/(1+k^2) \quad , \tag{24}$$

and this yields (20), which is thereby proven.

Integration of (20) yields

$$\varphi(t) = \varphi_0 + (1+k^2)^{1/2}\arctan(t) \quad , \tag{25}$$

where φ_0 is an (integration) constant.

The formulas (8), (19) and (25) provide, via (18), complete information on the motion of the monopole, which turns out, remarkably, to be given, by these formulas, quite explicitly and in terms of *elementary* functions.

The same results can be obtained by noticing that (again quite remarkably, although, now, not surprisingly), for $p=-3$ (which entails that (5.2.1-1,2) or (5.2.1-13) coincides with (1)) the third-order linear ODE (5.2.1-18) can be completely solved in terms of *elementary* functions.

Before proceeding, the diligent reader is advised to review the derivation of (5.2.1-18) and to understand the (minor) differences among the notations used in Sects. 5.2.1 and 5.2.2 (also beware of the very different meanings of the symbol φ in the two contexts).

For $p=-3$, the third-order ODE (5.2.1-18) reads

$$(1+t^2)^3\,\dddot{\psi}+3t(1+t^2)^2\,\ddot{\psi}+k^2(1+t^2)\dot{\psi}-k^2\,t\psi=0 \quad . \tag{26}$$

Here we employ a notation consistent with that used in (the latter part of) Sect. 5.2.2, and we have also written $\psi(t)$ in place of $\varphi(t)$ (and k in place of λ; see (5.2.1-19) and recall that we have chosen units of space and time such that $a=v=1$).

This third-order linear ODE has the remarkable property to possess solutions all of which are expressed in terms of *elementary* functions. Indeed it is easily seen that 3 independent solutions of (26) are provided by the formula

$$\psi_s(t)=(1+it)^{\frac{1}{2}(1+s\gamma)}\,(1-it)^{\frac{1}{2}(1-s\gamma)}=(1+t^2)\,\exp[\,is\arctan(t)\,],\qquad s=0,\pm1 \quad , \tag{27}$$

where

$$\gamma=(1+k^2)^{1/2} \quad . \tag{28}$$

Proof. From (27) one easily gets

$$\dot{\psi}_s(t)/\psi_s(t)=(is\gamma+t)(1+t^2)^{-1} \quad , \tag{29a}$$

$$\ddot{\psi}_s(t)/\psi_s(t) = (1-s^2\gamma^2)(1+t^2)^{-2} \quad , \tag{29b}$$

$$\dddot{\psi}_s(t)/\psi_s(t) = (1-s^2\gamma^2)(is\gamma-3t)\,(1+t^2)^{-3} \quad , \tag{29c}$$

and the insertion of these expressions in (26) yields

$$s\gamma[\ s^2\gamma^2-(1+k)\ ]=0 \tag{30a}$$

namely, via (28),

$$s(s^2-1)=0 \quad . \tag{30b}$$

Hence (27) satisfies (26) for $s=0$, $s=+1$ and $s=-1$.

Exercise 5.2.2-1. Show the consistency of these findings with those given above, in particular the consistency of (18) (with (8), (19) and (25)) with (5.2.1-12,17).

Let us conclude Sect. 5.2.2 by displaying the final results for the trajectories, reinstating the original variables (but maintaining the simplification to set the origin of time when the moving particle is closest to the origin). The results are better presented in spherical coordinates, in the reference system introduced above, whose origin coincides with the position of the fixed electric charge, and whose azimuthal axis (z-axis) is parallel to the Poincaré vector $\vec{J}$, see (11a) with (10) and (12):

$$r(t)=a[\ 1+(t/T)^2\ ]^{1/2} \quad , \tag{31a}$$

$$\theta = \mathrm{arccotan}(\kappa), \tag{31b}$$

$$\varphi(t)=\varphi_0+(1+\kappa^2)\ \arctan(t/T) \quad , \tag{31c}$$

where T is given by (5) and κ is the $\tilde{k}$ of (6),

$$T=a/v,\ \ \kappa=k/(av)\ . \tag{31d}$$

Let us emphasize here the difference among the Greek κ and the Latin k, the latter being of course the constant that appears in the right hand side of (1), the former being the $\tilde{k}$ in (6), $\kappa=\tilde{k}$; while a is again the distance of closest approach of the trajectory, and v is the (constant) velocity with which the moving particle travels.

Exercise 5.2.2-2. Show that asymptotically (in the remote past and future) all trajectories become straight lines, namely as $t \to \pm\infty$

$$\vec{r}(t) = \vec{v}_{\pm} t + \vec{r}_{\pm}^{(0)} + O(t^{-1}) \quad , \tag{32}$$

and obtain the following formulas that characterize their behavior,

$$b_{\pm} = a \quad , \tag{33}$$

$$\cos\alpha = 1 - 2(1+\kappa^2)\sin^2\left[(1+\kappa^2)^{1/2}\pi/2\right] \quad , \tag{34}$$

$$\Delta\varphi \equiv \varphi(\infty) - \varphi(-\infty) = \pi(1+\kappa^2)^{1/2} \quad , \tag{35}$$

where $b_{\pm}$ are the impact parameters, $b_{\pm} = r_{\pm}^{(0)} \equiv \left|\vec{r}_{\pm}^{(0)}\right|$ (namely the distance from the origin of the asymptotic straight lines, see (32)), of course $v_{\pm} \equiv \left|\vec{v}_{\pm}\right| = v$, $\cos\alpha = (\vec{v}_{+} \cdot \vec{v}_{-})/v^2$ is the "scattering angle" and $\Delta\varphi$, see (35), is the overall angular rotation of the trajectory.

Exercise 5.2.2-3. Using the results of the preceding *Exercise 5.2.2-2* (see in particular (34)) verify that, if $\kappa^2 << 1$, then $\alpha \approx \pi$, as well as $\Delta\varphi \approx \pi \bmod(2\pi)$, and discuss the paradoxical aspects of these findings (which originate from the singular, and long-range, dependence of the force on the distance r from the origin, see (1); the paradoxical aspect is that of course $\kappa = k = 0$ entails no force at all, hence free motion, hence $\alpha = \Delta\varphi = 0$).

Exercise 5.2.2-4. Draw a picture of a typical trajectory. *Hint*: see <S2000>.

5.2.3 Solvable cases of the (2×2)-matrix evolution equation $\ddot{\underline{U}} = 2a\dot{\underline{U}} + b\underline{U} + c[\,\dot{\underline{U}}, \underline{U}\,]$

In this Sect. 5.2.3 we discuss the *linearizable* matrix evolution equation (see (5.2-16))

$$\ddot{\underline{U}} = 2a\dot{\underline{U}} + b\underline{U} + c[\,\dot{\underline{U}}, \underline{U}\,], \tag{1}$$

with a,b and c three arbitrary "coupling constants" (c could of course be rescaled away), and we identify two cases in which, at least for (2×2)-

matrices, this equation can be explicitly solved in terms of known special functions. We set

$$\underline{U}(t)=\varphi(t)\,\underline{V}(\tau)\ ,\ \tau=\tau(t)\ , \tag{2}$$

and thereby obtain

$$\underline{V}''=\underline{V}'\left[2a\dot{\tau}-2\dot{\tau}(\dot{\varphi}/\varphi)-\ddot{\tau}\right]/\dot{\tau}^2+\underline{V}\left[b+2a(\dot{\varphi}/\varphi)-(\ddot{\varphi}/\varphi)\right]/\dot{\tau}^2$$

$$+c(\varphi/\dot{\tau})[\,\underline{V}',\underline{V}\,]\,. \tag{3}$$

Here, and throughout Sect. 5.2.3, primes denote derivatives with respect to τ.

A choice naturally suggested by this equation is

$$\dot{\tau}(t)=\varphi(t)\ , \tag{4a}$$

which entails that $\underline{V}(\tau)$ satisfies the matrix ODE

$$\underline{V}''=\varphi^{-2}\left[2a\varphi-3\dot{\varphi}\right]\underline{V}'+\varphi^{-3}\left[2a\dot{\varphi}+b\varphi-\ddot{\varphi}\right]\underline{V}+c[\,\underline{V}',\underline{V}\,]\,, \tag{4b}$$

which we report here for future memory. However, we prefer now to focus on two special cases of (1), which allow a completely explicit solution (at least in the case of (2×2)-matrices). Hence we set

$$\varphi(t)=\exp(\mu\, t)\ , \tag{5a}$$

$$\dot{\tau}(t)=\exp(\nu\, t)\ , \tag{5b}$$

so that (3) becomes

$$\underline{V}''=p\exp(-\nu t)\,\underline{V}'+q\exp(-2\nu t)\,\underline{V}+c\exp[(\mu-\nu)t][\,\underline{V}',\underline{V}\,]\,, \tag{5c}$$

with

$$p=2(a-\mu)-\nu\,, \tag{5d}$$

$$q=b+2a\,\mu-\mu^2\ . \tag{5e}$$

The first case we consider is characterized by the condition

$$b = -(8/9)\,a^2, \tag{6a}$$

to which we associate the choices $\mu = \nu = 2a/3$, which entail, via (5d), $p = 0$ and, via (5e) and (6a), $q = 0$ as well. Hence in this case the changes of variables

$$\underline{U}(t) = \exp(2at/3)\,\underline{V}(\tau), \tag{6b}$$

$$\tau(t) = [\exp(2at/3) - 1]/(2a/3)\ , \tag{6c}$$

yield

$$\underline{V}'' = c[\,\underline{V}', \underline{V}\,], \tag{6d}$$

which, up to trivial notational changes, is just the (2×2)-matrix evolution equation shown to be solvable in Sect. 5.2 (see (5-28b)). Hence a convenient prescription to solve (1) with (6a) is to solve instead (6d), and then perform the change of dependent and independent variables (6b) with (6c). Clearly, this implies that *all* solutions of (1) are completely periodic with period $T = 2\pi/\omega$ if

$$a = (3/2)i\omega,\ b = 2\omega^2\ , \tag{6e}$$

with ω an arbitrary (nonvanishing) *real* constant. Note however that in this case the matrix evolution equation (1) is *complex*.

The second case we consider is characterized by the restriction

$$b = 0\ . \tag{7a}$$

In this case we set $\mu = 0, \nu = 2a$, which again entails, via (5d), $p = 0$ and, via (5e) and (6a), $q = 0$ as well. Hence in this case the change of variable

$$\underline{U}(t) = \underline{V}(\tau), \tag{7b}$$

$$\tau(t) = [\exp(2at) - 1]/(2a)\ , \tag{7c}$$

yields

$$\underline{V}'' = c(1+2a\tau)^{-1}[\,\underline{V}',\underline{V}\,] \,, \tag{7d}$$

which, up to trivial notational changes, is the (2×2)-matrix evolution equation shown to be *solvable* at the end of Sect. 5.2. (see (5.2-60b)). Hence a convenient prescription to solve (1) with (7a) is (at least in the case of (2×2)-matrices), to solve instead (7d), and then perform the change of dependent and independent variables (7b) with (7c). Clearly, this again implies that *all* solutions of (1) are completely periodic with period $T = 2\pi / \omega$ if

$$a = i\omega / 2, \; b = 0 \,. \tag{7e}$$

Of course in this case as well the matrix evolution equation (1) is complex.

5.3 Association, complexification, multiplication: solvable few and many-body problems obtained from the previous ones

In Sect. 5.3 we indicate some rather elementary techniques whereby from a matrix evolution equation for one matrix one can obtain evolution equations for a few, or for many, matrices. These techniques are introduced by showing how they work in simple specific cases; we use for this purpose the solvable examples of the preceding two Sects. 5.1 and 5.2, see below. We also exhibit some solvable few- and many-body problems that correspond to these solvable matrix evolution equations via appropriate parametrizations of matrices in terms of 3-vectors (see for instance (5.2-18); other parametrizations, in terms of S-vectors with S an *arbitrary* positive integer, are also introduced below, at the end of Sect. 5.3, and more systematically in Sect. 5.5).

Let us emphasize that the techniques introduced herein can of course be used in more general contexts than the simple example used here to introduce them. Some other examples are exhibited below, mainly in Sect. 5.4. and its subsections. But there remains an ample scope for additional applications of these techniques, which the interested reader is advised to explore.

We use below the following two matrix evolution equations as basic examples:

$$\ddot{\underline{M}} = 2a\dot{\underline{M}} + b\underline{M} + c\,\dot{\underline{M}}\,\underline{M}^{-1}\,\dot{\underline{M}} \quad , \tag{1}$$

$$\ddot{\underline{U}} = 2a\dot{\underline{U}} + b\underline{U} + [\,\dot{\underline{U}}, F(\underline{U})\,] \quad . \tag{2}$$

Here $\underline{M} \equiv \underline{M}(t)$ and $\underline{U} \equiv \underline{U}(t)$ are square matrices of arbitrary rank, a, b, c, are 3 scalar constants and $F(\underline{U})$ is an (*a priori* arbitrary) function of the matrix $\underline{U}$ and of no other matrix (so that (5.2-2b) and (5.2-12) hold). Let us recall that the matrix evolution equation (1) is *solvable* (see Sect. 5.1), while the matrix evolution equation (2) is generally *linearizable*, sometimes also *solvable* (see Sects. 5.2 and 5.2.3).

Association. We use the term "association" to denote this technique, because it involves the association, to the matrix evolution equation under consideration, of another matrix evolution equation, see below. This simple technique is interesting because it allows to transform, as we now show, a matrix evolution equation which is *not* translation-invariant, i.e., not invariant under addition of a constant (matrix) to the dependent (matrix) variable, into a coupled set of 2 equations, featuring 2 (matrix) dependent variables, which are invariant under a (common, constant) translation of these (matrix) dependent variables.

To illustrate this technique let us focus, say, on (1), and *associate* to it the matrix evolution equation

$$\ddot{\underline{P}}(t) = \alpha\,\dot{\underline{P}}(t) \quad . \tag{3}$$

This matrix evolution equation is, of course, trivially solvable:

$$\underline{P}(t) = \underline{P}(0) + \dot{\underline{P}}(0)[\ \exp(\alpha t) - 1\ ]/\alpha \tag{4}$$

We then introduce the two matrices $\underline{M}^{(\pm)}(t)$ by setting

$$\underline{M}^{(\pm)}(t) = \underline{P}(t) \pm \underline{M}(t) \quad , \tag{5a}$$

so the

$$\underline{P}(t) = \frac{1}{2}[\ \underline{M}^{(+)}(t) + \underline{M}^{(-)}(t)\ ],\ \ \underline{M}(t) = \frac{1}{2}[\ \underline{M}^{(+)}(t) - \underline{M}^{(-)}(t)\ ] \quad , \tag{5b}$$

and we thereby get for these two matrices $\underline{M}^{(+)}(t)$ and $\underline{M}^{(-)}(t)$, the following two matrix evolution equations:

$$\underline{\ddot{M}}^{(\pm)} = (\frac{1}{2}\alpha \pm a)\underline{\dot{M}}^{(+)} + (\frac{1}{2}\alpha \mp a)\underline{\dot{M}}^{(-)} \pm \frac{1}{2}b[\underline{M}^{(+)} - \underline{M}^{(-)}]$$

$$\pm \frac{1}{2}c[\underline{\dot{M}}^{(+)} - \underline{\dot{M}}^{(-)}][\underline{M}^{(+)} - \underline{M}^{(-)}]^{-1}[\underline{\dot{M}}^{(+)} - \underline{\dot{M}}^{(-)}] \quad . \tag{6}$$

These equations of motion are obviously solvable (via (5)), and they are clearly invariant under the translation $\underline{M}^{(\pm)}(t) \to \underline{\tilde{M}}^{(\pm)}(t) = \underline{M}^{(\pm)}(t) + \underline{C}$, $\underline{\dot{C}} = 0$.

Exercise 5.3-1. Display the solution of the initial value problem for (6). *Hint*: see (5), (4) and (5.1-2).

Exercise 5.3-2. Display the solution of the initial-value problem for the (translation- and rotation-invariant) Newtonian equations of motion

$$\ddot{\vec{r}}^{(\pm)} = (\frac{1}{2}\alpha \pm a)\dot{\vec{r}}^{(+)} + (\frac{1}{2}\alpha \mp a)\dot{\vec{r}}^{(-)} \pm \frac{1}{2}b\vec{r} \pm c[\dot{\vec{r}}(\dot{\vec{r}}\cdot\vec{r}) - \frac{1}{2}\vec{r}(\dot{\vec{r}}\cdot\dot{\vec{r}})]/r^2 \quad , \tag{7a}$$

where

$$\vec{r} \equiv \vec{r}^{(+)} - \vec{r}^{(-)} \quad . \tag{7b}$$

Hint: note that the two equations of motion (7) correspond to (6) via (5.2-18), and use the solution of the preceding *Exercise 5.3-1*; or obtain these equations of motion by applying directly the *association* trick (appropriately modified) to (5.1-29).

Exercise 5.3-3. State conditions on the 4 parameters a,b,c,α in (7) which are sufficient to guarantee that all solutions of these Newtonian equations of motion, (7), are: *(i)* confined for all time (including the limits $t \to \pm\infty$), *(ii)* multiply periodic, *(iii)* completely periodic. *Hint*: allow the 4 parameters a,b,c,α to be complex, and use the solution of the preceding *Exercise 5.3-2* (see also *Propositions 5.1-7* and *5.1-10*, and *Exercise 5.1-14*).

Exercise 5.3-4. Display the equations of motion (more general than (7)) that correspond to (6) via (5.1-25) (rather than (5.2-18)), and discuss the behavior of their solutions.

Exercise 5.3-5. Apply the technique of *association*, as described above, to obtain the pair of coupled, translation-invariant, Newtonian equations of motion (for two 3-vectors, $\vec{r}^{(+)}(t)$ and $\vec{r}^{(-)}(t)$) that correspond to (5.2-19).

Complexification. The next technique we review in Sect. 5.3 consists merely in the process of *complexification* of the matrices, or of the 3-vectors, under consideration. For instance consider the (*linearizable*; see (5.2-19)) 3-vector equation of motion

$$\ddot{\vec{r}} = 2a\dot{\vec{r}} + b\vec{r} + c\dot{\vec{r}} \wedge \vec{r} \tag{8}$$

and set

$$a = \alpha + i\alpha', \ b = \beta + i\beta', \ c = \gamma + i\gamma' \ , \tag{9a}$$

as well as

$$\vec{r}(t) = \vec{r}_1(t) + i\vec{r}_2(t) \ , \tag{9b}$$

with the understanding that the 6 constants $\alpha, \alpha', \beta, \beta', \gamma, \gamma'$, as well as the 2 three-vectors $\vec{r}_1(t)$, $\vec{r}_2(t)$, are *real*. Then clearly (8) yields the following system of 2 coupled 3-vector Newtonian equations of motion:

$$\ddot{\vec{r}}_1 = 2\alpha\,\dot{\vec{r}}_1 - 2\alpha'\dot{\vec{r}}_2 + \beta\vec{r}_1 - \beta'\vec{r}_2 + \gamma(\dot{\vec{r}}_1 \wedge \vec{r}_1 - \dot{\vec{r}}_2 \wedge \vec{r}_2) - \gamma'(\dot{\vec{r}}_1 \wedge \vec{r}_2 + \dot{\vec{r}}_2 \wedge \vec{r}_1) \ , \tag{10a}$$

$$\ddot{\vec{r}}_2 = 2\alpha\,\dot{\vec{r}}_2 + 2\alpha'\dot{\vec{r}}_1 + \beta\vec{r}_2 - \beta'\vec{r}_1 + \gamma(\dot{\vec{r}}_1 \wedge \vec{r}_2 + \dot{\vec{r}}_2 \wedge \vec{r}_1) + \gamma'(\dot{\vec{r}}_1 \wedge \vec{r}_1 - \dot{\vec{r}}_2 \wedge \vec{r}_2) \ . \tag{10b}$$

Clearly this trick, as well as that illustrated above (*association*), results in doubling the number of (real) quantities under consideration. But such a duplication of the number of quantities evolving in time (be they matrices or vectors) is generally not the main motivation for employing these tricks. Indeed, as we emphasized above, the main motivation to introduce the *association* trick described above is to obtain, from equations of motion which are not invariant under translation, other equations of motion which do possess this invariance property. Likewise, the *complexification* trick we just illustrated, see (9), is generally introduced, not just to artificially duplicate the number of (real) dependent variables, but rather because more interesting models become available when the attention extends from real to complex variables; indeed we have already seen several instances in which the introduction of complex (perhaps imaginary) "coupling constants" is instrumental to generate models which feature remarkable behaviors, such as *completely periodic* motions.

Here we illustrated the complexification trick by showing how it works on a simple example, see (8) and (10). It is clear (in fact too obvious to warrant any additional elaboration) that it can be applied in much more general contexts, and of course to matrices as well as to vectors.

Multiplication. By this term we refer to the possibility to go from equations that involve only one, or a few, matrices, to equations that involve several (indeed, arbitrarily many) matrices. We now illustrate two techniques suitable to perform this trick.

The first technique is equally applicable to scalars, to vectors, to matrices. Here we illustrate it in a matrix context, taking as starting point the following *linearizable* matrix evolution equation,

$$\ddot{\underline{U}} = 2a\dot{\underline{U}} + b\underline{U} + c\left[\dot{\underline{U}},\underline{U}\right] \quad , \tag{11}$$

which is clearly the special case of (2) with $F(\underline{U}) = c\underline{U}$. Let us recall that of course a,b,c are 3 arbitrary scalar constants.

We now set

$$\underline{U}(t) = \sum_{j=1}^{J} \eta_j \underline{U}_j(t) \quad , \tag{12a}$$

as well as

$$a = \sum_{j=1}^{J} \eta_j a_j \, , \quad b = \sum_{j=1}^{J} \eta_j b_j \, , \quad c = \sum_{j=1}^{J} \eta_j c_j \quad , \tag{12b}$$

where the J quantities η_j are the elements of an Abelian algebra satisfying the multiplication law

$$\eta_j \eta_k = \eta_k \eta_j = \eta_{j+k} ; \; j,k = 1,...,J, \; \mathrm{mod}(J) \quad . \tag{13}$$

Here and below J is an arbitrary positive integer. Note that via (12) one has introduced J matrices $\underline{U}_j(t)$ as well as $3J$ arbitrary constants a_j, b_j, c_j, and that a standard representation of the Abelian algebra (13) is provided by the formula

$$\eta_j = \exp(2\pi i j / J) \quad . \tag{14}$$

It is then clear that insertion of the *ansaetze* (12) into (11) yields the following system of coupled equations for the J matrices $\underline{U}_j(t)$:

$$\ddot{\underline{U}}_j = \sum_{k=1}^{J} (2a_{j-k}\dot{\underline{U}}_k + b_{j-k}\underline{U}_k) + \sum_{k,\ell=1}^{J} c_{j-k-\ell}[\,\dot{\underline{U}}_k, \underline{U}_\ell\,] \quad , \tag{15}$$

of course with all indices defined $\mathrm{mod}(J)$.

Now assume that $\underline{U}[\,a,b,c;\underline{U}(0),\dot{\underline{U}}(0);t\,]$ is the solution of (11), corresponding to given constants a,b,c and initial data $\underline{U}(0), \dot{\underline{U}}(0)$; then the solution of (the J coupled matrix evolution equations) (15) with initial conditions $\underline{U}_j(0), \dot{\underline{U}}_j(0),\ j=1,\dots,J$, is given by the (rather explicit !) formula

$$\underline{U}_j(t) = J^{-1}\sum_{k=1}^{J} \exp(-2\pi i j k/J)\,\underline{U}[\,a^{(k)},b^{(k)},c^{(k)};\underline{U}^{(k)}(0),\dot{\underline{U}}^{(k)}(0);t\,] \quad , \tag{16a}$$

$$a^{(k)} \equiv \sum_{\ell=1}^{J} a_\ell \exp(2\pi i \ell k/J) \quad , \tag{16b}$$

$$b^{(k)} \equiv \sum_{\ell=1}^{J} b_\ell \exp(2\pi i \ell k/J) \quad , \tag{16c}$$

$$c^{(k)} \equiv \sum_{\ell=1}^{J} c_\ell \exp(2\pi i \ell k/J) \quad , \tag{16d}$$

$$\underline{U}^{(k)}(0) \equiv \sum_{k=1}^{J} \underline{U}_\ell(0)\exp(2\pi i \ell k/J) \quad , \tag{16e}$$

$$\dot{\underline{U}}^{(k)}(0) \equiv \sum_{k=1}^{J} \dot{\underline{U}}_\ell(0)\exp(2\pi i \ell k/J) \quad . \tag{16f}$$

Hence the system of J coupled matrix evolution equations (15) is as well *linearizable* as (11) (indeed *solvable* for (2×2)-matrices if all the constants a_j, b_j vanish, $a_j = b_j = 0$, see Sect. 5.2.3).

Proof. Let us start by proving a key formula. Assume $f(z)$ to be an analytic function of z, so that, at least for small enough $|z|$,

$$f(z) = \sum_{m=0}^{\infty} f^{(m)} z^m / m! \tag{17}$$

is well defined. Then clearly the formula

$$f(\sum_{k=1}^{J} \eta_k z_k) = \sum_{j=1}^{J} \eta_j f_j(\underline{z}) \quad , \tag{18}$$

with the η_j being elements of the Abelian algebra (13), defines uniquely the J functions $f_j(\underline{z})$. Here and below $\underline{z} \equiv (z_1, z_2, ..., z_J)$, and all indices are defined $\bmod(J)$. The key formula we now prove reads

$$f_j(\underline{z}) = J^{-1} \sum_{K=1}^{J} \exp(-2\pi i j K/J)\, f(z^{(K)}) \quad , \tag{19a}$$

$$z^{(K)} = \sum_{k=1}^{J} z_k \exp[2\pi i k K/J] \quad , \tag{19b}$$

hence it provides an explicit expression of the quantities $f_j(\underline{z})$ defined by (18).

To prove (19) we note that the algebra (13) admits the following J realizations:

$$\eta_j^{(K)} = \exp(2\pi i j K/J) \quad , \tag{20}$$

with $K = 1, ..., J$ (which reduce to (14) for $K = 1$). Hence (18) entails

$$f(\sum_{k=1}^{J} \eta_k^{(K)} z_k) = \sum_{j=1}^{J} \eta_j^{(K)} f_j(\underline{z}) \quad , \tag{21a}$$

namely

$$f(\sum_{k=1}^{J} z_k \exp(2\pi i k K/J)) = \sum_{j=1}^{J} f_j(\underline{z}) \exp(2\pi i j K/J) \quad . \tag{21b}$$

Multiplication by $\exp(-2\pi i j K/J)$ yields

$$\exp(-2\pi i k K/J)\, f(\sum_{\ell=1}^{J} z_\ell \exp(2\pi i \ell K/J)) = \sum_{j=1}^{J} f_j(\underline{z}) \exp[2\pi\, i(j-k)K/J] . \tag{22}$$

We now sum over K from 1 to J, and use the *identity*

$$\sum_{K=1}^{J} \exp(2\pi i j K/J) = J\,\delta_{j,J}, \quad j = 1, 2, ..., \ \bmod(J), \tag{23}$$

getting thereby (up to some renaming of dummy summation indices) (19), which is therefore proven.

The generalization of the formula (19) to the case of a matrix-valued function $f(z)$ is obvious, as well as the extension to the case when it depends on more than one argument, say $f \equiv f(x; y)$. Then by setting

$$f(\sum_{k=1}^{J} \eta_k x_k; \sum_{\ell=1}^{J} \eta_\ell y_\ell) = \sum_{j=1}^{J} \eta_j f_j(\underline{x}, \underline{y}) \tag{24}$$

one defines the J functions $f_j(\underline{x}; \underline{y})$, where of course $\underline{x} \equiv (x_1, x_2, ..., x_J)$, $\underline{y} \equiv (y_1, y_2, ..., y_J)$, and their explicit expressions are then given by the following obvious generalization of (19):

$$f_j(\underline{x}; \underline{y}) = J^{-1} \sum_{k=1}^{J} \exp(-2\pi i j K/J)\, f(\underline{x}^{(K)}; \underline{y}^{(K)}) \;, \tag{25a}$$

$$x^{(K)} \equiv \sum_{\ell=1}^{J} x_\ell \exp(2\pi i \ell K/J), \quad y^{(K)} = \sum_{\ell=1}^{J} y_\ell \exp(2\pi i \ell K/J) \;. \tag{25b}$$

It is now plain that these formulas, and their obvious extensions to the case of a function f that depends on several arguments rather than just on one or two, entail the validity of (16), which can therefore be considered as proven.

We have thus seen how from a single evolution equation, say (11), for a single matrix, say $\underline{U}(t)$, one can obtain a system of J coupled evolution equations, (15), for the J matrices $\underline{U}_j(t)$, defined by (12a). Let us however emphasize that this trick, by its very nature, generally yields coupled equations that can be decoupled by a linear transformation. For instance it is easily seen that the coupled equations (15) get transformed into the following decoupled equations,

$$\ddot{\underline{\tilde{U}}}_n = 2\tilde{a}_n \dot{\underline{\tilde{U}}}_n + \tilde{b}_n \underline{U}_n + \tilde{c}_n \left[\dot{\underline{\tilde{U}}}_n, \underline{\tilde{U}}_n \right] \;, \tag{26}$$

via the following *linear* transformation (applicable equally to scalars, to vectors and to matrices) among tilded and untilded variables:

$$\tilde{g}_n = J^{-1} \sum_{j=1}^{J} g_j \exp(2\pi i n j/J) \;, \tag{27a}$$

$$g_n = J^{-1} \sum_{k=1}^{J} \tilde{g}_k \exp(-2\pi i n k/J) \;. \tag{27b}$$

Proof. Note first of all the equivalence of (27a) with (27b): indeed by multiplying (27a) by $\exp(-2\pi i n k/J)$, then by summing over n from 1 to J, one obtains, via the identity (23), precisely (27b) (up to an exchange of the roles of the indices n and

k); and likewise one can obtain (27a) from (27b). Next, we multiply (15) by $\exp(2\pi i n j/J)$ and sum over j from 1 to J. Thereby we get

$$\ddot{\tilde{U}}_n = \sum_{j,k=1}^{J} \exp(2\pi i n j/J)\,(2a_{j-k}\,\dot{\underline{U}}_k + b_{j-k}\,\underline{U}_k)$$

$$+ \sum_{j,k,\ell=1}^{J} \exp(2\pi i n j/J)\, c_{j-k-\ell}\,[\,\dot{\underline{U}}_k, \underline{U}_\ell\,] \quad , \tag{28a}$$

$$\ddot{\tilde{U}}_n = \sum_{k,\ell=1}^{J} \exp[\,2\pi i n(\ell+k)/J)\,(2a_\ell\,\dot{\underline{U}}_k + b_\ell\,\underline{U}_k)$$

$$+ \sum_{k,\ell,m=1}^{J} \exp[\,2\pi i n(m+k+\ell)/J\,]\,c_m\,[\,\dot{\underline{U}}_k, \underline{U}_\ell\,] \quad , \tag{28b}$$

$$\ddot{\tilde{U}} = 2\,\tilde{a}_n\,\dot{\tilde{U}}_n + \tilde{b}_n\,\tilde{U}_n + \tilde{c}_n\,[\,\dot{\tilde{\underline{U}}}_n, \tilde{\underline{U}}_n\,] \quad . \tag{28c}$$

To obtain (28a) we used, in the left hand side, (27a); to go from (28a) to (28b) we first replaced, in the argument of the exponential in the first sum, j with $(j-k)+k$ and then we set $j-k=\ell$, likewise in the argument of the exponential in the second sum we replaced j with $(j-k-\ell)+k+\ell$ and then we set $j-k-\ell=m$ (this replacements do not affect the limits of the sums, since all indices are defined $\mathrm{mod}(J)$); finally, to go from (28b) to (28c) we used again (27a). And (28c) coincides with (26), which is thereby proven.

Exercise 5.3-6. Discuss the solvability, and the behavior, of the N-body system in 3-dimensional space characterized by the Newtonian equations of motion

$$\ddot{\vec{r}}_n = \sum_{m=1}^{N} [\,2\,a_{n-m}\,\dot{\vec{r}}_m + b_{n-m}\,\vec{r}_m\,] + 2\sum_{\ell,m=1}^{N} c_{n-m-\ell}\,\vec{r}_m \wedge \dot{\vec{r}}_\ell \quad . \tag{29}$$

Hint: see (15) and (5.2-1,19), and Sect. 5.2.3.

Exercise 5.3-7. Write out, and discuss, the Newtonian equations of motions that obtain from (5.2-19) by applying simultaneously (or rather, sequentially) all three the tricks (association, complexification, multiplication) discussed above (in Sect. 5.3)

Let us end Sect. 5.3 by introducing a second, perhaps more interesting although also rather trivial, *multiplication* trick, whereby from an evolution equation for a single matrix (of higher rank) one can obtain several evolution equations for several matrices (of lower rank). As already entailed by this description, and in contrast to the tricks described above, this technique of *multiplication* is specifically appropriate for matrices (it is not applicable to scalars).

The idea is to consider *block matrices*, namely matrices whose elements are themselves matrices, say

$$\underline{U} \equiv \begin{pmatrix} \underline{U}^{(11)} & \underline{U}^{(12)} & \cdots & \underline{U}^{(1M)} \\ \vdots & \vdots & \vdots\vdots\vdots & \vdots \\ \underline{U}^{(M1)} & \underline{U}^{(M2)} & \cdots & \underline{U}^{(MM)} \end{pmatrix} \quad . \tag{30}$$

Here we assume the block matrix $\underline{U}$ to be a square $(M \times M)$-matrix, whose overall order depends of course on the order of the matrices $\underline{U}^{(m_1 m_2)}$ (which themselves need not be square matrices, see below).

To illustrate this technique we now use the following two *linearizable* matrix evolution equations,

$$\ddot{\underline{U}} = 2a\dot{\underline{U}} + b\underline{U} + c[\dot{\underline{U}}, \underline{U}] \ , \tag{31}$$

respectively

$$\ddot{\underline{U}} = 2a\dot{\underline{U}} + b\underline{U} + c[\dot{\underline{U}}, \underline{U}^2] \ , \tag{32}$$

which correspond of course to (2) with $F(\underline{U}) = c\,\underline{U}$ respectively $F(\underline{U}) = c\,\underline{U}^2$.

It is clear that, via (30), these matrix evolution equations, (31) respectively (32), read

$$\ddot{\underline{U}}^{(m_1 m_2)} = 2a\,\dot{\underline{U}}^{(m_1 m_2)} + b\,\underline{U}^{(m_1 m_2)} + c\sum_{\ell=1}^{M} [\,\dot{\underline{U}}^{(m_1,\ell)}\,\underline{U}^{(\ell,m_2)} - \underline{U}^{(m_1,\ell)}\,\dot{\underline{U}}^{(\ell,m_2)}\,] \, , \tag{33}$$

respectively

$$\ddot{\underline{U}}^{(m_1 m_2)} = 2a\,U^{(m_1 m_2)} + b\,\underline{U}^{(m_1 m_2)}$$

$$+c \sum_{\ell_1,\ell_2=1}^{M} [\,\dot{\underline{U}}^{(m_1,\ell_1)}\,\underline{U}^{(\ell_1\ell_2)}\,\underline{U}^{(\ell_2 m_2)} - \underline{U}^{(m_1,\ell_1)}\,\underline{U}^{(\ell_1\ell_2)}\,\dot{\underline{U}}^{(\ell_2 m_2)}\,] \ , \tag{34}$$

with $m_1, m_2 = 1, ..., M$. We have thereby obtained, from a single matrix evolution equation, (31) respectively (32), for a single matrix

$\underline{U} \equiv \underline{U}(t)$, M^2 (coupled) matrix evolution equations for the M^2 matrices $\underline{U}^{(m_1m_2)} \equiv \underline{U}^{(m_1m_2)}(t)$, namely (33) respectively (34).

We have illustrated this trick by displaying how it works in the case of the two specific equations (31) and (32). It is quite obvious how it can be applied more generally, as well as the hunch that it is likely to be particularly useful for matrix evolution equations that feature polynomial (indeed, low-degree-polynomial) nonlinearities.

Exercise 5.3-8. Show that the following system of $N = M^2$ scalar, and $N = M^2$ three-vector (rotation-invariant) Newtonian equations of motion is *linearizable*:

$$\ddot{\rho}^{(m_1m_2)} = 2a\,\dot{\rho}^{(m_1m_2)} + b\,\rho^{(m_1m_2)}$$

$$+c\sum_{\ell=1}^{M} \left[\, \dot{\rho}^{(m_1\ell)}\,\rho^{(\ell m_2)} - \dot{\rho}^{(\ell m_2)}\,\rho^{(m_1\ell)} - \dot{\vec{r}}^{(m_1\ell)} \cdot \vec{r}^{(\ell m_2)} + \dot{\vec{r}}^{(\ell m_2)} \cdot \vec{r}^{(m_1\ell)} \,\right] , \tag{35a}$$

$$\dot{\vec{r}}^{(m_1m_2)} = 2\,a\,\dot{\vec{r}}^{(m_1m_2)} + b\,\vec{r}^{(m_1m_2)}$$

$$+c\sum_{\ell=1}^{M} \left[\, \dot{\rho}^{(m_1\ell)}\vec{r}^{(\ell m_2)} - \dot{\rho}^{(\ell m_2)}\,\vec{r}^{(m_1\ell)} + \rho^{(\ell m_2)}\,\dot{\vec{r}}^{(m_1\ell)} - \rho^{(m_1\ell)}\,\dot{\vec{r}}^{(\ell m_2)} \right.$$

$$\left. + \vec{r}^{(m_1\ell)} \wedge \dot{\vec{r}}^{(\ell m_2)} + \vec{r}^{(\ell m_2)} \wedge \dot{\vec{r}}^{(m_1\ell)} \,\right] . \tag{35b}$$

Hint: assume that all the M^2 matrices $\underline{U}^{(m_1m_2)}$ in (33) are (2×2)-matrices, and use for each of them a parametrization of type (5.1-25).

Exercise 5.3-9. Show that the following system of $N = M^2$ scalar, and $N = M^2$ three-vector (rotation-invariant) Newtonian equations of motion is *linearizable*:

$$\ddot{\rho}^{(m_1m_2)} = 2a\,\dot{\rho}^{(m_1m_2)} + b\,\rho^{(m_1m_2)}$$

$$+c\sum_{\ell_1,\ell_2=1}^{M} \left\{\, \dot{\rho}^{(m_1\ell_1)}\rho^{(\ell_1\ell_2)}\,\rho^{(\ell_2m_2)} - \dot{\rho}^{(\ell_2m_2)}\,\rho^{(\ell_1\ell_2)}\rho^{(m_1\ell_1)} \right.$$

$$-\dot{\rho}^{(m_1\ell_1)}\,(\vec{r}^{(\ell_1\ell_2)} \cdot \vec{r}^{(\ell_2m_2)}) + \dot{\rho}^{(\ell_2m_2)}\,(\vec{r}^{(\ell_1\ell_2)} \cdot \vec{r}^{(m_1l_1)})$$

$$+\rho^{(\ell_2m_2)}\,(\vec{r}^{(\ell_1\ell_2)} \cdot \dot{\vec{r}}^{(m_1\ell_1)}) - \rho^{(m_1\ell_1)}\,(\vec{r}^{(\ell_1\ell_2)} \cdot \dot{\vec{r}}^{(\ell_2m_2)})$$

$$-\rho^{(\ell_1\ell_2)}\left[(\dot{\vec{r}}^{(m_1\ell_1)} \cdot \vec{r}^{(\ell_2m_2)}) - (\vec{r}^{(m_1\ell_1)} \cdot \dot{\vec{r}}^{(\ell_2m_2)})\right]$$

$$+\dot{\vec{r}}^{(\ell_1 m_1)} \cdot (\vec{r}^{(\ell_1 \ell_2)} \wedge \vec{r}^{(\ell_2 m_2)}) - \vec{r}^{(m_1 \ell_1)} \cdot (\vec{r}^{(\ell_1 \ell_2)} \wedge \dot{\vec{r}}^{(\ell_2 m_2)}) \} \quad , \tag{36a}$$

$$\ddot{\vec{r}}^{(m_1 m_2)} = 2\, a\, \dot{\vec{r}}^{(m_1 m_2)} + b\, \vec{r}^{(m_1 m_2)}$$

$$+c \sum_{\ell_1, \ell_2 = 1}^{M} \{ -\dot{\vec{r}}^{(m_1 \ell_1)} (\vec{r}^{(\ell_1 \ell_2)} \cdot \vec{r}^{(\ell_2 m_2)}) + \dot{\vec{r}}^{(\ell_2 m_2)} (\vec{r}^{(\ell_1 \ell_2)} \cdot \vec{r}^{(m_1 \ell_1)})$$

$$+\vec{r}^{(m_1 \ell_1)} (\vec{r}^{(\ell_1 \ell_2)} \cdot \dot{\vec{r}}^{(\ell_2 m_2)}) - r^{(\ell_2 m_2)} (\vec{r}^{(\ell_1 \ell_2)} \cdot \dot{\vec{r}}^{(m_1 l_1)})$$

$$+\vec{r}^{(\ell_1 \ell_2)} [(\dot{\vec{r}}^{(m_1 \ell_2)} \cdot \vec{r}^{(\ell_2 m_2)}) - (r^{(m_1 \ell_1)} \cdot \dot{\vec{r}}^{(\ell_2 m_2)})]$$

$$-\dot{\rho}_1^{(m_1 \ell_1)}\, \vec{r}^{(\ell_1 \ell_2)} \wedge \vec{r}^{(\ell_2 m_2)} - \dot{\rho}^{(\ell_2 m_2)}\, \vec{r}^{(\ell_1 \ell_2)} \wedge \vec{r}^{(m_1 \ell_1)}$$

$$+\rho_1^{(m_1 \ell_1)}\, \vec{r}^{(\ell_1 \ell_2)} \wedge \dot{\vec{r}}^{(\ell_2 m_2)} + \rho^{(\ell_2 m_2)}\, \vec{r}^{(\ell_1 \ell_2)} \wedge \dot{\vec{r}}^{(m_1 \ell_1)}$$

$$-\rho^{(\ell_1 \ell_2)} [\, \dot{\vec{r}}^{(m_1 \ell_1)} \wedge \vec{r}^{(\ell_2 m_2)} - \vec{r}^{(m_1 \ell_1)} \wedge \dot{\vec{r}}^{(\ell_2 m_2)} \,]$$

$$+\dot{\rho}^{(m_1 \ell_1)} [\, \rho^{(\ell_1 \ell_2)}\, \vec{r}^{(\ell_2 m_2)} + \rho^{(\ell_2 m_2)}\, \vec{r}^{(\ell_1 \ell_2)} \,] - \dot{\rho}^{(\ell_2 m_2)}\, \rho^{(\ell_1 \ell_2)}\, \vec{r}^{(m_1 \ell_1)}$$

$$-\rho^{(m_1 \ell_1)} [\, \rho^{(\ell_1 \ell_2)}\, \dot{\vec{r}}^{(\ell_2 m_2)} + \rho^{(\ell_2 m_2)}\, \dot{\vec{r}}^{(\ell_1 \ell_2)} \,] - \rho^{(\ell_1 \ell_2)}\, \rho^{(\ell_2 m_2)}\, \dot{\vec{r}}^{(m_1 \ell_1)} \} \quad . \tag{36b}$$

Hint: as for the preceding *Exercise 5.3-8*, except for the replacement of (33) with (34).

Let us end Sect. 5.3 by adding to the multiplication trick we just described a further twist, whose relevance in yielding interesting many-body problems in multidimensional space will become clear later (see, for instance, Sect. 5.6.5), but is also illustrated here. To this end we focus hereafter on the matrix evolution equation (32), featuring a *cubic* nonlinearity.

We now set

$$M = 2K \quad , \tag{37a}$$

so that M is even, and we assume the (block) $(M \times M)$-matrix $\underline{U} \equiv \underline{U}(t)$ to have the following (block) structure:

$$
\underline{U}=\begin{pmatrix}
\underline{0} & \underline{W}^{(11)} & \underline{0} & \underline{W}^{(12)} & \cdots & \underline{0} & \underline{W}^{(1K)} \\
\underline{V}^{(11)} & \underline{0} & \underline{V}^{(12)} & \underline{0} & \cdots & \underline{V}^{(1K)} & \underline{0} \\
\underline{0} & \underline{W}^{(21)} & \underline{0} & \underline{W}^{(22)} & \cdots & \underline{0} & \underline{W}^{(2K)} \\
\underline{V}^{(21)} & \underline{0} & \underline{V}^{(22)} & \underline{0} & \cdots & \underline{V}^{(2K)} & \underline{0} \\
\vdots & \vdots & \vdots & \vdots & ⋮⋮⋮ & \vdots & \vdots \\
\underline{0} & \underline{W}^{(K1)} & \underline{0} & \underline{W}^{(K2)} & \cdots & \underline{0} & \underline{W}^{(KK)} \\
\underline{V}^{(K1)} & \underline{0} & \underline{V}^{(K2)} & \underline{0} & \cdots & \underline{V}^{(KK)} & \underline{0}
\end{pmatrix} . \tag{37b}
$$

The motivation for using a different notation for the matrices appearing in the even- and odd-numbered lines of this block-matrix $\underline{U}$ is because we like to keep open the option that these be (different) *rectangular* matrices; say, the matrices $\underline{W}^{(k_1,k_2)}$ $(L\times S)$-matrices, namely matrices with L lines and S columns, and, correspondingly, the matrices $\underline{V}^{(k_1,k_2)}$ be $(S\times L)$-matrices, namely matrices with S lines and L columns. Here L and S are two arbitrary positive integers. It is clear that these assumptions are compatible with the block structure (37b), provided we correspondingly assume the identically vanishing matrices appearing in the right hand side of (37b) to be square matrices, and more specifically, those appearing on the odd-numbered lines of the block-matrix (37b) to be $(L\times L)$-matrices, and those appearing on the even-numbered lines to be $(S\times S)$-matrices. The consistency of this block structure of the (sparse) square matrix $\underline{U}$ (which then features $(L+S)\cdot K$ lines and as many columns) is plain. Also plain is the consistency of this block structure with the matrix evolution equation (32), whereby one gets for the $2K^2$ matrices $\underline{W}^{(k_1,k_2)}\equiv\underline{W}^{(k_1,k_2)}(t)$ and $\underline{V}^{(k_1,k_2)}\equiv\underline{V}^{(k_1,k_2)}(t)$ the following (of course no less *linearizable* than (32)!) coupled evolution equations

$$
\ddot{\underline{W}}^{(k_1k_2)}=2\,a\,\dot{\underline{W}}^{(k_1k_2)}+b\,\underline{W}^{(k_1k_2)}
$$

$$
+c\sum_{k_1',k_2'=1}^{K}\left[\,\dot{\underline{W}}^{(k_1k_1')}\,\underline{V}^{(k_1'k_2')}\,\underline{W}^{(k_2'k_2)}-\underline{W}^{(k_1k_1')}\,\underline{V}^{(k_1'k_2')}\,\dot{\underline{W}}^{(k_2'k_2)}\,\right] , \tag{38a}
$$

$$
\ddot{\underline{V}}^{(k_1k_2)}=2\,a\,\dot{\underline{V}}^{(k_1k_2)}+b\,\underline{V}^{(k_1k_2)}
$$

$$
+c\sum_{k_1',k_2'=1}^{K}\left[\,\dot{\underline{V}}^{(k_1k_1')}\,\underline{W}^{(k_1'k_2')}\,\underline{V}^{(k_2'k_2)}-\underline{V}^{(k_1k_1')}\,\underline{W}^{(k_1'k_2')}\,\dot{\underline{V}}^{(k_2'k_2)}\,\right] . \tag{38b}
$$

Here $k_1,k_2=1,\dots,K$, and all matrix products are performed according to the standard "lines by columns" rule, namely (in self-evident notation: $\underline{W}$ is a generic ($L\times S$)-matrix, $\underline{V}$ a generic ($S\times L$)-matrix)

$$(\underline{W}\,\underline{V})_{\ell_1\ell_2}\equiv\sum_{s=1}^{S}W_{\ell_1 s}V_{s\ell_2}\ ,\quad \ell_1,\ell_2=1,\dots,L\quad , \tag{39a}$$

$$(\underline{V}\,\underline{W})_{s_1s_2}\equiv\sum_{\ell=1}^{L}V_{s_1\ell}W_{\ell s_2}\ ,\quad s_1,s_2=1,\dots,S\quad . \tag{39b}$$

Proof. The consistency of the block structure (37b) of the matrix $\underline{U}\equiv\underline{U}(t)$ with any evolution equation for $\underline{U}(t)$ whose nonlinear part only features products of an *odd* number of matrices $\underline{U}$ and $\dot{\underline{U}}$ (hence in particular with (32)) is entailed by the fact that the product of an *odd* number of matrices, all of them sharing the block structure (37b) but being otherwise arbitrary, still has the same block structure. Indeed let us define any element of a matrix as *odd* or *even* depending on the parity (*odd* or *even*) of the sum of its two indices (those identifying the line and column it belongs to). This definition is equally applicable to ordinary matrices (whose elements are numbers), and to block matrices (whose elements are themselves matrices). The block matrix (37b) is then characterized by the property that *all* its *even* elements vanish. Now consider the product of an *odd* number of matrices. It is clear that every *even* element of this product matrix is given by a sum of products of elements of the factor matrices (those entering as factors in the product), each of which must contain an *odd* number of (hence, at least *one*) *even* element. Hence, if *all even* elements of the factor matrices vanish, *all even* elements of the product matrix also vanish.

The consistency of the block structure (37b) with the evolution equation (32) is thereby proven. The derivation of (38) via (37) from (32) is then plain.

Because of its special structure, the matrix evolution equation (38) can be easily transformed into a set of covariant (hence, rotation-invariant) vector equations in S-dimensional space, for arbitrary S. Indeed it is easily seen that the parametrizations

$$\underline{W}^{(k_1k_2)}\equiv\begin{pmatrix} w_1^{(k_1k_2)(1)} & w_2^{(k_1k_2)(1)} & \cdots & w_S^{(k_1k_2)(1)}\\ w_1^{(k_1k_2)(2)} & w_2^{(k_1k_2)(2)} & \cdots & w_S^{(k_1k_2)(2)}\\ \vdots & \vdots & \vdots & \vdots\\ w_1^{(k_1k_2)(L)} & w_2^{(k_1k_2)(L)} & \cdots & w_S^{(k_1k_2)(L)}\end{pmatrix}\quad , \tag{40a}$$

$$\underline{V}^{(k_1k_2)} \equiv \begin{pmatrix} v_1^{(k_1k_2)(1)} & v_1^{(k_1k_2)(2)} & \cdots & v_1^{(k_1k_2)(L)} \\ v_2^{(k_1k_2)(1)} & v_2^{(k_1k_2)(2)} & \cdots & v_2^{(k_1k_2)(L)} \\ \vdots & \vdots & ::: & \vdots \\ v_S^{(k_1k_2)(1)} & v_S^{(k_1k_2)(2)} & \cdots & v_S^{(k_1k_2)(L)} \end{pmatrix} \quad , \tag{40b}$$

whereby we introduce the $2LK^2$ S-vectors

$$\vec{w}^{(k_1k_2)(\ell)} \equiv (w_1^{(k_1k_2)(\ell)}, w_2^{(k_1k_2)(\ell)}, ..., w_S^{(k_1k_2)(\ell)}),\ \ell = 1,...,L \quad , \tag{41a}$$

$$\vec{v}^{(k_1k_2)(\ell)} \equiv (v_1^{(k_1k_2)(\ell)}, v_2^{(k_1k_2)(\ell)}, ..., v_S^{(k_1k_2)(\ell)}),\ \ell = 1,...,L \tag{41b}$$

-- parametrizations which are clearly consistent with the assumed rectangular structures of the $(L \times S)$-matrices $\underline{W}^{(k_1k_2)}$ and of the $(S \times L)$-matrices $\underline{V}^{(k_1k_2)}$ -- entail that the matrix evolution equations (38) get re-written in the following covariant S-vector form:

$$\ddot{\vec{w}}^{(k_1k_2)(\ell)} = 2a\dot{\vec{w}}^{(k_1k_2)(\ell)} + b\vec{w}^{(k_1k_2)(\ell)}$$

$$+c \sum_{k_1',k_2'=1}^{K} \sum_{\ell'=1}^{L} [\ \vec{w}^{(k_2'k_2)(\ell')} (\dot{\vec{w}}^{(k_1k_1')(\ell)} \cdot \vec{v}^{(k_1'k_2')(\ell')}) - \dot{\vec{w}}^{(k_2'k_2)(\ell')} (\vec{w}^{(k_1k_1')(\ell)} \cdot \vec{v}^{(k_1'k_2')(\ell')})\ ] \quad , \tag{42a}$$

$$\ddot{\vec{v}}^{(k_1k_2)(\ell)} = 2a\dot{\vec{v}}^{(k_1k_2)(\ell)} + b\vec{v}^{(k_1k_2)(\ell)}$$

$$-c \sum_{k_1',k_2'=1}^{K} \sum_{\ell'=1}^{L} [\ \vec{v}^{(k_1k_1')(\ell')} (\dot{\vec{v}}^{(k_2'k_2)(\ell)} \cdot \vec{w}^{(k_1'k_2')(\ell')}) - \dot{\vec{v}}^{(k_1k_1')(\ell')} (\vec{v}^{(k_2'k_2)(\ell)} \cdot \vec{w}^{(k_1'k_2')(\ell')})\ ] \quad . \tag{42b}$$

Here of course $k_1, k_2 = 1,...,K$, and $\ell = 1,...,L$; hence the total number of S-vectors, and of equations, is $2LK^2$.

Let us reemphasize that here and below superimposed arrows denote S-vectors (with S an *arbitrary* positive integer), and of course a dot sandwiched between two S-vectors denotes the standard scalar product in S-dimensional space.

There are two natural reductions of these Newtonian equations of motion, each of which entails that the total number of S-vectors is reduced from $2LK^2$ to $2LK$. The first one obtains by setting

$$\vec{w}^{(k,k')(\ell)} = \vec{w}^{(k)(\ell)},\ k,k' = 1,...,K \quad , \tag{43a}$$

$$\vec{v}^{(k',k)(\ell)} = \vec{v}^{(k)(\ell)},\ k,k' = 1,...,K \quad ; \tag{43b}$$

it is easy to verify that this reduction is compatible with the equations of motion (42), which then take the following form:

$$\ddot{\vec{w}}^{(k)(\ell)} = 2a\dot{\vec{w}}^{(k)(\ell)} + b\vec{w}^{(k)(\ell)}$$

$$+C\sum_{k'=1}^{K}\sum_{\ell'=1}^{L}\left[\ \vec{w}^{(k')(\ell')}(\dot{\vec{w}}^{(k)(\ell)}\cdot\vec{v}^{(k')(\ell')}) - \dot{\vec{w}}^{(k')(\ell')}(\vec{w}^{(k)(\ell)}\cdot\vec{v}^{(k')(\ell')})\ \right]\ , \tag{44a}$$

$$\ddot{\vec{v}}^{(k)(\ell)} = 2a\dot{\vec{v}}^{(k)(\ell)} + b\vec{v}^{(k)(\ell)}$$

$$-C\sum_{k'=1}^{K}\sum_{\ell'=1}^{L}\left[\ \vec{v}^{(k')(\ell')}(\dot{\vec{v}}^{(k)(\ell)}\cdot\vec{w}^{(k')(\ell')}) - \dot{\vec{v}}^{(k')(\ell')}(\vec{v}^{(k)(\ell)}\cdot\vec{w}^{(k')(\ell')})\ \right]\ , \tag{44b}$$

with

$$C = cK\ . \tag{45}$$

The second reduction obtains by setting, instead of (43),

$$\vec{w}^{(k',k)(\ell)} = \vec{w}^{(k)(\ell)},\ k,k' = 1,\ldots,K\ , \tag{46a}$$

$$\vec{v}^{(k,k')(\ell)} = \vec{v}^{(k)(\ell)},\ k,k' = 1,\ldots,K\ , \tag{46b}$$

which is as well compatible with the equations of motion (42) and transforms them into the following form (different from (44)!):

$$\ddot{\vec{w}}^{(k)(\ell)} = 2a\dot{\vec{w}}^{(k)(\ell)} + b\vec{w}^{(k)(\ell)}$$

$$+C\sum_{k'=1}^{K}\sum_{\ell'=1}^{L}\left[\ \vec{w}^{(k)(\ell')}(\dot{\vec{w}}^{(k')(\ell)}\cdot\vec{v}^{(k')(\ell')}) - \dot{\vec{w}}^{(k)(\ell')}(\vec{w}^{(k')(\ell)}\cdot\vec{v}^{(k')(\ell')})\ \right]\ , \tag{47a}$$

$$\ddot{\vec{v}}^{(k)(\ell)} = 2a\dot{\vec{v}}^{(k)(\ell)} + b\vec{v}^{(k)(\ell)}$$

$$-C\sum_{k'=1}^{K}\sum_{\ell'=1}^{L}\left[\ \vec{v}^{(k)(\ell')}(\dot{\vec{v}}^{(k')(\ell)}\cdot\vec{w}^{(k')(\ell')}) - \dot{\vec{v}}^{(k)(\ell')}(\vec{v}^{(k')(\ell)}\cdot\vec{w}^{(k')(\ell')})\ \right]\ , \tag{47b}$$

again with (45).

Exercise 5.3-10. Verify !

Exercise 5.3-11. Verify that *partial* reductions, such as those entailed by (43a) *without* (43b), or (43b) *without* (43a), or (46a) *without* (46b), or (46b) *without* (46a), are also compatible with the equations of motion (42a); write down the corresponding equations of motion; and show that their solution can always be reduced to solving equations of type (44) or (47), up to the additional solution of linear equations (which therefore does not spoil the *linearizable* character of these equations of motion -- let us emphasize that this finding is valid even if the initial data for (42) are only compatible with a *partial* reduction, as defined above).

Exercise 5.3-12. Redo (with appropriate modifications) the treatment given above, under the assumption that M is *odd*, $M = 2K + 1$, rather than even, see (37).

5.4 A survey of matrix evolution equations amenable to exact treatments

In Sect. 5.4, or rather in the 3 subsections in which it is conveniently organized, we survey matrix evolution equations amenable to exact treatments. Specifically, in Sect. 5.4.1 we display a fairly large (perhaps excessively large!) class of *linearizable* matrix evolution equations, identified by a straightforward if notationally heavy generalization of the approach of Sect. 5.2; in Sect. 5.4.2, we introduce certain matrix evolution equations which are related to the so-called non-Abelian Toda lattice (an *integrable*, indeed *solvable*, system); and finally, in Sect. 5.4.3, we collect some other matrix evolution equations amenable to exact treatments.

As the reader will see, we do not try to provide a systematic presentation of *all* the treatable matrix ODEs that can be obtained by various techniques; our main focus below is to present a number of different techniques whereby such treatable equations can be uncovered (or, equivalently, manufactured), and to illustrate them via representative examples -- which are subsequently (see Sect. 5.6 and its subsections) employed to uncover (or, equivalently, to manufacture) treatable many-body problems in three-dimensional space. Hence the alert reader will find much instructive scope for additional experimentation of his/her own, with the possibility to obtain quite interesting new results, both in the guise of treatable matrix ODEs, as well via the subsequent derivation from these of many-body problems also amenable to exact treatments.

Throughout the following subsections matrices are denoted by underlined (upper or lower case) letters.

5.4.1 A class of linearizable matrix evolution equations

In Sect. 5.4.1 we present a methodology to manufacture *new* (generally nontrivial) *linearizable* matrix evolution equations from *known* (possibly trivially solvable) *linearizable* matrix evolution equations; this technique generalizes the approach already discussed in Sect. 5.2 (which the diligent reader is advised to review).

Assume that the $N+5$ square matrices $\underline{u}_n$, $\underline{f}$, $\underline{g}$, $\underline{h}$, $\underline{v}$, $\underline{y}$ satisfy the following $N+5$ matrix ODEs:

$$\underline{\tilde{U}}_n(\underline{u}_m^{(0)},\underline{u}_m^{(1)},...,m=1,...,N;\underline{f}^{(0)},\underline{f}^{(1)},...;\underline{g}^{(0)},\underline{g}^{(1)},...;\underline{h}^{(0)},\underline{h}^{(1)},...;t)=0\,, \tag{1a}$$

$$\underline{\tilde{F}}(\underline{u}_m^{(0)},\underline{u}_m^{(1)},...,m=1,...,N;\underline{f}^{(0)},\underline{f}^{(1)},...;\underline{g}^{(0)},\underline{g}^{(1)},...;\underline{h}^{(0)},\underline{h}^{(1)},...;t)=0\,, \tag{1b}$$

$$\underline{\tilde{G}}(\underline{u}_m^{(0)},\underline{u}_m^{(1)},...,m=1,...,N;\underline{f}^{(0)},\underline{f}^{(1)},...;\underline{g}^{(0)},\underline{g}^{(1)},...;\underline{h}^{(0)},\underline{h}^{(1)},..;t)=0\,, \tag{1c}$$

$$\underline{\tilde{H}}(\underline{u}_m^{(0)},\underline{u}_m^{(1)},...,m=1,...,N;\underline{f}^{(0)},\underline{f}^{(1)},...;\underline{g}^{(0)},\underline{g}^{(1)},...;\underline{h}^{(0)},\underline{h}^{(1)},..;t)=0\,, \tag{1d}$$

$$\sum_{j=0}^{J_1}\underline{\tilde{V}}_j(\underline{u}_m^{(0)},\underline{u}_m^{(1)},...,m=1,...,N;\underline{f}^{(0)},\underline{f}^{(1)},..;\underline{g}^{(0)},\underline{g}^{(1)},...;\underline{h}^{(0)},\underline{h}^{(1)},..;t)\,\underline{v}^{(j)}=0\,, \tag{1e}$$

$$\sum_{j=0}^{J_2}\underline{y}^{(j)}\,\underline{\tilde{Y}}_j(\underline{u}_m^{(0)},\underline{u}_m^{(1)},...,m=1,...,N;\underline{f}^{(0)},\underline{f}^{(1)},...;\underline{g}^{(0)},\underline{g}^{(1)},...;\underline{h}^{(0)},\underline{h}^{(1)},...;t)=0\,. \tag{1f}$$

Here $\underline{u}_n^{(j)}\equiv d^j\underline{u}_n/dt^j$, $j=0,1,2,...$, with analogous formulas for $\underline{f}^{(j)}$, $\underline{g}^{(j)}$, $\underline{h}^{(j)}$, $\underline{v}^{(j)}$, $\underline{y}^{(j)}$, and the functions $\underline{\tilde{U}}_n,\underline{\tilde{F}},\underline{\tilde{G}},\underline{\tilde{H}},\underline{\tilde{V}}_j,\underline{\tilde{Y}}_j$, are "scalar/matrix functions of matrices," namely matrices built only out of their matrix arguments (whose ordering is of course important) and of scalars (including, possibly, the independent variable t). Hence these square matrices satisfy the identity

$$\begin{aligned}&\underline{W}^{-1}\underline{\tilde{Z}}(\underline{u}_m^{(0)},\underline{u}_m^{(1)},...,m=1,...,N;\underline{f}^{(0)},\underline{f}^{(1)},...;\underline{g}^{(0)},\underline{g}^{(1)},...;\underline{h}^{(0)},\underline{h}^{(1)},...;t)\,\underline{W}=\\&\underline{\tilde{Z}}(\underline{W}^{-1}\underline{u}_m^{(0)}\underline{W},\underline{W}^{-1}\underline{u}_m^{(1)}\underline{W},...,m=1,...,N;\underline{W}^{-1}\underline{f}^{(0)}\underline{W},\underline{W}^{-1}\underline{f}^{(1)}\underline{W},...;\\&\qquad\underline{W}^{-1}\underline{g}^{(0)}\underline{W},\underline{W}^{-1}\underline{g}^{(1)}\underline{W},...;\underline{W}^{-1}\underline{h}^{(0)}\underline{W},\underline{W}^{-1}\underline{h}^{(1)}\underline{W},...;t)\,,\end{aligned} \tag{2}$$

where $\underline{\tilde{Z}}$ denotes any one of these matrix functions.

The notation employed to write these equations is meant to suggest (in a sense which will become clear below) that the matrix equation $\underline{\tilde{Z}}=0$ plays the primary role in determining the matrix $\underline{z}$, with the obvious correspondence (if $\underline{\tilde{Z}}=\underline{\tilde{U}}_n$, then $\underline{z}=\underline{u}_n$; if $\underline{\tilde{Z}}=\underline{\tilde{F}}$, then $\underline{z}=\underline{f}$; and so on). Note moreover the qualitative difference among the equations satisfied by the $N+3$ matrices $\underline{u}_n$, $\underline{f}$, $\underline{g}$, $\underline{h}$, (1a,b,c,d), and those satisfied by $\underline{v}$ and $\underline{y}$, (1e,f) (in these latter equations the parameters J_1 and J_2 are of course two nonnegative integers). The different role of the N matrices $\underline{u}_n$, and of the 3 matrices $\underline{f}$, $\underline{g}$, $\underline{h}$, will become clear below.

Note that here (contrary to what we wrote elsewhere) we indicate a matrix/scalar function of matrices by an *underlined* character, to emphasize its matrix character.

Consider now $N+5$ square matrices $\underline{U}_n, \underline{F}, \underline{G}, \underline{H}, \underline{V}, \underline{Y}$, related to the $N+5$ square matrices $\underline{u}_n$, $\underline{f}$, $\underline{g}$, $\underline{h}$, $\underline{v}$, $\underline{y}$ as follows:

$$\underline{u}_n=\underline{W}\,\underline{U}_n\,\underline{W}^{-1},\quad \underline{f}=\underline{W}\,\underline{F}\,\underline{W}^{-1},\quad \underline{g}=\underline{W}\,\underline{G}\,\underline{W}^{-1},\quad \underline{h}=\underline{W}\,\underline{H}\,\underline{W}^{-1}, \tag{3a}$$

$$\underline{U}_n=\underline{W}^{-1}\,\underline{u}_n\,\underline{W},\quad \underline{F}=\underline{W}^{-1}\,\underline{f}\,\underline{W},\quad \underline{G}=\underline{W}^{-1}\,\underline{g}\,\underline{W},\quad \underline{H}=\underline{W}^{-1}\,\underline{h}\,\underline{W}, \tag{3b}$$

$$\underline{v}=\underline{W}\,\underline{V},\quad \underline{y}=\underline{Y}\,\underline{W}^{-1}, \tag{4a}$$

$$\underline{V}=\underline{W}^{-1}\,\underline{v},\quad \underline{Y}=\underline{y}\,\underline{W}, \tag{4b}$$

with

$$\dot{\underline{W}}=\underline{g}\,\underline{v}+\underline{f}\,\underline{W}+\underline{W}\,\underline{y}\,\underline{h}\,\underline{W}, \tag{5a}$$

$$\dot{\underline{W}}=\underline{W}\,\underline{K}, \tag{5b}$$

where

$$\underline{K}=\underline{F}+\underline{G}\,\underline{V}+\underline{Y}\,\underline{H}. \tag{6}$$

Note that (5b) with (6) follows from (5a) via (3a) and (4a).

To complete the definition of the (invertible!) matrix $\underline{W}\equiv\underline{W}(t)$ one must supplement the matrix ODE (5) with an "initial condition", say $\underline{W}(0)=\underline{W}_0$. It will remain our privilege to make an appropriate choice for the (invertible!) matrix $\underline{W}_0$; generally it will be convenient to set simply $\underline{W}_0 = \underline{1}$.

Lemma 5.4.1-1. The $N+5$ square matrices $\underline{U}_n$, $\underline{F}$, $\underline{G}$, $\underline{H}$, $\underline{V}$, $\underline{Y}$ can be obtained from the $N+5$ square matrices $\underline{u}_n$, $\underline{f}$, $\underline{g}$, $\underline{h}$, $\underline{v}$, $\underline{y}$ by algebraic operations (inversion and multiplication of matrices) and by solving a *linear second-order matrix ODE* (which in some cases reduces to a *first-order matrix ODE*, see below). If the matrices $\underline{f}$, $\underline{g}$, $\underline{h}$, $\underline{v}$, $\underline{y}$ are time-dependent, this *linear ODE* is generally *nonautonomous.*

Proof. It follows easily from the relations (3) and (4), taking into account (5) and (6). Indeed the matrix Riccati equation (5a) is *linearizable* via the position

$$\underline{W} = -\left(\underline{y}\,\underline{h}\right)^{-1} \underline{\dot{M}}\,\underline{M}^{-1}, \tag{7}$$

which yields for $\underline{M}$ the *linear ODE*

$$\underline{\ddot{M}} = \left(\underline{\dot{y}}\,\underline{h} + \underline{y}\,\underline{\dot{h}} + \underline{y}\,\underline{h}\,\underline{f}\right)\left(\underline{y}\,\underline{h}\right)^{-1} \underline{\dot{M}} - \underline{y}\,\underline{h}\,\underline{g}\,\underline{v}\,\underline{M} \ . \tag{8}$$

Hence, if the 5 matrices $\underline{f}$, $\underline{g}$, $\underline{h}$, $\underline{v}$, $\underline{y}$ are known, the matrix $\underline{W}$ can be evaluated, via (7), by solving the *linear* (generally *nonautonomous*) matrix ODE (8). Then the $N+5$ matrices $\underline{U}_n$, $\underline{F}$, $\underline{G}$, $\underline{H}$, $\underline{V}$, $\underline{Y}$ can be obtained from the, assumedly known, $N+5$ matrices $\underline{u}_n$, $\underline{f}$, $\underline{g}$, $\underline{h}$, $\underline{v}$, $\underline{y}$ via (3b) and (4b).

Next, let us define matrices $\underline{U}_n^{[j]}$, $\underline{F}^{[j]}$, $\underline{G}^{[j]}$, $\underline{H}^{[j]}$ and $\underline{V}^{\{j\}}$, ${}^{\{j\}}\underline{Y}$, via the following recursive formulas:

$$\underline{Z}^{[0]} = \underline{Z},\ \underline{Z}^{[j+1]} = \underline{\dot{Z}}^{[j]} - [\ \underline{Z}^{[j]}, \underline{K}\ ]\,, \tag{9a}$$

$$\underline{Z}^{\{0\}} = \underline{Z},\ \underline{Z}^{\{j+1\}} = \underline{\dot{Z}}^{\{j\}} + \underline{K}\,\underline{Z}^{\{j\}}\,, \tag{9b}$$

$${}^{\{0\}}\underline{Z} = \underline{Z},\ {}^{\{j+1\}}\underline{Z} = {}^{\{j\}}\underline{\dot{Z}} - {}^{\{j\}}\underline{Z}\,\underline{K}\,, \tag{9c}$$

so that

$$\underline{Z}^{[1]} = \underline{\dot{Z}} - [\underline{Z}, \underline{K}],\ \ \underline{Z}^{[2]} = \underline{\ddot{Z}} - 2[\underline{\dot{Z}}, \underline{K}] - [\underline{Z}, \underline{\dot{K}}] + [\ [\underline{Z}, \underline{K}], \underline{K}\ ]\ ,..., \tag{10a}$$

$$\underline{Z}^{\{1\}} = \underline{\dot{Z}} + \underline{K}\,\underline{Z},\ \ \underline{Z}^{\{2\}} = \underline{\ddot{Z}} + 2\underline{K}\,\underline{\dot{Z}} + \underline{\dot{K}}\,\underline{Z} + \underline{K}^2\,\underline{Z}\ ,..., \tag{10b}$$

$$ {}^{\{1\}}\underline{Z}=\dot{\underline{Z}}+\underline{Z}\,\underline{K},\quad {}^{\{2\}}\underline{Z}=\ddot{\underline{Z}}-2\dot{\underline{Z}}\,\underline{K}-\underline{Z}\,\underline{K}+\underline{Z}\,\underline{K}^2\;,\ldots\,. \qquad (10c) $$

Here and throughout $[\underline{A},\underline{B}]$ denotes the *commutator* of the two matrices $\underline{A}$ and $\underline{B}$, $[\underline{A},\underline{B}]\equiv\underline{A}\,\underline{B}-\underline{B}\,\underline{A}$.

Lemma 5.4.1-2. If the $N+5$ square matrices $\underline{u}_n$, $\underline{f}$, $\underline{g}$, $\underline{h}$, $\underline{v}$, $\underline{y}$ satisfy the $N+5$ matrix ODEs (1), then the $N+5$ square matrices $\underline{U}_n$, $\underline{F}$, $\underline{G}$, $\underline{H}$, $\underline{V}$, $\underline{Y}$ satisfy the $N+5$ matrix ODEs

$$ \tilde{\underline{U}}_n(\underline{U}_m^{[0]},\underline{U}_m^{[1]},\ldots,m=1,\ldots,N;\underline{F}^{[0]},\underline{F}^{[1]},\ldots;\underline{G}^{[0]},\underline{G}^{[1]},\ldots;\underline{H}^{[0]},\underline{H}^{[1]},\ldots;t)=0\,, \qquad (11a) $$

$$ \tilde{\underline{F}}(\underline{U}_m^{[0]},\underline{U}_m^{[1]},\ldots,m=1,\ldots,N;\underline{F}^{[0]},\underline{F}^{[1]},\ldots;\underline{G}^{[0]},\underline{G}^{[1]},\ldots;\underline{H}^{[0]},\underline{H}^{[1]},\ldots;t)=0\,, \qquad (11b) $$

$$ \tilde{\underline{G}}(\underline{U}_m^{[0]},\underline{U}_m^{[1]},\ldots,m=1,\ldots,N;\underline{F}^{[0]},\underline{F}^{[1]},\ldots;\underline{G}^{[0]},\underline{G}^{[1]},\ldots;\underline{H}^{[0]},\underline{H}^{[1]},\ldots;t)=0\,, \qquad (11c) $$

$$ \tilde{\underline{H}}(\underline{U}_m^{[0]},\underline{U}_m^{[1]},\ldots,m=1,\ldots,N;\underline{F}^{[0]},\underline{F}^{[1]},\ldots;\underline{G}^{[0]},\underline{G}^{[1]},\ldots;\underline{H}^{[0]},\underline{H}^{[1]},\ldots;t)=0\,, \qquad (11d) $$

$$ \sum_{j=0}^{J_1}\tilde{\underline{V}}_j(\underline{U}_m^{[0]},\underline{U}_m^{[1]},\ldots,m=1,\ldots,N;\underline{F}^{[0]},\underline{F}^{[1]},\ldots;\underline{G}^{[0]},\underline{G}^{[1]},\ldots;\underline{H}^{[0]},\underline{H}^{[1]},\ldots;t)\,\underline{V}^{\{j\}}=0\;, \qquad (11e) $$

$$ \sum_{j=0}^{J_2}{}^{\{j\}}\underline{Y}\,\tilde{\underline{Y}}_j(\underline{U}_m^{[0]},\underline{U}_m^{[1]},\ldots,m=1,\ldots,N;\underline{F}^{[0]},\underline{F}^{[1]},\ldots;\underline{G}^{[0]},\underline{G}^{[1]},\ldots;\underline{H}^{[0]},\underline{H}^{[1]},\ldots;t)=0. \qquad (11f) $$

These formulas are obtained from (1) by replacing, wherever they appear, the matrices $\underline{u}_n^{(j)},\underline{f}^{(j)},\underline{g}^{(j)},\underline{h}^{(j)},\underline{v}^{(j)},\underline{y}^{(j)}$ respectively with the matrices $\underline{U}_n^{[j]},\underline{F}^{[j]},\underline{G}^{[j]},\underline{H}^{[j]},\underline{V}^{\{j\}},\,{}^{\{j\}}\underline{Y}$). Their validity is implied by the formulas

$$ \underline{z}^{(j)}=\underline{W}\,\underline{Z}^{[j]}\,\underline{W}^{-1}\;, \qquad (12a) $$

where $\underline{z}$ stands for $\underline{u}_n,\underline{f},\underline{g}$ or $\underline{h}$, and correspondingly $\underline{Z}$ stands for $\underline{U}_n,\underline{F},\underline{G}$ or $\underline{H}$, and

$$ \underline{v}^{(j)}=\underline{W}\,\underline{V}^{\{j\}}\;, \qquad (12b) $$

$$ \underline{y}^{(j)}={}^{\{j\}}\underline{Y}\,\underline{W}^{-1}, \qquad (12c) $$

which, via (3) and (4), clearly entail the equivalence of (1) and (11). As for the validity of these formulas, (12), they are an immediate consequence of the definitions (9) and of the formulas (3) and (4), as can be easily verified by time-differentiating them and using (5) and (6).

Note that we are implicitly assuming that the 2 matrices $\underline{W}$ and $y\underline{h}$ are invertible, see (3), (4) and (7). A breakdown of one of these two conditions at some time t_c will generally show up as a singularity of the corresponding solutions of (11). But note that the second condition (invertibility of $y\underline{h}$) is invalid but irrelevant in the special case in which $y\underline{h}$ vanishes identically, see the *Remark 5.4.1-7* below. Indeed, in this special case the (first-order) matrix evolution equation (5a) is already *linear*, hence there is no need to introduce via (7) the matrix $\underline{M}$.

We are now ready to formulate and prove the main result of Sect. 5.4.1.

Proposition 5.4.1-3. If the matrix ODEs (1) are *linearizable*, the matrix ODEs (11) are also *linearizable*.

Proof. *Lemma 5.4.1-2* entails that the solutions of (11) can be obtained by first solving the ODEs (1) to determine the $N+5$ matrices $\underline{u}_n, \underline{f}, \underline{g}, \underline{h}, \underline{v}, \underline{y}$ and by then obtaining from these, according to *Lemma 5.4.1-1*, the $N+5$ matrices $\underline{U}_n, \underline{F}, \underline{G}, \underline{H}, \underline{V}, \underline{Y}$.

Remark 5.4.1-4. In the context of the initial-value problem, the convenient choice $\underline{W}_0 = \underline{1}$, see above (paragraph after (6)), entails that the initial conditions for the $N+5$ matrices $\underline{u}_n, \underline{f}, \underline{g}, \underline{h}, \underline{v}, \underline{y}$ can be explicitly obtained from the initial conditions for the $N+5$ matrices $\underline{U}_n, \underline{F}, \underline{G}, \underline{H}, \underline{V}, \underline{Y}$ via (12) with (9,10).

Remark 5.4.1-5. Even if the equations (1) satisfied by the $N+5$ matrices $\underline{u}_n, \underline{f}, \underline{g}, \underline{h}, \underline{v}, \underline{y}$ are *linear* (and possibly quite trivial, see examples below), the equations (11) satisfied by the $N+5$ matrices $\underline{U}_n, \underline{F}, \underline{G}, \underline{H}, \underline{V}, \underline{Y}$ are generally *nonlinear*.

Remark 5.4.1-6. Not all the equations (1a,b,c,d,e,f) need be differential, for instance (1b) might read

$$\underline{f} = \underline{\tilde{f}}(\underline{u}_m^{(j)}, m=1,\ldots,N; \underline{g}^{(j)}; \underline{h}^{(j)}; j=0,1,2,\ldots) , \tag{13}$$

with $\underline{\tilde{f}}$ a given "scalar/matrix function" of its matrix arguments, yielding thereby an explicit definition of the matrix $\underline{f}$ in terms of the matrices $\underline{u}_n, \underline{g}, \underline{h}$ and their time-derivatives, and likewise an explicit definition of the matrix $\underline{F}$ in term of the matrices

$\underline{U}_n, \underline{G}, \underline{H}$ and their time-derivatives, see (11b) and (12a); and so on (see examples below). But this applies only to the equations (1a,b,c,d), and likewise to their counterparts (11a,b,c,d), not to (1e,f) and (11e,f) (except in the trivial cases discussed in the following *Remark 5.4.1-7*).

Remark 5.4.1-7. The linear homogeneous equations satisfied by $\underline{v}$ and $\underline{y}$, see (1e,f), and likewise the (generally nonlinear) equations satisfied by $\underline{V}$ and $\underline{Y}$, see (11e,f), clearly admit the trivial solutions $\underline{v}=\underline{V}=0$ or $\underline{y}=\underline{Y}=0$ (the vanishing of $\underline{v}$ entails the vanishing of $\underline{V}$, and viceversa, and likewise for $\underline{y}$ and $\underline{Y}$, see (4)). This corresponds to reduced versions of *Proposition 5.4.1-3*, which are obtained by setting to zero one of the pairs of functions $\underline{v}, \underline{V}$ and $\underline{y}, \underline{Y}$ (or possibly both pairs), as indeed entailed by the special cases of (1e) and correspondingly of (11e), respectively of (1f) and correspondingly of (11f), with $\tilde{V}_0=1, \tilde{V}_j=0$ for $j>0$, respectively $\tilde{Y}_0=1, \tilde{Y}_j=0$ for $j>0$. In the first case, $\underline{v}=\underline{V}=0$, the matrices $\underline{g}$ and $\underline{G}$ can also be ignored, see (6), and one can then ignore the equations (1e) and (11e) as well as (1c) and (11c). In the second case, $\underline{y}=\underline{Y}=0$, $\underline{h}$ and $\underline{H}$ can likewise be ignored, see (6), as well as the equations (1f), (11f) and (1d), (11d). If both conditions hold, $\underline{v}=\underline{V}=0$ and $\underline{y}=\underline{Y}=0$, then the only matrices that play a significant role are $\underline{u}_n$ and $\underline{f}$, and correspondingly $\underline{U}_n$ and $\underline{F}$, and the only relevant equations are (1a,b) and (11a,b); in this special case $\underline{K}=\underline{F}$, see (6).

We now exhibit two classes of *linearizable* second-order matrix ODEs satisfied by the $N+5$ matrices $\underline{U}_n, \underline{F}, \underline{G}, \underline{H}, \underline{V}, \underline{Y}$, obtained from *Proposition 5.4.1-3* by making specific choices for the functions $\underline{\tilde{U}}_n, \underline{\tilde{F}}, \underline{\tilde{G}}, \underline{\tilde{H}}, \underline{\tilde{V}}_j, \underline{\tilde{Y}}_j$, see (1) and (11). Once and for all, let us emphasize that these are representative examples, selected to display the type of *linearizable* matrix ODEs encompassed by our approach, and chosen with an eye to the manufacture of *linearizable many-body problems*, see Sect. 5.6 and its subsections; it will of course be easy (and instructive!) for the diligent reader to manufacture additional examples.

The *first class of examples* we consider is characterized by *linearizable* (in fact, *linear*) matrix ODEs of type (1) which read as follows:

$$\mu_n^{(u)} \ddot{\underline{u}}_n = \sum_{m=1}^{N} \left[a_{nm}^{(uu)} \dot{\underline{u}}_m + b_{nm}^{(uu)} \underline{u}_m\right] + a_n^{(uf)} \dot{\underline{f}} + b_n^{(uf)} \underline{f} + a_n^{(ug)} \dot{\underline{g}} + b_n^{(ug)} \underline{g} + a_n^{(uh)} \dot{\underline{h}} + b_n^{(uh)} \underline{h} , \tag{14a}$$

$$\mu^{(f)} \ddot{\underline{f}} = \sum_{m=1}^{N} \left[a_m^{(fu)} \dot{\underline{u}}_m + b_m^{(fu)} \underline{u}_m\right] + a^{(ff)} \dot{\underline{f}} + b^{(ff)} \underline{f} + a^{(fg)} \dot{\underline{g}} + b^{(fg)} \underline{g} + a^{(fh)} \dot{\underline{h}} + b^{(fh)} \underline{h} , \tag{14b}$$

$$\mu^{(g)}\,\underline{\ddot{g}}=\sum_{m=1}^{N}\left[a_m^{(gu)}\,\underline{\dot{u}}_m+b_m^{(gu)}\,\underline{u}_m\right]+a^{(gf)}\,\underline{\dot{f}}+b^{(gf)}\,\underline{f}+a^{(gg)}\,\underline{\dot{g}}+b^{(gg)}\,\underline{g}+a^{(gh)}\,\underline{\dot{h}}+b^{(gh)}\,\underline{h}\;, \tag{14c}$$

$$\mu^{(h)}\,\underline{\ddot{h}}=\sum_{m=1}^{N}\left[a_m^{(hu)}\,\underline{\dot{u}}_m+b_m^{(hu)}\,\underline{u}_m\right]+a^{(hf)}\,\underline{\dot{f}}+b^{(hf)}\,\underline{f}+a^{(hg)}\,\underline{\dot{g}}+b^{(hg)}\,\underline{g}+a^{(hh)}\,\underline{\dot{h}}+b^{(hh)}\,\underline{h}\;, \tag{14d}$$

$$\mu^{(v)}\,\underline{\ddot{v}}=\Big\{\sum_{m=1}^{N}\left[\tilde{a}_m^{(vu)}\,\underline{\dot{u}}_m^2+\tilde{b}_m^{(vu)}\,\underline{u}_m^2\right]$$

$$+\tilde{a}^{(vf)}\,\underline{\dot{f}}^2+\tilde{b}^{(vf)}\,\underline{f}^2+\tilde{a}^{(vg)}\,\underline{\dot{g}}^2+\tilde{b}^{(vg)}\,\underline{g}^2+\tilde{a}^{(vh)}\,\underline{\dot{h}}^2+\tilde{b}^{(vh)}\,\underline{h}^2\Big\}\,\underline{\dot{v}}\;,$$

$$+\Big\{\sum_{m=1}^{N}\left[a_m^{(vu)}\,\underline{\dot{u}}_m^2+b_m^{(vu)}\,\underline{u}_m^2\right]$$

$$+a^{(vf)}\,\underline{\dot{f}}^2+b^{(vf)}\,\underline{f}^2+a^{(vg)}\,\underline{\dot{g}}^2+b^{(vg)}\,\underline{g}^2+a^{(vh)}\,\underline{\dot{h}}^2+b^{(vh)}\,\underline{h}^2\Big\}\,\underline{v}\;, \tag{14e}$$

$$\mu^{(y)}\,\underline{\ddot{y}}=\underline{\dot{y}}\,\Big\{\sum_{m=1}^{N}\left[\tilde{a}_m^{(yu)}\,\underline{\dot{u}}_m^2+\tilde{b}_m^{(yu)}\,\underline{u}_m^2\right]$$

$$+\tilde{a}^{(yf)}\,\underline{\dot{f}}^2+\tilde{b}^{(yf)}\,\underline{f}^2+\tilde{a}^{(yg)}\,\underline{\dot{g}}^2+\tilde{b}^{(yg)}\,\underline{g}^2+\tilde{a}^{(yh)}\,\underline{\dot{h}}^2+\tilde{b}^{(yh)}\,\underline{h}^2\Big\}$$

$$+\underline{y}\,\Big\{\sum_{m=1}^{N}\left[a_m^{(yu)}\,\underline{\dot{u}}_m^2+b_m^{(yu)}\,\underline{u}_m^2\right]$$

$$+a^{(yf)}\,\underline{\dot{f}}^2+b^{(yf)}\,\underline{f}^2+a^{(yg)}\,\underline{\dot{g}}^2+b^{(yg)}\,\underline{g}^2+a^{(yh)}\,\underline{\dot{h}}^2+b^{(yh)}\,\underline{h}^2\Big\}\,. \tag{14f}$$

These equations contain $N+5$ quantities of type μ, and $2N^2+20N+42$ quantities of type a,b (variously decorated with lower indices and upper symbols of identification); our treatment would apply even if all these quantities were (arbitrarily!) given functions of time, but for simplicity we assume hereafter that they are (arbitrarily!) given constants. Then the $N+3$ matrices $\underline{u}_n$, $\underline{f}$, $\underline{g}$, $\underline{h}$ can be obtained from (14a,b,c,d) via purely algebraic operations, and the 2 matrices $\underline{v}$ and $\underline{y}$ can subsequently be obtained by solving the *linear nonautonomous* matrix ODEs (14e,f).

The corresponding (generally nonlinear!) *linearizable* matrix ODEs satisfied by the $N+5$ matrices $\underline{U}_n$, $\underline{F}$, $\underline{G}$, $\underline{H}$, $\underline{V}$, $\underline{Y}$ read as follows:

$$\mu_n^{(u)}\ddot{\underline{U}}_n = 2[\dot{\underline{U}}_n,\underline{K}]+[\underline{U}_n,\dot{\underline{K}}]-[[\underline{U}_n,\underline{K}],\underline{K}]+\sum_{m=1}^{N}\left[a_{nm}^{(uu)}\{\dot{\underline{U}}_m-[\underline{U}_m,\underline{K}]\}+b_{nm}^{(uu)}\underline{U}_m\right]$$

$$+a_n^{(uf)}\{\ \dot{\underline{F}}-[\underline{F},\underline{K}]\ \}+b_n^{(uf)}\underline{F}+a_n^{(ug)}\{\ \dot{\underline{G}}-[\underline{G},\underline{K}]\ \}+b_n^{(ug)}\underline{G}$$

$$+a_n^{(uh)}\{\ \dot{\underline{H}}-[\underline{H},\underline{K}]\ \}+b_n^{(uh)}\underline{H}, \qquad (15a)$$

$$\mu^{(f)}\ddot{\underline{F}} = 2[\dot{\underline{F}},\underline{K}]+[\underline{F},\dot{\underline{K}}]-[[\underline{F},\underline{K}],\underline{K}]+\sum_{m=1}^{N}\left[a_m^{(fu)}\{\dot{\underline{U}}_m-[\underline{U}_m,\underline{K}]\}+b_m^{(fu)}\underline{U}_m\right]$$

$$+a^{(ff)}\{\ \dot{\underline{F}}-[\underline{F},\underline{K}]\ \}+b^{(ff)}\underline{F}+a^{(fg)}\{\ \dot{\underline{G}}-[\underline{G},\underline{K}]\ \}+b^{(fg)}\underline{G}$$

$$+a^{(fh)}\{\ \dot{\underline{H}}-[\underline{H},\underline{K}]\ \}+b^{(fh)}\underline{H}\ , \qquad (15b)$$

$$\mu^{(g)}\ddot{\underline{G}} = 2[\dot{\underline{G}},\underline{K}]+[\underline{G},\dot{\underline{K}}]-[[\underline{G},\underline{K}],\underline{K}]+\sum_{m=1}^{N}\left[a_m^{(gu)}\{\dot{\underline{U}}_m-[\underline{U}_m,\underline{K}]\}+b_m^{(gu)}\underline{U}_m\right]$$

$$+a^{(gf)}\{\ \dot{\underline{F}}-[\underline{F},\underline{K}]\ \}+b^{(gf)}\underline{F}+a^{(gg)}\{\ \dot{\underline{G}}-[\underline{G},\underline{K}]\ \}+b^{(gg)}\underline{G}$$

$$+a^{(gh)}\{\ \dot{\underline{H}}-[\underline{H},\underline{K}]\ \}+b^{(gh)}\underline{H}\ , \qquad (15c)$$

$$\mu^{(h)}\ddot{\underline{H}} = 2[\dot{\underline{H}},\underline{K}]+[\underline{H},\dot{\underline{K}}]-[[\underline{H},\underline{K}],\underline{K}]+\sum_{m=1}^{N}\left[a_m^{(hu)}\{\dot{\underline{U}}_m-[\underline{U}_m,\underline{K}]\}+b_m^{(hu)}\underline{U}_m\right]$$

$$+a^{(hf)}\{\dot{\underline{F}}-[\underline{F},\underline{K}]\}+b^{(hf)}\underline{F}+a^{(hg)}\{\dot{\underline{G}}-[\underline{G},\underline{K}]\}+b^{(hg)}\underline{G}+a^{(hh)}\{\dot{\underline{H}}-[\underline{H},\underline{K}]\}+b^{(hh)}\underline{H}, \qquad (15d)$$

$$\mu^{(v)}\ddot{\underline{V}} = -2\underline{K}\dot{\underline{V}}-\dot{\underline{K}}\underline{V}-\underline{K}^2\underline{V}+\{\ \sum_{m=1}^{N}\left[\tilde{a}_m^{(vu)}\{\dot{\underline{U}}_m-[\underline{U}_m,\underline{K}]\}^2+\tilde{b}_m^{(vu)}\underline{U}_m^2\right]$$

$$+\tilde{a}^{(vf)}\{\ \dot{\underline{F}}-[\underline{F},\underline{K}]\ \}^2+\tilde{b}^{(vf)}\underline{F}^2+\tilde{a}^{(vg)}\{\ \dot{\underline{G}}-[\underline{G},\underline{K}]\ \}^2+\tilde{b}^{(vg)}\underline{G}^2$$

$$+\tilde{a}^{(vh)}\{\ \dot{\underline{H}}-[\underline{H},\underline{K}]\ \}^2+\tilde{b}^{(vh)}\underline{H}^2\ \}\{\dot{\underline{V}}+\underline{K}\underline{V}\}$$

$$+\{\ \sum_{m=1}^{N}\left[a_m^{(vu)}\{\dot{\underline{U}}_m-[\underline{U}_m,\underline{K}]\}^2+b_m^{(vu)}\underline{U}_m^2\right]$$

$$+a^{(vf)}\{\ \dot{\underline{F}}-[\underline{F},\underline{K}]\ \}^2+b^{(vf)}\underline{F}^2+a^{(vg)}\{\ \dot{\underline{G}}-[\underline{G},\underline{K}]\ \}^2+b^{(vg)}\underline{G}^2$$

$$+a^{(vh)}\{\ \underline{\dot{H}}-[\underline{H},\underline{K}]\ \}^2+b^{(vh)}\underline{H}^2\ \}\underline{V}, \tag{15e}$$

$$\mu^{(y)}\underline{\ddot{Y}}=2\underline{\dot{Y}}\underline{K}+\underline{Y}\underline{\dot{K}}-\underline{Y}\underline{K}^2+\{\underline{\dot{Y}}-\underline{Y}\underline{K}\}\{\ \sum_{m=1}^{N}\left[a_m^{(yu)}\{\underline{\dot{U}}_m-[\underline{U}_m,\underline{K}]\}^2+b_m^{(yu)}\underline{U}_m^2\right]$$

$$+\tilde{a}^{(yf)}\{\ \underline{\dot{F}}-[\underline{F},\underline{K}]\ \}^2+\tilde{b}^{(yf)}\underline{F}^2+\tilde{a}^{(yg)}\{\ \underline{\dot{G}}-[\underline{G},\underline{K}]\ \}^2+\tilde{b}^{(yg)}\underline{G}^2$$

$$+\tilde{a}^{(yh)}\{\ \underline{\dot{H}}-[\underline{H},\underline{K}]\ \}^2+\tilde{b}^{(yh)}\underline{H}^2\ \}$$

$$+\underline{Y}\{\ \sum_{m=1}^{N}\left[a_m^{(yu)}\{\underline{\dot{U}}_m-[\underline{U}_m,\underline{K}]\}^2+b_m^{(yu)}\underline{U}_m^2\right]$$

$$+a^{(yf)}\{\ \underline{\dot{F}}-[\underline{F},\underline{K}]\ \}^2+b^{(yf)}\underline{F}^2+a^{(yg)}\{\ \underline{\dot{G}}-[\underline{G},\underline{K}]\ \}^2+b^{(yg)}\underline{G}^2$$

$$+a^{(yh)}\{\ \underline{\dot{H}}-[\underline{H},\underline{K}]\ \}^2+b^{(yh)}\underline{H}^2\ \}. \tag{15f}$$

In these formulas $\underline{K}$ is of course defined by (6).

The *second class of examples* we consider is characterized by (much simpler!) *linearizable* (or *linear*, see below) matrix ODEs of type (1) which read as follows:

$$\underline{\ddot{u}}_n=\underline{\tilde{u}}(\underline{u}_m,\underline{\dot{u}}_m,\ m=1,...,N;t)\ , \tag{16a}$$

$$\underline{f}=\underline{\tilde{f}}(\underline{u}_m,\underline{\dot{u}}_m,\ m=1,...,N;t)\ . \tag{16b}$$

Here $\underline{\tilde{u}}$ is a given scalar/matrix function, such that (16a) is a *solvable* or *linearizable* system of ODEs (see for instance the preceding Sect. 5.3, or below), and $\underline{\tilde{f}}$ is an (arbitrarily) given scalar/matrix function. The corresponding ODEs of type (3.17) then read

$$\underline{\ddot{U}}_n=2[\ \underline{\dot{U}}_n,\underline{F}\ ]+[\ \underline{U}_n,\underline{\dot{F}}\ ]-[\ [\ \underline{U}_n,\underline{F}\ ],\ \underline{F}\ ]+\underline{\tilde{u}}\left(\underline{U}_m,\underline{\dot{U}}_m-[\underline{U}_m,\underline{F}],\ m=1,...,N;t\right), \tag{17a}$$

$$\underline{F}=\underline{\tilde{f}}\left(\underline{U}_m,\underline{\dot{U}}_m-[\underline{U}_m,\underline{F}],\ m=1,...,N;t\right)\ . \tag{17b}$$

Note that we are here in the special case mentioned in the *Remark 5.4.1-7,* see above, corresponding to the choice $\underline{v}=\underline{V}=0$ and $\underline{y}=\underline{Y}=0$. Also note that, if $\underline{f}$ depends on the matrices $\underline{\dot{u}}_m$ (besides the matrices $\underline{u}_m$), see (16b), hence $\underline{F}$ depends on the matrices $\dot{\underline{U}}_m$ (besides the matrices $\underline{U}_m$; see (17b)), then the ODE (17a) contains the second time-derivative of the matrices $\underline{U}_m$ in the right hand side as well (from the second term).

For the special choice

$$\tilde{\underline{u}}_n = \sum_{m=1}^{N} \left[2\, a_{nm}(t)\, \dot{\underline{u}}_m + b_{nm}(t)\, \underline{u}_m\right], \tag{18}$$

which entails of course that the ODEs (16a) are *linear*, the ODEs (17a) read

$$\ddot{\underline{U}}_n = 2\left[\dot{\underline{U}}_n, \underline{F}\right] + \left[\underline{U}_n, \dot{\underline{F}}\right] - \left[\left[\underline{U}_n, \underline{F}\right], \ \underline{F}\right] + \sum_{m=1}^{N} \left[2\, a_{nm}(t)\, \left\{\dot{\underline{U}}_m - \left[\underline{U}_n, \underline{F}\right]\right\} + b_{nm}(t)\, \underline{U}_m\right]. \tag{19}$$

The factor 2 in the right hand sides of the last two equations is introduced for notational convenience, see below. In the last equation, (19), the matrix $\underline{F}$ is of course always given by (17b), with an arbitrarily chosen $\tilde{\underline{f}}$. For instance, if we make for this function the simple choice

$$\tilde{\underline{f}}\left(\underline{u}_m, \dot{\underline{u}}_m, m = 1, \ldots, N; t\right) = \sum_{m=1}^{N} \left[c_m(t)\underline{u}_m\right], \tag{20a}$$

which entails

$$\underline{F} = \sum_{m=1}^{N} \left[c_m(t)\underline{U}_m\right], \tag{20b}$$

then the *linearizable* matrix ODEs satisfied by the N matrices $\underline{U}_n$ read as follows:

$$\ddot{\underline{U}}_n = \sum_{m=1}^{N} \{ \ 2a_{nm}(t)\dot{\underline{U}}_m + b_{nm}(t)\underline{U}_m$$

$$+ 2c_m(t)[\ \dot{\underline{U}}_n, \underline{U}_m \ ] + c_m(t)[\ \underline{U}_n, \dot{\underline{U}}_m \ ] + \dot{c}_m(t)[\ \underline{U}_n, \underline{U}_m \ ] \}$$

$$- \sum_{m_1, m_2 = 1}^{N} \{ \ 2\, a_{nm_1}(t)\, c_{m_2}(t)\, [\ \underline{U}_{m_1}, \underline{U}_{m_2} \ ] + c_{m_1}(t)\, c_{m_2}(t)\, [\ [\ \underline{U}_n, \underline{U}_{m_1} \ ], \underline{U}_{m_2} \ ] \}. \tag{21}$$

These equations deserve further elaboration, which for simplicity is hereafter restricted to the case with time-independent constants a_{nm}, b_{nm} and c_m. Then the above ODEs becomes

$$\ddot{\underline{U}}_n = \sum_{m=1}^{N} \left\{ 2\, a_{nm}\, \dot{\underline{U}}_m + b_{nm}\, \underline{U}_m + 2\, c_m \left[\dot{\underline{U}}_n, \underline{U}_m\right] + c_m \left[\underline{U}_n, \dot{\underline{U}}_m\right] \right\}$$

$$- \sum_{m_1, m_2=1}^{N} \left\{ 2 a_{nm_1}\, c_{m_2} \left[\underline{U}_{m_1}, \underline{U}_{m_2}\right] + c_{m_1} c_{m_2} \left[\left[\underline{U}_n, \underline{U}_{m_1}\right], \underline{U}_{m_2} \right] \right\}, \tag{22}$$

and the corresponding ODEs satisfied by the matrices $\underline{u}_n$ read

$$\ddot{\underline{u}}_n = \sum_{m=1}^{N} \left[2\, a_{nm}\, \dot{\underline{u}}_m + b_{nm}\, \underline{u}_m \right], \tag{23}$$

and can therefore be solved by purely algebraic operations. Particularly simple is the "diagonal case" characterized by the restrictions $a_{nm} = \delta_{nm} a_n, b_{nm} = \delta_{nm} b_n$, which entail that the ODEs (23) decouple and their solution reads

$$\underline{u}_n(t) = \exp(a_n t) \left[\underline{u}_n(0) \cosh(\Delta_n t) + \dot{\underline{u}}_n(0)\, \Delta_n^{-1} \sinh(\Delta_n t) \right], \tag{24a}$$

$$\Delta_n = (a_n{}^2 + b_n)^{1/2}. \tag{24b}$$

Even simpler is the case with $a_n = a, b_n = b$, hence $\Delta_n = \Delta = (a^2 + b)^{1/2}$. Then the evolution of the matrix $\underline{W}(t)$, see (5a), (16b) and (20a), (24), reads simply

$$\dot{\underline{W}}(t) = \exp(at) \left[\underline{A} \cosh(\Delta t) + \underline{B}\, \Delta^{-1} \sinh(\Delta t) \right] \underline{W}(t), \tag{25}$$

with $\underline{A} = \sum_{m=1}^{N} \left[c_m \underline{u}_m(0)\right]$, $\underline{B} = \sum_{m=1}^{N} \left[c_m \dot{\underline{u}}_m(0)\right]$ two *constant* matrices. But let us emphasize that the simplicity of this case, see (25), has a rather trivial origin: indeed the nonlinearity of the matrix ODEs (22) in this special case $a_{nm} = \delta_{nm} a,\ b_{nm} = \delta_{nm} b$ is in a way marginal, since in this case the ODEs (22), which can of course be rewritten as follows,

$$\ddot{\underline{U}}_n = 2a \dot{\underline{U}}_n + b \underline{U}_n + 2\left[\dot{\underline{U}}_n, \underline{F}\right] + \left[\underline{U}_n, \dot{\underline{F}}\right] - 2a\left[\underline{U}_n, \underline{F}\right] - \left[\left[\underline{U}_n, \underline{F}\right], \underline{F}\right], \tag{26a}$$

$$\underline{F} = \sum_{m=1}^{N} \left[c_m \underline{U}_m\right], \tag{26b}$$

are easily seen (by multiplying (26a) by c_n, summing over n and using (26b)) to imply a single (decoupled) ODE for the quantity $\underline{F}$,

$$\ddot{\underline{F}}=2a\dot{\underline{F}}+b\underline{F}+[\dot{\underline{F}},\underline{F}]. \tag{26c}$$

And after this equation has been solved for $\underline{F}$, the ODEs (26a) for the matrices $\underline{U}_n$ become *linear* (albeit *nonautonomous*). But the matrix ODE (26c) coincides (up to a trivial rescaling) with the special case of the system (22) corresponding to $N=1$. There is then a single matrix $\underline{U}_1 \equiv \underline{U}(t)$ which satisfies the matrix evolution equation

$$\ddot{\underline{U}}=2a\dot{\underline{U}}+b\underline{U}+c[\dot{\underline{U}},\underline{U}], \tag{27}$$

or, more generally,

$$\ddot{\underline{U}}=2a\dot{\underline{U}}+b\underline{U}+[\dot{\underline{U}},\underline{\tilde{f}}(\underline{U})], \tag{28}$$

if we retain the freedom to make an arbitrary choice (rather than the special choice (20)) for the scalar/matrix function $\underline{\tilde{f}}$, see (16b) and (17b). We have thereby returned to the cases discussed in Sect. 5.2 (for (28)) and in Sect. 5.2.3 (for (27)).

Finally, let us point out that the examples given so far in Sect. 5.4.1 focussed on *second-order ODEs* for the $N+5$ matrices $\underline{U}_n$, $\underline{F}$, $\underline{G}$, $\underline{H}$, $\underline{V}$, $\underline{Y}$ (or for a subset of them), obtained starting from *second-order ODEs* for the $N+5$ matrices $\underline{u}_n$, $\underline{f}$, $\underline{g}$, $\underline{h}$, $\underline{v}$, $\underline{y}$ (or for a subset of them). The motivation for doing so is because *second-order matrix ODEs* are a convenient starting point to obtain the Newtonian equations of motion which characterize many-body problems in three-dimensional space (our main interest in Chap. 5) -- as we saw in Sects. 5.1 and 5.2 and we shall see in Sect. 5.6 and its subsections. But it is also of interest to consider *first-order matrix ODEs*, both because of their possible applicative relevance, and because such equations may also be connected to our main goal in Chap. 5, namely to construct models of many-body problems in ordinary (three-dimensional) space. Indeed there are two ways in which such a goal may be realized also by starting from *first-order matrix ODEs* of the kind yielded by the technique introduced in Sect. 5.4.1: such equations may be eventually interpreted as Hamiltonian (rather than Newtonian) equations of motion for a many-body problem (see Sect. 5.6.4); or they may be used to obtain (by time-differentiation and appropriate substitutions) new *second-order matrix ODEs*, which can then be appropriately interpreted as Newtonian equations of motion for a many-body problem. We end Sect. 5.4.1 by providing two examples of matrix evolution equations which correspond -- as we shall show in Sects. 5.6.4 and 5.6.3 -- to these two possibilities.

Firstly let us show -- starting, as it were, again from first principle -- that the system of matrix evolution equations

$$\dot{\underline{U}}_n = \sum_{m=1}^{N} \left(a_{nm}\, \underline{U}_m + b_{nm}\, \underline{V}_m\right) + \left[\underline{U}_n, \underline{\tilde{f}}(\underline{U}_j, \underline{V}_j; t)\right] , \tag{29a}$$

$$\dot{\underline{V}}_n = \sum_{m=1}^{N} \left(c_{nm}\, \underline{U}_m + d_{nm}\, \underline{V}_m\right) + \left[\underline{V}_n, \underline{\tilde{f}}(\underline{U}_j, \underline{V}_j; t)\right] , \tag{29b}$$

is *linearizable*. Here the $4N^2$ quantities $a_{nm}, b_{nm}, c_{nm}, d_{nm}$ are arbitrary (they could also be time-dependent functions), and $\underline{\tilde{f}}$ is an arbitrary function of the $2N$ matrices $\underline{U}_m, \underline{V}_m$ and of the time t; note however that the same $\underline{\tilde{f}}$ enters in (29a) and (29b), that this quantity is independent of the index n, and that it is a scalar/matrix function of its arguments, namely it satisfies the property $\underline{W}\, \underline{\tilde{f}}(\underline{U}_j, \underline{V}_j; t)\, \underline{W}^{-1} = \underline{\tilde{f}}(\underline{W}\, \underline{U}_j\, \underline{W}^{-1}, \underline{W}\, \underline{V}_j\, \underline{W}^{-1}; t)$.

To prove that (29) is *linearizable* we proceed again as above, namely we set (see (3) and (5); beware of the notational changes!)

$$\underline{u}_n(t) = \underline{W}(t)\, \underline{U}_n(t) \left[\underline{W}(t)\right]^{-1}, \; \underline{U}_n(t) = \left[\underline{W}(t)\right]^{-1} \underline{u}_n(t)\, \underline{W}(t) , \tag{30a}$$

$$\underline{v}_n(t) = \underline{W}(t)\, \underline{V}_n(t) \left[\underline{W}(t)\right]^{-1}, \; \underline{V}_n(t) = \left[\underline{W}(t)\right]^{-1} \underline{v}_n(t)\, \underline{W}(t) , \tag{30b}$$

$$\dot{\underline{W}}(t) = \underline{W}(t)\, \underline{\tilde{f}}\left[\underline{U}_j(t), \underline{V}_j(t); t\right], \; \dot{\underline{W}}(t) = \underline{\tilde{f}}\left[\underline{u}_j(t), \underline{v}_j(t); t\right] \underline{W}(t) . \tag{31}$$

Time-differentiation of the first of the (30a,b) yields (using the first of the (31))

$$\dot{\underline{u}}_n = \underline{W} \{\, \dot{\underline{U}}_n - [\, \underline{U}_n, \underline{\tilde{f}}(\underline{U}_j, \underline{V}_j; t)\,]\, \} \underline{W}^{-1}, \tag{32a}$$

$$\dot{\underline{v}}_n = \underline{W} \{\, \dot{\underline{V}}_n - [\, \underline{V}_n, \underline{\tilde{f}}(\underline{U}_j, \underline{V}_j; t)\,]\, \} \underline{W}^{-1}, \tag{32b}$$

and from these equations and (29) , (30) we see that the matrices $\underline{u}_n(t), \underline{v}_n(t)$ satisfy the *linear* evolution equations

$$\dot{\underline{u}}_n = \sum_{m=1}^{N} \left(a_{nm}\, \underline{u}_m + b_{nm}\, \underline{v}_m\right) , \; \dot{\underline{v}}_n = \sum_{m=1}^{N} \left(c_{nm}\, \underline{u}_m + d_{nm}\, \underline{v}_m\right) . \tag{33}$$

The linearizability of (29) is thereby proven, since its solution can be achieved by solving firstly (33), then the second of the (31) (a linear equation for the matrix $\underline{W}(t)$), and then recovering $\underline{U}_n(t), \underline{V}_n(t)$ from the second of the (30a,b).

It is moreover easily seen that in some cases, see below, the system (29) is not only *linearizable*, it is in fact *solvable*. An obvious (and rather trivial) case is if the quantities $a_{nm}, b_{nm}, c_{nm}, d_{nm}$ all vanish, $a_{nm} = b_{nm} = c_{nm} = d_{nm} = 0$, and $\underline{\tilde{f}}$ does not depend explicitly on the time t, $\underline{\tilde{f}}(\underline{U}_j, \underline{V}_j; t) \equiv \underline{\tilde{f}}(\underline{U}_j, \underline{V}_j)$. In such a case (33) entails that the matrices $\underline{u}_n(t), \underline{v}_n(t)$ are in fact time-independent, $\underline{u}_n(t) = \underline{u}_n(0), \underline{v}_n(t) = \underline{v}_n(0)$, and the second of the (31) becomes explicitly solvable. Hence one concludes that the equations

$$\dot{\underline{U}}_n = \left[\underline{U}_n, \underline{\tilde{f}}(\underline{U}_j, \underline{V}_j)\right], \quad \dot{\underline{V}}_n = \left[\underline{V}_n, \underline{\tilde{f}}(\underline{U}_j, \underline{V}_j)\right], \tag{34a}$$

are explicitly *solvable*:

$$\underline{U}_n(t) = \exp\left[-t\underline{\tilde{f}}\left(\underline{U}_j(0), \underline{V}_j(0)\right) \right] \underline{U}_n(0) \exp\left[t\underline{\tilde{f}}\left(\underline{U}_j(0), \underline{V}_j(0)\right) \right], \tag{34b}$$

$$\underline{V}_n(t) = \exp\left[-t\underline{\tilde{f}}\left(\underline{U}_j(0), \underline{V}_j(0)\right) \right] \underline{V}_n(0) \exp\left[t\underline{\tilde{f}}\left(\underline{U}_j(0), \underline{V}_j(0)\right) \right]. \tag{34c}$$

Another, perhaps less trivial, case in which the equations (29) are in fact also *solvable* obtains if

$$\underline{\tilde{f}} = \sum_{j=1}^{N} \lambda_j \left[\underline{U}_j, \underline{V}_j\right], \tag{35a}$$

and the quantities $a_{nm}, b_{nm}, c_{nm}, d_{nm}, \lambda_n$ are all time-independent and satisfy the constraints

$$\lambda_n a_{nm} + \lambda_m d_{mn} = 0, \quad \lambda_n c_{nm} - \lambda_m c_{mn} = 0, \quad \lambda_n b_{nm} - \lambda_m b_{mn} = 0 \ . \tag{35b}$$

Indeed it is easily seen that these conditions are sufficient to guarantee, via (29), that $\underline{\tilde{f}}$ is time-independent, $\dot{\underline{\tilde{f}}} = 0$, so that there holds again the explicit solution (34b,c), of course with $\underline{\tilde{f}}$ given by (35a) (at $t = 0$). An interesting case (see below) is that with $\lambda_n = \lambda$, so that $d_{nm} = -a_{mn}$ and (29) read

$$\dot{\underline{U}}_n = \sum_{m=1}^{N} \left(a_{nm} \underline{U}_m + b_{nm} \underline{V}_m + \lambda [\underline{U}_n, [\underline{U}_m, \underline{V}_m]] \right), \tag{36a}$$

$$\dot{\underline{V}}_n=\sum_{m=1}^{N}\left(c_{nm}\,\underline{U}_m-a_{mn}\,\underline{V}_m+\lambda[\,\underline{V}_n,[\,\underline{U}_m,\underline{V}_m\,]]\right), \tag{36b}$$

with

$$b_{nm}=b_{mn},\quad c_{nm}=c_{mn}\;. \tag{36c}$$

We shall show in Sect. 5.6.4 how these *solvable* matrix evolution equations can be recast in the form of the Hamiltonian equations of motion of a many-body problem in three-dimensional (or indeed, in S-dimensional) space.

Secondly, and lastly, we illustrate, via a simple example, the possibility, in the context of the technique illustrated above, to restrict firstly attention to *first-order* matrix ODEs and to obtain subsequently *second-order* ODEs by appropriate additional steps. We only consider an illustrative, very simple, example; this allows, at very little cost in terms of repetitiveness, a completely self-contained presentation; but for the diligent reader interested in the connection with the treatment given above we note that the case considered below corresponds, up to a trivial notational change ($\underline{u}_1\rightarrow\underline{u},\underline{u}_2\rightarrow\underline{v}$), to the treatment given above with $N=2$, $\underline{\tilde{U}}_1=\dot{\underline{u}}_1-(\alpha\underline{u}_1+\beta\underline{u}_2)$, $\underline{\tilde{U}}_2=\dot{\underline{u}}_2-(\gamma\underline{u}_1+\delta\underline{u}_2)$, $\underline{\tilde{F}}=\underline{f}-(a\underline{u}_1+b\underline{u}_2+c[\underline{u}_1,\underline{u}_2])$ and $\underline{g}=\underline{h}=\underline{v}=\underline{y}=0$.

Let us set

$$\underline{u}=\underline{W}\,\underline{U}\,\underline{W}^{-1},\quad \underline{U}=\underline{W}^{-1}\,\underline{u}\,\underline{W}\;, \tag{37a}$$

$$\underline{v}=\underline{W}\,\underline{V}\,\underline{W}^{-1},\quad \underline{V}=\underline{W}^{-1}\,\underline{v}\,\underline{W}\;, \tag{37b}$$

$$\dot{\underline{W}}=\underline{W}\left\{a\,\underline{U}+b\,\underline{V}+c\left[\underline{U},\underline{V}\right]\right\},\quad \dot{\underline{W}}=\left\{a\,\underline{u}+b\,\underline{v}+c\left[\underline{u},\underline{v}\right]\right\}\underline{W}\;, \tag{38}$$

$$\dot{\underline{u}}=\alpha\,\underline{u}+\beta\,\underline{v},\quad \dot{\underline{v}}=\gamma\,\underline{u}+\delta\,\underline{v}\;, \tag{39}$$

with $a,b,c,\alpha,\beta,\gamma,\delta$ arbitrary constants.

This entails for the matrices $\underline{U}$ and $\underline{V}$ the first-order nonlinear ODEs

$$\dot{\underline{U}}=\alpha\underline{U}+\beta\underline{V}+b[\underline{U},\underline{V}]+c[\underline{U},[\underline{U},\underline{V}]]\;, \tag{40a}$$

$$\dot{\underline{V}}=\gamma\underline{U}+\delta\underline{V}+a[\underline{V},\underline{U}]+c[\underline{V},[\underline{V},\underline{U}]]\;. \tag{40b}$$

These nonlinear matrix ODEs are of course *linearizable*, since their solution can be achieved via the following steps: *(i)* set (for simplicity) $\underline{W}(0)=\underline{1}$ as initial condition to complement (38); *(ii)* note that this entails $\underline{u}(0)=\underline{U}(0)$, $\underline{v}(0)=\underline{V}(0)$ (see (37a,b)); *(iii)* evaluate $\underline{u}(t)$ and $\underline{v}(t)$ from the (explicitly solvable) *linear* evolution equation (39), taking into account the appropriate initial conditions, see *(ii)*; *(iv)* evaluate $\underline{W}(t)$ by solving the second of the (38), with initial condition $\underline{W}(0)=\underline{1}$, see *(i)* (note that this is a *linear nonautonomous* matrix ODE, entailing the solutions of M analogous systems of M linear first-order coupled nonautonomous ODEs -- assuming we are dealing with ($M\times M$)-matrices); *(v)* finally evaluate $\underline{U}(t)$ and $\underline{V}(t)$ from the second of the (37a,b).

Let us now derive, from the 2 *first-order* ODEs (40) satisfied by the 2 matrices $\underline{U}(t)$ and $\underline{V}(t)$, a single *second-order* ODE for one of these two matrices, say for $\underline{U}(t)$. This is easily obtained by time-differentiating (40a), thereby obtaining (using (40b)) the *second-order linearizable ODE*

$$\ddot{\underline{U}}=\alpha\dot{\underline{U}}+\beta\gamma\underline{U}+\beta\delta\underline{V}+(b\delta-a\beta)[\underline{U},\underline{V}]+b[\dot{\underline{U}},\underline{V}]+(c\delta-a\beta)[\underline{U},[\underline{U},\underline{V}]]$$

$$+c\{[\underline{U},[\dot{\underline{U}},\underline{V}]]+[\dot{\underline{U}},[\underline{U},\underline{V}]]-\beta[[\underline{U},\underline{V}],\underline{V}]\}-c\{a[\underline{U},[\underline{U},[\underline{U},\underline{V}]]]+b[\underline{U},[[\underline{U},\underline{V}],\underline{V}]]\}$$

$$-c^2[\underline{U},[\underline{U},[[\underline{U},\underline{V}],\underline{V}]]]\,, \qquad (41a)$$

where the matrix $\underline{V}$ should be expressed in terms of $\underline{U}$ and $\dot{\underline{U}}$ by solving for $\underline{V}$ the (non differential) linear matrix equation (40a), namely

$$\beta\underline{V}-b[\underline{V},\underline{U}]+c[[\underline{V},\underline{U}],\underline{U}]=\dot{\underline{U}}-\alpha\underline{U}\ . \qquad (41b)$$

At the end of Sect. 5.6.1 we shall display a (highly nonlinear) *linearizable one-body problem* in three-dimensional space which corresponds to this *second-order linearizable matrix ODE*, (41).

5.4.2 Some integrable matrix evolution equations related to the non Abelian Toda lattice

In Sect. 5.4.2 we consider various (systems of) matrix evolution ODEs which coincide, or are closely related, with the *integrable* evolution equations of the so-called *non Abelian Toda lattice*. A version of the (matrix) ODES of this model reads as follows:

$$\ddot{\underline{G}}_n = \dot{\underline{G}}_n [\underline{G}_n]^{-1} \dot{\underline{G}}_n + \gamma \{ \underline{G}_{n+1} - \underline{G}_n [\underline{G}_{n-1}]^{-1} \underline{G}_n \} . \tag{1}$$

Here $\underline{G}_n \equiv \underline{G}_n(t)$ is a time-dependent square matrix, labeled by the index n, and γ is a ("coupling") constant (possibly complex, see below), which could be eliminated via the scale transformation $\underline{G}_n \to \gamma^n \underline{G}_n$, or, if it is positive, via the time rescaling $t \to \gamma^{-1/2} t$.

In this book we do not discuss the actual solution of this matrix evolution equation: it suffices for us to know that this is an *integrable* (in fact, *solvable*) equation (see Sect. 5.N). But let us emphasize that this matrix evolution ODE, (1), to the extent it is meant to hold for $n = 1, \ldots, N$ (as we generally assume hereafter), should be completed by prescriptions "at the n-boundaries," such as, say, $\underline{G}_0(t) = \underline{G}_{N+1}(t) = 0$ ("free ends") or $\underline{G}_0(t) = \underline{G}_N(t)$, $\underline{G}_1(t) = \underline{G}_{N+1}(t)$ ("periodic"); these prescriptions are of course relevant to determine the solution, but we generally ignore them hereafter (except in some cases in which we assume periodic boundary conditions, see below).

Another, perhaps more interesting, version of this *integrable* matrix model reads

$$\ddot{\underline{Q}}_n = \dot{\underline{Q}}_n [\underline{Q}_{n+1} - \underline{Q}_n] - [\underline{Q}_n - \underline{Q}_{n-1}] \dot{\underline{Q}}_n . \tag{2a}$$

Here of course $\underline{Q}_n \equiv \underline{Q}_n(t)$ is again a square matrix.

The *integrability* of this matrix evolution equation is entailed by its relation to the *integrable* equation (1). Indeed it is clear that, by setting

$$\underline{A}_n(t) = [\underline{G}_n(t)]^{-1} \, \underline{G}_{n+1}(t) , \tag{3a}$$

$$\underline{B}_n(t) = [\underline{G}_n(t)]^{-1} \, \dot{\underline{G}}_n(t) , \tag{3b}$$

(1) can be rewritten as

$$\dot{\underline{A}}_n = \underline{A}_n \underline{B}_{n+1} - \underline{B}_n \underline{A}_n , \tag{4a}$$

$$\dot{\underline{B}}_n = \gamma [\underline{A}_n - \underline{A}_{n-1}] . \tag{4b}$$

Exercise 5.4.2-1. Verify!

We now set

$$\underline{Q}_n = \sum_{m=m_0}^{n} \underline{B}_m , \qquad (5a)$$

entailing

$$\underline{B}_n = \underline{Q}_n - \underline{Q}_{n-1} , \qquad (5b)$$

hence, via (4b),

$$\gamma \underline{A}_n = \dot{\underline{Q}}_n . \qquad (5c)$$

Insertion of (5c,b) in (4a) yields precisely (2a).

Exercise 5.4.2-2. Verify!

Of course also this matrix evolution ODE, (2), must eventually be completed with "end-point" prescriptions, namely with appropriate definitions for $\underline{Q}_0(t)$ and $\underline{Q}_{N+1}(t)$ (remember: the index n in (2a) runs from 1 to N); consistently with this ambiguity we left undefined the lower limit m_0 of the sum in the right-hand-side of (5a).

Remark 5.4.2-3. An arbitrary ("coupling") constant c can be reintroduced in (2a) via the rescaling $\underline{Q} \rightarrow c\,\underline{Q}$, so that it read

$$\ddot{\underline{Q}}_n = c\{\, \dot{\underline{Q}}_n [\, \underline{Q}_{n+1} - \underline{Q}_n \,] - [\, \underline{Q}_n - \underline{Q}_{n-1} \,] \dot{\underline{Q}}_n \,\} . \qquad (2b)$$

Remark 5.4.2-4. The version (2) of the non Abelian Toda lattice system of ODEs is *translation-invariant*, namely it is invariant under the *translation* $\underline{Q}_n(t) \rightarrow \tilde{\underline{Q}}_n = \underline{Q}_n(t) + \underline{Q}_0$, $\dot{\underline{Q}}_0 = 0$. Of course this invariance property could be destroyed by the "end-point" conditions; it is, however, compatible with *periodic* boundary conditions, $\underline{Q}_0(t) = \underline{Q}_N(t)$, $\underline{Q}_1(t) = \underline{Q}_{N+1}(t)$.

Remark 5.4.2-5. The *ansatz* $\underline{Q}_n(t) = n\, \underline{U}(t)$ is compatible with this evolution equation, (2), and it yields for $\underline{U}(t)$ the simplest one of the evolution equations discussed in the preceding Sect. 5.4.1, see (5.4.1-27) and see as well Sect. 5.2.3 (in both cases with $a = b = 0$).

Proposition 5.4.2-6. The following simple matrix evolution ODE is *integrable*:

$$\ddot{\underline{U}} = c^2 \left(2\underline{U}^3 + \underline{C}\,\underline{U} + \underline{U}\,\underline{C}\right) , \tag{6}$$

where c is an arbitrary scalar constant (which could of course be rescaled away, as well as the factor 2), and the arbitrary matrix $\underline{C}$ is also constant ($\dot{\underline{C}} = 0$; one could of course set $\underline{C} = 0$, or $\underline{C} = C\,\underline{1}$).

This important finding can indeed be considered a special case of the *integrable* equation (2b). Consider indeed the following special ("periodic") solution of (2b):

$$\underline{Q}_{2n}(t) = \underline{A}(t),\ \underline{Q}_{2n+1}(t) = \underline{B}(t) , \tag{7}$$

so that the 2 matrices $\underline{A}(t)$, $\underline{B}(t)$ satisfy the equations

$$\ddot{\underline{A}} = c\{\ \dot{\underline{A}}\,\underline{B} + \underline{B}\,\dot{\underline{A}} - \dot{\underline{A}}\,\underline{A} - \underline{A}\,\dot{\underline{A}}\ \},\ \ \ddot{\underline{B}} = c\{\ \dot{\underline{B}}\,\underline{A} + \underline{A}\,\dot{\underline{B}} - \dot{\underline{B}}\,\underline{B} - \underline{B}\,\dot{\underline{B}}\ \}. \tag{8}$$

Now set

$$\underline{S} = \underline{A} + \underline{B},\ \underline{D} = \underline{A} - \underline{B} , \tag{9a}$$

$$\underline{A} = (\underline{S} + \underline{D})/2,\ \underline{B} = (\underline{S} - \underline{D})/2 , \tag{9b}$$

so that

$$\ddot{\underline{S}} = c\{\ \underline{\dot{A}B} + \underline{A\dot{B}} + \underline{\dot{B}A} + \underline{B\dot{A}} - \underline{\dot{A}A} - \underline{A\dot{A}} - \underline{\dot{B}B} - \underline{B\dot{B}}\ \}, \tag{10a}$$

$$\ddot{\underline{D}} = c\{\ \underline{\dot{A}B} - \underline{A\dot{B}} - \underline{\dot{B}A} + \underline{B\dot{A}} - \underline{\dot{A}A} - \underline{A\dot{A}} + \underline{\dot{B}B} + \underline{B\dot{B}}\ \}. \tag{10b}$$

It is now clear that (10a) can be integrated once to yield

$$\dot{\underline{S}} = -c\{\ \underline{D}^2 + \underline{C}\ \}, \tag{11a}$$

while (10b) can be rewritten as follows:

$$\ddot{\underline{D}} = -c\{\ \underline{\dot{S}D} + \underline{D\dot{S}}\ \}. \tag{11b}$$

But, via (11a) and the identification

$$\underline{D}(t) = \underline{U}(t) , \tag{11c}$$

this last equation yields precisely (6).

Note that the initial data for (6), $\underline{U}(0)$ and $\dot{\underline{U}}(0)$, determine, via (11c), $\underline{D}(0)$ and $\dot{\underline{D}}(0)$; moreover (11a), with $\underline{C}$ assigned (arbitrarily!) and $\underline{U}(0)$ (hence $\underline{D}(0)$, see (11c)) given, determines $\dot{\underline{S}}(0)$, while $\underline{S}(0)$ can be assigned arbitrarily, consistently with the translation-invariance of (11a). From the initial data for $\underline{D}(t)$ and $\underline{S}(t)$ one obtains, via (11b), the initial data for $\underline{A}(t)$ and $\underline{B}(t)$, namely $\underline{A}(0)$, $\dot{\underline{A}}(0)$, $\underline{B}(0)$, $\dot{\underline{B}}(0)$; one then solves (8) (or equivalently, via (7), one solves (2b)), and in this manner one finally gets $\underline{A}(t)$, $\underline{B}(t)$, hence, via (the second of the) (9a) and (11c), the solution $\underline{U}(t)$ of (6) (with c an arbitrarily assigned scalar constant and $\underline{C}$ an arbitrarily assigned constant matrix).

This finding, *Proposition 5.4.2-6*, is important: it is one of the very few nontrivial examples (see Sect. 5.4.4) of *second-order* matrix ODEs that does not contain the first derivative of the dependent variable and which is amenable to exact treatment for arbitrary initial data; hence it shall yield many-body problems of the more standard type, with velocity-independent forces (see Sect. 5.6.5). Let us re-emphasize that its *integrability* -- indeed, its *solvability* -- is predicated upon the possibility to deal with the *periodic non Abelian Toda lattice*, a result which we take for granted (see Sect. 5.N). There is however a subclass of initial data for which (6) is more directly *solvable*, as entailed by the following two *exercises*.

Exercise 5.4.2-7. Show that the following *first-order* matrix ODE is *solvable*:

$$\dot{\underline{U}} = c\,\underline{C} + c\,\underline{U}^2, \tag{12}$$

with c respectively $\underline{C}$ arbitrary scalar respectively matrix constants. *Hint*: set

$$\underline{U} = -c^{-1}\,\dot{\underline{V}}\,\underline{V}^{-1}. \tag{13}$$

Exercise 5.4.2-8. Show that, if the matrix $\underline{U} \equiv \underline{U}(t)$ satisfies (12), it also satisfies (6). *Hint*: time-differentiate (12), and use it again to eliminate $\dot{\underline{U}}$ and thereby obtain (6).

5.4.3 Some other matrix evolution equations amenable to exact treatments

In Sect. 5.4.3 we present some other *solvable* and/or *integrable* and/or *linearizable* nonlinear matrix evolution equations.

A *solvable* second-order matrix ODE reads as follows:

$$\ddot{\underline{U}}=\alpha\underline{1}+\beta\underline{U}+\gamma(\dot{\underline{U}}+c\underline{U}^2)-c(\dot{\underline{U}}\underline{U}+2\underline{U}\dot{\underline{U}}+c\underline{U}^3)\,, \tag{1}$$

with α,β,γ,c arbitrary constants ($c\neq 0$).

Exercise 5.4.3-1. Show that (1) is explicitly *solvable. Hint*: set

$$c\underline{U}=\underline{V}^{-1}\dot{\underline{V}}\,, \tag{2}$$

and obtain thereby the following *third-order, linear, constant-coefficient* (hence *solvable)* ODE for the matrix $\underline{V}\equiv\underline{V}(t)$:

$$\dddot{\underline{V}}=c\alpha\underline{V}+\beta\dot{\underline{V}}+\gamma\ddot{\underline{V}}. \tag{3}$$

Exercise 5.4.3-2. Show that, if the 4 constants α,β,γ,c are real, a necessary and sufficient condition to guarantee that *all* solutions of (1) be *completely periodic*, is validity of the equalities $\alpha=\gamma=0$, together with the inequality $\beta<0$. *Hint*: see (3).

A *linearizable* system of matrix ODEs reads

$$\ddot{\underline{U}}_n=\dot{\underline{U}}_n\left[\underline{U}_n\right]^{-1}\dot{\underline{U}}_n+\underline{U}_n\sum_{m=1}^{N}a_{nm}\,\dot{\underline{U}}_m\left[\underline{U}_m\right]^{-1}\,, \tag{4}$$

where the time-dependent square matrices $\underline{U}_n\equiv\underline{U}_n(t)$ are labeled by the index n and the (scalar) quantities a_{nm} are arbitrary constants (there are N^2 of them).

Indeed by setting

$$\underline{V}_n=\dot{\underline{U}}_n\left[\underline{U}_n\right]^{-1}, \tag{5a}$$

$$\dot{\underline{U}}_n=\underline{V}_n\underline{U}_n\,, \tag{5b}$$

one gets from (4) for $\underline{V}_n(t)$ the *linear*, explicitly *solvable*, matrix ODE

$$\dot{\underline{V}}_n = \sum_{m=1}^{N} a_{nm} \underline{V}_m \,. \tag{6}$$

Exercise 5.4.3-3. Verify!

Hence to solve (4) one first solves, explicitly, this *linear* equation with constant coefficients, and then the *linear nonautonomous ODE* (5b).

A *solvable* matrix evolution ODE reads as follows:

$$\ddot{\underline{U}} = 2a\left(\underline{U}\dot{\underline{U}} + \dot{\underline{U}}\underline{U} \right) - 2a^2 \underline{U}^3 - 4b\underline{U}^2 + 3b\dot{\underline{U}} - 2b^2 \underline{U}$$

$$-2a\left(a\underline{U}^2 + b\underline{U} - \dot{\underline{U}} \right)^{1/2} \underline{U} \left(a\underline{U}^2 + b\underline{U} - \dot{\underline{U}} \right)^{1/2} , \tag{7}$$

where a, b are 2 arbitrary scalar constants.

To demonstrate the solvability of this matrix ODE, we start from the matrix evolution equation

$$\dot{\underline{M}} = a\,\underline{M}^2 + b\,\underline{M} \,, \tag{8}$$

whose solution reads (as the diligent reader will verify)

$$\underline{M}(t) = \left\{ \left[\underline{M}(0)\right]^{-1} \exp(-bt) + (b/a)\left[\exp(-bt) - 1\right] \right\}^{-1} . \tag{9}$$

Now set

$$\underline{M}(t) = \begin{pmatrix} \underline{A}(t) & \underline{B}(t) \\ \underline{B}(t) & \underline{A}(t) \end{pmatrix} , \tag{10}$$

a position which is clearly compatible with (8) and (9), and which yields for the 2 matrices $\underline{A}(t)$ and $\underline{B}(t)$ the equations

$$\dot{\underline{A}} = a(\underline{A}^2 + \underline{B}^2) + b\,\underline{A}, \quad \dot{\underline{B}} = a(\underline{A}\,\underline{B} + \underline{B}\,\underline{A}) + b\,\underline{B} \,, \tag{11}$$

the first of which can be solved for $\underline{B}$, yielding

$$\underline{B} = \left(\dot{\underline{A}}/a - \underline{A}^2 - b\,\underline{A}/a\right)^{1/2} . \tag{12}$$

The matrix ODE (7) is then easily obtained, via the identification $\underline{U}(t)=\underline{A}(t)$, by time-differentiating the first, and using the second, of the (11), as well as (12).

Exercise 5.4.3-4. Verify, and discuss the solution of the initial-value problem for the matrix ODE (7).

Exercise 5.4.3-5. Ponder on the relation of the technique used above, see (10), to the second *multiplication* technique discussed in Sect. 5.3.

Next, we report another *integrable* first-order matrix evolution equation of Toda ("nearest neighbor") type, which is the simplest instance of a class of *integrable* nonlinear matrix evolution equations <BRL81>. It reads (see eq. (5.1) of <BRL81>)

$$\dot{\underline{A}}_n = c\left\{\underline{A}_{n-1}\,\underline{A}_n - \underline{A}_n\,\underline{A}_{n+1}\right\}, \tag{13a}$$

and it yields, by positing

$$\underline{A}_n = \dot{\underline{U}}_n\left[\underline{U}_n\right]^{-1},\quad \dot{\underline{U}}_n = \underline{A}_n\,\underline{U}_n\,, \tag{13b}$$

the second-order *linearizable* matrix evolution equation

$$\ddot{\underline{U}}_n = \dot{\underline{U}}_n\left[\underline{U}_n\right]^{-1}\dot{\underline{U}}_n + c\left\{\,\dot{\underline{U}}_{n-1}\left[\underline{U}_{n-1}\right]^{-1}\underline{U}_n - \underline{U}_n\left[\underline{U}_n\right]^{-1}\dot{\underline{U}}_{n+1}\left[\underline{U}_{n+1}\right]^{-1}\underline{U}_n\,\right\}. \tag{13c}$$

Exercise 5.4.3-6. Verify!

This *second-order* matrix ODE, (13c), is *linearizable* because, to solve it, one must solve firstly the *integrable* ODE (13a) and then a *linear nonautonomous* matrix ODE (see the second of the (13b)).

The structure of this evolution equation, (13c), entails that, if $\underline{U}_n(t)$ satisfies it, then $\underline{\tilde{U}}_n(t)=\underline{U}_n(\tau)$ with $\tau=\left[\exp(a\,t)-1\right]/a$ satisfies the more general evolution equation

$$\ddot{\underline{\tilde{U}}}_n = a\dot{\underline{\tilde{U}}}_n + \dot{\underline{\tilde{U}}}_n\left[\underline{\tilde{U}}_n\right]^{-1}\dot{\underline{\tilde{U}}}_n + c\left\{\,\dot{\underline{\tilde{U}}}_{n-1}\left[\underline{\tilde{U}}_{n-1}\right]^{-1}\dot{\underline{\tilde{U}}}_n - \dot{\underline{\tilde{U}}}_n\left[\underline{\tilde{U}}_n\right]^{-1}\dot{\underline{\tilde{U}}}_{n+1}\left[\underline{\tilde{U}}_{n+1}\right]^{-1}\underline{\tilde{U}}_n\,\right\}. \tag{13d}$$

Exercise 5.4.3-7. Verify!

This suggests that, if $a=\pm i\omega,\ \omega>0$, the generic solution of this (complex) second order ODE, (13d), will feature some properties of *periodicity*.

Exercise 5.4.3-8. Show that the following second-order matrix ODE (somewhat analogous to (7)) is *integrable*:

$$\ddot{\underline{U}} = 2c\left\{c\underline{U}^3 + \left[c\underline{U} + \dot{\underline{U}}\right]^{1/2} \underline{U}\left[c\underline{U} + \dot{\underline{U}}\right]^{1/2}\right\} . \tag{14}$$

Hint: start from the *integrable equation* (13a), in the special ("periodic") case with $\underline{A}_{n\pm3} = \underline{A}_n$, and perform the following steps: *(i)* consider the special case of this integrable equation characterized by the additional restriction $\underline{A}_0 + \underline{A}_1 + \underline{A}_2 = 0$, whose compatibility with (13a) is easily verified; *(ii)* use this restriction to eliminate, say, $\underline{A}_2$; *(iii)* set $\underline{A}_0 + \underline{A}_1 = \underline{S}$, $\underline{A}_0 - \underline{A}_1 = \underline{D}$; *(iv)* express $\underline{S}$ via $\underline{D}$ and $\dot{\underline{D}}$, thereby obtaining a second-order ODE for $\underline{D}$ that does not contain $\underline{S}$; *(v)* finally make the identification $\underline{D} = -2\underline{U}$.

Next, we report the *solvable* system of matrix evolution ODEs

$$\ddot{\underline{U}}_n = \sum_{m=1}^{N} \left\{ (\tilde{d}_{nm} a_m \alpha_m + b_{nm})\, \dot{\underline{U}}_m + \tilde{d}_{nm} c_m \left[\dot{\underline{U}}_m, \underline{V}_m\right] + \tilde{d}_{nm} c_m \alpha_m \left[\underline{U}_m, \dot{\underline{U}}_m\right] \right\}, \tag{15a}$$

where the matrices $\underline{V}_n$ are obtained, in terms of the matrices $\underline{U}_m$ and their time-derivatives $\dot{\underline{U}}_m$, by solving the (linear, non differential) matrix equations

$$a_n \underline{V}_n + c_n \left[\underline{U}_n, \underline{V}_n\right] = \sum_{m=1}^{N} \left\{ d_{nm} \left[\dot{\underline{U}}_m - \sum_{l=1}^{N} b_{ml} \underline{U}_l\right]\right\} . \tag{15b}$$

Here the $N(3+2N)$ constants $a_n, \alpha_n, c_n, \tilde{d}_{nm}, b_{nm}$ are essentially arbitrary, while the N^2 constants d_{nm} are the matrix elements of the matrix $\underline{D}$ which is the inverse of the matrix $\underline{\tilde{D}}$ having matrix elements $\tilde{d}_{nm}$ (we are of course assuming this matrix to be invertible).

To demonstrate the solvability of this system of matrix ODEs, (15), we start from the system of matrix ODEs

$$\dot{\underline{U}}_n = \sum_{m=1}^{N} \left\{ a_{nm} \underline{V}_m + b_{nm} \underline{U}_m + c_{nm} \left[\underline{U}_m, \underline{V}_m\right]\right\} , \tag{16a}$$

$$\dot{\underline{V}}_n = \alpha_n \dot{\underline{U}}_n . \tag{16b}$$

For our purposes it is sufficient to assume the $N(3N+1)$ quantities $a_{nm}, b_{nm}, c_{nm}, \alpha_n$ to be *time-independent* "coupling constants," although the more general case in which they are given functions of the time t could be easily treated as well.

Exercise 5.4.3-9. Treat this more general case!

We now obtain from these ODEs, (16), the *second-order* evolution equations satisfied by the N matrices $\underline{U}_n \equiv \underline{U}_n(t)$, and show that they coincide (indeed, in a special case) with (15); and then we show that the ODEs (16) are *solvable*.

Time-differentiation of (16a) yields, using (16b),

$$\ddot{\underline{U}}_n = \sum_{m=1}^{N} \left\{ \lambda_{nm} \dot{\underline{U}}_m + c_{nm} \left[\dot{\underline{U}}_m, \underline{V}_m \right] + c_{nm} \, \alpha_m \left[\underline{U}_m, \dot{\underline{U}}_m \right] \right\} , \tag{17a}$$

where we have introduced the convenient notation

$$\lambda_{nm} = a_{nm} \, \alpha_m + b_{nm} \, . \tag{17b}$$

In (17a), the matrices $\underline{V}_n$ are supposed to be expressed in terms of $\underline{U}_m$ and $\dot{\underline{U}}_m$ by solving the (non differential) equations (16a), not the (differential) equations (16b). This can always be done algebraically (up to obvious restrictions, see below), but hereafter we restrict attention to the simpler case characterized by the restriction

$$a_{nm} = \tilde{d}_{nm} \, a_m, \quad c_{nm} = \tilde{d}_{nm} \, c_m \, , \tag{18a}$$

which expresses the $2N^2$ constants a_{nm}, c_{nm} in terms of the $N(N+2)$ (arbitrary) constants $\tilde{d}_{nm}, a_n, c_n$. We moreover assume that the ($N \times N$)-matrix $\underline{\tilde{D}}$, with matrix elements $\tilde{d}_{nm}$, is *invertible*, and we term d_{nm} the matrix elements of the inverse matrix $\underline{\tilde{D}}^{-1} \equiv \underline{D}$:

$$\tilde{d}_{nm} \equiv (\underline{\tilde{D}})_{nm}, \quad d_{nm} \equiv (\underline{\tilde{D}}^{-1})_{nm} \equiv (\underline{D})_{nm} \, . \tag{18b}$$

Then clearly from (16a) we get (15b). The solution of this linear algebraic equation, (15b), is of course, in principle, a trivial task; for a discussion of a methodology to obtain it explicitly in matrix form the interested reader is referred to <BR83>.

Let us now show how to solve (17), or rather, equivalently, (16). From (16b) we get

$$\underline{V}_n(t) = \alpha_n \, \underline{U}_n(t) + \underline{C}_n \, , \tag{19a}$$

with the *constant* matrices $\underline{C}_n$ given in terms of the initial data as follows:

$$\underline{C}_n = \underline{V}_n(0) - \alpha_n \, \underline{U}_n(0) \, . \tag{19b}$$

Insertion of (19a) in (16a) yields the following set of *linear ODEs with constant coefficients* for the matrices $\underline{U}_n(t)$, which can of course be explicitly solved by purely algebraic operations:

$$\dot{\underline{U}}_n = \sum_{m=1}^{N} \{ a_{nm} [\alpha_m \underline{U}_m + \underline{C}_m] + b_{nm} \underline{U}_m + c_{nm} [\underline{U}_m, \underline{C}_m] \} . \tag{20}$$

Next, we report the first-order *integrable* "Nahm equations":

$$\dot{\underline{M}}_n = c [\underline{M}_{n+1}, \underline{M}_{n+2}], \quad n = 1,2,3 \bmod(3) , \tag{21}$$

where the constant c could of course be rescaled away (see below). A simple way to obtain from these equations a set of 3 coupled *linearizable* second-order matrix ODEs is by setting

$$c \underline{M}_n = \mu_n \dot{\underline{U}}_n + \sum_{m=1}^{3} a_{n,m} \underline{U}_m, \quad n = 1,2,3 \bmod(3) , \tag{22}$$

which transform (21) into

$$\mu_n \ddot{\underline{U}}_n = -\sum_{m=1}^{3} a_{n,m} \dot{\underline{U}}_m + \left[\mu_{n+1} \dot{\underline{U}}_{n+1} + \sum_{m=1}^{3} a_{n+1,m} \underline{U}_m, \ \mu_{n+2} \dot{\underline{U}}_{n+2} + \sum_{m=1}^{3} a_{n+2,m} \underline{U}_m \right],$$

$$n = 1,2,3 \bmod(3) . \tag{23}$$

Clearly we are assuming here, for simplicity, that the 3 quantities μ_n, as well as the 9 quantities $a_{n,m}$, are (arbitrary) constants. These equations, (25), are categorized as *linearizable*, since to solve them one must first solve the *integrable* ODEs (21) for $\underline{M}_n(t)$ and then the *linear* (generally nonautonomous) ODEs (22) for $\underline{U}_n(t)$.

Finally, let us outline some other (well known) techniques to manufacture *solvable* and *linearizable* matrix ODEs. Most of these are systems of matrix ODEs of "nearest-neighbor" type; as we generally do in this book, we ignore in these contexts the question of the boundary conditions to be assigned at the extremal values of n (say, for $n = 0$ and $n = N+1$). But before delving in the derivation let us display two *solvable* and one *linearizable* matrix evolution ODEs which are yielded by these developments (see the fine print treatment below).

Solvable matrix ODEs:

$$\ddot{\underline{U}}=c+a\alpha\underline{U}+(\alpha-a)\dot{\underline{U}}+b\alpha\underline{U}^2-2b\dot{\underline{U}}\underline{U}\ ; \tag{24}$$

$$\ddot{\underline{U}}_n=(a_n-a_{n+1})\tilde{c}_n+(a_{n+1}-a_n)(a_n-\tilde{a}_n)\underline{U}_n+\tilde{c}_n(b_n\underline{U}_n-b_{n+1}\underline{U}_{n+1})$$

$$+(a_n-\tilde{a}_n)b_{n+1}\underline{U}_{n+1}\underline{U}_n-2b_n\dot{\underline{U}}_n\underline{U}_n$$

$$+(\tilde{a}_n+a_{n+1}-2a_n+b_{n+1}\underline{U}_{n+1}-b_n\underline{U}_n)(\dot{\underline{U}}_n+b_n\underline{U}_n^2). \tag{25}$$

Linearizable matrix ODE:

$$\ddot{\underline{M}}_n=(a_{n+1}-a_n)\dot{\underline{M}}_n+\{(1-b_n)\dot{\underline{M}}_n[\underline{M}_n]^{-1}+b_{n+1}\dot{\underline{M}}_{n+1}[\underline{M}_{n+1}]^{-1}\}\dot{\underline{M}}_n$$
$$+c_{n+1}\underline{M}_n-c_n\dot{\underline{M}}_n[\underline{M}_n]^{-1}\underline{M}_{n-1}[\dot{\underline{M}}_{n-1}]^{-1}\underline{M}_n\ . \tag{26}$$

Let us take as starting point the following *solvable* linear matrix evolution equation with constant (time-independent, matrix) coefficients:

$$\dot{\underline{W}}_n(t)=\underline{A}_n\underline{W}_n(t)+\underline{B}_n\underline{W}_{n+1}(t)+\underline{C}_n\underline{W}_{n-1}(t), \tag{27}$$

and let us set

$$\underline{V}_n=\underline{W}_{n+1}\underline{W}_n^{-1} \tag{28}$$

(here and below we often omit, for notational simplicity, the explicit indication of the time dependence -- as we generally did above). Then the matrix $\underline{V}_n(t)$ evolves according to the nonlinear equation

$$\dot{\underline{V}}_n=\underline{A}_{n+1}\underline{V}_n-\underline{V}_n\underline{A}_n+\underline{B}_{n+1}\underline{V}_{n+1}\underline{V}_n-\underline{V}_n\underline{B}_n\underline{V}_n+\underline{C}_{n+1}-\underline{V}_n\underline{C}_n[\underline{V}_{n-1}]^{-1}. \tag{29}$$

The special case of this equation with $\underline{A}_n=\underline{A}$, $\underline{B}_n=\underline{B}$ and $\underline{C}_n=0$ is the first nontrivial evolution equation of the so-called discrete Burger's hierarchy <LRB83>.

There are now various ways to derive, from this first-order *solvable* matrix evolution equation, *solvable* or *linearizable* second-order matrix evolution equations. We describe two of them.

A first trick is to separate the odd/even labeled matrices, by setting, say, $\underline{V}_{2m}=\underline{U}_m$, $\underline{V}_{2m+1}=\tilde{\underline{U}}_m$, $\underline{A}_{2m}=\hat{\underline{A}}_m$, $\underline{A}_{2m+1}=\tilde{\underline{A}}_m$, and so on. This yields

$$\dot{\underline{U}}_n=\tilde{\underline{A}}_n\underline{U}_n-\underline{U}_n\hat{\underline{A}}_n+\tilde{\underline{B}}_n\tilde{\underline{U}}_n\underline{U}_n-\underline{U}_n\hat{\underline{B}}_n\underline{U}_n+\tilde{\underline{C}}_n-\underline{U}_n\hat{\underline{C}}_n[\tilde{\underline{U}}_{n-1}]^{-1}, \tag{30a}$$

$$\dot{\underline{\tilde{U}}}_n = \underline{\hat{A}}_{n+1}\underline{\tilde{U}}_n - \underline{\tilde{U}}_n\underline{\tilde{A}}_n + \underline{\hat{B}}_{n+1}\underline{U}_{n+1}\underline{\tilde{U}}_n - \underline{\tilde{U}}_n\underline{\tilde{B}}_n\underline{\tilde{U}}_n + \underline{\hat{C}}_{n+1} - \underline{\tilde{U}}_n\underline{\tilde{C}}_n\left[\underline{U}_n\right]^{-1}. \qquad (30b)$$

We then time-differentiate the first of these two equations, use the second to eliminate $\dot{\underline{\tilde{U}}}_n$, and use the first (undifferentiated) to eliminate $\underline{\tilde{U}}_n$ after having set, for simplicity's sake, $\underline{\hat{C}}_n = 0$.

Exercise 5.4.3-10. Treat the more general case!

We thus get:

$$\ddot{\underline{U}}_n = \underline{\tilde{A}}_n\dot{\underline{U}}_n - \dot{\underline{U}}_n\underline{\hat{A}}_n - \dot{\underline{U}}_n\underline{\hat{B}}_n\underline{U}_n - \underline{U}_n\underline{\hat{B}}_n\dot{\underline{U}}_n$$

$$+\underline{\tilde{B}}_n\left[\underline{\hat{A}}_{n+1} + \underline{\hat{B}}_{n+1}\underline{U}_{n+1}\right]\left[\underline{\tilde{B}}_n\right]^{-1}\left[\dot{\underline{U}}_n - \underline{\tilde{A}}_n\underline{U}_n + \underline{U}_n\underline{\hat{A}}_n + \underline{U}_n\underline{\hat{B}}_n\underline{U}_n - \underline{\tilde{C}}_n\right]$$

$$-\left[\dot{\underline{U}}_n - \underline{\tilde{A}}_n\underline{U}_n + \underline{U}_n\underline{\hat{A}}_n + \underline{U}_n\underline{\hat{B}}_n\underline{U}_n - \underline{\tilde{C}}_n\right]\left[\underline{\hat{A}}_n + \underline{\hat{B}}_n\underline{U}_n\right]. \qquad (31)$$

We now restrict attention to the simpler case in which the constant matrices are replaced by scalars, namely we set $\underline{A}_n = a_n\underline{1}$, $\underline{\tilde{A}}_n = \tilde{a}_n\underline{1}$, and so on. This yields (25). A special case of this *solvable* system of matrix evolution ODEs, (25), corresponds to the assignment (which is easily seen to be consistent with this system of evolution matrix ODEs, (25)) $\underline{U}_n = \underline{U}$, $\tilde{c}_n = -c/\alpha$, $b_n = b$, $a_n = \tilde{a} + \alpha\, n$, $\tilde{a}_n = \tilde{a} - a + \alpha\, n$. This yields (24).

Exercise 5.4.3-11. Provide a more straightforward derivation of (24).

A second method to obtain a *second-order* equation from the *solvable first-order* equation (29) is by setting

$$\underline{V}_n(t) = \dot{\underline{M}}_n(t)\left[\underline{M}_n(t)\right]^{-1},\ \dot{\underline{M}}_n(t) = \underline{V}_n(t)\underline{M}_n(t). \qquad (32)$$

One obtains thereby the following evolution equation for the matrix $\underline{M}_n(t)$:

$$\ddot{\underline{M}}_n = \dot{\underline{M}}_n\left[\underline{M}_n\right]^{-1}\dot{\underline{M}}_n + \underline{A}_{n+1}\dot{\underline{M}}_n - \dot{\underline{M}}_n\left[\underline{M}_n\right]^{-1}\underline{A}_n\underline{M}_n$$

$$+\underline{B}_{n+1}\dot{\underline{M}}_{n+1}\left[\underline{M}_{n+1}\right]^{-1}\dot{\underline{M}}_n - \dot{\underline{M}}_n\left[\underline{M}_n\right]^{-1}\underline{B}_n\dot{\underline{M}}_n$$

$$+\underline{C}_{n+1}\underline{M}_n - \dot{\underline{M}}_n\left[\underline{M}_n\right]^{-1}\underline{C}_n\underline{M}_{n-1}\left[\dot{\underline{M}}_{n-1}\right]^{-1}\underline{M}_n. \qquad (33)$$

This should be categorized as a *linearizable* system of matrix evolution equations, since to solve it one must firstly solve the system of *linear constant-coefficient* (hence *solvable*) matrix ODEs (27) (to get $\underline{W}_n(t)$ and then, via (28), $\underline{V}_n(t)$), and then a *linear nonautonomous* matrix evolution equation (to get $\underline{M}_n(t)$; see the second of the (32)).

Remark 5.4.3-12. Note the presence, in this equation (33), of a time-differentiated matrix *in the denominator* (see the last term in the right-hand-side).

Remark 5.4.3-13. As an obvious consequence of the way this equation, (33), has been derived, see (32), it is invariant under the transformation $\underline{M}_n \rightarrow \underline{M}_n \underline{D}_n$ with $\underline{D}_n$ arbitrary *constant* matrices, $\dot{\underline{D}}_n = 0$.

When all the constant matrices are replaced by scalars, namely if we set in (33) $\underline{A}_n = a_n \underline{1}$, $\underline{B}_n = b_n \underline{1}$, $\underline{C}_n = c_n \underline{1}$, we obtain (26).

Remark 5.4.3-14. The special case of (26) which obtains by setting $\underline{M}_n(t) = \underline{M}(t)$, $a_n = 2an$, $b_n = (c-1)n$, $c_n = nb$ (an assignment which is easily seen to be compatible with this evolution equation, (26)) yields for the matrix $\underline{M}(t)$ the *solvable* evolution equation (5.1-1).

Exercise 5.4.3-15. Do an analogous treatment to that given above, based on (27) with (28), but replacing (27) with the following *solvable* system of ODEs:

$$\dot{\underline{W}}_n(t) = \underline{A}_n \underline{W}_n(t) + \underline{B}_n \underline{W}_{n+1}(t) + \underline{C}_n \underline{W}_{n+2}(t). \tag{34}$$

Hint: show that (34) with (28) yield

$$\dot{\underline{V}}_n(t) = \underline{A}_{n+1} \underline{V}_n(t) - \underline{V}_n(t) \underline{A}_n + \underline{B}_{n+1} \underline{V}_{n+1}(t) \underline{V}_n(t) - \underline{V}_n(t) \underline{B}_n \underline{V}_n(t)$$

$$+ \underline{C}_{n+1} \underline{V}_{n+2}(t) \underline{V}_{n+1}(t) \underline{V}_n(t) - \underline{V}_n(t) \underline{C}_n \underline{V}_{n+1}(t) \underline{V}_n(t). \tag{35}$$

5.4.4 On the integrability of the matrix evolution equation $\ddot{\underline{U}} = \underline{f}(\underline{U})$

In Sect. 5.4.4 we discuss some properties of the matrix evolution equation

$$\ddot{\underline{U}} = \underline{f}(\underline{U}), \tag{1}$$

in particular we show that the following two matrix evolution ODEs are *integrable* indeed *solvable*:

$$\ddot{\underline{U}} = \underline{U}^2 + c\underline{1}, \tag{2}$$

$$\ddot{\underline{U}} = 2\underline{U}^3 + \underline{C}\,\underline{U} + \underline{U}\,\underline{C}. \tag{3}$$

Here and below c is an arbitrary scalar constant, $\underline{C}$ an arbitrary scalar matrix, and of course underlined letters denote matrices (generally

($N\times N$)-matrices; but sometimes also larger matrices, as it will be clear from the context).

Remark 5.4.4-1. The *first-order* matrix evolution equation

$$\dot{\underline{U}}=\underline{U}^2+\underline{C} \tag{4}$$

is *integrable* indeed *solvable* (see *Exercise 5.4.2-7*). Time-differentiation of (4) yields (3) (see *Exercise 5.4.2-8*).

Remark 5.4.4-2. This argument (see the preceding *Remark 5.4.4-1*) demonstrates that the *second-order* matrix evolution ODE (3) is *partially solvable*: indeed, any solution of (4) yields a solution of (3); but these solutions are only a subclass of the solutions of (3) (in terms of the initial-value problem, for these solutions one can assign arbitrarily the initial value of $\underline{U}(t)$, say $\underline{U}(0)$, but not the initial value of $\dot{\underline{U}}(t)$, say $\dot{\underline{U}}(0)$, which is fixed by (4), namely $\dot{\underline{U}}(0)=[\underline{U}(0)]^2+\underline{C}$).

The fact that (3) is *partially solvable (*see *Remark 5.4.4-2*) does not entail that it is *integrable*. But *integrable* (indeed *solvable*) it is. We demonstrated this in Sect. 5.4.2, by showing that this matrix evolution ODE, (3), is related to a special (periodic) case of the so-called non-Abelian Toda lattice, a matrix evolution system which was solved two decades ago (see Sect. 5.N). Below we also provide Lax pairs which correspond to (3). This *integrable* matrix evolution ODE, (3), plays a key role in identifying *integrable* systems of quartic oscillators (see Sect. 5.6.5).

Before reporting some Lax pairs for (3), let us interject some other *remarks*.

Remark 5.4.4-3. For $N=1$, namely in the *scalar* case, the solution of (4) reads

$$U(t)=C^{1/2}\tan\left(C^{1/2}\,t\right). \tag{5a}$$

This solution is the "separatrix" among the solutions of (3), whose general solution reads

$$U(t)=A\,\mathrm{sn}(\lambda t+a,k),\ A^2=-(\lambda^2+2C),\ k^2=-(1+2C/\lambda^2), \tag{5b}$$

where λ and a are two arbitrary constants (to be determined by the initial conditions) and the function $\mathrm{sn}(u,k)$ is the Jacobian elliptic function (see Appendix A).

Remark 5.4.4-4. The factor 2 in the right hand side of (3) could of course be eliminated by rescaling the dependent variable, but we shall not do so, as the "canonical" form (3) (see also (5.4.2-6)) seems the most convenient one to work with. Of course a generalized version of (3), and of (2) as well, can be obtained by setting, say

$$\underline{U}(t)=\underline{K}_1\,\underline{\tilde{U}}(t)\,\underline{K}_2+\underline{K}\;, \tag{6}$$

with $\underline{K}_1,\underline{K}_2,\underline{K}$ 3 constant matrices (*a priori* arbitrary, and which can be chosen at one's convenience), and by then looking at the evolution of $\underline{\tilde{U}}(t)$ rather than $\underline{U}(t)$.

Remark 5.4.4-5. The matrix evolution ODEs discussed in this paper are generally Hamiltonian. In particular it is easily seen that (3) obtains in the standard manner from the Hamiltonian function

$$h(\underline{U},\underline{P})=\text{trace}\left[\frac{1}{2}\underline{P}^2-\underline{U}\,\underline{C}\,\underline{U}-\frac{1}{2}\underline{U}^4\right], \tag{7a}$$

where the N^2 canonical variables u_{jk} are the matrix elements of the matrix $\underline{U}$, and the corresponding N^2 canonical momenta p_{jk} are the matrix elements of the matrix $\underline{P}$, so that the Hamiltonian equations entailed by (7a) read

$$\dot{u}_{jk}=\partial h/\partial p_{jk}=p_{kj}, \tag{7b}$$

$$\dot{p}_{kj}=-\partial h/\partial u_{kj}=\sum_{s=1}^{n}\left(c_{js}u_{sk}+u_{js}c_{sk}\right)+2\sum_{r,s=1}^{n}u_{jr}u_{rs}u_{sk}\;. \tag{7c}$$

Here of course c_{jk} are the N^2 elements of the $N\times N$ matrix $\underline{C}$. The fact that t-differentiation of (7b) yields (3) (via (7c)) is plain.

Exercise 5.4.4-6. Verify!

Let us reemphasize that the *integrability* indeed *solvability* of (3) has already been established via its connection with the periodic non Abelian Toda lattice, see Sects. 3.4.2 and 5.N. But we now back this result by exhibiting various Lax pairs, indeed various hierarchies of Lax pairs, such that the standard Lax equation (see Sect. 2.1),

$$\dot{\underline{L}}=[\underline{L},\underline{M}]\;, \tag{8}$$

correspond to (3).

The first reads

$$\underline{L}=\begin{pmatrix} \dot{\underline{U}} & \underline{U}^2+\underline{C} \\ -\left(\underline{U}^2+\underline{C}\right) & -\dot{\underline{U}} \end{pmatrix}, \quad \underline{M}=\begin{pmatrix} \underline{0} & \underline{U} \\ \underline{U} & \underline{0} \end{pmatrix}, \tag{9a}$$

as well as

$$\underline{L}=\begin{pmatrix} \dot{\underline{U}} & \underline{0} & \underline{0} & \underline{U}^2+\underline{C} \\ \underline{0} & -\dot{\underline{U}} & -\left(\underline{U}^2+\underline{C}\right) & \underline{0} \\ \underline{0} & \underline{U}^2+\underline{C} & \dot{\underline{U}} & \underline{0} \\ -\left(\underline{U}^2+\underline{C}\right) & \underline{0} & \underline{0} & -\dot{\underline{U}} \end{pmatrix}, \quad \underline{M}=\begin{pmatrix} \underline{0} & \underline{0} & \underline{0} & \underline{U} \\ \underline{0} & \underline{0} & \underline{U} & \underline{0} \\ \underline{0} & \underline{U} & \underline{0} & \underline{0} \\ \underline{U} & \underline{0} & \underline{0} & \underline{0} \end{pmatrix}, \tag{9b}$$

and so on. Here we are displaying block-matrices, namely matrices of matrices; hence the square matrices $\underline{L}$ and $\underline{M}$ in (9a) are of order $2N$, those in (9b) of order $2^2N=4N$. These matrices are the first two specimens of a sequence of square matrices of order 2^pN, $p=1,2,3,....$, whose structure is, we trust, self-evident from (9a,b). And it is as well plain that the insertion of (9) into (8) yields (3).

Exercise 5.4.4-7. Verify!

Another sequence of Lax pairs that also yield (3) reads as follows:

$$\underline{L}=\begin{pmatrix} \underline{U} & \underline{1}-\left(\underline{U}^2+\underline{C}+\dot{\underline{U}}\right)/2 \\ \underline{1}-\left(\underline{U}^2+\underline{C}-\dot{\underline{U}}\right)/2 & -\underline{U} \end{pmatrix},$$

$$\underline{M}=\begin{pmatrix} \underline{0} & -\left(\underline{U}^2+\underline{C}+\dot{\underline{U}}\right)/2 \\ -\left(\underline{U}^2+\underline{C}+\dot{\underline{U}}\right)/2 & \underline{0} \end{pmatrix}; \tag{10a}$$

$$\underline{L}=\begin{pmatrix} \underline{U} & -\left(\underline{U}^2+\underline{C}+\dot{\underline{U}}\right)/2 & \underline{0} & \underline{1} \\ \underline{1} & -\underline{U} & -\left(\underline{U}^2+\underline{C}-\dot{\underline{U}}\right)/2 & \underline{0} \\ \underline{0} & \underline{1} & \underline{U} & -\left(\underline{U}^2+\underline{C}+\dot{\underline{U}}\right)/2 \\ -\left(\underline{U}^2+\underline{C}-\dot{\underline{U}}\right)/2 & \underline{0} & \underline{1} & -\underline{U} \end{pmatrix},$$

$$\underline{M}=\begin{pmatrix} \underline{0} & -\left(\underline{U}^2+\underline{C}+\dot{\underline{U}}\right)/2 & \underline{0} & \underline{0} \\ \underline{0} & \underline{0} & -\left(\underline{U}^2+\underline{C}+\dot{\underline{U}}\right)/2 & \underline{0} \\ \underline{0} & \underline{0} & \underline{0} & -\left(\underline{U}^2+\underline{C}+\dot{\underline{U}}\right)/2 \\ -\left(\underline{U}^2+\underline{C}+\dot{\underline{U}}\right)/2 & \underline{0} & \underline{0} & \underline{0} \end{pmatrix}, \tag{10b}$$

and so on.

Exercise 5.4.4-8. Verify!

Let us now discuss tersely the second-order matrix evolution ODE (1), assuming $\underline{f}(\underline{U})$ to be an arbitrary function of the matrix U and of no other matrix ($\underline{f}(\underline{U})$ can of course depend on an arbitrary number of *scalar* coefficients), so that

$$[\ \underline{U}, \underline{f}(\underline{U})\ ] = 0\ . \tag{11}$$

Then (1) entails that the commutator

$$\underline{\overline{C}} = [\ \underline{U}(t), \dot{\underline{U}}(t)\ ] \tag{12}$$

is constant:

$$\dot{\underline{\overline{C}}} = \underline{0}\ . \tag{13}$$

Exercise 5.4.4-9. Verify!

Remark 5.4.4-10. This property, (13), is independent of the functional form of $\underline{f}(\underline{U})$ (as long as (11) holds), and for its validity it is *not* required that (1) be autonomous, namely that $\underline{f}(\underline{U})$ *not* depend explicitly on the time t.

Remark 5.4.4-11. This property, (13), generally yields $N^2 - N$ (scalar) constants of motion; but of course N^2 are needed for the *complete integrability* of the ($N \times N$)-matrix evolution equation (1).

Indeed the definition (12) of the ($N \times N$)-matrix $\underline{\overline{C}}$ clearly entails that

$$\text{trace}[\ \underline{U}^p \underline{\overline{C}}\ ] = 0, p = 0,1,2,\ldots\ . \tag{14}$$

And these conditions, of which of course only N are independent, reduce the number of (scalar) conserved quantities entailed by the time-independence of the ($N \times N$)-matrix $\underline{\overline{C}}$, see (13), from N^2 (the number of matrix elements of $\underline{\overline{C}}$, see (12)) to $N^2 - N$.

Let us end our discussion of the matrix evolution equation (1) (with (11)) by exhibiting a Lax pair associated with it, which however does not yield additional constants of motion. It reads:

$$\underline{L}=\begin{pmatrix}\dot{\underline{U}}\,\underline{U} & \dot{\underline{U}}^2 \\ -\dot{\underline{U}}^2 & -\dot{\underline{U}}\,\underline{U}\end{pmatrix}, \quad \underline{M}=\begin{pmatrix}0 & \underline{U}^{-1}\underline{f}(\underline{U}) \\ \underline{1} & 0\end{pmatrix}. \tag{15}$$

Note that this matrix $\underline{L}$ is independent of the function $\underline{f}(\underline{U})$. Consistently with this fact, this Lax pair does not yield any additional constant of motion besides those entailed by the time independence of the $(N\times N)$-matrix $\overline{\underline{C}}$, see (12): indeed, the traces of the powers of this matrix $\underline{L}$, which are the (scalar) constants of motion implied by the Lax equation (8), reproduce the series of constants of motion entailed by the time-independence of the traces of the powers of the (time-independent!) matrix $\overline{\underline{C}}$, see (12) and (13).

Exercise 5.4.4-12. Verify!

Finally let us focus on the simplest nontrivial instance of (1), namely on (2). In this case we provide another Lax pair (additional to those exhibited above), which yields the N additional (scalar) constants of motion required to reach the total number of constants needed for *complete integrability*. Hence we conclude that (2), as well as (3), is *completely integrable.*

The additional Lax pair reads as follows:

$$\underline{L}=\begin{pmatrix}\dot{\underline{U}} & \lambda(\underline{U}^2+3c\underline{1}) \\ -2\underline{U}/(3\lambda) & -\dot{\underline{U}}\end{pmatrix}, \quad \underline{M}=\begin{pmatrix}\underline{0} & \lambda\underline{U} \\ \underline{1}/(3\lambda) & \underline{0}\end{pmatrix}. \tag{16}$$

Exercise 5.4.4-13. Verify that the Lax equation (8) with this Lax pair, (16) (which features the arbitrary constant parameter λ), corresponds to (2), that the constants of motion given by the traces of the powers of this Lax matrix $\underline{L}$, see (16), are different from those yielded by the constancy of the matrix $\overline{\underline{C}}$, see (12) and (13), and that one can thereby get all the N^2 scalar constants required for the complete integrability of the matrix evolution ODE (2).

We complete Sect. 5.4.4 with some simple but interesting results and conjectures.

Exercise 5.4.4-14. Show that the following 2 systems of 2 coupled matrix ODEs,

$$\ddot{\underline{U}} = 6(\lambda^2 - \omega^2)\underline{U} - 12\lambda\omega\underline{V} + 5\lambda\dot{\underline{U}} - 5\omega\dot{\underline{V}} + \underline{U}^2 - \underline{V}^2, \tag{17a}$$

$$\ddot{\underline{V}} = 6(\lambda^2 - \omega^2)\underline{V} + 12\lambda\omega\underline{U} + 5\lambda\dot{\underline{V}} + 5\omega\dot{\underline{U}} + \underline{U}\,\underline{V} + \underline{V}\,\underline{U}, \tag{17b}$$

respectively

$$\ddot{\underline{U}} = 2(\lambda^2 - \omega^2)\underline{U} - 4\lambda\omega\underline{V} + 3\lambda\dot{\underline{U}} - 3\omega\dot{\underline{V}} + 2(\underline{U}^3 - \underline{U}\,\underline{V}^2 - \underline{V}\,\underline{U}\,\underline{V} - \underline{V}^2\underline{U}), \tag{18a}$$

$$\ddot{\underline{V}} = 2(\lambda^2 - \omega^2)\underline{V} + 4\lambda\omega\underline{U} + 3\lambda\dot{\underline{V}} + 3\omega\dot{\underline{U}} + 2(-\underline{V}^3 + \underline{V}\,\underline{U}^2 + \underline{U}\,\underline{V}\,\underline{U} + \underline{U}^2\underline{V}), \tag{18b}$$

are as *integrable* as (2) (with $c = 0$) respectively as (3) (with $\underline{C} = \underline{0}$). *Hint*: firstly set, in (2) (with $c = 0$) respectively in (3) (with $\underline{C} = \underline{0}$),

$$\underline{W}(t) = \exp(p\eta t)\,\underline{U}(\tau), \quad \tau = [\exp(\eta t) - 1]/\eta, \tag{19}$$

with $p = 2$ respectively $p = 1$, then complexify:

$$\underline{W}(t) = \underline{U}(t) + i\,\underline{V}(t), \quad \eta = \lambda + i\omega. \tag{20}$$

Conjecture 5.4.4-15. If $\lambda = 0, \omega \neq 0$ *all* nonsingular solutions of the system of matrix ODEs (17) are *completely periodic.*

Conjecture 5.4.4-16. If $\lambda = 0, \omega \neq 0$ *all* nonsingular solutions of the system of matrix ODEs (18) are *completely periodic.*

5.5 Parametrization of matrices via three-vectors

Consistently with the strategy to identify treatable many-body problems in three-dimensional space outlined at the beginning of Chap. 5 and illustrated by the examples treated in Sects. 5.1 and 5.2, in Sect. 5.5 we review various convenient parametrizations of matrices in terms of 3-vectors (and, if need be, scalars). Obviously some of these results could be trivially extended to vectors of higher, or lower, dimensionality than 3; but here we prefer to focus (almost) exclusively on 3-dimensional (ordinary!) space. However, at the end of Sect. 5.5 we also outline a convenient parametrization in terms of vectors of arbitrary dimension S, which is particularly convenient to treat cases involving only certain "alternating" products of a (finite) *odd* number of matrices (we already introduced this technique in the latter part of Sect. 5.3, and we shall utilize it again in Sect. 5.6.5; in both these cases, the odd number of matrices in question is just 3).

We always denote matrices (whose rank will be specified on a case-by-case basis) by underlining their symbols, and 3-vectors by superimposed arrows.

Because of the structure of the matrix evolution equations of Sect. 5.4, we are particularly interested in parametrizations (of matrices in terms of one or more 3-vectors, and possibly of some scalars as well) which belong to one (or more) of the following three categories.

Definition 5.5-1. We term parametrizations of *type (i)*, for *invertible* matrices, those which are preserved under the operation $\underline{\tilde{M}}\,\underline{M}^{-1}\underline{\tilde{M}}$, namely are such that, if both $\underline{M}$ and $\underline{\tilde{M}}$ are so parametrized in terms of one or more 3-vectors, the combination $\underline{\tilde{M}}\,\underline{M}^{-1}\underline{\tilde{M}}$ admits the *same* parametrization in terms of (appropriately defined) 3-vectors.

Definition 5.5-2. We term parametrizations of *type (ii)* those which are preserved for commutators, namely are such that, if both $\underline{M}$ and $\underline{\tilde{M}}$ are so parametrized in terms of one or more 3-vectors, the commutator $\left[\underline{M},\underline{\tilde{M}}\right]\equiv\underline{M}\,\underline{\tilde{M}}-\underline{\tilde{M}}\,\underline{M}$ admits the *same* parametrization in terms of (appropriately defined) 3-vectors.

Definition 5.5-3. We term parametrizations of *type (iii)* those which are preserved under the product operation, namely are such that, if both $\underline{M}$ and $\underline{\tilde{M}}$ are so parametrized in terms of one or more 3-vectors, the product $\underline{M}\,\underline{\tilde{M}}$ admits the *same* parametrization in terms of (appropriately defined) 3-vectors.

Remark 5.5-4. Obviously parametrizations of *type (iii)* are also of *type (ii)*, and, for *invertible* matrices (the inverses of which preserve the same parametrization), of *type (i)* as well.

We indicate with the symbol $\doteq$ the one-to-one correspondence that the parametrization under consideration institutes among matrices and 3-vectors. For instance the most common parametrization we use, for (2×2)-matrices, reads

$$\underline{M} = \rho \underline{1} + i\vec{r}\cdot\vec{\underline{\sigma}} \ , \tag{1a}$$

where ρ is a scalar and the 3 matrices $\underline{\sigma}_x, \underline{\sigma}_y, \underline{\sigma}_z$ are the standard Pauli matrices,

$$\underline{\sigma}_x = \begin{pmatrix} 0 & 1 \\ 1 & 0 \end{pmatrix},\ \underline{\sigma}_y = \begin{pmatrix} 0 & i \\ -i & 0 \end{pmatrix},\ \underline{\sigma}_z = \begin{pmatrix} 1 & 0 \\ 0 & -1 \end{pmatrix}. \tag{2}$$

(Note that in the following the unit matrix $\underline{1} = \begin{pmatrix} 1 & 0 \\ 0 & 1 \end{pmatrix}$ is often omitted). So in this case, in correspondence to (1a), we write

$$\underline{M} \doteq (\rho, \vec{r}), \tag{1b}$$

and, via standard calculations (see Appendix H, where we report for convenience a number of standard formulas involving $\underline{\sigma}$-matrices), we also have

$$\underline{M}^{-1} \doteq (\rho, -\vec{r}) / (\rho^2 + r^2) \ , \tag{1c}$$

$$\underline{M}\,\tilde{\underline{M}} \doteq \left(\rho\,\tilde{\rho} - \vec{r}\cdot\tilde{\vec{r}}, \rho\tilde{\vec{r}} + \tilde{\rho}\vec{r} - \vec{r}\wedge\tilde{\vec{r}}\right) \ , \tag{1d}$$

$$\tilde{\underline{M}}\,\underline{M}^{-1}\tilde{\tilde{\underline{M}}} \doteq (\rho^2 + r^2)^{-1} \left(\tilde{\rho}\rho\tilde{\tilde{\rho}} + \tilde{\rho}(\vec{r}\cdot\tilde{\tilde{\vec{r}}}) - \rho(\tilde{\vec{r}}\cdot\tilde{\tilde{\vec{r}}}) + \tilde{\tilde{\rho}}(\vec{r}\cdot\tilde{\vec{r}}) + \vec{r}\cdot\left(\tilde{\vec{r}}\wedge\tilde{\tilde{\vec{r}}}\right), \right.$$

$$\left. \tilde{\vec{r}}\left[\rho\tilde{\tilde{\rho}} + (\vec{r}\cdot\tilde{\tilde{\vec{r}}})\right] - \vec{r}\left[\tilde{\rho}\tilde{\tilde{\rho}} + (\tilde{\vec{r}}\cdot\tilde{\tilde{\vec{r}}})\right] + \tilde{\tilde{\vec{r}}}\left[\tilde{\rho}\rho + (\tilde{\vec{r}}\cdot\vec{r})\right] + \tilde{\rho}\vec{r}\wedge\tilde{\tilde{\vec{r}}} - \rho\tilde{\vec{r}}\wedge\tilde{\tilde{\vec{r}}} + \tilde{\tilde{\rho}}\tilde{\vec{r}}\wedge\vec{r} \ \right). \tag{1e}$$

The formula (1d) shows that this parametrization (which was already introduced in Sect. 5.1) belongs to *type (iii)*.

Exercise 5.5-5. Verify these formulas, (1c,d,e). *Hint*: see *Exercise 5.1-13*.

Next, let us restrict consideration to the class of *traceless* (2×2)-matrices, that admit the parametrization

$$\underline{M}=i\,\vec{r}\cdot\underline{\vec{\sigma}} \tag{3a}$$

(namely to the special case of (1) with $\rho=0$). The formulas written above entail that this parametrization belongs both to *type (i)* and to *type (ii)*, but not to *type (iii)*. The relevant formulas read:

$$\underline{M}\doteq\vec{r}\ , \tag{3b}$$

$$\underline{M}^{-1}\doteq-\vec{r}\,/\,r^{2}\ , \tag{3c}$$

$$\underline{\tilde{M}}\,\underline{M}^{-1}\,\underline{\tilde{M}}\doteq\left[2\tilde{\vec{r}}\,(\tilde{\vec{r}}\cdot\vec{r})-\vec{r}\,(\tilde{\vec{r}}\cdot\tilde{\vec{r}})\right]/\,r^{2}\ , \tag{3d}$$

$$\left[\underline{M},\underline{\tilde{M}}\right]\doteq-2\,\vec{r}\wedge\tilde{\vec{r}}\ . \tag{3e}$$

Exercise 5.5-6. Verify these formulas, (3c,d,e).

The next parametrization we consider is, in terms of 2 three-vectors, for (invertible) antisymmetrical (4×4)-matrices. It reads

$$\underline{M}=\begin{pmatrix} 0 & x^{(1)} & y^{(1)} & z^{(1)} \\ -x^{(1)} & 0 & z^{(2)} & -y^{(2)} \\ -y^{(1)} & -z^{(2)} & 0 & x^{(2)} \\ -z^{(1)} & y^{(2)} & -x^{(2)} & 0 \end{pmatrix}, \tag{4a}$$

which entails (in self-evident notation)

$$\underline{M}\doteq\left(\vec{r}^{(1)},\vec{r}^{(2)}\right). \tag{4b}$$

It is then easy to verify that

$$\underline{M}^{-1}\doteq-\left(\vec{r}^{(2)},\,\vec{r}^{(1)}\right)/\,(\vec{r}^{(1)}\cdot\vec{r}^{(2)})\ , \tag{4c}$$

$$\underline{\tilde{M}}\;\underline{M}^{-1}\,\underline{\tilde{M}}\doteq\Big(\ \tilde{\vec{r}}^{(1)}\left[(\tilde{\vec{r}}^{(1)}\cdot\vec{r}^{(2)})+(\tilde{\vec{r}}^{(2)}\cdot\vec{r}^{(1)})\right]-\vec{r}^{(1)}(\tilde{\vec{r}}^{(1)}\cdot\tilde{\vec{r}}^{(2)})\ ,$$

$$\tilde{\vec{r}}^{(2}\left[(\tilde{\vec{r}}^{(1)}\cdot\vec{r}^{(2)})+(\tilde{\vec{r}}^{(2)}\cdot\vec{r}^{(1)})\right]-\vec{r}^{(2)}(\tilde{\vec{r}}^{(1)}\cdot\tilde{\vec{r}}^{(2)})\ \Big)/\,(\vec{r}^{(1)}\cdot\vec{r}^{(2)})\quad . \tag{4d}$$

The last formula shows that this is a parametrization of *type (i)* (iff the 2 three-vectors $\vec{r}^{(1)}$, $\vec{r}^{(2)}$ are not orthogonal).

Exercise 5.5-7. Verify these formulas, (4c,d,e).

The special case of this parametrization with $\vec{r}^{(1)} = \vec{r}, \vec{r}^{(2)} = \lambda\,\vec{r}$,

$$\underline{M} = \begin{pmatrix} 0 & x & y & z \\ -x & 0 & \lambda z & -\lambda y \\ -y & -\lambda z & 0 & \lambda x \\ -z & \lambda y & -\lambda x & 0 \end{pmatrix}, \tag{5a}$$

with λ an arbitrary (nonvanishing) constant, is also of *type (i)*, yielding

$$\underline{M} \doteq \vec{r}\ , \tag{5b}$$

$$\underline{\tilde{M}}\,\underline{M}^{-1}\,\underline{\tilde{M}} \doteq \left[2\vec{\tilde{r}}\,(\vec{\tilde{r}}\cdot\vec{r}) - \vec{r}\,(\vec{\tilde{r}}\cdot\vec{\tilde{r}})\right]/\,r^2\ ; \tag{5c}$$

but these two formulas merely reproduce (3b) and (3d).

Exercise 5.5-8. Verify!

A parametrization of (3×3)-matrices in terms of 3-vectors takes the natural form

$$\underline{M} = \begin{pmatrix} x^{(1)} & y^{(1)} & z^{(1)} \\ x^{(2)} & y^{(2)} & z^{(2)} \\ x^{(3)} & y^{(3)} & z^{(3)} \end{pmatrix}, \tag{6a}$$

which we write as

$$\underline{M} = \begin{pmatrix} \vec{r}^{(1)} \\ \vec{r}^{(2)} \\ \vec{r}^{(3)} \end{pmatrix}. \tag{6b}$$

Then clearly

$$\underline{M}^{-1} = \begin{pmatrix} u^{(1)}{}_x & u^{(2)}{}_x & u^{(3)}{}_x \\ u^{(1)}{}_y & u^{(2)}{}_y & u^{(3)}{}_y \\ u^{(1)}{}_z & u^{(2)}{}_z & u^{(3)}{}_z \end{pmatrix}, \tag{6c}$$

which we can also write

$$\underline{M}^{-1} = \left(\vec{u}^{(1)}, \vec{u}^{(2)}, \vec{u}^{(3)}\right) , \tag{6d}$$

where the 3 three-vectors $\vec{u}^{(j)}$ are defined, in terms of the 3 three-vectors $\vec{r}^{(k)}$, so that

$$\vec{u}^{(j)} \cdot \vec{r}^{(k)} = \delta_{jk}; \; j,k = 1,2,3 , \tag{6e}$$

which also entail

$$\sum_{j=1}^{3} u_x^{(j)} x^{(j)} = 1 , \tag{6f}$$

$$\sum_{j=1}^{3} u_y^{(j)} x^{(j)} = \sum_{j=1}^{3} u_z^{(j)} x^{(j)} = 0 , \tag{6g}$$

as well as the analogous 6 equalities obtained by cyclic permutations of the x, y, z components of the 3-vectors $\vec{u}^{(j)} \equiv (u_x^{(j)}, u_y^{(j)}, u_z^{(j)})$ and $\vec{r}^{(j)} \equiv (x^{(j)}, y^{(j)}, z^{(j)})$.

An explicit definition of the 3-vectors $\vec{u}^{(j)}$ reads therefore

$$\vec{u}^{(j)} = \vec{r}^{(j+1)} \wedge \vec{r}^{(j+2)} / \Delta; \; j = 1,2,3, \; \mathrm{mod}(3) , \tag{6h}$$

$$\Delta = \vec{r}^{(1)} \cdot \vec{r}^{(2)} \wedge \vec{r}^{(3)} . \tag{6i}$$

Note that Δ coincides (up to a factor $1/6$, and possibly a sign) with the volume of the tetrahedron of vertices $\vec{0}, \vec{r}^{(1)}, \vec{r}^{(2)}, \vec{r}^{(3)}$.

Hence for this parametrization we can write

$$\underline{M} \doteq \left(\vec{r}^{(j)}, \; j = 1,2,3\right) , \tag{6l}$$

and

$$\underline{\tilde{M}}\, \underline{M}^{-1}\, \underline{\tilde{\tilde{M}}} \doteq \left(\vec{v}^{(j)}, \; j = 1,2,3\right) , \tag{6m}$$

with

$$\vec{v}^{(j)} = \sum_{k=1,2,3,\mathrm{mod}(3)} \left[\vec{\tilde{r}}^{(j)} \cdot \vec{r}^{(k+1)} \wedge \vec{\tilde{\tilde{r}}}^{(k+2)} \right] \vec{\tilde{\tilde{r}}}^{(k)} / \Delta , \tag{6n}$$

with Δ defined by (4.6i).

The formula (4.6m) with (4.6n) holds *a fortiori* if $\underline{\tilde{\tilde{M}}} = \underline{\tilde{M}}$, hence it is clear that this parametrization is of *type (i)*.

Exercise 5.5-9. Verify these formulas, (6c,d,e,f,g,h,i,m,n).

A special case of this parametrization is obtained by replacing the 3 three-vectors $\vec{r}^{(j)}$ as follows:

$$\vec{r}^{(j)} \to \bar{\vec{r}}^{(j)} \equiv \vec{r}^{(j)} - \frac{1}{3}\sum_{k=1}^{3} \vec{r}^{(k)} . \tag{7a}$$

Then the treatment given above remains applicable, with the constraint on the (new) input vectors $\bar{\vec{r}}^{(j)}$ to have zero sum,

$$\sum_{j=1}^{3} \bar{\vec{r}}^{(j)} = 0 . \tag{7b}$$

It is then clear that the (new) vectors $\bar{\vec{v}}^{(j)}$, see (6m,n),

$$\bar{\vec{v}}^{(j)} = \sum_{k=1,2,3,\mathrm{mod}(3)} \left[\tilde{\bar{\vec{r}}}^{(j)} \cdot \bar{\vec{r}}^{(k+1)} \wedge \tilde{\tilde{\bar{\vec{r}}}}^{(k+2)} \right] \tilde{\tilde{\bar{\vec{r}}}}^{(k)} / \Delta , \tag{7c}$$

also satisfy the condition to have zero sum,

$$\sum_{j=1}^{3} \bar{\vec{v}}^{(j)} = 0 . \tag{7d}$$

Hence this parametrization is also of *type (i)*. It has the advantage to yield, in terms of the original 3-vectors $\vec{r}^{(j)}$, *translation-invariant* equations.

Exercise 5.5-10. Verify!

The next parametrization we consider is, for (4×4)-matrices, in terms of 4 three-vectors and one scalar. It reads

$$\underline{M} = \begin{pmatrix} \rho & x^{(1)} & y^{(1)} & z^{(1)} \\ \rho & x^{(2)} & y^{(2)} & z^{(2)} \\ \rho & x^{(3)} & y^{(3)} & z^{(3)} \\ \rho & x^{(4)} & y^{(4)} & z^{(4)} \end{pmatrix} , \tag{8a}$$

which we denote as follows:

$$\underline{M} \doteq \left(\rho, \vec{r}^{(j)}, j=1,\ldots,4\right) . \tag{8b}$$

It is a matter of standard vector and matrix algebra to obtain the corresponding formula for the (4×4)-matrix $\underline{\tilde{M}}\,\underline{M}^{-1}\,\underline{\tilde{M}}$:

$$\underline{\tilde{M}}\,\underline{M}^{-1}\,\underline{\tilde{M}} \doteq \left(\tilde{\rho}^2 \rho^{-1}, \vec{v}^{(j)}, j=1,\ldots,4\right) , \tag{8c}$$

$$\vec{v}^{(j)} = \sum_{k=1,2,3,4,\mathrm{mod}(4)} (-)^k\, \vec{\tilde{r}}^{(k)} \left\{ \tilde{\rho}\,\rho^{-1}\,\vec{r}^{(k+1)} \cdot \left(\vec{r}^{(k+2)} \wedge \vec{r}^{(k+3)}\right)\right.$$

$$\left. + \vec{\tilde{r}}^{(j)} \cdot \left[\left(\vec{r}^{(k+1)} - \vec{r}^{(k+2)}\right) \wedge \left(\vec{r}^{(k+2)} - \vec{r}^{(k+3)}\right)\right] \right\} / \Delta , \tag{8d}$$

$$\Delta = \left(\vec{r}^{(2)} - \vec{r}^{(1)}\right) \cdot \left[\left(\vec{r}^{(3)} - \vec{r}^{(1)}\right) \wedge \left(\vec{r}^{(4)} - \vec{r}^{(1)}\right)\right] . \tag{8e}$$

Note that the quantity Δ defined by this formula, (8e), is *translation-invariant* and coincides, up to a factor $1/6$ and possibly a sign, with the volume of the tetrahedron of vertices $\vec{r}^{(j)}, j=1,2,3,4$.

These formulas entail that this parametrization is of *type (i)*.

Exercise 5.5-11. Verify these formulas, (8c,d,e).

The next parametrization we consider is applicable to *antisymmetric* (3×3)-matrices. It reads

$$\underline{M} = \begin{pmatrix} 0 & x & y \\ -x & 0 & z \\ -y & -z & 0 \end{pmatrix} , \tag{9a}$$

and we denote it as follows:

$$\underline{M} \doteq \vec{r} . \tag{9b}$$

The following formulas are then easy to verify:

$$\underline{M}\,\underline{\tilde{M}} = \begin{pmatrix} -x\tilde{x} - y\tilde{y} & -y\tilde{z} & x\tilde{z} \\ -z\tilde{y} & -x\tilde{x} - z\tilde{z} & -x\tilde{y} \\ z\tilde{x} & -y\tilde{x} & -y\tilde{y} - z\tilde{z} \end{pmatrix} , \tag{9c}$$

$$[\underline{M},\tilde{\underline{M}}] \doteq -\vec{r} \wedge \tilde{\vec{r}} \;, \tag{9d}$$

$$\underline{M}^{2m+1} \doteq (-)^m \, r^{2m} \, \vec{r}, m = 0,1,2,... \;. \tag{9e}$$

These formulas entail that this parametrization is of *type (ii)* (note, however, that (9e) entails that this parametrization is also preserved for any *odd* power of the matrix $\underline{M}$).

Exercise 5.5-12. Verify these formulas, (9c,d,e).

The last parametrization was already introduced at the end of Sect. 5.3. We report it here for completeness.

Let $\underline{V}$ respectively $\underline{W}$ be a $(S \times L)$-matrix, namely a matrix with S rows and L columns, respectively a $(L \times S)$-matrix, namely a matrix with L rows and S columns (hereafter L and S denote two, *a priori* arbitrary, positive integers); and let us parametrize them in terms of the $2L$ S-vectors $\vec{v}^{(l)}$ respectively $\vec{w}^{(l)}$,

$$\vec{v}^{(l)} \equiv \left(v_1^{(l)}, v_2^{(l)}, ..., v_S^{(l)}\right), \quad \vec{w}^{(l)} \equiv \left(w_1^{(l)}, w_2^{(l)}, ..., w_S^{(l)}\right), \quad l = 1,2,...,L \,, \tag{10a}$$

in the following manner:

$$\underline{V} \equiv \begin{pmatrix} v_1^{(1)} & v_1^{(2)} & \cdots & v_1^{(L)} \\ v_2^{(1)} & v_2^{(2)} & \cdots & v_2^{(L)} \\ \vdots & \vdots & \vdots\vdots\vdots & \vdots \\ v_S^{(1)} & v_S^{(2)} & \cdots & v_S^{(L)} \end{pmatrix}, \qquad \underline{W} \equiv \begin{pmatrix} w_1^{(1)} & w_2^{(1)} & \cdots & w_S^{(1)} \\ w_1^{(2)} & w_2^{(2)} & \cdots & w_S^{(2)} \\ \vdots & \vdots & \vdots\vdots\vdots & \vdots \\ w_1^{(L)} & w_2^{(L)} & \cdots & w_S^{(L)} \end{pmatrix} . \tag{10b}$$

These formulas can be denoted via the notation

$$\underline{V} \doteq \left(\vec{v}^{(1)}, \vec{v}^{(2)}, ..., \vec{v}^{(L)}\right), \;\; \underline{W} \doteq \left(\vec{w}^{(1)}, \vec{w}^{(2)}, ..., \vec{w}^{(L)}\right). \tag{10c}$$

Then (and this is the interesting point, as we shall see in particular in Sect. 5.6.5, but we already saw at the end of Sect. 5.3) the $(S \times L)$-matrix $\tilde{\tilde{\underline{V}}}$ respectively the $(L \times S)$-matrix $\tilde{\tilde{\underline{W}}}$, defined by the "alternating" triple products

$$\tilde{\tilde{\underline{V}}} = \underline{V}\,\underline{W}\,\tilde{\underline{V}} \;, \quad \tilde{\tilde{\underline{W}}} = \underline{W}\,\underline{V}\,\tilde{\underline{W}} \;, \tag{10d}$$

admit an analogous representation in terms of the $2L$ S-vectors $\tilde{\tilde{\vec{v}}}^{(l)}$ respectively $\tilde{\tilde{\vec{w}}}^{(l)}$,

$$\underline{\tilde{\tilde{V}}} \doteq \left(\vec{\tilde{\tilde{v}}}^{(1)}, \vec{\tilde{\tilde{v}}}^{(2)}, \ldots, \vec{\tilde{\tilde{v}}}^{(L)}\right), \quad \underline{\tilde{\tilde{W}}} \doteq \left(\vec{\tilde{\tilde{w}}}^{(1)}, \vec{\tilde{\tilde{w}}}^{(2)}, \ldots, \vec{\tilde{\tilde{w}}}^{(L)}\right), \tag{10e}$$

with these $2L$ S-vectors expressed in terms of the S-vectors associated with the matrices that enter in the triple products by the following *covariant* formulas:

$$\vec{\tilde{\tilde{v}}}^{(l)} = \sum_{\lambda=1}^{L} \vec{\tilde{v}}^{(\lambda)} \left(\vec{\tilde{w}}^{(\lambda)} \cdot \vec{\tilde{v}}^{(l)}\right), \quad \vec{\tilde{\tilde{w}}}^{(l)} = \sum_{\lambda=1}^{L} \vec{\tilde{w}}^{(\lambda)} \left(\vec{\tilde{v}}^{(\lambda)} \cdot \vec{\tilde{w}}^{(l)}\right). \tag{10f}$$

Exercise 5.5-13. Verify!

Exercise 5.5-14. Formulate and prove analogous results for multiple alternating products. *Hint*: if

$$\underline{\tilde{V}} = \underline{V}^{(1)} \underline{W}^{(1)} \underline{V}^{(2)} \underline{W}^{(2)} \cdots \underline{V}^{(p)} \underline{W}^{(p+1)} \underline{V}^{(p+1)}, \tag{11a}$$

then

$$\vec{\tilde{v}}^{(l)} = \sum_{\lambda_1, \lambda_2, \ldots, \lambda_p = 1}^{L} \vec{v}^{(\lambda_1)} \left(\vec{w}^{(\lambda_1)} \cdot \vec{v}^{(\lambda_2)}\right) \cdots \left(\vec{w}^{(\lambda_p)} \cdot \vec{v}^{(l)}\right); \tag{11b}$$

if

$$\underline{\tilde{W}} = \underline{W}^{(1)} \underline{V}^{(1)} \underline{W}^{(2)} \underline{V}^{(2)} \cdots \underline{W}^{(p-1)} \underline{V}^{(p-1)} \underline{W}^{(p)}, \tag{12a}$$

then

$$\vec{\tilde{w}}^{(l)} = \sum_{\lambda_1, \lambda_2, \ldots, \lambda_p = 1}^{L} \vec{w}^{(\lambda_1)} \left(\vec{v}^{(\lambda_1)} \cdot \vec{w}^{(\lambda_2)}\right) \cdots \left(\vec{v}^{(\lambda_p)} \cdot \vec{w}^{(l)}\right). \tag{12b}$$

5.6 A survey of N-body systems in three-dimensional space amenable to exact treatments

In Sect. 5.6, or rather in its subsections, we display a number of few- and many-body problems amenable to exact treatments (*solvable* and/or *integrable* and/or *linearizable*) -- including some already discussed in preceding sections. The environment for almost all these models (the exceptions being mainly in Sect. 5.6.5) is ordinary (three-dimensional) space:

accordingly, unless otherwise specified, hereafter vectors, denoted by superimposed arrows, say $\vec{r}$, indicate 3-vectors,

$$\vec{r} \equiv (x, y, z) , \tag{1}$$

and of course the standard formulas hold for the scalar and vector products,

$$\vec{r}_1 \cdot \vec{r}_2 = x_1 x_2 + y_1 y_2 + z_1 z_2 , \tag{2}$$

$$\vec{r}_1 \wedge \vec{r}_2 = (y_1 z_2 - z_1 y_2, z_1 x_2 - x_1 z_2, x_1 y_2 - y_1 x_2). \tag{3}$$

Let us also report, for convenience, the corresponding formulas for spherical (rather than Cartesian) co-ordinates:

$$\vec{r} = r(\cos\theta \, \cos\varphi, \, \cos\theta \, \sin\varphi, \, \sin\theta) , \tag{4}$$

$$\vec{r}_1 \cdot \vec{r}_2 = r_1 \, r_2 \left[\cos(\theta_1 - \theta_2) + \cos\theta_1 \, \cos\theta_2 \, \left[\cos(\varphi_1 - \varphi_2) - 1\right]\right] = r_1 \, r_2 \cos\theta_{12} , \tag{5}$$

$$\left|\vec{r}_1 \wedge \vec{r}_2 = r_1 \, r_2 \cos\theta_{12}\right| . \tag{6}$$

In the last two formulas, (5) and (6), θ_{12} denotes of course the angle among the vectors $\vec{r}_1$ and $\vec{r}_2$.

Remark 5.6-1. Essentially all formulas are written below in *covariant* form, in terms of 3-vectors and occasionally of scalars. This guarantees their *rotation-invariance* (except when they feature constant vectors -- unless these are also assumed to transform themselves as 3-vectors under rotations).

The presentation below is conveniently split in several parts, the contents of which are clearly indicated by the titles of the following subsections. The clarity thereby gained overcompensates for the minor repetitions entailed by this compartmentalization.

5.6.1 Few-body problems of Newtonian type

In Sect. 5.6.1 we display a representative set of few-body problems in 3-dimensional space, and we indicate how to treat them. As it will be immediately clear to the alert reader, many more examples could be manufactured by analogous techniques -- the interested readers will profitably try their brains at this game (in addition to solving all the *exercises* proposed below).

A simple *solvable one-body problem* is obtained by applying the parametrization (5.5-3) to the *solvable* matrix evolution equation (5.1-1). Its equation of motion reads (see (5.1-29))

$$\ddot{\vec{r}} = 2a\dot{\vec{r}} + b\vec{r} + c[\ 2\dot{\vec{r}}\,(\dot{\vec{r}}\cdot\vec{r}) - \vec{r}\,(\dot{\vec{r}}\cdot\dot{\vec{r}})\ ]/r^2\ . \tag{1a}$$

Its general solution reads (see (5.1-2))

$$\vec{r}(t) = \vec{r}(0)\left[\varphi^{(+)}(t) + i\,B\varphi^{(-)}(t)\right] - i\,\dot{\vec{r}}(0)\varphi^{(-)}(t)/C\ , \tag{1b}$$

$$\varphi^{(\pm)}(t) = \exp(a\,\gamma\,t)\{\ \left[\cosh(\Delta t) + (A\Delta + i\,C)\Delta^{-1}\sinh(\Delta t)\right]^{\gamma}$$

$$\pm\left[\cosh(\Delta t) + (A\Delta - iC)\Delta^{-1}\sinh(\Delta t)\right]^{\gamma}\ \}/2\,, \tag{1c}$$

$$\gamma \equiv 1/(1-c)\ , \tag{1d}$$

$$\Delta = \left[a^2 + b(1-c)\right]^{1/2}\ , \tag{1e}$$

$$A = \{\ -a + (1-c)\left[\dot{\vec{r}}(0)\cdot\vec{r}(0)\right]/\left[r(0)\right]^2\ \}/\Delta\ , \tag{1f}$$

$$B = \left[\dot{\vec{r}}(0)\cdot\vec{r}(0)\right]/\left|\dot{\vec{r}}(0)\wedge\vec{r}(0)\right|\ , \tag{1g}$$

$$C = \left|\dot{\vec{r}}(0)\wedge\vec{r}(0)\right|/\left[r(0)^2\right]\ . \tag{1h}$$

Here the symbol $|\vec{v}|$ indicates the modulus of the 3-vector $\vec{v}$, so that $|\vec{v}|^2 \equiv {v_x}^2 + {v_y}^2 + {v_z}^2$ (irrespective of whether the 3-vector $\vec{v}$ is real or complex).

Exercise 5.6.1-1. Verify! *Hint*: see (5.1-2) and (5.5-3).

The behavior of this system can be read from this explicit formula; see also the detailed analysis in Sect. 5.1. Of course for this model, (1), to admit of a "physical" interpretation the constants a,b,c must be *real*, as well as the initial conditions $\vec{r}(0)$, $\dot{\vec{r}}(0)$; then $\varphi^{(+)}(t)$ is *real*, $\varphi^{(-)}(t)$ is *imaginary* (see (1c)), and of course $\vec{r}(t)$ is *real*. Note that all nonsingular solutions of this problem are periodic, if $a=0$ and b,c are given by (5.1-24b,c).

A *solvable 2-body problem* obtains from (1) by complexification:

$$\vec{r}(t)=\vec{r}^{(1)}(t)+i\,\vec{r}^{(2)}(t),\ \ a=\alpha+i\,\tilde{\alpha},\ \ b=\beta+i\,\tilde{\beta},\ \ c=\gamma+i\,\tilde{\gamma}\ . \tag{2a}$$

Its equations of motion read as follows:

$$\ddot{\vec{r}}^{(1)}=2\,\alpha\,\dot{\vec{r}}^{(1)}-2\,\tilde{\alpha}\,\dot{\vec{r}}^{(2)}+\beta\,\vec{r}^{(1)}-\tilde{\beta}\,\vec{r}^{(2)}+\frac{\gamma\,\vec{R}^{(1)}-\tilde{\gamma}\,\vec{R}^{(2)}}{(\vec{r}^{(1)}\cdot\vec{r}^{(1)}-\vec{r}^{(2)}\cdot\vec{r}^{(2)})^2+4\,(\vec{r}^{(1)}\cdot\vec{r}^{(2)})^2}\ , \tag{2b}$$

$$\ddot{\vec{r}}^{(2)}=2\,\alpha\,\dot{\vec{r}}^{(2)}+2\,\tilde{\alpha}\,\dot{\vec{r}}^{(1)}+\beta\,\vec{r}^{(2)}+\tilde{\beta}\,\vec{r}^{(1)}+\frac{\gamma\,\vec{R}^{(2)}+\tilde{\gamma}\,\vec{R}^{(1)}}{(\vec{r}^{(1)}\cdot\vec{r}^{(1)}-\vec{r}^{(2)}\cdot\vec{r}^{(2)})^2+4\,(\vec{r}^{(1)}\cdot\vec{r}^{(2)})^2}\ , \tag{2c}$$

where

$$\vec{R}^{(1)}=(\vec{r}^{(1)}\cdot\vec{r}^{(1)}-\vec{r}^{(2)}\cdot\vec{r}^{(2)})\,\vec{\rho}^{(1)}+2\,(\vec{r}^{(1)}\cdot\vec{r}^{(2)})\,\vec{\rho}^{(2)}\ , \tag{2d}$$

$$\vec{R}^{(2)}=(\vec{r}^{(1)}\cdot\vec{r}^{(1)}-\vec{r}^{(2)}\cdot\vec{r}^{(2)})\,\vec{\rho}^{(2)}-2\,(\vec{r}^{(1)}\cdot\vec{r}^{(2)})\,\vec{\rho}^{(1)}\ , \tag{2e}$$

$$\vec{\rho}^{(1)}=2\left(\dot{\vec{r}}^{(1)}\cdot\vec{r}^{(1)}-\dot{\vec{r}}^{(2)}\cdot\vec{r}^{(2)}\right)\dot{\vec{r}}^{(1)}-2\left(\dot{\vec{r}}^{(1)}\cdot\vec{r}^{(2)}-\dot{\vec{r}}^{(2)}\cdot\vec{r}^{(1)}\right)\dot{\vec{r}}^{(2)}$$

$$-\left(\dot{\vec{r}}^{(1)}\cdot\dot{\vec{r}}^{(1)}-\dot{\vec{r}}^{(2)}\cdot\dot{\vec{r}}^{(2)}\right)\vec{r}^{(1)}+2\,(\dot{\vec{r}}^{(1)}\cdot\dot{\vec{r}}^{(2)})\,\vec{r}^{(2)}, \tag{2f}$$

$$\vec{\rho}^{(2)}=2\left(\dot{\vec{r}}^{(1)}\cdot\vec{r}^{(1)}-\dot{\vec{r}}^{(2)}\cdot\vec{r}^{(2)}\right)\dot{\vec{r}}^{(2)}+2\left(\dot{\vec{r}}^{(1)}\cdot\vec{r}^{(2)}-\dot{\vec{r}}^{(2)}\cdot\vec{r}^{(1)}\right)\dot{\vec{r}}^{(1)}$$

$$-\left(\dot{\vec{r}}^{(1)}\cdot\dot{\vec{r}}^{(1)}-\dot{\vec{r}}^{(2)}\cdot\dot{\vec{r}}^{(2)}\right)\vec{r}^{(2)}-2\,(\dot{\vec{r}}^{(1)}\cdot\dot{\vec{r}}^{(2)})\,\vec{r}^{(1)}. \tag{2g}$$

An "unphysical" aspect of these equations of motion is the appearance of certain components of the force which are independent of the coordinate and velocity of the particle on which the force acts (we refer for instance to the terms $-2\,\tilde{\alpha}\,\dot{\vec{r}}^{(2)}$ and $-\tilde{\beta}\,\vec{r}^{(2)}$ in the right hand side of (2b)). *This phenomenon is characteristic of several equations considered below, and will not be highlighted again in the following.*

Exercise 5.6.1-2. (i) Discuss the explicit solution of the initial-value problem for (2); *(ii)* give conditions which guarantee that *all* solutions of (2) are periodic. *Hint*: see Sect. 5.1 (in particular, *Propositions 5.1-7* and *5.1-10*).

A *solvable translation-invariant 2-body problem* is obtained by applying the reduction (5.5-3) to the coupled matrix evolution equations (5.3-6) (instead of (5.1-1)). It reads

$$\ddot{\vec{r}}^{(\pm)} = [(\alpha/2) \pm a]\dot{\vec{r}}^{(+)} + [(\alpha/2) \mp a]\dot{\vec{r}}^{(-)} \pm \{b\vec{r} + c[2\dot{\vec{r}}(\dot{\vec{r}}\cdot\vec{r}) - \vec{r}(\dot{\vec{r}}\cdot\dot{\vec{r}})]/r^2\}/2 \ , \tag{3a}$$

where

$$\vec{r}(t) = \vec{r}^{(+)}(t) - \vec{r}^{(-)}(t) \ . \tag{3b}$$

These equations of motion are *translation-invariant.* Their solution is given, via (3b) and

$$\vec{s}(t) = \vec{r}^{(+)}(t) + \vec{r}^{(-)}(t) \ , \tag{3c}$$

by (1) and (see (5.3-4))

$$\vec{s}(t) = \vec{s}(0) + \dot{\vec{s}}(0)[\exp(\alpha t) - 1]/\alpha \ , \tag{3d}$$

which correspond to the trivially *solvable* equation of motion satisfied by $\vec{s}(t)$,

$$\ddot{\vec{s}} = \alpha \dot{\vec{s}} \tag{3e}$$

see (5.3-3).

Exercise 5.6.1-3. Display the equations of motion of the *solvable translation-invariant 4-body problem* that obtains from the previous one by complexification (see (2a), and add, say, $\alpha = \eta + i\tilde{\eta}$), and analyze the behavior of their solutions (in particular identify restrictions on the "coupling constants" which guarantee that *all* solutions remain confined or are completely periodic). *Hint*: as for *Exercise 5.6.1-2*.

Exercise 5.6.1-4. Generalize all the models presented above by using the parametrization (5.5-1) rather than (5.5-3). *Hint*: every 3-vector $\vec{r}(t)$ gets then associated with a scalar $\rho(t)$.

Another *solvable 2-body problem* is obtained by applying the parametrization (5.5-4) (rather than (5.5-1) or (5.5-3)) to the (same) matrix evolution equation (5.1-1). Its equations of motion read

$$\ddot{\vec{r}}^{(1)} = 2a\dot{\vec{r}}^{(1)} + b\vec{r}^{(1)}$$

$$+c\,\{\,\dot{\vec{r}}^{(1)}[\,(\dot{\vec{r}}^{(1)}\cdot\vec{r}^{(2)})+(\dot{\vec{r}}^{(2)}\cdot\vec{r}^{(1)})\,]-\vec{r}^{(1)}(\dot{\vec{r}}^{(1)}\cdot\dot{\vec{r}}^{(2)})\,\}/(\vec{r}^{(1)}\cdot\vec{r}^{(2)})\,, \tag{4a}$$

$$\ddot{\vec{r}}^{(2)} = 2a\dot{\vec{r}}^{(2)} + b\vec{r}^{(2)}$$

$$+c\,\{\,\dot{\vec{r}}^{(2)}[\,(\dot{\vec{r}}^{(2)}\cdot\vec{r}^{(1)})+(\dot{\vec{r}}^{(1)}\cdot\vec{r}^{(2)})\,]-\vec{r}^{(2)}(\dot{\vec{r}}^{(2)}\cdot\dot{\vec{r}}^{(1)})\,\}/(\vec{r}^{(2)}\cdot\vec{r}^{(1)})\,. \tag{4b}$$

Exercise 5.6.1-5. Obtain and discuss the solution (of the initial value problem) for these Newtonian equations of motion. *Hint*: see (5.1-2) and (5.5-4), and the discussion in Sect. 5.1.

Remark 5.6.1-6. The reduction $\vec{r}^{(1)}=\vec{r}$, $\vec{r}^{(2)}=\lambda\vec{r}$ is clearly compatible with these equations of motion, and it yields back (1).

Exercise 5.6.1-7. Display the equations of motion of the *solvable 4-body problem* that obtains from (4) via the complexification (2b), and identify the cases in which all solutions are confined, multiply periodic or completely periodic. *Hint*: again, as for *Exercise 5.6.1-2*.

Exercise 5.6.1-8. Display the equations of motion of the *solvable translation-invariant 4-body problem* that obtains by applying the parametrization (5.5-4) to the coupled matrix evolution equations (5.3-6), as well as the equations of motion of the *solvable translation-invariant 8-body problem* that can be subsequently obtained by complexification; and analyze the corresponding motions, at least to the extent of identifying restrictions on the "coupling constants" sufficient to guarantee that *all* solutions are *(i)* confined, *(ii)* multiply periodic or *(iii)* completely periodic. *Hint*: see Sect. 5.1.

Next, let us consider the *solvable 3-body problem* that is obtained by applying the parametrization (5.5-6) to the matrix evolution equation (5.1-1). The corresponding equations of motion read

$$\ddot{\vec{r}}^{(j)} = 2a\,\dot{\vec{r}}^{(j)} + b\,\vec{r}^{(j)} + c\sum_{k=1,2,3,\mathrm{mod}(3)}\left\{\dot{\vec{r}}^{(k)}\left[\dot{\vec{r}}^{(j)}\cdot\vec{r}^{(k+1)}\wedge\vec{r}^{(k+2)}\right]\right\}/\Delta\,,\ j=1,2,3, \tag{5a}$$

with

$$\Delta \equiv \vec{r}^{(1)}\cdot\vec{r}^{(2)}\wedge\vec{r}^{(3)}\,. \tag{5b}$$

Via the simple transformation

$$\vec{r}^{(j)}(t) \rightarrow \exp(\lambda t)\,\vec{r}^{(j)}(t) \tag{6a}$$

this model takes the more general form

$$\ddot{\vec{r}}^{(j)} = [\ 2a + \lambda(c-2)\ ]\,\dot{\vec{r}}^{(j)} + [\ b + 2\lambda a + \lambda^2 (c-1)\ ]\,\vec{r}^{(j)}$$

$$+c \sum_{k=1,2,3,\mathrm{mod}(3)} \{\ (\dot{\vec{r}}^{(k)} + \lambda \vec{r}^{(k)})[\ \dot{\vec{r}}^{(j)} \cdot \vec{r}^{(k+1)} \wedge \vec{r}^{(k+2)}\ ]\ \}/\Delta\ ,\quad j=1,2,3, \tag{6b}$$

with Δ always defined by (5b).

Exercise 5.6.1-9. Obtain the solution of this 3-body problem, and compare it with that given (for a marginally less general model) in the literature <CJX94>. *Hint*: use (5.1-2) and (5.5-6), and the discussion of Sect. 5.1.

Exercise 5.6.1-10. Display the Newtonian equations of motion of the *solvable 6-body problem* that obtains from (6b) by complexification, and find conditions on the "coupling constants" a, b, c, λ which guarantee that *all* its *nonsingular* solutions are *periodic* (and find the periods). *Hint*: see the preceding *Exercise 5.6.1-9*.

Exercise 5.6.1-11. Display the Newtonian equations of motion of the *solvable translation-invariant 12-body problem* that obtains from that of the preceding *Exercise 5.6.1-9* by applying to it appropriately the technique of *association*, and determine conditions on the (10, real) coupling constants of this model sufficient to guarantee that all its solutions are *completely periodic*. *Hint*: ponder on the relation among (3) and (1).

Remark 5.6.1-12. The equations of motion (6b) (as well as, *a fortiori*, (5)) are clearly consistent with the restriction that the "center of mass" $\vec{R} = (1/3)\sum_{j=1}^{3} \vec{r}^{(j)}$ stay put at the origin, $\vec{R} = 0$ .The special solutions of (6b), or (5), that fulfill this constraint, correspond to the solutions of the *solvable translation-invariant 3-body problem* that is obtained by formally replacing in (6b) or (5) every 3-vector $\vec{r}^{(j)}$ with $\vec{\tilde{r}}^{(j)} \equiv \vec{r}^{(j)} - \vec{R}$.

Next we consider the *solvable translation-invariant 4-body problem* that is obtained by applying the parametrization (5.5-8) with $\rho = 1$ to the matrix evolution equation (5.1-1) with $b=0$. Its Newtonian equations of motion read

$$\ddot{\vec{r}}^{(j)} = 2a\dot{\vec{r}}^{(j)} + c \sum_{k=1,2,3,4,\text{mod}(4)} (-)^k \{ \dot{\vec{r}}^{(k)} [\dot{\vec{r}}^{(j)} \cdot (\vec{r}^{(k+1)} - \vec{r}^{(k+2)}) \wedge (\vec{r}^{(k+2)} - \vec{r}^{(k+3)})] \} / \Delta, \tag{7a}$$

$$\Delta \equiv (\vec{r}^{(2)} - \vec{r}^{(1)}) \cdot (\vec{r}^{(3)} - \vec{r}^{(1)}) \wedge (\vec{r}^{(4)} - \vec{r}^{(1)}) . \tag{7b}$$

Exercise 5.6.1-13. Obtain the solution of this 4-body problem, and compare it with that given (for a marginally less general model) in the literature <CJX94>. *Hint*: use (5.1-2) and (5.5-8), and the discussion of Sect. 5.1.

Exercise 5.6.1-14. Display and discuss (with particular attention to *periodic* motions) the Newtonian equations of motion of the *solvable translation-invariant 8-body problem* that obtains from (7) by *complexification*. *Hint*: see *Exercise 5.6.1-9*.

Exercise 5.6.1-15. Display the Newtonian equations of motion that obtain by applying the parametrization (5.5-8) to the matrix evolution equation (5.1-1) (without assuming $\rho=1$ nor $b=0$; hence these equations of motion generalize (7)).

Next, let us exhibit the *solvable 2-body problem* that is obtained by applying the simple parametrization (5.5-3) to the *solvable* system of 2 coupled matrix evolution equations that itself obtains by applying the multiplication trick (5.3-30) (with $M=2$ and $\underline{U}^{(11)} = \underline{U}^{(22)}$, $\underline{U}^{(12)} = \underline{U}^{(21)}$) to the *solvable* matrix evolution ODE (5.1-1):

$$\ddot{\vec{r}}^{(j)} = 2a\dot{\vec{r}}^{(j)} + b\vec{r}^{(j)} + c\{ \dot{\vec{r}}^{(j)} [q\dot{q} - 4p\dot{p}] + 2\dot{\vec{r}}^{(j+1)} [q\dot{p} - p\dot{q}]$$

$$+\vec{r}^{(j)} [2p\tilde{p} - q\tilde{q}/2] + \vec{r}^{(j+1)} [p\tilde{q} - q\tilde{p}] \}/d, \; j=1,2,\text{mod}(2) , \tag{8a}$$

$$q \equiv (\vec{r}^{(1)})^2 + (\vec{r}^{(2)})^2, p \equiv \vec{r}^{(1)} \cdot \vec{r}^{(2)}, \tilde{q} \equiv 2[\dot{\vec{r}}^{(1)} \cdot \dot{\vec{r}}^{(1)} + \dot{\vec{r}}^{(2)} \cdot \dot{\vec{r}}^{(2)}], \; \tilde{p} \equiv 2\dot{\vec{r}}^{(1)} \cdot \dot{\vec{r}}^{(2)}, \quad \text{8b)}$$

$$d \equiv q^2 - 4p^2 \equiv (\vec{r}^{(1)} + \vec{r}^{(2)})^2 (\vec{r}^{(1)} - \vec{r}^{(2)})^2 . \tag{8c}$$

Exercise 5.6.1-16. Derive these equations of motion and find their solution. *Hint*: follow the instructions given above, then use (5.1-2) and (5.5-3), and see Sect. 5.1.

This model, (8), is presented as another simple example (besides those already reported in Sect. 5.3) of the kind of results that can be obtained by using the *multiplication* (in fact, in this case, only *duplication*) technique described in that Sect. 5.3; clearly many more models (which, however, become more and more complicated) can be obtained by iterated uses of these techniques, which can be moreover combined with those used above (*complexification*, and *association* with other solvable models, in particular the latter to manufacture *translation-invariant* models).

Next we consider the scalar/vector *solvable one-body problem* which is obtained by applying the parametrization (5.5-1) to the solvable matrix evolution equation (5.4.3-1). It is characterized by the following equations of motion:

$$\ddot{\rho}=\alpha+\beta\rho+\gamma\left[\dot{\rho}+c(\rho^2-r^2)\right]-c\left[3\rho\dot{\rho}-3(\vec{r}\cdot\dot{\vec{r}})+c\rho(\rho^2-3r^2)\right], \tag{9a}$$

$$\ddot{\vec{r}}=\beta\vec{r}+\gamma\left[\dot{\vec{r}}+2c\rho\vec{r}\right]-c\left[3\dot{\rho}\vec{r}+3\rho\dot{\vec{r}}-\vec{r}\wedge\dot{\vec{r}}+c\vec{r}(3\rho^2-r^2)\right]. \tag{9b}$$

Note that in this case the 3-vector equation of motion (9b) is coupled to the scalar equation (9a).

Exercise 5.6.1-17. Assuming the 4 constants α,β,γ,c to be real, find restrictions on their values which guarantee that *all* solutions of (9) are *completely periodic*. *Hint*: see *Exercise 5.4.3-2*.

Exercise 5.6.1-18. Obtain from (9), by *complexification*, the Newtonian equations of motion of a scalar/vector *solvable 2-body problem*. *Hint*: set, say,

$$\vec{r}=\vec{r}^{(1)}+i\vec{r}^{(2)},\ \rho=\rho^{(1)}+i\rho^{(2)},\ \alpha=a+i\tilde{a},\ \beta=b+i\tilde{b},\ \gamma=c+i\tilde{c},\ c=C+i\tilde{C}. \tag{10}$$

Exercise 5.6.1-19. Obtain from (9), by appropriate *association*, the Newtonian equations of motion of a scalar/vector *translation/invariant solvable 2-body problem*. *Hint*: see (3b,c,d,e).

Exercise 5.6.1-20. Apply the parametrization (5.5-1) to the matrix evolution equation (5.4.3-7), and discuss the (*solvable*!) scalar/vector Newtonian equations of motion obtained in this manner; in particular, determine conditions on the, assumedly real, "coupling constants" sufficient to guarantee that *all* solutions of this Newtonian equations of motion are *completely periodic*.

Next, we report the equations of motion of the scalar/vector *solvable one-body problem* that is obtained by applying the parametrization (5.5-1) to the *solvable* matrix evolution equation (5.4.3-24):

$$\ddot{\rho}=c+(\alpha-a)\dot{\rho}+a\alpha\rho+b\alpha(\rho^2-r^2)-2b(\dot{\rho}\rho-\dot{\vec{r}}\cdot\vec{r}) \ , \tag{11a}$$

$$\ddot{\vec{r}}=(\alpha-a)\dot{\vec{r}}+a\alpha\vec{r}+2b\alpha\rho\vec{r}-2b(\dot{\rho}\vec{r}+\rho\dot{\vec{r}})-2b\vec{r}\wedge\dot{\vec{r}} \ . \tag{11b}$$

Note the similarity of these equations of motion to (9).

Exercise 5.6.1-21. Assuming the 4 constants α,β,γ,c to be real, find restrictions on their values which guarantee that *all* solutions of (11) are *completely periodic*. *Hint*: see *Exercise 5.6.1-17*.

Application of the parametrization (5.5-1) to the *integrable* matrix evolution equation (5.4.2-6) (with the position $\underline{C}=\gamma\underline{1}+i\vec{C}\cdot\vec{\underline{\sigma}}$) yields the following scalar/vector *integrable one-body problem* :

$$\ddot{\rho}=2c^2\left[\rho(\rho^2-3r^2)+\gamma\rho-(\vec{C}\cdot\vec{r})\right] , \tag{12a}$$

$$\ddot{\vec{r}}=2c^2\left[-\vec{r}(r^2-3\rho^2)+\gamma\vec{r}+\rho\vec{C}\right] . \tag{12b}$$

Of course only in the case with $\vec{C}=0$ is *rotation-invariance* preserved.

Exercise 5.6.1-22. Apply to these equations of motion, (12), the techniques of *complexification* and/or *association*, and display the scalar/vector Newtonian equations of motion of the *integrable 2- and 4-body problems* obtained in this manner.

Exercise 5.6.1-23. Display the scalar/vector Newtonian equations of motion entailed by application of the parametrization (5.5-1) to the *integrable* matrix evolution equation (5.4.3-14).

Next let us recall that the *linearizable one-body problem* characterized by the Newtonian equation of motion (5.2-19) has been treated in Sect. 5.2.

Exercise 5.6.1-24. Display the Newtonian equations of motion of the *solvable 2-body translation-invariant problems* which obtain via the *as-*

sociation trick from (5.2-58c), (5.2-59d) and (5.2.2-1). *Hint*: see (3b,c,d,e).

Exercise 5.6.1-25. Show that the following Newtonian equation of motion is *linearizable*:

$$\ddot{\vec{r}} = 2a\dot{\vec{r}} + b\vec{r} + C\vec{r} \wedge \dot{\vec{r}} \ . \tag{13}$$

Hint: apply (5.5-3) to (5.2.3-1) (with $C = 2c$).

Complexification of equation (13) via the positions

$$\vec{r} = \vec{r}^{(1)} + i\vec{r}^{(2)},\ a = \alpha + i\tilde{\alpha},\ b = \beta + i\tilde{\beta},\ C = c + i\tilde{c}, \tag{14a}$$

yields the *integrable 2-body problem* characterized by the equations of motion

$$\ddot{\vec{r}}^{(1)} = 2(\alpha\dot{\vec{r}}^{(1)} - \tilde{\alpha}\dot{\vec{r}}^{(2)}) + \beta\vec{r}^{(1)} - \tilde{\beta}\vec{r}^{(2)}$$

$$+ c(\vec{r}^{(1)} \wedge \dot{\vec{r}}^{(1)} - \vec{r}^{(2)} \wedge \dot{\vec{r}}^{(2)}) - \tilde{c}(\vec{r}^{(1)} \wedge \dot{\vec{r}}^{(2)} + \vec{r}^{(2)} \wedge \dot{\vec{r}}^{(1)}), \tag{14b}$$

$$\ddot{\vec{r}}^{(2)} = 2(\alpha\dot{\vec{r}}^{(2)} + \tilde{\alpha}\dot{\vec{r}}^{(1)}) + \beta\vec{r}^{(2)} + \tilde{\beta}\vec{r}^{(1)}$$

$$+ c(\vec{r}^{(1)} \wedge \dot{\vec{r}}^{(2)} + \vec{r}^{(2)} \wedge \dot{\vec{r}}^{(1)}) + \tilde{c}(\vec{r}^{(1)} \wedge \dot{\vec{r}}^{(1)} - \vec{r}^{(2)} \wedge \dot{\vec{r}}^{(2)}). \tag{14c}$$

Exercise 5.6.1-26. Show that, at least in the 2 cases characterized by the restrictions $\alpha = \tilde{\beta} = 0,\ \tilde{\alpha} = 3\omega/2, \beta = 2\omega^2$ or $\alpha = \beta = \tilde{\beta} = 0,\ \tilde{\alpha} = \omega/2$, with ω an arbitrary (real, nonvanishing) constant, this model, (14), is *solvable* and *all* its solutions are completely periodic with period $T = 2\pi/\omega$. *Hint*: see Sect. 5.2.3.

Association of (14) with an appropriate, trivially solvable, model (see (3b,c,d,e), with α replaced by γ) yields the *linearizable translation-invariant 4-body problem* characterized by the Newtonian equations of motion

$$\ddot{\vec{r}}^{(1,\pm)} = \{\ \gamma\dot{\vec{s}}^{(1)} - \tilde{\gamma}\dot{\vec{s}}^{(2)} \pm [\ 2(\alpha\dot{\vec{r}}^{(1)} - \tilde{\alpha}\dot{\vec{r}}^{(2)}) + \beta\vec{r}^{(1)} - \tilde{\beta}\vec{r}^{(2)}$$

$$+ c(\vec{r}^{(1)} \wedge \dot{\vec{r}}^{(1)} - \vec{r}^{(2)} \wedge \dot{\vec{r}}^{(2)}) - \tilde{c}(\vec{r}^{(1)} \wedge \dot{\vec{r}}^{(2)} + \vec{r}^{(2)} \wedge \dot{\vec{r}}^{(1)})\ ]\ \}/2\ , \tag{15a}$$

$$\ddot{\vec{r}}^{(2,\pm)} = \{\ \gamma\dot{\vec{s}}^{(2)} + \tilde{\gamma}\dot{\vec{s}}^{(1)} \pm [\ 2(\alpha\dot{\vec{r}}^{(2)} + \tilde{\alpha}\dot{\vec{r}}^{(1)}) + \beta\vec{r}^{(2)} + \tilde{\beta}\vec{r}^{(1)}$$

$$+c(\vec{r}^{(1)}\wedge\dot{\vec{r}}^{(2)}+\vec{r}^{(2)}\wedge\dot{\vec{r}}^{(1)})+\tilde{c}(\vec{r}^{(1)}\wedge\dot{\vec{r}}^{(1)}-\vec{r}^{(2)}\wedge\dot{\vec{r}}^{(2)})\;]\;\}/2\;, \tag{15b}$$

$$\vec{r}^{(1)}\equiv\vec{r}^{(1,+)}-\vec{r}^{(1,-)},\;\vec{r}^{(2)}\equiv\vec{r}^{(2,+)}-\vec{r}^{(2,-)},\;\;\vec{s}^{(1)}\equiv\vec{r}^{(1,+)}+\vec{r}^{(1,-)},\;\vec{s}^{(2)}\equiv\vec{r}^{(2,+)}+\vec{r}^{(2,-)}\;. \tag{15c}$$

Exercise 5.6.1-27. Show that this model is *solvable* and that *all* its solutions are completely periodic with period $T=2\pi/\omega$ if there holds either one of the 2 sets of restrictions on the 4 coupling constants $\alpha,\tilde{\alpha},\beta,\tilde{\beta}$ detailed in the preceding *Exercise 5.6.1-26*, and in addition there hold the 2 constraints $\gamma=0,\,\tilde{\gamma}=m\,\omega$, with m an arbitrary integer ($m\neq0$; for $m=0$, namely $\gamma=\tilde{\gamma}=0$, (15) reduces to (14)). *Hint*: use (3b,c,d,e) (with α replaced by γ), and see Sect. 5.2.3.

Remark 5.6.1-28. Recall that the *one-body problem* characterized by the Newtonian equation of motion

$$\ddot{\vec{r}}=f(r)\,\vec{r}\wedge\dot{\vec{r}}\quad, \tag{16}$$

is *linearizable* (for arbitrary $f(r)$, see (5.2.1-1); this equation of motion, (16), is actually *solvable* if $f(r)=k\,r^{p}$ with $p=0,p=-2$ or $p=-3$, see Sects. 5.2.1 and 5.2.2).

Next, let us report two *linearizable three-body problems* that are obtained from the "Nahm equations" (5.4.3-21). The first one is obtained by applying the parametrization (5.5-3) to the *linearizable* matrix equation (5.4.3-23) and it reads:

$$\ddot{\vec{r}}_n=(2/\mu_n)\,\{\;\mu_{n+1}\,\mu_{n+2}\,\dot{\vec{r}}_{n+2}\wedge\dot{\vec{r}}_{n+1}$$

$$+\sum_{m=1}^{3}\left[-\frac{1}{2}a_{nm}\,\dot{\vec{r}}_m+\left(a_{n+1,m}\,\mu_{n+2}\,\dot{\vec{r}}_{n+2}-a_{n+2,m}\,\mu_{n+1}\,\dot{\vec{r}}_{n+1}+\sum_{k=1}^{3}a_{n+1,m}a_{n+2,k}\vec{r}_k\right)\wedge\vec{r}_m\right]\}. \tag{17}$$

In these equations, (17), as well as in the following ones, (18) and (19), all indices are defined mod(3).

Exercise 5.6.1-29. Display the Newtonian equations of motion of the *translation-invariant linearizable 6-body problem* that obtains from (17) by applying the *association* trick (see (3b,c,d,e)).

The second *linearizable three-body problem* related to the Nahm equations (5.4.3-21) is characterized by the Newtonian equations of motions

$$\ddot{\vec{r}}_n=\left[a_n+c^2\left(r_{n+1}^2+r_{n+2}^2\right)\right]\vec{r}_n-\left[b_{n,n+1}+c^2\left(\vec{r}_n\cdot\vec{r}_{n+1}\right)\right]\vec{r}_{n+1}-\left[b_{n,n+2}+c^2\left(\vec{r}_n\cdot\vec{r}_{n+2}\right)\right]\vec{r}_{n+2}\ , \quad (18a)$$

where:

$$a_n=\{\ u_{n+1}^2+u_{n+2}^2+\left(\gamma_{n,n}-\gamma_{n+1,n+1}-\gamma_{n+2,n+2}\right)\left[\left(\vec{r}_{n+1}\cdot\vec{u}_{n+1}\right)+\left(\vec{r}_{n+2}\cdot\vec{u}_{n+2}\right)\right]$$

$$-\gamma_{n,n+1}\left[\left(\vec{r}_n\cdot\vec{u}_{n+1}\right)+\left(\vec{r}_{n+1}\cdot\vec{u}_n\right)\right]-\gamma_{n,n+2}\left[\left(\vec{r}_n\cdot\vec{u}_{n+2}\right)+\left(\vec{r}_{n+2}\cdot\vec{u}_n\right)\right]\ \}/\Delta^2\ , \quad (18b)$$

$$b_{n,m}=\{\ \vec{u}_n\cdot\vec{u}_m+\frac{1}{2}\left(\gamma_{m,m}-\gamma_{m+1,m+1}-\gamma_{m+2,m+2}\right)\left[\left(\vec{r}_n\cdot\vec{u}_m\right)+\left(\vec{r}_m\cdot\vec{u}_n\right)\right]$$

$$-\gamma_{m,n}\left[\left(\vec{r}_{n+1}\cdot\vec{u}_{n+1}\right)+\left(\vec{r}_{n+2}\cdot\vec{u}_{n+2}\right)\right]+\gamma_{m,2n-m}\left[\left(\vec{r}_n\cdot\vec{u}_{2n-m}\right)+\left(\vec{r}_{2n-m}\cdot\vec{u}_n\right)\right]\ \}/\Delta^2\ , \quad (18c)$$

$$\gamma_{n,m}=\dot{\vec{r}}_n\cdot\vec{r}_m\ , \quad (18d)$$

$$\vec{u}_n=\frac{1}{3}\left(\gamma_{n,n}-\gamma_{n+1,n+1}-\gamma_{n+2,n+2}\right)\vec{r}_n+\gamma_{n+1,n}\,\vec{r}_{n+1}+\gamma_{n+2,n}\,\vec{r}_{n+2}\ , \quad (18e)$$

$$\Delta=\vec{r}_1\cdot\vec{r}_2\wedge\vec{r}_3. \quad (18f)$$

These equations of motion, (18), are obtained from (5.4.3-21) via matrix *duplication* and by again applying the parametrization (5.5-3). The relevant formulas read:

$$\underline{M}_n=\begin{pmatrix}\underline{A}_n & \underline{B}_n\\ \underline{B}_n & \underline{A}_n\end{pmatrix}\ , \quad (19a)$$

$$\underline{A}_n=(i/c)\,\vec{u}_n\cdot\vec{\sigma}_n,\quad \underline{B}_n=i\,\vec{r}_n\cdot\vec{\sigma}_n\ . \quad (19b)$$

Note that an apparent paradox arises, namely, setting $c=0$ in the equations of motion (18), the "free motion" equation $\ddot{\vec{r}}=0$ is *not* recovered (in contrast to what seems suggested by (5.4.3-21)): this is due to the definition of $\underline{A}_n$, see (19b), which features the coupling constant c in the denominator.

We end Sect. 5.6.1 by displaying the *linearizable one-body problem* that corresponds, via the parametrization (5.5-3), to the linearizable matrix ODE (5.4.1-41):

$$\ddot{\vec{r}}=\alpha\dot{\vec{r}}+\beta\gamma\vec{r}+\beta\delta\vec{\vec{r}}+2(\beta a-b\delta)\vec{r}\wedge\vec{\vec{r}}+2b\vec{\vec{r}}\wedge\dot{\vec{r}}$$

$$+4(c\delta-a\beta)\vec{r}\wedge(\vec{r}\wedge\vec{\vec{r}})+4c\left\{\vec{r}\wedge(\dot{\vec{r}}\wedge\vec{\vec{r}})+\dot{\vec{r}}\wedge(\vec{r}\wedge\vec{\vec{r}})\right\}+4c\{\ \beta(\vec{\vec{r}}\wedge\vec{r})\wedge\vec{r}$$

$$+2ac\left[\vec{r}\wedge(\vec{r}\wedge\vec{\vec{r}})\right]\wedge\vec{r}+2b\vec{r}\wedge\left[\vec{\vec{r}}\wedge(\vec{r}\wedge\vec{\vec{r}})\right]+16c^2\vec{r}\wedge\left[\vec{r}\wedge\left(\vec{\vec{r}}\wedge(\vec{\vec{r}}\wedge\vec{r})\right)\right]\},\qquad(20a)$$

$$\vec{\vec{r}}=\left[(\beta-4cr^2)^2+4b^2r^2\right]^{-1}\left\{2b\vec{r}\wedge\dot{\vec{r}}+(\beta-4cr^2)\dot{\vec{r}}\right.$$

$$\left.+\left[4b^2(\beta-4cr^2)^{-1}\left(\vec{r}\cdot\dot{\vec{r}}+(4c\chi-\alpha)r^2\right)+(\beta-4cr^2)(4c\chi-\alpha)\right]\vec{r}\right\},\qquad(20b)$$

$$\chi=\left[(\beta-4cr^2)^3+8bcr^2(\beta-4cr^2)^2+4b^2\beta r^2\right]^{-1}\cdot$$

$$\cdot\left[(\beta-4cr^2)^2+4b^2r^2\right]\left(\alpha r^2-\vec{r}\cdot\dot{\vec{r}}\right).\qquad(20c)$$

Exercise 5.6.1-30. Verify! *Hint*: use (H.11).

Exercise 5.6.1-31. Clearly (20) can be rewritten in the form

$$\ddot{\vec{r}}=f^{(1)}\dot{\vec{r}}+f^{(2)}\vec{r}+f^{(3)}\vec{r}\wedge\dot{\vec{r}}\ ,\qquad(20d)$$

with $f^{(1)},f^{(2)},f^{(3)}$ scalar functions of the 3 scalars $r^2,\left|\dot{\vec{r}}\right|^2,\vec{r}\cdot\dot{\vec{r}}$, and as well of course of the 7 "coupling constants" $\alpha,\beta,\gamma,\delta,a,b,c$. Find $f^{(1)},f^{(2)},f^{(3)}$.

5.6.2 Few-body problems of Hamiltonian type

In this short Sect. 5.6.2 we report two simple cases of few-body systems, whose Newtonian equations of motion are Hamiltonian.

The first case is characterized by the scalar/vector *integrable* equations of motion (5.6.1-12), which we rewrite here to please the lazy reader:

$$\ddot{\rho}=2c^2\left[\rho(\rho^2-3r^2)+\gamma\rho-(\vec{C}\cdot\vec{r})\right]\ ,\qquad(1a)$$

$$\ddot{\vec{r}}=2c^2\left[-\vec{r}(r^2-3\rho^2)+\gamma\vec{r}+\rho\vec{C}\right]\ .\qquad(1b)$$

It is easily seen that they are yielded by the following Hamiltonian, of normal type (but note the *negative* sign in front of the second "kinetic energy term" in the right hand side):

$$H(\vec{p},\pi;\vec{r},\rho)=(p^2-\pi^2)/2+c^2[\ 2\gamma(\rho^2-r^2)-4\rho\vec{C}\cdot\vec{r}+r^4+\rho^4-6r^2\rho^2\ ]/2. \tag{1c}$$

Exercise 5.6.2-1. Verify!

These Newtonian equations of motion, (1), are *rotation-invariant* only if $\vec{C}=0$. If this condition holds, one can moreover set $\rho=0$ (consistently with (1a), which is then trivially satisfied). Then (1b) becomes the standard Newtonian equation of motion of a single unharmonic (quartic) oscillator,

$$\ddot{\vec{r}}=a\vec{r}-br^2\vec{r}, \tag{2a}$$

where we set, for notational simplicity, $c^2=b/2, 2c^2\gamma=a$. This is of course the one-body Newtonian equation of motion yielded by the standard Hamiltonian

$$H(\vec{p},\vec{r})=p^2/2+V(r), \tag{2b}$$

with

$$V(r)=-ar^2/2+br^4/4, \tag{2c}$$

whose integrability is well-known (the standard one-body Hamiltonian (2b) is of course integrable for any spherically symmetrical potential energy $V(r)$). Clearly all (real) solutions of this model are confined if the coupling constant b is positive, $b>0$.

A more complete analysis of *integrable* unharmonic oscillators (including also several *few-body* cases) is given in Sect. 5.6.5.

A *solvable* (but rather trivially so) *Hamiltonian one-body problem* is characterized by the Hamiltonian

$$H(\vec{p},\vec{q})=a\vec{q}\cdot\vec{p}+bp^2/2-cq^2/2+(\lambda/2)(\vec{q}\wedge\vec{p})^2, \tag{3a}$$

entailing the (Hamiltonian) equations of motion

$$\dot{\vec{q}}=a\vec{q}+b\vec{p}+\lambda[\vec{q}\wedge[\vec{q}\wedge\vec{p}]],\ \dot{\vec{p}}=c\vec{q}-a\vec{p}+\lambda[\vec{p}\wedge[\vec{q}\wedge\vec{p}]]. \tag{3b}$$

The solvability of these equations of motion, (3b), is entailed by their correspondence, via the parametrization (5.5-3), with the $N=1$ case of the *solvable* matrix evolution equations (5.4.1-36), with $a_{11}=a$, $b_{11}=b$, $c_{11}=c$ and λ replaced by $\lambda/4$; the trivial character of this property of solvability is entailed by the fact that the "angular momentum" $\vec{q}\wedge\vec{p}$ is a constant of the motion, so that the equations of motion (3b) are effectively linear.

Exercise 5.6.2-2. Verify the correspondence of (3a) with (3b), as well as the validity of the two assertions stated immediately above.

Exercise 5.6.2-3. Investigate the behavior of the Hamiltonian system (3).

5.6.3 Many-body problems of Newtonian type

In Sect. 5.6.3 we report a representative list of many-body problems featuring Newtonian equations of motion amenable to exact treatments. Let us reemphasize that our main purpose here is to illustrate the kind of systems that can be obtained and the techniques to get them. Of course many more such models can be manufactured: all readers are welcome to try their brains at this instructive and interesting sport.

A *linearizable N-body problem* (featuring $3N$ arbitrary coupling constants):

$$\ddot{\vec{r}}_n=\sum_{n_1=1}^{N}\left(2a_{n-n_1}\dot{\vec{r}}_{n_1}+b_{n-n_1}\vec{r}_{n_1}\right)+\sum_{n_1,n_2=1}^{N}\left(c_{n-n_1-n_2}\vec{r}_{n_1}\wedge\dot{\vec{r}}_{n_2}\right). \tag{1}$$

Here, and in analogous equations below, all indices are defined $\mathrm{mod}(N)$. If $a_n=b_n=0$ the model is not only *linearizable*: it is *solvable*.

These equations of motion may be obtained from (5.6.1-13) via the first *multiplication* technique of Sect. 5.3, or equivalently by applying the parametrization (5.5-3) to (5.3-15).

Exercise 5.6.3-1. Verify! (But beware of the need, in the second case, to rescale the coefficients c_n by a factor 2).

For instance for $N=3$ these equations of motion, when displayed in longhand, read

$$\ddot{\vec{r}}_1 = 2\,(a_1\dot{\vec{r}}_3 + a_2\dot{\vec{r}}_2 + a_3\dot{\vec{r}}_1) + b_1\vec{r}_3 + b_2\vec{r}_2 + b_3\vec{r}_1 + 2\left[\; c_1(\vec{r}_3 \wedge \dot{\vec{r}}_3) + c_2(\vec{r}_1 \wedge \dot{\vec{r}}_1) + c_3(\vec{r}_2 \wedge \dot{\vec{r}}_2)\;\right]$$

$$+ c_1(\vec{r}_1 \wedge \dot{\vec{r}}_2 + \vec{r}_2 \wedge \dot{\vec{r}}_1) + c_2(\vec{r}_2 \wedge \dot{\vec{r}}_3 + \vec{r}_2 \wedge \dot{\vec{r}}_3) + c_3(\vec{r}_1 \wedge \dot{\vec{r}}_3 + \vec{r}_3 \wedge \dot{\vec{r}}_1)\;, \tag{2a}$$

$$\ddot{\vec{r}}_2 = 2\,(a_1\dot{\vec{r}}_1 + a_2\dot{\vec{r}}_3 + a_3\dot{\vec{r}}_2) + b_1\vec{r}_1 + b_2\vec{r}_3 + b_3\vec{r}_2 + 2\left[\; c_1(\vec{r}_2 \wedge \dot{\vec{r}}_2) + c_2(\vec{r}_3 \wedge \dot{\vec{r}}_3) + c_3(\vec{r}_1 \wedge \dot{\vec{r}}_1)\;\right]$$

$$+ c_1(\vec{r}_1 \wedge \dot{\vec{r}}_3 + \vec{r}_3 \wedge \dot{\vec{r}}_1) + c_2(\vec{r}_1 \wedge \dot{\vec{r}}_2 + \vec{r}_1 \wedge \dot{\vec{r}}_2) + c_3(\vec{r}_2 \wedge \dot{\vec{r}}_3 + \vec{r}_3 \wedge \dot{\vec{r}}_2)\;, \tag{2b}$$

$$\ddot{\vec{r}}_3 = 2\,(a_1\dot{\vec{r}}_2 + a_2\dot{\vec{r}}_1 + a_3\dot{\vec{r}}_3) + b_1\vec{r}_2 + b_2\vec{r}_1 + b_3\vec{r}_3 + 2\left[\; c_1(\vec{r}_1 \wedge \dot{\vec{r}}_1) + c_2(\vec{r}_2 \wedge \dot{\vec{r}}_2) + c_3(\vec{r}_3 \wedge \dot{\vec{r}}_3)\;\right]$$

$$+ c_1(\vec{r}_2 \wedge \dot{\vec{r}}_3 + \vec{r}_2 \wedge \dot{\vec{r}}_3) + c_2(\vec{r}_1 \wedge \dot{\vec{r}}_3 + \vec{r}_3 \wedge \dot{\vec{r}}_1) + c_3(\vec{r}_1 \wedge \dot{\vec{r}}_2 + \vec{r}_2 \wedge \dot{\vec{r}}_1)\;. \tag{2c}$$

Let us however recall that the equations obtained via this technique can generally be decoupled via a linear transformation (see (5.3-27)).

A *translation-invariant linearizable (2N)-body problem* (featuring $4N$ arbitrary coupling constants):

$$\ddot{\vec{r}}^{(\pm)}{}_n = \left\{\; \alpha_n \dot{\vec{s}}_n \pm \sum_{n_1=1}^{N} \left(2\,a_{n-n_1}\dot{\vec{r}}_{n_1} + b_{n-n_1}\vec{r}_{n_1}\right) + \sum_{n_1,n_2=1}^{N} \left(c_{n-n_1-n_2}\vec{r}_{n_1} \wedge \dot{\vec{r}}_{n_2}\right) \right\}/2\;, \tag{3a}$$

$$\vec{s}_n = \vec{r}_n^{(+)} + \vec{r}_n^{(-)},\ \vec{r}_n = \vec{r}_n^{(+)} - \vec{r}_n^{(-)}\;,\ \vec{r}_n^{(\pm)} = (\vec{s}_n \pm \vec{r}_n)/2\;. \tag{3b}$$

Again, if $a_n = b_n = 0$, the model is actually *solvable*.

Here the $2N$ "particle coordinates" are of course the 3-vectors $\vec{r}_n^{(\pm)} \equiv \vec{r}_n^{(\pm)}(t)$, $n=1,\ldots N$.

Exercise 5.6.3-2. Show that these equations of motion, (3), are indeed *linearizable* (and *solvable* if $a_n = b_n = 0$). *Hint*: apply the *association* technique to (1), or equivalently, and more directly, note that (3a) entail that the 3-vectors $\vec{s}_n \equiv \vec{s}_n(t)$, see (3b), satisfy the trivially solvable equations of motion $\ddot{\vec{s}}_n = \alpha_n\,\dot{\vec{s}}_n$, and the 3-vectors $\vec{r}_n \equiv \vec{r}_n(t)$, see (3b), satisfy (1).

A scalar/vector *solvable N-body problem* (featuring $4N$ arbitrary coupling constants):

$$\ddot{\rho}_n = \alpha_n + \sum_{n_1=1}^{N} \{\beta_{n-n_1}\rho_{n_1} + \gamma_{n-n_1}\dot{\rho}_{n_1}\} - 3\sum_{n_1,n_2=1}^{N} \{c_{n-n_1-n_2}[\rho_{n_1}\dot{\rho}_{n_2} - (\vec{r}_{n_1}\cdot\dot{\vec{r}}_{n_2})]\}$$

$$+ \sum_{n_1,n_2,n_3=1}^{N} \{c_{n-n_1-n_2-n_3}\gamma_{n_1}[\rho_{n_2}\rho_{n_3} - (\vec{r}_{n_2}\cdot\vec{r}_{n_3})]\}$$

$$- \sum_{n_1,n_2,n_3,n_4=1}^{N} \{c_{n-n_1-n_2-n_3-n_4}c_{n_1}\rho_{n_2}[\rho_{n_3}\rho_{n_4} - 3(\vec{r}_{n_3}\cdot\vec{r}_{n_4})]\}, \qquad (4a)$$

$$\ddot{\vec{r}}_n = \sum_{n_1=1}^{N} \{\beta_{n-n_1}\vec{r}_{n_1} + \gamma_{n-n_1}\dot{\vec{r}}_{n_1}\} - \sum_{n_1,n_2=1}^{N} \{c_{n-n_1-n_2}[3\dot{\rho}_{n_1}\vec{r}_{n_2} + 3\rho_{n_1}\dot{\vec{r}}_{n_2} - (\vec{r}_{n_1}\wedge\dot{\vec{r}}_{n_2})]\}$$

$$+2\sum_{n_1,n_2,n_3=1}^{N} \{c_{n-n_1-n_2-n_3}\gamma_{n_1}\rho_{n_2}\vec{r}_{n_3}\} - \sum_{n_1,n_2,n_3,n_4=1}^{N} \{c_{n-n_1-n_2-n_3-n_4}c_{n_1}\vec{r}_{n_2}[3\rho_{n_3}\rho_{n_4} - (\vec{r}_{n_3}\cdot\vec{r}_{n_4})]\}. \qquad (4b)$$

These equations of motion, (4), are obtained by applying to (5.6.1-9) the (first) *multiplication* technique described in Sect. 5.3. Hence these equations of motion can be decoupled by using a linear transformation of type (5.3-27).

Exercise 5.6.3-3. Verify!

A *solvable N-body problem* (featuring N^2+3N arbitrary coupling constants):

$$\ddot{\vec{r}}_n = a_n\dot{\vec{r}}_n - \alpha_n\gamma_n\vec{r}_n\wedge\dot{\vec{r}}_n$$

$$+[a_n(a_n^{\,2}+\gamma_n^{\,2}r_n^{\,2})]^{-1}\{\,-\gamma_n\dot{\vec{r}}_n\wedge[\gamma_n^{\,2}(\vec{r}_n\cdot\dot{\vec{r}}_n)\vec{r}_n + a_n\gamma_n\vec{r}_n\wedge\dot{\vec{r}}_n]$$

$$-\sum_{m=1}^{N}\{\,b_{nm}\{\,\dot{\vec{r}}_m + \gamma_n\dot{\vec{r}}_n\wedge[\,\gamma_n^{\,2}(\vec{r}_n\cdot\vec{r}_m)\vec{r}_n + a_n^{\,2}\vec{r}_m + a_n\gamma_n\vec{r}_n\wedge\vec{r}_m\,]\,\}\,\}\,\}. \qquad (5)$$

These equations of motion, (5), are obtained by applying the parametrization (5.5-3) to the special case of the *solvable* matrix evolution equation (5.4.3-15) with $\underline{D}=\underline{\tilde{D}}=\underline{1}$.

Exercise 5.6.3-4. Verify, and analyze the motions entailed by (5). *Hint*: do first the one-body case, $N=1$.

A scalar/vector *solvable N-body problem* (featuring $4N$ arbitrary coupling constants, and "nearest-neighbor" interactions):

$$\ddot{\rho}_n = (a_n - a_{n+1})\tilde{c}_n + (a_n - a_{n+1})(\tilde{a}_n - a_n)\rho_n + \tilde{c}_n(b_n\rho_n - b_{n+1}\rho_{n+1})$$

$$-(\tilde{a}_n - a_n)b_{n+1}(\rho_n\rho_{n+1} - \vec{r}_n\cdot\vec{r}_{n+1}) - 3b_n(\dot{\rho}_n\rho_n - \dot{\vec{r}}_n\cdot\vec{r}_n)$$

$$+(\tilde{a}_n - 2a_n + a_{n+1})[\,\rho_n + b_n(\rho_n^{\,2} - r_n^{\,2})\,] + b_{n+1}(\dot{\rho}_n\rho_{n+1} - \dot{\vec{r}}_n\cdot\vec{r}_{n+1})$$

$$+b_n b_{n+1}[\,\rho_{n+1}(\rho_n^{\,2} - r_n^{\,2}) - 2\rho_n\vec{r}_n\cdot\vec{r}_{n+1}\,] - b_n^{\,2}\rho_n(\rho_n^{\,2} - 3r_n^{\,2}), \tag{6a}$$

$$\ddot{\vec{r}}_n = (a_n - a_{n+1})(\tilde{a}_n - a_n)\vec{r}_n + \tilde{c}_n(b_n\vec{r}_n - b_{n+1}\vec{r}_{n+1})$$

$$-(\tilde{a}_n - a_n)b_{n+1}(\rho_n\vec{r}_{n+1} + \rho_{n+1}\vec{r}_n + \vec{r}_n\wedge\vec{r}_{n+1}) - b_n(3\,\dot{\rho}_n\vec{r}_n + 3\,\rho_n\dot{\vec{r}}_n - \dot{\vec{r}}_n\wedge\vec{r}_n)$$

$$+(\tilde{a}_n - 2\,a_n + a_{n+1})(\dot{\vec{r}}_n + 2\,b_n\rho_n\vec{r}_n) + b_{n+1}(\rho_{n+1}\dot{\vec{r}}_n + \dot{\rho}_n\vec{r}_n + \dot{\vec{r}}_n\wedge\vec{r}_{n+1})$$

$$+b_n b_{n+1}[\,2\,\rho_{n+1}\rho_n\vec{r}_n + (\rho_n^{\,2} - r_n^{\,2})\vec{r}_{n+1} + 2\,\rho_n\vec{r}_n\wedge\vec{r}_{n+1}\,] - b_n^{\,2}(3\,\rho_n^{\,2} - r_n^{\,2})\vec{r}_n\,. \tag{6b}$$

These equations of motion, (6), are obtained by applying the parametrization (5.5-1) to the matrix evolution ODE (5.4.3-25).

Exercise 5.6.3-5. Verify, and analyze the motions entailed by (6). *Hint*: do first the one-body case, $N=1$.

A *linearizable N-body problem* (featuring $2N^2+N$ arbitrarily assigned functions of time):

$$\ddot{\vec{r}}_n = \sum_{m=1}^{N}\left\{2\,a_{nm}(t)\dot{\vec{r}}_m + b_{nm}(t)\vec{r}_m + C_m(t)\left[2(\vec{r}_m\wedge\dot{\vec{r}}_n) - (\vec{r}_n\wedge\dot{\vec{r}}_m)\right] - \dot{C}_m(t)(\vec{r}_n\wedge\vec{r}_m)\right\}$$

$$+\sum_{l,m=1}^{N}\left\{2\,a_{nm}C_l(t)(\vec{r}_m\wedge\vec{r}_l) - C_m(t)\,C_l(t)\,[(\vec{r}_n\wedge\vec{r}_m)\wedge\vec{r}_l]\right\}. \tag{7}$$

These equations of motion, (7), are obtained by applying the parametrization (5.5-3) to the *linearizable* system of matrix ODEs (5.4.1-21) (with $C_n = 2c_n$). (Verify!). Note that, as in (5.4.1-21), we are allowing in this case all the $N(2N+1)$ quantities $a_{nm}(t)$, $b_{nm}(t)$, $C_n(t)$ to be *time-dependent*, since this does not negate the property of these equations of motion, (7), to be *linearizable*. Of course one may limit consideration to the case in which all the coefficients a_{nm}, b_{nm}, C_m are time-independent.

Exercise 5.6.3-6. Obtain the (of course more general, and as well *linearizable*) equations of motion that obtain from (5.4.1-21) via the parametrization (5.5-1).

An *integrable N-body problem* (with "nearest-neighbor" interactions):

$$\ddot{\vec{r}}_n = \left[2\dot{\vec{r}}_n(\dot{\vec{r}}_n \cdot \vec{r}_n) - \vec{r}_n(\dot{\vec{r}}_n \cdot \dot{\vec{r}}_n)\right]/r_n^2 + \gamma\left\{\vec{r}_{n+1} - \left[2\vec{r}_n(\vec{r}_n \cdot \vec{r}_{n-1}) - \vec{r}_{n-1}(\vec{r}_n \cdot \vec{r}_n)\right]/r_{n-1}^2\right\}. \tag{8}$$

These equations of motion, (8), are obtained by applying the parametrization (5.5-3) to (5.4.2-1). (Verify!). Note their "nearest-neighbor" character. We do not repeat here, nor below in analogous cases, the comments on the need that these equations of motion be supplemented by "boundary conditions at the n-ends" (see Sect. 5.4.2).

Exercise 5.6.3-7. Obtain the (more general) equations of motion that obtain from (5.4.2-1) via the parametrization (5.5-1).

A scalar/vector *translation-invariant integrable N-body problem* (with "nearest-neighbor" interactions):

$$\ddot{\rho}_n = c\left[\dot{\rho}_n(\rho_{n+1} - 2\rho_n + \rho_{n-1}) - \dot{\vec{r}}_n \cdot (\vec{r}_{n+1} - 2\vec{r}_n + \vec{r}_{n-1})\right], \tag{9a}$$

$$\ddot{\vec{r}}_n = c\left[\dot{\vec{r}}_n(\rho_{n+1} - 2\rho_n + \rho_{n-1}) + \dot{\rho}_n(\vec{r}_{n+1} - 2\vec{r}_n + \vec{r}_{n-1}) + \dot{\vec{r}}_n \wedge (\vec{r}_{n+1} - \vec{r}_{n-1})\right]. \tag{9b}$$

These Newtonian equations of motion, (9), are obtained by applying the parametrization (5.5-1) to (5.4.2-2b) (in this case the parametrization (5.5-3) is not applicable!).

Exercise 5.6.3-8. Verify!

Note the *translation-invariant* character of these equations of motion, (9): they are indeed invariant under $\rho_n \to \tilde{\rho}_n = \rho_n + \rho_0$, $\vec{r}_n \to \tilde{\vec{r}}_n = \vec{r}_n + \vec{r}_0$ with $\rho_0, \vec{r}_0$ time-independent but otherwise arbitrary.

Exercise 5.6.3-9. Obtain from (9) more general equations of motion by applying the (first) multiplication technique of Sect. 5.3.

Exercise 5.6.3-10. Show that the following Newtonian equations of motion (which feature the 4 arbitrary real coupling constants $\lambda, \omega, \gamma, \tilde{\gamma}$) are just as *integrable* as (9) (to which they reduce back for $\lambda = \omega = \tilde{\gamma} = 0$, $\gamma = c$ and $\rho_n^{(2)} = 0, \vec{r}_n^{(2)} = 0$).

$$\ddot{\rho}_n^{(1)} = 2(\omega^2 - \lambda^2)\,\rho_n^{(1)} + 4\lambda\omega\,\rho_n^{(2)} + 3\lambda\dot{\rho}_n^{(1)} - 3\omega\dot{\rho}_n^{(2)}$$

$$+\gamma A_n - \tilde{\gamma}\tilde{A}_n - (\lambda\gamma - \omega\tilde{\gamma})\,B_n + (\lambda\tilde{\gamma} + \omega\gamma)\,\tilde{B}_n, \tag{10a}$$

$$\ddot{\rho}_n^{(2)} = 2(\omega^2 - \lambda^2)\,\rho_n^{(2)} - 4\lambda\omega\,\rho_n^{(1)} + 3\lambda\,\dot{\rho}_n^{(2)} + 3\omega\,\dot{\rho}_n^{(1)}$$

$$+\tilde{\gamma} A_n + \gamma\tilde{A}_n - (\lambda\tilde{\gamma} + \omega\gamma)\,B_n - (\lambda\gamma - \omega\tilde{\gamma})\,\tilde{B}_n\ , \tag{10b}$$

$$\ddot{\vec{r}}_n^{(1)} = 2(\omega^2 - \lambda^2)\vec{r}_n^{(1)} + 4\lambda\omega\vec{r}_n^{(2)} + 3\lambda\,\dot{\vec{r}}_n^{(1)} - 3\omega\,\dot{\vec{r}}_n^{(2)}$$

$$+\gamma\,\vec{C}_n - \tilde{\gamma}\,\tilde{\vec{C}}_n - (\lambda\gamma - \omega\tilde{\gamma})\,\vec{D}_n + (\lambda\tilde{\gamma} + \omega\gamma)\,\tilde{\vec{D}}_n, \tag{10c}$$

$$\ddot{\vec{r}}_n^{(2)} = 2(\omega^2 - \lambda^2)\vec{r}_n^{(2)} - 4\lambda\omega\vec{r}_n^{(1)} + 3\lambda\,\dot{\vec{r}}_n^{(2)} + 3\omega\,\dot{\vec{r}}_n^{(1)}$$

$$+\tilde{\gamma}\,\vec{C}_n + \gamma\,\tilde{\vec{C}}_n - (\lambda\tilde{\gamma} + \omega\gamma)\,\vec{D}_n - (\lambda\gamma - \omega\tilde{\gamma})\,\tilde{\vec{D}}_n, \tag{10d}$$

$$A_n \equiv A_n^{(1)} - A_n^{(2)},\ \tilde{A}_n \equiv \tilde{A}_n^{(1)} + \tilde{A}_n^{(2)};\ B_n \equiv B_n^{(1)} - B_n^{(2)},\ \tilde{B}_n \equiv \tilde{B}_n^{(1)} + \tilde{B}_n^{(2)}, \tag{10e}$$

$$A_n^{(j)} \equiv \dot{\rho}_n^{(j)}(\rho_{n+1}^{(j)} - 2\rho_n^{(j)} + \rho_{n-1}^{(j)}) - \dot{\vec{r}}_n^{(j)} \cdot (\vec{r}_{n+1}^{(j)} - 2\vec{r}_n^{(j)} + \vec{r}_{n-1}^{(j)}), j = 1,2\ , \tag{10f}$$

$$\tilde{A}_n^{(j)} \equiv \dot{\rho}_n^{(j+1)}(\rho_{n+1}^{(j)} - 2\rho_n^{(j)} + \rho_{n-1}^{(j)}) - \dot{\vec{r}}_n^{(j+1)} \cdot (\vec{r}_{n+1}^{(j)} - 2\vec{r}_n^{(j)} + \vec{r}_{n-1}^{(j)}), j = 1,2,\text{mod}(2), \tag{10g}$$

$$B_n^{(j)} \equiv \rho_n^{(j)}(\rho_{n+1}^{(j)} - 2\rho_n^{(j)} + \rho_{n-1}^{(j)}) - \vec{r}_n^{(j)} \cdot (\vec{r}_{n+1}^{(j)} - 2\vec{r}_n^{(j)} + \vec{r}_{n-1}^{(j)}), j = 1,2\ , \tag{10h}$$

$$\tilde{B}_n^{(j)} \equiv \rho_n^{(j+1)}({\rho^{(j)}}_{n+1} - 2\rho_n^{(j)} + \rho_{n-1}^{(j)}) - \vec{r}_n^{(j+1)} \cdot (\vec{r}_{n+1}^{(j)} - 2\vec{r}_n^{(j)} + \vec{r}_{n-1}^{(j)}), j = 1,2,\text{mod}(2), \tag{10i}$$

$$\vec{C}_n \equiv \vec{C}_n^{(1)} - \vec{C}_n^{(2)},\ \tilde{\vec{C}}_n \equiv \tilde{\vec{C}}_n^{(1)} + \tilde{\vec{C}}_n^{(2)};\ \vec{D}_n \equiv \vec{D}_n^{(1)} - \vec{D}_n^{(2)},\ \tilde{\vec{D}}_n \equiv \tilde{\vec{D}}_n^{(1)} + \tilde{\vec{D}}_n^{(2)}\ , \tag{10j}$$

$$\vec{C}_n^{(j)} \equiv \dot{\vec{r}}_n^{(j)}\,(\rho_{n+1}^{(j)} - 2\,\rho_n^{(j)} + \rho_{n-1}^{(j)}) + \dot{\rho}_n^{(j)}(\vec{r}_{n+1}^{(j)} - 2\,\vec{r}_n^{(j)} + \vec{r}_{n-1}^{(j)})$$

$$+\dot{\vec{r}}_n^{(j)} \wedge (\vec{r}_{n+1}^{(j)} - \vec{r}_{n-1}^{(j)}), j = 1,2\ , \tag{10k}$$

$$\widetilde{C}_n^{(j)} \equiv \dot{\vec{r}}_n^{(j+1)} (\rho_{n+1}^{(j)} - 2\,\rho_n^{(j)} + \rho_{n-1}^{(j)}) + \dot{\rho}_n^{(j+1)} (\vec{r}_{n+1}^{(j)} - 2\vec{r}_n^{(j)} + \vec{r}_{n-1}^{(j)})$$

$$+\dot{\vec{r}}_n^{(j+1)} \wedge (\vec{r}_{n+1}^{(j)} - \vec{r}_{n-1}^{(j)}),\ j = 1,2,\mathrm{mod}(2)\ , \tag{10l}$$

$$\vec{D}_n^{(j)} \equiv \vec{r}_n^{(j)} (\rho_{n+1}^{(j)} - 2\,\rho_n^{(j)} + \rho_{n-1}^{(j)}) + \rho_n^{(j)} (\vec{r}_{n+1}^{(j)} - 2\vec{r}_n^{(j)} + \vec{r}_{n-1}^{(j)})$$

$$+\vec{r}_n^{(j)} \wedge (\vec{r}_{n+1}^{(j)} - \vec{r}_{n-1}^{(j)}),\ j = 1,2\ , \tag{10m}$$

$$\widetilde{\vec{D}}_n^{(j)} \equiv \vec{r}_n^{(j+1)} (\rho_{n+1}^{(j)} - 2\,\rho_n^{(j)} + \rho_{n-1}^{(j)}) + \rho_n^{(j+1)} (\vec{r}_{n+1}^{(j)} - 2\vec{r}_n^{(j)} + \vec{r}_{n-1}^{(j)})$$

$$+\vec{r}_n^{(j+1)} \wedge (\vec{r}_{n+1}^{(j)} - \vec{r}_{n-1}^{(j)}),\ j = 1,2,\mathrm{mod}(2)\ . \tag{10n}$$

Hint: replace formally in (9) $\rho_n(t), \vec{r}_n(t)$ with, say, $\widetilde{\rho}_n(\tau), \widetilde{\vec{r}}_n(\tau)$ (also replacing, of course, derivatives with respect to t with derivatives with respect to τ), then set $\widetilde{\rho}_n(\tau) = \exp(-a\,t)\,\rho_n(t)$, $\widetilde{\vec{r}}_n(\tau) = \exp(-a\,t)\,\vec{r}_n(t)$, $\tau = [\exp(a\,t) - 1]/a$, and finally complexify ($\rho_n = \rho_n^{(1)} + i\,\rho_n^{(2)}$, $\vec{r}_n = \vec{r}_n^{(1)} + i\,\vec{r}_n^{(2)}$, $c = \gamma + i\,\widetilde{\gamma}$, $a = \lambda + i\,\omega$).

Conjecture 5.6.3.11. The model (10) features many *completely periodic* solutions, with period $T = 2\pi/\omega$ (or a multiple of it), if $\lambda = 0, \omega > 0$.

A scalar-vector *linearizable N-body problem* (with 2 arbitrary coupling constants, and "nearest-neighbor" interactions):

$$\ddot{\rho}_n = a\,\dot{\rho}_n + \widetilde{\rho}_n \dot{\rho}_n - \widetilde{\vec{r}}_n \cdot \dot{\vec{r}}_n + c\left\{ \widetilde{\rho}_{n-1}\dot{\rho}_n - \widetilde{\vec{r}}_{n-1} \cdot \dot{\vec{r}}_n - \rho_n \widetilde{\widetilde{\rho}}_{n+1} + \vec{r}_n \cdot \widetilde{\widetilde{\vec{r}}}_{n+1} \right\}, \tag{11a}$$

$$\ddot{\vec{r}}_n = a\,\dot{\vec{r}}_n + \widetilde{\rho}_n \dot{\vec{r}}_n + \dot{\rho}_n \widetilde{\vec{r}}_n - \widetilde{\vec{r}}_n \wedge \dot{\vec{r}}_n$$

$$+ c\left\{ \widetilde{\rho}_{n-1}\dot{\vec{r}}_n + \dot{\rho}_n \widetilde{\vec{r}}_{n-1} - \widetilde{\vec{r}}_{n-1} \wedge \dot{\vec{r}}_n - \rho_n \widetilde{\widetilde{\vec{r}}}_{n+1} - \widetilde{\widetilde{\rho}}_{n+1}\vec{r}_n - \vec{r}_n \wedge \widetilde{\widetilde{\vec{r}}}_{n+1} \right\}, \tag{11b}$$

$$\widetilde{\rho}_n \equiv \left(\rho_n \dot{\rho}_n + \vec{r}_n \cdot \dot{\vec{r}}_n\right) / \left(\rho_n^{\ 2} + r_n^{\ 2}\right), \tag{11c}$$

$$\widetilde{\vec{r}}_n \equiv \left(\rho_n \dot{\vec{r}}_n - \dot{\rho}_n \vec{r}_n - \vec{r}_n \wedge \dot{\vec{r}}_n\right) / \left(\rho_n^{\ 2} + r_n^{\ 2}\right), \tag{11d}$$

$$\widetilde{\widetilde{\rho}}_{n+1} \equiv \widetilde{\rho}_n \widetilde{\rho}_{n+1} - \widetilde{\vec{r}}_n \cdot \widetilde{\vec{r}}_{n+1}\ , \tag{11e}$$

$$\widetilde{\widetilde{\vec{r}}}_{n+1} \equiv \widetilde{\rho}_n \widetilde{\vec{r}}_{n+1} + \widetilde{\rho}_{n+1} \widetilde{\vec{r}}_n - \widetilde{\vec{r}}_n \wedge \widetilde{\vec{r}}_{n+1}\ . \tag{11f}$$

These Newtonian equations of motion, (11), are obtained by applying the parametrization (5.5-1) to the *linearizable* matrix ODE (5.4.3-13d).

Exercise 5.6.3-12. Verify!

Exercise 5.6.3-13. Complexify!

Conjecture 5.6.3-14. If in (the complexified version of) these equations of motion, (11), $a = \pm i\,\omega, \omega > 0$, the model features a lot of *completely periodic* solutions, with period $T = 2\pi / \omega$ (or a multiple of it).

Exercise 5.6.3-15. Obtain the Newtonian equations of motion (involving only 3-vectors, no scalars) yielded by (5.4.3-13d) via the parametrization (5.5-6).

A *linearizable N-body problem* (with $3N$ arbitrary coupling constants, and "nearest-neighbor" interactions):

$$\ddot{\rho}_n = \left(a_{n+1} - a_n - b_n\right)\dot{\rho}_n + \tilde{\rho}_n\dot{\rho}_n - \vec{\tilde{r}}_n \cdot \dot{\vec{r}}_n + b_{n+1}\left(\tilde{\rho}_{n+1}\dot{\rho}_n - \vec{\tilde{r}}_{n+1} \cdot \dot{\vec{r}}_n\right)$$

$$+ c_{n+1}\rho_n - c_n\left(\rho_n \tilde{\tilde{\rho}}_{n-1} - \vec{r}_n \cdot \vec{\tilde{\tilde{r}}}_{n-1}\right), \tag{12a}$$

$$\ddot{\vec{r}}_n = \left(a_{n+1} - a_n - b_n\right)\dot{\vec{r}}_n + \tilde{\rho}_n\dot{\vec{r}}_n + \dot{\rho}_n\vec{\tilde{r}}_n - \vec{\tilde{r}}_n \wedge \dot{\vec{r}}_n + b_{n+1}\left(\tilde{\rho}_{n+1}\dot{\vec{r}}_n + \dot{\rho}_n\vec{\tilde{r}}_{n+1} - \vec{\tilde{r}}_{n+1} \wedge \dot{\vec{r}}_n\right)$$

$$+ c_{n+1}\vec{r}_n - c_n\left(\rho_n \vec{\tilde{\tilde{r}}}_{n-1} + \tilde{\tilde{\rho}}_{n-1}\vec{r}_n + \vec{r}_n \wedge \vec{\tilde{\tilde{r}}}_{n-1}\right), \tag{12b}$$

with $\tilde{\rho}_n, \vec{\tilde{r}}_n$ given again by (11c), (11d) and

$$\hat{\rho}_n \equiv \left(\rho_n\dot{\rho}_n + \vec{r}_n \cdot \dot{\vec{r}}_n\right) / \left(\dot{\rho}_n^{\,2} + \dot{\vec{r}}_n \cdot \dot{\vec{r}}_n\right), \tag{12c}$$

$$\vec{\hat{r}}_n \equiv \left(\dot{\rho}_n\vec{r}_n - \rho_n\dot{\vec{r}}_n + \vec{r}_n \wedge \dot{\vec{r}}_n\right) / \left(\dot{\rho}_n^{\,2} + \dot{\vec{r}}_n \cdot \dot{\vec{r}}_n\right), \tag{12d}$$

$$\tilde{\tilde{\rho}}_{n-1} \equiv \tilde{\rho}_n\hat{\rho}_{n-1} - \vec{\tilde{r}}_n \cdot \vec{\hat{r}}_{n-1}\ , \tag{12e}$$

$$\vec{\tilde{\tilde{r}}}_{n-1} \equiv \tilde{\rho}_n\vec{\hat{r}}_{n-1} + \hat{\rho}_{n-1}\vec{\tilde{r}}_n - \vec{\tilde{r}}_n \wedge \vec{\hat{r}}_{n-1}\ . \tag{12f}$$

These Newtonian equations of motion, (12), are obtained by applying the parametrization (5.5-1) to the system of evolution matrix ODEs (5.4.3-26).

Exercise 5.6.3-16. Verify!

Remark 5.6.3-17. These equations of motion are highly nonlinear: they feature a "velocity-dependence" in the denominator, see (12c,d,e,f).

Exercise 5.6.3-18. Obtain the Newtonian equations of motion (which only involve 3-vectors, no scalars) yielded by the application to the (same) system of evolution matrix ODEs, (5.4.3-26), of the parametrization (5.5-6) (rather than (5.5-1)).

Next, we display the scalar/vector *linearizable many-body problem* that is obtained by applying the parametrizations (5.5-3) and (5.5-1) to the *linearizable* matrix ODEs (5.4.1-15), via the positions $\underline{U}_n = i\vec{r}_n \cdot \underline{\vec{\sigma}}$, $\underline{F} = i\vec{f}\cdot\vec{\sigma}$, $\underline{G} = i\vec{g}\cdot\vec{\sigma}$, $\underline{H} = i\vec{h}\cdot\vec{\sigma}$, $\underline{V} = \eta + i\vec{v}\cdot\underline{\vec{\sigma}}$, $\underline{Y} = \theta + i\vec{y}\cdot\underline{\vec{\sigma}}$:

$$\mu_n^{(u)}\ddot{\vec{r}}_n = 4\vec{k}\wedge\dot{\vec{r}}_n + 2\dot{\vec{k}}\wedge\vec{r}_n - 4\vec{k}\wedge(\vec{k}\wedge\vec{r}_n) + \sum_{m=1}^{N}\left[a_{nm}^{(uu)}\left\{\dot{\vec{r}}_m - 2\vec{k}\wedge\vec{r}_m\right\} + b_{nm}^{(uu)}\vec{r}_m\right]$$

$$+a_n^{(uf)}\left\{\dot{\vec{f}} - 2\vec{k}\wedge\vec{f}\right\} + b_n^{(uf)}\vec{f} + a_n^{(ug)}\left\{\dot{\vec{g}} - 2\vec{k}\wedge\vec{g}\right\} + b_n^{(ug)}\vec{g} + a_n^{(uh)}\left\{\dot{\vec{h}} - 2\vec{k}\wedge\vec{h}\right\} + b_n^{(uh)}\vec{h} \tag{13a}$$

$$\mu^{(f)}\ddot{\vec{f}} = 4\vec{k}\wedge\dot{\vec{f}} + 2\dot{\vec{k}}\wedge\vec{f} - 4\vec{k}\wedge(\vec{k}\wedge\vec{f}) + \sum_{m=1}^{N}\left[a_m^{(fu)}\left\{\dot{\vec{r}}_m - 2\vec{k}\wedge\vec{r}_m\right\} + b_m^{(fu)}\vec{r}_m\right] +$$

$$+a^{(ff)}\left\{\dot{\vec{f}} - 2\vec{k}\wedge\vec{f}\right\} + b^{(ff)}\vec{f} + a^{(fg)}\left\{\dot{\vec{g}} - 2\vec{k}\wedge\vec{g}\right\} + b^{(fg)}\vec{g} + a^{(fh)}\left\{\dot{\vec{h}} - 2\vec{k}\wedge\vec{h}\right\} + b^{(fh)}\vec{h} \tag{13b}$$

$$\mu^{(g)}\ddot{\vec{g}} = 4\vec{k}\wedge\dot{\vec{g}} + 2\dot{\vec{k}}\wedge\vec{g} - 4\vec{k}\wedge(\vec{k}\wedge\vec{g}) + \sum_{m=1}^{N}\left[a_m^{(gu)}\left\{\dot{\vec{r}}_m - 2\vec{k}\wedge\vec{r}_m\right\} + b_m^{(gu)}\vec{r}_m\right]$$

$$+a^{(gf)}\left\{\dot{\vec{f}} - 2\vec{k}\wedge\vec{f}\right\} + b^{(gf)}\vec{f} + a^{(gg)}\left\{\dot{\vec{g}} - 2\vec{k}\wedge\vec{g}\right\} + b^{(gg)}\vec{g} + a^{(gh)}\left\{\dot{\vec{h}} - 2\vec{k}\wedge\vec{h}\right\} + b^{(gh)}\vec{h} \tag{13c}$$

$$\mu^{(h)}\ddot{\vec{h}} = 4\vec{k}\wedge\dot{\vec{h}} + 2\dot{\vec{k}}\wedge\vec{h} - 4\vec{k}\wedge(\vec{k}\wedge\vec{h}) + \sum_{m=1}^{N}\left[a_m^{(hu)}\left\{\dot{\vec{r}}_m - 2\vec{k}\wedge\vec{r}_m\right\} + b_m^{(hu)}\vec{r}_m\right]$$

$$+a^{(hf)}\left\{\dot{\vec{f}}-2\vec{k}\wedge\vec{f}\right\}+b^{(hf)}\,\vec{f}+a^{(hg)}\left\{\dot{\vec{g}}-2\vec{k}\wedge\vec{g}\right\}+b^{(hg)}\,\vec{g}+a^{(hh)}\left\{\dot{\vec{h}}-2\vec{k}\wedge\vec{h}\right\}+b^{(hh)}\,\vec{h} \tag{13d}$$

$$\mu^{(v)}\,\ddot{\vec{v}}=2\vec{k}\wedge\dot{\vec{v}}+\dot{\vec{k}}\wedge\vec{v}+k^2\,\vec{v}-\left\{\dot{\vec{v}}-\vec{k}\wedge\vec{v}\right\}\{\ \sum_{m=1}^{N}\left[\widetilde{a}_m^{(vu)}\left\{\dot{\vec{r}}_m-2\vec{k}\wedge\vec{r}_m\right\}^2+\widetilde{b}_m^{(vu)}\,{r_m}^2\right]$$

$$+\widetilde{a}^{(vf)}\left\{\dot{\vec{f}}-2\vec{k}\wedge\vec{f}\right\}^2+\widetilde{b}^{(vf)}\,f^2+\widetilde{a}^{(vg)}\left\{\dot{\vec{g}}-2\vec{k}\wedge\vec{g}\right\}^2+\widetilde{b}^{(vg)}\,g^2$$

$$+\widetilde{a}^{(vh)}\left\{\dot{\vec{h}}-2\vec{k}\wedge\vec{h}\right\}^2+\widetilde{b}^{(vh)}\,h^2\ \}-\vec{v}\{\ \sum_{m=1}^{N}\left[a_m^{(vu)}\left\{\dot{\vec{r}}_m-2\vec{k}\wedge\vec{r}_m\right\}^2+b_m^{(vu)}\,{r_m}^2\right]$$

$$+a^{(vf)}\left\{\dot{\vec{f}}-2\vec{k}\wedge\vec{f}\right\}^2+b^{(vf)}\,f^2+a^{(vg)}\left\{\dot{\vec{g}}-2\vec{k}\wedge\vec{g}\right\}^2+b^{(vg)}\,g^2$$

$$+a^{(vh)}\left\{\dot{\vec{h}}-2\vec{k}\wedge\vec{h}\right\}^2+b^{(vh)}\,h^2\ \}, \tag{13e}$$

$$\mu^{(v)}\,\ddot{\vec{y}}=-2\vec{k}\wedge\dot{\vec{y}}-\dot{\vec{k}}\wedge\vec{y}+k^2\,\vec{y}-\left\{\dot{\vec{y}}+\vec{k}\wedge\vec{y}\right\}\{\ \sum_{m=1}^{N}\left[\widetilde{a}_m^{(vy)}\left\{\dot{\vec{r}}_m-2\vec{k}\wedge\vec{r}_m\right\}^2+\widetilde{b}_m^{(yu)}\,{r_m}^2\right]$$

$$+\widetilde{a}^{(yf)}\left\{\dot{\vec{f}}-2\vec{k}\wedge\vec{f}\right\}^2+\widetilde{b}^{(yf)}\,f^2+\widetilde{a}^{(yg)}\left\{\dot{\vec{g}}-2\vec{k}\wedge\vec{g}\right\}^2+\widetilde{b}^{(yg)}\,g^2$$

$$+\widetilde{a}^{(yh)}\left\{\dot{\vec{h}}-2\vec{k}\wedge\vec{h}\right\}^2+\widetilde{b}^{(yh)}\,h^2\ \}-\vec{y}\{\ \sum_{m=1}^{N}\left[a_m^{(yu)}\left\{\dot{\vec{r}}_m-2\vec{k}\wedge\vec{r}_m\right\}^2+b_m^{(yu)}\,{r_m}^2\right]$$

$$+a^{(yf)}\left\{\dot{\vec{f}}-2\vec{k}\wedge\vec{f}\right\}^2+b^{(yf)}\,f^2+a^{(yg)}\left\{\dot{\vec{g}}-2\vec{k}\wedge\vec{g}\right\}^2+b^{(yg)}\,g^2$$

$$+a^{(yh)}\left\{\dot{\vec{h}}-2\vec{k}\wedge\vec{h}\right\}^2+b^{(yh)}\,h^2\ \}, \tag{13f}$$

$$\mu^{(v)}\ddot{\eta}=-2\left[\gamma\dot{\eta}-(\vec{k}\cdot\dot{\vec{v}})\right]-\left[\dot{\gamma}\eta-(\dot{\vec{k}}\cdot\vec{v})\right]+(k^2-\gamma^2)\eta$$

$$-\dot{\eta}\{\ \sum_{m=1}^{N}\left[\widetilde{a}_m^{(vu)}\left\{\dot{\vec{r}}_m-2\vec{k}\wedge\vec{r}_m\right\}^2+\widetilde{b}_m^{(vu)}{r_m}^2\right]+\widetilde{a}^{(vf)}\left\{\dot{\vec{f}}-2\vec{k}\wedge\vec{f}\right\}^2+\widetilde{b}^{(vf)}f^2$$

$$+\widetilde{a}^{(vg)}\left\{\dot{\vec{g}}-2\vec{k}\wedge\vec{g}\right\}^2+\widetilde{b}^{(vg)}g^2+\widetilde{a}^{(vh)}\left\{\dot{\vec{h}}-2\vec{k}\wedge\vec{h}\right\}^2+\widetilde{b}^{(vh)}h^2\ \}$$

$$-\eta\{\ \sum_{m=1}^{N}\left[a_m^{(vu)}\left\{\dot{\vec{r}}_m-2\vec{k}\wedge\vec{r}_m\right\}^2+b_m^{(vu)}{r_m}^2\right]+a^{(vf)}\left\{\dot{\vec{f}}-2\vec{k}\wedge\vec{f}\right\}^2+b^{(vf)}f^2,$$

$$+a^{(vg)}\left\{\dot{\vec{g}}-2\vec{k}\wedge\vec{g}\right\}^2+b^{(vg)}g^2+a^{(vh)}\left\{\dot{\vec{h}}-2\vec{k}\wedge\vec{h}\right\}^2+b^{(vh)}h^2\ \}, \qquad (13g)$$

$$\mu^{(v)}\ddot{\theta}=2\left[\gamma\dot{\theta}-(\vec{k}\cdot\dot{\vec{y}})\right]+\left[\dot{\gamma}\theta-(\dot{\vec{k}}\cdot\vec{y})\right]+(k^2-\gamma^2)\theta-\dot{\theta}\{\ \sum_{m=1}^{N}\left[\tilde{a}_m^{(vy)}\left\{\dot{\vec{r}}_m-2\vec{k}\wedge\vec{r}_m\right\}^2+\tilde{b}_m^{(yu)}r_m^{\ 2}\right]$$

$$+\tilde{a}^{(yf)}\left\{\dot{\vec{f}}-2\vec{k}\wedge\vec{f}\right\}^2+\tilde{b}^{(yf)}f^2+\tilde{a}^{(yg)}\left\{\dot{\vec{g}}-2\vec{k}\wedge\vec{g}\right\}^2+\tilde{b}^{(yg)}g^2$$

$$+\tilde{a}^{(yh)}\left\{\dot{\vec{h}}-2\vec{k}\wedge\vec{h}\right\}^2+\tilde{b}^{(yh)}h^2\ \}-\theta\{\ \sum_{m=1}^{N}\left[a_m^{(yu)}\left\{\dot{\vec{r}}_m-2\vec{k}\wedge\vec{r}_m\right\}^2+b_m^{(yu)}r_m^{\ 2}\right]$$

$$+a^{(yf)}\left\{\dot{\vec{f}}-2\vec{k}\wedge\vec{f}\right\}^2+b^{(yf)}f^2+a^{(yg)}\left\{\dot{\vec{g}}-2\vec{k}\wedge\vec{g}\right\}^2+b^{(yg)}g^2$$

$$+a^{(yh)}\left\{\dot{\vec{h}}-2\vec{k}\wedge\vec{h}\right\}^2+b^{(yh)}h^2\ \}, \qquad (13h)$$

where

$$\vec{k}\equiv\vec{f}+\eta\vec{g}+\theta\vec{h}-\vec{g}\wedge\vec{f}+\vec{h}\wedge\vec{y},\quad \gamma\equiv\vec{g}\cdot\vec{v}+\vec{h}\cdot\vec{y}. \qquad (13i)$$

Exercise 5.6.3-19. Verify!

These equations of motion determine the evolution of the $N+5$ three-vectors $\vec{r}_n(t),\ \vec{f}(t),\ \vec{g}(t),\ \vec{h}(t),\ \vec{v}(t),\vec{y}(t)$ and of the 2 scalars $\eta(t),\ \theta(t)$. They feature $2N^2+20N+42$ arbitrary "coupling constants", hence they include many special cases, corresponding to appropriate choices of these coupling constants, many of which could be set to zero to obtain simpler systems: the exploration of these special cases provides ample ground to hunt for new interesting Newtonian many-body problems amenable to exact treatment (*linearizable*).

The diligent reader may write the more general equations of motion involving $N+5$ scalar quantities, as well as $N+5$ three-vectors, which obtain by using the parametrization (5.5-1) for *all* the $N+5$ matrices that evolve according to (5.4.1-15), rather than for only those two, $\underline{V}$ and $\underline{Y}$, for which this parametrization, (5.5-1) rather than (5.5-3), is mandated by the very structure of (5.4.1-15).

Next, we report the scalar/vector *linearizable many-body problem* characterized by the equations of motion

$$\ddot{\rho}_n=\left[\dot{\rho}_n^{\ 2}\rho_n+2\dot{\rho}_n(\dot{\vec{r}}_n\cdot\vec{r}_n)-\rho_n(\dot{\vec{r}}_n\cdot\dot{\vec{r}}_n)\right]/(\rho_n^{\ 2}+r_n^{\ 2})$$

$$+\sum_m \{ a_{nm}[\dot{\rho}_m\rho_m\rho_n+\dot{\rho}_m(\vec{r}_m\cdot\vec{r}_n)-\rho_m(\dot{\vec{r}}_m\cdot\vec{r}_n)+\rho_n(\dot{\vec{r}}_m\cdot\vec{r}_m)]/(\rho_m^{\,2}+r_m^{\,2})\}, \quad (14a)$$

$$\ddot{\vec{r}}_n=\{\dot{\vec{r}}_n[\dot{\rho}_n\rho_n+(\dot{\vec{r}}_n\cdot\vec{r}_n)]-\vec{r}_n[\dot{\rho}_n^{\,2}+(\dot{\vec{r}}_n\cdot\dot{\vec{r}}_n)]+\dot{\vec{r}}_n[\dot{\rho}_n\rho_n+(\dot{\vec{r}}_n\cdot\vec{r}_n)]\}/(\rho_n^{\,2}+r_n^{\,2})$$

$$+\sum_m [a_{nm}\{\dot{\vec{r}}_m[\rho_m\rho_n+(\vec{r}_m\cdot\vec{r}_n)]-\vec{r}_m[\dot{\rho}_m\rho_n+(\dot{\vec{r}}_m\cdot\vec{r}_n)]+\vec{r}_n[\dot{\rho}_m\rho_m+(\dot{\vec{r}}_m\cdot\vec{r}_m)]$$

$$+[\dot{\rho}_m\vec{r}_m\wedge\vec{r}_n-\rho_m\dot{\vec{r}}_m\wedge\vec{r}_n+\rho_n\dot{\vec{r}}_m\wedge\vec{r}_m]\}/(\rho_m^{\,2}+r_m^{\,2})]. \quad (14b)$$

Exercise 5.6.3-20. Obtain these equations of motion, (14), by applying the parametrization (5.5-1) to the matrix ODEs (5.4.3-4).

Exercise 5.6.3-21. Obtain the Newtonian equations of motion (which only involve 3-vectors, no scalars) yielded by the application to the (same) system of evolution matrix ODEs, (5.4.3-4), of the parametrization (5.5-6) (rather than (5.5-1)).

A scalar/vector *solvable N^2-body problem* (featuring $4N^2$ arbitrary coupling constants):

$$\ddot{\rho}_{nm}=a_{nm}+\sum_{m_1=1}^{N}[b_{m_1m}\rho_{nm_1}+c_{m_1m}\dot{\rho}_{nm_1}]$$

$$-\sum_{m_1,m_2=1}^{N} d_{m_1m_2}[\dot{\rho}_{nm_1}\rho_{m_2n}-(\dot{\vec{r}}_{nm_1}\cdot\vec{r}_{m_2n})+2\rho_{nm_1}\dot{\rho}_{m_2n}-2(\vec{r}_{nm_1}\cdot\dot{\vec{r}}_{m_2n})]$$

$$+\sum_{m_1,m_2,m_3=1}^{N} d_{m_1m_2}c_{m_3m}[\rho_{nm_1}\rho_{m_2m_3}-(\vec{r}_{nm_1}\cdot\vec{r}_{m_2m_3})]-\sum_{m_1,m_2,m_3,m_4=1}^{N} d_{m_1m_2}d_{m_3m_4}[\rho_{nm_1}\rho_{m_2m_3}\rho_{m_4m}$$

$$-\rho_{nm_1}(\vec{r}_{m_2m_3}\cdot\vec{r}_{m_4m})-\rho_{m_2m_3}(\vec{r}_{nm_1}\cdot\vec{r}_{m_4m})-\rho_{m_4m}(\vec{r}_{nm_1}\cdot\vec{r}_{m_2m_3})+(\vec{r}_{nm_1}\wedge\vec{r}_{m_2m_3})\cdot\vec{r}_{m_4m}], \quad (15a)$$

$$\ddot{\vec{r}}_{nm}=\sum_{m_1=1}^{N}[b_{m_1m}\vec{r}_{nm_1}+c_{m_1m}\dot{\vec{r}}_{nm_1}]+\sum_{m_1,m_2=1}^{N} d_{m_1m_2}[\dot{\vec{r}}_{nm_1}\wedge\vec{r}_{m_2n}+2\vec{r}_{nm_1}\wedge\dot{\vec{r}}_{m_2n}]$$

$$-\sum_{m_1,m_2,m_3=1}^{N} d_{m_1m_2}c_{m_3m}\vec{r}_{nm_1}\wedge\vec{r}_{m_2m_3}-\sum_{m_1,m_2,m_3,m_4=1}^{N} d_{m_3m_4}d_{m_1m_2}[\vec{r}_{nm_1}\{\rho_{m_2m_3}\rho_{m_4m}-(\vec{r}_{m_2m_3}\cdot\vec{r}_{m_4m})\}$$

$$+\vec{r}_{m_2m_3}\{\rho_{nm_1}\rho_{m_4m}-(\vec{r}_{nm_1}\cdot\vec{r}_{m_4m})\}+\vec{r}_{m_4m}\{\rho_{nm_1}\rho_{m_2m_3}-(\vec{r}_{nm_1}\cdot\vec{r}_{m_2m_3})\}]. \quad (15b)$$

These Newtonian equations of motion, (15), describe the evolution of the N^2 scalars $\rho_{nm}(t)$ and of the N^2 three-vectors $\vec{r}_{nm}(t)$; they feature $4N^2$ arbitrary "coupling constants". Their solvable character is demonstrated by noting that they are obtained by applying the parametrization (5.5-1) to the following *solvable* matrix ODEs:

$$\ddot{\underline{U}}_{nm} = a_{nm}\underline{1} + \sum_{m_1=1}^{N} \left[b_{m_1 m} \underline{U}_{nm_1} + c_{m_1 m} \dot{\underline{U}}_{nm_1} \right] + \sum_{m_1,m_2,m_3=1}^{N} d_{m_1 m_2} c_{m_3 m} \underline{U}_{nm_1} \underline{U}_{m_2 m_3}$$

$$- \sum_{m_1,m_2=1}^{N} d_{m_1 m_2} \left[\dot{\underline{U}}_{nm_1} \underline{U}_{m_2 m} + 2\underline{U}_{nm_1} \dot{\underline{U}}_{m_2 m} \right] + \sum_{m_1,m_2,m_3,m_4=1}^{N} d_{m_1 m_2} d_{m_3 m_4} \underline{U}_{nm_1} \underline{U}_{m_2 m_3} \underline{U}_{m_4 m} . \quad (16)$$

Exercise 5.6.3-22. Verify!

There remains to show that (16) is *solvable.* Indeed it corresponds to the block-matrix evolution ODE

$$\ddot{\underline{M}} = \underline{A} + \underline{M}\,\underline{B} + \dot{\underline{M}}\,\underline{C} + \underline{M}\,\underline{D}\,\underline{M}\,\underline{C} - \dot{\underline{M}}\,\underline{D}\,\underline{M} - 2\underline{M}\,\underline{D}\,\dot{\underline{M}} - \underline{M}\,\underline{D}\,\underline{M}\,\underline{D}\,\underline{M} , \quad (17)$$

with $\underline{M} \equiv \underline{M}(t)$ a block-matrix whose nm-element is the matrix $\underline{U}_{nm} \equiv \underline{U}_{nm}(t)$ and with the 4 constant block-matrices $\underline{A}, \underline{B}, \underline{C}, \underline{D}$ all having their nm-elements proportional to the unit matrix, namely (in self-evident notation)

$$\underline{A}_{nm} = \alpha_{nm}\underline{1},\ \underline{B}_{nm} = \beta_{nm}\underline{1},\ \underline{C}_{nm} = \gamma_{nm}\underline{1},\ \underline{D}_{nm} = c_{nm}\underline{1} . \quad (18)$$

Exercise 5.6.3-23. Verify!

On the other hand the matrix ODE (17) is *solvable* (for any arbitrary assignment of the 4 matrices $\underline{A}, \underline{B}, \underline{C}, \underline{D}$ -- with $\underline{D}$ invertible, see below), since via the position

$$\underline{M} = \underline{D}^{-1}\underline{V}^{-1}\dot{\underline{V}} \quad (19)$$

it gets transformed into the following *linear constant-coefficient* (hence obviously *solvable*) third-order ODE for the block-matrix $\underline{V} \equiv \underline{V}(t)$:

$$\dddot{\underline{V}} = \underline{V}\,\underline{D}\,\underline{A} + \dot{\underline{V}}\,\underline{B} + \ddot{\underline{V}}\,\underline{C} . \quad (20)$$

Exercise 5.6.3-24. Verify!

Remark 5.6.3-25. Up to trivial notational changes, the matrix ODE (17) reduces to (5.4.3-1) in the special case $\underline{A} = \alpha\,\underline{1}$, $\underline{B} = \beta\,\underline{1}$, $\underline{C} = \gamma\,\underline{1}$, $\underline{D} = c\underline{1}$.

Let us end Sect. 5.6.3 by pointing out that the scalar/vector Newtonian equations of motion of two other *linearizable* N^2*-body problems* are exhibited in Sect. 5.3, see (5.3-35,36), and that several other *integrable many-body problems* are exhibited in Sect. 5.6.5.

5.6.4 Many-body problems of Hamiltonian type

In this short Sect. 5.6.4 we report two analogous, and rather trivial, *solvable* many-body problems of Hamiltonian type.

The first model is characterized by the following *solvable* equations of motion:

$$\dot{\vec{q}}_n = \sum_{m=1}^{N} \left(a_{nm}\vec{q}_m + b_{nm}\vec{p}_m + \lambda\left[\vec{q}_n \wedge \left[\vec{q}_m \wedge \vec{p}_m\right]\right]\right), \tag{1a}$$

$$\dot{\vec{p}}_n = \sum_{m=1}^{N} \left(c_{nm}\vec{q}_m - a_{mn}\vec{p}_m + \lambda\left[\vec{p}_n \wedge \left[\vec{q}_m \wedge \vec{p}_m\right]\right]\right), \tag{1b}$$

with $b_{nm} = b_{mn}$, $c_{nm} = c_{mn}$, which are yielded by the Hamiltonian

$$H(\vec{p}_1,\ldots,\vec{p}_N;\vec{q}_1,\ldots,\vec{q}_N) = \sum_{n,m=1}^{N} \left\{a_{nm}\vec{q}_n\cdot\vec{p}_m + b_{nm}\vec{p}_n\cdot\vec{p}_m/2 - c_{nm}\vec{q}_n\cdot\vec{q}_m/2\right\}$$

$$+(\lambda/2)\left[\sum_{n=1}^{N}(\vec{q}_n \wedge \vec{p}_n)\right]^2 . \tag{1c}$$

These equations of motion, (1a,b), correspond to the *solvable* matrix ODEs (5.4.1-36) (with λ replaced, for notational simplicity, by $\lambda/4$) via the parametrization (5.5-3): $\underline{U}_n = i\,\vec{q}_n\cdot\vec{\sigma}$, $\underline{V}_n = i\,\vec{p}_n\cdot\vec{\sigma}$. They feature $2N^2+N+1$ (arbitrary) constants. But the solvability of this Hamiltonian system has a rather trivial origin: indeed, the evolution equations (1a,b) are hardly nonlinear, since they entail (as it is of course implied by the treatment that led to them, see Sect. 5.4.1) that the 3-vector $\sum_{m=1}^{N}(\vec{q}_m \wedge \vec{p}_m)$ is a constant of motion.

Exercise 5.6.4-1. Verify all the above assertions, namely the fact that the Hamiltonian (1c) yields the Hamiltonian equations (1a,b), that these equations of motion, (1a,b), correspond to (5.4.1-36) via (5.5-3), and that they entail that the 3-vector $\sum_{m=1}^{N}(\vec{q}_m \wedge \vec{p}_m)$ is a constant of motion.

Remark 5.6.4-2. For $N=1$ this model, (1), becomes the model (5.6.2-3).

A second *solvable Hamiltonian many-body problem* (with $4N$ arbitrary coupling constants) is characterized by the Hamiltonian

$$H(\vec{p}_1,...,\vec{p}_N;\vec{q}_1,...,\vec{q}_N)$$

$$=\sum_{n_1,n_2=1}^{N}\left[a_{N-n_1-n_2}\left(\vec{p}_{n_1}\cdot\vec{q}_{n_2}\right)+b_{N-n_1-n_2}\left(\vec{p}_{n_1}\cdot\vec{p}_{n_2}\right)/2-c_{N-n_1-n_2}\left(\vec{q}_{n_1}\cdot\vec{q}_{n_2}\right)/2\right]$$

$$+(1/2)\sum_{n_1,n_2,n_3,n_4=1}^{N}\lambda_{N-n_1-n_2-n_3-n_4}\left(\vec{p}_{n_1}\wedge\vec{q}_{n_2}\right)\cdot\left(\vec{p}_{n_3}\wedge\vec{q}_{n_4}\right), \tag{2a}$$

and by the corresponding Hamiltonian equations of motion

$$\dot{\vec{q}}_n=\sum_{n_1=1}^{N}\left(a_{n-n_1}\vec{q}_{n_1}+b_{n-n_1}\vec{p}_{n_1}\right)+\sum_{n_1,n_2,n_3=1}^{N}\left[\lambda_{n-n_1-n_2-n_3}(\vec{p}_{n_1}\wedge\vec{q}_{n_2})\wedge\vec{q}_{n_3}\right], \tag{2b}$$

$$\dot{\vec{p}}_n=\sum_{n_1=1}^{N}\left(c_{n-n_1}\vec{q}_{n_1}-a_{n-n_1}\vec{p}_{n_1}\right)+\sum_{n_1,n_2,n_3=1}^{N}\left[\lambda_{n-n_1-n_2-n_3}(\vec{p}_{n_1}\wedge\vec{q}_{n_2})\wedge\vec{p}_{n_3}\right]. \tag{2c}$$

Here of course all indices are defined $\mathrm{mod}(N)$.

Let us exhibit in longhand the form (2b) takes for $N=2$:

$$\dot{\vec{q}}_1=b_1\vec{p}_2+b_2\vec{p}_1+a_1\vec{q}_2+a_2\vec{q}_1$$

$$+\lambda_1[(\vec{p}_1\wedge\vec{q}_1)\wedge\vec{q}_2+(\vec{p}_1\wedge\vec{q}_2)\wedge\vec{q}_1+(\vec{p}_2\wedge\vec{q}_1)\wedge\vec{q}_1+(\vec{p}_2\wedge\vec{q}_2)\wedge\vec{q}_2]$$

$$+\lambda_2[(\vec{p}_1\wedge\vec{q}_1)\wedge\vec{q}_1+(\vec{p}_1\wedge\vec{q}_2)\wedge\vec{q}_2+(\vec{p}_2\wedge\vec{q}_1)\wedge\vec{q}_2+(\vec{p}_2\wedge\vec{q}_2)\wedge\vec{q}_1], \tag{2d}$$

$$\dot{\vec{q}}_2=b_1\vec{p}_1+b_2\vec{p}_2+a_1\vec{q}_1+a_2\vec{q}_2$$

$$+\lambda_1[(\vec{p}_1\wedge\vec{q}_1)\wedge\vec{q}_1+(\vec{p}_1\wedge\vec{q}_2)\wedge\vec{q}_2+(\vec{p}_2\wedge\vec{q}_1)\wedge\vec{q}_2+(\vec{p}_2\wedge\vec{q}_2)\wedge\vec{q}_1]$$

$$+\lambda_2[(\vec{p}_1\wedge\vec{q}_1)\wedge\vec{q}_2+(\vec{p}_1\wedge\vec{q}_2)\wedge\vec{q}_1+(\vec{p}_2\wedge\vec{q}_1)\wedge\vec{q}_1+(\vec{p}_2\wedge\vec{q}_2)\wedge\vec{q}_2]. \tag{2e}$$

These results are obtained by applying the first *multiplication* trick of Sect. 5.3 to the one-body ($N=1$) version of the model treated above, (1), or rather, equivalently, to the model (5.6.2-3) (see the *Remark 5.6.4-2*). Let us go through this development.

We set

$$\vec{p}=\sum_{n=1}^{N}\eta_n\,\vec{p}_n,\quad \vec{q}=\sum_{n=1}^{N}\eta_n\,\vec{q}_n\,, \tag{3}$$

with analogous formulas for the "coupling constants" a,b,c,λ (see (5.6.2-3)), as well as

$$H(\vec{p},\vec{q})=\sum_{n=1}^{N}\eta_n H_n(\vec{p}_1,...,\vec{p}_N;\vec{q}_1,...,\vec{q}_N)\;, \tag{4a}$$

with

$$H_n(\vec{p}_1,...,\vec{p}_N;\vec{q}_1,...,\vec{q}_N)=\sum_{n_1,n_2=1}^{N}\,[\;a_{n-n_1-n_2}\left(\vec{p}_{n_1}\cdot\vec{q}_{n_2}\right)+b_{n-n_1-n_2}\left(\vec{p}_{n_1}\cdot\vec{p}_{n_2}\right)/2$$
$$-c_{n-n_1-n_2}\left(\vec{q}_{n_1}\cdot\vec{q}_{n_2}\right)/2\;]+(1/2)\sum_{n_1,n_2,n_3,n_4=1}^{N}\lambda_{n-n_1-n_2-n_3-n_4}\left(\vec{p}_{n_1}\wedge\vec{q}_{n_2}\right)\cdot\left(\vec{p}_{n_3}\wedge\vec{q}_{n_4}\right). \tag{4b}$$

Here and below all the indices $n,n_1,\ldots,n_4$ are of course defined mod(N).

Then the N Hamiltonians H_n are in involution, and the Hamiltonian H_N coincides with $H, H_N\equiv H$.

But of course, as emphasized in Sect. 5.3, these coupled equations of motion, (2b,c), can be decoupled by linear transformations of type (5.3-27).

Exercise 5.6.4-3. Verify all the above assertions.

Exercise 5.6.4-4. Show *(i)* that the scalar/vector many-body problem characterized by the Newtonian equations of motion

$$\ddot{\rho}_{nm}=a\,\delta_{nm}+\sum_{l=1}^{N}\;(\rho_{nl}\,\rho_{lm}-\vec{r}_{nl}\cdot\vec{r}_{lm})\;, \tag{5a}$$

$$\ddot{\vec{r}}_{nm}=\sum_{l=1}^{N}\;(\rho_{nl}\,\vec{r}_{lm}+\rho_{lm}\,\vec{r}_{nl}-\vec{r}_{nl}\wedge\vec{r}_{lm})\;, \tag{5b}$$

is *integrable*, and *(ii)* verify that these Newtonian equations of motion, (5), are yielded by the following Hamiltonian (of normal type; but note the negative sign in front of the first "kinetic energy" term):

$$H(\underline{\pi},\underline{\vec{p}};\underline{\rho},\underline{\vec{r}})=\frac{1}{2}\sum_{n,m=1}^{N}\left(-\pi_{nm}\pi_{mn}+\vec{p}_{nm}\cdot\vec{p}_{mn}\right)+a\sum_{n=1}^{N}\rho_{nn}$$

$$+\frac{1}{3}\sum_{n,m,l=1}^{N}\left[\rho_{nm}\rho_{ml}\rho_{ln}-3\rho_{nm}(\vec{r}_{ml}\cdot\vec{r}_{ln})+(\vec{r}_{nm}\cdot\vec{r}_{ml}\wedge\vec{r}_{ln})\right]. \tag{6}$$

Hint (for *(i)*): apply the second *multiplication* technique of Sect. 5.3, see (5.3-30), to the *integrable* matrix ODE (5.4.4-2), then use the parametrization (5.5-1) (for (2×2)-matrices).

Remark 5.6.4-5. These Newtonian equations of motion, (5), involve N^2 scalars, $\rho_{nm}\equiv\rho_{nm}(t)$, and N^2 3-vectors, $\vec{r}_{nm}\equiv\vec{r}_{nm}(t)$; they are compatible with the symmetrical reduction, $\rho_{nm}=\rho_{mn}$, $\vec{r}_{nm}=\vec{r}_{mn}$, whereby the number of scalar respectively 3-vector dependent variables is reduced from N^2 to $N(N+1)/2$.

Remark 5.6.4-6. These Newtonian equations of motion, (5), are of course *invariant under rotations*, but they are *not invariant under reflections* (except in the special "scalar" case, characterized by the vanishing of all 3-vectors, $\vec{r}_{nm}\equiv 0$).

Exercise 5.6.4-7. Show that the scalar/vector many-body problem characterized by the Newtonian equations of motion

$$\ddot{\rho}_{nm}=6(\lambda^2-\omega^2)\rho_{nm}-12\lambda\omega\eta_{nm}+5\lambda\dot{\rho}_{nm}-5\omega\dot{\eta}_{nm}$$

$$+\sum_{l=1}^{N}\left(\rho_{nl}\rho_{lm}-\vec{r}_{nl}\cdot\vec{r}_{lm}-\eta_{nl}\eta_{lm}+\vec{q}_{nl}\cdot\vec{q}_{lm}\right), \tag{7a}$$

$$\ddot{\eta}_{nm}=6(\lambda^2-\omega^2)\eta_{nm}+12\lambda\omega\rho_{nm}+5\lambda\dot{\eta}_{nm}+5\omega\dot{\rho}_{nm}$$

$$+\sum_{l=1}^{N}\left(\rho_{nl}\eta_{lm}+\eta_{nl}\rho_{lm}+\vec{r}_{nl}\cdot\vec{q}_{lm}+\vec{q}_{nl}\cdot\vec{r}_{lm}\right), \tag{7b}$$

$$\ddot{\vec{r}}_{nm}=6(\lambda^2-\omega^2)\vec{r}_{nm}-12\lambda\omega\vec{q}_{nm}+5\lambda\dot{\vec{r}}_{nm}-5\omega\dot{\vec{q}}_{nm}$$

$$\sum_{l=1}^{N}\left(\rho_{nl}\vec{r}_{lm}+\rho_{lm}\vec{r}_{nl}-\eta_{nl}\vec{q}_{lm}-\eta_{lm}\vec{q}_{nl}-\vec{r}_{nl}\wedge\vec{r}_{lm}+\vec{q}_{nl}\wedge\vec{q}_{lm}\right), \tag{7c}$$

$$\ddot{\vec{q}}_{nm}=6(\lambda^2-\omega^2)\vec{q}_{nm}+12\lambda\omega\vec{r}_{nm}+5\lambda\dot{\vec{q}}_{nm}+5\omega\dot{\vec{r}}_{nm}$$

$$\sum_{l=1}^{N} (\rho_{nl}\,\vec{q}_{lm} + \eta_{lm}\,\vec{r}_{nl} + \rho_{nl}\,\vec{q}_{lm} + \eta_{lm}\,\vec{r}_{nl} - \vec{r}_{nl} \wedge \vec{q}_{lm} - \vec{q}_{nl} \wedge \vec{r}_{lm}) \ , \tag{7d}$$

is *integrable*. *Hint*: same as for *Exercise 5.6.4-4*, but with (5.4.4-2) replaced by (5.4.4-17).

Conjecture 5.6.4-8. *All* the nonsingular solutions of the many-body problem (7) are *completely periodic*, with period $T = |2\pi/\omega|$, if $\lambda = 0$ and ω is a *nonvanishing real* constant. *Hint*: see (5.4.4-19,20).

5.6.5 Many-body problems in multidimensional space with velocity-independent forces: integrable unharmonic ("quartic") oscillators, and nonintegrable oscillators with lots of completely periodic motions

In Sect. 5.6.5 several *completely integrable*, indeed *solvable*, Hamiltonian many-body problems are exhibited, characterized by Newtonian equations of motion, with linear and cubic forces, in S-dimensional space (S = arbitrary positive integer, with special attention to S=3). As usual the equations of motion are always written (see below) in covariant form ("S-vector equal S-vector"), entailing their *rotational invariance*. All these many-body problems are Hamiltonian: the corresponding Hamiltonian functions are of normal type, with the kinetic energy *quadratic* in the canonical momenta, and the potential energy *quadratic* and *quartic* in the canonical coordinates (see below).

The investigation of quartic oscillators is, since the time of the Fermi-Pasta-Ulam numerical experiment <FPU55>, at the origin (see for instance <Kr77>) of the revolution that has occurred, over the last three/four decades, in the understanding of the behavior and relevance of integrable systems. It has moreover an obvious and ubiquitous applicative interest, inasmuch as it generally provides the first nonlinear ("unharmonic") correction to the behavior of linear ("harmonic") oscillators, the physical relevance of which is of course universal.

The foundation of the following results is the *integrable* matrix evolution equation (5.4.4-3), which we conveniently write now in the following guise:

$$\ddot{\underline{U}} = \frac{1}{2}\left(\underline{A}\,\underline{U} + \underline{U}\,\underline{A}\right) + b\,\underline{U}^{3}\ . \tag{1}$$

Here $\underline{U} \equiv \underline{U}(t)$ is a square matrix of arbitrary rank, $\underline{A}$ is a constant matrix and b is a constant scalar (which could of course be rescaled away).

Exercise 5.6.5-1. Compare the 3 versions, (1), (5.4.4-3) and (5.4.2-6), of (essentially) the same matrix ODE, and detail the changes of variables that transform each of them into each other. *Hint*: rescale both the dependent and the independent variables.

In Sect. 5.6.5 we denote as usual matrices by (upper case) underlined characters, while superimposed arrows, say $\vec{r}$, denote S-vectors. The actual dimensionality of vectors and matrices will, we hope, always be clear from the context. A dot sandwiched among two S-vectors (see below) denotes the standard scalar product in S-dimensional space. (This is the same notation used at the end of Sect. 5.3, which the diligent reader is advised to revisit before proceeding any further).

And let us emphasize, once and for all, that we generally report below the *integrable* equations of motion in their neatest form. Generalizations are easily obtainable by standard changes of variables (see for instance the *Remark 5.4.4-4)* and/or by "multiplication of variables" techniques such as the first one described in Sect. 5.3.

A rather general class of *integrable* Newtonian equations of motion in S-dimensional space, with arbitrary $S=1,2,3...$, reads as follows:

$$\ddot{\vec{r}}^{\,(m_1,m_2)(l)} = \sum_{\mu=1}^{M}\left[a_{m_1,\mu}\,\vec{r}^{\,(\mu,m_2)(l)} + a_{\mu,m_2}\,\vec{r}^{\,(m_1,\mu)(l)}\right]$$

$$+\,b\sum_{\mu_1,\mu_2=1}^{M}\sum_{\lambda=1}^{L}\left\{P_{m_1}\left[\vec{r}^{\,(m_1,\mu_1)(\lambda)}\left(\vec{r}^{\,(\mu_2,m_2)(l)}\cdot\vec{r}^{\,(\mu_1,\mu_2)(\lambda)}\right)\right] + P_{m_2}\left[\vec{r}^{\,(\mu_2,m_2)(\lambda)}\left(\vec{r}^{\,(m_1,\mu_1)(l)}\cdot\vec{r}^{\,(\mu_1,\mu_2)(\lambda)}\right)\right]\right\}. \tag{2a}$$

Here $m_1, m_2 = 1,...,M$; $l=1,...,L$ (M, L being arbitrary positive integers) and

$$P_m = 1 \text{ if } m \text{ is even},\ P_m = 0 \text{ if } m \text{ is odd},\ m = 1,2,...,M\ . \tag{2b}$$

Moreover

$$a_{nm}=0 \text{ if } n+m \text{ is odd}, \ \vec{r}^{(n,m)(l)}=0 \text{ if } n+m \text{ is even;} \tag{2c}$$

hence, for given M and L, the number N of S-vectors $\vec{r}^{(n,m)(l)}$ involved in (2) is $N=L(M^2-1+P_M)/2$, while the number of (nonvanishing) arbitrary constants $a_{n,m}$ is $(M^2+1-P_M)/2$. The additional arbitrary constant b in (2a) could of course be rescaled away.

Exercise 5.6.5-2. Check this simple arithmetic!

We show below that these Newtonian equations of motion, (2), are *integrable*, and we exhibit Hamiltonian functions of standard type that entail them. But firstly let us display a number of reductions of (2a).

A first reduction yields the Newtonian equations of motion

$$\ddot{\vec{r}}_k^{(l)}=\sum_{k'=1}^{K} \alpha_{k,k'}\,\vec{r}_{k'}^{(l)}+c\sum_{l'=1}^{L}\sum_{k'=1}^{K} \vec{r}_k^{(l')}\left(\vec{r}_{k'}^{(l')}\cdot\vec{r}_{k'}^{(l)}\right), \tag{3a}$$

where $l=1,\dots,L$ and $k=1,\dots,K$.

These equations of motion, (3a), obtain from (2) by setting

$$M=2K,\ \vec{r}^{(2k,2k_1-1)(l)}=\vec{r}^{(2k_2-1,2k)(l)}=\vec{r}_k^{(l)}, \tag{3b}$$

as well as

$$a_{2k_1-1,2k_2-1}=0, \tag{3c}$$

$$a_{2k_1,2k_2}=a_{2k_2,2k_1}=\alpha_{k_1,k_2}=\alpha_{k_2,k_1}. \tag{3d}$$

Here all k-indices range from 1 to K, while the index l ranges of course from 1 to L. Note the independence of the right hand side of (3b) from the indices k_1 and k_2; this entails a substantial reduction, which is easily seen to be compatible with (2) (with (3c,d)).

Exercise 5.6.5-3. Verify; and also check that the reduction (3b) would be compatible with the evolution (2) even if the right hand side of (3c) contained a nonvanishing constant, independent of the indices k_1 and k_2; but this would not entail any additional generality for the equations of motion (3a).

The Newtonian equations of motion (3a) involve now the $N=LK$ S-vectors $\vec{r}_k^{(l)}$, and they contain, in addition to the arbitrary constant $c=bK$ (that could of course be rescaled away), the $K(K+1)/2$ arbitrary constants $\alpha_{k_1,k_2}=\alpha_{k_2,k_1}$.

Exercise 5.6.5-4. Replace (3b) with

$$M=2K,\quad \vec{r}^{(2k,2k_1-1)(l)}=\vec{r}_k^{(l)}\ ,\quad \vec{r}^{(2k_2-1,2k)(l)}=\vec{q}_k^{(l)}, \tag{4}$$

and verify that the corresponding reduction (with (3c,d)) is also compatible with (2); and write the corresponding Newtonian equations of motion (more general than (3a), and which reduce to (3a) for $\vec{r}_k^{(l)}=\vec{q}_k^{(l)}$).

A different reduction yields the equations of motion

$$\ddot{\vec{r}}_k^{(l)}=\sum_{k'=1}^{K}\alpha_{k,k'}\,\vec{r}_{k'}^{(l)}+c\sum_{l'=1}^{L}\sum_{k'=1}^{K}\vec{r}_{k'}^{(l')}\left(\vec{r}_{k'}^{(l')}\cdot\vec{r}_k^{(l)}\right), \tag{5a}$$

where $l=1,\ldots,L$ and $k=1,\ldots,K$.

These equations of motion, (5a), obtain from (2) by setting

$$M=2K,\quad \vec{r}^{(2k_1,2k-1)(l)}=\vec{r}^{(2k-1,2k_2)(l)}=\vec{r}_k^{(l)}\ , \tag{5b}$$

as well as

$$a_{2k_1,2k_2}=0\ , \tag{5c}$$

$$a_{2k_1-1,2k_2-1}=a_{2k_2-1,2k_1-1}=\alpha_{k_1,k_2}=\alpha_{k_2,k_1}\ . \tag{5d}$$

Here all k-indices range again from 1 to K, while the index l ranges of course from 1 to L. Note again the independence of the right hand side of (5b) from the indices k_1 and k_2; this substantial reduction is again easily seen to be compatible with (2) (with (5c,d)).

Exercise 5.6.5-5. Verify; and also check that the reduction (5b) would be compatible with the evolution (2) even if the right hand side of (5c) contained a nonvanishing constant, independent of the indices k_1 and k_2; but this would not entail any additional generality for the equations of motion (5a).

The Newtonian equations of motion (5a) involve again $N=LK$ S-vectors $\vec{r}_k^{(l)}$, and they contain, in addition to the arbitrary constant $c=bK$ (that could of course be rescaled away), $K(K+1)/2$ arbitrary constants $\alpha_{k_1,k_2}=\alpha_{k_2,k_1}$.

Exercise 5.6.5-6. Replace (5b) with

$$M=2K,\ \vec{r}^{(2k_1,2k-1)(l)}=\vec{r}_k^{(l)},\ \vec{r}^{(2k-1,2k_2)(l)}=\vec{q}_k^{(l)}, \tag{6}$$

and verify that the corresponding reduction (with (5c,d)) is also compatible with (2); and write the corresponding Newtonian equations of motion (more general than (5a), and which reduce to (5a) for $\vec{r}_k^{(l)}=\vec{q}_k^{(l)}$).

Exercise 5.6.5-7. Show that a completely analogous treatment to that given above (from (3) onward) is applicable for odd $M=2K-1$ (rather than even $M=2K$), and that it leads to the same results.

Obvious special cases of (3a) and (5a) obtain for $L=1$ and for $K=1$; the former case, $L=1$, is richer than the latter, $K=1$, because the number of arbitrary constants α_{k_1,k_2} is always $K(K+1)/2$, as indicated above.

Additional and/or different reductions are also possible. Let us illustrate this point by displaying some of them, in specific *few-body* cases.

The simplest case involves just a single S-vector; it obtains from both (3a) and (5a) for $L=K=1$, and it reads

$$\ddot{\vec{r}}=\left(a+c\,r^2\right)\vec{r} \tag{7}$$

(which coincides, up to a trivial notational change, with (5.6.2-2a)). The *integrability* of this rotation-invariant equation for a single S-vector is of course rather trivial: indeed (7) can be reduced to an equation for the (scalar) radius r, which can then be solved by quadratures.

Next, let us report 4 cases involving 2 S-vectors. The first one obtains from (3a) with $K=2$, $L=1$, $\alpha_{1,1}=\alpha_1$, $\alpha_{1,2}=\alpha_{2,1}=\alpha_2$ and $\alpha_{2,2}=\alpha_3$. It reads

$$\ddot{\vec{r}}_1=\alpha_1\vec{r}_1+\alpha_2\vec{r}_2+c\,\vec{r}_1\left(r_1^2+r_2^2\right),\quad \ddot{\vec{r}}_2=\alpha_2\vec{r}_1+\alpha_3\vec{r}_2+c\,\vec{r}_2\left(r_1^2+r_2^2\right) \tag{8}$$

(here and below r_k^2 denotes of course the squared-modulus of the S-vector $\vec{r}_k$, namely $r_k^2=\left|\vec{r}_k\right|^2=\vec{r}_k\cdot\vec{r}_k$).

The second one obtains from (5a) with $K=2$, $L=1$, $\alpha_{1,1}=\alpha_1$, $\alpha_{1,2}=\alpha_{2,1}=\alpha_2$ and $\alpha_{2,2}=\alpha_3$. It reads

$$\ddot{\vec{r}}_1=\alpha_1\vec{r}_1+\alpha_2\vec{r}_2+c\left[\vec{r}_1 r_1^2+\vec{r}_2\left(\vec{r}_1\cdot\vec{r}_2\right)\right],\quad \ddot{\vec{r}}_2=\alpha_2\vec{r}_1+\alpha_3\vec{r}_2+c\left[\vec{r}_1\left(\vec{r}_1\cdot\vec{r}_2\right)+\vec{r}_2 r_2^2\right]. \tag{9}$$

The third one obtains from (3a) with $K=2$, $L=2$, $\alpha_{1,1}=\alpha_{2,2}=\alpha_1$, $\alpha_{1,2}=\alpha_{2,1}=\alpha_2$ and $\vec{r}_1^{(1)}=\vec{r}_2^{(2)}=\vec{r}_1$, $\vec{r}_2^{(1)}=\vec{r}_1^{(2)}=\vec{r}_2$. It reads

$$\ddot{\vec{r}}_1=\alpha_1\vec{r}_1+\alpha_2\vec{r}_2+c[\ \vec{r}_1\left(r_1^2+r_2^2\right)+2\vec{r}_2(\vec{r}_1\cdot\vec{r}_2)\ ], \tag{10a}$$

$$\ddot{\vec{r}}_2=\alpha_2\vec{r}_1+\alpha_1\vec{r}_2+c[\ 2\vec{r}_1(\vec{r}_1\cdot\vec{r}_2)+\vec{r}_2\left(r_1^2+r_2^2\right)\ ]. \tag{10b}$$

The fourth one obtains directly from (2a) with $M=3$, $L=1$ by setting $\vec{r}^{(1,2)(1)}=\vec{r}^{(2,3)(1)}=\vec{r}_1$, $\vec{r}^{(2,1)(1)}=\vec{r}^{(3,2)(1)}=\vec{r}_2$, $a_{1,1}+a_{2,2}=a_{2,2}+a_{3,3}=\alpha_1$, $a_{1,3}=\alpha_2$, $a_{3,1}=\alpha_3$. It reads

$$\ddot{\vec{r}}_1=\alpha_1\vec{r}_1+\alpha_2\vec{r}_2+c[\ \vec{r}_1(\vec{r}_1\cdot\vec{r}_2)+\vec{r}_2\,r_1^2\ ], \tag{11a}$$

$$\ddot{\vec{r}}_2=\alpha_3\vec{r}_1+\alpha_1\vec{r}_2+c[\ \vec{r}_1\,r_2^2+\vec{r}_2(\vec{r}_1\cdot\vec{r}_2)\ ]. \tag{11b}$$

Newtonian equations of motion involving 3 S-vectors are obviously obtained, as mentioned above, from (3a) and (5a) by setting $K=3, L=1$ or $K=1, L=3$. Let us display a different case that obtains from (3a) with $K=2, L=3$ by setting $\vec{r}_1^{(1)}=\vec{r}_2^{(3)}=\vec{r}_1$, $\vec{r}_1^{(2)}=\vec{r}_2^{(2)}=\vec{r}_2$, $\vec{r}_1^{(3)}=\vec{r}_2^{(1)}=\vec{r}_3$ and $a_{1,1}=a_{2,2}=\alpha_1$, $a_{1,2}=a_{2,1}=\alpha_2$; it reads

$$\ddot{\vec{r}}_1=\alpha_1\vec{r}_1+\alpha_2\vec{r}_3+c\left\{\vec{r}_1\left(r_1^2+r_3^2\right)+\vec{r}_2\left(\vec{r}_1\cdot\vec{r}_2+\vec{r}_2\cdot\vec{r}_3\right)+2\vec{r}_3\left(\vec{r}_1\cdot\vec{r}_3\right)\right\}, \tag{12a}$$

$$\ddot{\vec{r}}_2=\left(\alpha_1+\alpha_2\right)\vec{r}_2+c\left\{\vec{r}_1\left(\vec{r}_1\cdot\vec{r}_2+\vec{r}_2\cdot\vec{r}_3\right)+2\vec{r}_2\left(r_2^2\right)+\vec{r}_3\left(\vec{r}_1\cdot\vec{r}_2+\vec{r}_2\cdot\vec{r}_3\right)\right\}, \tag{12b}$$

$$\ddot{\vec{r}}_3=\alpha_2\vec{r}_1+\alpha_1\vec{r}_3+c\left\{2\vec{r}_1\left(\vec{r}_1\cdot\vec{r}_3\right)+\vec{r}_2\left(\vec{r}_1\cdot\vec{r}_2+\vec{r}_2\cdot\vec{r}_3\right)+\vec{r}_3\left(r_1^2+r_3^2\right)\right\}. \tag{12c}$$

Let us now indicate how the Newtonian S-vector equations of motion (2) are obtained from the *integrable* matrix evolution equation (1). To this end we consider $\underline{U}$ and $\underline{A}$ as ($M\times M$)-block-matrices, i. e. matrices whose elements are themselves (possibly rectangular) matrices, with the self-evident notation $\underline{U}_{m_1 m_2}=\underline{U}^{(m_1,m_2)}$, $\underline{A}_{m_1 m_2}=\underline{A}^{(m_1,m_2)}$, so that (1) yields

$$\ddot{\underline{U}}^{(m_1,m_2)}=\frac{1}{2}\sum_{\mu=1}^{N}\left(\underline{A}^{(m_1,\mu)}\underline{U}^{(\mu,m_2)}+\underline{U}^{(m_1,\mu)}\underline{A}^{(\mu,m_2)}\right)+b\sum_{\mu_1,\mu_2=1}^{N}\underline{U}^{(m_1,\mu_1)}\underline{U}^{(\mu_1,\mu_2)}\underline{U}^{(\mu_2,m_2)}, \tag{13}$$

where the M^2 matrices $\underline{U}^{(m_1,m_2)}$ are of course time-dependent, $\underline{U}^{(m_1,m_2)} \equiv \underline{U}^{(m_1,m_2)}(t)$, while the M^2 matrices $\underline{A}^{(m_1,m_2)}$ are constant. Of course if the matrices $\underline{U}^{(m_1,m_2)}$ and $\underline{A}^{(m_1,m_2)}$ are rectangular, their dimensions must be chosen appropriately, so that the matrix products in (13) make sense. Indeed we set (with S and L arbitrary positive integers):

(i) if m_1 and m_2 are both even then $\underline{U}^{(m_1,m_2)}=\underline{0}$, where $\underline{0}$ is the null (square) ($S \times S$)-matrix, and $\underline{A}^{(m_1,m_2)}=a_{m_1,m_2}\underline{1}$, where $\underline{1}$ is the identity (square) ($S \times S$)-matrix;

(ii) if m_1 and m_2 are both odd then $\underline{U}^{(m_1,m_2)}=\underline{0}$, where $\underline{0}$ is the null (square) ($L \times L$)-matrix, and $\underline{A}^{(m_1,m_2)}=a_{m_1,m_2}\underline{1}$, where $\underline{1}$ is the identity (square) ($L \times L$)-matrix;

(iii) if m_1 is even and m_2 is odd then $\underline{U}^{(m_1,m_2)}=\underline{V}^{(m_1,m_2)}$, where $\underline{V}^{(m_1,m_2)}$ is a (rectangular) ($S \times L$)-matrix, and $\underline{A}^{(m_1,m_2)}=\underline{0}$, where $\underline{0}$ is the null (rectangular) ($S \times L$)-matrix;

(iv) if m_1 is odd and m_2 is even then $\underline{U}^{(m_1,m_2)}=\underline{W}^{(m_1,m_2)}$, where $\underline{W}^{(m_1,m_2)}$ is a (rectangular) ($L \times S$)-matrix, and $\underline{A}^{(m_1,m_2)}=\underline{0}$, where $\underline{0}$ is the null (rectangular) ($L \times S$)-matrix.

Note that now the sparse block ($M \times M$)-matrices $\underline{U}$ and $\underline{A}$ are in fact square $\tilde{M} \times \tilde{M}$ matrices, with $\tilde{M}=[(M+P_M-1)S+(M-P_M+1)L]/2$.

Exercise 5.6.5-8. Check this simple arithmetic!

We now introduce a representation of matrices in terms of S-vectors by identifying the S elements of the L rows of the matrices $\underline{W}^{(m_1,m_2)}$ as the S components of L different S-vectors, and, conversely, by identifying the S elements of the L columns of the matrices $\underline{V}^{(m_1,m_2)}$ as the S components of L (different) S-vectors:

$$\left(\underline{W}^{(m_1,m_2)}\right)_{l,s}=x_s^{(m_1,m_2)(l)} \; , \; l=1,\ldots,L \; ; \; \left(\underline{V}^{(m_1,m_2)}\right)_{s,l}=x_s^{(m_1,m_2)(l)} \; , \; l=1,\ldots,L \; . \tag{14}$$

Note that we have thereby introduced the following S-vectors (altogether, $N=(M^2-1+P_M)L/2$ of them -- check this arithmetic!) :

$$\vec{r}^{(m_1,m_2)(l)} \equiv \left(x_1^{(m_1,m_2)(l)}, x_2^{(m_1,m_2)(l)}, \ldots, x_S^{(m_1,m_2)(l)}\right) . \tag{15}$$

It is now a matter of elementary algebra to verify that the matrix evolution equation (1), taking into account the above assignments, yields the Newtonian equations of motion (2).

Exercise 5.6.5-9. Do verify. *Hint*: see the analogous treatment of (5.3-32), as given in Sect. 5.3, from (5.3-37) onward; as well as the following formulas, (16).

To help visualization, we end this discussion by displaying the matrices $\underline{U}$ and $\underline{A}$ for the case $M=2, S=3, L=2$:

$$\underline{U}=\begin{pmatrix} \underline{0} & \underline{W}^{(1,2)} \\ \underline{V}^{(2,1)} & \underline{0} \end{pmatrix}=\begin{pmatrix} 0 & 0 & x_1^{(1,2)(1)} & x_2^{(1,2)(1)} & x_3^{(1,2)(1)} \\ 0 & 0 & x_1^{(1,2)(2)} & x_2^{(1,2)(2)} & x_3^{(1,2)(2)} \\ x_1^{(2,1)(1)} & x_1^{(2,1)(2)} & 0 & 0 & 0 \\ x_2^{(2,1)(1)} & x_2^{(2,1)(2)} & 0 & 0 & 0 \\ x_3^{(2,1)(1)} & x_3^{(2,1)(2)} & 0 & 0 & 0 \end{pmatrix}, \tag{16a}$$

$$\underline{A}=\begin{pmatrix} \underline{A}^{(1,1)} & \underline{0} \\ \underline{0} & \underline{A}^{(2,2)} \end{pmatrix}=\begin{pmatrix} a_{1,1} & 0 & 0 & 0 & 0 \\ 0 & a_{1,1} & 0 & 0 & 0 \\ 0 & 0 & a_{2,2} & 0 & 0 \\ 0 & 0 & 0 & a_{2,2} & 0 \\ 0 & 0 & 0 & 0 & a_{2,2} \end{pmatrix}. \tag{16b}$$

Note that more general choices for the constant matrix $\underline{A}$ are possible (with the nonvanishing matrices $\underline{A}^{(n,m)}$ no more proportional to the identity matrix; see above), but they then yield non-rotation-invariant many-body systems.

Exercise 5.6.5-10. Convince yourself of this, by constructing some such models.

The results displayed thus far have been obtained by applying to the *integrable* matrix evolution equation (1) the technique of *multiplication* "with a further twist", as described at the end of Sect. 5.3. But what if we apply to (1) this technique of *multiplication* in its simpler ("untwisted") form? That clearly amounts to focusing on (13), which we rewrite here for notational convenience as follows:

$$\ddot{\underline{U}}_{nm}=\frac{1}{2}\sum_{j=1}^{N}\left(\underline{A}_{nj}\underline{U}_{jm}+\underline{U}_{nj}\underline{A}_{jm}\right)+b\sum_{j,k=1}^{N}\underline{U}_{nj}\underline{U}_{jk}\underline{U}_{km}. \tag{17}$$

Here the N^2 matrices $\underline{U}_{nm}$ are of course again time-dependent, $\underline{U}_{nm}\equiv\underline{U}_{nm}(t)$, while the N^2 matrices $\underline{A}_{nm}$ are constant; and we now like to treat them not as block matrices (as we instead did after (13)), but rather to parametrize them directly in terms of 3-vectors (note that we now restrict attention to

$S=3$). A convenient way is to assume that all these matrices have rank 2, and moreover that the matrices $\underline{A}_{nm}$ are just multiples of the unit matrix,

$$\underline{A}_{nm}=a_{nm}\underline{1}=a_{nm} \; , \tag{18a}$$

where the a_{nm}'s are N^2 *scalar* (*a priori* arbitrary) constants, while the matrices $\underline{U}_{nm}$ take the standard form (see (5.5-1))

$$\underline{U}_{nm}(t)=\rho_{nm}(t)+i\vec{r}_{nm}(t)\cdot\underline{\vec{\sigma}} \; . \tag{18b}$$

We have thereby introduced the N^2 scalars (or rather pseudoscalars, see below) $\rho_{nm}\equiv\rho_{nm}(t)$ as well as the N^2 3-vectors $\vec{r}_{nm}\equiv\vec{r}_{nm}(t)$, and we get for them from (17) the following system of scalar/vector Newtonian equations of motion:

$$\ddot{\rho}_{nm}=\frac{1}{2}\sum_{j=1}^{N}\left(a_{nj}\rho_{jm}+a_{jm}\rho_{nj}\right)+b\sum_{j,k=1}^{N}\left\{\rho_{nj}\rho_{jk}\rho_{km}+\left[\left(\vec{r}_{nj}\wedge\vec{r}_{jk}\right)\cdot\vec{r}_{km}\right]\right.$$

$$\left.-\left[\,\rho_{nj}\left(\vec{r}_{jk}\cdot\vec{r}_{km}\right)+\rho_{jk}\left(\vec{r}_{nj}\cdot\vec{r}_{km}\right)+\rho_{km}\left(\vec{r}_{nj}\cdot\vec{r}_{jk}\right)\,\right]\right\} , \tag{19a}$$

$$\ddot{\vec{r}}_{nm}=\frac{1}{2}\sum_{j=1}^{N}\left(a_{nj}\vec{r}_{jm}+a_{jm}\vec{r}_{nj}\right)$$

$$+b\sum_{j,k=1}^{N}\left\{\vec{r}_{nj}\left[\rho_{jk}\rho_{km}-\left(\vec{r}_{jk}\cdot\vec{r}_{km}\right)\right]+\vec{r}_{jk}\left[\rho_{nj}\rho_{km}+\left(\vec{r}_{nj}\cdot\vec{r}_{km}\right)\right]+\vec{r}_{km}\left[\rho_{jk}\rho_{jk}-\left(\vec{r}_{nj}\cdot\vec{r}_{jk}\right)\right]\right.$$

$$\left.-\left[\,\rho_{nj}\left(\vec{r}_{jk}\wedge\vec{r}_{km}\right)+\rho_{jk}\left(\vec{r}_{nj}\wedge\vec{r}_{km}\right)+\rho_{km}\left(\vec{r}_{nj}\wedge\vec{r}_{jk}\right)\,\right]\right\} . \tag{19b}$$

These equations of motion feature the N^2 arbitrary scalar constants a_{nm}. Of course here and below the "wedge" respectively "dot" symbols sandwiched among two 3-vectors indicate the standard vector respectively scalar products for 3-vectors.

Exercise 5.6.5-11. Verify!

Exercise 5.6.5-12. Write out the (more general, but not rotation-invariant) set of Newtonian equations of motion that obtain if the simplification (18a) is forsaken by setting instead (see (18b) and (5.5-1))

$$\underline{A}_{nm} = a_{nm} + i\vec{c}_{nm} \cdot \underline{\vec{\sigma}} \,. \tag{20}$$

Hint: these equations, more general than (19), may also be written in covariant form, but they lack rotation-invariance because they contain the N^2 constant 3-vectors $\vec{c}_{nm}$ that introduce N^2 privileged directions.

Remark 5.6.5-13. The equations of motion (19), which are obviously *rotation-invariant*, are also *invariant under reflections* if the N^2 3-vectors $\vec{r}_{nm}$ behave as ordinary vectors and the N^2 scalars ρ_{nm} behave as pseudoscalars (or viceversa).

A reduction obviously consistent with (19) is to the case when all the 3-vectors $\vec{r}_{nm}$ vanish (while the converse, when all the scalars ρ_{nm} vanish, is not consistent with (19), except in some special cases, see below). No further elaboration of this "one-dimensional" case is reported below.

If the N^2 "coupling constants" a_{nm} depend symmetrically on their 2 indices,

$$a_{nm}=a_{mn} \,, \tag{21}$$

so that there are effectively only $N(N+1)/2$ of them, these evolution equations, (19), are consistent with the reductions

$$\rho_{nm}(t)=\eta\, \rho_{mn}(t),\ \vec{r}_{nm}(t) =-\eta\, \vec{r}_{mn}(t) \,, \tag{22}$$

with $\eta=1$, respectively $\eta=-1$. Via these reductions the N^2 scalars ρ_{nm} and the N^2 three-vectors $\vec{r}_{nm}$ are effectively reduced to $N(N+1)/2$ scalars and to $N(N-1)/2$ three-vectors, respectively to $N(N-1)/2$ scalars and to $N(N+1)/2$ three-vectors. For instance for $N=2$ and $\eta=1$, by setting $a_{11}=a_1,\, a_{22}=a_2,\, a_{12}=a_{21}=a_3,\ \rho_{11}=\rho_1,\, \rho_{22}=\rho_2,\, \rho_{12}=\rho_{21}=\rho_3,\ \vec{r}_{12}=-\vec{r}_{21}=\vec{r}$, we get

$$\ddot{\rho}_1=a_1\rho_1+a_3\rho_3+b[\ \rho_1^{\ 3}+2\rho_1\rho_3^{\ 2}+\rho_2\rho_3^{\ 2}+(2\rho_1+\rho_2)r^2\ ] \,, \tag{23a}$$

$$\ddot{\rho}_2=a_2\rho_2+a_3\rho_3+b[\ \rho_2^{\ 3}+2\rho_2\rho_3^{\ 2}+\rho_1\rho_3^{\ 2}+(2\rho_2+\rho_1)r^2\ ] \,, \tag{23b}$$

$$\ddot{\rho}_3 = \frac{1}{2}(a_1 + a_2)\rho_3 + \frac{1}{2}a_3(\rho_1 + \rho_2) + b\rho_3\left[\rho_1^{\,2} + \rho_2^{\,2} + \rho_3^{\,2} + \rho_1\rho_2 + r^2\right], \tag{23c}$$

$$\ddot{\vec{r}} = \left[\frac{1}{2}(a_1 + a_2) + b\left[\rho_1^{\,2} + \rho_2^{\,2} + \rho_3^{\,2} + \rho_1\rho_2 + r^2\right]\right]\vec{r}\ , \tag{23d}$$

while for $N=2$ and $\eta=-1$, by setting $a_{11}=a_1$, $a_{22}=a_2$, $a_{12}=a_{21}=a_3$, $\rho_{12}=-\rho_{21}=\rho$, $\vec{r}_{11}=\vec{r}_1$, $\vec{r}_{22}=\vec{r}_2$, $\vec{r}_{12}=\vec{r}_{21}=\vec{r}_3$, we get

$$\ddot{\rho} = \frac{1}{2}(a_1 + a_2)\rho - b\left\{\rho^3 + (\vec{r}_1 \wedge \vec{r}_2)\cdot\vec{r}_3 + \rho\left[r_1^{\,2} + r_2^{\,2} + r_3^{\,2} + \vec{r}_1\cdot\vec{r}_2\right]\right\}, \tag{24a}$$

$$\ddot{\vec{r}}_1 = a_1\vec{r}_1 + a_3\vec{r}_3 - b\left[\ 2\,\rho(\vec{r}_2 \wedge \vec{r}_3) + \vec{r}_1\left(2\rho^2 + r_1^{\,2} + 2r_3^{\,2}\right) + \vec{r}_2\left(\rho^2 - r_3^{\,2}\right) + 2\vec{r}_3\,(\vec{r}_3\cdot\vec{r}_2)\ \right], \tag{24b}$$

$$\ddot{\vec{r}}_2 = a_2\vec{r}_2 + a_3\vec{r}_3 - b\left[\ 2\,\rho(\vec{r}_1 \wedge \vec{r}_3) + \vec{r}_2\left(2\rho^2 + r_2^{\,2} + 2r_3^{\,2}\right) + \vec{r}_1\left(\rho^2 - r_3^{\,2}\right) + 2\vec{r}_3\,(\vec{r}_3\cdot\vec{r}_1)\ \right], \tag{24c}$$

$$\ddot{\vec{r}}_3 = \frac{1}{2}(a_1 + a_2)\vec{r}_3 + \frac{1}{2}a_3(\vec{r}_1 + \vec{r}_2)$$
$$-b\left\{\rho(\vec{r}_1 \wedge \vec{r}_2) + \vec{r}_1\,(\vec{r}_2\cdot\vec{r}_3) + \vec{r}_2\,(\vec{r}_3\cdot\vec{r}_1) + \vec{r}_3\left[\ r_1^{\,2} + r_2^{\,2} + r_3^{\,2} - (\vec{r}_1\cdot\vec{r}_2)\ \right]\right\}. \tag{24d}$$

If $a_1=a_2=a$ an additional reduction, consistent with (23), respectively with (24), obtains by setting $\rho_1(t)=\rho_2(t)=\rho(t)$, respectively $\vec{r}_1(t)=\vec{r}_2(t)=\vec{r}(t)$. Then (23) read

$$\ddot{\rho} = a\rho + a_3\rho_3 + b\,\rho\left[\ \rho^2 + 3\rho_3^{\,2} + 3r^2\ \right], \tag{25a}$$

$$\ddot{\rho}_3 = a\rho_3 + a_3\rho + b\,\rho_3\left[\ 3\rho^2 + \rho_3^{\,2} + r^2\ \right], \tag{25b}$$

$$\ddot{\vec{r}} = \left[\ a + b\left(3\rho^2 + \rho_3^{\,2} + r^2\right)\ \right]\vec{r}\ , \tag{25c}$$

respectively (24) read

$$\ddot{\rho} = \rho\left[\ a - b\left(\rho^2 + r_3^{\,2} + 3r^2\right)\ \right], \tag{26a}$$

$$\ddot{\vec{r}} = a\vec{r} + a_3\vec{r}_3 - b\left[\ 2\rho(\vec{r} \wedge \vec{r}_3) + \vec{r}\left(3\rho^2 + r^2 + r_3^{\,2}\right) + 2\vec{r}_3\,(\vec{r}_3\cdot\vec{r})\ \right], \tag{26b}$$

$$\ddot{\vec{r}}_3 = a\vec{r}_3 + a_3\vec{r} - b\left\{2\vec{r}\,(\vec{r}\cdot\vec{r}_3) + \vec{r}_3\,(r_3^{\,2} + r^2)\right\}. \tag{26c}$$

These evolution equations, (26), are clearly consistent with the additional reduction $\rho(t)=0$. Then they read

$$\ddot{\vec{r}}=\vec{r}\,[\;a-b(r^2+r_3^2)\;]+\vec{r}_3\,[\;a_3-2b(\vec{r}_3\cdot\vec{r})\;]\,, \tag{27a}$$

$$\ddot{\vec{r}}_3=\vec{r}_3\,[\;a-b(r_3^2+r^2)\;]+\vec{r}\,[\;a_3-2b(\vec{r}\cdot\vec{r}_3)\;]\,, \tag{27b}$$

which, up to obvious notational changes, coincide with (10), and are of course consistent with the additional reduction $\vec{r}_3(t)=\vec{r}(t)$, which then yields (7) (up to obvious notational changes).

Let us now go back to (23) to point out that, if $a_3=0$, another consistent reduction obtains by setting $\rho_3(t)=0$. Then (23) read

$$\ddot{\rho}_1=a_1\rho_1+b[\;\rho_1^3+(2\rho_1+\rho_2)r^2\;]\,, \tag{28a}$$

$$\ddot{\rho}_2=a_2\rho_2+b[\;\rho_2^3+(2\rho_2+\rho_1)r^2\;]\,, \tag{28b}$$

$$\ddot{\vec{r}}=\vec{r}\left[\frac{1}{2}(a_1+a_2)+b(\rho_1^2+\rho_2^2+\rho_1\rho_2+r^2)\right]. \tag{28c}$$

If $a_3=0$ and moreover $a_1=a_2=a$, further reductions are possible. By setting $\rho(t)=0$ in (25) one gets

$$\ddot{\rho}_3=\rho_3[\;a+b(\rho_3^2+r^2)\;]\,,\;\ddot{\vec{r}}=[\;a+b(\rho_3^2+r^2)\;]\vec{r}\;; \tag{29}$$

by setting $\rho_3(t)=0$ in (25) (or, equivalently, $\rho_1(t)=\rho_2(t)=\rho(t)$ in (28)), one gets

$$\ddot{\rho}=\rho[\;a+b(\rho^2+3r^2)\;]\,,\;\ddot{\vec{r}}=\vec{r}\;[\;a+b(3\rho^2+r^2)\;]\;; \tag{30}$$

and by setting $\rho(t)=\rho_3(t)=0$, one gets again (7) (up to trivial notational changes).

Likewise, by setting $\vec{r}(t)=0$ in (26) with $a_3=0$, one gets

$$\ddot{\rho}=\rho[\;a-b(\rho^2+r_3^2)\;]\,,\;\ddot{\vec{r}}_3=(a-b\,r_3^2)\,\vec{r}_3\,, \tag{31}$$

while by setting $\vec{r}_3(t)=0$ in (26) with $a_3=0$ one gets

$$\ddot{\rho}=\rho[\;a-b(\rho^2+3r^2)\;]\,,\;\ddot{\vec{r}}=\vec{r}\,[\;a-b(3\rho^2+r^2)\;]\,. \tag{32}$$

Both these evolution equations, (31) and (32), are compatible with the further reduction $\rho(t)=0$, and they thereby both yield, up to trivial notational changes, the same evolution equation, namely (7).

Exercise 5.6.5-14. Check all these reductions and explore new ones.

All the Newtonian equations written above, both in the contexts of S-dimensional and 3-dimensional space, are Hamiltonian: this is not surprising, since they are all obtained as reductions (i. e., special cases) of the matrix evolution equation (1) which is itself Hamiltonian, see (5.4.4-7). In particular (2) respectively (19) obtain from the Hamiltonians (written in self-evident notation)

$$H=\frac{1}{2}\sum_{m_1,m_2=1}^{M}\sum_{l=1}^{L}\left(\vec{p}^{\,(m_1,m_2)(l)}\cdot\vec{p}^{\,(m_2,m_1)(l)}\right)$$

$$-\sum_{m_1,m_2,\mu=1}^{M}P_{m_1+m_2}\,a_{m_1,m_2}\sum_{l=1}^{L}(1-P_{m_1+\mu})(1-P_{m_2+\mu})\left(\vec{r}^{\,(\mu,m_1)(l)}\cdot\vec{r}^{\,(m_2,\mu)(l)}\right)$$

$$-\frac{b}{4}\sum_{m_1,m_2,\mu_1,\mu_2=1}^{M}\;\sum_{l,\lambda=1}^{L}\{\;P_{m_1}P_{m_2}(1-P_{\mu_1})(1-P_{\mu_2})\cdot$$

$$\cdot[\;\left(\vec{r}^{\,(\mu_2,m_1)(l)}\cdot\vec{r}^{\,(m_1,\mu_1)(\lambda)}\right)\left(\vec{r}^{\,(\mu_1,m_2)(\lambda)}\cdot\vec{r}^{\,(m_2,\mu_2)(l)}\right)+\left(\vec{r}^{\,(m_1,\mu_2)(l)}\cdot\vec{r}^{\,(\mu_1,m_1)(\lambda)}\right)\left(\vec{r}^{\,(m_2,\mu_1)(\lambda)}\cdot\vec{r}^{\,(\mu_2,m_2)(l)}\right)\;]\;\}$$

(33a)

respectively

$$H=\frac{1}{2}\sum_{i,j=1}^{N}\left[\left(\vec{p}_{ij}\cdot\vec{p}_{ji}\right)-\pi_{ij}\pi_{ji}\right]-\frac{1}{2}\sum_{i,j,k=1}^{N}a_{ij}\left[\vec{r}_{jk}\cdot\vec{r}_{ki}-\rho_{jk}\rho_{ki}\right]$$

$$-\frac{b}{4}\sum_{i,j,k,l=1}^{N}\{\;2[\;\rho_{ij}\rho_{kl}\left(\vec{r}_{jk}\cdot\vec{r}_{li}\right)+\rho_{ij}\rho_{li}\left(\vec{r}_{jk}\cdot\vec{r}_{kl}\right)+\rho_{ij}\rho_{jk}\left(\vec{r}_{kl}\cdot\vec{r}_{li}\right)\;]-\rho_{ij}\rho_{jk}\rho_{kl}\rho_{li}$$

$$+2[\;\left(\vec{r}_{ij}\cdot\vec{r}_{kl}\right)\left(\vec{r}_{jk}\cdot\vec{r}_{li}\right)-\left(\vec{r}_{ij}\cdot\vec{r}_{li}\right)\left(\vec{r}_{jk}\cdot\vec{r}_{kl}\right)-\left(\vec{r}_{ij}\cdot\vec{r}_{jk}\right)\left(\vec{r}_{kl}\cdot\vec{r}_{li}\right)\;]-\left(\vec{r}_{kl}\wedge\vec{r}_{li}\right)\cdot\left(\rho_{ij}\vec{r}_{jk}+\rho_{jk}\vec{r}_{ij}\right)\;\}.$$

(34a)

In (33a), the canonical coordinates are the S-vectors $\vec{r}^{\,(m_1,m_2)(l)}$, and the corresponding canonical momenta are the S-vectors $\vec{p}^{\,(m_1,m_2)(l)}$ (and of course P_m is defined by (2b)); note that this Hamiltonian, (33a), entails

$$\dot{\vec{r}}^{\,(m_1,m_2)(l)}=\vec{p}^{\,(m_2,m_1)(l)}\;,\;\;m_1,m_2=1,\dots M,\;\;m_1+m_2=\text{odd},\;\;l=1,\dots L\;,\qquad(33\text{b})$$

as well as the other set of Hamiltonian equations (which we do not display), from which the Newtonian equations (2) follow. Likewise in (34a) the canonical coordinates respectively momenta are the (pseudo) scalars and the 3-vectors ρ_{nm}, $\vec{r}_{nm}$ respectively π_{nm} , $\vec{p}_{nm}$, $n,m=1,...,N$, and this Hamiltonian, (34a), entails

$$\dot{\rho}_{nm}=-\pi_{mn} \ , \quad \dot{\vec{r}}_{nm}=\vec{p}_{mn} \ , \quad n,m=1,...,N \ , \tag{34b}$$

as well as the other set of Hamiltonian equations (which we do not display), from which the Newtonian equations (19) follow.

Exercise 5.6.5-15. Verify!

Exercise 5.6.5-16. Write the Hamiltonian functions for all the many- and few-body problems, in S-dimensional and in 3-dimensional space, displayed above.

Exercise 5.6.5-17. Show that the following 2-body and 3-body problems are *solvable*, that they are Hamiltonian, and that these Newtonian equations of motions, (35) and (36), can be decoupled via a linear reshuffling of the dependent variables:

$$\ddot{\vec{r}}_n = a_2\,\vec{r}_n + a_1\,\vec{r}_{n+1} + c_2\,\vec{r}_n\,(r_1^2+r_2^2) + 2\,c_1\,\vec{r}_{n+1}\,(\vec{r}_1\cdot\vec{r}_2), \quad n=1,2 \ \mathrm{mod}(2)\,; \tag{35}$$

$$\ddot{\vec{r}}_n = a_3\,\vec{r}_n + a_2\,\vec{r}_{n+1} + a_1\,\vec{r}_{n+2}$$

$$+2c_1\{\ \vec{r}_{n+2}[\ r_3^2+2(\vec{r}_1\cdot\vec{r}_2)\ ]+\vec{r}_n[\ r_1^2+2(\vec{r}_2\cdot\vec{r}_3)\ ]+\vec{r}_{n+1}[\ r_2^2+2(\vec{r}_1\cdot\vec{r}_3)\ ]\ \}$$

$$+2c_2\{\ \vec{r}_{n+1}[\ r_3^2+2(\vec{r}_1\cdot\vec{r}_2)\ ]+\vec{r}_{n+2}[\ r_1^2+2(\vec{r}_2\cdot\vec{r}_3)\ ]+\vec{r}_n[\ r_2^2+2(\vec{r}_1\cdot\vec{r}_3)\ ]\ \}$$

$$+2c_3\{\ \}\{\ \vec{r}_n[\ r_3^2+2(\vec{r}_1\cdot\vec{r}_2)\ ]+\vec{r}_{n+1}[\ r_1^2+2(\vec{r}_2\cdot\vec{r}_3)\ ]+\vec{r}_{n+2}[\ r_2^2+2(\vec{r}_1\cdot\vec{r}_3)\ ]\ \}\,,$$

$$n=1,2,3 \ \mathrm{mod}(3)\,. \tag{36}$$

Hint: note that application of the first *multiplication* trick of Sect. 5.3. to the *solvable* equation of motion (7) yields

$$\ddot{\vec{r}}_n = \sum_{m=1}^{N} a_{n-m}\vec{r}_m + \sum_{m_1,m_2,m_3=1,..,N \ \mathrm{mod}(N)} c_{m_3}\,\vec{r}_{n-m_1-m_2-m_3}\,(\vec{r}_{m_1}\cdot\vec{r}_{m_2})\,. \tag{37}$$

Exercise 5.6.5-18. Repeat all the developments given above (in Sect. 5.6.5), but taking as starting point, instead of the *integrable* matrix evolution ODE (1), the system of two coupled matrix ODEs (5.4.4-18), and formulate *conjectures* analogous to *Conjecture 5.4.4-16* (and perhaps also prove all these *conjectures*?!).

As promised by the title of Sect. 5.6.5, devoted to Newtonian equations of motion with *velocity-independent* forces, we complete it by presenting, via the following two *Exercises 5.6.5-19* and *5.6.5-20*, some remarkable *exact* results for *nonintegrable* systems of this kind.

Exercise 5.6.5-19. Consider the system of $2N$ unharmonic (real) oscillators characterized by the Newtonian equations of motion

$$\ddot{\vec{u}}_n + 3\Omega\dot{\vec{v}}_n - 2\Omega^2\vec{u}_n$$

$$= \sum_{m_1,m_2,m_3=1}^{N} \{ (a_{nm_1m_2m_3}\vec{u}_{m_1} - b_{nm_1m_2m_3}\vec{v}_{m_1})(\vec{u}_{m_2}\cdot\vec{u}_{m_3} - \vec{v}_{m_2}\cdot\vec{v}_{m_3})$$

$$-(a_{nm_1m_2m_3}\vec{v}_{m_1} + b_{nm_1m_2m_3}\vec{u}_{m_1})(\vec{u}_{m_2}\cdot\vec{v}_{m_3} + \vec{v}_{m_2}\cdot\vec{u}_{m_3}) \}, \qquad (38a)$$

$$\ddot{\vec{v}}_n - 3\Omega\dot{\vec{u}}_n - 2\Omega^2\vec{v}_n$$

$$= \sum_{m_1,m_2,m_3=1}^{N} \{ (a_{nm_1m_2m_3}\vec{u}_{m_1} - b_{nm_1m_2m_3}\vec{v}_{m_1})(\vec{u}_{m_2}\cdot\vec{v}_{m_3} + \vec{v}_{m_2}\cdot\vec{u}_{m_3})$$

$$+(a_{nm_1m_2m_3}\vec{v}_{m_1} + b_{nm_1m_2m_3}\vec{u}_{m_1})(\vec{u}_{m_2}\cdot\vec{u}_{m_3} - \vec{v}_{m_2}\cdot\vec{v}_{m_3}) \}, \qquad (38b)$$

where N is of course an *arbitrary* positive integer, superimposed arrows denote S-vectors with S also an *arbitrary* positive integer, superimposed dots denote of course time-differentiations, dots sandwiched among vectors denote the standard scalar product, and the $2N^4+1$ constants $a_{nm_1m_2m_3}, b_{nm_1m_2m_3}, \Omega$ are also *arbitrary* (real, $\Omega \neq 0$). Prove that there exist then, in the neighborhood of the equilibrium configuration $\vec{u}_n = \vec{v}_n = \dot{\vec{u}}_n = \dot{\vec{v}}_n = 0$, a ball of initial data $\vec{u}_n(0), \vec{v}_n(0), \dot{\vec{u}}_n(0), \dot{\vec{v}}_n(0)$, of non-vanishing volume in the initial-data space of $4NS$ dimensions, such that *all* the corresponding trajectories are *completely periodic* with period $T = 2\pi/|\Omega|$. *Hint:* start from the system

$$\vec{r}_n'' = \sum_{m_1,m_2,m_3=1}^{N} c_{nm_1m_2m_3} \vec{r}_{m_1} (\vec{r}_{m_2} \cdot \vec{r}_{m_3}) \ , \tag{39}$$

with $\vec{r}_n \equiv \vec{r}_n(\tau)$ and where the primes denote of course differentiations with respect to τ; then note that the solutions of these (complex) equations of motion, considered as functions of the complex variable τ, are certainly holomorphic in a disk of the complex τ-plane centered at $\tau = 0$ and of any given radius, say ρ, provided the initial data, $\vec{r}_n(0), \vec{r}_n'(0)$, are all sufficiently small (the degree of smallness required depends of course on the assigned radius ρ, and on the coupling constants $c_{nm_1m_2m_3}$); then set

$$\vec{w}_n(t) = \exp(i\Omega t)\, \vec{r}_n(\tau), \quad \tau = [\exp(i\Omega t) - 1]/(i\Omega) \ , \tag{40a}$$

which of course entails

$$\vec{r}_n(0) = \vec{w}_n(0), \quad \vec{r}_n'(0) = \dot{\vec{w}}_n(0) - i\Omega \vec{w}_n(0) \tag{40b}$$

(so that if all the quantities $\vec{w}_n(0), \vec{w}_n'(0)$ are small, the initial data $\vec{r}_n(0), \vec{r}_n'(0)$ for (39) are also small); then note that if $\vec{r}_n(\tau)$ is holomorphic in a disk of the complex τ-plane centered at $\tau = 0$ and of radius $\rho > 2/|\Omega|$, the corresponding $\vec{w}_n(t)$, see (40a), is periodic in t with period $T = 2\pi/|\Omega|$; and finally set $\vec{w}_n(t) = \vec{u}_n(t) + i\vec{v}_n(t)$, $c_{nm_1m_2m_3} = a_{nm_1m_2m_3} + ib_{nm_1m_2m_3}$, whereby (39) with (40a) becomes (38).

The results given in the preceding *Exercise 5.6.5-19* refer to Newtonian multi-oscillator type equations written in covariant form in S-dimensional space and characterized by *cubic* nonlinearities, see (38). Neither of these two restrictions are however essential for the validity of this finding, as entailed by the following

Exercise 5.6.5-20. Consider the system of N (complex, coupled) Newtonian equations of motion of oscillator type

$$\ddot{w}_n - i(2 + 2/p)\Omega \dot{w}_n - (1 + p/2)\Omega^2 w_n = F_n(\underline{w}) \ , \tag{41a}$$

where the N complex quantities $w_n \equiv w_n(t)$ are the dependent variables, Ω is a *real nonvanishing* constant, p is a *positive integer*, and the N functions $F_n(\underline{w})$ of the N dependent variables w_m are *arbitrary* except for the requirements *(i)* that in the neighborhood of $w_m = 0$ they be analytic and *(ii)* that they satisfy the scaling property

$$F_n(\lambda \underline{w}) = \lambda^{p+1} F_n(\underline{w}) \ . \tag{41b}$$

Prove that there exist then, in the neighborhood of the equilibrium configuration $w_n = \dot{w}_n = 0$, a ball of initial data $w_n(0), \dot{w}_n(0)$, of nonvanishing volume in the space (having $4N$ real dimensions) of initial data, such that *all* the corresponding solutions of (41) are *completely periodic*, with period $T = 2\pi/|\Omega|$ if p is even, with period $\tilde{T} = 2T = 4\pi/|\Omega|$ if p is odd. (Note that the results of the preceding *Exercise 5.6.5-19* are a special case of those of this *Exercise 5.6.5-20* with $p = 2$). *Hint*: see the *hint* for the preceding *Exercise 5.6.5-19*, but start from the system

$$z_n'' = F_n(\underline{z}), \ z_n \equiv z_n(\tau) \tag{42}$$

(rather than from (39)), and set

$$w_n(t) = \exp(i\Omega t)\ z_n(\tau), \quad \tau = [\exp(i\omega t) - 1]/(i\omega), \tag{43}$$

(instead of (40a)), with $\omega = p\Omega/2$.

Note the analogy (but also the difference) of the treatment that yields the results given in the last two *Exercises 5.6.5-19* and *5.6.5-20*, to that given in Sect. 4.5 (see in particular the proof of *Proposition 4.5-9*). Of course these findings are also applicable if the nonlinear parts of these equations of motion are missing: indeed the general solution of the *linear* part of (41a) is clearly the linear superposition of two periodic solutions, one with period $T = 2\pi/|\Omega|$, the other with period $\tilde{T}_p = 2T/(2+p)$ (verify!), hence it is periodic with period T if p is even, with period $2T$ if p is odd (verify!). But let us re-emphasize that in the nonlinear case these findings identify systems that are generally *not* integrable, but nevertheless do behave in a very simple (*completely periodic*!) manner for a certain set (of *nonvanishing* measure!) of initial data.

5.7 Outlook

The results presented in Chap. 5 have been obtained quite recently. Clearly the techniques introduced herein could be exploited much more systematically and extensively than it has been done up to now, and one could thereby obtain a much larger collection of exactly treatable many-body problems in 3-dimensional (and also in S-dimensional) space than have been exhibited herein: indeed a motivation to write this book was just the hope to stimulate such a development. The amusing task to study in detail the behavior of these systems remains moreover largely undone, and it offers an ample prospect of interesting investigations.

5.N Notes to Chapter 5

The results presented in Chap. 5 are mainly based on <BC2000a> (which we often followed *verbatim*, but correcting several misprints -- hopefully without introducing new ones!); but see also the references quoted, and the credits given, there (let us in particular mention that a computational tool quite useful in this context is provided by <BR83>). There are however also some new results, for instance the many-body problems (5.3-44) and (5.3-47).

The main idea on which the developments of Sects. 5.2 and 5.4.1 are based is probably not new, see for instance <SS96>, <GS97>.

For the treatment of the magnetic monopole problem (Sects. 5.2.1 and 5.2.2) see also <P1896>, <SMTDC76> and <S2000>, and of course the references quoted in these papers. Our treatment follows closely <ABC2001>.

The *integrability* (indeed, *solvability*) of the periodic non Abelian Toda lattice (Sects. 5.4.2 and 5.4.4) has been proven by I. Krichever <K81> (see also <BMRL80>, <BRL81>, <RLB83>).

For the integrable matrix Nahm equations (5.4.3-21), see <N82> (I wish to thank Mark Ablowitz for bringing these *integrable* matrix ODEs to my attention and for providing this reference).

For the treatment of unharmonic ("quartic") oscillators, see <BC2000b,c> (which we often followed *verbatim* in Sect. 5.6.5). The fundamental observation that the matrix evolution equation (5.4.4-3) or, equivalently, (5.6.5-1), is *integrable* generalizes the previous finding by V. I. Inozemtsev <I90>, that (5.4.4-3) respectively (5.6.5-1) are *integrable* when the matrices $\underline{A}$ respectively $\underline{C}$ are multiples of the unit matrix.

The approach of the *Exercises 5.6.5-19* and *5.6.5-20* is fully discussed and exploited in <CF2001>.

Appendix A: Elliptic functions

In this Appendix, for the convenience of the reader and also to define our notation (there exist variations in the standard literature) we collect (without any commentary) some formulas for the elliptic functions associated with the names of Jacobi and Weierstrass.

Jacobian elliptic functions.

$$s \equiv \mathrm{sn}(u,k),\ c \equiv \mathrm{cn}(u,k),\ d \equiv \mathrm{dn}(u,k); \tag{1}$$

$$0 \le k \le 1,\ k' = \left(1-k^2\right)^{1/2},\ 0 \le k' \le 1; \tag{2}$$

$$s^2+c^2=1,\ k^2\ s^2+d^2=1,\ d^2-k^2\ c^2=k'^2=1-k^2; \tag{3}$$

$$s' \equiv \partial[\mathrm{sn}(u,k)]/\partial u = c\,d, \tag{4a}$$

$$c' \equiv \partial[\mathrm{cn}(u,k)]/\partial u = -s\,d, \tag{4b}$$

$$d' \equiv \partial[\mathrm{dn}\,(u,k)]/\partial u = -k^2\,s\,c; \tag{4c}$$

$$(s')^2 = \left(1-s^2\right)\left(1-k^2\,s^2\right), \tag{5a}$$

$$(c')^2 = \left(1-c^2\right)\left(k^2\,c^2+1-k^2\right), \tag{5b}$$

$$(d')^2 = \left(1-d^2\right)\left(d^2+1-k^2\right); \tag{5c}$$

$$(c/s)' = -d/s^2, \tag{6a}$$

$$(d/s)' = -c/s^2, \tag{6b}$$

$$s'' = -s\left(d^2+k^2\ c^2\right) = -(1+k^2)s+2k^2\,s^3\ , \tag{7a}$$

$$c'' = -c\left(d^2-k^2\ s^2\right) = -(1-2k^2)s-2k^2\,c^3\ , \tag{7b}$$

$$d'' = -k^2\ d\left(c^2-s^2\right) = (2-k^2)d-d^3\ ; \tag{7c}$$

F. Calogero: LNPm 66, pp. 663 - 674, 2001

$$\mathrm{sn}(-u,k) = -\mathrm{sn}(u,k),\ \ \mathrm{cn}(-u,k) = \mathrm{cn}(u,k),\ \ \mathrm{dn}(-u,k) = \mathrm{dn}(u,k); \tag{8}$$

$$\mathrm{sn}(u,k) = u - (1+k^2)u^3/3! + (1+14k^2+k^4)\,u^5/5! + \ldots, \tag{9a}$$

$$\mathrm{cn}(u,k) = 1 - u^2/2 + (1+4k^2)u^4/4! - (1+44k^2+16k^4)u^6/6! + \ldots, \tag{9b}$$

$$\mathrm{dn}(u,k) = 1 - k^2u^2/2 + k^2(4+k^2)u^4/4! - k^2(16+44k^2+k^4)u^6/6! + \ldots, \tag{9c}$$

Addition formulas of Jacobian functions.

$$s_j \equiv \mathrm{sn}(u_j,k),\ \ c_j \equiv \mathrm{cn}(u_j,k),\ \ d_j \equiv \mathrm{dn}(u_j,k),\ \ j = 1,2, \tag{10a}$$

$$\mathrm{sn}(u_1+u_2,k) = \left(s_1\,c_2\,d_2 + c_1\,d_1\,s_2\right)/D, \tag{10b}$$

$$\mathrm{cn}(u_1+u_2,k) = \left(c_1\,c_2 - s_1\,d_1\,s_2\,d_2\right)/D, \tag{10c}$$

$$\mathrm{dn}(u_1+u_2,k) = \left(d_1\,d_2 - k^2\,s_1\,c_1\,s_2\,c_2\right)/D, \tag{10d}$$

$$D = 1 - k^2\ s_1^2\ s_2^2. \tag{10e}$$

Degenerate cases of Jacobian function.

$$k = 0,\ \ k' = 1;\ \ \mathrm{sn}(u,0) = \sin(u),\ \ \mathrm{cn}(u,0) = \cos(u),\ \ \mathrm{dn}(u,0) = 1, \tag{11a}$$

$$k = 1,\ \ k' = 0;\ \ \mathrm{sn}(u,1) = \tanh(u),\ \ \mathrm{cn}(u,1) = \mathrm{dn}(u,1) = \left[\sinh(u)\right]^{-1}. \tag{11b}$$

Doubly periodic Weierstrass functions.

$$\wp(z) \equiv \wp(z\,|\,\omega,\omega) = z^{-2} + \sum{}'\left\{(z-w)^{-2} - w^{-2}\right\} \tag{12}$$

Here and throughout Appendix A: $w \equiv w_{mn} \equiv 2\omega\, m + 2\omega' n$, $\sum{}' f(w)$ is the sum over all (positive, vanishing and negative) integers m,n, excluding only the single term with both m and n vanishing, $m = n = 0$ (likewise for the product $\prod{}' f(w)$).

$$\wp(-z) = \wp(z), \tag{13}$$

$$\wp(z + 2m\omega + 2n\omega') = \wp(z). \tag{14}$$

ω, ω': "semiperiods". $\mathrm{Im}(\omega'/\omega) \neq 0$ (typically: ω' imaginary, ω real, so that $\wp(z)$ be real).

$$\wp'(z) = \partial \wp(z|\omega,\omega')/\partial z = -2z^{-3} - 2\sum{}'(z-w)^{-3}; \tag{15}$$

$$\wp'(\omega) = 0, \tag{16a}$$

$$\wp'(\omega') = 0; \tag{16b}$$

$$\wp(z; g_2, g_3) \equiv \wp(z|\omega,\omega') \equiv \wp(z); \tag{17}$$

$$\omega_1 = \omega,\ \omega_2 = -\omega - \omega',\ \omega_3 = \omega',\ \omega_1 + \omega_2 + \omega_3 = 0, \tag{18}$$

$$\wp(\omega_j) = e_j,\ j = 1,2,3, \tag{19}$$

$$\omega_j = \int_{-\infty}^{e_j} dx(4x^3 - g_2 x - g_3)^{-1/2},\ j = 1,2,3, \tag{20}$$

e_j are the 3 roots of the cubic equation $4x^3 - g_2 x - g_3 = 0$, hence

$$e_1 + e_2 + e_3 = 0,\ -4(e_1 e_2 + e_2 e_3 + e_3 e_1) = g_2,\ 4e_1 e_2 e_3 = g_3; \tag{21}$$

$$\wp(z) = z^{-2} + \sum_{k=2}^{\infty} c_k z^{2(k-1)}, \tag{22a}$$

$$c_2 = g_2/20,\ \ c_3 = g_3/28,\ \ c_k = 3[(2k+1)(k-3)]^{-1} \sum_{j=2}^{k-2} c_j c_{k-j},\ k \geq 4. \tag{22b}$$

Differential equations.

$$[\wp'(z)]^2 = 4[\wp(z)]^3 - g_2 \wp(z) - g_3, \tag{23a}$$

$$[\wp'(z)]^2 = 4[\wp(z) - e_1][\wp(z) - e_2][\wp(z) - e_3]; \tag{23b}$$

$$\wp''(z) = 6[\wp(z)]^2 - g_2/2, \tag{24}$$

$$\wp'''(z) = 12\wp'(z)\,\wp(z). \tag{25}$$

The last two equations imply, by induction, that the z-derivative of order $2n$ of $\wp(z)$ is a polynomial (with z-independent coefficients) of degree $n+1$ in $\wp(z)$, while the z-derivative of order $2n+1$ of $\wp(z)$ equals $\wp'(z)$ times a polynomial (with z-independent coefficients) in $\wp(z)$ of degree n.

Addition formulas.

$$\wp(z_1+z_2)=\left\{\frac{1}{2}\left[\wp'(z_1)-\wp'(z_2)\right]/\left[\wp(z_1)-\wp(z_2)\right]\right\}^2-\wp(z_1)-\wp(z_2), \tag{26}$$

$$\begin{vmatrix} 1 & \wp(z_1) & \wp'(z_1) \\ 1 & \wp(z_2) & \wp'(z_2) \\ 1 & \wp(z_1+z_2) & -\wp'(z_1+z_2) \end{vmatrix}=0, \tag{27}$$

$$\wp(z_1+z_2)=\wp(z_1)-\frac{1}{2}\frac{\partial}{\partial z_1}\{\left[\wp'(z_1)-\wp'(z_2)\right]/\left[\wp(z_1)-\wp(z_2)\right]\}, \tag{28}$$

$$\wp(z_1+z_2)=\wp(z_2)-\frac{1}{2}\frac{\partial}{\partial z_2}\{\left[\wp'(z_1)-\wp'(z_2)\right]/\left[\wp(z_1)-\wp(z_2)\right]\}, \tag{29}$$

$$\wp(z_1+z_2)+\wp(z_1-z_2)=2\wp(z_1)-\frac{\partial^2}{\partial z_1^2}\{\log\left[\wp(z_1)-\wp(z_2)\right]\}. \tag{30}$$

Other formulas.

$$\wp(z+\omega_j)=e_j+(e_j-e_k)\,(e_j-e_\ell)/[\;\wp(z)-e_j\;], \tag{31}$$

where j,k,ℓ are any permutation of $1,2,3$;

$$\wp(2z)=-2\wp(z)+\left[\frac{1}{2}\wp''(z)/\wp'(z)\right]^2; \tag{32}$$

$$\wp(z/2)=\wp(z)+\sum_{j=1,2,3 \bmod(3)}\{\left[\wp(z)-e_j\right]\left[\wp(z)-e_{j+1}\right]\}^{1/2}\;. \tag{33}$$

Rescalings.

$$\wp(\lambda z|\lambda\omega,\lambda\omega')=\lambda^{-2}\,\wp(z|\omega,\omega'),\quad \lambda\neq 0, \tag{34a}$$

$$\wp'(\lambda z|\lambda\omega,\lambda\omega') = \lambda^{-3}\,\wp'(z|\omega,\omega'),\quad \lambda\neq 0, \tag{34b}$$

$$\wp(\lambda z;\lambda^{-4}g_2,\lambda^{-6}g_3) = \lambda^{-2}\,\wp(z;g_2,g_3),\quad \lambda\neq 0, \tag{35a}$$

$$\wp'(\lambda z;\lambda^{-4}g_2,\lambda^{-6}g_3) = \lambda^{-3}\,\wp'(z;g_2,g_3),\quad \lambda\neq 0\,. \tag{35b}$$

Hence $\wp(z|\omega,\omega')$ depends effectively on 2, rather than 3, parameters: for instance, it could be considered a function of z/ω and z/ω' (or ω'/ω). Likewise, of course, for $\wp(z;g_2,g_3)$.

Degenerate cases.

$$\omega=\infty, \omega'=i\pi/(2a),\ e_1=e_2=a^2/3,\ e_3=-2a^2/3, \tag{36a}$$

$$\wp(z)=a^2/3+a^2[\sinh(az)]^{-2}\ ; \tag{36b}$$

$$\omega=\infty,\ \omega'=i\infty,\ e_1=e_2=e_3=0, \tag{37a}$$

$$\wp(z)=z^{-2}\,. \tag{37b}$$

"Sigma" and "zeta" Weierstrass functions.

$$\sigma(z)\equiv\sigma(z|\omega,\omega')=z\prod{}'\left\{(1-z/w)\exp\left[(z/w)+(z/w)^2/2\right]\right\}, \tag{38a}$$

$$\zeta(z)\equiv\zeta(z|\omega,\omega')=z^{-1}+\sum{}'\left[(z-w)^{-1}+w^{-1}+zw^{-2}\right]\ ; \tag{38b}$$

$$\zeta(z)\equiv\zeta(z|\omega,\omega')=\sigma'(z)/\sigma(z)\,; \tag{39}$$

$$\zeta'=-\wp(z)=\left\{\sigma''(z)\sigma(z)-[\sigma'(z)]^2\right\}/[\sigma(z)]^2\,; \tag{40}$$

$$\sigma(-z)=-\sigma(z),\ \ \zeta(-z)=-\zeta(z)\,; \tag{41}$$

$$\eta\equiv\zeta(\omega),\ \eta'\equiv\zeta(\omega');\ \eta_j\equiv\zeta(\omega_j),\ j=1,2,3, \tag{42}$$

$$\eta\,\omega'-\eta'\,\omega=i\pi/2\,; \tag{43}$$

$$\sigma(z+2m\omega+2n\omega')=(-)^{m+n+mn}\,\sigma(z)\exp\left[(z+2m\omega+2n\omega')(2m\eta+2n\eta')\right], \tag{44}$$

$$\zeta(z+2m\omega+2n\omega')=\zeta(z)+2m\eta+2n\eta'; \tag{45}$$

$$\sigma(z)=\sum_{m,n=0}^{\infty} a_{m,n}\,(g_2/2)^m\,(2g_3)^n\,z^{(4m+6n+1)}\Big/(4m+6n+1)!, \tag{46a}$$

$$a_{0,0}=1,\; a_{m,n}=0 \text{ if } m<0 \text{ or } n<0, \tag{46b}$$

$$a_{m,n}=(3m+1)\,a_{m+1,n-1}+(16/3)(n+1)\,a_{m-2,n+1}-(m+n-1/3)(4m+6n-1)\,a_{m-1,n}, \tag{46c}$$

$$\sigma(0)=0, \sigma'(0)=1, \sigma''(0)=\sigma'''(0)=\sigma^{(4)}(0)=0\;; \tag{46d}$$

$$\zeta(z)=z^{-1}-\sum_{k=2}^{\infty} c_k\,z^{2k-1}/(2k-1). \tag{47}$$

Relations among Weierstrass and Jacobian functions:

$$\sigma_j(z|\omega,\omega')\equiv\exp(-\eta_j z)\,\sigma(z+\omega_j|\omega,\omega')/\sigma(\omega_j|\omega,\omega'),\; j=1,2,3, \tag{48}$$

$$u=(e_1-e_3)^{1/2}\,z, \tag{49}$$

$$k^2=(e_2-e_3)/(e_1-e_2), \tag{50}$$

$$\mathrm{sn}(u,k)=(e_1-e_3)^{1/2}\,\sigma(z|\omega,\omega')/\sigma_3(z|\omega,\omega'), \tag{51a}$$

$$\mathrm{cn}(u,k)=\sigma_1(z|\omega,\omega')/\sigma_3(z|\omega,\omega'), \tag{51b}$$

$$\mathrm{dn}(u,k)=\sigma_2(z|\omega,\omega')/\sigma_3(z|\omega,\omega'); \tag{51c}$$

$$[\mathrm{sn}(u,k)]^2=(e_1-e_3)/[\wp(z|\omega,\omega')-e_3], \tag{52a}$$

$$[\mathrm{cn}(u,k)]^2=[\wp(z|\omega,\omega')-e_1]/[\wp(z|\omega,\omega')-e_3], \tag{52b}$$

$$[\mathrm{dn}(u,k)]^2=[\wp(z|\omega,\omega')-e_2]/[\wp(z|\omega,\omega')-e_3], \tag{52c}$$

$$[\mathrm{cn}(u,k)/\mathrm{sn}(u,k)]^2=(e_1-e_3)^{-1}[\wp(z|\omega,\omega')-e_1], \tag{53a}$$

$$[\mathrm{dn}(u,k)/\mathrm{sn}(u,k)]^2=(e_1-e_3)^{-1}[\wp(z|\omega,\omega')-e_2]. \tag{53b}$$

Degenerate cases of Weierstrass functions.

$$\omega=\infty,\ \omega'=i\pi/(2a),\ e_1=e_2=a^2/3,\ e_3=-2a^2/3, \tag{54a}$$

$$\sigma(z)=a^{-1}\sinh(az)\exp(-a^2z^2/6), \tag{54b}$$

$$\zeta(z)=-a^2z/3+a\ \mathrm{cotanh}(az), \tag{54c}$$

and see (36);

$$\omega=\infty,\ \omega'=i\infty,\ e_1=e_2=e_3=0, \tag{55a}$$

$$\sigma(z)=z,\ \zeta(z)=1/z, \tag{55b}$$

and see (37).

Duplication formulas of Weierstrass functions.

$$\sigma(2z)=-\wp'(z)\,[\sigma(z)]^4, \tag{56a}$$

$$\zeta(2z)=2\zeta(z)+\frac{1}{2}\wp''(z)/\wp(z), \tag{56b}$$

and see (32).

Additional relations satisfied by Weierstrass functions.

$$\begin{vmatrix} 1 & \wp(z_1) & \wp'(z_1) & \wp''(z_1)\cdot & \cdot & \cdot & \wp^{(N-2)}(z_1) \\ 1 & \wp(z_2) & \wp'(z_2) & \wp''(z_2)\cdot & \cdot & \cdot & \wp^{(N-2)}(z_2) \\ \cdot & \cdot & \cdot & \cdot & \cdot & \cdot & \cdot \\ 1 & \wp(z_N) & \wp'(z_N) & \wp''(z_N)\cdot & \cdot & \cdot & \wp^{(N-2)}(z_N) \end{vmatrix}$$

$$=(-)^{(N-1)(N-2)/2}\left[\prod_{n=1}^{N-1}n!\right]\sigma\Big(\sum_{m=1}^{N}z_m\Big)\left\{\prod_{n=1}^{N}[\sigma(z_n)]\right\}^{-N}\left[\prod_{l,m=1;l>m}^{N}\sigma(z_m-z_l)\right]. \tag{57}$$

Here N is a positive integer larger than 1, and the N numbers z_n are of course arbitrary (but different among themselves, and different from zero, both properties being of course valid $\mathrm{mod}(2\omega,2\omega')$). Note that the right hand side vanishes if $\sum_{n=1}^{N}z_n=0$; indeed, for $N=3$ and $z_3=-(z_1+z_2)$, this formula, (57), coincides with (27). For $N=2$ it yields

$$\sigma(z_1+z_2)\,\sigma(z_1-z_2)=[\sigma(z_1)]^2\,[\sigma(z_2)]^2\,[\wp(z_2)-\wp(z_1)]\,, \tag{58a}$$

whose logarithmic derivative (with respect to z_1) yields

$$\zeta(z_1+z_2)+\zeta(z_1-z_2)-2\zeta(z_1)=\wp'(z_1)/[\wp(z_1)-\wp(z_2)]\,. \tag{58b}$$

$$[\zeta(z_1)+\zeta(z_2)-\zeta(z_1+z_2)]^2+\zeta'(z_1)+\zeta'(z_2)+\zeta'(z_1+z_2)=0\,, \tag{59a}$$

$$\zeta(z_1+z_2)[\zeta(z_1)+\zeta(z_2)]+\zeta(z_1-z_2)[\zeta(z_1)-\zeta(z_2)]$$

$$=\zeta'(z_1)+\zeta^2(z_1)+\zeta'(z_2)+\zeta^2(z_2)$$

$$+[\zeta'(z_1+z_2)+\zeta^2(z_1+z_2)+\zeta'(z_1-z_2)+\zeta^2(z_1-z_2)]/2\,, \tag{59b}$$

$$\zeta(z_1)\zeta(z_2)=\zeta(z_1-z_2)[\zeta(z_2)-\zeta(z_1)]$$

$$+[\zeta'(z_1)+\zeta^2(z_1)+\zeta'(z_2)+\zeta^2(z_2)+\zeta'(z_1-z_2)+\zeta^2(z_1-z_2)]/2\,, \tag{59c}$$

$$\zeta(z_1+z_2)-\zeta(z_1)-\zeta(z_2)=\frac{1}{2}\,[\wp'(z_1)-\wp'(z_2)]/[\wp(z_1)-\wp(z_2)], \tag{60}$$

$$\wp(z_1+z_2)+\wp(z_1)+\wp(z_2)=[\zeta(z_1+z_2)-\zeta(z_1)-\zeta(z_2)]^2; \tag{61}$$

$$\zeta(z_1)+\zeta(z_2)+\zeta(z_3)-\zeta(z_1+z_2+z_3)$$

$$=\sigma(z_1+z_2)\sigma(z_2+z_3)\,\sigma(z_3+z_1)/[\sigma(z_1)\sigma(z_2)\sigma(z_3)\sigma(z_1+z_2+z_3)]\;; \tag{62}$$

$$\sigma(z_0+z_1)\sigma(z_0-z_1)\sigma(z_2+z_3)\sigma(z_2-z_3)$$

$$+\sigma(z_0+z_2)\sigma(z_0-z_2)\sigma(z_3+z_1)\sigma(z_3-z_1)$$

$$+\sigma(z_0+z_3)\sigma(z_0-z_3)\sigma(z_1+z_2)\sigma(z_1-z_2)=0\,. \tag{63}$$

This last formula, (63), contains 4 arbitrary variables, hence many other relations can be obtained from it, for instance by assigning special values (such as zero) to one or more of these variables, perhaps after having performed some differentiations. Some such formulas are displayed in <BC90>.

$$\det\left[\sigma(x_n - y_m + \alpha)/\sigma(x_n - y_m)\right]$$

$$\equiv \begin{vmatrix} \sigma(x_1-y_1+\alpha)/\sigma(x_1-y_1) & \sigma(x_1-y_2+\alpha)/\sigma(x_1-y_2) & \cdots & \sigma(x_1-y_N+\alpha)/\sigma(x_1-y_N) \\ \sigma(x_2-y_1+\alpha)/\sigma(x_2-y_1) & \sigma(x_2-y_2+\alpha)/\sigma(x_2-y_2) & \cdots & \sigma(x_2-y_N+\alpha)/\sigma(x_2-y_N) \\ \vdots & \vdots & \vdots\vdots\vdots & \vdots \\ \sigma(x_N-y_1+\alpha)/\sigma(x_N-y_1) & \sigma(x_N-y_2+\alpha)/\sigma(x_N-y_2) & \cdots & \sigma(x_N-y_N+\alpha)/\sigma(x_N-y_N) \end{vmatrix}$$

$$= \sigma\left[\alpha + \sum_{j=1}^{N} (x_j - y_j)\right]\left[\sigma(\alpha)\right]^{N-1} \prod_{n,m=1,n>m}^{N} \left[\sigma(x_n - x_m)\sigma(y_m - y_n)\right] / \prod_{n,m=1}^{N} \sigma(x_n - y_m). \tag{64}$$

Here (and below) $\sigma(z) \equiv \sigma(z|\omega,\omega')$, α is an arbitrary constant and the $2N$ variables x_n, y_n are also arbitrary (but different: $x_n \neq x_m$ and $y_n \neq y_m$ for $n \neq m$, and $x_n \neq y_m$; otherwise appropriate limits must be taken). The diligent reader will check that, for $N=2$, this formula reproduces (63) (*hint*: set $\alpha = z_1 + z_2, x_1 = z_0, x_2 = z_3, y_1 = z_2, y_2 = z_1$).

$$\sum_{n=1}^{N} \left\{\left[\sigma(z + \sum_{\ell=1,\ell\neq n}^{N} y_\ell - \sum_{j=1}^{N} x_j)\right]\left[\prod_{m=1}^{N} \sigma(y_n - x_m)\right] / \left[\prod_{\ell=1,\ell\neq n}^{N} \sigma(y_n - y_\ell)\right]\right\}$$

$$= \sigma\left[\sum_{j=1}^{N} (y_j - x_j)\right] \prod_{n=1}^{N} \left[\sigma(z - x_n)/\sigma(z - y_n)\right] \quad . \tag{65a}$$

Here the $2N+1$ variables z, x_n, y_n are all arbitrary, except for the usual requirement that they be *different*, and it is moreover required that

$$\sum_{j=1}^{N} (y_j - x_j) \neq 0, \quad \mathrm{mod}(2\omega, 2\omega') \quad . \tag{65b}$$

If instead this condition does *not* hold, (65a) is replaced by the following N identities:

$$\sum_{m=1}^{N} \left[\zeta(z - y_m) - \zeta(x_n - y_m)\right]\left[\prod_{j=1}^{N} \sigma(y_m - x_j)\right] / \left[\prod_{\ell=1,\ell\neq m}^{N} \sigma(y_m - y_\ell)\right]$$

$$= \prod_{m=1}^{N} \left[\sigma(z - x_m)/\sigma(z - y_m)\right], \quad n = 1,\ldots,N \quad , \tag{66a}$$

$$\sum_{j=1}^{N} (y_j - x_j) = 0, \quad \mathrm{mod}(2\omega, 2\omega') \quad . \tag{66b}$$

In the left-hand side of (66a) $\zeta(x) \equiv \zeta(x|\omega,\omega')$ is the Weierstrass zeta function, as defined above. Note that the right-hand side of (66a) does not depend on the index n, while the left-hand side does; this of course entails a number of additional identities.

$$\zeta(z)+\sum_{n=1}^{N}\left[\zeta(x_n)+\zeta(y_n)\right]-\zeta(z+s)$$

$$=\sum_{m=1}^{N}\frac{\sigma(z+x_m)}{\sigma(z)\,\sigma(x_m)}\frac{\sigma(z+s-x_m)}{\sigma(z+s)}\left\{\prod_{\ell=1,\ell\neq m}^{N}\left[\frac{\sigma(x_\ell)}{\sigma(x_\ell-x_m)}\right]\right\}\left\{\prod_{j=1}^{N}\left[\frac{\sigma(x_m+y_j)}{\sigma(y_j)}\right]\right\}, \quad (67a)$$

$$s\equiv\sum_{n=1}^{N}(x_n+y_n)\ . \quad (67b)$$

Here N is an arbitrary positive integer, and z, x_n, y_n are $2N+1$ arbitrary (complex) numbers. For $N=1$, up to trivial notational changes, this formula coincides with (62).

$$\text{cotanh}(z)+\sum_{n=1}^{N}\left[\text{cotanh}(x_n)+\text{cotanh}(y_n)\right]-\text{cotanh}(z+s)$$

$$=\sum_{m=1}^{N}\frac{\sinh(z+x_m)}{\sinh(z)\sinh(x_m)}\frac{\sinh(z+s-x_m)}{\sinh(z+s)}\cdot$$

$$\cdot\left\{\prod_{\ell=1,\ell\neq m}^{N}\left[\frac{\sinh(x_\ell)}{\sinh(x_\ell-x_m)}\right]\right\}\left\{\prod_{j=1}^{N}\left[\frac{\sinh(x_m+y_j)}{\sinh(y_j)}\right]\right\}\ . \quad (68)$$

$$z^{-1}+\sum_{n=1}^{N}\left[x_n^{-1}+y_n^{-1}\right]-(z+s)^{-1}$$

$$=\sum_{m=1}^{N}\frac{(z+x_m)}{z\,x_m}\frac{(z+s-x_m)}{(z+s)}\left\{\prod_{\ell=1,\ell\neq m}^{N}\left[\frac{x_\ell}{(x_\ell-x_m)}\right]\right\}\left\{\prod_{j=1}^{N}\left[\frac{(x_m+y_j)}{y_j}\right]\right\}\ . \quad (69)$$

These formulas, (68) and (69), are degenerate cases of (67), see (54) and (55); here of course s is given by (67b), and z, x_n, y_n are $2N+1$ arbitrary (complex) numbers.

$$\wp^{(k+1)}(x_n)=\left[\zeta\Big(\sum_{j=1}^{N}x_j\Big)-N\,\zeta(x_n)+\sum_{l=1,l\neq n}^{N}\zeta(x_n-x_l)\right]\wp^{(k)}(x_n)$$

$$- \sum_{m=1,m\neq n}^{N} \{ [\sigma(x_m)/\sigma(x_n)]^N \prod_{l=1,l\neq n,l\neq m}^{N} [\sigma(x_n - x_l)/\sigma(x_m - x_l)] \cdot$$

$$\cdot \left\{ \sigma(x_n - x_m + \sum_{j=1}^{N} x_j) / \left[\sigma(\sum_{j=1}^{N} x_j)\, \sigma(x_n - x_m) \right] \right\} \} \wp^{(k)}(x_m),\ k = 1,2,\ldots,N-2 \ . \tag{70a}$$

Here N is an arbitrary integer larger than 2, $N > 2$, the N numbers x_n are arbitrary (except for the requirement that their sum not vanish, $\sum_{n=1}^{N} x_n \neq 0$), of course $\wp(x) \equiv \wp(x|\omega,\omega'), \sigma(x) \equiv \sigma(x|\omega,\omega'), \zeta(x) \equiv \zeta(x|\omega,\omega')$ are the usual Weierstrass functions, and we use the notation

$$\wp^{(k)}(x) \equiv d^k \wp(x)/dx^k \ . \tag{70b}$$

$$(k+2)(x_n)^{-(k+3)} = \left[(N/x_n) - \left(\sum_{j=1}^{N} x_j \right)^{-1} - \sum_{l=1,l\neq n}^{N} (x_n - x_l)^{-1} \right] (x_n)^{-(k+2)}$$

$$+ \sum_{m=1,m\neq n}^{N} \{ [x_m/x_n]^N \prod_{l=1,l\neq n,l\neq m}^{N} [(x_n - x_l)/(x_m - x_l)] \left[\left(\sum_{j=1}^{N} x_j \right)^{-1} + (x_n - x_m)^{-1} \right] \} (x_m)^{-(k+2)},$$

$$k = 1,2,\ldots,N-2 \ . \tag{71}$$

This last formula is the completely degenerate case of (70), see (37b) and (55b).

A.N Notes to Appendix A

Most of the formulas reported in Appendix A are standard, and can be found in any compilation of mathematical formulas, see for instance <E53> (which we mainly followed), <MT56> or <GRJ94> (but see also <WW27> and <BC90>). We could not find the very useful identities (A-59b,c) in the literature, and therefore we have provided a proof of them in Sect. 2.3.6.2 (see the last part of the proof of *Proposition 2.3.6.2-7*). The determinantal identity (A-57), as well as the sum rule (A-65) can be found in the classic textbook by E. T. Whittaker and G. N. Watson <WW27> (see pp. 458 and 451). The determinantal identity (A-64) is due to G. Frobenius <F1892>, and (A-66) is taken from a recent paper by F.

W. Nijhoff and G.D. Pang <NP96>; I am grateful to Frank Nijhoff for providing these references. We show in Sect. 3.1.2.1 (see *Exercise 3.1.2.1-9*) how to prove the sum rule (A-67); we are not aware of its having being displayed elsewhere (not even in the degenerate trigonometric/hyperbolic, or rational, cases, see (A-68) and (A-69)). Likewise, the procedure to prove (A-70) is indicated in the *hint* which goes with *Exercise 3.1.2.1-10*. For several other identities involving the Weierstrass sigma and zeta elliptic functions see the last part of Appendix D.

Appendix B: Functional equations

In Appendix B we review the results on functional equations obtained elsewhere in this book, and we also report and introduce some new findings and conjectures.

Let us emphasize that, throughout, we focus on *analytic* solutions of the functional equations we consider. *Nonanalytic* solutions may also have an important role to play, see for instance Sect. 2.1.16, but they are not discussed in this Appendix B.

The first functional equation appears in Sect. 2.1.1, see (2.1.1-16); we write it here as follows:

$$\alpha(x+y)\left[\beta(x)-\beta(y)\right]=\alpha(x)\alpha'(y)-\alpha(y)\alpha'(x)\,. \tag{1}$$

The unknown functions are $\alpha(z)$ and $\beta(z)$; as for the latter, in Sect. 2.1.1 attention is restricted to *even* functions, $\beta(-z)=\beta(z)$. The *general* solution (with this restriction) of this functional equation is provided and discussed in Sect. 2.1.4; it involves *elliptic* functions (and their degenerate versions: *trigonometric, hyperbolic, rational*).

The functional equation

$$\left[\alpha(x+y)-\alpha(x)\alpha(y)\right]\left[\eta(x)-\eta(y)\right]=\alpha(x)\,\alpha'(y)-\alpha(y)\,\alpha'(x)\,, \tag{2a}$$

appears in Sect. 2.1.8, see (2.1.8-19); its *general* solution is provided and discussed in Sect. 2.1.11; it also involves *elliptic* functions (as well as their degenerate versions: *trigonometric, hyperbolic, rational*).

As entailed by the contexts in which these two functional equations, (1) and (2), have been introduced, as well as from their structure, (1) can be considered a limiting case of (2) (in the same sense as *nonrelativistic* equations are the limit, as the speed of light goes to infinity, of *relativistic* equations of motion), although, in the functional equation context, the limiting procedure is not entirely trivial (it corresponds to different singular behaviors of the solutions at the origin, see *Exercise B-2*).

F. Calogero: LNPm 66, pp. 675 - 684, 2001

Note that (2a) can also be rewritten in the form

$$\alpha(x+y)\,[\eta(x)-\eta(y)]=\alpha(x)\,\mu(y)-\alpha(y)\,\mu(x)\,, \tag{2b}$$

$$\mu(z)\equiv\alpha'(z)-\alpha(z)\eta(z)\,. \tag{2c}$$

Moreover, in the process of solving (2), another equivalent functional equation has been introduced:

$$\alpha(x+y)=\alpha(x)\;\alpha(y)+\varphi(x)\;\varphi(y)\;\psi(x+y) \tag{3a}$$

(see (2.1.11-23)). In contrast to (1) and (2), this functional equation, (3a), features 3 *a priori* unknown functions rather than only 2 and, more importantly, it does not feature any derivative; it is indeed, obtained by integrating the functional equation (2). The general solution of this functional equation, (3a), is also provided in Sect. 2.1.11; of course it involves *elliptic* functions, and their degenerate versions (*trigonometric*, *hyperbolic*, *rational*).

Likewise, the following nondifferential functional equation can be obtained by integrating (1):

$$\alpha(x+y)=\alpha(x)\;\theta(y)+\alpha(y)\;\theta(x)+\alpha(x)\;\alpha(y)\,\chi(x+y)\,; \tag{4a}$$

its *general* solution is, of course, also known (*elliptic* functions, and their degenerate versions).

Many other avatars of these functional equations, (3a) respectively (4a), can be obtained by appropriate redefinitions of the dependent variables. For instance the following functional equations are equivalent to (3a):

$$\alpha(x+y)/[\alpha(x)\alpha(y)]-\psi(x+y)/[\omega(x)\omega(y)]=1, \tag{3b}$$

$$\Psi(x)\Psi(y)/\Psi(x+y)-\Phi(x)\Phi(y)/\Omega(x+y)=1\,, \tag{3c}$$

$$\log[\alpha(x+y)-\alpha(x)\alpha(y)]=f(x)+f(y)+g(x+y)\,, \tag{3d}$$

$$\log[1-\alpha(x)\alpha(y)/\alpha(x+y)]=f(x)+f(y)+h(x+y)\,; \tag{3e}$$

likewise, the following functional equations are equivalent to (4a):

$$\alpha(x+y)/[\alpha(x)\alpha(y)]=\rho(x)+\rho(y)+\chi(x+y)\,, \tag{4b}$$

$$b(x+y)-b(x)b(y)=\log[\rho(x)+\rho(y)+\chi(x+y)], \tag{4c}$$

$$\exp\{\alpha(x+y)/[\alpha(x)\,\alpha(y)]\}=G(x)\,G(y)\,H(x+y). \tag{4d}$$

The keys to these transformations read as follows: for (3b), $\omega(z)=\alpha(z)/\varphi(z)$; for (3c), $\Psi(z)=1/\alpha(z)$, $\Phi(z)=\varphi(z)/\alpha(z)$, $\Omega(z)=\psi(z)$; for (3d) and (3e) $f(z)=\log[\varphi(z)]$, $g(z)=\log[\psi(z)]$, $h(z)=\log[\psi(z)/\alpha(z)]$; for (4b) and (4c), $\rho(z)=\theta(z)/\alpha(z)$, $b(z)=\log[\alpha(z)]$; for (4d) $G(z)=\exp[\theta(z)/\alpha(z)]$, $H(z)=\exp[\chi(z)]$.

A functional equation that seems more general than (1) and (2a), since it features 3 dependent variables (functions of a single argument) rather than 2, reads

$$\alpha(x+y)\,[\beta(x)-\beta(y)]=\alpha(x)\,\gamma(y)-\alpha(y)\,\gamma(x); \tag{5a}$$

but, in fact, this functional equation, (5a), is hardly more general than (1) and (2), since it essentially reduces to one or the other of these two functional equations (see *Exercise B-2* below).

Via appropriate changes of (dependent) variables, this functional equation, (5a), can assume other avatars, for instance

$$u(x+y)/[u(x)u(y)]=[\beta(x)-\beta(y)]/[v(x)-v(y)] \tag{5b}$$

(via $u(z)=1/\alpha(z)$, $v(z)=-\gamma(z)/\alpha(z)$), or

$$u(x+y)\,[u(x)\,w(y)-u(y)\,w(x)]=\psi(x)\,u^2(y)-\psi(y)\,u^2(x) \tag{5c}$$

(via $u(z)=1/\alpha(z)$, $w(z)=\gamma(z)/\alpha^2(z)$, $\psi(z)=\beta(z)/\alpha^2(z)$). Hereafter we refer for definiteness to the version (5a).

This functional equation, (5a), admits the following *trivial* solutions: $\alpha(z)=0$ with $\beta(z)=-\gamma(z)$ arbitrary; $\alpha(z)=A$, $\beta(z)=-\gamma(z)$, with A an arbitrary constant and $\gamma(z)$ an arbitrary function; $\beta(z)=B$ and $\alpha(z)=\gamma(z)$ with B an arbitrary constant and $\gamma(z)$ an arbitrary function. Hereafter we ignore this kind of *trivial* solutions, as well as those obtained from these by transformations such as those discussed immediately below. This kind of neglect of *trivial* solutions extends to *all* the functional equations discussed in this Appendix B, even though we do not bother to emphasize it in every case.

This functional equation, (5a), is clearly invariant under the following transformation:

$$\tilde{\alpha}(z) = A\,\alpha(a\,z)\,\exp(b\,z)\,, \tag{6a}$$

$$\tilde{\beta}(z) = B\,\beta(a\,z) + C\,, \tag{6b}$$

$$\tilde{\gamma}(z) = B\,\gamma(a\,z)\,\exp(b\,z) + D\,\tilde{\alpha}(z)\,, \tag{6c}$$

where the 6 constants A, B, C, D, a, b are *arbitrary*.

Exercise B-1. Verify that, if $\alpha(z)$, $\beta(z)$, $\gamma(z)$ satisfy (5), so do $\tilde{\alpha}(z)$, $\tilde{\beta}(z)$, $\tilde{\gamma}(z)$, as given by (6).

Exercise B-2. Show that, as $z \to 0$, the only possible behaviors of the *analytic* solutions of (5a) are (of course up to the transformations (6))

$$\alpha(z) = z^{-1} + O(1)\,, \quad \beta(z) = z^{-2} + O(z^{-1})\,, \quad \gamma(z) = -z^{-2} + O(z^{-1})\,, \tag{7a}$$

$$\alpha(z) = 1 + O(z)\,, \quad \beta(z) = -z^{-1} + O(1)\,, \quad \gamma(z) = z^{-1} + O(1)\,, \tag{7b}$$

and that, in the first case, (7a),

$$\gamma(z) = \alpha'(z) \tag{8a}$$

(so that (5a) becomes (1)), while in the second, (7b),

$$\gamma(z) = \alpha'(z) - \alpha(z)\,\beta(z) \tag{8b}$$

(so that (5a) becomes (2), up to trivial notational changes). *Hint*: firstly set $y = -x + \delta$, $\delta \to 0$, in (5a), to establish (7); then set $y = \delta$, $\delta \to 0$, in (5a), with (7a) respectively (7b), and thereby obtain (8a) respectively (8b) by equating the terms of order δ^{-p} with $p = 2,1$ respectively $p = 1,0$.

Remark B-3. The condition that $\beta(z)$ be even, $\beta(-z) = \beta(z)$, selects automatically the behavior (7a), hence the functional equation (5a) with this condition corresponds to (1).

The third functional equation we report was introduced in Sect. 2.3.6.1, and its *general* solution is provided in Sect. 2.3.6.2; it involves *elliptic* functions, including of course their degenerate versions (*trigonometric, hyperbolic, rational*). We write it here in the form (2.3.6.2-1e):

$$f(x-y)\,[g(x)-g(y)] = \frac{1}{2}\left\{[g(x)-g(y)]^2 + g'(x) + g'(y)\right\} + h(x-y)\,. \tag{9}$$

The 3 dependent variables are of course $f(z)$, $g(z)$ and $h(z)$; the treatment of Sect. 2.3.6.2 is restricted to functions $f(z)$ that are *odd*, $f(-z)=-f(z)$, and to functions $h(z)$ that are *even*, $h(-z)=h(z)$; note that either one of these two assumptions entails, via (9), the other one.

The fourth functional equation we review here was introduced in Sect. 2.1.16.1; it reads as follows (see (2.1.16.1-3)):

$$\alpha'(x)\,\alpha(x+y+z)-\alpha'(x+y+z)\,\alpha(x)+\alpha'(y)\,\alpha(z)+\alpha'(z)\,\alpha(y)$$

$$=\alpha(y+z)\,[\beta_1(x)+\beta_2(y)+\beta_3(z)-\beta_4(x+y+z)]. \tag{11}$$

Note that this functional equation features 3 independent variables and 5 dependent variables (functions of a single argument). However, its known *analytic* solutions are rather trivial:

$$\alpha(u)=A\cos(au)\,;\quad \beta_s(u)=b_s\,,\quad b_1+b_2+b_3-b_4=0\,, \tag{11a}$$

$$\alpha(u)=A\sin(au)\,;\quad \beta_s(u)=b_s\,,\quad b_1+b_2+b_3-b_4=2\,, \tag{11b}$$

with A, a and b_s arbitrary constants (5 altogether, since the sum of the 4 constants b_s is fixed).

Exercise B-4. Verify!

Remark B-5. Any nontrivial solution of the functional equation (10), with the additional restriction

$$\beta_s(-u)=-\beta_s(u)\,,\qquad s=1,2,3,4\,, \tag{12}$$

would be of great interest, since to it there corresponds an *integrable* dynamical system, see *Proposition 2.1.16.1-1*. (The solution (11a) with $b_s=0$, $s=1,2,3,4$, is of this type, and the corresponding *integrable*, indeed *solvable*, dynamical system is given by the Hamiltonian (2.1.15-16); the *nonanalytic* solution (2.1.16-1) with (2.1.16.1-5) is also of this type, and the corresponding *integrable* dynamical system is given by the Hamiltonian (2.1.16-12)).

Exercise B-6. Prove that, if $\alpha(u)$ is *even*, $\alpha(-u)=\alpha(u)$, and the functions $\beta_s(u)$ are all *odd*, see (12), then for every *analytic* solution of the functional equation (10) (if any exists!) the 4 functions $\beta_s(u)$ are *all* equal,

$$\beta_s(u) = \beta(u), \qquad s = 1,2,3,4 \ . \tag{13}$$

Hint: firstly set $y = -x$, and then $z = -x$, in (10).

The fifth functional equation we review here was introduced in Sect. 2.1.14. It reads

$$2\alpha'(x+y)\,[f(x)-f(y)] - \alpha(x+y)\,[f'(x)-f'(y)]$$

$$= \alpha(x)\,\gamma(y) - \alpha(y)\,\gamma(x) \tag{14}$$

(see (2.1.14-8)). This functional equation features 2 independent variables and 3 dependent variables.

Clearly if $\alpha(z)$, $f(z)$, $\gamma(z)$ satisfy this functional equation (14), so do

$$\tilde{\alpha}(z) = A\alpha(a\,z) \ , \tag{15a}$$

$$\tilde{f}(z) = B\,f(a\,z) + C\,, \tag{15b}$$

$$\tilde{\gamma}(z) = B\,\gamma(a\,z) + D\tilde{\alpha}(z)\,, \tag{15c}$$

with A, B, C, D, arbitrary constants.

Exercise B-7. Verify!

Two *analytic* solutions of (14) are known:

$$\alpha(z) = \sin(z+c), \qquad f(z) = \cos(2\,z+c)\,, \qquad \gamma(z) = 0\,, \tag{16a}$$

$$\alpha(z) = \sin(z+c), \qquad f(z) = \sin(2\,z+c), \qquad \gamma(z) = 4\cos(z+4c)\,, \tag{16b}$$

with c an arbitrary constant.

Exercise B-8. Verify!

Exercise B.9. Show that $\alpha(x) = \mu + x$, $f(x) = x\,(\nu + x)$, $\gamma(x) = 2\,(\nu - \mu)$ with μ, ν arbitrary constants, is, up to the transformation (15), the most general polynomial solution of (14) and verify that it can be obtained from (16) via (15) and an appropriate limiting process.

Conjecture B-10. Up to the transformation (15) (including its limiting cases), (16) provide *all* the *analytic* solutions of (14).

Remark B-11. There exist, however, also *nonanalytic* solutions of (14), see (2.1.16-1,2,3,4).

An interesting functional equation that generalizes (14) reads as follows

$$\lambda\, \alpha'(x+y)\,[f(x)-f(y)]-\alpha(x+y)\,[f'(x)-f'(y)]=\alpha(x)\,\gamma(y)-\alpha(y)\,\gamma(x). \quad (17)$$

It features, in addition to the 3 (dependent) functions $\alpha(z)$, $f(z)$, $\gamma(z)$, the "eigenvalue" λ.

It is easily seen that this functional equation, (17), is invariant (as well as (14)) under the transformation (15).

Exercise B-12. Verify!

Remark B-13. For $\lambda=2$, the functional equation (17) reduces to (14), and it therefore possesses the solutions (16); for $\lambda=0$, (17) reduces to (5a) (up to the notational change $f'(z)=-\beta(z)$), and it therefore possesses the solutions of (1) and of (2a), see *Exercise B-2* (this solutions involve generally *elliptic* functions; they are displayed in Sects. 2.1.4 and 2.1.11).

This functional equation, (17), possesses, for arbitrary λ, the solution

$$\alpha(z)=\sin(z+b), \qquad f(z)=[\sin(z)]^{\lambda}, \qquad \gamma(z)=\lambda\,[\sin(z)]^{\lambda-1}. \quad (18)$$

Exercise B-14. Verify!

Conjecture B-15. For $\lambda\neq 0$ and $\lambda\neq 2$, (18) is (up to the transformation (15)), the *general* solution of the functional equation (17).

Remark B-16. The additional requirement that *all* the functions, $\alpha(z)$, $f(z)$, $\gamma(z)$, see (18), that satisfy the functional equation (17) be *entire* entails that the "eigenvalues" λ are positive *integers*,

$$\lambda_n=n, \qquad n=1,2,3,\ldots; \quad (19a)$$

the requirements that these solutions, (18), of the functional equation (17), all be *meromorphic* functions entails that the "eigenvalues" λ are *integers*,

$$\lambda_n = n, \qquad n = 0, \pm 1, \pm 2, \pm 3, \ldots . \tag{19b}$$

The notion of "eigenvalue" introduced here in the context of functional equations refers of course to the existence, when λ is an eigenvalue, of *nontrivial* solutions of the functional equation, satisfying the additional specific requirement that characterizes the specific eigenvalue problem: in this instances, that (all the 3 functions that constitute) the solutions of the functional equation be *entire* respectively *meromorphic.*

Finally, let us consider a functional equation that clearly generalizes (17):

$$\alpha'(x+y)\,[g(x)-g(y)]+\alpha(x+y)\,[h(x)-h(y)] = \alpha(x)\,\gamma(y)-\alpha(y)\,\gamma(x). \tag{20a}$$

This functional equation features 2 independent, and 4 dependent, variables (functions of a single argument). Many other avatars of this functional equation are obtained by changes of (dependent) variables, for instance

$$\alpha'(x+y)\,[\varphi(x)\,\eta(y)-\varphi(y)\,\eta(x)]+\alpha(x+y)\,[\psi(x)\,\eta(y)-\psi(y)\,\eta(x)]$$

$$= \alpha(x)\,\eta(x)-\alpha(y)\,\eta(y), \tag{20b}$$

(via $\varphi(z)=g(z)/\gamma(z)$, $\psi(z)=h(z)/\gamma(z)$, $\eta(z)=1/\gamma(z)$), and

$$\alpha'(x+y)\,[\mu(x)\,\rho(y)-\mu(y)\,\rho(x)]+\alpha(x+y)\,[\nu(x)\,\rho(y)-\nu(y)\,\rho(x)]$$

$$= \alpha(x)\,\alpha(y)\,[\rho(x)-\rho(y)], \tag{20b}$$

(via $\mu(z)=\alpha(z)\,g(z)/\gamma(z)$, $\nu(z)=\alpha(z)h(z)/\gamma(z)$, $\rho(z)=\alpha(z)/\gamma(z)$).

Exercise B-17. Verify that, if $\alpha(z)$, $g(z)$, $h(z)$, $\gamma(z)$, satisfy the functional equation (20a), so do

$$\tilde{\alpha}(z) = A\alpha(az)\,\exp(bz), \tag{21a}$$

$$\tilde{g}(z) = B\, g(a\, z) + C\,, \tag{21b}$$

$$\tilde{h}(z) = B\big[a\, h(a\, z) - b\, g(a\, z)\big] + D$$

$$\tilde{\gamma}(z) = a\, B\, \gamma(a\, z)\, \exp(b\, z) + E\, \tilde{\alpha}(z)\,, \tag{21c}$$

where A, B, C, D, E, a, b are 7 *arbitrary* constants.

It is plain that this functional equation, (20a), possesses the following solutions: $\alpha(z) = 0$, no restriction on $g(z)$, $h(z)$, $\gamma(z)$; $\alpha'(z) = 0$, $h(z) = -\gamma(z)$, no restriction on $g(z)$, $\gamma(z)$; $g'(z) = h'(z) = 0$, $\alpha(z) = \gamma(z)$, no restriction on $\gamma(z)$; as well as the solutions that obtain from these via (21). These *trivial* solutions are hereafter ignored.

It is also plain that, if

$$g'(z) = 0\,, \tag{22}$$

the functional equation (20a) coincides, up to trivial notational changes, with (5a), whose nontrivial solutions, as we saw above, are the union of the solutions of (1) and (2), see *Exercise B-2*. Let us recall that these solutions involve *elliptic* functions, see Sect. 2.1.4 and 2.1.11, and of course as well their degenerate versions: *trigonometric, hyperbolic, rational* functions. In the following we also exclude from consideration these solutions of (20a) with (22), as well as all those obtained from these via the transformation (21) (which preserves (22), see (21b)).

Conjecture B-18. Up to the transformation (21), and excluding the trivial solutions detailed above as well as those associated with the condition (22), *all* analytic solutions of the functional equation (20a) read as follows:

$$\alpha(z) = \sin(z + c)\,, \qquad g(z) = \sin(z)\, \gamma(z)\,, \qquad h(z) = -\cos(z)\, \gamma(z)\,, \tag{23a}$$

with $\gamma(z)$ an *arbitrary* (analytic) function;

$$\alpha(z) = \sin(z + c)\,, \quad g(z) = \cos(2\, z + c)\,, \quad h(z) = \sin(2\, z + c)\,, \quad \gamma(z) = 0\,; \tag{23b}$$

$$\alpha(z) = \sin(z + c)\,, \; g(z) = \sin(2\, z + c)\,, \; h(z) = -\cos(2\, z + c)\,, \; \gamma(z) = 2\cos(z + c)\,. \tag{23c}$$

In all these expressions, (23), c is an *arbitrary* constant.

Exercise B-19. Verify that (23) satisfy (20a).

Remark B-20. The additional solutions (23b) and (23c) are *not* special cases of (23a).

Remark B-21. For the solutions (23b,c) the functional equations (14) and (20a) *coincide*, as demonstrated by the substitution

$$f(z) \Rightarrow g(z)\,, \qquad f'(z) \Rightarrow -h(z)\,, \qquad \gamma(z) \Rightarrow 2\gamma(z)\,, \tag{24}$$

which also entail that (16a,b) correspond to (23b,c).

Exercise B-22. Reobtain the solution (18) of (17) as a special case of the solution (23a) of (20a). *Hint*: use the relation among $g(z)$ and $h(z)$, which reduces (20a) to (17).

B.N Notes to Appendix B

The functional equation (B-1) is the first one to have appeared <C75> in the context of (the Lax matrix approach to) classical (i.e., non quantal) *integrable* systems; its general solution was exhibited in the same paper <C75>, and discussed in <C76a>, and also, more or less simultaneously, by A. M. Perelomov (see Appendix A of <OP76b>) and by S. I. Pydkuyko and A. M. Stepin <PS76>.

The functional equation (B-2) was introduced and solved in <BC87>; a more detailed discussion of the *general* solution of this functional equation, and as well of (B-3) and (B-4), is given in <BC90>.

The functional equation (B-5) is a special case of the more general (but in fact rather closely related) functional equation

$$\varphi_1(x+y)=\begin{vmatrix}\varphi_2(x) & \varphi_2(y)\\ \varphi_3(x) & \varphi_3(y)\end{vmatrix} / \begin{vmatrix}\varphi_4(x) & \varphi_4(y)\\ \varphi_5(x) & \varphi_5(y)\end{vmatrix}, \tag{1}$$

(set $\varphi_1(z)=\varphi_2(z)=\alpha(z)$, $\varphi_3(z)=\gamma(z)$, $\varphi_4(z)=\beta(z)$, $\varphi_5(z)=1$). This functional equation, (1), is fully treated in the monograph <BB97b>, where the interested reader will find additional references on functional equations of this type, such as <BP96>, <BK96>, BB97a>.

The functional equation (B-9) is, to the best of my knowledge, new.

The functional equations (B-10) and (B-14) were introduced in <CF96>.

The remaining material in Appendix B is, to the best of my knowledge, new (including the introduction of the functional equation (B-17) with *integer* "eigenvalues", see *Remark B-16*).

Appendix C:
Hermite polynomials:
zeros, determinantal representations

In Appendix C we collect a number of formulas for Hermite polynomials; this is only a representative sample, many other analogous results are available in the literature, not only for Hermite polynomials, but as well for all the classical polynomials, see Sect. C.N.

Hermite polynomials.

$$H_n(x) = n! \sum_{m=0}^{[[n/2]]} (-1)^m \left[m!(n-2m)!\right]^{-1} (2x)^{n-2m}, \tag{1a}$$

$$H_0(x)=1\,, \quad H_1(x)=2x, \quad H_2(x)=4x^2-2, \quad H_3(x)=8x^3-12x; \tag{1b}$$

$$H_n(-x) = (-1)^n H_n(x); \tag{1c}$$

$$H_n'(x) = 2n H_{n-1}(x) = 2x H_n(x) - H_{n+1}(x), \tag{1d}$$

$$H_n''(x) = 2x H_n'(x) + 2n H_n(x); \tag{1e}$$

$$\sum_{n=0}^{\infty} H_n(x) z^n / n! = \exp(2xz - z^2), \tag{1f}$$

$$\sum_{n=0}^{\infty} (z/2)^n H_n(x) H_n(y)/n!$$

$$= (1-z^2)^{-1/2} \exp\left\{\left[\ 2xyz - (x^2+y^2)z^2\ \right]/(1-z^2)\right\}. \tag{1g}$$

Sum rules for the N zeros of Hermite polynomials:

$$H_N(x_n^{(H)(N)}) = 0\,, \qquad z_n \equiv x_n^{(H)(N)}\,, \tag{2}$$

$$\sigma_n^{(p)} \equiv \sum_{m=1,\, m\neq n}^{N} (z_n - z_m)^{-p}, \tag{3a}$$

F. Calogero: LNPm 66, pp. 685 - 688, 2001

$$\sigma_n^{(1)} = z_n , \tag{3b}$$

$$\sigma_n^{(2)} = 2(N-1)/3 - z_n^2/3 , \tag{3c}$$

$$\sigma_n^{(3)} = z_n/2 , \tag{3d}$$

$$\sigma_n^{(4)} = [\ 2(N+2) - z_n^2\ ][\ 2(N-1) - z_n^2\ ]/45 , \tag{3e}$$

$$\sigma_n^{(5)} = z_n(2N+1-z_n^2)/18 ; \tag{3f}$$

$$\sum_{n=1}^{N} z_n^2 = N(N-1)/2 ; \tag{4}$$

$$\sum_{m,n=1;\, m\neq n}^{N} (z_n - z_m)^{-2} = N(N-1)/2 . \tag{5}$$

Remarkable matrices defined in terms of the N zeros z_n of the Hermite polynomial $H_n(z)$, see (2): the $(N \times N)$-matrices

$$N_{nm} = \delta_{nm} z_n^2 + (1-\delta_{nm}) z_n/(z_n - z_m) , \tag{6}$$

$$A_{nm} = \delta_{nm} \sum_{l=1,\, l\neq n}^{N} (z_n - z_l)^{-2} - (1-\delta_{nm})(z_n - z_m)^{-2} , \tag{7a}$$

$$A_{nm} = \delta_{nm} [\ 2(N-1)/3 - z_n^2/3\ ] - (1-\delta_{nm})(z_n - z_m)^{-2} , \tag{7b}$$

both have the first N nonnegative integers $0,1,\ldots,N-1$ as eigenvalues; the $(N \times N)$-matrix $\underline{B}$,

$$B_{nm} = 6\, \delta_{nm} \sum_{l=1,\, l\neq n}^{N} (z_n - z_l)^{-4} - (1-\delta_{nm})\, 6\, (z_n - z_m)^{-4} , \tag{8a}$$

$$B_{nm} = \delta_{nm}(2/15)[\ 2(N+2) - z_n^2\ ][\ 2(N-1) - z_n^2\ ] - (1-\delta_{nm})\, 6\, (z_n - z_m)^{-4} , \tag{8b}$$

is related to the matrix $\underline{A}$, see (7), as follows:

$$\underline{B} = \underline{A}(\underline{A} + 2) , \tag{9}$$

hence it has the eigenvalues $(p^2 - 1)$, $p = 1,2,\ldots,N$; the (Hermitian) $(N \times N)$-matrix

$$\underline{M} \equiv \underline{M}(\varphi),$$

$$M_{nm}(\varphi) = \delta_{nm}\, z_n \cos\varphi + (1-\delta_{nm})\, i\, (z_n - z_m)^{-1} \sin\varphi\,, \tag{10}$$

has the N zeros z_n, see (2), as its eigenvalues (for all values of the "angle" φ; this result is of course trivial for $\varphi = 0$).

It had been conjectured that, if one defined an $(N \times N)$-matrix $\underline{A}$ in terms of N *a priori arbitrary* numbers z_n via (7a) and then *required* that this matrix $\underline{A}$ have the first N nonnegative integers as its eigenvalues, then the N numbers z_n would be determined and would coincide, up to a common shift, with the N zeros of the Hermite polynomial of degree N. But this conjecture has been *disproved* <C82b>. It has been likewise *disproved* <C82b> that the requirement that the $(N \times N)$-matrix $\underline{N}$, defined by (6) in terms of N *a priori arbitrary* numbers z_n, have the first N nonnegative integers as its N eigenvalues, determines uniquely the N numbers z_n (which would then coincide with the N zeros of the Hermite polynomial of degree N; since if the N numbers z_n are so defined, then the $(N \times N)$-matrix $\underline{N}$ does indeed have the first N nonnegative integers as its N eigenvalues).

Determinantal representation of Hermite polynomials, in terms of N *arbitrary* numbers x_n:

$$H_n(x) = (N!)^{-1} \det\left[\, \underline{M}^{(H)}(x \mid \underline{x})\, \right], \tag{11a}$$

where the $(N \times N)$-matrix $\underline{M}^{(H)}(x \mid \underline{x})$ (a function of the variable x, and of the N-vector $\underline{x}$ whose N components are the N *arbitrary* numbers x_n, $\underline{x} \equiv (x_1, x_2, \ldots, x_N)$) reads as follows:

$$\underline{M}^{(H)}(x \mid \underline{x}) = 2(\underline{X} - \underline{D}) + (x - \underline{X})(\underline{D} - 2\underline{X})\underline{D} + 2N(x - \underline{X}), \tag{11b}$$

$$\underline{X} = \text{diag}[x_n], \qquad X_{nm} = \delta_{nm}\, x_n\,, \tag{11c}$$

$$D_{nm} = \delta_{nm} \sum_{l=1,\, l \neq n}^{N} (x_n - x_l)^{-1} + (1-\delta_{nm})(x_n - x_m)^{-1}. \tag{11d}$$

Other determinantal representations of Hermite polynomials – indeed, of *all* classical polynomials, and even more generally, of any polynomial characterized as the solution either of a linear differential equation or of a linear recursion relation – can be easily manufactured, see Sect. 2.4.5.5 (for explicit examples see the literature quoted in Sect. C.N).

Let us end this Appendix C by re-emphasizing that, although here we only reported results for Hermite polynomials, analogous, and also more general, results are as well available for *all* the classical polynomials, see Sect. C.N.

C.N Notes to Appendix C

The results reported (only for Hermite polynomials) in Appendix C are a representative, but incomplete, sample of those that can be found in the literature (for all the classical polynomials: Jacobi, Laguerre, Gegenbauer, Lagrange, besides Hermite); see firstly the standard compilations covering classical polynomials, for instance <S39>, <H65>, Vol. II of <E53>, and <GRJ94>, for the standard formulas (definitions of the classical polynomials, differential and recursion relations, generating functions); then see (in addition, again, to <S39> for some key properties of the zeros) the following papers which introduced the main new ideas <C78a> and which provide overviews on the "new" results (zeros, remarkable matrices, determinantal representations): <C78b>, <ABCOP79>, <C80a>, <C81a>, <C81c>, <C82c>, <C84b>; and finally, if need be, see the following original papers: <C77a>, <C77b>, <C77d>, <ABC78>, <BC79>, <C80b>, <C82a>, <C82b>, <C85a>, <C85d>.

For certain relations among classical polynomials in the limit in which certain of their parameters diverge see <C78d>, <C78e>.

For certain results related with the limit in which the degrees of the polynomials diverge, so that their zeros fill a continuum distribution, and (the analogs of) the "remarkable matrices" become integral operators, see: <CP78b>, <CP78c>, <C79a>, <C79b>.

For some results analogous to (some of) those presented in Appendix C for the zeros of Hermite polynomials, but featuring instead the zeros of combinations of Hermite polynomials, see <ABC78>.

Finally, for some analogous results featuring the zeros of Bessel functions (of which there are an infinite number, in contrast to the polynomial case), see <C77c>, <C77e>, <AC78a>, <AC78b>, <AC78c>.

Appendix D: Remarkable matrices and related identities

By *remarkable matrices* we mean matrices, generally defined by rather neat expressions containing many arbitrary parameters, which feature simple properties, typically an explicitly known spectrum given by a very neat rule, and often as well explicitly known eigenvectors also given by neat expressions. There generally follows the validity of *identities*, obtained for instance by writing in long-hand the eigenvalue equation satisfied by the remarkable matrix, or by evaluating the traces of its powers, or its determinant, in terms of its eigenvalues. In Appendix D we report, with minimal commentary, a representative sample of such formulas. The developments that led to these formulas have been described in various places throughout this book, mainly in the parts treating Lagrangian interpolation, see Sect. 2.4 and its subsections (in particular, of course, Sect. 2.4.5 and its subsections), as well as Sect. 3.1 and its subsections.

Hereafter, unless otherwise specified, indices run as usual from 1 to N, and numbers denoted as θ_n, x_n and so on are *arbitrary* (possibly even *complex*) but *distinct* ($\theta_n \neq \theta_m$ if $n \neq m$; $\varphi_n \neq \varphi_m$ if $n \neq m$; and so on; most formulas remain valid even if this condition is dropped, but in such cases suitable limits may be required).

Proposition D-1. Define the $(N \times N)$-matrix $\underline{C}(\underline{\theta})$ in terms of the N *arbitrary* "angles" θ_n by the neat rule

$$C_{nm}(\underline{\theta}) = i \sum_{l=1, l \neq n}^{N} \operatorname{cotan}(\theta_n - \theta_l), \qquad \text{if } n = m, \tag{1a}$$

$$C_{nm}(\underline{\theta}) = i \left[\sin(\theta_n - \theta_m)\right]^{-1}, \qquad \text{if } n \neq m. \tag{1b}$$

Then the N eigenvalues c_n and the N (right) eigenvectors $\underline{u}^{(n)}(\underline{\theta})$ of this $(N \times N)$-matrix $\underline{C}(\underline{\theta})$,

$$\underline{C}(\underline{\theta})\underline{u}^{(n)}(\underline{\theta}) = c_n \underline{u}^{(n)}(\underline{\theta}) \ , \tag{1c}$$

F. Calogero: LNPm 66, pp. 689 - 702, 2001

are given by the simple rule

$$c_n = 2n - N - 1 \ , \qquad n = 1,2,\dots,N \ , \tag{1d}$$

$$\underline{u}_m^{(n)}(\underline{\theta}) = \exp\left[i(N+1-2n)\theta_m\right]\left[\prod_{l=1,l\neq m}^{N} \sin(\theta_m - \theta_l)\right]^{-1} . \tag{1e}$$

Proposition D-2. Define the $(N \times N)$-matrix $\underline{\tilde{C}}(\underline{\theta})$ in terms of the N *arbitrary* "angles" θ_n by the neat rule

$$\tilde{C}_{nm}(\underline{\theta}) = C_{nm}(\underline{\theta}) = i \sum_{l=1,l\neq n}^{N} \text{cotan}(\theta_n - \theta_l) \ , \qquad \text{if } n = m, \tag{2a}$$

$$\tilde{C}_{nm}(\underline{\theta}) = i \ \text{cotan}(\theta_n - \theta_m) \ , \qquad \text{if } n \neq m. \tag{2b}$$

Then the $N-1$ (hence, all but, at most, one) eigenvalues $\tilde{c}_n$ of this $(N \times N)$-matrix $\underline{\tilde{C}}(\underline{\theta})$, and the corresponding right eigenvectors $\underline{\tilde{u}}^{(n)}(\underline{\theta})$,

$$\underline{\tilde{C}}(\underline{\theta}) \ \underline{\tilde{u}}^{(n)}(\underline{\theta}) = \tilde{c}_n \ \underline{\tilde{u}}^{(n)}(\underline{\theta}) \ , \qquad n = 1,2,\dots,N-1, \tag{2c}$$

are given by the simple rule

$$\tilde{c}_n = 2n - N \ , \qquad n = 1,2,\dots,N-1, \tag{2d}$$

$$\underline{\tilde{u}}_m^{(n)}(\underline{\theta}) = \exp\left[i(N-2n)\theta_m\right]\left[\prod_{l=1,l\neq m}^{N} \sin(\theta_m - \theta_l)\right]^{-1} , \qquad n = 1,2,\dots,N-1; \tag{2e}$$

and these $N-1$ eigenvectors $\underline{\tilde{u}}_m^{(n)}(\underline{\theta})$ are also eigenvectors, *all* of them with *zero* eigenvalue, of the matrix $\underline{J}$ defined by the simple rule that *all* its elements are *unity*:

$$J_{nm} = 1 \ , \tag{2f}$$

$$\underline{J} \ \underline{\tilde{u}}_m^{(n)}(\underline{\theta}) = 0 \ , \qquad n = 1,2,\dots,N-1, \tag{2g}$$

which also entails the (obvious) matrix identity (see(2a,b))

$$\underline{J} \ \underline{\tilde{C}}(\underline{\theta}) = 0 \ . \tag{2h}$$

Moreover the $(N \times N)$-matrix $\underline{\tilde{C}}(\theta)$ possesses obviously the eigenvalue 0, with the left eigenvector $\underline{u}$ characterized by the simple rule to have *all* its components equal to *unity*:

$$u_m = 1, \tag{2i}$$

$$\underline{u}\,\underline{\tilde{C}}(\theta) = 0; \tag{2l}$$

while of course $\underline{u}$ is also an eigenvector of $\underline{J}$ (of course both right and left, since $\underline{J}$ is symmetrical), with eigenvalue N:

$$\underline{J}\,\underline{u} = \underline{u}\,\underline{J} = N\,\underline{u}. \tag{2m}$$

Hence if N is *odd*, the matrix $\underline{\tilde{C}}(\theta)$ possesses the N distinct eigenvalues $-(N-2), -(N-4), ..., -1, 0, 1, ..., N-4, N-2$, and it is therefore diagonalizable; while if N is *even*, it possesses the $N-1$ eigenvalues $-(N-2), -(N-4), ..., -2, 0, 2, ..., N-4, N-2$, and, *iff* it is diagonalizable, the eigenvalue 0 has multiplicity 2 (but, for N even, $\underline{\tilde{C}}(\theta)$ need *not* be diagonalizable; for instance for $N = 2$

$$\underline{\tilde{C}}(\theta) = i \operatorname{cotan}(\theta_1 - \theta_2) \begin{pmatrix} 1 & 1 \\ -1 & -1 \end{pmatrix}, \tag{2n}$$

and the only case when this (2×2)-matrix is diagonalizable is when it vanishes identically, namely if $\theta_1 - \theta_2 = \pi/2 \mod(\pi)$); while the matrix

$$\underline{A}(\underline{\theta}; \alpha, \beta, \gamma) = \alpha \underline{1} + \beta \underline{J} + \gamma \underline{\tilde{C}}(\theta), \tag{2r}$$

has, in addition to the $N-1$ eigenvalues

$$a_n = \alpha + \gamma(2n - N), \qquad n = 1, 2, ..., N-1, \tag{2p}$$

the eigenvalue

$$a_n = \alpha + \beta N; \tag{2q}$$

and if these N eigenvalues are distinct, it is of course diagonalizable.

Remark D-3. The similarities and differences among the results of *Propositions D-1* and *D-2* should be emphasized; note in particular that, for any given N, their eigenvalues and eigenvectors are *different*, for instance for *odd* N the N eigenvalues c_n of $\underline{C}(\underline{\theta})$ are *even integers*, see (1d), while the $N-1$ nonvanishing eigenvalues of $\underline{\tilde{C}}(\underline{\theta})$ are *odd integers*, see (2d); and this in spite of the fact that their diagonal elements coincide, see (1a) and (1b), hence their traces also coincide; indeed both their traces vanish,

$$\text{trace}[\ \underline{C}(\underline{\theta})\ ] = \text{trace}[\ \underline{\tilde{C}}(\underline{\theta})\ ] = 0\ , \tag{3a}$$

consistently with the symmetrical location of their (real!) eigenvalues to the left and the right of *zero*, and also consistently with the trivial *identity*

$$\sum_{n,m=1;n\neq m}^{N} \text{cotan}(\theta_n - \theta_m) = 0\,. \tag{3b}$$

Proposition D-4. Define the $(N \times N)$-matrix $\underline{M}(\underline{\theta})$ in terms of the N *arbitrary* "angles" θ_n by the neat rule

$$M_{nm}(\underline{\theta}) = -\sum_{l=1,l\neq n}^{N} \cos(\theta_n)\sin(\theta_l)/\sin(\theta_n - \theta_l)\,, \qquad \text{if } n = m, \tag{4a}$$

$$M_{nm}(\underline{\theta}) = -\cos(\theta_n)\sin(\theta_n)/\sin(\theta_n - \theta_m)\,, \qquad \text{if } n \neq m\,. \tag{4b}$$

Then the N eigenvalues μ_n and the N eigenvectors $\underline{v}^{(n)}(\underline{\theta})$ of this $(N \times N)$-matrix $\underline{M}(\underline{\theta})$,

$$\underline{M}(\underline{\theta})\ \underline{v}^{(n)}(\underline{\theta}) = \mu_n\ \underline{v}^{(n)}(\underline{\theta})\ , \tag{4c}$$

are given by the simple rule

$$\mu_n = n - 1\ , \qquad n = 1,2,\ldots,N\,, \tag{4d}$$

$$\underline{v}_m^{(n)}(\underline{\theta}) = [\cos\theta_m]^{n-1}[\sin\theta_m]^{N-n}\left[\prod_{l=1,l\neq m}^{N} \sin(\theta_m - \theta_l)\right]^{-1}. \tag{4e}$$

Remark D-5. Note the possibility to generalize/reformulate these results, see (4), by shifting the *arbitrary* "angles" θ_n (which are all obviously defined $\text{mod}(2\pi)$) by an arbitrary (common!) amount θ, $\theta_n \to \theta_n + \theta$ (and then perhaps setting $\theta = \pi/2$ to get a neater result). Also note the isospectral character of all these three $(N \times N)$-matrices, $\underline{C}(\underline{\theta})$, $\underline{\tilde{C}}(\underline{\theta})$ and

$\underline{M}(\theta)$, as manifested by the independence of their spectra, see (1d), (2d) and (4d), from the N parameters θ_n.

Proposition D-6. In addition to the *trigonometric identities* implied by (1c) with (1a,b,d,e), by (2c) with (2a,b,d,e) and by (4c) with (4a,b,d,e), there also hold the following *sum rules*:

$$\sum_{n,m,l=1;n\neq m,m\neq l,l\neq n}^{N} \cot\mathrm{an}(\theta_n-\theta_m)\cot\mathrm{an}(\theta_n-\theta_l)=-N(N-1)(N-2)/3 \ , \quad (5a)$$

$$\sum_{n,m,l=1;n\neq m,m\neq l,l\neq n}^{N} \cos^2\theta_n \sin\theta_m \sin\theta_l \left[\sin(\theta_n-\theta_m)\sin(\theta_n-\theta_l)\right]^{-1}=N(N-1)(N-2)/3, \quad (5b)$$

$$\sum_{n,m,l=1,n\neq m,m\neq l,l\neq n}^{N} \sin(2\theta_n)\sin(\theta_m+\theta_l)\left[\sin(\theta_n-\theta_m)\sin(\theta_n-\theta_l)\right]^{-1}=-2N(N-1)(N-2)/3, \quad (5c)$$

$$\sum_{n,m,l=1;n\neq m,m\neq l,l\neq n}^{N} \sin(\theta_n+\theta_m)\sin(\theta_n+\theta_l)\left[\sin(\theta_n-\theta_m)\sin(\theta_n-\theta_l)\right]^{-1}=N(N-1)(N-2)/3, \quad (5d)$$

$$\sum_{n,m,l=1;n\neq m,m\neq l,l\neq n}^{N} \sin(2\theta_n+\theta_m+\theta_l)\left[\sin(\theta_n-\theta_m)\sin(\theta_n-\theta_l)\right]^{-1}=0 \ . \quad (5e)$$

Remark D-7. Other identities can be obtained from these by shifting (*all* the) arbitrary quantities θ_n, namely by replacing θ_n with $\theta_n+\theta$ (possibly with $\theta=\pi/2$ or $\theta=\pi/4$ or $\theta=\pi/8$, to get neater results); in this manner one can, for instance, replace the sines with cosines and the cosines with sines in the numerator in the left hand side of (5b,c,d,e). And of course other identities may be obtained by combining those displayed above with one another as well as with those obtained by such shifts.

Proposition D-8. Define the $(N\times N)$-matrix $\underline{R}(\underline{\varphi},\underline{\theta})$ and $\underline{B}(\underline{\theta},\alpha)$, and the N N-vectors $\underline{w}^{(n)}(\underline{\theta})$, in terms of the $2N+1$ "angles" θ_n, φ_n, α, by the neat rules

$$R_{nm}(\underline{\varphi},\underline{\theta})=\prod_{l=1,l\neq m}^{N}\left[\sin(\varphi_n-\theta_l)/\sin(\theta_m-\theta_l)\right] \ , \quad (6a)$$

$$\underline{B}(\underline{\theta},\alpha)=\underline{R}(\underline{\theta}+\alpha,\underline{\theta}), \quad B_{nm}(\underline{\theta},\alpha)=\prod_{l=1,l\neq m}^{N}\left[\sin(\theta_n-\theta_l+\alpha)/\sin(\theta_m-\theta_l)\right], \tag{6b}$$

$$\underline{w}_m^{(n)}(\underline{\theta})=\exp\left[i(2n-N-1)\,\theta_m\right]; \tag{6c}$$

there hold then the following *identities*:

$$\underline{R}(\underline{\theta},\underline{\theta})=\underline{1}; \tag{6d}$$

$$\underline{R}(\underline{\varphi},\underline{\eta})\;\underline{R}(\underline{\eta},\underline{\theta})=\underline{R}(\underline{\varphi},\underline{\theta}), \tag{6e}$$

$$\underline{R}(\underline{\varphi},\underline{\eta}_1)\;\underline{R}(\underline{\eta}_1,\underline{\eta}_2)\cdots\underline{R}(\underline{\eta}_p,\underline{\theta})=\underline{R}(\underline{\varphi},\underline{\theta}), \qquad p=1,2,3,\ldots, \tag{6f}$$

$$\left[\underline{R}(\underline{\varphi},\underline{\theta})\right]^{-1}=\underline{R}(\underline{\theta},\underline{\varphi}); \tag{6g}$$

$$\underline{w}^{(n)}(\underline{\varphi})=\underline{R}(\underline{\varphi},\underline{\theta})\;\underline{w}^{(n)}(\underline{\theta}); \tag{6h}$$

$$\underline{B}(\underline{\theta},0)=\underline{1}; \tag{6i}$$

$$\underline{B}(\underline{\theta},\alpha)\;\underline{B}(\underline{\theta},\beta)=\underline{B}(\underline{\theta},\beta)\;\underline{B}(\underline{\theta},\alpha)=\underline{B}(\underline{\theta},\alpha+\beta), \tag{6l}$$

$$\prod_{s=1}^{p}\underline{B}(\underline{\theta},\alpha_s)=\underline{B}(\underline{\theta},\sum_{s=1}^{p}\alpha_s), \qquad p=1,2,3,\ldots, \tag{6m}$$

$$\left[\underline{B}(\underline{\theta},\alpha)\right]^{-1}=\underline{B}(\underline{\theta},-\alpha); \tag{6n}$$

$$\underline{B}(\underline{\theta},\alpha)\;\underline{w}^{(n)}(\underline{\theta})=\beta_n(\alpha)\;\underline{w}^{(n)}(\underline{\theta}), \tag{6o}$$

$$\beta_n(\alpha)=\exp\left[i\,(2n-N-1)\,\alpha\right]; \tag{6p}$$

$$\mathrm{trace}\left[\underline{B}(\underline{\theta},\alpha)\right]=\sin(N\alpha)/\sin(\alpha), \tag{6q}$$

$$\mathrm{trace}\left[\prod_{s=1}^{p}\underline{B}(\underline{\theta},\alpha_s)\right]=\sin(N\sum_{s=1}^{p}\alpha_s)/\sin(\sum_{s=1}^{p}\alpha_s); \tag{6r}$$

$$\det\left[\underline{B}(\underline{\theta},\alpha)\right]=\underline{1}\,. \tag{6s}$$

Remark D-9. The N eigenvalues $\beta_n(\alpha)$ of the $(N\times N)$-matrix $\underline{B}(\underline{\theta},\alpha)$ are independent of the N parameters θ_m, and its N eigenvectors $\underline{w}^{(n)}(\underline{\theta})$ are independent of the parameter α (see (6o), (6p), (6c)); hence any change of the parameters θ_m entails for $\underline{B}(\underline{\theta},\alpha)$ an *isospectral* deformation, as displayed by the formula

$$\underline{B}(\underline{\varphi},\alpha)=\underline{R}(\underline{\varphi},\underline{\theta})\ \underline{B}(\underline{\theta},\alpha)\ \left[\underline{R}(\underline{\varphi},\underline{\theta})\right]^{-1}, \tag{6t}$$

$$\underline{B}(\underline{\varphi},\alpha)=\underline{R}(\underline{\varphi},\underline{\theta})\ \underline{B}(\underline{\theta},\alpha)\ \underline{R}(\underline{\theta},\underline{\varphi}). \tag{6u}$$

Remark D-10. The following *trigonometric identities* are merely explicit versions of some of the formulas written above (specifically: (6q), (6r), (6o,p), (6g), and (6f) an (6e)):

$$\sum_{n=1}^{N}\ \prod_{m=1,m\neq n}^{N}\left[\sin(\theta_n-\theta_m+\alpha)/\sin(\theta_n-\theta_m)\right]=\sin(N\alpha)/\sin(\alpha), \tag{7a}$$

$$\prod_{s=1}^{p}\left\{\sum_{n_s=1}^{N}\ \prod_{m_s=1,m_s\neq n_s}^{N}\left[\sin(\theta_{n_{s-1}}-\theta_{m_s}+\alpha_s)/\sin(\theta_{n_s}-\theta_{m_s})\right]\right\}$$

$$=\sin(N\sum_{s=1}^{p}\alpha_s)/\sin(\sum_{s=1}^{p}\alpha_s)\ ,\qquad n_0\equiv n_p,\qquad p=1,2,3,...\ ; \tag{7b}$$

$$\sum_{l=1}^{N}\ \cos\left[m(\theta_n-\theta_l+\alpha)\right]\ \prod_{k=1,k\neq l}^{N}\left[\sin(\theta_n-\theta_k+\alpha)/\sin(\theta_l-\theta_k)\right]=1,$$

$$m=N-1,N-3,N-5,...,1\ \text{or}\ 0;\quad n=1,2,...,N; \tag{7c}$$

$$\sum_{l=1}^{N}\ \sin\left[m(\theta_n-\theta_l+\alpha)\right]\ \prod_{k=1,k\neq l}^{N}\left[\sin(\theta_n-\theta_k+\alpha)/\sin(\theta_l-\theta_k)\right]=0,$$

$$m=N-1,N-3,N-5,...,1\ \text{or}\ 0;\quad n=1,2,...,N; \tag{7d}$$

$$\sum_{l=1}^{N}\left[\left\{\prod_{j=1,j\neq l}^{N}\left[\sin(\varphi_n-\theta_j)/\sin(\theta_l-\theta_j)\right]\right\}\left\{\prod_{k=1,k\neq m}^{N}\left[\sin(\theta_l-\varphi_k)/\sin(\varphi_m-\varphi_k)\right]\right\}\right]=\delta_{nm}, \tag{7e}$$

$$\sum_{l=1}^{N}\left[\left\{\prod_{j=1,j\neq l}^{N}\left[\sin(\varphi_n-\eta_j)/\sin(\eta_l-\eta_j)\right]\right\}\left\{\prod_{k=1,k\neq m}^{N}\left[\sin(\eta_l-\theta_k)/\sin(\varphi_n-\theta_k)\right]\right\}\right]=1. \tag{7f}$$

Several other *trigonometric identities* are implied by the other formulas written above.

Proposition D-11. The $(N \times N)$-matrix $\underline{Q}(\underline{\theta},\gamma)$, defined in terms of the $N+1$ *arbitrary* "angles" θ_n, γ by the neat rule

$$Q_{nm}(\underline{\theta},\gamma) = \cos(\theta_n - \theta_m + \gamma), \tag{8a}$$

has the 2 eigenvalues

$$q^{(\pm)}(\underline{\theta},\gamma) = (N/2)\{ \cos\gamma \pm [-\sin^2\gamma + \gamma_2(\underline{\theta})]^{1/2} \}, \tag{8b}$$

$$q^{(\pm)}(\underline{\theta},\gamma) = (N/2)\{ \cos\gamma \pm [-\cos^2\gamma - 2\gamma_1(\underline{\theta})]^{1/2} \}, \tag{8c}$$

$$\gamma_2(\underline{\theta}) \equiv N^{-2} \sum_{n,m=1}^{N} \cos[2\,(\theta_n - \theta_m)] = 1 - 2\gamma_1(\underline{\theta}), \tag{8d}$$

$$\gamma_1(\underline{\theta}) \equiv N^{-2} \sum_{n,m=1}^{N} \sin^2(\theta_n - \theta_m) = [1 - 2\gamma_2(\underline{\theta})]/2, \tag{8e}$$

and *all* its other eigenvalues (if any, namely if $N>2$) vanish; the 4 $[(2N)\times(2N)]$-matrices $\underline{Q}(\underline{\theta},\gamma;\sigma,\tau)$ with $\sigma=+1,-1$ and $\tau=0,1$, defined in terms of the matrix $\underline{Q}(\underline{\theta},\gamma)$, see (8a), by the neat "block-matrix" rule

$$\underline{Q}(\underline{\theta},\gamma;\sigma,\tau) = \begin{pmatrix} \underline{Q}(\underline{\theta},\gamma) & i^{1-\tau}\,\sigma\,\underline{Q}(\underline{\theta},\gamma+\pi/2) \\ i^{1+\tau}\,\sigma\,\underline{Q}(\underline{\theta},\gamma+\pi/2) & \underline{Q}(\underline{\theta},\gamma) \end{pmatrix}, \quad \sigma=+1,-1,\ \tau=0,1, \tag{9a}$$

have the 2 eigenvalues $\exp(\pm i\gamma)$, and *all* their other $2N-2$ eigenvalues *vanish*; and they satisfy the neat relations

$$\underline{Q}(\underline{\theta},\alpha;\sigma,\tau)\ \underline{Q}(\underline{\theta},\beta;\sigma,\tau) = \underline{Q}(\underline{\theta},\alpha+\beta;\sigma,\tau). \tag{9b}$$

Remark D-12. The 2 nonvanishing eigenvalues of the $(N \times N)$-matrix $\underline{Q}(\underline{\theta},\gamma)$, see (8a), depend on the N parameters θ_n only via the *single* quantity $\gamma_2(\underline{\theta})$ (or $\gamma_1(\underline{\theta})$), see (8b,d) (or (8c,e)); the 2 nonvanishing eigenvalues of the 4 $[(2N)\times(2N)]$-matrices $\underline{Q}(\underline{\theta},\gamma;\sigma,\tau)$, see (9a) with (8a), are independent of the N parameters θ_n, hence their variation when these parameters change is *isospectral*.

Proposition D-13. There holds the following *trigonometric identities*:

$$\text{trace}\{\,[\underline{Q}(\underline{\theta},\gamma)]^p\,\} = \sum_{n_1,\ldots,n_p=1}^{N}\ \prod_{s=1,[n_{p+1}\equiv n_1]}^{p} \cos(\theta_{n_s}-\theta_{n_{s+1}}+\gamma) \tag{10a}$$

$$= 2^{1-p}\,N^p \sum_{r=0}^{[[p/2]]} \binom{p}{2r} (\cos\gamma)^{p-2r}\left[-\sin^2\gamma+\gamma_2(\underline{\theta})\right]^r \tag{10b}$$

$$= 2^{1-p}\,N^p \sum_{r=0}^{[[p/2]]} \binom{p}{2r} (\cos\gamma)^{p-2r}\left[\cos^2\gamma-2\gamma_1(\underline{\theta})\right]^r\ , \tag{10c}$$

where p is an *arbitrary positive* integer, $p=1,2,3,\ldots$, $[[p/2]]$ is the integer part of $p/2$ (namely $[[p/2]]=p/2$ if p is even, $[[p/2]]=(p-1)/2$ if p is odd), and the quantities $\gamma_2(\underline{\theta})$, $\gamma_1(\underline{\theta})$ are defined by (8d,e) in terms of the N *arbitrary* "angles" θ_n.

Proposition D-14. Let the $(N\times N)$-matrix $\underline{M}\equiv\underline{M}(a_1,a_2,\ldots,a_p;b_1,b_2,\ldots,b_r;c;q;z)$ be defined as follows:

$$M_{nm} = \frac{c^N(c^{-1};q)_N}{(q;q)_{N-1}\,(cq^{n-m}-1)}\ \frac{(cq;q)_{n-1}\,(q^{-N+1};q)_{m-1}}{(cq^{-N+1};q)_{n-1}\,(q;q)_{m-1}}\ {}_{p+2}\Phi_{r+2}\left(\begin{matrix} cq^n,\, cq^{n-m},\, a_1,\, a_2,\, \ldots,\, a_p;\\ cq^{m-N}, cq^{n-m+1}, b_1, b_2,\ldots, b_r;\end{matrix}\, q,z\right), \tag{11a}$$

where

$$(a;q)_l = \prod_{s=0}^{l-1}(1-aq^s) = (1-a)(1-aq)(1-aq^2)\cdots(1-aq^{l-1}),\quad (a;q)_0=1, \tag{11b}$$

and ${}_p\Phi_r$ is the *basic hypergeometric function,*

$${}_p\Phi_r\left(\begin{matrix} a_1,a_2,\ldots,a_p;\\ b_1,b_2,\ldots,b_r;\end{matrix}\, q,z\right) = \sum_{j=0}^{\infty} \frac{(a_1;q)_j\,(a_2;q)_j\cdots(a_p;q)_j\; z^j}{(b_1;q)_j\,(b_2;q)_j\cdots(b_r;q)_j\,(q;q)_j}\ . \tag{11c}$$

The N eigenvalues $\mu_n\equiv\mu_n(a_1,a_2,\ldots,a_p;b_1,b_2,\ldots,b_r;c;q;z)$ of this matrix, and the corresponding eigenvectors $\underline{w}^{(n)}(q)$,

$$\underline{M}\,\underline{w}^{(n)}(q) = \mu_n\,\underline{w}^{(n)}(q), \tag{11d}$$

are then given by the following neat rules:

$$\mu_n = c^{n-1}\ {}_p\Phi_q\left(\begin{matrix} a_1,a_2,\ldots,a_p;\\ b_1,b_2,\ldots,b_r;\end{matrix}\, q,zq^{n-1}\right), \tag{11e}$$

$$w_m^{(n)} = q^{(m-1)(n-1)} . \tag{11f}$$

Here p and r are two *arbitrary nonnegative integers,* and $a_1, a_2, \ldots, a_p; b_1, b_2, \ldots, b_r; c; q; z$ are $p+r+3$ arbitrary complex numbers (up to the obvious restrictions required by the definitions given above).

Remark D-15. For $N=2$, by equating the trace respectively the determinant of $\underline{M}$, see (11a), to the sum respectively the product of its 2 eigenvalues, see (11e), one gets the following (rather trivial) *linear* respectively (perhaps less trivial) *quadratic identities* relating "contiguous" *basic hypergeometric functions* (see (11c)):

$$(c-q)\ {}_{p+1}\Phi_{r+1}\left(\begin{matrix} c,\ \underline{a}; \\ cq^{-1}, \underline{b}; \end{matrix}\ q, z\right) + (1-cq)\ {}_{p+1}\Phi_{r+1}\left(\begin{matrix} cq^2, \underline{a}; \\ cq,\ \underline{b}; \end{matrix}\ q, z\right)$$

$$= (1-q)\left[{}_p\Phi_r\left(\begin{matrix} \underline{a}; \\ \underline{b}; \end{matrix}\ q, z\right) + c\ {}_p\Phi_r\left(\begin{matrix} \underline{a}; \\ \underline{b}; \end{matrix}\ q, zq\right)\right], \tag{12a}$$

respectively

$$(c-q)(1-cq)\ {}_{p+1}\Phi_{r+1}\left(\begin{matrix} c,\ \underline{a}; \\ cq^{-1}, \underline{b}; \end{matrix}\ q, z\right) {}_{p+1}\Phi_{r+1}\left(\begin{matrix} cq^2, \underline{a}; \\ cq,\ \underline{b}; \end{matrix}\ q, z\right) + q(1-c)^2\left[{}_{p+1}\Phi_{r+1}\left(\begin{matrix} cq,\ \underline{a}; \\ c, \underline{b}; \end{matrix}\ q, z\right)\right]^2$$

$$= c(1-q)^2\ {}_p\Phi_r\left(\begin{matrix} \underline{a}; \\ \underline{b}; \end{matrix}\ q, z\right) {}_p\Phi_r\left(\begin{matrix} \underline{a}; \\ \underline{b}; \end{matrix}\ q, zq\right) . \tag{12b}$$

Here of course $\underline{a}$ denotes the arbitrary p-vector of components a_s, and likewise $\underline{b}$ denotes the arbitrary r-vector of components b_s (with p and r arbitrary nonnegative integers).

Proposition D-16. Let the $(N \times N)$-matrix $\underline{\tilde{M}} \equiv \underline{\tilde{M}}(\alpha_1, \alpha_2, \ldots, \alpha_p; \beta_1, \beta_2, \ldots, \beta_r; \gamma; z)$ be defined as follows:

$$\tilde{M}_{nm} = \frac{(-\gamma)_N\, (\gamma)_{n-1}\, (1-N)_{m-1}}{(N-1)!(n-m-\gamma)(\gamma-N+1)_{n-1}\, (m-1)!}\ {}_{p+2}F_{r+2}\left(\begin{matrix} \gamma+n, \gamma+n-m, \alpha_1, \alpha_2, \ldots, \alpha_p; \\ \gamma+n-N, \gamma+n-m+1, \beta_1, \beta_2, \ldots, \beta_r; \end{matrix}\ z\right), \tag{13a}$$

where

$$(\alpha)_l=\prod_{s=0}^{l-1}\ (\alpha+s)=\alpha(\alpha+1)(\alpha+2)\cdots(\alpha+l-1)\,,\quad (\alpha)_0=1\,, \tag{13b}$$

and ${}_pF_r$ is the standard *hypergeometric function,*

$${}_pF_r\begin{pmatrix}\alpha_1,\alpha_2,\ldots,\alpha_p; \\ \beta_1,\beta_2,\ldots,\beta_r; \end{pmatrix} z=\sum_{j=0}^{\infty}\frac{(\alpha_1)_j(\alpha_2)_j\cdots(\alpha_p)_j\ z^j}{(\beta_1)_j(\beta_2)_j\cdots(\beta_r)_j\ j!}\,. \tag{13c}$$

Then *all* the N eigenvalues of this (nondiagonalizable) $(N\times N)$-matrix coincide with the hypergeometric function ${}_pF_r\begin{pmatrix}\alpha_1,\alpha_2,\ldots,\alpha_p; \\ \beta_1,\beta_2,\ldots,\beta_r;\end{pmatrix} z$, see (13c). Here p and r are two *arbitrary nonnegative integers,* and $\alpha_1,\alpha_2,\ldots,\alpha_p;\beta_1,\beta_2,\ldots,\beta_r;\gamma;\ z$ are $p+r+2$ *arbitrary* complex numbers (up to the obvious restrictions required by the definitions given above).

Remark D-17. For $N=2$, by equating the trace respectively the determinant of $\underline{\tilde{M}}$, see (13a), to the sum respectively the product of its 2 (equal) eigenvalues, see (13c), one gets the following (relatively trivial) *linear* respectively (perhaps less trivial) *quadratic identities* relating "contiguous" *hypergeometric functions* (see (13c)):

$$(1-\gamma)\ {}_{p+1}F_{r+1}\begin{pmatrix}\gamma,\underline{\alpha}; \\ \gamma-1,\underline{\beta};\end{pmatrix} z+(1+\gamma)\ {}_{p+1}F_{r+1}\begin{pmatrix}\gamma+2,\underline{\alpha}; \\ \gamma+1,\underline{\beta};\end{pmatrix} z=2\ {}_pF_r\begin{pmatrix}\underline{\alpha}; \\ \underline{\beta};\end{pmatrix} z\,, \tag{14a}$$

$$(\gamma^2-1)\ {}_{p+1}F_{r+1}\begin{pmatrix}\gamma,\underline{\alpha}; \\ \gamma-1,\underline{\beta};\end{pmatrix} z\ {}_{p+1}F_{r+1}\begin{pmatrix}\gamma+2,\underline{\alpha}; \\ \gamma+1,\underline{\beta};\end{pmatrix} z+\gamma^2\left[{}_{p+1}F_{r+1}\begin{pmatrix}\gamma+1,\underline{\alpha}; \\ \gamma,\underline{\beta};\end{pmatrix} z\right]^2=\left[{}_pF_r\begin{pmatrix}\gamma,\underline{\alpha}; \\ \gamma,\underline{\beta};\end{pmatrix} z\right]^2. \tag{14b}$$

Here of course $\underline{\alpha}$ denotes the arbitrary p-vector of components α_s, and likewise $\underline{\beta}$ denotes the arbitrary r-vector of components β_s (with p and r arbitrary nonnegative integers).

Proposition D-18. Let the $(N\times N)$-matrix $\underline{W}^{(p)}\equiv\underline{W}^{(p)}(\underline{\theta};\theta_0;\omega_1,\omega_2;\omega)$ be defined as follows:

$$W_{nm}^{(p)}=(-1)^p\exp\left\{-2(p/N)\eta\left[\theta_0+p\omega+\sum_{j=1}^{N}\ (\theta_n-\theta_j)\right]\right\}\cdot$$

$$\cdot\{\sigma[\theta_n-\theta_m+\theta_0+2(p/N)\omega]/\sigma(\theta_0)\}\cdot$$

$$\cdot \prod_{l=1, l\neq m}^{N} \{\sigma[\theta_n - \theta_l + 2(p/N)\omega] / \sigma(\theta_m - \theta_l)\}, \tag{15a}$$

where $\sigma(z) \equiv \sigma(z \mid \omega_1, \omega_2)$ is the Weierstrass sigma function, see Appendix A, ω coincides with one of the two semiperiods, ω_1, ω_2, of these sigma functions, η is the corresponding complementary quantity, see (A-42), the $N+1$ numbers θ_n, θ_0 are *arbitrary* and p is a *positive integer* defined mod(N). Then the ($N \times N$)-matrix $\underline{W}^{(p)}$ satisfies the matrix formulas characteristic of a shift operator,

$$\underline{W}^{(p)} = \left[\underline{W}^{(1)}\right]^p, \tag{15b}$$

$$\underline{W}^{(p_1)} \underline{W}^{(p_2)} = \underline{W}^{(p_1+p_2)}, \tag{15c}$$

$$\mathrm{trace}[\ \underline{W}^{(p)}\ ] = N\,\delta_{0p}\ , \qquad p = 0, \pm 1, \pm 2, \ldots \mathrm{mod}(N)\ , \tag{15d}$$

$$\det[\ \underline{W}^{(p)}\ ] = (-1)^{p(N+1)}, \tag{15e}$$

and its N eigenvalues are given by the simple expression

$$\exp(2i\pi\, p\, n/N)\ , \qquad n = 1, 2, \ldots, N\,. \tag{15f}$$

Remark D-19. The relations (15d) and (15c) yield via (15a) the following *identities*:

$$\sum_{n=1}^{N} \exp(-2p\eta\theta_n) \prod_{m=1, m\neq n}^{N} [\sigma(\theta_n - \theta_m + 2\omega p/N)/\sigma(\theta_n - \theta_m)]$$

$$= \delta_{0p}\, N \exp[\ 2(p/N)\eta(\theta_0 + p\omega - \sum_{j=1}^{N}\theta_j)\ ], \tag{16a}$$

$$\sum_{l=1}^{N} \exp[2p\eta(\theta_n - \theta_l)]\,[\sigma(\theta_n - \theta_l + \theta_0 + 2\omega q/N)/\sigma(\theta_n - \theta_l + 2\omega(p+q)/N)] \cdot$$

$$\cdot [\sigma(\theta_l - \theta_m + \theta_0 + 2\omega p/N)/\sigma(\theta_l - \theta_m + 2\omega p/N)] \cdot$$

$$\cdot \prod_{k=1, k\neq l}^{N} \{\sigma(\theta_n - \theta_k + 2\omega q/N)\, \sigma(\theta_l - \theta_k + 2\omega p/N) \cdot$$

$$\cdot [\ \sigma(\theta_m - \theta_k + 2\omega(p+q)/N)\, \sigma(\theta_l - \theta_k)\ ]^{-1}\ \}$$

$$= \exp(-4\eta\omega\, p\, q/N)\left[\sigma(\theta_0)/\sigma(2\omega\, p/N)\right] . \qquad (16b)$$

Here N is an *arbitrary positive integer*, $N \geq 2$; p and q are two arbitrary integers, defined $\mathrm{mod}(N)$; the $N+1$ quantities θ_n, θ_0 are arbitrary (possibly complex) numbers, up to the restrictions required to make proper sense of these formulas; $\sigma(z) \equiv \sigma(z \mid \omega_1, \omega_2)$ is the Weierstrass sigma function, see Appendix A, with semiperiods ω_1, ω_2, while ω coincides with one of these two semiperiods and η is the complementary quantity to ω, see (A-42). The indices n, m in (16b) can take N integer values from 1 to N; and since this is as well true for p and q, the formula (16b) entails in fact N^4 identities.

Remark D-20. For $N = 2$, the formula (15e) with (15a) and $p=1$ (and $\theta = \theta_1 - \theta_2$) yields the identity

$$\sigma(\theta + \theta_0 + \omega)\, \sigma(\theta - \theta_0 - \omega)\, \sigma^2(\omega) - \sigma(\theta + \omega)\, \sigma(\theta - \omega)\, \sigma^2(\theta_0 + \omega)$$

$$= \sigma^2(\theta)\, \sigma^2(\theta_0)\, \exp[2\eta(\theta_0 + \omega)] . \qquad (17)$$

Proposition D-21. There hold the N *identities*

$$\sum_{m=1}^{N} \{ [\prod_{l=1, l\neq n}^{N} (a_m c_l - b_m d_l)] / [(a_m - x b_m) \prod_{j=1, j\neq m}^{N} (a_m b_j - b_m a_j)] \}$$

$$= [\prod_{l=1, l\neq n}^{N} (d_l - x c_l)] / [\prod_{k=1}^{N} (a_k - x b_k)] , \quad n = 1,2,\ldots,N . \qquad (18)$$

Here a_n, b_n, c_n, d_n, x are $4N+1$ *arbitrary* numbers (up to the obvious restrictions required to make good sense of (18)).

Remark D-22. The *identities* (18), whose left hand side corresponds merely to the "partial fractions" decomposition of the product in the right hand side, provides a convenient tool to identify "sums which can be transformed into products".

D.N Notes to Appendix D

The original idea to identify "remarkable matrices", in the specific sense used herein, should be perhaps traced to <C78a>; see also <C77a>, <CP79>, <ABCOP79>, <C80b>, <C80c>, <C81b>, <BC81>. A more complete fruition of these ideas came via the connection with the standard theory of Lagrangian interpolation; for an overview see <C84b>, for explicit applications (also involving basic hypergeometric functions) see <C86b> and <C88>; for a more elementary treatment (aimed at computer applications), see <C97a>, <C99a> and <C95d>. A final boost to this approach came from the connection with the generalized theory of Lagrangian interpolation <C93a>, see <C98b> and some of the findings in this book (mainly in Sect. 3.1.2.1; see also some of the identities reported in the latter part of Appendix A).

The remarkable matrices, and related trigonometric identities involving angles that are rational fractions of π, obtained in <CP79>, are reported in Sect. 15.823 of <GRJ94> (where two different notations, $\lambda_s^{(b)}$ and b_s, and likewise $\lambda_s^{(c)}$ and c_s, are used for the *same* quantities); some of these findings are special cases of results reported in Appendix D, in particular the matrix $\underline{A}$ defined in Sect. 15.823 of <GRJ94> coincides with $\underline{A}(\underline{\theta};\alpha,\beta,\gamma)$, see (D-2o), in the special case $\theta_n = \pi n / N$, $\alpha = -1$, $\beta = 1$, $\gamma = 1$, thanks to the (obvious) identities

$$\sum_{m=1, m\neq n}^{N} \text{cotan}\left[\pi(n-m)/N\right] = 0 \ . \qquad (1)$$

Not all the formulas obtained in the papers quoted above are reported in Appendix D; but we trust the selection reported there, which comes essentially, and in this order, from <C97a>, <C84b>, <C99a>, <C95d> (whose results are reformulations and simple generalizations of those given in <C85c>, <C86b>, <C98b>), is sufficiently representative to provide a fair idea of the formulas that may be found (with their proofs!) in the papers quoted above, and, perhaps more importantly, of the kind of identities that the alert reader may uncover by using the techniques described in those papers and in the relevant sections of this book. And, of course, the material contained in the preceding Appendix C, and in the references quoted in Sect. C.N, provides additional examples of *remarkable matrices* and of *related identities*.

Appendix E: Lagrangian approximation for eigenvalue problems in one and more dimensions

In the last part of Chap. 2, and in the first part of Chap. 3, a finite-dimensional representation of the operator of differentiation is introduced and its relation with the technique of Lagrangian interpolation is elucidated. In Appendix E we tersely outline the possibility to exploit such a representation in the context of numerical analysis, in particular to evaluate the eigenvalues of differential operators. Our purpose here is merely to introduce the main idea in the simplest context; the readers who are interested in pursuing this approach are referred to the literature (see Sect. E.N), although it should be made immediately clear that much more can probably be done than has been done up to now. This is likely to be especially true for applications in the multidimensional context, based on the results reported in the first part of Chap. 3; while our presentation here is, for simplicity's sake, mainly focussed on the one-dimensional case (Sturm-Liouville eigenvalue problems). Let us also mention that the techniques of Chap. 3, including in particular the consideration of time-dependent nodes and of their time evolution, might have other interesting applications in numerical analysis, for instance in the context of fluid mechanics; but we are then talking of (possible) *future* developments.

Consider the Sturm-Liouville eigenvalue problem

$$c_2(x)\,\psi''_m(x)+c_1(x)\,\psi'_m(x)+c_0(x)\,\psi_m(x)=\lambda_m\,\psi_m(x)\,, \tag{1a}$$

in the finite interval $a \le x \le b$, with boundary conditions

$$\psi_m(a)=\psi_m(b)=0\,. \tag{1b}$$

Assume that all the eigenvalues λ_m are *real* and that they are bounded below, so that they can be ordered as an increasing sequence,

$$\lambda_m \le \lambda_{m+1} \quad , \qquad m=1,2,\dots\,. \tag{2}$$

Generally this ordering corresponds to the property of the eigenfunction $\psi_m(x)$ to posses $m-1$ zeros inside the interval (a,b).

F. Calogero: LNPm 66, pp. 703 - 707, 2001

We now set

$$\psi_m(x) = (x-a)(x-b)\varphi_m(x)\,, \tag{3}$$

transforming thereby the eigenvalue problem (1) into the equivalent problem

$$\mathsf{A}\varphi_m(x) = \lambda_m\,\varphi_m(x)\,, \tag{4a}$$

$$\varphi_m(x) \text{ regular at } x=a \text{ and } x=b\,, \tag{4b}$$

where A is the differential operator (singular at $x=a$ and $x=b$) defined as follows:

$$\mathsf{A} = \alpha_2(x)\,(d/dx)^2 + \alpha_1(x)\,(d/dx) + \alpha_0(x) \tag{5a}$$

where

$$\alpha_2(x) = c_2(x)\,, \tag{5b}$$

$$\alpha_1(x) = c_1(x) + 2\,[(x-a)^{-1} + (x-b)^{-1}]\,c_2(x)\,, \tag{5c}$$

$$\alpha_0(x) = c_0(x) + \left[(x-a)^{-1} + (x-b)^{-1}\right]c_1(x) + 2\left[(x-a)(x-b)\right]^{-1}c_2(x)\,. \tag{5d}$$

Exercise E-1. Verify!

Let us moreover assume that the functions $c_2(x)$, $c_1(x)$ and $c_0(x)$, see (1a), are *entire* and that $c_2(x)$ has no zeros (even for complex x). This implies that any solution $\psi(x)$ of (1a), hence as well the eigenfunctions $\psi_m(x)$ of the eigenvalue problem (1), are *entire* functions of x. This is, of course, not the case for the generic solution $\varphi(x)$ of (4a) with (5), that generally has simple poles at $x=a$ and at $x=b$, see (3). But the eigenfunctions $\varphi_m(x)$ of (4) with (5) are *entire*; indeed they, and the corresponding eigenvalues λ_m, are determined by the requirement that the *singular* Sturm-Liouville equation (4a) with (5) possess an *entire* solution $\varphi_m(x)$ (see (4b), itself implied by (3) with (1b)).

Let us now associate to the differential operator A, see (5), an $(N\times N)$-matrix $\underline{A}$, via the replacement $x \Rightarrow \underline{X}$, $d/dx \Rightarrow \underline{D}$,

$$\underline{A} = \alpha_2(\underline{X})\,\underline{D}^2 + \alpha_1(\underline{X})\,\underline{D} + \alpha_0(\underline{X})\,, \tag{6}$$

with $\underline{X}$ respectively $\underline{D}$ defined by (2.4.1-1) respectively (2.4.1-2),

$$\underline{X} = \mathrm{diag}(x_n; n = 1, 2, \ldots, N), \tag{7a}$$

$$(\underline{D})_{nm} = \delta_{nm} \sum_{l=1}^{N} (x_n - x_l)^{-1} + (1 - \delta_{nm})(x_n - x_m)^{-1}, \tag{7b}$$

in terms of N distinct, but otherwise arbitrary, numbers x_n. Let a_n be the N eigenvalues of this ($N \times N$)-matrix $\underline{A}$:

$$\underline{A}\, \underline{v}^{(n)} = a_n\, \underline{v}^{(n)}, \quad n = 1, 2, \ldots, N. \tag{8}$$

These eigenvalues, a_m, need not be (all) *real*, since the matrix $\underline{A}$ need not be Hermitian (see, however, below). Indicate by $\hat{\lambda}_m$ those eigenvalues a_m which are real,

$$\hat{\lambda}_m = a_m, \qquad \hat{\lambda}_m \text{ real}, \tag{9}$$

and order them in increasing order,

$$\hat{\lambda}_m \leq \hat{\lambda}_{m+1}, \tag{10}$$

(see (2)).

It is now plausible to conjecture that, for sufficiently large N, and small m, the m-th real eigenvalue $\hat{\lambda}_m$ of the matrix $\underline{A}$ provides an approximation to the m-th (real) eigenvalue λ_m of the Sturm-Liouville problem (1),

$$\hat{\lambda}_m \approx \lambda_m \qquad (m \text{ "small", } N \text{ "large"}). \tag{11}$$

Indeed, if the eigenfunction $\varphi_m(x)$ is a polynomial of degree *less* than N, λ_m *coincides* with $\hat{\lambda}_m$ (or at least it coincides with an eigenvalue of (1); and the ordering conventions (2) and (10) support then the conjecture (11)).

Exercise E-2. Review the results that imply the validity of this statement. *Hint*: see *Corollary 2.4.1-4.*

But the requirement that identifies the eigenfunction $\varphi_m(x)$ of (4a) is the condition that this function be *entire*. An *entire* function is generally well approximated by a *polynomial*, the more so the higher the degree of the polynomial is (note that, presumably, the accuracy of the approximation in question refers to a comparison limited to the real interval $a \le x \le b$, or perhaps to its neighborhood in the complex x-plane). Hence it is reasonable to expect that, if $\varphi_m(x)$ is generally (for large enough N) well approximated by a polynomial (of degree less than N), since $\hat{\lambda}_m$ would *exactly* coincide with λ_m if $\varphi_m(x)$ were exactly a polynomial (of degree less than N), than, for large enough N, $\hat{\lambda}_m$ will generally approximate λ_m well.

The conjecture (11) is reinforced and made quantitative by the (non-rigorous but plausible <C83b> <C83c>) estimate

$$\left|\lambda_m - \hat{\lambda}_m(N)\right| / \left|\lambda_m\right| \approx \left[(\pi/2)(m/N)\right]^{N-2}. \tag{12}$$

Note that the convergence at large N of $\hat{\lambda}_m(N)$ to λ_m entailed by this formula is quite fast.

For a justification of this formula, (12), the interested reader is referred to the literature <C83b>, <C83c>, <C84b>.

Note that, in (12), we wrote $\hat{\lambda}_m(N)$ in place of $\hat{\lambda}_m$, to underline the dependence of this number on N. This quantity, $\hat{\lambda}_m$, depends moreover on the choice of the N, *a priori* arbitrary, numbers x_n that enter in the definition of the two matrices $\underline{X}$ and $\underline{D}$, see (7) (indeed to obtain the estimate (12) the assumption is made that the nodes x_n are equispaced in the interval (a,b)). The dependence on the values of these parameters x_n is however expected to be weak: indeed when $\varphi_m(x)$ is a polynomial (of degree *less* than N), the exact result for the eigenvalue is obtained for an *arbitrary* choice of the N numbers x_n. But in any case the freedom in the choice of the N nodes x_n can sometimes be taken advantage of, as it were *a priori*: for instance sometimes an appropriate choice of these N parameters x_n can guarantee that the ($N \times N$)-matrix $\underline{A}$ be Hermitian, hence that *all* its eigenvalues a_n be *real*, clearly a desirable feature.

The advantage of this method to evaluate numerically the (first few) eigenvalues of a Sturm-Liouville type problem reside in its simplicity and efficiency. The simplicity is evidenced by the fact that the problem to compute the eigenvalues of a differential operator gets transformed into the task to compute the eigenvalues of an ($N \times N$)-matrix, without the need to perform any integration (as is instead the case in most other methods, for instance in variational ones). The efficiency is suggested by the estimate (12), and it is indeed confirmed by numerical tests <D83>, <CF85>.

The general philosophy of this approach is based on the ideas of *collocation* and *discretization*; one evaluates the continuum problem under consideration at a number of discrete points, the N nodes, and uses the information from *all* these points to approximate the values of the derivatives of the unknown function at these points. The fact that one is effectively using the values of the function at *all* the nodes, to evaluate the derivative of the unknown function at each node, is a main cause of the fast convergence at large N evidenced by (12). On the other hand one has in this approach to cope with *full* ($N \times N$)-matrices, rather than only three-diagonal ones (or few-diagonal ones), as it is instead the case in approaches, based on discretization, where the derivatives of the unknown function at one point are expressed via the values of that function in the immediate, or close, neighborhood of that point.

Finally let us reiterate that, so far, the additional potentialities of the generalized approach described in the first part of Chap. 3 have been hardly used, to the best of my (most imperfect) knowledge, in numerical analysis, at least from the point of view emphasized in that Chap. 3: finite dimensional representations of the differential operator, with large flexibility in the choice of the nodes, especially important in the multidimensional context; including the possibility to let them evolve in time, in the context of problems which study the time-evolution of (discrete, or continuous) systems. There seems therefore to be much scope for further work in these directions.

E.N Notes to Appendix E

The idea on which the approach outlined in Appendix E is based was introduced in <C83b>, and pursued in <C83c>, <C84a>, <C85c>; see also <D83>, <D85>, <Ca86> and, for numerical applications, <CF85> and <BCP90>. The presentation of Appendix E follows rather closely Sect. 6 of <C84b>.

While some of the numerical examples discussed in <CF85> are multidimensional, they do not employ the more advanced technique, see the first part of Chap. 3, that extends the Lagrangian interpolation approach to a multidimensional environment. The main advantage of such a technique is the great flexibility it entails in the choice of nodes. The possibility of applications in numerical analysis are mentioned (but without subsequent follow up, so far) in (section VI of) the paper <C93a> where this multidimensional extension of the Lagrangian interpolation approach was first introduced.

Appendix F: Some theorems of elementary geometry in multidimensions

In Appendix F we report several *theorems* of elementary (Euclidean) geometry, whose common origin is in some basic properties of *determinants*. The justification for including such a topic in this book is because a first result of this kind emerged naturally in the context of the treatment of the generalized Lagrangian approximation technique, see *Exercise 3.1.2.2-4*. We discuss below this nice result <CK96> firstly, and we then proffer several other *theorems*, mainly using the format of proposing *exercises* (equipped with appropriate *hints*). It will be clear to the alert reader who studies these findings, how lots of other, analogous, results could be *manufactured* (or should one say *discovered*?).

Define the following ($N \times N$)-determinant:

$$\Delta(\underline{\vec{r}}) = \begin{vmatrix} 1 & x_1^{(1)} & x_2^{(1)} & \dots & x_S^{(1)} & r^2(\underline{x}^{(1)}) \\ 1 & x_1^{(2)} & x_2^{(2)} & \dots & x_S^{(2)} & r^2(\underline{x}^{(2)}) \\ \cdot & \cdot & \cdot & \dots & \cdot & \cdot \\ 1 & x_1^{(N)} & x_2^{(N)} & \dots & x_S^{(N)} & r^2(\underline{x}^{(N)}) \end{vmatrix}, \tag{1a}$$

of course with

$$N = S + 2 , \tag{1b}$$

and with

$$r^2(\underline{x}) = \sum_{j=1}^{S} x_s^2 . \tag{1c}$$

Here S is an arbitrary positive integer, the determinant (1a) has $N = S + 2$ lines and of course as many columns, see (1b), $\underline{\vec{r}}$ stands for the set of N S-vectors $\vec{r}^{(n)}$, $n = 1,2,...,N$, and $x_j^{(n)}$, $j = 1,2,...,S$ are the S coordinates of the S-vector $\vec{r}^{(n)}$ in some Cartesian coordinate system.

If one *translates* or *rotates* the Cartesian reference frame, the coordinates $x_j^{(n)}$ of course change accordingly; but the value of the determinant (1) does *not* change.

F. Calogero: LNPm 66, pp. 708 - 722, 2001

Exercise F-1. Prove this fact. *Hint*: remember that a determinant does not change if to a column one *adds* any other column times an *arbitrary* constant (the *addition* to be of course made element by element), and that if every element of a column gets *multiplied* by an *arbitrary* constant, c, the value of the determinant gets *multiplied* by the *same* constant, c (note that it is quite easy, using the first of the two properties of determinants, to prove invariance under *translations*; the proof of invariance under *rotations* is more cumbersome; it is actually implied by the following developments, see *Remark F-3* below).

These invariance properties of $\Delta(\vec{r})$, see (1), suggest that this quantity have an intrinsic, *geometrical*, significance, namely that it can be defined in terms of the N points $\vec{r}_n$ in S-space (as it were, *independently* of the specific values of their coordinates in some specific Cartesian reference frame). Indeed, we find below that it can be defined geometrically in several different ways, and the equalities among these different definitions entail nontrivial *theorems* of elementary geometry.

A first approach goes as follows. Select one of the N vectors $\vec{r}^{(n)}$, say $\vec{r}^{(p)}$, and define the simplex Σ_p in S-space having as its $S+1=N-1$ vertices the $N-1$ points $\vec{r}^{(n)}$, $n=1,2,\ldots,p-1,p+1,\ldots,N$, as well as the (unique!) hypersphere S_p in S-space that goes through these $N-1$ points. Now *translate* the Cartesian system so that its origin coincide with the center of the hypersphere S_p. In this new system of coordinates the determinant (1) (whose value has not changed, since we only performed a *translation*), reads

$$\Delta(\vec{r}) = \begin{vmatrix} 1 & x_1^{(1)} & x_2^{(1)} & \cdots & x_S^{(1)} & R_p^2 \\ 1 & x_1^{(2)} & x_2^{(2)} & \cdots & x_S^{(2)} & R_p^2 \\ \cdot & \cdot & \cdot & \cdots & \cdot & \cdot \\ 1 & x_1^{(p-1)} & x_2^{(p-1)} & \cdots & x_S^{(p-1)} & R_p^2 \\ 1 & x_1^{(p)} & x_2^{(p)} & \cdots & x_S^{(p)} & r_p^2 \\ 1 & x_1^{(p+1)} & x_2^{(p+1)} & \cdots & x_S^{(p+1)} & R_p^2 \\ \cdot & \cdot & \cdot & \cdots & \cdot & \cdot \\ 1 & x_1^{(N)} & x_2^{(N)} & \cdots & x_S^{(N)} & R_p^2 \end{vmatrix}, \tag{2}$$

where R_p is now the hyperradius of the hypersphere S_p (see (1c)), and r_p is the Euclidean distance in S-space (see (1c)) of the point $\vec{r}^{(p)}$ from the center of coordinates, or, equivalently, from the center of the hypersphere S_p (equivalently, r_p is the hyperradius of the, uniquely defined, hypersphere s_p, *concentric* to S_p, on which $\vec{r}^{(p)}$ lies).

Now subtract the first column of the determinant (2a), multiplied by r_p^2, from the last. This does not change the value of the determinant, which now reads

$$\Delta(\bar{r}) = \begin{vmatrix} 1 & x_1^{(1)} & x_2^{(1)} & \dots & x_S^{(1)} & 0 \\ 1 & x_1^{(2)} & x_2^{(2)} & \dots & x_S^{(2)} & 0 \\ . & . & . & \dots & . & . \\ 1 & x_1^{(p-1)} & x_2^{(p-1)} & \dots & x_S^{(p-1)} & 0 \\ 1 & x_1^{(p)} & x_2^{(p)} & \dots & x_S^{(p)} & r_p^2 - R_p^2 \\ 1 & x_1^{(p+1)} & x_2^{(p+1)} & \dots & x_S^{(p+1)} & 0 \\ . & . & . & \dots & . & . \\ 1 & x_1^{(N)} & x_2^{(N)} & \dots & x_S^{(N)} & 0 \end{vmatrix} . \tag{3}$$

Hence

$$\Delta(\bar{r}) = (-1)^{N+p} (r_p^2 - R_p^2)\, W_p, \tag{4}$$

where W_p is the $(N-1)\times(N-1)$ determinant that obtains from $\Delta(\bar{r})$, see (2) or (3), by eliminating the p-th line and the last column,

$$W_p = \begin{vmatrix} 1 & x_1^{(1)} & x_2^{(1)} & \dots & x_S^{(1)} \\ 1 & x_1^{(2)} & x_2^{(2)} & \dots & x_S^{(2)} \\ . & . & . & \dots & . \\ 1 & x_1^{(p-1)} & x_2^{(p-1)} & \dots & x_S^{(p-1)} \\ 1 & x_1^{(p+1)} & x_2^{(p+1)} & \dots & x_S^{(p+1)} \\ . & . & . & \dots & . \\ 1 & x_1^{(N)} & x_2^{(N)} & \dots & x_S^{(N)} \end{vmatrix} . \tag{5}$$

But it is well known that this determinant, W_p, coincides (up to a sign, and the numerical factor $(N-1)!$) with the volume $V^{(N-1)}$ of the *simplex* Σ_p^{N-1}. Hence we arrive at the following

Proposition F-2. Up to its sign, and the factor $(N-1)!/\pi$, the determinant $\Delta(\bar{r})$, see (1), is the product of the volume V of the *simplex* $\Sigma^{(N-1)}$, in S-dimensional space, having as vertices $N-1=S+1$ of the N points $\bar{r}^{(n)}$, times the area A of the plane (2-dimensional) annulus whose two radii are given by the following prescription: one of them is the hyperradius R of the hypersphere $S^{(N-1)}$ in which the simplex $\Sigma^{(N-1)}$ is inscribed (namely, the hypersphere which goes through the $N-1=S+1$ vertices of the simplex $\Sigma^{(N-1)}$), the other one is the hyperradius r of the *concentric* hypersphere s on which lies the extra point (that one of the N points $\vec{r}^{(n)}$ which is *not* used to define the symplex $\Sigma^{(N-1)}$).

It is clear that this *Proposition F-2* is implied by the above treatment and that it provides the desired geometrical interpretation of the determinant (1).

Remark F-3. To prove *Proposition F-2* the invariance of (1) under *translations* of the Cartesian coordinate system was utilized, but its invariance under *rotations* (of the Cartesian coordinate system) was *not* invoked; this property is now implied by *Proposition F-2,* which provides a purely *geometrical* definition of the value of $\Delta(\vec{\underline{r}})$, clearly independent of the choice of the Cartesian system.

The validity of *Proposition F-2* provides N different geometrical interpretations for the determinant (1), corresponding to the arbitrary choice that must be made of one point, say $\vec{r}^{(p)}$, which is singled out and treated differently from all others, $\vec{r}^{(n)}$ with $n=1,2,...,p-1,p+1,...,N$; the identity of these N geometrical interpretations gives rise to a (purely geometrical) *theorem*. While we leave the formulation of this result in the context of a space with an arbitrary number S of dimensions ($S>3$) as an *exercise* (unnumbered!) for the diligent reader (*solution*: see <CK96>), we now report the formulations <CK96> of this finding for $S=1,2,3$ (consistently with the title of this book!).

Theorem F-4. Let $x^{(1)}$, $x^{(2)}$ and $x^{(3)}$ indicate 3 points on a straight line. Choose any one of them, say $x^{(3)}$, and let $X^{(3)}=(x^{(1)}+x^{(2)})/2$ be the center of the segment $[x^{(1)},x^{(2)}]$, of length $2R_3=\left|x^{(1)}-x^{(2)}\right|$; also let $r_3=\left|x^{(3)}-X^{(3)}\right|$ be the distance of the point $x^{(3)}$ from $X^{(3)}$. Let C_3 respectively c_3 be the two coplanar concentric circles centered at $X^{(3)}$, having respectively radii R_3 (so that $x^{(1)}$ and $x^{(2)}$ lie on C_3) and r_3 (so that $x^{(3)}$ lies on c_3). Let $A_3=\pi(r_3^2-R_3^2)$ be the area of the plane *annulus* comprised between c_3 and C_3. Let $P_3=R_3\cdot A_3$ be the volume of the *annular cylinder* characterized by the height R_3 and the radii r_3 and R_3. Let P_1 respectively P_2 be the analogous quantities, corresponding to the choice of $x^{(1)}$ respectively $x^{(2)}$ in place of $x^{(3)}$. Then

$$P_1=P_2=P_3. \tag{6}$$

This *Theorem F-4* could of course be interpreted as an elementary result in 3-dimensional geometry (since the annular cylinders of volume P_j are 3-dimensional objects), or in 1-dimensional geometry (since the starting point of the treatment are 3 points on a straight line). It is of course implied by *Proposition F-2* with $S=1$. The diligent reader will draw appropriate diagrams to display graphically the geometrical significance of this elementary finding.

Theorem F-5. Let $\vec{r}^{(n)}$, $n=1,2,3,4$, indicate 4 points *in the plane*. Select any one of them, say $\vec{r}^{(4)}$. Let T_4 be the area of the *triangle* with vertices $\vec{r}^{(1)}$, $\vec{r}^{(2)}$, $\vec{r}^{(3)}$; C_4 be the circle on which these 3 points lie; and c_4 be the *concentric* circle on which $\vec{r}^{(4)}$ lies (draw the diagram!). Let A_4 be the area of the *annulus* comprised between the two circles, and $P_4=T_4\cdot A_4$. Let P_n, $n=1,2,3$ be analogously defined, by replacing the role of $\vec{r}^{(4)}$ with $\vec{r}^{(n)}$, $n=1,2,3$. Then

$$P_1=P_2=P_3=P_4. \tag{7}$$

Theorem F-6. Let $\vec{r}^{(n)}$, $n=1,2,3,4,5$, indicate 5 points *in 3-dimensional space*. Select any one of them, say $\vec{r}^{(5)}$. Let T_5 be the area of the *tetrahedron* with vertices $\vec{r}^{(1)}$, $\vec{r}^{(2)}$, $\vec{r}^{(3)}$, $\vec{r}^{(4)}$; C_5 be the sphere on which these 4 points lie; and c_5 be the concentric sphere, of radius r_5, on which $\vec{r}^{(5)}$ lies. Let A_5 be the area of the plane *annulus* comprised among the two concentric circles of radii r_5 and R_5. Let $P_5=T_5\cdot A_5$. Let P_n, $n=1,2,3,4$ be analogously defined, by replacing the role of $\vec{r}^{(5)}$ with $\vec{r}^{(n)}$, $n=1,2,3,4$. Then

$$P_1=P_2=P_3=P_4=P_5. \tag{8}$$

These two *Theorems, F-5* respectively *F-6,* are clearly immediate consequences of *Proposition F-2* with $S=2$ respectively $S=3$.

In the formulation of all these results we have always tacitly understood that the points under consideration are *generic*. All results remain of course valid as well for *nongeneric* configurations (for instance, in the case of *Theorem F-5*, if 3 of the 4 points in the plane are aligned, so that they form a triangle of vanishing area), but in such cases one must assign an appropriate (limiting) value to indeterminate products (of type $0\cdot\infty$).

Exercise F-7. Generalize all the above results by replacing *hyperspheres* with *quadrics* (in S-dimensional space). *Hint*: replace the definition (1c) with

$$r^2(\underline{x}) = \sum_{j,k=1}^{S} a_{jk} x_j x_k \ . \qquad (9)$$

Solution: see <CK96>.

Many more *theorems* of elementary geometry can be obtained by analogous techniques; the examples given below, mainly in the guise of *exercises*, are restricted to spaces of one and two dimensions.

Exercise F-8. Consider 4 points on a straight line (embedded in a plane), and indicate by $x^{(n)}$, $n = 1,2,3,4$, their coordinates (on the line). Select any (unordered) pair of them, say $x^{(1)}$, $x^{(2)}$ (there are of course 6 possible choices). Construct the circle, C_{12}, that has the segment $[\, x^{(1)}, x^{(2)} \,]$ as its *diameter*, and the two *concentric* circles (in the same plane), $c_{12,3}$ respectively $c_{12,4}$, on which $x^{(3)}$ respectively $x^{(4)}$ lie. Let $A_{12,3}$ respectively $A_{12,4}$ be the areas of the two *annuli* comprised between C_{12} and $c_{12,3}$ respectively $c_{12,4}$; let B_{12} be the area of the *rectangle* (in the plane) of sides $\left|x^{(1)} - x^{(2)}\right|$ and $\left|x^{(3)} - x^{(4)}\right|$; and define the product $P_{12} = B_{12} \cdot A_{12,3} \cdot A_{12,4}$. Let P_{nm} be the quantity analogous to P_{12}, but with the pair $x^{(1)}$, $x^{(2)}$ replaced by $x^{(n)}$, $x^{(m)}$. Prove that the 6 quantities P_{nm} are all equal,

$$P_{12} = P_{13} = P_{14} = P_{23} = P_{24} = P_{34} \ . \qquad (10)$$

Hint: show that $P_{nm} = \pi^2 \left|V_4(\underline{x})\right|$, where $V_4(\underline{x})$ is the (Vandermonde) determinant

$$V_4(\underline{x}) = \begin{vmatrix} 1 & x_1 & x_1^2 & x_1^3 \\ 1 & x_2 & x_2^2 & x_2^3 \\ 1 & x_3 & x_3^2 & x_3^3 \\ 1 & x_4 & x_4^2 & x_4^3 \end{vmatrix} \qquad (11)$$

(and to evaluate this determinant, rather than using the standard Vandermonde formula, exploit its translation invariance, setting the origin of coordinates in the middle of two points, say at the center of the pair $x^{(1)}$, $x^{(2)}$; then subtract the first column, multiplied by a suitable factor, from the third, and likewise the second from the fourth, ...).

Exercise F-9. Consider 5 points on a straight line (embedded in a plane), and partition them into a *pair* and a *trio* (both unordered; this can be done in 10 different ways). Now draw (in the plane) the circle C that has as *diameter* the segment joining the selected *pair* of points, and the 3 *concentric* circles, c_1, c_2, c_3, on which the other 3 points lie. Let A be the area of the circle C, and A_1, A_2 respectively A_3 be the 3 areas of the 3 circular *annuli* (in the plane) comprised between C and c_1, c_2 respectively c_3. Next select *one* point from the *trio* (this can be done of course in 3 ways), and draw (in the plane) the circle $\tilde{C}$ that has as *diameter* the segment joining the other two points, as well as the *concentric* circle $\tilde{c}$ on which the chosen point lies, and let $\tilde{A}$ be the area of the circle $\tilde{C}$, and $\tilde{A}'$ the area of the annulus comprised between these two circles, $\tilde{C}$ and $\tilde{c}$ (draw diagram!). Finally let

$$P=(A\tilde{A})^{1/2}\, A_1\, A_2\, A_3\, \tilde{A}' \; . \tag{12}$$

Given 5 generic points on a straight line, there are 30 different constructions that lead to as many, *a priori* different, evaluations of P. Prove that *all* these values of P are equal, and that they indeed coincide with $\pi^5 |V_5(\underline{x})|/4$, where $V_5(\underline{x})$ is the (5×5) Vandermonde determinant corresponding to the 5 points,

$$V(\underline{x})=\begin{vmatrix} 1 & x_1 & x_1^2 & x_1^3 & x_1^4 \\ 1 & x_2 & x_2^2 & x_2^3 & x_2^4 \\ 1 & x_3 & x_3^2 & x_3^3 & x_3^4 \\ 1 & x_4 & x_4^2 & x_4^3 & x_4^4 \\ 1 & x_5 & x_5^2 & x_5^3 & x_5^4 \end{vmatrix} \; . \tag{13}$$

Hint: to evaluate $V_5(\underline{x})$, exploit its translation invariance by setting the origin of coordinates in the middle of a pair of particles; then subtract from the third, fourth respectively fifth columns the first, second respectively third columns multiplied by an appropriate factor; then proceed as in the proof of *Proposition F-2* (with $S=1$, $N=3$).

Exercise F-10. Let $\vec{r}^{(n)}$ be 4 generic points *in the plane*. Select any (unordered) pair of them, say $\vec{r}^{\,(1)}$ and $\vec{r}^{\,(2)}$; this can be done in 6 different ways. Let C_{12} be a circle that goes through these two points; there are of course an infinity of such circles, characterized by their radii R, with $\left|r^{(1)}-r^{(2)}\right|\le 2R<\infty$. Let $C_{12,3}$ respectively $C_{12,4}$ be the two circles, both *concentric* to C_{12}, on which the other points, $\vec{r}^{(3)}$ respectively $\vec{r}^{(4)}$, lie; once C_{12} has been chosen, these two circles are uniquely defined (draw diagram!). Let $A_{12,3}$ respectively $A_{12,4}$ be the areas of the circular *annuli* comprised among C_{12} and $C_{12,3}$ respectively $C_{12,4}$; and let $T_{12,3}$ respectively $T_{12,4}$ be the areas of the two *triangles* with vertices $\vec{r}^{(1)}$, $\vec{r}^{(2)}$ and $\vec{r}^{(3)}$ respectively $\vec{r}^{(4)}$. Then set

$$P = \left| A_{12,3}\, T_{12,4} + s\, A_{12,4}\, T_{12,3} \right| \ , \tag{14}$$

with the sign s determined as follows: $s = \sigma\sigma'$, where $\sigma = +$ if the circle C_{12} falls between $C_{12,3}$ and $C_{12,4}$ and $\sigma = -$ otherwise, and $\sigma' = +$ if the points $\vec{r}^{(3)}$, $\vec{r}^{(4)}$ lie on the same side of the straight line going through $\vec{r}^{(1)}$ and $\vec{r}^{(2)}$, $\sigma' = -$ otherwise (note that, in the borderline cases, the value of the sign s becomes irrelevant, since one of the addenda in the right hand side of (14) vanishes). Prove that the value of P is then independent, not only of the initial selection of the first pair of points (out of 6 possibilities), but as well of the value of the parameter R (which can vary continuously within its allowed range), and that in fact

$$P = (\pi/2)\left|\Delta(\underline{\vec{r}})\right| , \tag{15}$$

with $\Delta(\underline{\vec{r}})$ defined by (1) (with $S=2$, $N=4$). *Hint*: to evaluate $\Delta(\underline{\vec{r}})$ take again advantage of its translation invariance, but now in a different manner than that used to prove *Proposition F-2*: choose the origin of coordinates at a point at the (same!) distance R from the two points of the selected pair; then subtract the first column, multiplied by R^2, from the last ...

Exercise F-11. Let $\vec{r}^{(n)}$, $n=1,2,3$, be 3 (distinct) *aligned* points in the *plane*, that lie on a common straight line, and $\vec{r}^{(4)}$ be a fourth, *nonaligned*, point in the plane, that does *not* lie on that straight line. Let, say, $\vec{r}^{(2)}$ be the (uniquely defined) middle one of the 3 aligned points, and set $\alpha = \left|\vec{r}^{(2)} - \vec{r}^{(1)}\right| / \left|\vec{r}^{(3)} - \vec{r}^{(1)}\right|$, hence $1-\alpha = \left|\vec{r}^{(3)} - \vec{r}^{(2)}\right| / \left|\vec{r}^{(3)} - \vec{r}^{(1)}\right|$ with $0<\alpha<1$. (Check!). Now let $p=1,2$ or 3 and indicate with A_p the area of the *annulus* comprised among the circle C_p that goes through the 3 points $\vec{r}^{(4)}$ and $\vec{r}^{(n)}$ with n taking the two values, in the set $(1,2,3)$, *different* from p, and the *concentric* circle c_p on which $\vec{r}^{(p)}$ lies (draw diagram!). Prove that

$$(1-\alpha)\, A_1 = A_2 = \alpha\, A_3 \ . \tag{16}$$

Hint: see the previous *Exercise F-10*, for the special configuration considered here.

Exercise F-12. Let $\vec{r}^{(n)}$ be 5 generic points *in the plane* (the requirement of genericity entails that no 3 points are aligned). Select any one of them, say $\vec{r}^{(5)}$ (there are of course 5 possible choices). Now *translate* and *rotate* the Cartesian reference frame so that in the new coordinate system, call it K_5, the *hyperbola* characterized by the equation $xy = H_5$, with an appropriately chosen value of the parameter H_5, go through the 4 points $\vec{r}^{(1)}$, $\vec{r}^{(2)}$, $\vec{r}^{(3)}$, $\vec{r}^{(4)}$. *(i)* Prove that these requirements determine uniquely the *translation*, the *rotation* (up to a minor ambiguity, see below), the modulus of the quantity H_5, and the corresponding *hyperbola*, as well as the *new* reference frame K_5 (up, as regard K_5, to the trivial symmetries corresponding to a rotation by $\pm\pi/2$ and a change of sign of H_5, or a rotation by $\pm\pi$ without change of H_5; hereafter we stick to one specific choice, K_5, among these, essentially equivalent, Cartesian frames of reference). Set then $h_5 = x^{(5)} y^{(5)}$, where $x^{(5)}$, $y^{(5)}$ are the Cartesian coordinates of $\vec{r}^{(5)}$ in the *new* (translated and rotated) reference frame K_5; hence the *hyperbola* defined, in the *new* reference frame K_5, by the formula $xy = h_5$ goes through the point $\vec{r}^{(5)}$. Next, select one of the 4 points $\vec{r}^{(1)}$, $\vec{r}^{(2)}$, $\vec{r}^{(3)}$, $\vec{r}^{(4)}$, say $\vec{r}^{(4)}$ (there are of course 4 possibilities). Then *translate* (without any *rotation*) the reference frame so that, in the *new* reference frame (call it K_{54}), the *hyperbola* defined by the equation $x^2 - y^2 = H_{54}$, with an appropriate choice of the parameter H_{54}, go through the 3 points $\vec{r}^{(1)}$, $\vec{r}^{(2)}$, $\vec{r}^{(3)}$. *(ii)* Prove that there is a unique translation that does this, with a corresponding unique value of H_{54} (note that only the *translation* can be adjusted in this case, not the *rotation*, in contrast to the previous case; indeed, in this case the *hyperbola* is required to go through 3 points, above it was required to go through 4 points). Set now $h_{54} = [x^{(4)}]^2 - [y^{(4)}]^2$, where $x^{(4)}$, $y^{(4)}$ are the *new* coordinates of $\vec{r}^{(4)}$; hence the *hyperbola* defined, in the new coordinate system, by the formula $x^2 - y^2 = h_{54}$, goes through the point $\vec{r}^{(4)}$. Finally evaluate the product $P_{54} = |H_5 - h_5||H_{54} - h_{54}|T_{123}$ where T_{123} is the area of the *triangle* whose 3 vertices are the 3 points $\vec{r}^{(1)}$, $\vec{r}^{(2)}$, $\vec{r}^{(3)}$; and let P_{nm} be the quantity analogously defined, with 5 replaced by n and 4 replaced by m. There clearly are 20 *a priori* different determinations of P_{nm}, corresponding to 5 choices for n and 4 for m ($m \neq n$, of course). Prove that they *all* have the same value,

$$P_{nm} = \left|\Delta_5^{(2)}(\underline{\vec{r}})\right| / 2 \,, \tag{17}$$

with the (5×5)-determinant $\Delta_5^{(2)}(\underline{\vec{r}})$ defined, in terms of the Cartesian coordinates $x^{(n)}$, $y^{(n)}$ of the 5 points $\vec{r}^{(n)}$ in the plane, by the formula

$$\Delta_5^{(2)} = \begin{vmatrix} 1 & x^{(1)} & y^{(1)} & [x^{(1)}]^2 - [y^{(1)}]^2 & x^{(1)}\, y^{(1)} \\ \vdots & \vdots & \vdots & \vdots & \vdots \\ 1 & x^{(5)} & y^{(5)} & [x^{(5)}]^2 - [y^{(5)}]^2 & x^{(5)}\, y^{(5)} \end{vmatrix} \,. \tag{18}$$

Hint: prove first of all that this determinant, (18), is invariant under *translations* $(\vec{r}^{(n)} \to \tilde{\vec{r}}^{(n)} = \vec{r}^{(n)} + \vec{r}^{(0)})$ and *rotations* $(x^{(n)} \to \tilde{x}^{(n)} = x^{(n)} \cos\theta - y^{(n)} \sin\theta \,,\; y^{(n)} \to \tilde{y}^{(n)} = x^{(n)} \sin\theta + y^{(n)} \cos\theta)$; then take advantage of this fact to show that, in an appropriately rotated and translated reference frame (as suggested by the formulation above)

$$\Delta_5^{(2)} = -(h_5 - H_5)\, \Delta_4^{(2)} \,, \tag{19a}$$

with

$$\Delta_4^{(2)} = \begin{vmatrix} 1 & x^{(1)} & y^{(1)} & [x^{(1)}]^2 - [y^{(1)}]^2 \\ \vdots & \vdots & \vdots & \vdots \\ 1 & x^{(4)} & y^{(4)} & [x^{(4)}]^2 - [y^{(4)}]^2 \end{vmatrix} \,; \tag{19b}$$

then note that this determinant, (19b), is invariant under translations (not rotations), and take advantage of this fact to show that, in an appropriately translated reference frame (as suggested by the formulation above)

$$\left|\Delta_4^{(2)}\right| = 2\left| h_{54} - H_{54}\right| T_{123} \,. \tag{19c}$$

Exercise F-13. Drop, in the preceding *Exercise F-12*, the condition that the 5 points $\vec{r}^{(n)}$ be *generic*; can you obtain thereby some new geometrical *theorems*? *Hint*: see *Exercise F-11*.

Finally let $\vec{r}^{(n)}$, $n = 1,...,6$, be 6 *generic* points *in the plane*, and let us focus on the following (6×6)-determinant defined in terms of their Cartesian coordinates $x^{(n)}$, $y^{(n)}$:

$$\Delta_6^{(2)}(\vec{r})=\begin{vmatrix}1 & x^{(1)} & y^{(1)} & [x^{(1)}]^2-[y^{(1)}]^2 & x^{(1)}y^{(1)} & [x^{(1)}]^2+[y^{(1)}]^2\\ \vdots & \vdots & \vdots & \vdots & \vdots & \vdots\\ 1 & x^{(6)} & y^{(6)} & [x^{(6)}]^2-[y^{(6)}]^2 & x^{(6)}y^{(6)} & [x^{(6)}]^2+[y^{(6)}]^2\end{vmatrix}. \tag{20a}$$

Exercise F-14. Prove that this determinant, (20a), is invariant under *translations* $(\vec{r}^{(n)}\to\tilde{\vec{r}}^{(n)}=\vec{r}^{(n)}+\vec{r}^{(0)})$ and *rotations* $(x^{(n)}\to\tilde{x}^{(n)}=x^{(n)}\cos\theta-y^{(n)}\sin\theta,\ y^{(n)}\to\tilde{y}^{(n)}=x^{(n)}\sin\theta+y^{(n)}\cos\theta)$. *Hint*: use the basic properties of determinants (see *hint* in *Exercise F-1*).

Exercise F-15. Show that there are several equivalent definitions of the determinant (20a), for instance

$$\Delta_6^{(2)}(\vec{r})=4\begin{vmatrix}1 & x^{(1)} & y^{(1)} & [x^{(1)}]^2 & x^{(1)}y^{(1)} & [y^{(1)}]^2\\ \vdots & \vdots & \vdots & \vdots & \vdots & \vdots\\ 1 & x^{(6)} & y^{(6)} & [x^{(6)}]^2 & x^{(6)}y^{(6)} & [y^{(6)}]^2\end{vmatrix}, \tag{20b}$$

and

$$\Delta_6^{(2)}(\vec{r})=\begin{vmatrix}1 & x^{(1)} & y^{(1)} & [x^{(1)}]^2-[y^{(1)}]^2 & x^{(1)}y^{(1)} & (1+\alpha)[x^{(1)}]^2+\beta x^{(1)}y^{(1)}+(1-\alpha)[y^{(1)}]^2+\gamma x^{(1)}+\delta y^{(1)}+\eta\\ \vdots & \vdots & \vdots & \vdots & \vdots & \vdots\\ 1 & x^{(6)} & y^{(6)} & [x^{(6)}]^2-[y^{(6)}]^2 & x^{(6)}y^{(6)} & (1+\alpha)[x^{(6)}]^2+\beta x^{(6)}y^{(6)}+(1-\alpha)[y^{(6)}]^2+\gamma x^{(6)}+\delta y^{(6)}+\eta\end{vmatrix} \tag{20c}$$

where α, β, γ, δ, η are 5 arbitrary coefficients. *Hint*: use the basic properties of determinants.

Exercise F-16. Show that, given 5 *generic* points $\vec{r}^{(n)}$ *in the plane*, there is a *unique* quadric (defined, in a given Cartesian frame, by the equation

$$(1+\alpha)x^2+\beta xy+(1-\alpha)y^2+\gamma x+\delta y+\eta=0, \tag{21}$$

with the 5 parameters α, β, γ, δ, η determined in terms of the Cartesian coordinates of these points) which goes through these 5 points. *Hint*: obtain 5 linear nonhomogeneous algebraic equations for the 5 parameters α, β, γ, δ, η by setting $x=x^{(n)}$, $y=y^{(n)}$, $n=1,\ldots,5$ in (21).

Exercise F-17. Show that

$$\Delta_6^{(2)}=\Delta_5^{(2)}\cdot D^{(6)}, \tag{22}$$

with $\Delta_6^{(2)}$ defined by (2), $\Delta_5^{(2)}$ defined by (18) and $D^{(6)} \equiv D^{(6)}(\underline{\vec{r}})$ defined as follows:

$$D^{(6)}(\underline{\vec{r}}) = d^{(6)}(\vec{r}^{(6)}) =$$

$$= (1+\alpha_6)[x^{(6)}]^2 + \beta_6\, x^{(6)}\, y^{(6)} + (1-\alpha_6)[y^{(6)}]^2 + \gamma_6\, x^{(6)} + \delta_6\, y^{(6)} + \eta_6 \,, \tag{23}$$

where the 5 parameters α_6, β_6, γ_6, δ_6, η_6 are (uniquely) defined by the requirement that the quadric uniquely characterized (see *Exercise F-16*) by the equation

$$d^{(6)}(\vec{r}) = 0 \,, \tag{24}$$

(see (21) and (23)) goes through the 5 points $\vec{r}^{(n)}$, $n = 1,...,5$. *Hint*: set $\alpha = \alpha_6$, $\beta = \beta_6$, $\gamma = \gamma_6$, $\delta = \delta_6$, $\eta = \eta_6$ in (20c), and use (23) and (24) (this latter formula entails of course $d^{(6)}(\vec{r}^{(n)}) = 0$, for $n = 1,...,5$).

Let us end *Appendix F* by formulating the following

Theorem F-18. Let $\vec{r}^{(n)}$, $n = 1,...,6$, be 6 generic points *in the plane*, and $x^{(n)}$, $y^{(n)}$ their Cartesian coordinates (in some reference frame). Select any one of them, say $\vec{r}^{(p)}$, $1 \le p \le 6$ (this can of course be done in 6 different ways). Then define the following two quantities: the (5×5)-determinant

$$\Delta_5^{(2)}(\vec{r}^{(1)},...,\vec{r}^{(p-1)},\vec{r}^{(p+1)},...,\vec{r}^{(6)}) =$$

$$= \begin{vmatrix} 1 & x^{(1)} & y^{(1)} & [x^{(1)}]^2 - [y^{(1)}]^2 & x^{(1)}\, y^{(1)} \\ \cdot & \cdot & \cdot & \cdot & \cdot \\ 1 & x^{(p-1)} & y^{(p-1)} & [x^{(p-1)}]^2 - [y^{(p-1)}]^2 & x^{(p-1)}\, y^{(p-1)} \\ 1 & x^{(p+1)} & y^{(p+1)} & [x^{(p+1)}]^2 - [y^{(p+1)}]^2 & x^{(p+1)}\, y^{(p+1)} \\ \cdot & \cdot & \cdot & \cdot & \cdot \\ 1 & x^{(6)} & y^{(6)} & [x^{(6)}]^2 - [y^{(6)}]^2 & x^{(6)}\, y^{(6)} \end{vmatrix} , \tag{25}$$

and the quantity

$$D^{(p)}(\underline{\vec{r}}) = (1+\alpha_p)[x^{(p)}]^2 + \beta_p\, x^{(p)}\, y^{(p)} + (1-\alpha_p)[y^{(p)}]^2 + \gamma_p\, x^{(p)} + \delta_p\, y^{(p)} + \eta_p \,, \tag{26}$$

where $x^{(p)}$, $y^{(p)}$ are of course the Cartesian components of $\vec{r}^{(p)}$ and the 5 coefficients α_p, β_p, γ_p, δ_p, η_p in (26) are uniquely defined by the requirement that the *quadric* defined by the formula

$$(1+\alpha_p)x^2+\beta_p xy+(1-\alpha_p)y^2+\gamma_p x+\delta_p y+\eta_p=0, \tag{27}$$

go through the 5 points $\vec{r}^{(1)},...,\vec{r}^{(p-1)},\vec{r}^{(p+1)},...,\vec{r}^{(6)}$. These two quantities, $\Delta_5^{(2)}$ and $D^{(p)}$, see (25) and (26), are defined in terms of the Cartesian components of the 6 2-vectors $\vec{r}^{(n)}$, but their values are independent of any rigid motion (translation or rotation) of the coordinate system, hence they should be both considered as *geometrical* objects having an intrinsic significance, independent of the coordinate system used to evaluate them: for $\Delta_5^{(2)}$, this has already been shown above, see *Exercise F-12*; as for the value of $D^{(p)}$, it clearly provides an intrinsic measure of the *failure* of the point $\vec{r}^{(p)}$ to lie on the (unique!) quartic that goes through the 5 points $\vec{r}^{(1)},...,\vec{r}^{(p-1)},\vec{r}^{(p+1)},...,\vec{r}^{(6)}$, since (26) and (27) entail that $D^{(p)}$ vanishes if the point $\vec{r}^{(p)}$ also lies on this quadric. Generally, given 6 generic points, to every one of the 6 possible different selections of $\vec{r}^{(p)}$ from the set $\{\vec{r}^{(n)}$, $n=1,...,6\}$, there correspond *different* values of $\Delta_5^{(2)}$ and $D^{(p)}$, see (25) and (26); but the product of these two quantities is always the same, for these 6 choices of p, indeed

$$\Delta_5^{(2)}(\vec{r}^{(1)},...,\vec{r}^{(p-1)},\vec{r}^{(p+1)},...,\vec{r}^{(6)})\cdot D^p(\underline{\vec{r}})=\Delta_6^{(2)}(\underline{\vec{r}}), \tag{28}$$

with $\Delta_6^{(2)}(\underline{\vec{r}})$ defined by (20).

Proof. The *uniqueness* of the quadric (27) is proven in *Exercise F-16*; the *intrinsic* (coordinate-independent) nature of $\Delta_5^{(2)}$ is proven in *Exercise F-12*; the *intrinsic* nature of $D^{(p)}$ is entailed by (28) and by the intrinsic nature of $\Delta_5^{(2)}$ (see above) as well as $\Delta_6^{(2)}$ (see *Exercise F-14*); the result (28) is proven in *Exercise F-17* (with 6 replaced by p).

Remark F-19. The value of the determinant $\Delta_6^{(2)}$, see (20), is an *intrinsic* measure of the *failure* of the 6 points $\vec{r}^{(n)}$, $n=1,...,6$, to lie on one and the same *quadric* (of course, if they do, $\Delta_6^{(2)}$ vanishes, see (28), (26) and (27)).

Exercise F-20. In the formulation of *Theorem F-18* the adjective *generic* was used to qualify the 6 points $\vec{r}^{(n)}$ in the plane; but this *Theorem*

F-18 is of course valid for *any* configuration of the 6 points $\vec{r}^{(n)}$, although for some special configuration it may entail assigning appropriately a value to an indeterminate product of type $0 \cdot \infty$. Is it possible to obtain interesting *theorems* by focusing on special configurations of the 6 oints $\vec{r}^{(n)}$ (in analogy to what is done above in *Exercise F-11*)?

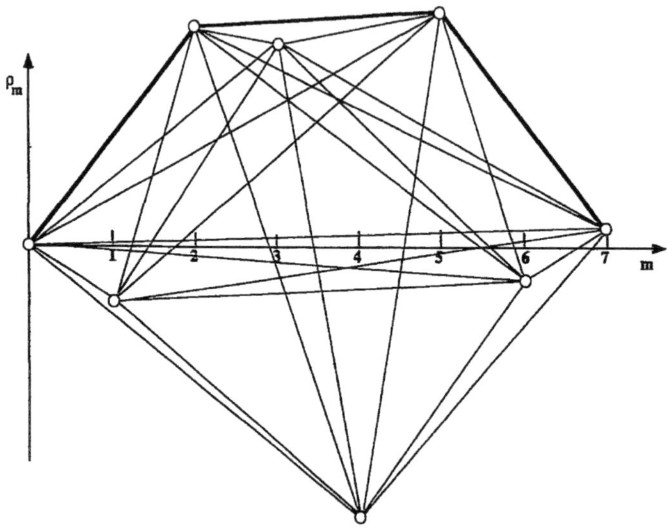

Fig. G. - 1. The "upper curve" is bolded (see text for explanantion).

Appendix G:
Asymptotic behavior of the zeros of a polynomial whose coefficients diverge exponentially

Let $z_n(t)$, $n=1,2,\ldots,N$, be the N zeros of a (monic) polynomial of degree N in z whose N coefficients depend exponentially on the real parameter t:

$$\left[z_n(t)\right]^N + \sum_{m=1}^{N} c_m(t)\left[z_n(t)\right]^{N-m} = 0, \quad n=1,2,\ldots,N, \tag{1}$$

$$c_m(t) = \bar{c}_m \exp\left[(\rho_m + i\gamma_m)\,t\right], \quad m=1,2,\ldots,N \quad . \tag{2}$$

The constants $\bar{c}_m$ are N arbitrary (nonvanishing) *complex* numbers, and the constants ρ_m, γ_m are $2N$ arbitrary *real* numbers (N is an arbitrary positive integer, $N \geq 2$).

We now formulate, and then prove, *Proposition G-1*, that details the behavior of the zeros $z_n(t)$ of (1) with (2) as $t \to \pm\infty$. Since the formulation of this *Proposition G-1* is fairly involved, the reader is warned that it might be easier to appreciate its significance fully, by proceeding, immediately after a first cursory reading of it, to understand the strategy of its proof, as detailed below.

Proposition G-1. As the real parameter t tend to (positive) infinity, $t \to \infty$,

$$z_n(t) = z_n^{(+)}(t)\left\{1 + O\left[\exp(-p_n t)\right]\right\} \quad , \tag{3a}$$

$$z_n^{(+)}(t) = \tilde{z}_n(t)\,\exp(q_n t) \quad , \tag{3b}$$

$$\tilde{z}_n(t) = \bar{z}_n \exp(i\, r_n t) \quad . \tag{3c}$$

Here the superscript "plus" attached to $z_n^{(+)}(t)$ serves to distinguish this quantity from $z_n(t)$ (clearly $z_n^{(+)}(t)$ is the "dominant part" of $z_n(t)$ as $t \to \infty$), and also as a reminder that we are investigating the behavior as t tends to positive infinity (an analogous superscript should be attached to

F. Calogero: LNPm 66, pp. 723 - 731, 2001

$\tilde{z}_n(t)$ and to $\bar{z}_n$; it is omitted to simplify the notation). The N complex constants $\bar{z}_n$, and the $3N$ real numbers $p_n > 0$, q_n and r_n are given by the following prescriptions.

Identify on a Cartesian plane the N points with integer abscissas $m = 1,2,...,N$ and ordinates ρ_m, and in addition the origin (abscissa $m = 0$, ordinate $\rho_0 = 0$). Draw the (clearly unique and continuous, if generally segmented) curve, which is the *upper envelope* of the $N(N+1)/2$ segments that connect pairwise these $N+1$ points (see the example with $N = 7$ in Figure G-1); hereafter we refer to this segmented curve as the *upper curve*. Associate to each segment of the upper curve the following numbers (the labels s identifies subsequent segments of this curve, from left to right): $m_s^{(-)}$ and $m_s^{(+)}$ are the values of m that correspond to the beginning and to the end of the s-th segment (so that $m_1^{(-)} = 0$, $m_s^{(+)} = m_{s+1}^{(-)}$, $m_S^{(+)} = N$, where S is the number of segments that make up the upper curve);

$$n_s = m_s^{(+)} - m_s^{(-)}, \quad s = 1,...,S, \tag{4}$$

(hence n_s is the number of points that lie below the s-th segment, increased by one, see Figure G-1 ; of course the n_s add up to N, $\sum_{s=1}^{S} n_s = N$);

$$z^{(s)} = [\ -\bar{c}_{m_s^{(+)}} / \bar{c}_{m_s^{(-)}}\ ]^{1/n_s}, \quad s = 1,...,S \tag{5}$$

(with the convention $\bar{c}_0 = 1$);

$$q^{(s)} = [\ \rho_{m_s^{(+)}} - \rho_{m_s^{(-)}}\ ]/[\ m_s^{(+)} - m_s^{(-)}\ ]\ ; \tag{6}$$

$$r^{(s)} = [\ \gamma_{m_s^{(+)}} - \gamma_{m_s^{(-)}}\ ]/[\ m_s^{(+)} - m_s^{(-)}\ ]\ ; \tag{7}$$

$$p^{(s)} = \min_{m=1,2,...,N;\, m \neq m_s^{(+)}, m_s^{(-)}} [\ (\{[m - m_s^{(-)}]\rho_{m_s^{(+)}} + [m_s^{(+)} - m]\rho_{m_s^{(-)}}\}/[m_s^{(+)} - m_s^{(-)}]) - \rho_m\ ]. \tag{8}$$

Note that the last formula implies that $p^{(s)}$ is positive, $p^{(s)} > 0$, since the straight line defined, as a function of the variable m, by the expression

$$f(m) = \{[m - m_s^{(-)}]\rho_{m_s^{(+)}} + [m_s^{(+)} - m]\rho_{m_s^{(-)}}\}/[m_s^{(+)} - m_s^{(-)}] \tag{9}$$

is, by construction, *above* all other points (m, ρ_m) with $m \neq m_s^{(-)}$ and $m \neq m_s^{(+)}$. (See Figure G-1, and note that we assume here to be in the *generic case*, thereby excluding that three or more of the $N+1$ points identified above lie on the same straight line; the exceptional cases when this instead happens are discussed below).

Then, to each segment s, are associated n_s asymptotic values $z_n^{(+)}(t)$, see (3a), with the following identification of the parameters in (3):

$$\bar{z}_n = z^{(s)} \exp(2\pi i j / n_s) , \quad j = 1, \ldots, n_s , \tag{10a}$$

$$p_n = p^{(s)} , \; q_n = q^{(s)} , \; r_n = r^{(s)} . \tag{10b}$$

For instance, in the case of Figure G-1, $S = 3$; $n_1 = 2, n_2 = 3, n_3 = 2$, $z^{(1)} = (-\bar{c}_2)^{1/2}$, $z^{(2)} = (-\bar{c}_5 / \bar{c}_2)^{1/3}$, $z^{(3)} = (-\bar{c}_7 / \bar{c}_5)^{1/2}$; $q^{(1)} = \rho_2 / 2$, $q^{(2)} = (\rho_5 - \rho_2)/3$, $q^{(3)} = (\rho_7 - \rho_5)/2$; $r^{(1)} = \gamma_2 / 2$, $r^{(2)} = (\gamma_5 - \gamma_2)/3$, $r^{(3)} = (\gamma_7 - \gamma_5)/2$; and

$$p^{(1)} = \min_{m=1,\ldots,7;\, m \neq 0,2} \left[m\,\rho_2 / 2 - \rho_m \right] , \tag{11a}$$

$$p^{(2)} = \min_{m=1,\ldots,7;\, m \neq 2,5} \left\{ \left[(m-2)\,\rho_5 + (5-m)\rho_2 \right] / 3 - \rho_m \right\} , \tag{11b}$$

$$p^{(3)} = \min_{m=1,\ldots,7;\, m \neq 5,7} \left\{ \left[(m-5)\,\rho_7 + (7-m)\rho_5 \right] / 2 - \rho_m \right\} . \tag{11c}$$

Hence

$$\bar{z}_n = \exp(2\pi i n/2)\, (-\bar{c}_2)^{1/2} , \; p_n = p^{(1)} , \; q_n = q^{(1)} , \; r_n = r^{(1)} , \; n = 1,2, \tag{12a}$$

$$\bar{z}_n = \exp\left[2\pi i (n-2)/3\right] (-\bar{c}_5 / \bar{c}_2)^{1/3} , \; p_n = p^{(2)} , \; q_n = q^{(2)} , \; r_n = r^{(2)} , \; n = 3,4,5, \tag{12b}$$

$$\bar{z}_n = \exp\left[2\pi i (n-5)/2\right] (-\bar{c}_7 / \bar{c}_5)^{1/2} , \; p_n = p^{(3)} , \; q_n = q^{(3)} , \; r_n = r^{(3)} , \; n = 6,7 . \tag{12c}$$

To sum up: as $t \to \infty$, the dominant terms $z_n^{(+)}(t)$, giving the asymptotic behaviors of the N zeros $z_n(t)$ (see (3a)) are divided into S families, where S is the number of segments that compose the *upper curve*. Each family includes n_s values (see (4)) which, in the complex plane, lie equispaced on a circle centered at the origin (see (3) and (10)), whose radius evolves proportionally to $\exp[q^{(s)} t]$ (see (3), (10) and (6)), and which rotate with constant velocity as entailed by the factor $\exp[i r^{(s)} t]$ (see (3c) and (10b)). The radius of the s-th circle diverges to infinity or converges to zero (in either case, exponentially), depending on whether the s-th segment has positive or negative slope (see (3b), (10b) and (6)); it is constant if the s-th segment is horizontal (so that $q^{(s)}$ vanishes, see (6)). In the case of *positive* slope $(q^{(s)} > 0)$, namely when the radius of the circle *diverges* exponentially as $t \to \infty$, hence the corresponding n_s zeros spiral to infinity, they may, or may not, approach their dominant parts $z_n^{(+)}(t)$, see (3); this depends on the behavior of the *difference* $z_n(t) - z_n^{(+)}(t)$, which is $O(\exp\{[q^{(s)} - p^{(s)}]t\})$ (see (3) and (10)), hence vanishes or diverges depending on the sign of the difference $q^{(s)} - p^{(s)}$. The zero $z_n(t)$ approaches of course its dominant part $z_n^{(+)}(t)$ if this does not diverge, namely when $q_n \le 0$ (see (3b), (6) and (10b)).

As mentioned above, this outcome describes the situation in the generic case in which no segment of the upper curve contains one additional point besides the two extremal ones. In the special cases when a segment of the upper curve contains one, or more, additional points (as it might for instance be the case in the example of Figure G-1 if ρ_3 were a bit larger, so that the point $(3, \rho_3)$ lie on the segment joining $(2, \rho_2)$ and $(5, \rho_5)$), then the formula (3) remains valid with the same definitions of q_n and also (essentially; but see below) of p_n (see (3a,b) and (10b)), while the definition (3c) of $\tilde{z}_n(t)$ is instead replaced by a new one, as we now explain.

But firstly let us note that the n_s quantities $\tilde{z}_n(t)$ defined by (3c), (10a) and (5) are the n_s finite (i.e., nonvanishing and nondivergent) roots of the following algebraic equation in $\tilde{z}$:

$$\bar{c}_{m_s^{(-)}} \exp[i \gamma_{m_s^{(-)}} t] \, \tilde{z}^{N-m_s^{(-)}} + \bar{c}_{m_s^{(+)}} \exp[i \gamma_{m_s^{(+)}} t] \, \tilde{z}^{N-m_s^{(+)}} = 0 \;, \tag{13a}$$

or equivalently

$$\bar{c}_{m_s^{(-)}} \exp[i \gamma_{m_s^{(-)}} t] \, \tilde{z}^{m_s^{(+)}} + \bar{c}_{m_s^{(+)}} \exp[i \gamma_{m_s^{(+)}} t] \, \tilde{z}^{m_s^{(-)}} = 0 \tag{13b}$$

(see (4), (3c), (7) and (10)). This is the equation whose roots determine the quantities $\tilde{z}_n(t)$ in the generic case considered above. To also cover the exceptional cases with additional points on some segments of the *upper curve*, the following supplementary rule applies: if the s-th segment of the *upper curve* contains $\Sigma_s > 2$ points $(m_s^{(\sigma)}, \rho_{m_s^{(\sigma)}})$, $\sigma = 1,\ldots,\Sigma_s$, of course with

$$m_s^{(-)} = m_s^{(1)} < m_s^{(\sigma)} < m_s^{(\Sigma_s)} = m_s^{(+)}\ ,\quad \sigma = 2,\dots,\Sigma_s - 1\ ,\tag{14}$$

then the n_s dominant values $z_n^{(+)}(t)$, see (3), belonging to the family associated with the s-th segment, have parameters p_n and q_n still defined by (10b) with (6) and (8) (except that the minimum in the right hand side of (8) must now be taken over all values of m different from *all* the values $m_s^{(\sigma)}$, $\sigma = 1,\dots,\Sigma_s$); but the quantities $\tilde{z}_n(t)$, see (3b), instead of being given by (10a) and (5), or equivalently as the n_s roots of (13), are now the n_s roots of the following algebraic equation in $\tilde{z}$,

$$\sum_{\sigma=1}^{\Sigma_s} \bar{c}_{m_s^{(\sigma)}} \exp[i\,\gamma_{m_s^{(\sigma)}}\, t\,]\, \tilde{z}^{\,m_s^{(+)} - m_s^{(\sigma)}} = 0\ .\tag{15}$$

The fact that this equation has indeed n_s roots is implied by (14) and (4).

This completes the formulation of *Proposition G-1*. Since clearly the shape of the upper curve is largely determined by the value ρ_+ of the largest ρ_m,

$$\rho_+ = \operatorname*{Max}_{m=1,2,..,N} [\rho_m]\ ,\tag{16}$$

Proposition G-1 entails the following

Corollary G-2. The behavior as $t\to\infty$ of the zeros $z_n(t)$, see (1) and (2), is largely determined by the parameter ρ_+, see (16), and, if ρ_+ is not negative, $\rho_+ \ge 0$, also by the value m_+ (or the values $m_+^{(\sigma)}$, see below) at which ρ_m attains its maximal value ρ_+.

Indeed, if ρ_+ is negative,

$$\rho_+ < 0\ ,\tag{17a}$$

then as $t\to\infty$ all N zeros $z_n(t)$ converge exponentially fast to zero,

$$z_n(t) \underset{t\to\infty}{\longrightarrow} 0\ .\tag{17b}$$

If instead ρ_+ vanishes

$$\rho_+ = 0\ ,\tag{18a}$$

then as $t\to\infty$ only some (if any) of the zeros $z_n(t)$ converge to zero, while the remaining ones neither converge to zero nor escape to infinity.

Specifically, if $\rho_m = 0$ for $m = m_+^{(\sigma)}$ and $\rho_m < 0$ for $m \neq m_+^{(\sigma)}$ with $\sigma = 1,...,\Sigma$ and $m_+^{(\sigma)} \leq m_+^{(\Sigma)} \equiv m_+$, then, as $t \to \infty$, $N - m_+$ of the N zeros $z_n(t)$ converge (exponentially fast) to zero, and m_+ of them approach (exponentially fast) the m_+ roots of the following algebraic equation in $\tilde{z}$:

$$\tilde{z}^{m_+} + \sum_{\sigma=1}^{\Sigma} \bar{c}_{m_+}^{(\sigma)} \exp\left[i \gamma_{m_+^{(\sigma)}} t\right] \tilde{z}^{m_+ - m_+^{(\sigma)}} = 0 \ . \tag{18b}$$

Note that, if $\Sigma = 1$, entailing $m_+^{(1)} = m_+$, these m_+ roots are given by the formula

$$\tilde{z}_j(t) = \exp(2\pi i j / m_+)\, (-c_{m_+})^{1/m_+} \exp(i \gamma_{m_+} t / m_+), \quad j = 1,...,m_+ \ . \tag{18c}$$

Finally, if ρ_+ is positive,

$$\rho_+ > 0 \ , \tag{19a}$$

as $t \to \infty$ some of the N zeros $z_n(t)$ escape to infinity, while the others converge to zero. Specifically, if $\rho_m = \rho_+ > 0$ for $m = m_+$ and $\rho_m < \rho_+$ for $m \neq m_+$ (namely, if the maximal, positive, value ρ_+ is attained only at the single value m_+ of the index $m = 1,...,N$), then, as $t \to \infty$, m_+ of the N zeros $z_n(t)$ escape (exponentially fast) to infinity and $N - m_+$ converge (exponentially fast) to zero. If instead $\rho_m = \rho_+ > 0$ for $m = m_+^{(\sigma)}$, and $\rho_m < \rho_+$ for $m \neq m_+^{(\sigma)}$, with $\sigma = 1,..,\Sigma$ and $m_+^{(-)} \equiv m_+^{(1)} \leq m_+^{(2)} \leq ... \, m_+^{(\Sigma)} \equiv m_+^{(+)}$, then, as $t \to \infty$, $m_+^{(-)}$ of the N zeros $z_n(t)$ escape (exponentially fast) to infinity, $N - m_+^{(+)}$ converge (exponentially fast) to zero, and $m_+^{(+)} - m_+^{(-)}$ approach (exponentially fast) the $m_+^{(+)} - m_+^{(-)}$ roots of the following algebraic equation in $\tilde{z}$:

$$\sum_{\sigma=1}^{\Sigma} \bar{c}_{m_\sigma^{(+)}} \exp\left[i \gamma_{m_+^{(\sigma)}} t\right] \tilde{z}^{m_+^{(+)} - m_+^{(\sigma)}} = 0 \ . \tag{19b}$$

This case differs from the previous one iff $\Sigma \geq 2$; if $\Sigma = 2$ (so that $m_+^{(1)} = m_+^{(-)}$, $m_+^{(2)} = m_+^{(+)}$) the $m_+^{(+)} - m_+^{(-)}$ roots of this equation are given by the explicit formula

$$\tilde{z}_j(t) = \exp\left[2\pi i j / (m_+^{(+)} - m_+^{(-)})\right] \left[\, \bar{c}_{m_+^{(+)}} / \bar{c}_{m_+^{(-)}} \,\right]^{1/(m_+^{(+)} - m_+^{(-)})} \cdot$$

$$\cdot \exp\left[i (\gamma_{m_+^{(+)}} - \gamma_{m_+^{(-)}}) t / (m_+^{(+)} - m_+^{(-)})\right] . \tag{19c}$$

Proofs. The proof of *Corollary G-2* requires no elaboration: it is an immediate consequence of *Proposition G-1*, combined with the topology of the *upper curve* under the various instances considered (except for the specific formulas (18b) and (19b), which are entailed by the proof of *Proposition G-1*, see below).

The basic idea to prove *Proposition G-1* is that, to find the zeros $z_n(t)$ of (1) with (2), one should focus on *two* of the $N+1$ terms of the polynomial equation

$$\sum_{m=0}^{N} c_m(t)\left[z(t)\right]^{N-m}=0 \ , \tag{20}$$

identifying the behavior of $z=z(t)$ as $t\to\infty$ so that these *two* terms are of the *same* order and *dominate* over all other terms. Note that, for notational convenience, we have replaced here (1) with (20); these two equations, (1) and (20), of course coincide since we also set

$$c_0(t)=1 \ , \tag{21a}$$

which is consistent with the validity of (2) also for $m=0$, with

$$\bar{c}_0=1, \ \rho_0=\gamma_0=0 \ . \tag{21b}$$

Hence our proof proceeds through the identification of such pairs, the demonstration that they indeed dominate, and the derivation, via the requirement that they cancel against each other, of the results detailed in the above formulation of *Proposition G-1*.

Let us then assume that the two terms with, say, $m=m_1$ and $m=m_2$ (with $m_2>m_1$) are of the same order and dominate over all others as $t\to\infty$, so that by setting

$$z=\tilde{z}\exp(q\,t) \tag{22}$$

with

$$\rho_{m_1}+(N-m_1)q=\rho_{m_2}+(N-m_2)q \ , \tag{23}$$

we can conveniently rewrite (20) as follows:

$$\bar{c}_{m_1}\exp(i\gamma_{m_1}t)\,\tilde{z}^{N-m_1}+\bar{c}_{m_2}\exp(i\gamma_{m_2}t)\,\tilde{z}^{N-m_2}$$

$$=-\sum_{m=0;m\neq m_1,m_2}^{N}\bar{c}_m\exp(i\gamma_m t)\exp(-p_m t)\,\tilde{z}^{N-m} \ ,$$

with

$$p_m=\rho_{m_1}+q(m-m_1)-\rho_m \ . \tag{25}$$

From (23), which corresponds to the requirement that the two selected terms be of the *same* order as $t \to \infty$, we get

$$q = (\rho_{m_2} - \rho_{m_1}) / (m_2 - m_1) \quad , \tag{26}$$

hence, via (25),

$$p_m = \{ [(m - m_1)\rho_{m_2} + (m_2 - m)\rho_{m_1}] / (m_2 - m_1) \} - \rho_m \quad . \tag{27}$$

It is now clear that the expression in the left hand side of (24) dominates, as $t \to \infty$, over every term in the right hand side, provided the quantities p_m, see (27), are positive,

$$p_m > 0 \quad , \tag{28}$$

for all values of $m \neq m_1, m_2$. Since the term inside the curly bracket in the right hand side of (27) represents, as a function of m, the *straight line* that goes, in the Cartesian (m, ρ_m) plane, through the two points (m, ρ_{m_1}) and (m_2, ρ_{m_2}), it is clear that this condition is satisfied, in the *generic case* (as defined in the formulation of *Proposition G-1*, see above), iff $m_1 = m_s^{(-)}$ and $m_2 = m_s^{(+)}$. With such a choice we clearly get (6) from (26), as well as, from (24),

$$\bar{c}_{m_s^{(-)}} \exp[i\gamma_{m_s^{(-)}} t]\, \tilde{z}^{N-m_s^{(-)}} + \bar{c}_{m_s^{(+)}} \exp[i\gamma_{m_s^{(+)}} t]\, \tilde{z}^{N-m_s^{(+)}} = O[\exp(-p^{(s)} t)] \tag{29}$$

with (8). Clearly this last formula, via (4), entails, in the asymptotic $t \to \infty$ limit, (13) hence (3c) with (7), (10) and (5).

Proposition G-1 is thereby proven in the generic case. Extending this proof to the *general case*, see above, is, we trust, sufficiently straightforward, to justify leaving this as a task for the diligent reader.

Remark G-3. In the formulation of *Proposition G-1* and of its *Corollary G-2*, we have assumed that none of the coefficients $c_m(t)$ in (1) vanish identically (see the first sentence after (2)). It is easy to extend *Proposition G-1* and *Corollary G-2* so that they also hold if one or more of the coefficients $c_m(t)$ (namely, one or more of the constants $\bar{c}_m$, see (2)) vanishes. Then the corresponding points (m, ρ) must simply be ignored in the construction leading to the definition of the *upper curve*, as well as in the definition of ρ_+, see (16), hence of m_+, and so on.

Remark G-4. Another easy extension of *Proposition G-1* and *Corollary G-2* deals with the other limit, $t \to -\infty$. Then the role of the *upper curve* is taken over by the, analogously defined (but with all inequalities reversed), *lower curve* and all formulas apply without changes (other than the obvious ones). In this case the quantities $p^{(s)}$ (defined by (8) with min replaced by Max) are *negative* (rather than positive), and the zeros associated with segments of the *lower curve* having *negative* slope go asymptotically to (or rather, in the remote past, come from) *infinity*, while those associated with segments of the *lower curve* having *positive* slope converge to (or rather come from) *zero*. As for the results of *Corollary G-2*, the key role to determine the behavior of the N zeros $z_n(t)$ as $t \to -\infty$ is played by the quantity

$$\rho_- = \min_{m=1,\dots,N} [\rho_m] \; ; \qquad (30)$$

and the formulation of the modified version of *Corollary G-2* detailing the behavior of the zeros as $t \to -\infty$ coincides essentially with that given above for the $t \to +\infty$ case, with ρ_+ replaced by ρ_- and a reversal of some of the inequalities, as obviously appropriate.

As an example, let us look at the instance illustrated by *Figure G-1*. In this (generic) case, as $t \to +\infty$ the 7 zeros $z_n(t)$ get separated into 3 families, one $(s=1,\, m_1^{(-)}=0,\, m_1^{(+)}=2)$ containing 2 members and spiraling to infinity, a second one $(s=2,\, m_2^{(-)}=2,\, m_2^{(+)}=5)$ containing 3 members and also spiraling to infinity (albeit less fast), and a third one $(s=3,\, m_3^{(-)}=5,\, m_3^{(+)}=7)$ containing 2 members which spiral to the origin; while only 2 families emerge in the $t \to -\infty$ limit, one containing 4 members $(s=1,\, m_1^{(-)}=0,\, m_2^{(+)}=4)$ which spiral out to (or rather in from) infinity in the remote past, the other containing 3 members $(s=2,\, m_2^{(-)}=4,\, m_2^{(+)}=7)$ which in the remote past spiral to (or rather from) the origin.

In conclusion we emphasize that, while these results provide an easy technique to predict the asymptotic behavior, as $t \to \pm\infty$, of the zeros $z_n(t)$ of the polynomial (1) with (2), there is instead no easy way to identify *which* zero behaves *how*, and in particular no easy way to connect the behavior of a particular zero as $t \to -\infty$ to its behavior as $t \to +\infty$.

G.N Notes to Appendix G

The treatment in this Appendix G follows closely Appendix A of <C96b>; in particular Figure G-1 coincides with the Figure 1 given there.

Appendix H:
Some formulas for Pauli matrices and three-vectors

In this Appendix we display some standard formulas for the $\underline{\sigma}$-matrices, and a useful 3-vector identity.

$$\underline{\sigma}_x = \begin{pmatrix} 0 & 1 \\ 1 & 0 \end{pmatrix}, \quad \underline{\sigma}_y = \begin{pmatrix} 0 & i \\ -i & 0 \end{pmatrix}, \quad \underline{\sigma}_z = \begin{pmatrix} 1 & 0 \\ 0 & -1 \end{pmatrix} . \tag{1}$$

$$\left[\rho + i(\vec{r}\cdot\underline{\vec{\sigma}})\right]^{-1} = \left[\rho - i(\vec{r}\cdot\underline{\vec{\sigma}})\right]/(\rho^2+r^2) , \tag{2}$$

$$(\vec{r}^{(1)}\cdot\underline{\vec{\sigma}})(\vec{r}^{(2)}\cdot\underline{\vec{\sigma}}) = (\vec{r}^{(1)}\cdot\vec{r}^{(2)}) - i(\vec{r}^{(1)}\wedge\vec{r}^{(2)})\cdot\underline{\vec{\sigma}} , \tag{3}$$

$$\left[(\vec{r}^{(1)}\cdot\underline{\vec{\sigma}}),(\vec{r}^{(2)}\cdot\underline{\vec{\sigma}})\right] = -2i(\vec{r}^{(1)}\wedge\vec{r}^{(2)})\cdot\underline{\vec{\sigma}} , \tag{4}$$

$$(\vec{r}^{(1)}\cdot\underline{\vec{\sigma}})(\vec{r}^{(2)}\cdot\underline{\vec{\sigma}})(\vec{r}^{(3)}\cdot\underline{\vec{\sigma}}) = -i(\vec{r}^{(1)}\wedge\vec{r}^{(2)})\cdot\vec{r}^{(3)}$$

$$+\left[\vec{r}^{(1)}(\vec{r}^{(2)}\cdot\vec{r}^{(3)}) - \vec{r}^{(2)}(\vec{r}^{(1)}\cdot\vec{r}^{(3)}) + \vec{r}^{(3)}(\vec{r}^{(1)}\cdot\vec{r}^{(2)})\right]\cdot\underline{\vec{\sigma}} , \tag{5}$$

$$(\vec{r}^{(1)}\cdot\underline{\vec{\sigma}})(\vec{r}^{(2)}\cdot\underline{\vec{\sigma}})(\vec{r}^{(3)}\cdot\underline{\vec{\sigma}}) + (\vec{r}^{(3)}\cdot\underline{\vec{\sigma}})(\vec{r}^{(2)}\cdot\underline{\vec{\sigma}})(\vec{r}^{(1)}\cdot\underline{\vec{\sigma}})$$

$$= 2\left[\vec{r}^{(1)}(\vec{r}^{(2)}\cdot\vec{r}^{(3)}) - \vec{r}^{(2)}(\vec{r}^{(1)}\cdot\vec{r}^{(3)}) + \vec{r}^{(3)}(\vec{r}^{(1)}\cdot\vec{r}^{(2)})\right]\cdot\underline{\vec{\sigma}} , \tag{6}$$

$$(\vec{r}^{(1)}\cdot\underline{\vec{\sigma}})(\vec{r}^{(2)}\cdot\underline{\vec{\sigma}})(\vec{r}^{(1)}\cdot\underline{\vec{\sigma}}) = \left[2\vec{r}^{(1)}(\vec{r}^{(1)}\cdot\vec{r}^{(2)}) - \vec{r}^{(2)}(\vec{r}^{(1)}\cdot\vec{r}^{(1)})\right]\cdot\underline{\vec{\sigma}} , \tag{7}$$

$$\left(\rho + i\vec{r}\cdot\underline{\vec{\sigma}}\right)^{1/2} \equiv \tilde{\rho} + i\tilde{\vec{r}}\cdot\underline{\vec{\sigma}}, \quad \tilde{\rho} = \left[(\rho+ir)^{1/2} + (\rho-ir)^{1/2}\right]/2 ,$$

$$\tilde{\vec{r}} = -i\,\vec{r}\left[(\rho+ir)^{1/2} - (\rho-ir)^{1/2}\right]/(2r) , \tag{8a}$$

$$\left(i\vec{r}\cdot\underline{\vec{\sigma}}\right)^{1/2} = (r/2)^{1/2}\left(1 + i\,r^{-1}\vec{r}\cdot\underline{\vec{\sigma}}\right) , \tag{8b}$$

$$\exp(i\vec{r}\cdot\underline{\vec{\sigma}}) = \cos(r) + i(\vec{r}\cdot\underline{\vec{\sigma}})r^{-1}\sin(r) , \tag{9}$$

$$f(\vec{r}\cdot\underline{\vec{\sigma}}) = \left\{f(r) + f(-r) + \left[f(r) - f(-r)\right]\vec{r}\cdot\underline{\vec{\sigma}}/r\right\}/2 . \tag{10}$$

F. Calogero: LNPm 66, pp. 732 - 733, 2001

The 3-vector formula

$$a\vec{r}+\vec{r}\wedge\vec{b}+(\vec{r}\cdot\vec{d})\vec{c}=\vec{f} \tag{11a}$$

entails

$$\vec{r}=(a^2+b^2)^{-1}[\ a\vec{f}+\vec{b}\wedge\vec{f}+\gamma\vec{b}\wedge\vec{c}+a\gamma\vec{c}+a^{-1}(\vec{f}\cdot\vec{b}+\gamma\vec{c}\cdot\vec{b})\,\vec{b}\ ], \tag{11b}$$

$$\gamma=-[\ a^2\vec{f}\cdot\vec{d}+a(\vec{f}\wedge\vec{d})\cdot\vec{b}+(\vec{f}\cdot\vec{b})(\vec{d}\cdot\vec{b})\ ]\cdot$$

$$\cdot[\ a(a^2+b^2)+a^2\vec{c}\cdot\vec{d}+a(\vec{c}\wedge\vec{d})\cdot\vec{b}+(\vec{c}\cdot\vec{b})(\vec{d}\cdot\vec{b})\ ]^{-1}. \tag{11c}$$

The subcases $\vec{b}=0$, $\vec{c}=0$ (or $\vec{d}=0$), $a=0$, while easily obtainable from the above formulas, deserve separate display.
Case $\vec{b}=0$:

$$\vec{r}=a^{-1}\left[\vec{f}-(\vec{f}\cdot\vec{d})(a+\vec{c}\cdot\vec{d})^{-1}\vec{c}\right]. \tag{11d}$$

Case $\vec{c}=0$ (or $\vec{d}=0$):

$$\vec{r}=(a^2+b^2)^{-1}\left[a\vec{f}+\vec{b}\wedge\vec{f}+a^{-1}(\vec{f}\cdot\vec{b})\vec{b}\right]. \tag{11e}$$

Case $a=0$ (note that it requires $\vec{b}\cdot\vec{c}\neq 0$ as well as $\vec{b}\cdot\vec{d}\neq 0$):

$$\vec{r}=b^{-2}\{\ \vec{b}\wedge\vec{f}-(\vec{b}\cdot\vec{c})^{-1}(\vec{b}\cdot\vec{f})\vec{b}\wedge\vec{c}$$

$$+(\vec{b}\cdot\vec{c})^{-1}(\vec{b}\cdot\vec{d})^{-1}[\ b^2(\vec{b}\cdot\vec{f})-(\vec{b}\cdot\vec{c})(\vec{b}\wedge\vec{f})\cdot\vec{d}+(\vec{b}\cdot\vec{f})(\vec{b}\wedge\vec{c})\cdot\vec{d}\ ]\vec{b}\ \}. \tag{11f}$$

References

The indications in square brackets after each reference identify the place in the book where it is referred to (possibly more than once): whether in the Foreword [For], in a Section or Appendix, or among the References [Ref]. Note that the format of this book entails that most of the call to references are located in the "Notes" sections attached to Chapters and to the Appendices.

<A74> V. Arnold, *Mathematical methods of classical mechanics,* Nauka, Moscow, 1974 (in Russian; translations in several languages, including English, were subsequently published by Mir, Moscow). [For, 1.N]

<A76> M. Adler, "Some finite-dimensional integrable systems", in <FG76>, pp. 237-244. [2.N]

<A77> M. Adler, "Some finite-dimensional integrable systems and their scattering behavior", Commun. Math. Phys. **55**, 195-230 (1977). [2.N]

<ABC78> S. Ahmed, M. Bruschi and F. Calogero, "On the zeros of combinations of Hermite polynomials", Lett. Nuovo Cimento **21**, 447-452 (1978). [C.N]

<ABC2001> E. Abadoglu, M. Bruschi and F. Calogero, "ABC of magnetic monopole dynamics", (to be published). [5.N]

<ABCOP79> S. Ahmed, M. Bruschi, F. Calogero, M. A. Olshanetsky and A. M. Perelomov, "Properties of the zeros of the classical polynomials and of Bessel functions", Nuovo Cimento **49B**, 173-199 (1979). [C.N., D.N]

<AC78a> S. Ahmed and F. Calogero, "On the zeros of Bessel functions. III". Lett. Nuovo Cimento **21**, 311-314 (1978). [C.N]

<AC78b> S. Ahmed and F. Calogero, "On the zeros of Bessel functions. IV", Lett. Nuovo Cimento **21**, 531-534 (1978). [C.N]

<AC78c> S. Ahmed and F. Calogero, "On the zeros of Bessel functions. V", Lett. Nuovo Cimento **21**, 535-536 (1978). [C.N]

<AM78> R. Abraham and J. E. Marsden, *Foundations of Mechanics,* second edition, Benjamin, 1985. [1.N]

<AMM77>H. Airault, H. P. McKean and J. Moser, "Rational and elliptic solutions of the Korteweg-de Vries equation and a related many-body problem", Commun. Pure Appl. Math. **30**, 95-148 (1977). [2.N]

<APS85> E. Arbarello, C. Procesi, E. Strickland (editors), *Geometry Today* (Giornate di Geometria, Roma 1984), Birkhauser, Boston, 1985. [Ref]

<APS99> Quoteworthy Science, APS News **8.4**, 5 (April 1999). [For]

<AS65> M. Abramowitz and I. Stegun, *Handbook of mathematical functions with formulas, graphs, and mathematical tables*, National Bureau of Standards, Applied Mathematics Series **55**, third printing, 1995. [Ref]

<AS97> V. E. Adler and A. B. Shabat, "Generalized Legendre Transformations," Theor. Math. Phys. **112**, 935-948 (1997). <Russian version: Teor. Mat. Fys. **112**, 179-194 (1997)>. [4.N]

<B78> A. O. Barut (editor), *Nonlinear Equations in Physics and Mathematics*, Proceedings of the Nato Advanced Study Institute held in Istanbul in August 1977, Reidel, Dordrecht, 1978. [Ref]

<BB80> C. Bardos and D. Bessis (editors), *Bifurcation phenomena in mathematical physics and related topics*, Reidel, Dordrecht, 1980. [Ref]

<BB97a> H. W. Braden and V. M. Buchstaber, "The general analytic solution of a functional equation of addition type", SIAM J. Math. Anal. **28**, 903-923 (1997). [B.N]

<BB97b> H. W. Braden and V. M. Buchstaber, "Integrable systems with pairwise interactions and functional equations", Rev. Math. & Math. Phys. **10**, 3-120 (1997). [2.N, B.N]

<BC79> M. Bruschi and F. Calogero, "Eigenvectors of a matrix related to the zeros of Hermite polynomials", Lett. Nuovo Cimento **24**, 601-604 (1979). [C.N]

<BC81> M. Bruschi and F. Calogero, "Finite-dimensional matrix representation of the operator of differentiation through the algebra of raising and lowering operators: general properties and explicit examples", Nuovo Cimento **62B**, 337-351 (1981). [2.N, D.N]

<BC87> M. Bruschi and F. Calogero, "The Lax representation for an integrable class of relativistic dynamical systems", Commun. Math. Phys. **109**, 481-492 (1987). [2.N, B.N]

<BC90> M. Bruschi and F. Calogero, "General analytic solution of certain functional equations of addition type", SIAM J. Math. Anal. **21**, 1019-1030 (1990). [2.N, A.N, B.N]

<BC2000a> M. Bruschi and F. Calogero, "Solvable and/or integrable and/or linearizable *N*-body problems in ordinary (three-dimensional) space. I", J. Nonlinear Math. Phys. **7**, 303-386 (2000). [5.N]

<BC2000b> M. Bruschi and F. Calogero, "On the integrability of certain matrix evolution equations", Physics Lett. **A 273**, 167-172 (2000). [5.N]

<BC2000c> M. Bruschi and F. Calogero, "Integrable systems of quartic oscillators", Physics Lett. **A 273**, 173-182 (2000). [5.N]

<BCP90> M. Bruschi, R. G. Campos and E. Pace, "On a method for computing eigenvalues and eigenfunctions of linear differential operators", Nuovo Cimento **B105**, 131-163 (1990). [2.N, E.N]

<BCS98> A. J. Bordner, E. Corrigan and R. Sasaki, "Calogero-Moser models. I - A new formulation", Prog. Theor. Phys. **100**, 1107-1129 (1998). [2.N]

<BCS99> A. J. Bordner, E. Corrigan and R. Sasaki, "Generalized Calogero-Moser models and universal Lax operators", Prog. Theor. Phys. **102**, 499-529 (1999). [2.N]

<BDMMR85> C. Brezinski, A. Draux, A. Magnus, P. Maroni, A. Ronveaux (editors), *Polynomes Orthogonaux et Applications*, Proceedings Bar-Le-Duc 1984, Lecture Notes in Mathematics **1171**, Springer, Heidelberg, 1985. [Ref]

<BK96> V. Buchstaber and I. Krichever, "Multidimensional vector addition theorems and the Riemann theta functions", Int. Math. Res. Notices **10**, 505-513 (1996). [B.N]

<BMP92> M. Boiti, L. Martina and F. Pempinelli (editors), *Nonlinear Evolution Equations and Dynamical Systems* (Proceedings NEEDS '91), World Scientific, 1992. [Ref]

<BMRL80> M. Bruschi, S. V. Manakov, O. Ragnisco and D. Levi, "The nonabelian Toda lattice - Discrete analogue of the matrix Schroedinger spectral problem," J. Math. Phys. **21**, 2749-2753 (1980). [5.N]

<BMS2000> A. J. Bordner, N. S. Manton and R. Sasaki, "Calogero-Moser models. V - Supersymmetry and quantum Lax pairs", Prog. Theor. Phys. **103**, 463-487 (2000). [2.N]

<BP96> V. M. Buchstaber and A. M. Perelomov, "On the functional equation related to the quantum three-body problem", Amer. Math. Soc. Transl. **175**, 15-34 (1996). [B.N]

<BR83> M. Bruschi and O. Ragnisco, "On the inversion of the commutation operator," Lett. Nuovo Cimento **38**, 41-44 (1983). There is a misprint in formula (7) of this paper, which should read: $\widetilde{M}_{ij} = A^i M A^j + \mathrm{sgn}(j-i) A^j M A^i$. [5.N]

<BR89> M. Bruschi and O. Ragnisco, "On a new integrable Hamiltonian system with nearest neighbor interaction", Inverse Problems **5**, 983-998 (1989). [2.N]

<BRL81> M. Bruschi, O. Ragnisco and D. Levi, "Evolution equations associated with the discrete analog of the matrix Schroedinger spectral problem solvable via the inverse spectral transform," J. Math. Phys. **22**, 2463-2471 (1981). [5.N]

<BS97> H. W. Braden and R. Sasaki, "The Ruijsenaars-Schneider model", Prog. Theor. Phys. **97**, 1003-1017 (1997). [2.N]

<BS99> A. J. Bordner and R. Sasaki, "Calogero-Moser models. III - Elliptic potentials and twisting", Prog. Theor. Phys. **101**, 799-829 (1999). [2.N]

<BST99> A. J. Bordner and R. Sasaki, "Calogero-Moser models. II - Symmetries and folding", Prog. Theor. Phys. **101**, 487-518 (1999). [2.N]

<C71> F. Calogero, "Solution of the one-dimensional *N*-body problems with quadratic and/or inversely quadratic pair potentials", J. Math. Phys. **12**, 419-436 (1971); Erratum, *ibidem* **37**, 3646 (1996). [For., 2.N]

<C75> F. Calogero, "Exactly solvable one-dimensional many-body problems", Lett. Nuovo Cimento **13**, 411-416 (1975). [2.N, B.N]

<C76a> F. Calogero, "On a functional equation connected with integrable many-body problems", Lett. Nuovo Cimento **16**, 77-80 (1976). [2.N, B.N]

<C76b> F. Calogero, "A sequence of Lax matrices for certain integrable Hamiltonian systems", Lett. Nuovo Cimento **16**, 22-24 (1976). [2.N]

<C77a> F. Calogero, "On the zeros of the classical polynomials", Lett. Nuovo Cimento **19**, 505-508 (1977). [C.N, D.N]

<C77b> F. Calogero, "Equilibrium configuration of the one-dimensional *N*-body problem with quadratic and inversely-quadratic pair potentials", Lett. Nuovo Cimento **20**, 251-253 (1977). [2.N, C.N]

<C77c> F. Calogero, "On the zeros of Bessel functions", Lett. Nuovo Cimento **20**, 254-256 (1977). [C.N]

<C77d> F. Calogero, "On the zeros of Hermite polynomials", Lett. Nuovo Cimento **20**, 489-490 (1977). [C.N]

<C77e> F. Calogero, "On the zeros of Bessel functions. II", Lett. Nuovo Cimento **20**, 476-478 (1977). [C.N]

<C78a> F. Calogero, "Motion of poles and zeros of special solutions of nonlinear and linear partial differential equations, and related "solvable" many-body problems", Nuovo Cimento **43B**, 177-241 (1978). [2.N, C.N, D.N]

<C78b> F. Calogero, "Integrable many-body problems", in <B78>, pp.3-53. [2.N, C.N]

<C78c> F. Calogero (editor), *Nonlinear evolution equations solvable by the spectral transform*, Pitman, London, 1978. [Ref]

<C78d> F. Calogero, "Asymptotic behavior of the zeros of the (generalized) Laguerre polynomial $L^a{}_n(x)$ as the index $a \to \infty$ and limiting formula relating Laguerre polynomials of large index and large argument to Hermite polynomials", Lett. Nuovo Cimento **23**, 101-102 (1978). [C.N]

<C78e> F. Calogero, "Asymptotic behavior of the zeros of the Jacobi polynomials $P_n^{(at,bt)}(x)$ as $t \to \infty$ and limit relations of these polynomials with Hermite polynomials", Lett. Nuovo Cimento **23**, 167-168 (1978). [C.N]

<C79a> F. Calogero, "Singular integral operators with integral eigenvalues and polynomial eigenfunctions", Nuovo Cimento **51B**,1-14 (1979); **53B**, 463 (1979). [C.N]

<C79b> F. Calogero, "Integral representation and generating function for the polynomials $U_n^{(a,b)}(x)$", Lett. Nuovo Cimento **24**, 595-600 (1979). [C.N]

<C80a> F. Calogero, "Solvable many-body problems and related mathematical findings (and conjectures)", in <BB80>, pp.371-384. [2.N, C.N]

<C80b> F. Calogero, "Isospectral matrices and polynomials", Nuovo Cimento **58B**, 169-180 (1980). [2.N, C.N, D.N]

<C80c> F. Calogero, "Finite transformations of certain isospectral matrices", Lett. Nuovo Cimento **28**, 502-504 (1980). [2.N, D.N]

<C81a> F. Calogero, "Matrices, differential operators and polynomials", J. Math. Phys. **22**, 919-932 (1981). [2.4.4, 2.4.5.1, 2.N, C.N]

<C81b> F. Calogero, "Additional identities for certain isospectral matrices", Lett. Nuovo Cimento **30**, 342-344 (1981). [2.N, D.N]

<C81c> F. Calogero, "Integrable many-body problems and related mathematical results", in <Co81>, pp.143-150. [2.N, C.N]

<C82a> F. Calogero, "Isospectral matrices and classical polynomials", Linear Algebra Appl. **44**, 55-60 (1982). [2.N, C.N]

<C82b> F. Calogero, "Disproof of a conjecture", Lett. Nuovo Cimento **35**, 181-185 (1982). [C.N]

<C82c> F. Calogero, "Integrable dynamical systems and related mathematical results". In <W83>, pp. 47-109. [2.N, C.N] K. B. Wolf (editor), *Nonlinear Phenomena,* (Proceeding of the CIFMO School and Workshop held at Oaxtepec, Mexico, Nov. 29–Dec. 17, 1982), Lecture Notes in Physics **189**, Springer, 1983. [Ref.]

<C83a> F. Calogero, "Lagrangian interpolation and differentiation", Lett. Nuovo Cimento **35**, 273-278 (1983) & **36**, 447 (1983). [2.N]

<C83b> F. Calogero, "Computation of Sturm-Liouville eigenvalues via Lagrangian interpolation", Lett. Nuovo Cimento **37**, 9-16 (1983). [2.N, E, E.N]

<C83c> F. Calogero, "Interpolation, differentiation and solution of eigenvalue problems in more than one dimension", Lett. Nuovo Cimento **38**, 453-459 (1983). [2.N, E, E.N]

<C84a> F. Calogero, "Interpolation, differentiation and solution of eigenvalue problems for periodic functions", Lett. Nuovo Cimento **39**, 305-311 (1984). [2.N, E.N]

<C84b> F. Calogero, "Some applications of a convenient finite-dimensional matrix representation of the differential operator", Proceedings of the International Conference on "Special Functions: Theory and Computation", Rend. Sem. Mat., Univ. & Polit. Torino, October 1984 (special issue), pp. 23-61. [2.4.5.5, 2.N, C.N, D.N, E, E.N]

<C85a> F. Calogero, "Determinantal representations of the classical polynomials", Bollettino U.M.I. (6) **4-A**, 407-414 (1985). [2.4.5.5, 2.N, C.N]

<C85b> F. Calogero, "Some solvable dynamical systems", in <APS85>, pp. 39-45. [2.N]

<C85c> F. Calogero, "Interpolation and differentiation for periodic functions", Lett. Nuovo Cimento **42**, 106-110 (1985). [2.N, D.N, E.N]

<C85d> F. Calogero, "Determinantal representation of polynomials satisfying recurrence relations", in <BDMMR85>, pp. 568-570. [2.4.5.5, 2.N, C.N]

<C85e> F. Calogero, "A class of integrable dynamical systems", Inverse Problems **1**, L21-L24 (1985). [2.5, 2.N]

<C86a> F. Calogero, "A class of solvable dynamical systems", Proceedings of the International Conference on Solitons and Coherent Structures, Santa Barbara, 11-16.1.1985, Physica **18D**, 280-302 (1986). [2.5, 2.N]

<C86b> F. Calogero, "Integrable dynamical systems and some other mathematical results (remarkable matrices, identities, basic hypergeometric functions)", Notes of lectures presented at the summer school held at the Université de Montréal, July 29- August 16, 1985, in <W86>, pp.40-70. [2.N, D.N]

<C86c> F. Calogero, "Integrable many-body problems in more than one dimension", Reports Math. Phys. **24**, 141-143 (1986). [2.N]

<C88> F. Calogero, "A remarkable matrix", in <HG88>, pp. 975-980. [2.N, D.N]

<C92> F. Calogero, "Some recent findings on nonlinear evolution equations and dynamical systems", in <BMP92>, pp. 253-259. [2.N]

<C93a> F. Calogero, "Interpolation in multidimensions, a convenient finite-dimensional matrix representation of the (partial) differential operators, and some applications", J. Math. Phys. **34**, 4704-4724 (1993). [3.2.2, 3.N, D.N, E.N]

<C93b> F. Calogero, "Remarks on certain integrable one-dimensional many-body problems", Phys. Letters **A183**, 85-88 (1993). [2.N]

<C94> F. Calogero, "A class of C-integrable PDEs in multidimensions", Inverse Problems **10**, 1231-1234 (1994). [2.3.4.2]

<C95a> F. Calogero, "Integrable nonlinear evolution equations and dynamical systems in multidimensions", in <HCJ95>, pp.229-244. [2.N]

<C95b> F. Calogero, "An integrable Hamiltonian system", Physics Letters **A201**, 306-310 (1995). [2.N]

<C95c> F. Calogero, "A solvable Hamiltonian system", J. Math. Phys. **36**, 4832-4840 (1995). [2.N]

<C95d> F. Calogero, "Trigonometric identities. III", (unpublished). [D.N]

<C96a> F. Calogero, "Solvable (nonrelativistic, classical) *n*-body problems on the line. II", J. Math. Phys. **37**, 1253-1258 (1996). [3.N]

<C96b> F. Calogero, "A solvable *n*-body problem in the plane. I", J. Math. Phys. **37**, 1735-1759 (1996). [3.N, 4.N, G.N]

<C97a> F. Calogero, "Remarkable matrices and trigonometric identities", J. Comput. Appl. Math. **83**, 127-130 (1997). [D.N]

<C97b> F. Calogero, "Motion of strings in the plane: a solvable model", J. Math. Phys. **38**, 821-829 (1997). [4.N]

<C97c> F. Calogero, "A class of integrable Hamiltonian systems whose solutions are (perhaps) all completely periodic", J. Math. Phys. **38**, 5711-5719 (1997). [2.N, 4.N]

<C97d> F. Calogero, "Tricks of the trade: relating and deriving solvable and integrable dynamical systems", in <vDV2000>, pp. 93-116. [2.N, 3.N, 4.N]

<C98a> F. Calogero, "Three solvable many-body problems in the plane", Acta Applicandae Mathematicae **51**, 93-111 (1998). [4.N]

<C98b> F. Calogero, "Generalized Lagrangian interpolation, finite-dimensional representations of shift operators, remarkable matrices, trigonometric and elliptic identities", in <RJVV98>, pp. 50-59. [D.N]

<C98c> F. Calogero, "Integrable and solvable many-body problems in the plane via complexification", J. Math. Phys. **39**, 5268-5291 (1998). [4.4.1, 4.N]

<C98d> F. Calogero, "A solvable many-body problem in the plane", J. Nonlinear Mat. Phys. **5**, 289-293 (1998). [4.N]

<C99a> F. Calogero, "Remarkable matrices and trigonometric identities. II", Commun. Appl. Anal. **3**, 267-270 (1999). [D.N]

<C99b> F: Calogero, "The 'neatest' many-body problem amenable to exact treatments (a 'goldfish'?)", *Proceedings* of the Workshop on Solitons, Collapses and Turbulence, to celebrate V. E. Zakharov's 60th birthdate, Chernogolovka near Moscow, Russia, 3-10 August, 1999; Physica D (in press). [2.N, 4.N]

<Ca86> R. G. Campos, "A non-perturbative method for the $kx^2+\beta x^4$ interaction", Rev. Mex. Fis. **32**, 379-400 (1986). [2.N, E.N]

<CD82> F. Calogero and A. Degasperis, *Spectral Transform and Solitons: Tools to Solve and Investigate Nonlinear Evolution Equations*. Volume One. North Holland, Amsterdam, 1982, pp.514. [2.N, 4.2.4]

<CDM95> M. Costato, A. Degasperis and M. Milani (editors), *National Workshop on Nonlinear Dynamics*, Conference Proceedings vol. **48**, Pavullo nel Frignano (Modena), Italy, 19-22 May 1994; Società Italiana di Fisica, Bologna, 1995. [Ref]

<CF85> F. Calogero and E. Franco, "Numerical tests of a novel technique to compute the eigenvalues of differential operators", Nuovo Cimento **89B**, 161-208 (1985). [2.N, E.N]

<CF92> F. Calogero and J.-P. Françoise, "Integrable dynamical systems obtained by duplications", Ann. Ins. H. Poincaré **57**, 167-181 (1992). [2.N]

<CF96> F. Calogero and J.-P. Françoise, "A completely integrable Hamiltonian system", J. Math. Phys. **37**, 2863-2871 (1996). [2.N, B.N]

<CF97> F. Calogero and J.-P. Françoise, "Hamiltonian character of the motion of the zeros of a polynomial whose coefficients oscillate over time", J. Phys. A: Math. Gen. **30**, 211-218 (1997). [2.3.4.2, 4.2.3]

<CF2000a> F. Calogero and J.-P. Françoise, "Solution of certain integrable dynamical systems of Ruijsenaars-Schneider type with completely periodic trajectories", Ann. Henri Poincaré **1**, 173-191 (2000). [2.N, 4.N]

<CF2000b> F. Calogero and J.-P. Françoise, "A novel solvable many-body problem with elliptic interactions", Int. Math. Res. Notices **15**, 775-786 (2000). [2.N]

<CF2000c> F. Calogero and J.-P. Françoise, "Periodic solutions of a many-rotator problem in the plane", Inverse Problems (in press). [4.5, 4.N]

<CF2001> F. Calogero and J.-P. Françoise, "Periodic motions galore: how to modify nonlinear evolution equations so that they feature a lot of periodic solutions", (to be published). [4.N, 5.N]

<CFS2000> R. Caseiro, J.-P. Françoise an R. Sasaki, "Algebraic linearization of dynamics of Calogero type for any Coxeter group", J. Math. Phys. **41**, 4679-4686 (2000). [2.N]

<CH37> R. Courant and D. Hilbert, *Methoden der mathematischen Physik*, vol. 2, Springer, Berlin, 1937. [For]

<CH93> R. Camassa and D. D. Holm, "An integrable shallow water equation with peaked solitons", Phys. Rev. Lett. **71**, 1661-1664 (1993). [2.N]

<CHH94> R. Camassa, D. D. Holm and J. M. Hyman, "A new integrable shallow water equation", Adv. Appl. Mech. **31**, 1-33 (1994). [2.N]

<CJX93> F. Calogero and Ji Xiaoda, "Solvable (nonrelativistic, classical) *n*-body problems on the line. I", J. Math. Phys. **34**, 5659-5670 (1993). [3.2.2, 3.N]

<CJX94> F. Calogero and Ji Xiaoda, "Solvable (nonrelativistic, classical) *n*-body problems in multidimensions. I", J. Math. Phys. **35**, 710-733 (1994). [3.2.3, 3.2.4, 3.N, 5.6.1]

<CJX95> F. Calogero and Ji Xiaoda, "Solvable (nonrelativistic, classical) *n*-body problems in multidimensions. II", in <CDM95>, pp. 21-32 . [3.N]

<CK96> F. Calogero and M. D. Kruskal, "An elementary theorem in plane geometry and its multidimensional extension", in <MTV96>, pp. 37-41. [F]

<CMR75> F. Calogero, C. Marchioro and O. Ragnisco, "Exact solution of the classical and quantal one-dimensional many-body problems with the two-body potential $V_a(x) = g^2a^2 / \sinh^2(ax)$", Lett. Nuovo Cimento **13**, 383-387 (1975).[2.N]

<CN91> F. Calogero and M. C. Nucci, "Lax pairs galore", J. Math. Phys. **32**, 72-74 (1991). [2.N]

<Co81> E. G. D. Cohen (editor), *Proceedings of the 1980 Summer School on Fundamental Problems in Statistical Mechanics*, North Holland, 1981. [Ref]

<CP78a> F. Calogero and A. M. Perelomov, "Properties of certain matrices related to the equilibrium configuration of the one-dimensional many-body problems with the pair potentials $V_1(x) = -\ln|\sin x|$ and $V_2(x) = 1/\sin^2(x)$", Comm. Math. Phys. **59**, 109-116 (1978). [2.3.5]

<CP78b> F. Calogero and A. M. Perelomov, "Asymptotic density of the zeros of Hermite polynomials of diverging order, and related properties of certain singular integral operators", Lett. Nuovo Cimento **23**, 650-652 (1978). [C.N]

<CP78c> F. Calogero and A. M. Perelomov, "Asymptotic density of the zeros of Laguerre and Jacobi polynomials", Lett. Nuovo Cimento **23**, 653-656 (1978). [C.N]

<CP79> F. Calogero and A. M. Perelomov, "Some diophantine relations involving circular functions of rational angles", Linear Algebra Appl. **25**, 91-94 (1979). [2.3.5, D.N]

<CvD95> F. Calogero and J. F. van Diejen, "An exactly solvable Hamiltonian system: quantum version", Physics Letters **A205**, 143-148 (1995). [2.N]

<CvD96> F. Calogero and J. F. van Diejen, "Solvable quantum version of an integrable Hamiltonian system", J. Math. Phys. **37**, 4243-4251 (1996). [2.N]

<D81> B. A. Dubrovin, "Theta functions and non-linear equations", Russian Mathematical Surveys **36**, 11-92 (1981) [Russian original: Uspekhi Mat. Nauk **36**, 11-80 (1981)]. [Ref]

<D83> L. Durand, "Lagrangian differentiation, integration and eigenvalue problems", Lett. Nuovo Cimento **38**, 311-317 (1983). [2.N., E.N]

<D85> L. Durand, "Lagrangian differentiation, Gauss-Jacobi integration, and Sturm-Liouville eigenvalue problems", in <BDMMR85>, pp. 331-339. [2.N, E.N]

<DHP98> E. D'Hoker and D. H. Phong, "Calogero-Moser Lax pairs with spectral parameter for general Lie algebras", Nucl. Phys. **B 530**, 537-610 (1998). [2.N]

<DM73> B. P. Demidovich and I. A. Maron, *Computational Mathematics*, Mir, Moscow, 1973 [English translation by G. Yankovsky, Mir, 1976]. [4.2.1]

<E53> A. Erdelyi (editor), *Higher transcendental functions,* McGraw Hill, New York, 1953. [2.3.3, A.N, C.N]

<F1892> G. Frobenius, "Über die elliptischen Funktionen zweiter Art", J. Reine Angew. Math. **93**, 53-68 (1892). [A.N]

<F74a> H. Flaschka, "The Toda lattice. I: Existence of integrals", Phys. Rev. **B9**, 1924-1925 (1974). [2.N]

<F74b> H. Flaschka, "On the Toda lattice. II: Inverse scattering solution", Prog. Theor. Phys. **51**, 703-716 (1974). [2.N]

<FG76> H. Flaschka and D. W. McLaughlin (editors), *Proceedings* of the Conference on the Theory and Application of Solitons (Tucson, Arizona, January 1976), Rocky Mountain J. Math. **8**, no. 1 & 2 (1978). [2.N, Ref]

<FPU55> E. Fermi, J. R. Pasta and S. M. Ulam, "Studies of nonlinear problems", Los Alamos Sci. Lab. Rep. LA-1940, 1955 [reprinted in: *Collected Works of Enrico Fermi,* University of Chicago Press, Chicago, 1965, vol. II, p. 978; and also in *Nonlinear Wave Motion* (edited by A. C. Newell), Lect. Appl. Math. **15**, American Mathematical Society, providence, R. I., 1974, pp. 143-156]. [5.6.5]

<G83> G. Gallavotti, *The Elements of Mechanics*, Springer, 1983. [1.N]

<GP99> L. Gavrilov and A. M. Perelomov, "On the explicit solutions of the elliptic Calogero system", J. Math. Phys. **40**, 6339-6352 (1999). [2.N]

<GRJ94> I. S. Gradshteyn, I. M. Ryzhik and (as editor) A. Jeffrey, *Tables of integrals, series and products,* Academic Press, New York, fifth edition, 1994. [2.4.5, 5.2, A.N, C.N, D.N]

<GS97> I. Z. Golubchik and V. V. Sokolov, "On some generalizations of the factorization method", Theor. Math. Phys. **110**, 267-276 (1997). <Russian version: Teor. Mat. Fyz. **110**, 339-350 (1997)>. [5.N]

<H65> U. W. Hochstrasser, "Orthogonal polynomials", in <AS65>, pp. 773-802. [C.N]

<H74> M. Henon, "Integrals of the Toda lattice", Phys. Rev. **B9**, 1921-1923 (1974). [2.N]

<H92> J. Hoppe, *Lectures on integrable systems*, Lectures Notes in Physics **M10**, Springer, Berlin, 1992. [2.N]

<HCJ95> M. Hazewinkel, H. W. Capel and E. M. de Jager (editors), *KdV '95*, Proceedings of the International Symposium held in Amsterdam, The Netherlands, April 23-26, 1995; reprinted from Acta Applicandae Mathematicae, vol. **39**, 1995. [Ref]

<HG88> M. Hazewinkel and M. Gerstenhaber (editors), *Deformation Theory of Algebras and Structure and Applications*, Kluwer Academic Publishers, 1988. [Ref]

<I90> V. I. Inozemtsev, "Matrix analogues of elliptic functions" , Funct. Anal. Appl. **23**, 323-325 (1990) [Russian original : Funct. Anal. Pril. **23**, 81-82 (1989)]. [5.N]

<J1866> C. Jacobi, "Problema trium corporum mutuis attractionibus cubis distantiarum inverse proportionalibus recta linea se moventium", in *Gesammmelte Werke*, vol. **4**, Berlin, 1866, pp. 533-539. [2.N]

<K77> P. P. Kulish, "Factorization of scattering characteristics and integrals of motion", in <C78c>, pp. 252-257. [2.N]

<K78> I. M. Krichever, "Rational solutions of the Kadomtsev-Petriashvili equation and integrable systems of *n* particles on the line", Funct. Anal. Appl. **12**, 59-61 (1978). [2.N]

<K80> I. M. Krichever, "Elliptic solutions of the Kadomtsev-Petriashvili equation and integrable systems of particles", Funct. Anal. Appl. **14**, 282-290 (1980). [2.N]

<K81> I. M. Krichever, "The periodic non-Abelian Toda chain and its two-dimensional generalization", published as Appendix (pp. 82-89) to <D81>. [5.N]

<Kr77> M. Kruskal, "The birth of the soliton", in <C78c>, pp. 1-8. [5.6.5]

<K90> B. Kupershmidt (editor), *Integrable and superintegrable systems*, World Scientific, Singapore, 1990. [Ref]

<KKS78> D. Kazhdan, B. Konstant and S. Sternberg, "Hamiltonian group actions and dynamical systems of Calogero type", Commun. Pure Appl. Math. **31**, 481-507 (1978). [2.N]

<KL72> D. C. Khandekar and S. V. Lawande, "Solution of a one-dimensional three-body problem in classical mechanics", Amer. J. Phys. **40**, 458-462 (1972). [2.N]

<KST99> S. P. Khastgir, R. Sasaki and K. Takasaki, "Calogero-Moser models. IV - Limits to Toda theory", Prog. Theor. Phys. **102**, 749-776 (1999). [2.N]

<L68> P. D. Lax, "Integrals of nonlinear equations of evolution and solitary waves", Commun. Pure Appl. Math. **21**, 467-490 (1968). [2.N]

<LRB83> D. Levi, O. Ragnisco and M. Bruschi, "Continuous and Discrete Matrix Burger's Hierarchies", Nuovo Cimento **74B**, 33-51 (1983).[5.N]

<LVW96> D. Levi, L. Vinet and P. Winternitz (editors), *Symmetries and Integrability of Difference Equations,* CRM Lecture Notes and Proceedings, vol. **9**, Montréal, 1996. [Ref]

<Man74> S. V. Manakov, "Complete integrability and stochastization of discrete dynamical systems", Sov. Phys. JETP **40**, 269-274 (1975) <Russian original: Zh. Eksp. Teor. Fiz. **67**, 543-555 (1974)>. [2.N]

<Mar70> C. Marchioro, "Solution of a three-body scattering problem in one dimension", J. Math. Phys. **11**, 2193-2196 (1970). [2.N]

<Mo75> J. Moser, "Three integrable Hamiltonian systems connected with isospectral deformations", Adv. Math. **16**, 197-220 (1975). [2.N]

<Mo80> J. Moser, "Various aspects of integrable Hamiltonian systems", in: *Dynamical Systems*, Progress in Mathematics **8**, Birkhauser, Basel, 1980, pp. 233-289. [2.N]

<MPP93> V. Makhankov, I. Puzynin and O. Pashaev (editors), *Nonlinear Evolution Equations and Dynamical Systems - NEEDS '92*, World Scientific, Singapore, 1993, pp. 423-431. [Ref]

<MT65> L. M. Milne-Thomson, "Jacobian elliptic functions and theta functions", in <AS65>, pp. 567-585. [A.N]

<MTV96> P. Marcellini, G. T. Talenti and E. Vesentini (editors), *Partial Differential Equations and Applications* (Collected papers in honor of Carlo Pucci), Marcel Dekker, New York, 1996. [Ref]

<N82> W. Nahm, "The Algebraic Geometry of Multipoles", in: *Group Theoretical Methods in Physics,* edited by M. Serdaroglu and E. Inonu, Lecture Notes in Physics **180**, Springer, 1982, pp. 456-466. [5.N]

<NP96> F. W. Nijhoff and G.N. Pang, "Discrete-time Calogero-Moser model and lattice KP equations", in <LVW96>, pp. 253-264. [A.N]

<OP76a> M. A. Olshanetsky and A. M. Perelomov, "Explicit solution of the Calogero model in the classical case and geodesic flows on symmetric spaces of zero curvature", Lett. Nuovo Cimento **16**, 333-339 (1976). [2.N]

<OP76b> M. A. Olshanetsky and A. M. Perelomov, "Completely integrable Hamiltonian systems connected with semisimple Lie algebras", Invent. Math. **37**, 93-108 (1976). [2.N., B.N]

<OP76c> M. A. Olshanetsky and A. M. Perelomov, "Explicit solutions of some completely integrable systems", Lett. Nuovo Cimento **17**, 97-101 (1976). [2.N]

<OP81> M. A. Olshanetsky and A. M. Perelomov, "Classical integrable finite-dimensional systems related to Lie algebras", Phys. Rep. **71**, 313-400 (1981). [2.N]

<OR78> M. A. Olshanetsky and V.-B. K. Rogov, "Bound states in completely integrable systems with two types of particles", Ann. Inst. H. Poincaré **29**, 169-177 (1978). [2.N]

<P1896> H. Poincaré, "Remarques sur une expérience de M. Birkeland", Compt. Rendus Acad. Sci. **123**, 530-533 (1896). [5.N]

<P76> A. M. Perelomov, "Completely integrable classical systems connected with semisimple Lie algebras. II", ITEF preprint **27** (1976). [2.N]

<P77> A. M. Perelomov, "Completely integrable classical systems connected with semisimple Lie algebras. III", Lett. Math. Phys. **1**, 531-534 (1977). [2.N]

<P78> A. M. Perelomov, "The simple relation between certain dynamical systems", Commun. Math. Phys. **63**, 9-11 (1978). [2.N]

<P90> A. M. Perelomov, *Integrable systems of classical mechanics and Lie algebras*, Birkhauser, Basel, 1990. [2.N]

<PS76> S. I. Pydkuyko and A. M. Stepin, "On the solution of a functional-difference equation", Funct. Anal. Appl. **10**, 84-85 (1976) (in Russian). [2.N., B.N]

<R87> S. N. M. Ruijsenaars, "Complete integrability of relativistic Calogero-Moser systems and elliptic function identities", Commun. Math. Phys. **110**, 191-213 (1987). [2.N]

<R88> S. N. M. Ruijsenaars, "Action-angle maps and scattering theory for some finite-dimensional integrable systems. I. The pure soliton case", Commun. Math. Phys. **115**, 127-165 (1988). [2.N]

<R90> S. N. M. Ruijsenaars, "Finite-dimensional soliton systems", in <K90>, pp. 165-206. [2.N]

<R94a> S. N. M. Ruijsenaars, "Systems of Calogero-Moser type", in <SV99>, pp. 251-352. [2.N]

<R94b> S. N. M. Ruijsenaars, "Action-angle maps and scattering theory for some finite-dimensional integrable systems. II. Solitons, antisolitons, and their bound states", Publ. RIMS, Kyoto Univ. **30**, 865-1008 (1994). [2.N]

<R95> S. N. M. Ruijsenaars, "Action-angle maps and scattering theory for some finite-dimensional integrable systems. III. Sutherland type systems and their duals", Publ. RIMS, Kyoto Univ. **31**, 247-353 (1995). [2.N]

<R97> S. N. M. Ruijsenaars, "Integrable particle systems vs solutions to the KP and 2D Toda equations", Ann. Phys. **256**, 226-301 (1997). [2.N]

<RB96> O. Ragnisco and M. Bruschi, "Peakons, r-matrix and Toda lattice", Physica **A228**, 150-159 (1996). [2.N]

<RJVV98>K. Srinivasa Rao, R. Jagannathan, G. Van den Berghe and J. Van der Jeugt (editors), *Special Functions and Differential Equations*, Proceedings of a Workshop held at The Institute of Mathematical Sciences, Madras, India, January 13-24, 1997. Institute of Mathematical Sciences, Madras, 1998. [Ref]

<RS86> S. N. M. Ruijsenaars and H. Schneider, "A new class of integrable systems and its relation to solitons", Ann. Phys. (NY) **170**, 370-405 (1986). [2.N]

<S39> G. Szego, *Orthogonal Polynomials*, AMS Colloquium Publications XXIII, AMS, Providence, R. I., 1939. [C.N]

<S71> B. Sutherland, "Exact results for a quantum many-body problem in one dimension", Phys. Rev. **A4**, 2019-2021 (1971). [2.N]

<S72> B. Sutherland, "Exact results for a quantum many-body problem in one dimension. II", Phys. Rev. **A5**, 1372-1376 (1972). [2.N]

<S90> P. C. Sabatier (editor), *Inverse methods in action*, Springer, Berlin, 1990. [Ref]

<S97> Yu. B. Suris, "New integrable systems related to the relativistic Toda lattice", J. Phys. A: Math. Gen. **30**, 1745-1761 (1997). [4.N]

<S2000> J. Sivardière, "On the classical motion of a charge in the field of a magnetic monopole", Eur. J. Phys. **21**, 183-190 (2000). [5.2.2,5.N]

<SMTDC76> J. Schwinger, K. A. Milton, W.-Y. Tsai, L. L. De Raad Jr and D. C. Clark, "Nonrelativistic dyon-dyon scattering", Ann. Phys. **101**, 451-495 (1976). [5.N]

<SS96> S. I. Svinolupov and V. V. Sokolov, "Deformations of Triple-Jordan Systems and Integrable Equations", Theor. Math. Phys. **108**, 1160-1163 (1996). <Russian version: Teor. Mat. Fyz. **108**, 388-392 (1996)>. [5.N]

<SV99> G. Semenoff and L. Vinet (editors), *Proceedings of the 1994 Banff summer school on "Particles and Fields"*, CRM series in mathematical physics, Springer, New York, 1999. [Ref]

<T67> M. Toda, "Vibration of a chain with nonlinear interaction", J. Phys. Soc. Japan **22**, 431-436 (1967). [2.N]

<T81> M. Toda, *Theory of nonlinear lattices*, Springer Series in Solid-State Sciences **20**, Springer, Berlin, 1981. [2.N]

<vD94> J. F. van Diejen, *Families of commuting difference operators*, PhD Thesis, University of Amsterdam, 1994. [2.N]

<vDV2000> J. F. van Diejen and L. Vinet (editors), *Calogero-Moser-Sutherland Models,* Proceedings of the Workshop on Calogero-Moser-Sutherland Models, Montreal, 10-15 March 1997, CRM Series in Mathematical Physics, Springer, 2000. [Ref]

<W38> H. Weyl, Bull. Amer. Math. Soc. **44**, 602-604 (1938) (book review of <CH37>). [For]

<W77> S. Wojciechowski, "New completely integrable Hamiltonian systems of N particles on the real line", Phys. Lett. **A59**, 84-86 (1977). [2.N]

<W83> K. B. Wolf (editor), *Nonlinear Phenomena,* (Proceeding of the CIFMO School and Workshop held at Oaxtepec, Mexico, Nov. 29–Dec. 17, 1982), Lecture Notes in Physics **189**, Springer, 1983. [Ref.]

<W84> S. Wojciechowski, "An explicit solution for the one-dimensional many-body system in an external potential", Phys. Lett. **A185**, 188-190 (1984). [2.N]

<W86> P. Winternitz (editor), *Systèmes Dynamiques Non Linéaires: Integrabilité et Comportement Qualitatif,* SMS **102**, Presses de l' Université de Montréal, 1986. [Ref]

<WW27> E. T. Whittaker and G. N. Watson, *A course of modern analysis,* Cambridge University Press, Cambridge, 1963. [2.3.6.2, 2.3.6.3, A.N]

<Y92> R. I. Yamilov, "Classification of Toda type scalar lattices", in <MPP93>. [4.N]

<Z90> V. E. Zakharov, "On the dressing method", in <S90>, pp. 602-623 (see p. 622). [For]

MIX
Papier aus verantwortungsvollen Quellen
Paper from responsible sources
FSC® C105338

If you have any concerns about our products,
you can contact us on
ProductSafety@springernature.com

In case Publisher is established outside the EU,
the EU authorized representative is:
Springer Nature Customer Service Center GmbH
Europaplatz 3, 69115 Heidelberg, Germany

Printed by Libri Plureos GmbH
in Hamburg, Germany